# THE ATLAS OF THE
# SOLAR SYSTEM

## PATRICK MOORE
## GARRY HUNT
## IAIN NICOLSON
## PETER CATTERMOLE

**Foreword by Sir Francis Graham-Smith**
**Astronomer Royal**

CHANCELLOR
PRESS

# CONTENTS

**The Atlas of the Solar System**
First published by Mitchell Beazley 1983. Revised 1990
All rights reserved.
No part of this work may be reproduced or utilized in any form by any means, electronic or mechanical, including photocopying, recording or by any information storage or retrieval system, without the prior written permission of the publisher.

This edition published in 1997 by Chancellor Press
an imprint of Reed Consumer Books Limited
Michelin House, 81 Fulham Road
London SW3 6RB

# CONTENTS

**Printed in Slovenia**
ISBN 0 75370 014 X

**Editor** Judy Garlick
**Art Editor and Designer** Wolfgang Mezger
**Executive Editor** Lawrence Clarke
**Executive Art Editor** John Ridgeway

# FOREWORD

Over three centuries ago Copernicus, Kepler and Newton led a revolution in man's understanding of the universe, when the Solar System was revealed as a simple dynamic entity. Astronomy at that time was concerned with positions and motions, and with the universal law of gravity. The nature of the stars and the planets was unknown, awaiting revelation in the age of telescopes and space probes.

Now we see a second revolution. Consider the following dates: in 1957 Sputniks I and II were placed in Earth orbit; in 1959 Lunik 3 showed the first view of the back of the Moon; in 1965 Mariner 4, the first planetary probe, reached Mars; in 1969 Neil Armstrong stood on the surface of the Moon; in 1976 Viking I made a soft landing on Mars; in 1979 Pioneer II reached Jupiter, followed by Voyagers 1 and 2 to Jupiter and Saturn. Voyager 2 passed by Uranus in 1986 and Neptune in 1989. In just over three decades we have been presented with an overwhelming quantity of photographs and other information about the Solar System, which has transformed our understanding of the Sun, the Moon, the planets, and interplanetary space. There will never again be such a period of expansion. Even if the Grand Tour of the planets undertaken by the Voyagers did not depend on a rare configuration of the planets, it now seems unlikely that the spectacular and expensive explorations of the more distant planets will be repeated for many years to come.

It would be unreasonable to expect that all the new data on the planets and their satellites would fall into a simple and neat pattern. In fact, what we now see is an astonishing diversity. This Atlas does much more than display the new photographs in all their variety, with their amazing details of the surfaces of bodies that were previously only known as points of light. There is here a synthesis of all our knowledge of the Solar System, including particularly the Sun. The atlas will be a source book for all working astronomers; it is at the same time easily accessible to anyone with amateur interest.

It is important and gratifying to note that the text has been approved by the Royal Astronomical Society. Like all the best atlases, this one is both authoritative and good to look at.

Astronomer Royal

# PREFACE

Astronomy is the oldest of the sciences; it probably began in the long-ago days of the cave-dwellers. With the Greeks, astronomy became a true science and, when telescopes were invented in the early part of the seventeenth century, it became possible to make real progress in understanding the nature of the stellar objects.

The Solar System is our home, and it is natural for us to regard it as most important—even though it makes up only a very small part of the universe as a whole. Telescopes have been able to show fine detail on the Moon, and also features on some of the planets, notably Mars, Jupiter and Saturn; but until modern times our knowledge was very incomplete. Even today we cannot pretend that we have solved all the problems of the Solar System, but at least we know more than would have seemed remotely possible a few decades ago.

It is, of course, the coming of rocket flight that has made all the difference. Men have been to the Moon, and we can now analyze samples of lunar materials in our laboratories. Unmanned probes have made controlled landings on Venus and Mars; previously we had no idea whether or not Venus was an ocean-covered planet or the furnace-like environment that it has proved to be. Flyby vehicles have rendezvoused with Mercury, and all the giant planets, so that in the Solar System all the planets have been contacted apart from Pluto—which seems to be in a class of its own, and may not be worthy of true planetary status. Voyager 2, launched in 1977, made its pass of Neptune in 1989, sending back images of very high quality even though it had been traveling in space for 12 years. Probes were also sent to Halley's Comet in 1986.

Many of the revelations have been unexpected, and some have even been unwelcome. Venus, in particular, offers no prospects for a future colony, even though in size and mass it is so similar to the Earth, while Mars has a tenuous carbon dioxide atmosphere and a bitterly cold climate. But there have been discoveries which nobody could have foreseen. In some ways the greatest surprises have been sprung by Jupiter and Saturn, whose satellites have proved to be remarkable worlds, ranging from the smooth billiard-ball surface of Europa to the sulphurous volcanoes of Io, and the strange environment on Titan, where orange clouds in a thick nitrogen atmosphere hide the true surface completely.

Quite apart from these individual discoveries, the whole outlook of Solar System astronomy has altered. We are able to look back into the distant past and make informed speculations about how the system has evolved. And we have learned much more about the Sun, without which the Earth and other planets could never have come into being. It is a normal star, but not as constant and unchanging as used to be thought.

Voyager 2's flight passed Neptune in 1989 after 12 years traveling through space effectively concluded the first stage of our exploration of the universe. The *Atlas of the Solar System* has been revised to complete its portrait gallery of the planets and to include the most recent space activities. All the new nomenclature has been incorporated. (It will be found that in some cases this new nomenclature, introduced by the controlling body of world astronomy—the International Astronomical Union—differs from that given in older books.)

Of the contributors to the *Atlas*, Patrick Moore, who is past President of the British Astronomical Association and a member of the I.A.U., has been actively engaged in lunar cartography as well as planetary studies. Garry Hunt is a Principal Scientific Investigator for NASA, and was formerly Head of Atmospheric Physics at Imperial College, London. Iain Nicolson is an astronomer at the Hatfield Polytechnic Observatory and an expert on the Sun and stars, and Peter Cattermole is a lecturer in geology at the University of Sheffield and whose main field is volcanology and comparative planetology.

It is hoped that this atlas will present a balanced view of Solar System research today. This is a fortunate time to produce it, because the major discoveries have been made, and the coming years will quite possibly be devoted to consolidation and augmentation of the discoveries and theories of the first period of intensive space research described here.

# Scale of the Universe

We live in an immense universe. The Earth is of great importance to man, but in the universe as a whole it is utterly insignificant. So, too, is the Sun, without which life on Earth would cease to exist. The Sun is a very ordinary star and only appears so glorious because of its relative proximity to the Earth. The distance between the Sun and Earth is on average 150,000,000 km (one astronomical unit). The nearest stars are millions of millions of kilometers away.

The star system to which the Sun belongs is known as the Galaxy (or Milky Way). It contains at least 100,000 million stars, which are of very different types. Some are much more powerful than the Sun, others are feeble by comparison. They have a wide range of surface temperatures: very hot stars such as Vega in the constellation of Lyra and Rigel in Orion glow white or blue; the Sun, with a surface temperature of about 6,000° C, is yellow; while cooler stars such as Betelgeux in Orion are orange or orange-red. There are stars that are large enough to contain the orbit of the Earth around the Sun—and stars that are smaller than the Earth itself.

## A universe of galaxies
Stars only exist in galaxies. Throughout space there are many millions of galaxies, each with its quota of suns. Some galaxies can be seen with the naked eye—notably the Nubeculae or Magellanic Clouds in the southern sky, and the Andromeda Galaxy seen from the northern hemisphere. Photographs of Andromeda taken with large optical telescopes show that it is spiral in form, although it lies at an unfavorable angle to us and the beauty of the spiral is lost. Its distance is 2,200,000 light years, so we are viewing the galaxy as it was more than two million years ago. The light year is a convenient unit with which to express large distances: it is equal to the distance travelled by light during one year. Since light travels at 300,000 km

per second, one light year is equal to 9.46 million million kilometers. The closest star beyond the Sun (Proxima Centauri) lies at a distance of 4.2 light years.

Galaxies rarely exist alone. They are usually found in groups that vary in number from two or three to 1,000 or more. In all groups the member galaxies are distributed at random and held together by gravity. Any object with mass, however small, exerts a gravitational force field, and it is gravitational attraction that gives an object its weight. The force of gravity is always positive and it can only be a force of attraction. In the case of a group of galaxies, the force of gravity is just as if all the galaxies' mass was concentrated at the center of the system (even if there is no object located there).

A cluster of galaxies can measure tens of millions of light years across, and there is increasing evidence from studies of distant space that groups of galaxies are further grouped into what might be termed "superclusters". Our Galaxy is part of a sparse group of galaxies known as the Local Group. The Milky Way is the second- or third-largest member of this group, which comprises about 30 members including the Andromeda Galaxy, another spiral called the Triangulum, and the irregularly shaped Magellanic Clouds. All the other galaxies in the Local Group are much smaller systems of relatively low mass.

The Milky Way is about 100,000 light years in diameter. It is a flattened system with a central bulge. When we look along the main plane of the Galaxy we see many stars in almost the same line of sight, which accounts for the familiar band of light across the night sky. The Milky Way stars are not genuinely crowded together and appearances can be very deceptive. The real center of the Galaxy is hidden by obscuring clouds of interstellar dust, but is known to be about 30,000 light years away. The Sun, with its attendant planets,

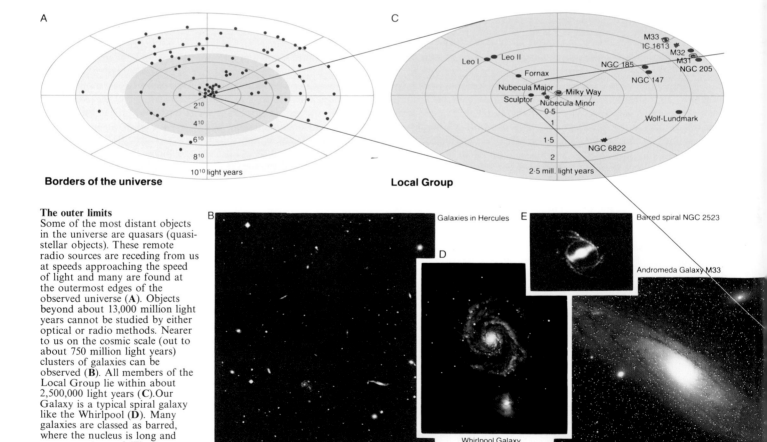

A
**Borders of the universe**

$2^{10}$
$4^{10}$
$6^{10}$
$8^{10}$
$10^{10}$ light years

C
**Local Group**

M33
IC 1613
M32
Leo I  Leo II
NGC 185   M31
NGC 205
Fornax
NGC 147
Nubecula Major   Milky Way
Sculptor   Nubecula Minor
0·5
1
Wolf-Lundmark
1·5
NGC 6822
2
2·5 mill. light years

## The outer limits
Some of the most distant objects in the universe are quasars (quasi-stellar objects). These remote radio sources are receding from us at speeds approaching the speed of light and many are found at the outermost edges of the observed universe (**A**). Objects beyond about 13,000 million light years cannot be studied by either optical or radio methods. Nearer to us on the cosmic scale (out to about 750 million light years) clusters of galaxies can be observed (**B**). All members of the Local Group lie within about 2,500,000 light years (**C**). Our Galaxy is a typical spiral galaxy like the Whirlpool (**D**). Many galaxies are classed as barred, where the nucleus is long and narrow, not round (**E**). Our spectacular neighbor Andromeda (**F**) is a spiral like the Milky Way.

B
Galaxies in Hercules   E   Barred spiral NGC 2523
D
Andromeda Galaxy M33
Whirlpool Galaxy NGC 5194 (M51)
F

6

lies close to the edge of one of the spiral arms. The Galaxy is known to rotate, and the Sun takes about 225,000,000 years to complete one journey around the galactic center.

Up to 1,000 million galaxies lie within the range of modern telescopes. They display a variety of forms, the most familiar of which are probably the spirals—some loose and some tightly coiled. Other galaxies are elliptical and range from long, narrow systems to almost circular. There are also irregular galaxies within the classification of star systems, and these have no symmetry at all. Most dwarf galaxies are irregular.

## Stars and planets
A normal star is a globe of intensely hot gas: it is self-luminous and produces energy by means of nuclear processes that occur deep within it. In the case of the Sun, the core temperature is in the order of 14,000,000° C—perhaps even more. With a body whose inner temperature is less than 10,000,000° C, no nuclear reactions can be sustained and the body cannot become luminous.

Stars are classified according to their spectral type. The absorption or emission of radiation by the constituents of a star produces tell-tale lines on the spectrum of light received from the star. The constituents of a star are also related to its surface temperature. Within this spectral classification, stars are said to be of a particular color. Their surface temperature is linked to the stage of stellar evolution they have reached. The course of stellar evolution depends largely upon the initial mass of the original nebular material from which it forms: a massive star evolves differently from a star of much less mass.

The Sun can be said to be a fairly average star. It is of course accompanied by a planetary system comprising nine planets and various bodies of lesser importance. The planets move in orbit around the Sun and planetary satellites orbit parent planets in a similar fashion. Bodies move in orbit at great speed. The force of gravity between an orbiting body and the parent body provides the centripetal force towards the center of motion that keeps a body in orbit and produces its centripetal acceleration. The speed of rotation is almost constant and it is the continuously changing direction of the orbiting body which implies acceleration. If the inward gravitational force were removed, the orbiting body would continue on in a straight line.

## Basic measurements
Information about the bodies of the Solar System has been derived from the analysis of the light received from them. The interpretation of basic measurements enables us to calculate physical dimensions and properties as well as chemical composition.

By careful observations of the positions of the planets it is possible to determine the dimensions of their orbits very accurately. Three positional measurements enable the preliminary orbit to be found. The physical size of a planetary body is calculated from the angular size of the disc that is presented to us. From any one apparent angular size, if the distance of the object is known, the size can be determined. Mass can best be calculated if the planet possesses its own satellite: the gravitational interaction implies the mass. If there is no satellite then the effects on the body's nearest neighbors can be used instead. From a knowledge of the mass and size of a body, its mean density can be obtained. The value of acceleration due to gravity on the surface of a body is determined by the body's mass and its radius. This determines the velocity of escape.

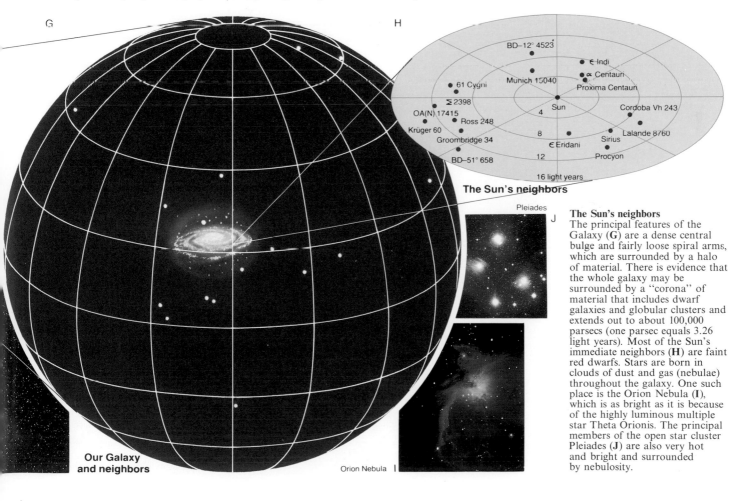

The Sun's neighbors

Our Galaxy and neighbors

Orion Nebula I

Pleiades J

**The Sun's neighbors**
The principal features of the Galaxy (**G**) are a dense central bulge and fairly loose spiral arms, which are surrounded by a halo of material. There is evidence that the whole galaxy may be surrounded by a "corona" of material that includes dwarf galaxies and globular clusters and extends out to about 100,000 parsecs (one parsec equals 3.26 light years). Most of the Sun's immediate neighbors (**H**) are faint red dwarfs. Stars are born in clouds of dust and gas (nebulae) throughout the galaxy. One such place is the Orion Nebula (**I**), which is as bright as it is because of the highly luminous multiple star Theta Orionis. The principal members of the open star cluster Pleiades (**J**) are also very hot and bright and surrounded by nebulosity.

# Scale of the Solar System

The Solar System comprises the Sun, the nine major planets (Mercury, Venus, Earth, Mars, Jupiter, Saturn, Uranus, Neptune and Pluto) together with their satellites, a host of minor bodies (asteroids, comets and meteoroids) and a certain amount of interplanetary gas and dust. The Sun is by far the dominant body, its gravitational field controlling the motions of the other bodies in the Solar System. The solar magnetic field and solar radiation play an important role, too, particularly with regard to the gas and dust.

With a diameter of 1,392,000 km, the Sun is 109 times larger than the Earth and nearly 10 times the diameter of Jupiter, the largest of the planets. Its mass—$1.9891 \times 10^{30}$ kg—is some 330,000 times that of the Earth, and more than 1,000 times that of Jupiter: indeed, the Sun contains more than 99.8 percent of the mass of the Solar System. When its volume, 1,303,000 times that of the Earth, is compared with its mass, it is clear that the mean density of the Sun is appreciably less than the Earth's density. In fact, at 1,410 kg m$^{-3}$, it is less than a quarter of the value for terrestrial material, reflecting the difference in structure and composition of the two bodies.

**Planetary orbits and distances**

It was Johannes Kepler in 1609 who discovered that the planets travelled around the sun in ellipses, with the Sun located at one focus of the ellipse. This principle is the first of the three "laws" governing planetary motion that bear Kepler's name (*see* page 438). Strictly speaking, the Sun and planets should be considered to revolve around the center of mass or 'barycenter" of the entire solar system, a point which does not coincide exactly with the center of the Sun. The second law states that for each planet the line between

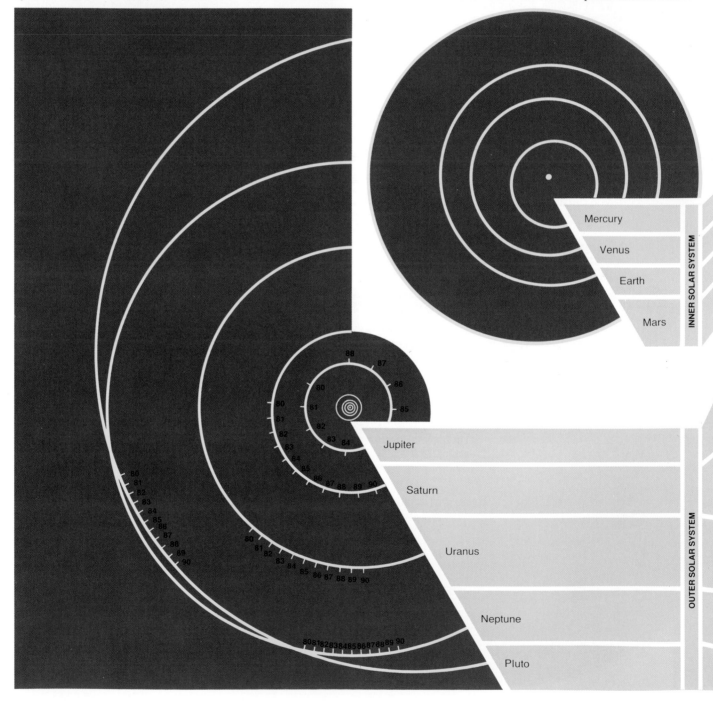

the planet and the Sun (the "radius vector") sweeps out equal areas in equal times. When a planet's orbit takes it nearer to the Sun it must therefore move faster than when it is far away. The point of closest approach to the Sun is called "perihelion", and the most distant point is "aphelion". The third of Kepler's laws defines a simple relationship between a planet's period of revolution and its mean distance from the Sun. Taking the year as the unit of time and the astronomical unit as the unit of distance, $P^2 = a^3$ where P is the period and a is the "mean distance".

The size and shape of an ellipse are determined by two quantities: the length of the "semi-major axis" and the "eccentricity". These are two of six orbital elements (see page 439) that have to be found in order to define the size, shape and orientation of a planet's orbit and also the planet's position in the orbit at any time.

**Planetary configurations**

An "inferior" planet is one whose orbit lies inside that of the Earth: the others are said to be "superior". A superior planet is in "opposition" when it lies directly opposite the Sun in the sky as viewed from Earth. It is in "quadrature" when the angle measured at the Earth between the planet and the Sun (called "elongation") is 90°. When either a superior or an inferior planet is directly in line with the Sun it is said to be in "conjunction". In the case of an inferior planet, the conjunction may be either inferior (when it is between the Earth and the Sun) or superior (when it is on the opposite side of the Sun).

Phenomena occurring as a result of the positional configurations of the planets relative to each other, their satellites and the Sun include phases (see pages 78, 100, 146–7) and eclipses (148–9).

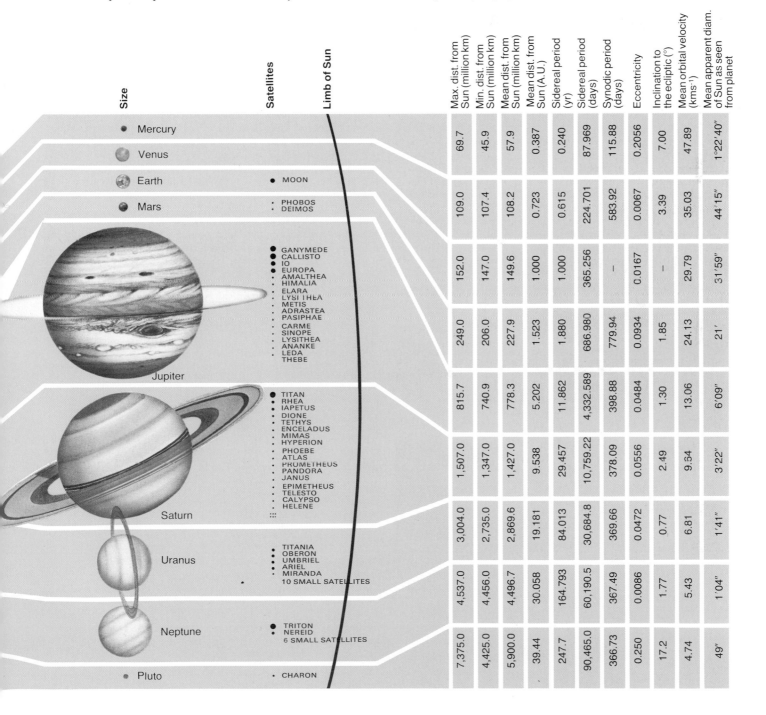

| Size | Satellites | Max. dist. from Sun (million km) | Min. dist. from Sun (million km) | Mean dist. from Sun (million km) | Mean dist. from Sun (A.U.) | Sidereal period (yr) | Sidereal period (days) | Synodic period (days) | Eccentricity | Inclination to the ecliptic (°) | Mean orbital velocity (kms⁻¹) | Mean apparent diam. of Sun as seen from planet |
|---|---|---|---|---|---|---|---|---|---|---|---|---|
| Mercury | | 69.7 | 45.9 | 57.9 | 0.387 | 0.240 | 87.969 | 115.88 | 0.2056 | 7.00 | 47.89 | 1°22'40" |
| Venus | | 109.0 | 107.4 | 108.2 | 0.723 | 0.615 | 224.701 | 583.92 | 0.0067 | 3.39 | 35.03 | 44'15" |
| Earth | MOON | 152.0 | 147.0 | 149.6 | 1.000 | 1.000 | 365.256 | – | 0.0167 | – | 29.79 | 31'59" |
| Mars | PHOBOS, DEIMOS | 249.0 | 206.0 | 227.9 | 1.523 | 1.880 | 686.980 | 779.94 | 0.0934 | 1.85 | 24.13 | 21' |
| Jupiter | GANYMEDE, CALLISTO, IO, EUROPA, AMALTHEA, HIMALIA, ELARA, LYSITHEA, METIS, ADRASTEA, PASIPHAE, CARME, SINOPE, LYSITHEA, ANANKE, LEDA, THEBE | 815.7 | 740.9 | 778.3 | 5.202 | 11.862 | 4,332.589 | 398.88 | 0.0484 | 1.30 | 13.06 | 6'09" |
| Saturn | TITAN, RHEA, IAPETUS, DIONE, TETHYS, ENCELADUS, MIMAS, HYPERION, PHOEBE, ATLAS, PROMETHEUS, PANDORA, JANUS, EPIMETHEUS, TELESTO, CALYPSO, HELENE | 1,507.0 | 1,347.0 | 1,427.0 | 9.538 | 29.457 | 10,759.22 | 378.09 | 0.0556 | 2.49 | 9.64 | 3'22" |
| Uranus | TITANIA, OBERON, UMBRIEL, ARIEL, MIRANDA, 10 SMALL SATELLITES | 3,004.0 | 2,735.0 | 2,869.6 | 19.181 | 84.013 | 30,684.8 | 369.66 | 0.0472 | 0.77 | 6.81 | 1'41" |
| Neptune | TRITON, NEREID, 6 SMALL SATELLITES | 4,537.0 | 4,456.0 | 4,496.7 | 30.058 | 164.793 | 60,190.5 | 367.49 | 0.0086 | 1.77 | 5.43 | 1'04" |
| Pluto | CHARON | 7,375.0 | 4,425.0 | 5,900.0 | 39.44 | 247.7 | 90,465.0 | 366.73 | 0.250 | 17.2 | 4.74 | 49" |

# Origins of the Solar System

The modern view is that the universe as we know it is about 13,000 million years old. The Solar System by comparison is about 4,560 million years old and forms but a tiny fraction of the huge volume we call space. It is believed that within the visible boundaries of the universe there are at least $10^{11}$ star systems (galaxies). There is currently convincing support for the view that everything in the universe commenced with a "Big Bang": particularly strong evidence is provided by the 2.7 K background radiation detected in interstellar space. Within the universe hydrogen and helium predominate: heavier elements are believed to have been generated within stars and then returned to the interstellar medium by means of both supernova explosions and the loss of mass experienced by stars in the red giant phase of their evolution.

## The solar nebula

The Solar System originated in what is termed the "solar nebula", and until very recently the general opinion was that its overall composition was akin to that of certain types of meteorite (but excluding hydrogen, helium and rare gases). Recent studies of certain isotopes, particularly $^{16}O$, have, however, revealed anomalies that strongly suggest that the nebula was not isotopically homogeneous. One suggestion is that the activity of a nearby supernova supplied material to produce the anomalies, and also triggered the condensation of the solar nebula and eventually the formation of the Solar System itself. As an alternative to this notion, it has been proposed that the isotopic anomalies might have been locked into "pre-solar grains" which were produced in expanding supernova shells. The most favored of this type of theory includes the activity of not one but two such supernovae, which eventually interacted.

Whichever theory proves to be ultimately the most acceptable, it is true to say that a majority of cosmologists have returned to the kind of notion first considered by Pierre Laplace and his colleagues, that an initially rotating gas cloud (nebula) existed, whose shape and internal movements were determined by both gravitational and rotational forces. It is thought that as time passed gravitational

attraction became the dominant force, whereupon contraction set in and, as a result, the rate at which the cloud rotated increased, thus conserving angular momentum. With time the cloud flattened out into a disc and material within the cloud slowly drifted towards the center, eventually accreting into the "proto-Sun". The latter eventually collapsed under its own gravitation, becoming dense and opaque as more and more matter was compressed into a smaller and smaller volume. Collapse had the effect of raising the internal temperature of the proto-Sun to the point at which thermonuclear reactions began. The star then took on a form similar to that with which we are familiar.

How hot did the nebula become during its evolution? Was it, for example, so hot (say, 1,800 K) that dust and gas particles were completely vaporized, or did it simply condense to produce a series of condensates such as are observed in some meteorites? Was the nebula never very hot at all (only about 300 K)? That the solar nebula was not well mixed is strongly suggested from the study of meteorites, which often consist of an agglomeration of particles, some of which clearly were once molten, but others of which evidently were never very hot at all. The most primitive meteorites (C1 carbonaceous chondrites) have a chemistry with roughly solar elemental abundances, but while the more refractory examples suggest melting and/or metamorphism within their parent bodies, others appear to have been heated since they aggregated. Anomalies in the distribution of $^{16}O$ would tend to have been ironed out if the whole nebula had once been very hot. The fact that they have not, and the meteorite evidence together suggest that the solar nebula never did become very hot and that the anomalous particles were added to the nebula in solid phases.

## Protoplanet versus planetesimal theories

The processes that occurred next are not at all well understood, but it is clear that, at some undefined stage, nuclei developed which eventually accreted into the planets as we know them. Many theories have been proposed but currently only two are really worth discussing; these are the proto-planet theory and the planetesimal

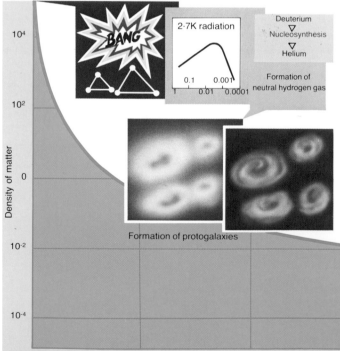

In a simple model universe all matter is created in a hot "Big Bang", prior to which all matter existed in a state of infinite density at a finite time in the past—known as the cosmic singularity. From this hot beginning follows the cosmic expansion. After about 25 sec this model universe has a temperature of about 4 billion degrees and a density of about 2 tons per liter. At this time the principal constituent is radiation. Today a relic radiation remains. At an age of 200 sec the temperature of the universe has fallen sufficiently for the formation of deuterium. This nucleosynthesis initiates a chain of nuclear reactions that produce helium and, after a million years and the combination of electrons

and protons, neutral hydrogen. This marks the end of the radiation era and the beginning of galaxy formation. The force of gravity now becomes important, and slight variations in density result in gravitational instabilities that finally produce the collapse of lumps (or areas of contrasting density) into protogalaxies. The Solar System itself formed from the collapse of a molecular gas cloud. The planets formed by hot, heterogeneous accretion. The protoplanet theory suggests the collapse of large and massive protoplanets due to gravitational instabilities in the nebula. The planetesimal theory starts with dust grains from which larger aggregations develop until planet-sized bodies remain.

theory. The gist of the former is that the collapse of the nebula into large massive protoplanets was caused by gravitational instability within it. The second theory sees growth as having started with dust grains which gradually collided to form larger aggregations, growing in size from millimeter-, through meter- to kilometer-sized particles, until eventually they reached asteroid proportions and finally developed into planetary-sized bodies.

It is probably true to say that for the very large gaseous planets, Jupiter and Saturn in particular, parts of the former theory may well have applied, but for the terrestrial worlds the weight of evidence favors the planetesimal idea. This theory is supported by the widespread evidence of large-scale impacts within the inner Solar System and out at least as far as Saturn. Major impacts could well account for the observed obliquities of the planets.

**Evidence from meteorites**
Weak electrostatic forces are generally believed to have been sufficient to accomplish the transformation from dust-sized particles into centimeter-sized grains. It is here that meteorites provide us with a vital clue in our search for the next step. Most meteorites contain at least 20 minerals that are in a state of disequilibrium with each other; that is, they could not have been produced as condensation products in equilibrium. This strongly favors a complex cycle of growth, disruption and re-aggregation until quite late in the accretion process. When we examine the overall composition of meteorites of all types, it is also clear that parent bodies of varying chemistries are required to account for the diversity seen. Such diversity is also observed among the asteroids. Diversity seems to have been the rule, and the evidence strongly indicates that very early on in the evolution of the solar nebula there were inhomogeneities, the various constituents being accreted in varying proportions in the growing planetesimals. Subsequently differences developed in the distribution of the different kinds of elements and these gave each planet its distinctive characteristics.

Another factor, of considerable importance in this context, is what was happening at the center of the Solar System, for at some point the Sun began to form and radiate energy. Both Jupiter and Saturn have retained near solar proportions of hydrogen and helium, and furthermore they have satellite systems containing a large proportion of ice. Evidently in this part of the young Solar System the overall temperature never exceeded 300 K. Closer to the Sun, however, hydrogen, helium and the rare gases have long since vanished, the conclusion being that they were removed before they could be enmeshed in the fabric of Earth- and Moon-sized bodies. One idea is that the Sun went through an early "T-Tauri" stage, whereupon a very strong "solar wind" swept the inner portion of the Solar System clear of these elements and perhaps removed the early atmospheres of the inner planets as well. The main period of planetary growth probably followed this phase of the Sun's evolutionary development.

While gravitational effects became extremely important during the growth of the planets from dust-sized to larger bodies, other factors played a vital role in the subsequent individual evolution of each planet. Radioactive elements, in particular $^{26}$Al, locked into the fabric of the planets would quickly have started to decay, producing heat energy which built up gradually until in some cases melting commenced. These bodies thus experienced the chemical differentiation of silicate-rich mantles and iron/nickel-enriched cores. These processes probably occurred during the first 100 million years of the planets' lives. Immense amounts of hydrogen at this stage would have created highly "reducing" conditions, but as the solar nebula "cleared", so oxidizing conditions would have come to dominate and there would have been a gradual depletion in the more volatile constituents.

During the final stages of the accretionary process, large-scale impacts on the young planets would affect considerable redistribution of material. Initially cratering was endemic to the Solar System. Subsequently melting and chemical differentiation became important in the development of planetary crusts and the arrangement of chemical compounds within the interiors of the planets, thus producing the kind of internal structure that scientists believe exists within the planets today.

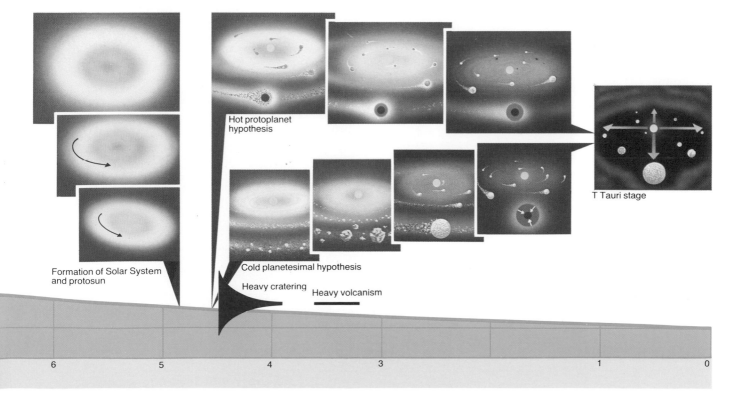

Hot protoplanet hypothesis

T Tauri stage

Formation of Solar System and protosun

Cold planetesimal hypothesis

Heavy cratering    Heavy volcanism

6    5    4    3    1    0

11

# Evolution of the Planets

All the planets have changed since they first accreted, the changes having been wrought by a variety of dynamic processes including cratering, chemical differentiation, melting, volcanism, degassing and structural deformation. We know more about modification of the inner planets than of the outer giants.

### Cratering
Cratering was endemic early in the history of the Solar System. The most complete record is found on the Moon, where the effects of erosion are minimal. The lunar maria, which were flooded by basaltic lava about 3,500 million years ago, provide a suitable place to study cratering. The number of craters of any one size is proportional to the inverse square of the diameter. The objects responsible for the creation of the craters also follow a pattern of distribution, and it emerges that the range of sizes of meteorites and other bodies that have impacted the planets and Moon is essentially the same as that of the asteroids in the Asteroid Belt. This is taken as strong confirmatory evidence that intense cratering originated in the infall of asteroid- and meteorite-sized bodies.

Impact cratering (1) occurs when a missile strikes the surface (1A), sending shock waves into the target and the projectile itself.

The resulting stresses cause the projectile to vaporize and material is ejected at high velocities. At the next stage material is thrown out at lower velocities, excavating the crater (1B) as a plume of material is ejected from the center (1C). Finally, ejected material falls back to the surface and the rim of the crater forms by uplift (1D).

Studies of cratering on the Moon's older highland crust show that the cratering rate changed over time. About 4,000 million years ago the rate was hundreds (and possibly thousands) of times greater than at present, but then fell off abruptly about 3,000 million years ago. This has led to speculation about a "Great Bombardment", a sort of cosmic sweeping-up operation, by means of which debris left over from the primordial solar nebula was cleaned up by the planets. Further calculations indicate that if all the inner planets were cratered as intensely as the Moon, about half the total number of bodies present in the Asteroid Belt today would suffice to produce the observed cratering. Since this number must represent but a small fraction of the total amount of debris left in the early stages of the Solar System, there is clearly no problem in explaining the production of craters in this way.

The Earth must also have been heavily scarred at this time and some scientists have suggested that the intense bombardment was

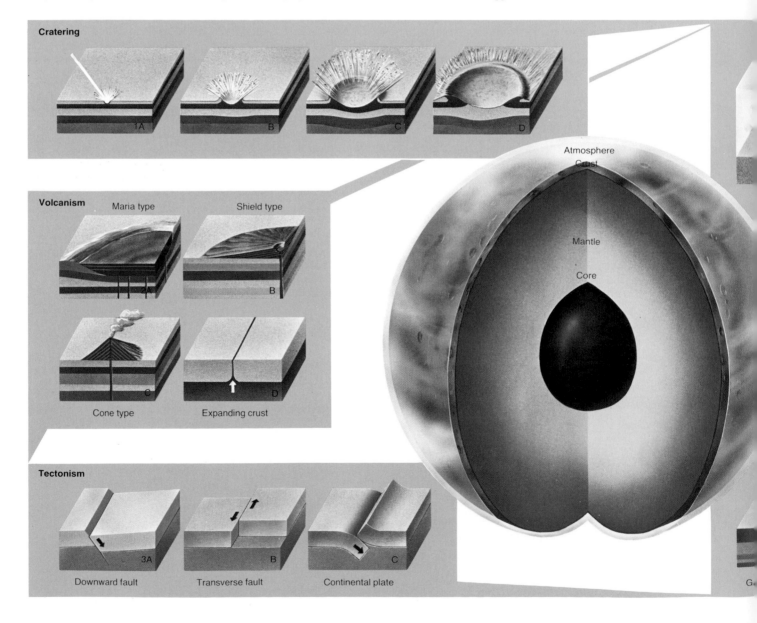

**Cratering**

1A    B    C    D

**Volcanism**

Maria type    Shield type

2A    B

Cone type    Expanding crust

**Tectonism**

3A    B    C

Downward fault    Transverse fault    Continental plate

Atmosphere
Crust
Mantle
Core

at least partly responsible for inhomogeneities that subsequently developed in the Earth's crust. However, dynamic processes like erosion and tectonics have long since removed most of the cratering record from this planet. Mercury, Mars and the moons of Jupiter and Saturn also bear the scars of this early cratering, while radar mapping of the cloud-shrouded Venus has revealed both volcanic structures and craters that may be of impact origin. Differences in detail from one planet to another probably reflect variations in the speed with which each evolved and whether or not erosion, fracturing and volcanism have affected the initial record.

## Chemical differentiation

After the accretionary stage, other internal processes took over and generated dramatic changes, the effects of which vary from planet to planet. In the case of the Moon, accretional energy and the decay of radioactive elements such as $^{26}Al$ was sufficient to melt at least half the body. Subsequently a small core formed and it seems that very soon any volatiles present within were soon lost to space. The presence of material molten within the Moon was instrumental in widespread volcanism (2), involving the uprising of basaltic magma onto the surface. Considerable thicknesses of lava filled in some of the ancient lunar basins (2A), producing the lunar maria. In general, lava can be released in a number of ways: it can be ejected by means of a single cone (2B), in which case stratified layers of volcanic rock build up, or, where the lava is more fluid, through a shield volcano (2C). Alternatively, it can seep out of fissures (2D) where, for example, the crust is expanding.

On Earth the sequence of events was rather different. During the melting and segregation of the terrestrial core, there was a depletion of potassium by a factor of four compared with the supposed proportions in the primordial nebula. This and other changes, such as the separation of iron, nickel and cobalt and their drift towards the core, meant that the Earth's interior underwent a process of chemical differentiation that was much more marked than that of the Moon. During this process a large proportion was melted, molten magma eventually rising back towards the surface (since it was less dense than its surroundings) to form the primitive crust.

In the early stages of planetary evolution, the outflow of heat from each body was mainly by means of conduction but, subsequently, where melting occurred, stirrings within the liquid core may have induced convective motions which speeded up the outflow of heat, thus cooling the planets more quickly. This convection resulted in crustal modification and the tectonic features (3) such as folding and faulting—either downward (3A) or transverse (3B)—and the movement of continental plates (3C).

The spinning of a planet, coupled with a liquid core, may have resulted in the generation of a magnetic field. The Earth's interior is believed to behave rather like a vast dynamo that is self-perpetuating once the generation of the initial current begins. A similar effect is observed on Jupiter and Saturn and, to a much lesser extent, on Mercury, but not on Mars or Venus.

## Planetary atmospheres

Exactly how and when the planets developed their atmospheres is difficult to establish, but it is clear that primitive atmospheres were quite unlike modern ones. Gases that accumulated were supplied by a kind of internal "drying out" process known as degassing. Most of these gases were undoubtedly provided through volcanic eruptions. In the case of the Earth early gases probably included water vapor, hydrogen, hydrogen chloride, carbon monoxide, carbon dioxide and nitrogen. Most of the $H_2$ would have escaped, as it did from all the inner planets, and any $O_2$ would not have remained in a free state for long since it would have combined with gases like methane and carbon monoxide and the crystalline materials of the crust. The production of significant amounts of free $O_2$ had to await the development of life forms such as green algae.

The position of a planet within the Solar System also influenced the fate of an atmosphere. Venus has a choking, carbon dioxide-rich mantle of air that evolved as a result of the planet having been subjected to much higher temperatures than existed on Earth, where $CO_2$ re-combined to form crustal materials such as limestone, coal and petroleum. Mercury was too near the Sun to retain any gases at all, and Mars was too small to hold on to many of the original volatiles. The giant planets have retained hydrogen atmospheres largely because of their large masses.

Atmospheres are further modified (4) by the release of volatiles during cratering (4A), outgassing from volcanoes (4B), the reaction of chemicals in the air with those in rocks (4C) and atmospheric circulation via evaporation and condensation (4D). Present-day planetary spectra (5) reveal the different evolutionary paths followed by the inner (5A) and the outer (5B) planets.

Finally, rocks are eroded, redistributed and eventually re-deposited, perhaps being totally recycled and remelted, thus changing the appearance of the crust of a planet (6). Material falls or slides under the influence of gravity (6A); wind erodes material and redeposits it elsewhere (6B); and flowing water carves many distinctive surface features (6C).

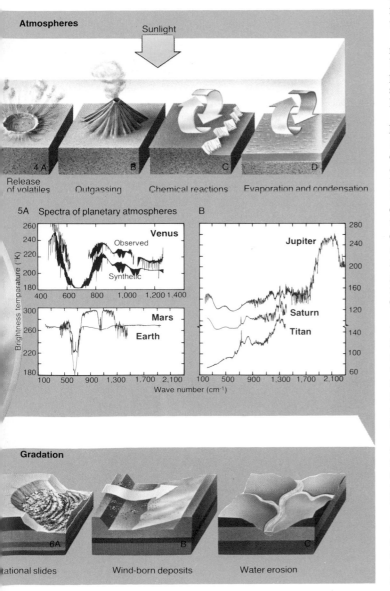

**Atmospheres**

Sunlight

4A     B     C     D

Release of volatiles    Outgassing    Chemical reactions    Evaporation and condensation

5A   Spectra of planetary atmospheres    B

Brightness temperature (K)

Venus
Observed
Synthetic
400   600   800   1,000   1,200 1,400

Jupiter

Mars
Earth
Saturn
Titan

100   500   900   1,300   1,700   2,100    100   500   900   1,300   1,700   2,100

Wave number (cm⁻¹)

**Gradation**

6A     B     C

Gravitational slides    Wind-born deposits    Water erosion

# Exploration of the Solar System

The idea of reaching other worlds is far from new. As long ago as the second century AD a Greek satirist, Lucian of Samosata, wrote a story about a journey to the Moon. Another early space-travel story was written by the German astronomer Johannes Kepler and was published in 1634. In Kepler's version the "astronaut" was transported to the Moon by an obliging demon. Since then the twentieth century has seen a rapid transformation of fictional fantasy into scientific reality. In less than three decades our scientific horizons have encompassed the whole Solar System.

## Trajectories

On 4 October 1957 the first artificial satellite, the Soviet Sputnik 1, was launched, marking the real start of the Space Age. The subsequent development of highly sophisticated Earth-orbiting satellites was paralleled by the exploration of the Moon, which culminated in the first man walking on the Moon in July 1969. This technical achievement can never be underestimated. Sending a probe to another planet requires both precise timing and accuracy. Even to reach the Moon a vehicle has to enter a 16 km diameter corridor within 26 km/h of the desired injection velocity. To reach beyond Earth's orbit the heaviest payload that can be sent is one

that travels in an elliptical orbit that touches the Earth's orbit at departure and that of its destination planet at arrival. Known as a Hohmann transfer orbit, it must be timed so that after leaving Earth, a probe meets its destination planet half way round the Sun from where it started. This period is known as the launch window.

The Hohmann transfer orbit was used to reach Venus in 1962, Mars in 1965 and Jupiter in 1973. Reaching more distant planets, however, requires a special trajectory. Because of the immense distances involved and the limited supply of fuel that can be carried on board, the "gravity-assist" trajectory is used to boost the vehicle on its journey. The principle is to use the gravitational field during the flyby of a closer planet to accelerate the spacecraft onto the next. The first vehicle to achieve this was Mariner 10 in 1974 (see pages 428–9). In escaping from Earth's gravity Mariner was slowed down enough to alter its orbit to an ellipse reaching onto the orbit of Venus at perihelion. Without the interaction with Venus it would have continued this elliptical path back out to the Earth's orbit. By approaching Venus from the outside and swinging across from behind the planet, the gravity field of Venus was able to slow Mariner down enough for it to enter a new "closed" ellipse and reach farther into the orbit of Mercury. Once this ellipse was

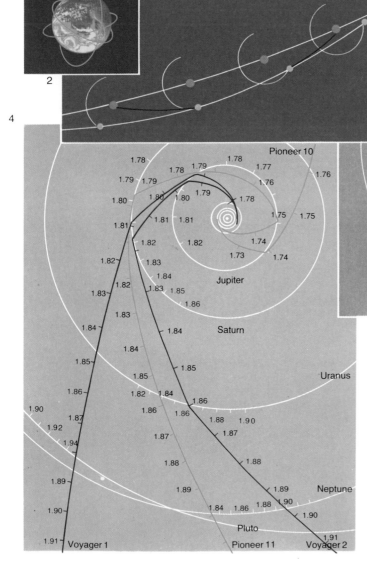

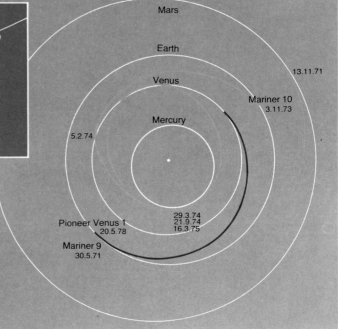

## Probes to the planets

Alongside the monitoring of our planet from orbiting space-probes (1), exploration of other planets and satellites in the Solar System started with the Earth's closest neighbor, the Moon. Here the principles of trajectory and communication were able to be fully tested; the intricacies of a return trip (2) involved both unmanned soft-landers and man himself. Successful exploration of the terrestrial planets (3) started with Mariner 2's flyby of Venus in 1962; eight years later the first soft-lander was transmitting from the surface. In 1973 Mariner 10's "gravity assist" trajectory took it from Venus to three close encounters with Mercury. In 1975,

four years on from the success of Mariner 9, Viking 1 soft-landed on Mars. Exploration of the outer planets (4) has depended on four missions, the Pioneers and Voyagers. Pioneer 10 made its closest approach to Jupiter in 1973, and Pioneer 11 exactly one year later. Pioneer 11's course was adjusted to enable it to rendezvous with Saturn in 1979. Both Pioneers are now traveling out of the Solar System. Although Voyager 2 was launched first, Voyager 1 traveled by a more economical route, and reached Jupiter first, in 1979, and Saturn in 1980. Voyager 2's trajectory took it on to Uranus in 1986 and Neptune in 1989.

established, the Sun's gravitational field automatically brought Mariner back to Mercury's orbit. Its period around the Sun was then adjusted to synchronize its return with Mercury.

Having tested the principle, the most ambitious voyage of the twentieth century was attainable. In 1977 a rare alignment of the outer planets, which occurs only once in 176 years, offered the possibility of sending probes on a "Grand Tour", taking them past Jupiter, Saturn, Uranus and possibly Neptune as well. On 20 August 1977 Voyager 2 was launched. It made its pass of Jupiter in July 1979 and Saturn in August 1981, after which it passed Uranus in January 1986 and Neptune in August 1989 (*see* pages 434–5).

### Communications

The technical achievements of landing a man on the Moon (*see* pages 426–7) or soft-landing vehicles on Venus (*see* page 430) or Mars (*see* page 433) were peculiar to each mission. However, the considerable problems of tracking, controlling and communicating with space-probes are common to all. Thus the power of onboard transmitters is severely limited by the restriction on weight: those on the Voyager spacecraft operated on a maximum of less than 30 W, and the power of the received signal over 1 m² of the Earth's surface was in the range of only $10^{-18}$ W. Moreover, because of the vast distances, there is a delay in reception of about 40 minutes.

Also, a certain amount of interference and distortion is inevitable. Some types of information, such as command signals, demand a high degree of accuracy; others, such as video data, are relatively tolerant, since errors can be removed by computer-processing on the ground. Sophisticated encoding techniques are used to protect the most error-sensitive data, while at the same time more robust data can be transmitted more economically. On Voyager the communications system operated on two different frequencies, one near 0.13 m in the so-called "S-band" and another near 0.4 m—the "X-band". The X-band was used exclusively for "downlink" transmissions, while the S-band carried both "uplink" and "downlink" data. Like all other probes the tracking depended on an international network of stations sited in Canberra, Madrid and Goldstone, at such longitudes that the spacecraft is within range of one at all times. The most ambitious probe to date, Voyager is the last major planetary flyby mission until the mid-1990s. In the coming decade, studies of satellites and planetary atmospheres will depend on the effectiveness of the Space Telescope (operative in 1990)—the world's first orbiting telescope.

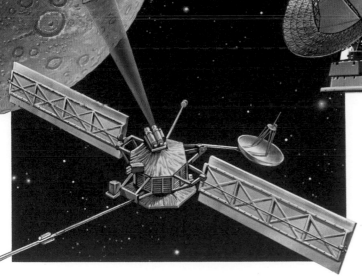

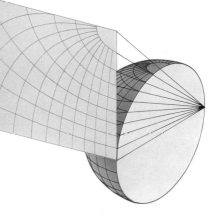

### Imaging
Most probes employ shuttered television type-cameras. Each image is analyzed into 640,000 picture elements ("pixels"); the brightness of each pixel is measured by sensors and its value expressed as an eight-digit binary number. The numbers are transmitted to a receiving station and translated back into the levels of light intensity to form the image. Full-color images are built up from monochromatic images taken through different filters.

### Mapping
The orthographic projection (**A**) represents the globe as a flattened disc (as the Moon appears from Earth). The cylindrical Mercator projection (**B**) minimizes distortion with a latitude scale that increases from the equator. Stereographic projections are used to avoid distortion at the poles when they are centered on the opposite pole (**C**), or to map exactly one hemisphere when they are centered on the equator (**D**), as in the maps of Mercury.

A 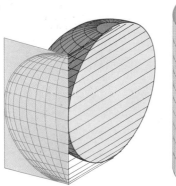 B 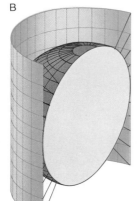 C 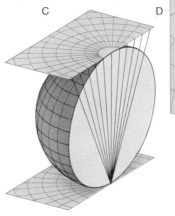 D

# THE SUN

# Characteristics

Looked at in cross-section, the Sun may be divided into a number of concentric layers, or shells, each of which will be treated in more detail in later pages. The visible surface is known as the "photosphere" (meaning "sphere of light"), but it is not a solid surface like that of the Earth; rather it is a shallow shell of gas, some 400 km thick, from which all but a tiny fraction of the Sun's light is emitted. Below the photosphere, where the temperature and density increase rapidly, the solar material is opaque to visible light, while above it the solar atmosphere is essentially transparent. The solar radius, measured from the center of the Sun to the photosphere, is 696,265 km; since most of the visible light emanates from a region within the photosphere only about 100 km thick, it is hardly surprising that the edge of the solar disc appears hard and sharp.

The Sun shines as a result of nuclear reactions, releasing vast quantities of energy, which take place in the hot, dense central core, a region extending to about 0.25 of the solar radius, a distance of some 175,000 km from the center (this value is not known with certainty). The central temperature is about 15 million K and the density is estimated to be 160,000 kg m$^{-3}$. Material outside the core is at lower temperatures, pressures and densities, and nuclear reactions do not take place at significant rates.

Energy is released in the core primarily in the form of $\gamma$-ray and X-ray radiation or high-energy, short wavelength "photons" (*see* pages 70–71). These photons cannot travel far before being absorbed: deep in the solar interior the mean distance travelled is about 1 cm. The photons are absorbed and reemitted a multitude of times, and their energies decline as they move out through the solar globe. Their paths take the form of a "random walk", in which the direction and energy change at each absorption and emission. As a result, it can take something like $10^4$ to $10^5$ years for a given packet of energy to make its way to the surface. This process of absorption, scattering and emission goes on through a region called the "radiative zone", which extends from about 0.25 solar radii ($R_S$) to over 0.8 $R_S$ (a value of 0.86 $R_S$ is commonly quoted). Photons, absorbed and emitted many times, are progressively converted from $\gamma$-ray and X-ray radiation into extreme ultraviolet and ultraviolet, finally emerging from the photosphere as visible light (low-energy or longish wavelength photons).

Beyond about 0.8 $R_S$ the lower temperature allows electrons to be captured by nuclei—particularly those of heavier elements—to form partially ionized atoms, which are very effective at capturing photons. As a result, the solar material becomes very much more opaque than in the layers below (by a factor of about 20) and there is much greater resistance to the flow of radiation out through the Sun. A steep temperature gradient is established, and hot bubbles of gas rising through this gradient are accelerated to form convective cells. Below about 0.85 $R_S$ convection is not believed to be significant, and the transport of energy from the interior is mainly by radiation. Above this level, however, where blocking of radiative transfer takes place, convection sets in and masses of hot gas are carried bodily to the surface, where they radiate their heat. In the outermost 15 to 20 percent of the solar radius, energy transport is primarily by convection; this layer is known as the "convective zone".

At the top of the convective zone lies the photosphere. Under ideal observing conditions, the photosphere may be seen to be made up of a large number of bright "granules" (*see* pages 26–7). These granules are often polygonal in shape, and are separated by thin dark lines. They are evidence of the bubbling motion of hot gases in the outer parts of the solar globe in the process of transporting heat outward from the interior by convection. On a much larger scale there is a network of "supergranular" cells, each of which contains hundreds of individual granules. Being larger patterns of circulation, they extend deeper into the convective zone.

The photosphere is overlaid by the "chromosphere", a more rarefied layer a few thousand kilometers thick. The temperature at

the base of the chromosphere is about 4,300 K, but it increases with altitude, rising very sharply through a thin transition region, where it merges with the "corona", an even more rarefied region, having a density at base of about $10^{-11}$ kg m$^{-3}$ and extending outward to a distance of several solar radii. The height of the transition region above the photosphere varies quite markedly over the solar surface, but is typically about 2,000 km.

The corona has a very high temperature of between 1 million and 5 million K but, because its density is so low, the amount of heat it contains is very small compared with the photosphere. Indeed, only about one millionth of the Sun's visible light comes from the corona, and even this is photospheric light scattered in the corona rather than light emitted by the corona. Under normal circumstances the chromosphere and corona cannot be seen without the aid of specialized instrumentation, but when a total eclipse occurs these layers may be seen directly. Beyond the corona there extends the solar "magnetosphere" and the "solar wind", the stream of atomic particles (mainly electrons and protons) which flows away from the Sun into interplanetary space (*see* pages 68–9).

The above is a summary of the structural properties of the quiet Sun. However, the Sun is an active star. Violent events occur on it, the most dramatic being explosive flares, while "prominences" (huge flame-like clouds of luminous gas) may be observed above the chromosphere. Aspects of solar activity and variability will be treated in later pages.

**Physical characteristics of the Sun**

| | |
|---|---|
| Diameter | 1,392,530 km |
| Volume | $1.414 \times 10^{27}$ m³ |
| Mass | $1.9891 \times 10^{30}$ kg |
| Magnetic field strengths (typical): | |
| Sunspots | 3,000 G |
| Polar field | 5 G |
| Bright, chromospheric network | 25 G |
| Chromospheric plages | 200 G |
| Prominences | 10 to 100 G |

Chemical composition of photosphere

| Element | % weight | Element | % weight |
|---|---|---|---|
| Hydrogen | 73.46 | Nitrogen | .09 |
| Helium | 24.85 | Silicon | .07 |
| Oxygen | .77 | Magnesium | .05 |
| Carbon | .29 | Sulphur | .04 |
| Iron | .16 | Other | .10 |
| Neon | .12 | | |

| | |
|---|---|
| Density (water = 1): | |
| Mean density of entire Sun | 1,410 kg m$^{-3}$ |
| Interior (center of Sun) | $1.6 \times 10^5$ kg m$^{-3}$ |
| Surface (photosphere) | $10^{-6}$ kg m$^{-3}$ |
| Chromosphere | $10^{-9}$ kg m$^{-3}$ |
| Low corona | $10^{-13}$ kg m$^{-3}$ |
| Solar radiation: | |
| Entire Sun | $3.86 \times 10^{23}$ kW |
| Unit area of surface of Sun | $6.29 \times 10^4$ kW m$^{-2}$ |
| Received at top of Earth's atmosphere | 1,368 W m$^{-2}$ |
| Surface brightness of the Sun (photosphere): | |
| Compared to full Moon | 398,000 times |
| Compared to inner corona | 300,000 times |
| Compared to outer corona | $10^{10}$ times |
| Temperature: | |
| Interior (center) | 15,000,000 K |
| Surface (photosphere) | 6,050 K |
| Sunspot umbra (typical) | 4,240 K |
| Penumbra (typical) | 5,680 K |
| Chromosphere | 4,300 to 50,000 K |
| Corona | 800,000 to 5,000,000 K |
| Rotation (as seen from Earth): | |
| Of solar equator | 26.8 days |
| At solar latitude 30° | 28.2 days |
| At solar latitude 60° | 30.8 days |
| At solar latitude 75° | 31.8 days |

# Observational Background

## Location of the Sun: geocentric and heliocentric theories

All early civilizations believed that the Earth lay at the center of the universe. The geocentric view was developed to its fullest extent by the Greeks, who conceived a system in which the Sun, the Moon, the stars and planets revolved around the Earth in perfect circular motions. It was apparent to them that the observed motions could not be explained simply in terms of uniform motion on a circle centered on the Earth, and elaborate schemes involving combinations of circular motions were devised to obtain better agreement between theory and observation. Best known of these devices was the epicycle, a circle whose center travels round the circumference of another circle. The Greek view of the universe was synthesized by Claudius Ptolemaeus (Ptolemy) in his book the *Almagest*, published about the middle of the second century AD. This view remained largely unchallenged for over a millenium.

However, long before Ptolemy's time, a few philosophers had questioned the geocentric theory. In particular, Aristarchus (310–230 BC) had suggested that the Earth and planets traveled around the Sun, but this view was strenuously rejected at that time by his contemporaries. The heliocentric (Sun-centered) theory did not receive further serious attention until the sixteenth century. In 1543 the Polish astronomer Nicolaus Copernicus (1473–1543) published a detailed account of such a theory—*De Revolutionibus Orbium Coelestium*—which, despite considerable opposition, came to be espoused by many leading thinkers, including Galileo Galilei (1564–1642). The final blow to the old geocentric theory was dealt by Johannes Kepler (1571–1630) when he established his laws of planetary motion based on the premise that all the planets travel around the Sun in elliptical orbits.

## Nature, distance and size of the Sun

In one of the earliest attempts to describe the physical nature of the Sun the Greek philospher Anaxagoras (c.499–c.427 BC) argued that the Sun was a mass of fiery hot stone about the size of the Peloponnesus (the southern peninsula of Greece). The idea that the Sun might be an ordinary star began to emerge in the sixteenth century, but there is some reason to suppose that this possibility had been mooted by Aristarchus.

About 270 BC Aristarchus attempted to find the distance of the Sun by a geometrical method based upon measuring the angle between Sun and Moon at the times of first and last quarter. Difficulties in making the angular measurements led him to a gross underestimate of the Sun's distance; his result indicated that the Sun was about 19 times further away than the Moon, an underestimate by a factor of about 20. Hipparchus (second century BC) improved upon Aristarchus' work and arrived at a distance equivalent to 10 million kilometers. This implied that the Sun must be at least seven times the size of the Earth.

By the seventeenth century astronomers had realized that if they could establish the distance of the planets from the Earth they could use this information to work out the distance to the Sun. Thus in 1672 the Italian astronomer G. D. Cassini used the method of parallax to find the distance of Mars and from this deduced the Sun's distance to be 138, 370, 000 km. On those rare occasions when Venus passes *directly* between Earth and Sun it is visible as a tiny black dot "in transit" across the solar disc; parallax measurements of Venus while in transit may be used to infer the solar distance. The transits of 1761, 1769, 1874 and 1882 were used for this purpose, and from the latter pair Sir George Airy obtained a value of 150,162,000 km.

Parallax measurements of some of those asteroids which approach the Earth closely further refined the value, and in recent years radar techniques—whereby signals are bounced back from planetary surfaces—have led to the currently accepted value of 149,597,892 km. The resultant value for the solar diameter is thus calculated at 1,392,530 km.

1A

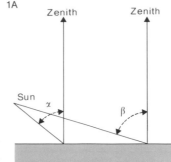

### 1. Anaxagoras' estimate of the distance of the Sun
The Greek philospher Anaxagoras (c.499–c.427 BC) argued that the Sun was a mass of fiery hot stone about the size of the southern peninsula of Greece. On the assumption that the Earth was flat, he probably measured the angles α and β from two places a known distance apart and calculated what he took to be the distance of the Sun by triangulation. In reality most of the difference in angles results from the Earth's curvature.

3A

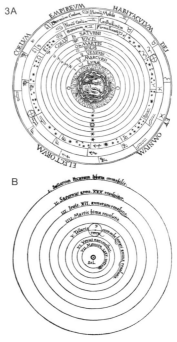

B

C

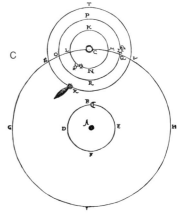

### 2. Early cosmological systems
In the Ptolemaic universe (**A**) the Earth was surrounded by a system of concentric spheres to which the planets and stars were fixed. The Copernican system (**B**) placed the Sun at the center of the universe. The Tychonic system (**C**) was a hybrid, with the Earth at the center but with the planets revolving round the Sun.

### 3. Transits of Venus
As Venus passes between the Sun and the Earth (**A**), observers at different latitudes, X and Y, see Venus follow different tracks, XX¹ and YY¹ across the solar disc (**B**). By timing the instants at which Venus enters and leaves the solar disc (X, Y, and X¹, Y¹, respectively) the precise lengths of the lines XX¹ and YY¹ may be found. From this

information the angular separation between the two lines can be determined. Knowing, from Kepler's laws, the *relative* distances from the Sun of the Earth and of Venus, and having measured by this method the parallax of Venus from sites X and Y separated by a known distance, the distance of the Sun can be found from geometry.

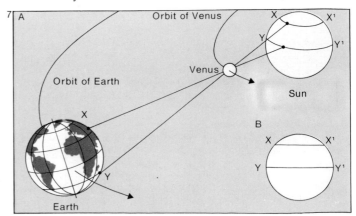

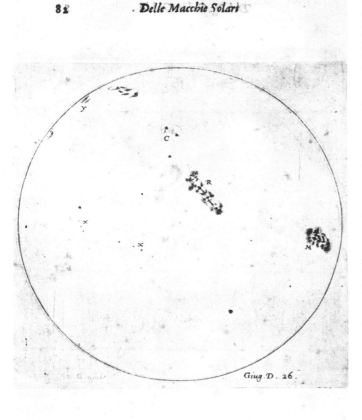

8$     *Delle Macchie Solari*

*Giug. D. 26.*

**4. Sunspot observations by Galileo**
The division of sunspots into umbral and penumbral regions is clearly apparent in this drawing from Galileo's *Istoria* (1613). Galileo first claimed to have observed sunspots three years earlier in November 1610.

**5. Early photography**
The quality of photographs of the Sun taken in the nineteenth century is often remarkably high. This example was taken by L. M. Rutherfurd on 21 September 1870. There are numerous clearly defined sunspot groups.

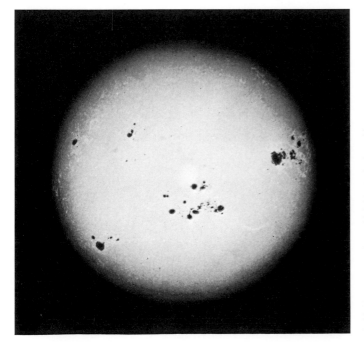

## Sunspots
Although the Sun was long believed to be a pure unblemished orb, naked eye reports of dark spots on its face date back some 2,000 years. Chinese records of such sightings date back to 28 BC (and possibly to 165 BC) but the spots were generally considered to be foreground objects.

Following the invention of the telescope a number of astronomers began to make telescopic observations of the Sun. Galileo claimed to have seen sunspots in this way in November 1610 and from subsequent observations deduced that the Sun rotates on its axis in a period of about a month. Christopher Scheiner (1575–1650) made a long series of observations between 1611 and 1625 from which he obtained a rotation period of 27 days, deduced the tilt of the solar axis to be between 6° and 8°, and examined the structure of spots and faculae.

The corona is believed first to have been mentioned in the writings of Plutarch (AD 46–120) and there are definite records of its having been seen at the eclipse of 22 December 968. Of other notable solar phenomena, prominences are thought first to have been studied in detail by the Swedish observer Vassenius at the total eclipse of 1733; they were certainly recorded at the eclipse of 1842. That most violent of events—the solar flare—was first seen as "two patches of intense bright and white light" on 1 September 1859 by the English amateur, Richard Carrington.

## The spectroscope
That sunlight could be split up by a prism into a rainbow band of its constituent colors was clearly demonstrated by Isaac Newton in 1666. In 1802 the English chemist William Wollaston discovered that the rainbow band was crossed by a number of dark lines, and in 1814 the German physicist Joseph von Fraunhofer used a spectroscope (an instrument for examining the spectrum in detail) to measure the wavelengths of 324 of these lines. As a result of work carried out by Gustav Kirchhoff and Robert Bunsen in 1859 it came to be realized that these lines were produced by the absorption of light in the solar atmosphere by the different chemical elements which were present; each chemical element giving its own characteristic set of lines

By this means the chemical composition of the Sun's outer layers could be determined. More detailed analysis of the spectrum reveals a tremendous range of additional information, such as temperature, density, velocity, rotation and the presence of magnetic fields. The importance of spectroscopy to astronomy can scarcely be over-rated.

## Photography and other techniques
The advent of photography offered a means of recording a solar image in a fraction of a second. The first good solar photograph was obtained on 2 April 1845 by H. Fizeau and L. Foucault in France, using the daguerrotype process, while in 1851 Berkowski made the first successful photograph of a total eclipse.

Another major step forward came in 1892 with the invention by George Ellery Hale of the "spectroheliograph", an instrument which allows the whole disc of the Sun to be studied at any selected wavelength of light, and permits the observation—at any time, rather than only during total eclipses—of phenomena such as prominences and filaments. The coronagraph, invented in 1930 by the Frenchman Bernard Lyot, allowed astronomers based at high altitude observatories to study some details of the inner corona at times other than eclipses.

That the Sun is a source of radio waves was deduced by J. S. Hey as a result of noise signals picked up in 1942 by British Army radar observers. With the coming of the Space Age and the ability to place instruments above our atmosphere, the whole range of solar radiations has become available for study, and our understanding of the Sun has leapt forward as a result.

# Solar Spectrum

Most of what is known about the Sun is derived from analysis of its spectrum. The Sun's visible spectrum, like that of most other stars, consists of a continuum (continuous spectrum) with dark absorption lines. An oversimplified view is that the continuum is emitted from the photosphere, and that the absorption lines are produced as this radiation passes through the more rarefied solar atmosphere above the base of the photosphere. Indeed, it used to be thought that there was one particular layer, called the "reversing layer" where these lines were formed. In fact emission and absorption take place throughout the solar atmosphere (photosphere and chromosphere), but as far as visible light is concerned the effects of absorption become more important with increasing height above the base of the photosphere.

### The electromagnetic spectrum
Visible light spans a range of wavelengths from about 400 to about 700 nm, but the visible or optical range of wavelengths represents only a tiny band in a much wider range of wavelengths making up the entire spectrum of electromagnetic radiation. Electromagnetic waves of all kinds travel through a vacuum at a constant speed, the velocity of light, equal to 299,792.5 km s$^{-1}$. They range in wavelength from a microscopic fraction of 1 nm to many kilometers; the complete range of wavelengths has been divided arbitrarily (from shortest to longest wavelength) into γ-ray, X-ray, ultraviolet, visible, infrared, microwave and radio. Other subdivisions encountered in solar astronomy include extreme ultraviolet (EUV), which spans from 10 nm to about 120 nm, soft X-ray (0.1 to 10 nm), and hard X-ray (less than 0.1 nm), while at wavelengths longer than visible (10 μm to 1 mm) the term "far infrared" is used, and the term "radio" is often used for any wavelength longer than 1 mm.

### Black body radiation
A "black body" is the name given to an idealized object that is a theoretically perfect emitter of radiation. Such a body absorbs all radiation falling on its surface and reflects nothing; it will also emit, in a characteristic way, all the energy that is supplied to it; in other words, a good absorber is also a good emitter. A black body will emit all kinds of radiation, but the amount of energy emitted at different wavelengths depends upon its temperature.

A graph of the amount of energy plotted against wavelength follows a distinctive curve called a "black body distribution curve" (see diagram 1); for any given temperature there is one particular wavelength of peak emission, and the intensity of emitted radiation drops off sharply at shorter wavelengths and more gradually at longer ones. According to the Wien displacement law, the wavelength of peak emission is inversely proportional to temperature: the higher the temperature the shorter the wavelength of peak emission. The color of a hot body, which depends on the wavelength of peak emission, is therefore determined by its temperature: in everyday terms, when a lump of metal is heated it first glows dull red, then as the temperature increases it becomes progressively orange, yellow and white hot. A body at room temperature (roughly 300 K) emits most strongly in the infrared at a wavelength of about 10 μm while a 6,000 K black body would radiate mainly in the visible, peaking at about 550 nm in the yellow-green part of the spectrum. A body at 30,000 K would have its peak of emission at around 100 nm—well into the ultraviolet.

Although stars are not ideal black bodies, their continuum emissions can broadly be described by black body curves. The "effective temperature" of a star is the temperature that a black body of the same radius as that star would need to have in order to emit the same *total* quantity of energy as the star. The effective temperature of the Sun, for example, is 5,780 K. As a result, the Sun emits most strongly in the middle of the visible range. Observed from a distance, the Sun would appear as a yellow star.

### The absorption lines
The solar spectrum reaching the Earth includes thousands of dark lines originating in the solar atmosphere. It is further complicated by additional absorption lines produced in the Earth's atmosphere called "telluric lines".

The way in which absorption lines are produced depends upon the physics of the atom. The simplest atom, hydrogen, may be visualized as consisting of a central proton (a heavy particle of

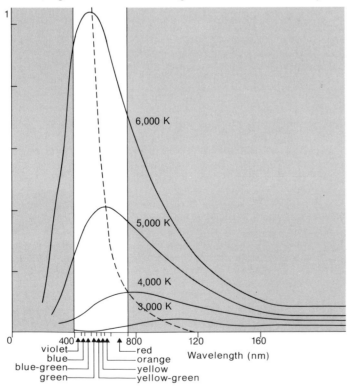

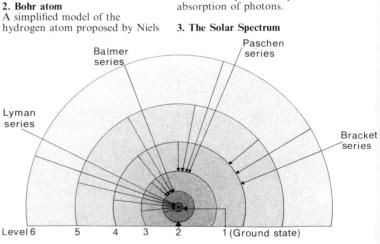

**1. Black-body radiation**
The distribution of energy flux against wavelength of a radiating black body follows a distinctive curve: for any given temperature the curve is unique. As the temperature of the body rises the total amount of energy emitted increases rapidly, and the wavelength of peak emission shifts towards the blue (as shown by the dotted line).

**2. Bohr atom**
A simplified model of the hydrogen atom proposed by Niels

Bohr helps to account for the production of spectral lines. The atom's electron is represented as orbiting the nucleus; only certain orbits are possible, each with its own energy level. When an electron makes a transition from a higher to a lower level, a photon (whose energy corresponds to the difference between the two levels) is emitted, producing an emission feature; conversely, absorption features are produced by the absorption of photons.

**3. The Solar Spectrum**

positive electrical charge) around which there circulates a single electron (a lighter particle of negative charge). According to quantum theory, this electron may exist only in one of a number of orbits, each of which corresponds to a certain energy level of the atom; the atom has least energy when the electron is in the lowest orbit—the "ground state". In some respects light behaves like a wave but in others it behaves like a beam of particles; a "particle" of light, or "photon", has a certain amount of energy corresponding to the wavelength of the light. In fact the energy is inversely proportional to wavelength, so that the shorter the wavelength, the higher the energy of the photon.

If the energy of an incoming photon corresponds to the energy difference between one electron orbit and another, then the photon may be absorbed by an atom. When a photon is absorbed in this way, the electron jumps (makes an "upward transition") to a higher orbit. On the other hand, an electron in a high orbit can fall down to a lower one, and the difference in energy is released in the form of a photon with a wavelength that again corresponds to the energy difference between the two orbits. In essence, then, absorption of light occurs when electrons make upward transitions, and emission occurs when they make downward transitions.

Many different transitions can take place in a hydrogen atom (see diagram 2). For example, an electron falling from level 3 to level 2 releases red light at a wavelength of 656.3 nm, while an electron falling from level 4 to level 2 emits at a wavelength of 486.1 nm. The greater energy difference corresponds to a shorter wavelength. The complete series of lines corresponding to all possible transitions to level 2 is known as the Balmer series, the lines being labelled $H_\alpha$ (at 656.3 nm), $H_\beta$ (486.1 nm) and so on. Other series occur in different regions of the spectrum. The Lyman series, for example, involving transitions to the ground state, is found in the ultraviolet, while the Paschen and Brackett series occur in the infrared.

Each chemical element has its own characteristic series of lines which allows its "fingerprint" to be recognized in the solar spectrum. The detailed analysis of the spectrum, however, is complicated by many factors. Heavier atoms have more electrons and can exist in various states of "ionization" (an atom is said to be ionized when it loses or gains one or more electrons). The state of ionization is indicated by means of Roman numerals; for example, ordinary neutral hydrogen is denoted by HI while ionized hydrogen, which has lost its one electron, is represented by HII, and iron which had lost, say, 13 electrons would be symbolized by FeXIV. The level of ionization affects the pattern of absorption features so that it is possible to determine from an examination of the spectrum not only which elements are present, but also their state of ionization. This information in turn can be used to measure the temperature of the gas because extremely high levels of ionization are known to be produced at very high temperatures. Molecules, moreover, have complex spectra.

**Doppler shift**
In practice a spectral line is not a perfectly black line at one precise wavelength. If the intensity of its absorption of the continuum is mapped against wavelength, it is usually found that the line has a shape, or "profile", dipping down slowly at first and then sharply as the central wavelength of the line is approached; with increasing wavelength the amount of absorption drops off sharply at first and then more slowly. The line appears to have a central core with "wings" on either side (see diagram 2). One of the factors responsible for spreading out the line over a range of wavelengths is the "Doppler effect": the movement of a source relative to an observer causes a shift in the observed wavelength. In a gas, the motions of the absorbing atoms, due essentially to the temperature of the gas, cause small changes in the wavelengths at which absorption is observed. An approaching atom will be seen to absorb (or emit) at a shorter wavelength than a stationary atom, while a receding atom will absorb or emit at a longer wavelength. The random motion of atoms in a hot gas ensures a spread of wavelengths on either side of the mean value.

Other factors also play a part. Pressure broadening, whereby collisions between atoms and electrons change the perceived wavelength, the presence of electric and magnetic fields (which can lead to the splitting of the line into two or more components) and other factors all contribute to the line profile (see pages 30–31).

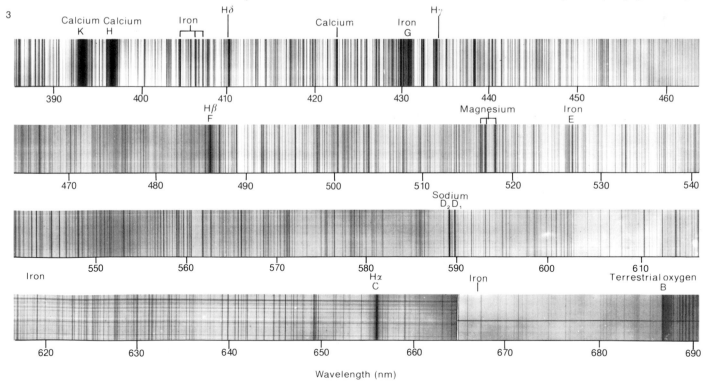

Wavelength (nm)

# Solar Radiation

Analysis of the solar spectrum can reveal a wealth of information about the Sun and its physical properties. By investigating the spectrum, for example, it is possible to deduce facts about the conditions at different heights above the base of the photosphere: at any given wavelength it is possible to "see" down into the solar atmosphere only as far as the level at which it becomes opaque to that particular radiation. Bearing in mind that a good absorber is also a good emitter, this implies that the greatest part of the radiation received at a given wavelength originates from the level at which the solar atmosphere becomes opaque to that wavelength (the "emission level"). From below that level, the radiation is absorbed too effectively to reach Earth, while above that level the solar atmosphere is transparent, absorbing and emitting very little.

At the central wavelength of a very dark absorption line the solar atmosphere is highly opaque, and so any radiation received must originate from a relatively high level; towards the "wings" the absorption is less and the radiation received originates at a lower level. The continuum radiation comes from the low photosphere—at least in the visible part of the spectrum. Ultraviolet radiation is received primarily from the chromosphere, the photosphere being opaque to most of it. At the shortest radio wavelengths (in the millimeter and centimeter ranges) radiation is received from just above the photosphere, but at longer wavelengths the solar atmosphere is opaque at higher levels; at meter wavelengths the radiation comes from the low corona.

## Radio emissions from the Sun

Although the black body curve for a body with a temperature of around 6,000 K, continued into the radio region, would suggest that the Sun ought to be a very weak radio emitter, in fact observations show that the radio output of the Sun is a highly variable quantity that can flare up at certain wavelengths by factors as large as 10,000.

Solar radio emission consists of three main components:
(1) The "quiet Sun" emission is the total background output of the Sun, excluding discrete (localized) sources, and takes the form of "thermal" (black body) radiation emitted from the randomly moving particles in a hot gas. The apparent "brightness temperature" of this radiation ranges from about 6,000 K at millimeter wavelengths to over $10^6$ K at meter wavelengths, the different temperatures relating to different levels in the atmosphere. (The brightness temperature associated with radiation of a particular wavelength is the temperature a black body of the same size as the Sun would need to have in order to emit the measured quantity of radiation at that wavelength.)
(2) The slowly varying component (S-component) is again thermal radiation, but is emitted from localized regions of the solar atmosphere, the total quantity of radiation emitted by the whole Sun seldom much exceeding the "quiet Sun" level, but being locally intense and depending on the level of solar activity. It is most prominent in the "decimeter" wavelength range between about 10 and 50 cm (frequencies of around 3 GHz to 600 MHz).
(3) Radio bursts can occur over the entire radio spectrum over timescales ranging from less than 1 sec up to several hours. Their power outputs can exceed the quiet Sun level by factors of 1,000 to 10,000, the radiation being mainly of a nonthermal nature: it is emitted by electrons which, instead of moving at random as in a hot gas, are directed in their motions, under the influence of, for example, magnetic fields. One important type of nonthermal process is "synchrotron radiation": electrons moving at a large fraction of the speed of light in strong magnetic fields are forced to spiral around the magnetic lines of force and, as a result, emit radiation in a narrow cone along the direction in which they are moving. Solar radio radiation is polarized, usually due to the role of the magnetic field.

Despite the spectacular nature of these events, the overall contribution of the radio bursts to the total radio output over a long period is quite small.

## Infrared radiation

At wavelengths from about 750 nm to just under 1 mm the observed radiation originates in the photosphere and lower chromosphere

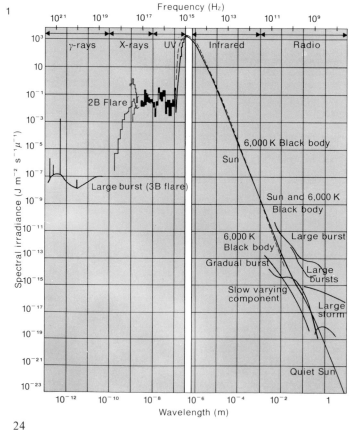

1

**1. Spectrum of solar radiation**
As can be seen from this graph, the peak intensity of solar emission is in the narrow band of visible wavelengths. At wavelengths between about 100 nm and 1 cm, radiation from the Sun corresponds quite closely to the radiation that would be emitted by a "black body" with a temperature of 6,000 K. (The curve for a black body at this temperature is shown as a dotted line in the graph.) At shorter wavelengths, extending from the ultraviolet to the shortest γ-rays, and at very long wavelengths, solar radiation deviates significantly from this pattern: the black body temperature to which these extremes of emission correspond is closer to $10^6$ K. Unlike the central portion of the spectrum, the intensity of radiation in the short and long wavelengths varies considerably, depending on the level of solar activity. At radio wavelengths, for example, there is a slowly varying component which changes with the solar cycle (*see* pages 36–7), while different types of flare (*see* pages 42–5) produce distinctive emissions; the strong emission lines that appear in the γ-ray region are shown in the graph.

**2. Absorption line profile**
If a graph is plotted of intensity against wavelength, it is found that the cut-off of an absorption feature is not abrupt but slightly spread out over a range of wavelengths. In the visible spectrum the center of the line is, generally, formed higher in the atmosphere than the "wings" of the line, which are formed at deeper regions where the temperature (and consequently the level of emission) is relatively high. Thus, by making monochromatic images of the Sun at wavelengths slightly off the centers of absorption lines it is possible to observe a range of depths into the solar atmosphere.

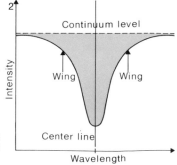

and follows quite closely the black body curve for a body in the temperature range 6,000 K to 4,000 K (the lower temperatures correspond to radiation coming from the region of the temperature minimum some 500 km above the base of the photosphere).

## Ultraviolet radiation
In the near ultraviolet (around 300 nm) the radiation originates, like visible light, in the photosphere, but at progressively shorter wavelengths the source of the ultraviolet continuum shifts to the chromosphere. At wavelengths shorter than about 140 nm the solar material is sufficiently opaque to radiation from lower levels for it to become possible to see emission lines from the chromosphere itself. Thus, below that wavelength, the spectrum changes from a bright continuum with dark-absorption lines to a faint continuum with bright emission lines. Studies of the Sun in the wavelengths of ultraviolet emission lines allow higher and higher layers of the chromosphere and transition region to be investigated. One particularly bright line is the hydrogen line Lyman-alpha ($L_\alpha$, the first line of the Lyman series) at 121.6 nm which emits more energy than the entire solar spectrum between 0 and 120 nm; this radiation exerts important effects on the Earth's atmosphere and on other Solar System phenomena.

## Extreme ultraviolet (EUV) and soft X-ray radiation
Radiation in the 1 nm to 120 nm range also has important terrestrial effects. It originates in the chromosphere, transition region and corona, and at wavelengths of 1 to 2 nm can show variations in intensity over the solar activity cycle of factors in excess of 100. The soft X-rays are emitted by hotter, denser concentrations of gas in the corona (with a temperature in excess of $10^6$ K) and a study of these radiations shows the detailed structure of the corona across and beyond the solar disc.

Hard X-rays (wavelengths less than 0.1 nm) originate primarily in solar flares, the intensity of such emission fluctuating widely by as large a factor as 10,000 depending on the level of solar activity. These outbursts have significant effects on the Earth's atmosphere.

## The solar constant
The total amount of energy per second over all wavelengths that would be received at the top of the Earth's atmosphere when the Earth is located at its mean distance from the Sun is called the "solar constant". Although this quantity is of the utmost importance for the existence of life on Earth, surprisingly its value is not yet known to an accuracy of better than 0.5 percent, and it is unclear by how much—if at all—the value of this "constant" varies with time. The difficulty arises mainly because, until comparatively recently, all measurements had to be made from the Earth's surface and allowance had to be made (often incorrectly) for the fact that the atmosphere obscures many wavelengths and exerts a variable absorption on the others. There is also a problem of calibration, since it is hard to find a reference standard for a source as bright as the Sun. One of the principal experiments on the Solar Maximum Mission was devoted to improving knowledge of this quantity and assessing variability (*see* pages 424–5).

The presently accepted value of the solar constant is about 1,370 W m$^{-2}$; in other words, a surface of area 1 m$^2$ held perpendicular to the Sun's rays should receive 1,370 W of power at the mean distance from the Sun to the Earth. There is an uncertainty of nearly $\pm$ 10 W on this figure—the mean of values measured for the first 153 days of the SMM was 1,368 W m$^{-2}$.

Overwhelmingly the greatest part of solar radiation is emitted in the visible and near infrared part of the spectrum, and so the major contribution to the solar constant is in those wavelengths. In fact, about 99 percent of solar radiation is emitted in the wavelength range 300 nm to 6,000 nm; just under 1 percent is contributed by ultraviolet radiation in the range 120 nm to 300 nm, and all other wavelengths contribute only a tiny fraction of 1 percent. The solar output in those wavelengths that contributes significantly to the solar constant does seem to be constant to within 1 percent. The radio and X-ray wavelengths, which show huge fluctuations, contribute insignificantly to the total. A variation of $\pm$ 3.5 percent from the average value arises annually, due to the elliptical form of the Earth's orbit.

**3. Radio map of the Sun**
This contour map shows the Sun at a wavelength of 3 cm. Active regions appear as the sources of enhanced radio emission.

**4. Synchrotron radiation**
When an electron moving at a "relativistic" velocity (a large fraction of the speed of light) spirals around magnetic field lines, it emits a narrow beam of strongly polarized radiation. The wavelength is dependent on the strength of the field and the velocity of the electron. This process is an important source of radio bursts emitted by the Sun as well as by many other astronomical radio sources.

**5. Contrasting solar images**
Spectroheliograms are used to show the Sun at various wavelengths, such as H$\alpha$ at 656.28 nm (**A**) or the K-line of doubly ionized calcium at 393.37 nm (**B**). (**C**) is an X-ray image, while (**D**) is a magnetogram.

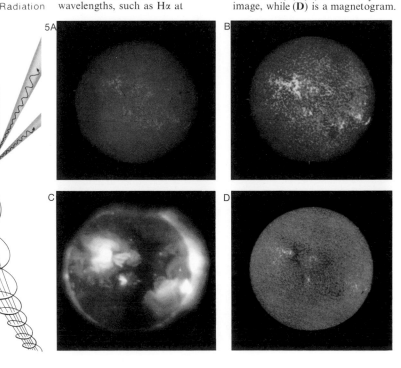

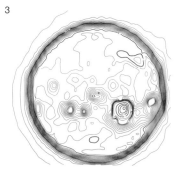

# Photosphere

The photosphere is the layer from which practically all the Sun's visible light is emitted, and as such represents what is normally called the "surface" of the Sun. Although the edge or "limb" of the solar disc appears to be quite sharp in a photograph or in the projected solar image as seen with a telescope (apart from the "boiling" effect due to turbulence in the atmosphere), it is fairly readily apparent that the brightness of the solar disc diminishes near its edge. This phenomenon of "limb darkening" is consistent with the Sun's being a gaseous body having a temperature that decreases with distance from its center.

Since the Sun is gaseous, its photosphere is to some extent transparent: it can be observed down to a depth of a few hundred kilometers before it becomes completely opaque. When an observer looks directly at the center of the solar disc, he is in fact looking through photospheric layers of increasing density and temperature towards the base of the photosphere, where the solar material becomes opaque. Most of the light he sees at the center comes from the hotter, and therefore more luminous, lower layers of the photosphere. However, looking towards the limb, an observer's line of sight is almost at a tangent to the edge of the solar disc, so that he is looking into the cooler, upper layers of the photosphere. Thus although it is again possible to penetrate a certain distance into the photosphere before the solar material becomes opaque, the region at which this occurs is higher above the base of the photosphere than at the center of the disc. The light from the region of the limb, therefore, comes from higher layers which, being at a lower temperature, emit less strongly than the deeper layers at the base of the photosphere: consequently the limb appears less bright than the center.

The degree of limb darkening is not the same for all wavelengths. It is most apparent in the blue and violet regions of the spectrum, because the intensity of blue light diminishes with decreasing temperature more rapidly than does the intensity of red light (*see* pages 22–3). At radio wavelengths and at shorter ultraviolet and X-ray wavelengths the opposite effect, "limb brightening", is observed. These types of radiation emerge primarily from layers in the solar atmosphere well above the photosphere, in the regions where temperature increases with height instead of decreasing as it does in the photosphere. As a result the radiation from the limb regions (which again become opaque at higher levels than the central regions) comes from *brighter* high-temperature layers. The limb therefore appears brighter than the center. In the case of X-ray sources in the corona, absorption is not important but there are more X-ray emitting sources to be seen along the line of sight when looking beyond the edge of the Sun than when looking towards the center; again, the limb regions appear brighter.

The pressure and density of the gas in the lower layers of the photosphere are, respectively, only about 10 percent and 0.01 percent of the values pertaining to the Earth's atmosphere at sea level. Terrestrial air is highly transparent to visible light; why should the photospheric material become opaque at such low densities and pressures? The *opacity* of a gas is a measure of its ability to absorb light, and the value of this quantity depends upon the state in which the gas exists. For example, a gas made up of neutral hydrogen atoms in their lowest energy states interacts very weakly with photons, and so is highly transparent; on the other hand, a hotter hydrogen gas is more ionized and excited, and interacts strongly with optical photons, absorbing them very effectively. The deep interior of the Sun is completely ionized and highly opaque.

The photospheric material is opaque mainly as a result of the presence of negative hydrogen ions ($H^-$) (hydrogen atoms that have captured an extra electron). A photon emerging from the interior which encounters an $H^-$ will be absorbed, and the extra electron will be ejected in the process. When at a later moment the resulting neutral hydrogen atom recaptures an extra electron a photon is

**1. Limb darkening**
The apparent fall in brightness away from the center of the Sun's disc is evident in this white light photograph taken on 9 April 1970. The solar atmosphere is partially transparent, and at the center of the disc it is possible to see through the upper layers down to a region where the temperature is higher and emission correspondingly brighter; at the limb, however, high temperature layers do not fall along the oblique line of sight and the limb therefore appears darker.

**2. Granulation**
The textured appearance of the photosphere, described as "solar granulation", results from the turbulent motion of hot gas rising and falling as heat is transported from the interior by means of convection. Each granule is about 1,000 km across and represents the top of a rising column of hot gas. Downward motion occurs in the dark lane between cells. The mean lifetime of a granular cell is approximately 10 min. This photograph was taken on 19 August 1975.

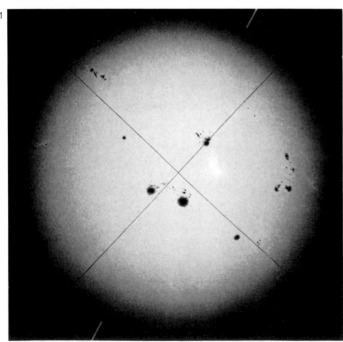

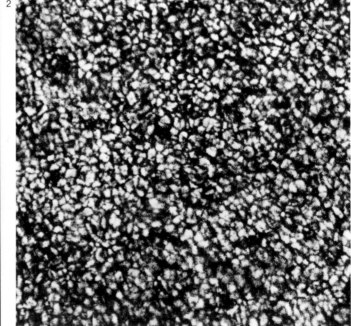

emitted, possibly with a different wavelength from the originally "captured" photon. In this way radiant energy from the interior is processed by the photosphere and emerges as visible light. The concentration of H⁻ drops off very rapidly with decreasing temperature, so that with increasing height above the base of the photosphere the opacity drops very sharply and emitted photons have increasing chances of escaping into space without being captured again. The drop in opacity is so rapid that the major part of the emitted light comes from a layer only about 100 km thick.

### Granulation and supergranulation
The whole photosphere is a seething mass of bright, shifting granules (*see* pages 18–19) whose average lifetime is about 8 min. The individual granules are typically about 1,000 km in diameter, corresponding to angular sizes, as seen from the Earth, of between 1 and 2 seconds of arc. Theoretically such features should be discernible in a telescope of about 10 cm aperture, but the turbulence of the atmosphere normally precludes this. Each granule represents a region in the center of which hot gas is rising from the interior (with a vertical velocity of about 0.5 km s⁻¹), then spreading horizontally outward at the top of the cell at speeds of the order of 0.25 km s⁻¹. Cooler gas descends at the edge of each cell in the intergranular lanes.

On a much larger scale there is a network of "supergranular" cells with typical cell diameters of about 30,000 km; each super-granular cell contains hundreds of individual granules. Being larger

patterns of circulation that extend deeper into the convective zone they survive longer than ordinary granules, with typical lifetimes of between 12 and 24 hr. The solar magnetic field is enhanced around the boundaries of supergranular cells, giving rise to the "chromo-spheric network" in the chromosphere (*see* page 28).

The pattern of supergranular cells is readily revealed in what are known as "velocity-cancelled spectroheliograms". By the Doppler effect, light from a source approaching the observer is blue-shifted (shortened in wavelength) while light from a receding source is red-shifted (lengthened in wavelength). Since supergranular cells exhibit a horizontal motion of gas out from their centers, it follows that some of the gas will be approaching the observer and matter in other parts of each cell will be receding. If one spectroheliogram is made at a wavelength a little to the red side of the center of a spectral line and another is made a little to the blue side, the red-shifted and blue-shifted regions of gas will be revealed. The pattern of motions is enhanced if the two spectroheliograms are combined photographically in an appropriate way, revealing the super-granular cells. Corrections are made to allow for solar rotation.

Telescopically, the most obvious signs of activity on the photo-sphere are the sunspots (*see* pages 32–7). Associated with sunspots, however, are "faculae", brighter than average patches of light in the upper regions of the photosphere which usually appear in the vicinity of a sunspot group before the sunspots themselves emerge, and which usually persist for several weeks after the disappearance of the spots.

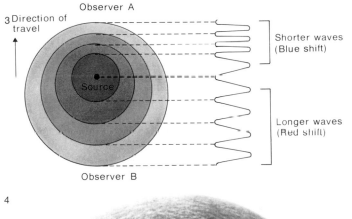

3 Direction of travel

Observer A

Source

Shorter waves (Blue shift)

Longer waves (Red shift)

Observer B

**3. Doppler effect**
The movement of a source of light relative to the observer produces an apparent shift in the wavelength (or color) of the light seen by the observer. If the source is moving towards observer A then it will appear that the distance between successive crests is compressed, producing a shift towards blue; if, instead, the source is moving away to observer B, the distance between crests will appear longer and the light will look redder than if the source had been stationary.

**4. Supergranulation**
Large-scale convective motion produces "supergranular" cells similar to granulation but about 180 times larger and with longer lifetimes. Supergranulation is shown up in this velocity-cancelled spectroheliogram, in which receding areas appear as dark regions and approaching areas as light regions.

**5. Faculae**
Bright patches in the photosphere, or faculae, can be seen near the limb of the Sun.

4

5

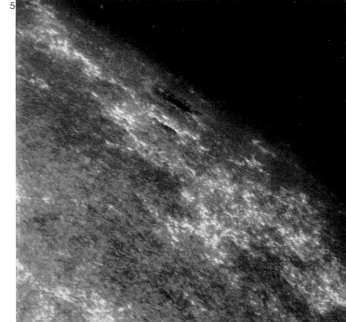

# Chromosphere I

The chromosphere is the tenuous layer of gas, a few thousand kilometers thick, which lies above the photosphere and which derives its name (literally, "color sphere") from the characteristic reddish-pink color it displays when it becomes visible for a few seconds at the onset and at the end of the total phase of a solar eclipse. Far too faint to be seen against the brilliance of the photosphere, the chromosphere may be observed visually only during a total eclipse, but it can be studied at any time with the aid of a spectroscope, the slit of which is placed at a tangent to the solar disc, when observations are made in one of the wavelengths at which chromospheric matter emits light. The chromosphere may be examined across the entire solar disc by means of a spectroheliograph or a monochromatic filter.

## Spicules

The chromosphere is not a homogeneous atmospheric layer, but has considerable structure and shows rapidly changing aspects both locally and on the large scale. In monochromatic light, the edge of the Sun is seen to be made up of large numbers of flame-like protrusions, known as "spicules", which rise and fall, looking like blades of grass blowing in the wind; their appearance has been described as a "burning prairie". Spicules are near-vertical cylinders of chromospheric gas which follow the directions of the local magnetic fields. Their temperatures typically range from 10,000 to 20,000 K; they are about 1,000 km broad and rise to heights of about 10,000 km. These jets of gas rise from the lower chromosphere with velocities of 20 to 30 km s⁻¹, most of their material eventually falling back and, possibly, contributing to the heating of the chromosphere in the process. Their average lifetimes are between 5 and 10 min.

According to one theory the energy stored in localized magnetic fields is responsible for throwing up spicular material; there is no doubt that local fields influence the paths along which this material moves. Spicules are seen as bright emission features at the limb, but it is possible to see them as dark absorbing features against the disc. At any one time there may be about 500,000 of them on the complete solar surface.

## Chromospheric network

The chromosphere exhibits a large-scale cellular structure known as the "chromospheric network". The cells making up the network coincide in position with the supergranules of the photosphere, their boundaries being determined by the enhanced magnetic fields that occur at the perimeters of supergranular cells. In the photosphere the density of matter is sufficiently great for the flow of hot plasma to sweep the solar magnetic fields along with it. The outward horizontal flow of material from the centers to the edges of supergranular cells thus sweeps the magnetic fields to the boundaries of the cells where they are concentrated and amplified; vertical magnetic fields of 1,000 to 2,000 G are found at the corners. The lower density plasma of the chromosphere is governed in its behavior by the magnetic fields acting upon it; hence the chromospheric network is determined by the underlying magnetic fields. Spicules rise up in bunches in the concentrated fields at the corners of the network.

The network takes the form of a pattern of bright emission features seen in monochromatic light. The pattern is not normally seen at altitudes greater than about 10,000 km above the photosphere (*see* page 51); beyond this level, the clumps of spicules at the cell boundaries spread out, blurring the pattern.

Skylab ultraviolet observations show that the chromosphere takes on a different appearance in polar regions. The chromospheric network cuts off at the edges of these regions, and darker polar caps are revealed coinciding with long-lasting "coronal holes" out of which solar wind particles are streaming (*see* pages 68–9). The height of the transition region jumps up by as much as

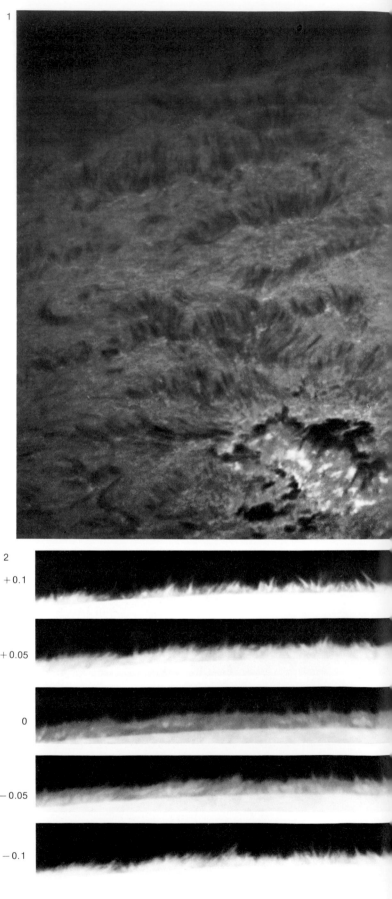

1

2

+0.1

+0.05

0

−0.05

−0.1

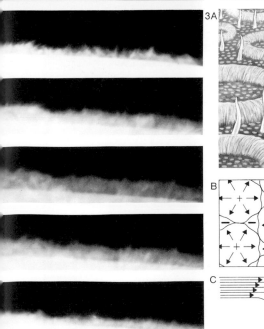

## 1. Spicules

When seen at the limb, spicules are visible as emission features, but against the relatively bright solar disc they show up as dark blade-like absorption features. Bunches of spicules at the edges of supergranular cells outline the chromospheric network. Some distance from the limb, the bunches of spicules make up patterns known as "rosettes"; the bases of these rosettes are marked by small bright mottles. Closer to the limb the rosettes are foreshortened so that the bunches of spicules look like "bushes". The material comprising spicules rises up from the lower chromosphere at speeds of some 20 to 30 km s$^{-1}$, reaching heights of about 10,000 km before falling back. This filtergram was taken on 13 February 1971 at a wavelength 0.08 nm longer than the central wavelength of Hα.

## 2. Spicule sequence

At the center of the Hα line many spicules overlap to give the impression that the chromosphere is a fairly uniform layer. At slightly longer and slightly shorter wavelengths, however, individual spicules are more readily seen. This sequence, taken at intervals of 0.05 nm, was made on 14 October 1970.

## 3. Structure of spicules

At the boundaries of supergranular cells (**A**) material from the solar interior rises, spreads out, then descends, just as it does in the ordinary small-scale granules. The direction of flow of material in these cells is illustrated (**B**)—the "+" signs indicate rising gas, the arrows indicate horizontal radial flow, and the

10,000 km as the solar atmosphere thins out and expands over these zones. Spicules reach greater heights over the poles, and giant spicules ("macrospicules") have been found in these regions. The macrospicules contain chromospheric material at temperatures in the region of 50,000 K and rise to heights in excess of 40,000 km. They endure typically for five to ten times the average lifetime of ordinary spicules (in other words for about 40 to 50 min).

It is thought that differential motion between adjacent supergranular cells is responsible for diffusing part of the localized magnetic fields towards the polar regions. North and south polarities show a net overall tendency to collect preferentially at opposite poles. In this way the impression is created that the Sun has a general magnetic field at the poles, broadly similar to the terrestrial field, and having a strength of about 5 or 6 G (compared to 0.6 G for the Earth); indeed, for a long time astronomers were convinced that this was the case. However, it now appears that the Sun does not have an overall "dipole" field like the Earth (the sort of field that would result from a bar magnet being buried inside its globe), but only a complex array of localized surface fields (*see* pages 46–8).

Other features of the chromosphere, visible in monochromatic light, include the "fibrils", short-lived (10 to 20 min) essentially horizontal strands of gas, typically 10,000 km long and 1,000 to 2,000 km wide, which show up as dark absorption features. The pattern of fibrils is dictated by the local magnetic fields, this being particularly obvious in and around active regions. The chromosphere in monochromatic light is seen to be the seat of many aspects of solar activity, including flares, prominences, filaments, and "plages" (bright patches at higher temperatures than their surroundings).

"−" signs indicate downflow. The downflow tends to be more strongly concentrated into funnels where cells meet; the direction of flow at such a boundary is indicated (**C**). The horizontal flow of the solar plasma sweeps magnetic field lines to the edges of cells, and from these regions the spicules arise. This pattern determines the structure of the chromospheric network.

## 4. Macrospicules

Macrospicules resemble spicules but are considerably larger; they are found in the polar regions. This sequence of ultraviolet brightness contours is based on readings made on 11 December 1973. It shows at intervals of a few minutes how macrospicules rise and fall, reaching heights of about 40,000 km, some four times greater than ordinary spicules.

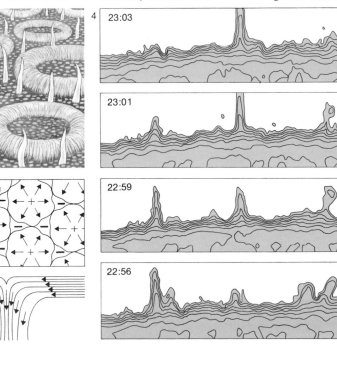

# Chromosphere II

During an eclipse, just as the photosphere disappears from view behind the Moon, the spectrum of the chromosphere flashes into view; because of the suddenness of its appearance, this phenomenon is known as the "flash spectrum" and lasts for only a few seconds before the chromospheric light is cut off. It was first observed by the American astronomer Charles Young in December 1870.

The visible spectrum of the chromosphere consists of a very weak continuum upon which is superimposed many bright emission lines; about 3,500 lines in all have been identified. It is weak because the chromospheric gas is highly transparent to most visible wavelengths (and, as a poor absorber, it is also a poor emitter). When the solar disc is looked at directly it is impossible to see the chromospheric emission lines against the intensely luminous photosphere. Instead, the pattern of dark absorption lines is seen, arising from the absorption that takes place in the photosphere and chromosphere. When, as at a total eclipse, chromospheric gas can be seen at the limb without the background photospheric light, its lines show up clearly in emission.

Many of the lines in the chromospheric spectrum coincide with familiar absorption lines in the photospheric spectrum. $H_\alpha$ at a wavelength of 656.3 nm in the red part of the spectrum is particularly prominent, and gives the chromosphere its characteristic color. Other lines are not present in the photospheric spectrum, such as, for example, the lines of neutral and ionized

helium, and of some ionized metals. The presence of these lines indicates that some parts of the chromosphere are at appreciably higher temperatures than the photosphere. Temperature and pressure are crucial factors in determining which lines are present, and in influencing their appearance (broad, narrow, faint, strong) (see pages 22–3). Since both temperature and pressure change rapidly with altitude, each level displays its own characteristic spectrum. Thus, different levels in the solar atmosphere may be sampled by examining a strong Fraunhofer line and looking at wavelengths a little longer or shorter than the center of the line (see page 24). By selecting lines produced at different temperatures, typical of different altitudes, it is also possible to scan through the solar atmosphere and to build up a picture of its vertical structure.

For example, light emitted in the ultraviolet by singly ionized helium (HeII) at a wavelength of 30.4 nm comes predominantly from a layer with a temperature in a region around 80,000 to 90,000 K. This layer is significantly higher than the level from which most of the $H_\alpha$ radiation is emitted. Triply ionized oxygen (OIV) at temperatures in excess of 100,000 K would radiate from even higher levels of the atmosphere.

It is found that the temperature *increases* very rapidly from a minimum of about 4,200 K in the lower reaches of the chromosphere to some 500,000 K towards the top of the transition zone, the average rate of increase being about 200 K km$^{-1}$. The density

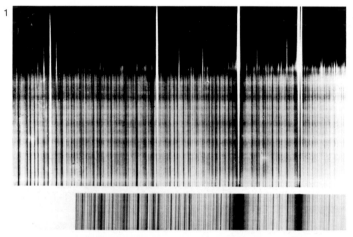

**1. Flash spectrum**
The flash spectrum of the chromosphere is shown with the photospheric spectrum (below) for comparison. During a total solar eclipse the normal photospheric spectrum—a continuous spectrum with dark absorption lines—is visible until the last sliver of the solar disc disappears behind the Moon. At that instant, the emission spectrum of the chromosphere, normally too faint to be seen against that of the photosphere, flashes into view for a few seconds until the chromosphere, too, is obscured by the Moon. Most of the emission lines have the same wavelengths as the dark lines in the photospheric spectrum.

**2. Layers of the chromosphere**
These spectroheliograms taken on 11 September 1961 show the different appearance of the chromosphere at four different wavelengths: $H_\alpha$ (**A**), $H_\alpha + 0.035$ nm (**B**), $H_\alpha + 0.07$ nm (**C**), and the K-line of calcium (**D**). At the center of the $H_\alpha$ line (**A**), areas of solar activity such as dark filaments and bright plages (also revealed in calcium K) are seen. The filaments curve towards the sunspots following the neutral line (see page 34). At slightly longer wavelengths (**B** and **C**) the filaments and plages fade from view. Structures at slightly higher levels of the chromosphere are shown up, since these features emit light over a relatively broad

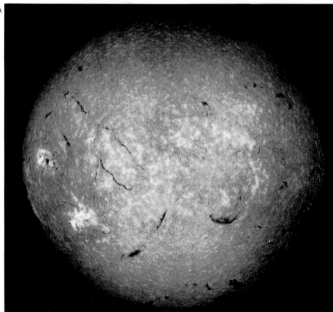

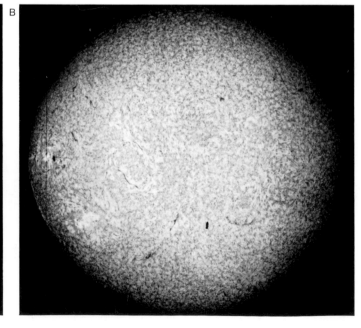

however *drops* equally dramatically from more than $10^{21}$ particles per cubic meter ($10^{21}$ m$^{-3}$) to about $10^{15}$ m$^{-3}$ (from about $10^{-5}$ to about $10^{-10}$ times the density of air at sea level).

The temperature in the solar atmosphere, then, drops from about 6,000 K at the base of the photosphere to the temperature minimum of some 4,200 K at an altitude of about 500 km, and thereafter increases sharply to well in excess of $10^6$ K in the lower corona. Why should this be so? According to the laws of thermodynamics, heat cannot flow from a cooler body to a hotter. It should not be possible for radiation emanating from a relatively cool body (the photosphere) to raise surrounding (coronal) material to a higher temperature than that of the emitting body. Solar energy is generated deep in the solar core and passes out through the Sun to be radiated from the photosphere at a temperature of about 6,000 K. How can the relatively low temperature of the photosphere be reconciled with the very much higher temperature of the chromosphere and corona?

In fact, the chromosphere and corona are *not* heated directly by radiation from the photosphere; the solar atmosphere above the photosphere is highly transparent to the optical and infrared radiations which comprise most of the radiation emitted from the photosphere, and as a result these radiations do not transfer significant quantities of energy to the chromosphere and corona. A popular view is that energy is transferred to the chromosphere (and to the corona) primarily by *mechanical* means: bulk kinetic energy is transmitted to the solar atmosphere and deposited there in the form of heat. What probably happens is that low frequency sound waves generated in the turbulent convective zone (rather like the rumble of thunder) move out through the photosphere and chromosphere, and the rapid decrease in density causes these waves to accelerate rapidly to become shock waves, rather like those generated by a supersonic aircraft. These shocks, moving through the chromosphere and corona, induce energetic collisions between particles with consequent heating effects.

Energy is also believed to be transmitted through the magnetic field in the form of "magnetohydrodynamic" (MHD) waves, which cause charged particles to oscillate violently. The resulting collisions contribute to raising the temperature. Another contributory effect is the downward conduction of heat from the corona, although this mechanism begs the question of why the corona is itself hotter than the chromosphere. Other mechanisms which may have a role to play include the effects of infalling material (most of which would have been hurled upwards prior to falling back), and magnetic effects, whereby currents flowing in the solar magnetic fields experience a degree of resistance and so dissipate heat in a fashion analogous to an electric bar heater. However, there is at present no completely satisfactory account of the mechanism of chromospheric heating.

waveband; at Hα+0.07 nm, for example, practically all that can be seen is the pattern of spicules outlining the chromospheric network. At the same time, however, the umbra and penumbra of the sunspots (which exist at deeper levels) can also be seen clearly, because the chromosphere is less opaque in the "wings" of the line than it is at the center of the Hα line. A band of plages appears concentrated near the equator (**D**).

**3. Solar limb**
The lower portions of these small prominences (*see* pages 40–41) are obscured by layers of the chromosphere, here seen edge-on in Hα light.

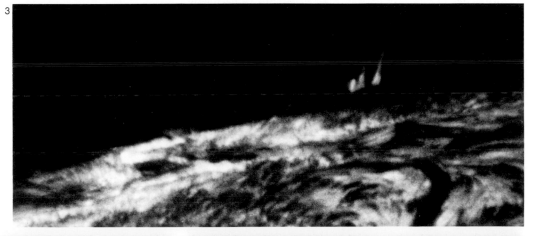

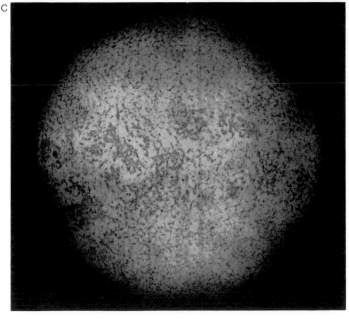

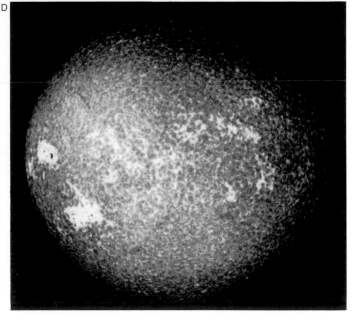

# Sunspots I

The existence of dark "sunspots" on the photosphere provides the most obvious evidence of activity on the Sun. Sunspots were first observed telescopically in late 1610. Although occasionally seen by naked-eye observers over the past two millenia, the fact that they were phenomena actually on the face of the Sun itself was not recognized until the era of the telescope began in the early years of the seventeenth century with the observations of Galileo, Scheiner and their contemporaries.

## Characteristics of sunspots

A typical sunspot consists of a darker central region—the "umbra"—surrounded by a lighter "penumbra". The diameter of the penumbra is, on average, about 2.5 times that of the umbra, and the penumbra can account for as much as 80 percent of the total sunspot area. The penumbra is made up of a pattern of lighter and darker filaments spreading out approximately radially from the umbra. Umbra and penumbra appear dark by contrast with the brighter photosphere because they are cooler than the average photospheric temperature; the central umbra has a temperature of around 4,000 K compared to about 5,600 K for the penumbra and about 6,000 K for the photospheric granules. Since the amount of radiation emitted by a hot body is proportional to the fourth power of its effective temperature, the umbra emits only about 30 percent of the light emitted by an equal area of photosphere, and the penumbra has a brightness of about 70 percent of the photospheric value. The darkness of a sunspot is merely an effect of contrast; if a typical sunspot, with an umbra about the size of the Earth, could be seen in isolation at the same distance as the Sun, it would shine about 50 times more brightly than the full Moon.

Sunspots range in size from tiny pores about the size of individual granules (about 1,000 km in diameter), which appear as dark spots within penumbras, to complex structures several tens of thousands of kilometers in diameter, covering areas of up to some $10^9$ km². A large group of sunspots may extend over a distance in excess of 100,000 km.

## Classification of sunspots

Various systems have been devised for the classification of sunspots; the one shown on this page (*see* diagram 2) is the "Zurich Spot Classification", which divides sunspots into nine classes designated by the letters A, B, C, D, E, F, G, H and J. Essentially the system relates to the evolutionary stages through which groups of spots pass, although not all spots go through the complete sequence. Starting as a small cluster (type A) or as a bipolar group (B), spots grow rapidly and within a period of 8 to 10 days they arrive at their maximum area (F) after passing through the intermediate stages (C, D and E). Decay takes slightly longer, so that the group spends most of its life in the final stage of the classification (G to J). A group of moderate size, however, might follow an abbreviated sequence of development, while a very small group might not develop beyond the first stage. Associated activity, such as flares, tends to reach a maximum fairly early in the evolution of the group, typically during stages D, E and F. The total lifetime of a large group may be as long as several weeks; the average is less than two weeks.

## Sunspots and solar rotation

The Sun rotates on its axis, but unlike a solid body different parts of the surface rotate in different periods of time. The mean rotation period of the Sun—or of its surface at least—is 25.380 days.

**1. Solar rotation**
This sequence shows the giant sunspot group of 1947. After passing round the far side of the Sun the group reappears to make a second crossing. This group, the largest ever recorded, reached a maximum area of $18 \times 10^9$ km².

**2. Zurich classification scheme**
This system of classification was developed by M. Waldmeier. Additional information may be added by means of a three-letter code (Zpc). The first letter denotes the basic Zurich class (A–J); the second denotes the penumbral type (large, small, etc.) of the largest spot in the group, and the third denotes the degree of compactness of the distribution of the spots.

However, from Earth it appears to be rotating more slowly, because the Earth is moving around the Sun in the same direction as the Sun is spinning on its axis; the apparent mean rotation period (known as the "synodic period") is 27.275 days.

The first means of deducing the period of solar rotation was provided by the motion of sunspots across the solar disc. Observations made on successive days reveal that sunspots change their positions on the visible disc, marching steadily across from east to west (from the eastern to the western limb of the Sun as seen in the sky) as they are carried around by the rotation of the Sun. The apparent paths taken by sunspots across the disc vary throughout the year because the solar equator is tilted to the plane of the ecliptic by an angle of about 7°25.

The position of features such as sunspots on the photosphere is specified by giving their "heliographic" latitudes and longitudes. These are coordinates essentially similar to latitude and longitude on Earth. Heliographic latitude is measured north or south of the solar equator (taking values from 0° to ±90°), while heliographic longitude is measured along or parallel to the solar equator from a standard meridian (an imaginary circle perpendicular to the equator and passing through the poles of the Sun). This meridian is defined as having crossed the center of the visible disc on 1 January 1854 at 12.00 U.T. and is assumed to have been rotating with a uniform period of 25.38 days ever since. The system was devised by Richard Carrington (*see* page 21) and does not relate to any specific visible feature of the solar disc, unlike terrestrial longitude, which operates with reference to the meridian passing through Greenwich; instead heliographic longitude is based on an imaginary meridian which revolves around the center of the Sun at the rate which Carrington assigned to the mean rotation of the Sun.

Moreover, the motion of sunspots at different heliographic latitudes reveals the "differential rotation" of the Sun: the photosphere rotates at different rates at different latitudes, with rotation periods ranging from about 25 days at the equator to about 27 days at latitudes ±30°; still longer periods are found at higher latitudes. Since sunspots are seldom found more than about 40° from the equator, the rotation periods of higher latitude regions are more difficult to determine.

However, the rotation rate of the photosphere may also be deduced from the Doppler effect (*see* page 23) in the wavelengths of spectral lines originating in the trailing and leading limbs at different latitudes. Light from the trailing limb will be blue-shifted, while light from the leading limb will be red-shifted. By measuring the Doppler shift in this way it has been shown that the rotation period ranges from 26 days at the equator to 37 days at the poles. The sunspots appear to rotate faster than the general photospheric background by about 4 to 5 percent.

**3. Differential rotation**
The rotation period of the photosphere increases with increasing latitude. In the idealized situation shown here, if a row of sunspots lay along the central meridian of the Sun (**A**) then, after one rotation, the spots would be spread out (**B**).

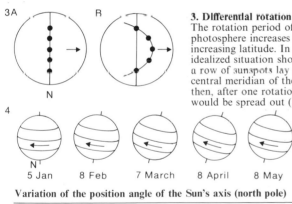

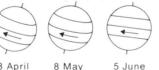

| 5 Jan | 8 Feb | 7 March | 8 April | 8 May | 5 June | | 7 July | 13 Aug | 8 Sept | 11 Oct | 9 Nov | 7 Dec |

**4. Apparent sunspot motions**
In the course of a year the angle between true north and the Sun's axis takes values between 0° and 26°3 east or west (*see* Table below) because of the combined effects of the tilts of the solar and terrestrial equators to the plane of the Earth's orbit.

**5. Rotation rate of the Sun**
In addition to the variations in the Sun's rate of rotation at different latitudes (*see* diagram 3), there are also considerable fluctuations about the mean equatorial value of 13°86 per day. The reason for these changes in the rotation rate are not clear.

**Variation of the position angle of the Sun's axis (north pole)**

| 5 Jan | 0° | 8 April | 26°3 | 7 July | 0° | 11 Oct | 26°3 |
|---|---|---|---|---|---|---|---|
| 16 Jan | 5° | 26 April | 25° | 19 July | 5° | 30 Oct | 25° |
| 27 Jan | 10° | 8 May | 23° | 30 July | 10° | 9 Nov | 23° |
| 8 Feb | 14° | 19 May | 20° | 13 Aug | 14° | 20 Nov | 20° |
| 23 Feb | 20° | 5 June | 14° | 28 Aug | 20° | 7 Dec | 14° |
| 7 March | 23° | 15 June | 10° | 8 Sept | 23° | 16 Dec | 10° |
| 18 March | 25° | 26 June | 5° | 21 Sept | 25° | 26 Dec | 5° |

**Variation of the heliographic latitude of the center of the Sun's disc**

| 1 Jan | 3°0 S | 21 Apr | 5°0 S | 1 July | 3°0 N | 25 Oct | 5°0 N |
|---|---|---|---|---|---|---|---|
| 10 Jan | 4°0 S | 2 May | 4°0 S | 11 July | 4°0 N | 4 Nov | 4°0 N |
| 19 Jan | 5°0 S | 11 May | 3°0 S | 21 July | 5°0 N | 13 Nov | 3°0 N |
| 1 Feb | 6°0 S | 20 May | 2°0 S | 3 Aug | 6°0 N | 21 Nov | 2°0 N |
| 18 Feb | 7°0 S | 28 May | 1°0 S | 24 Aug | 7°0 N | 24 Nov | 1°0 N |
| 7 Feb | 7°3 S | 6 June | 0° | 9 Sept | 7°3 N | 7 Dec | 0° |
| 21 Mar | 7°0 S | 14 June | 1°0 N | 22 Sept | 7°0 N | 15 Dec | 1°0 S |
| 8 Apr | 6°0 S | 22 June | 2°0 N | 12 Oct | 6°0 N | 23 Dec | 2°0 S |

**6. The Wilson effect**
A sunspot looks like a saucer-shaped depression when seen close to the limb. This sequence shows a small, almost circular spot having a symmetrical umbra and penumbra. The picture taken close to the limb shows that the apparent width of the umbra on the side further from the limb is less than the apparent width on the other side. Although this is consistent with the idea of a sunspot being a depression, the effect may be an illusion.

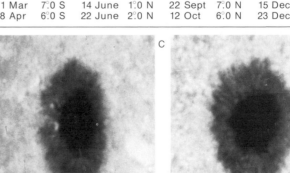

# Sunspots II

Sunspots generally occur in pairs or groups, isolated spots being relatively infrequent. Magnetograph observations clearly reveal that sunspots are the seats of intense magnetic fields, a typical sunspot group consisting of two spots of opposite magnetic polarity, one "north" and the other "south". The leading spot of a pair, in terms of the direction of solar rotation, is referred to as the "preceding" or p-spot while the other is known as the "following" or f-spot. With very infrequent exceptions, all the preceding spots in bipolar groups in one hemisphere have the same magnetic polarity, while all the preceding spots in the other hemisphere have the opposite polarity; similarly the following spots on either side of the equator will have opposite polarities.

**Magnetic fields in sunspots**

The vertical magnetic fields in the umbra of a spot are usually in the range 2,000–4,000 G, up to 10,000 times stronger than the field at the Earth's surface which ranges between about 0.3 G at the equator to a maximum of about 0.7 G at a pole. The pattern of magnetic field lines between spots of opposite polarity in a bipolar group is readily apparent in spectroheliograms taken in, for example, $H_\alpha$ or the K line of ionized calcium; the pattern of fibrils, aligned along the magnetic lines of force, closely resembles the pattern obtained when iron filings are scattered on a sheet of paper placed above an ordinary bar magnet (the lines of force delineating the direction of the magnetic field around the magnet loops, which curve round from the north magnetic pole to the south).

Solar material is highly ionized, the ions and electrons having electrical charge. A charged particle can move along a magnetic line of force without experiencing resistance, but if it tries to cross the lines a secondary magnetic field is induced. The resulting force opposes the direction of motion of the particle. Thus the flow of ionized material tends to take place along the lines of force of the local magnetic field. The high electrical conductivity of solar material, resulting from its high level of ionization, ensures that matter and magnetic fields in the Sun are thus closely coupled together; the fields are said to be "frozen in" to the matter. This implies that where the energy of the field is dominant, the flow of matter will be governed by the configuration of the local magnetic field, but where the kinetic energy of the matter is greater than the magnetic energy of the field, the field lines will be distorted and will follow the mass motions. Solar magnetism is, in fact, the controlling factor in a wide variety of solar phenomena.

The sunspot groups are divided on the basis of their magnetic properties into three principal classes, as follows:

$\alpha$: *unipolar* groups—single spots, or groups of spots having the same magnetic polarity;

$\beta$: *bipolar* groups, in which the p- and f-spots are of opposite magnetic polarity;

: *complex* groups, in which many spots of each magnetic polarity are jumbled together.

The regions of opposite polarity in a group are separated by a "neutral line" (or "line of inversion") where the vertical component of the magnetic field is zero. Dark absorbing filaments of material—which are prominences seen against the disc—are frequently found lying near neutral lines, and the violent explosive release of magnetic energy that produces flares also often occurs around these boundaries. Magnetograms show up clearly the regions of opposite polarity (*see* diagram 1B).

**Babcock-Leighton model**

At present there is no completely satisfactory theory of sunspots. However, the Babcock-Leighton model (*see* diagram 3) explains

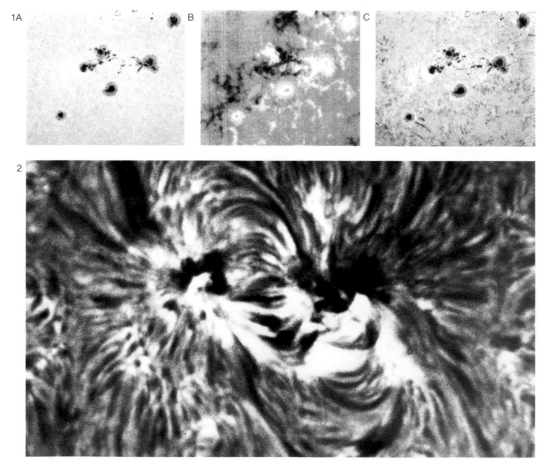

**1. Magnetic structure**
This sequence of photographs shows the intimate link between sunspot groups, photospheric magnetic fields and the structure of the chromosphere in Hα. Sunspot groups taken at a wavelength 0.4 nm longer than Hα are shown (**A**). The magnetogram (**B**) shows regions of positive magnetic polarity as bright patches and regions of negative polarity as dark patches, revealing the bipolar nature of sunspot groups. The influence of the magnetic field on the chromospheric structure at Hα is shown (**C**).

**2. Bipolar sunspot group**
This photograph taken in the light of Hα at a wavelength of 656.3 nm clearly outlines the magnetic field structure of a bipolar group. The dark filaments and fibrils follow the lines of force between the two major spots of opposite polarity, giving a pattern reminiscent of that obtained when iron filings are scattered on a sheet of paper placed over a bar magnet. The bright emission is a flare, a sudden release of energy often associated with complex sunspot groups (*see* pages 42–5).

most aspects of sunspots and their behavior. The theory assumes that a weak magnetic field permeates the outer solar layers and that differential rotation stretches the lines of force and wraps them round the solar globe like elastic threads. Where the concentrated lines breach the surface a pair or group of spots is formed. Since the field lines slope towards the equator, the preceding spot of an emerging pair should occur at a slightly lower latitude than the following one; this, again, is in qualitative agreement with observation.

As the bipolar magnetic region spreads out and declines, the magnetic flux is carried away and spread around by the formation and destruction of supergranular cells and by differential rotation. Because the f-spot is closer to the pole in each case, its polarity tends preferentially to be carried to the polar regions of the hemisphere in which it is located, building up sufficient strength by around the time of sunspot maximum to reverse the existing polarity there.

It is thought that as the polarity in a polar region changes, it begins to change the slope of the field lines in the locality; as the strength of the new polar field builds up (a result of the continued diffusion of f-spot polarity) so the zone in which slope reversal occurs moves closer to the equator. When the slope of the field lines changes from downwards (in the direction of solar rotation) to upwards, the effect of differential rotation is to unwind the field lines, rather than winding them up. The concentrated fields required for the production of sunspot groups decline and activity diminishes. Eventually a stage is reached where, once again, the lines of force run approximately from north to south and the winding-up process begins again. Since the field is in the opposite direction, the polarities of p- and f-spots will be reversed compared to the previous cycle. As the reversal begins to produce spots in higher latitudes, the residual field from the old cycle still gives rise to some spots of the old polarity close to the equator.

In this way, the Sun is thought to wind up and relax its magnetic fields in a periodic fashion. Magnetic energy built up during a cycle is released in various forms of solar activity, the most violent being the solar flare (*see* pages 42–5). Although this model has much to commend it, solar activity and its cyclic behavior is still far from being fully understood. Indeed, one key question is as yet unanswered: where does the magnetic field originate?

The most popular explanation is the "dynamo model", whereby the circulation of electrical currents (comprising ions and electrons) in the convective zone is responsible for generating the fields. A moving electrical charge generates a magnetic field: with powerful convective forces driving the highly conductive solar material, substantial fields can be generated and sustained. Similar processes in the hot liquid metallic core of the Earth are believed to be responsible for sustaining the terrestrial magnetic field.

### Why are sunspots cool and dark?
Although sunspots have been studied for centuries, there is still some debate about the precise reason why they are dark. It was believed that the intense localized magnetic field in the umbra reduced the flow of hot material to that region of the photosphere; as a result less radiant energy is transported to the photosphere in that area than in its surroundings. It has been suggested, however, that this theory of sunspot darkening is not altogether satisfactory since it should lead to a build-up of heat below the spot. An alternative view suggests that the strong magnetic field actually enhances the flow of heat, but converts 75 to 80 percent of this energy into "hydromagnetic waves", which propagate through the photospheric layer without dissipating significant amounts of energy; the wave energy contributes instead to the heating of the solar atmosphere higher up above the sunspot group.

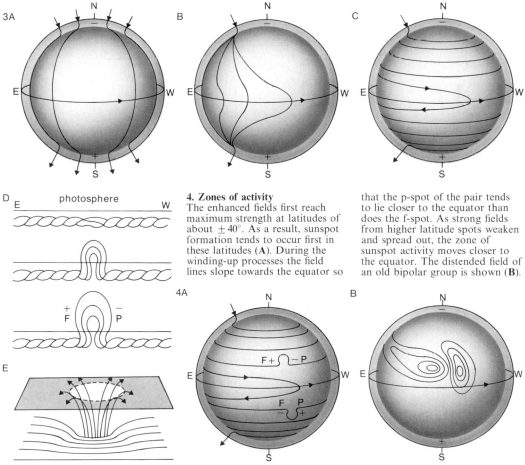

**3. Babcock-Leighton model**
According to this model the basic mechanism responsible for sunspot activity is the winding up of the solar magnetic field by the Sun's differential rotation. An idealized situation is represented (**A**) in which the north pole is taken to have negative magnetic polarity and the south pole, positive polarity. Lines of force lie along meridians in a north–south direction on and below the photosphere. The magnetic field lines are "frozen in" the solar material, and in time become distorted by the differential rotation (**B**). As the cycle progresses, the lines become stretched and wrapped round the Sun. By this means the field lines are drawn together and the field strength greatly enhanced (**C**). Field lines are brought together to form tubes of magnetic flux (**D**), which become twisted by the effects of convection. Bundles of tubes may be wound together into structures rather like ropes, this process amplifying still further the magnetic field strength. When the field strengths in the tubes or ropes becomes sufficiently great, they float to the surface, and where kinks in the bundles of tubes penetrate the photosphere, a sunspot group is formed. Below a sunspot (**E**) magnetic tubes of force are squeezed together— despite their natural tendency to push apart—by powerful convective currents.

**4. Zones of activity**
The enhanced fields first reach maximum strength at latitudes of about ± 40°. As a result, sunspot formation tends to occur first in these latitudes (**A**). During the winding-up processes the field lines slope towards the equator so that the p-spot of the pair tends to lie closer to the equator than does the f-spot. As strong fields from higher latitude spots weaken and spread out, the zone of sunspot activity moves closer to the equator. The distended field of an old bipolar group is shown (**B**).

# Sunspots III

The number of sunspots visible on the solar disc varies in a periodic way. This phenomenon was discovered by Heinrich Schwabe, who published his analysis of 17 years of observational data in 1843. At a maximum in the cycle there may be 100 or more spots on the disc, but at a minimum there are very few spots, and several weeks may pass without any at all being seen.

Schwabe's initial announcement was of a sunspot cycle of about 10 years, but later analysis by R. Wolf led to the more accurate value of 11 years, which is quoted conventionally today. However, over the past 50 years or so the period between successive maxima has averaged out at about 10.4 years. Since records began, the duration of individual cycles has varied from about 7 to about 17 years. The "11-year" cycle is readily apparent in a graph of sunspot numbers, which also shows some evidence for a longer term modulation of the height of maximum over a period of about 80 years. In 1893 E. W. Maunder of the Royal Greenwich Observatory concluded from his study of old solar records that for a period of about 70 years from about 1645 to about 1715 sunspot activity virtually vanished. Recent investigations have largely confirmed these conclusions, and have also shown that similar bouts of inactivity appear to have occurred in the more distant past. During the "Maunder Minimum" from 1645 to 1715 the appearance of a sunspot was regarded as a most noteworthy event; John Flamsteed, the first Astronomer Royal, at one stage scanned the Sun for fully seven years between sightings of a spot. Although absence of evidence of spots is not necessarily evidence of their absence, it seems unlikely that astronomers should not have maintained observations of the Sun. Sunspots were first studied telescopically in about 1610, but the 11-year cycle was not discovered until 1843: if the cycle had been behaving as regularly as it now does, it seems likely that it would have been noticed before Schwabe's time.

Another feature of the sunspot cycle is the cyclic change in the mean latitudes at which spots occur. At the beginning of a new cycle, spots tend to appear at latitudes of between 30° and 40° north and south of the equator, but as the cycle progresses they occur at successively lower latitudes. At maximum, spots tend to be found in zones at about $\pm 15°$, and by minimum the mean latitude of sunspots is between 5° and 7°; as the last spots of the old cycle are occurring in low latitudes, the first spots of the new cycle are beginning to emerge in higher latitudes once more. The progression of regions of spot activity towards the equator during each cycle is known as "Spörer's law"—the effect having been discovered in 1858 by R. Carrington and then investigated in more detail by G. Spörer—and is graphically represented in the so-called "butterfly diagram" (*see* diagram 3). In sunspot groups the preceding spot is usually found to lie at a slightly lower latitude than the following one and, whereas near the equator the line joining the two principal spots in a group may be inclined to the equator by as little as 1°, at high latitudes the inclination may be as great as 20°.

## Measurement of sunspot activity

The level of sunspot activity is assessed by the Zurich (or "Wolf") relative sunspot number, R, which is defined somewhat arbitrarily as $R = k(f + 10G)$, where f is the total number of spots, G is the number of groups and k is a factor that depends upon the idiosyncrasies of the individual observer and his telescope. For example, if there were three groups on the Sun, containing, respectively, 3, 6 and 10 spots, f would be 19, and G would be 3; assuming, for simplicity, k to be 1, the value of R would be $19 + (3 \times 10) = 49$. This means of measuring sunspot activity is not altogether satisfactory: for example, five groups of two spots each would give R a value of 50, compared to 49 for 19 spots in the

**1. Solar minima and maxima**
This pair of photoheliograms illustrates the contrast in the appearance of the Sun at solar minimum (**A**) and at solar maximum (**B**). The former was taken at 8 hr 32 min UT on 5 September 1974, near the time of a sunspot minimum; only a few small isolated spots may be seen. At minimum the solar disc may appear completely blank. The right-hand photograph was taken at 9 hr 46 min UT on 12 November 1979, during one of the more active solar maxima of the twentieth century. Large sunspot groups are seen to be confined to two bands on opposite sides of the equator.

**2. Historical record**
The sunspot cycle is displayed over a period of nearly four centuries in this graph of mean annual values of the Wolf Relative Sunspot Number. The basic 11-year cycle is readily apparent in the more recent data, although it is clear that there are variations in the lengths of the cycles and in the levels of activity at successive maxima. The period from 1645 to 1715 is the Maunder Minimum, when, it would seem, sunspots were almost completely absent. In addition to the 11-year cycle, there are indications of a longer-term modulation of the heights of successive maxima with a period of, perhaps, 80 years.

1A

B
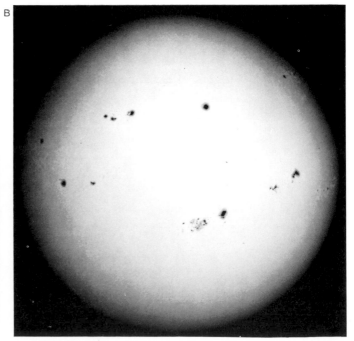

previous example; nevertheless, it has the merit that detailed records of R date back to 1848, when Rudolph Wolf introduced the system; Wolf himself by examining old observations, extended the sunspot record back to the time of Galileo and Scheiner. (A more precise measure of sunspot activity can be given by considering the total *areas* of sunspots.)

### Reversal of polarity

With very few exceptions, if all the p-spots in the northern hemisphere happen to have north magnetic polarity, all the p-spots in the southern hemisphere will have south polarity. At the end of each 11-year cycle the pattern of spot polarities reverses. Continuing the previous example, all the new spot pairs appearing around latitude 30° north would have south magnetic poles associated with their p-spots, while the p-spots of the corresponding new groups in the southern hemisphere would have north magnetic polarity. Because of the polarity reversal at the end of each 11-year cycle, 22 years elapse before the Sun returns completely to its original overall pattern.

Associated with this phenomenon is the periodic reversal of the weak "polar field". Although the Sun does not have an overall dipole field like that of the Earth, it has instead a field made up of large numbers of localized flux elements spread through the outer layers of its globe. Nevertheless there are large areas of net polarity at the polar regions, with the polarity at the north heliographic pole (usually) opposite to that at the south heliographic pole. Recent estimates of the net polar field are around 5 or 6 G at solar minimum. The polar magnetic polarities tend to reverse about a year or so after the sunspot maximum, in other words (about halfway through the cycle) since the rise to maximum activity tends to be more rapid than the decline to minimum (the average rise time is

4.6 years compared to the average decline of 6.7 years). The changeover at opposite poles generally does not occur simultaneously, and on occasions one polarity can change as much as a year or two before the other. For example, the south magnetic pole became a north magnetic pole in 1957, but the north magnetic pole did not become a south magnetic pole until the end of 1958; thus, for a time, the Sun may have two north or two south magnetic poles.

**Zurich yearly means of daily relative sunspot numbers**

| Year | No. | Year | No. | Year | No. | Year | No. | Year | No. |
|---|---|---|---|---|---|---|---|---|---|
| 1851 | 64.5 | 1877 | 12.3 | 1903 | 24.4 | 1929 | 65.0 | 1955 | 38.0 |
| 1852 | 54.2 | 1878 | 3.4 | 1904 | 42.0 | 1930 | 35.7 | 1956 | 141.7 |
| 1853 | 39.0 | 1879 | 6.0 | 1905 | 63.5 | 1931 | 21.2 | 1957 | 190.2 |
| 1854 | 20.6 | 1880 | 32.3 | 1906 | 53.8 | 1932 | 11.1 | 1958 | 184.6 |
| 1855 | 6.7 | 1881 | 54.3 | 1907 | 62.0 | 1933 | 5.6 | 1959 | 159.0 |
| 1856 | 4.3 | 1882 | 59.7 | 1908 | 48.5 | 1934 | 8.7 | 1960 | 112.3 |
| 1857 | 22.8 | 1883 | 63.7 | 1909 | 43.9 | 1935 | 36.0 | 1961 | 53.9 |
| 1858 | 54.8 | 1884 | 63.5 | 1910 | 18.6 | 1936 | 79.7 | 1962 | 37.5 |
| 1859 | 93.8 | 1885 | 52.2 | 1911 | 5.7 | 1937 | 114.4 | 1963 | 27.9 |
| 1860 | 95.7 | 1886 | 25.4 | 1912 | 3.6 | 1938 | 109.6 | 1964 | 10.2 |
| 1861 | 77.2 | 1887 | 13.1 | 1913 | 1.4 | 1939 | 88.8 | 1965 | 15.1 |
| 1862 | 59.1 | 1888 | 6.8 | 1914 | 9.6 | 1940 | 67.8 | 1966 | 47.0 |
| 1863 | 44.0 | 1889 | 6.3 | 1915 | 47.4 | 1941 | 47.5 | 1967 | 93.8 |
| 1864 | 47.0 | 1890 | 7.1 | 1916 | 57.1 | 1942 | 30.6 | 1968 | 105.9 |
| 1865 | 30.5 | 1891 | 35.6 | 1917 | 103.9 | 1943 | 16.3 | 1969 | 105.5 |
| 1866 | 16.3 | 1892 | 73.0 | 1918 | 80.6 | 1944 | 9.6 | 1970 | 104.5 |
| 1867 | 7.3 | 1893 | 84.9 | 1919 | 63.6 | 1945 | 33.1 | 1971 | 66.6 |
| 1868 | 37.3 | 1894 | 78.0 | 1920 | 37.6 | 1946 | 92.5 | 1972 | 68.9 |
| 1869 | 73.9 | 1895 | 64.0 | 1921 | 26.1 | 1947 | 151.5 | 1973 | 38.0 |
| 1870 | 139.1 | 1896 | 41.8 | 1922 | 14.2 | 1948 | 136.2 | 1974 | 34.5 |
| 1871 | 111.2 | 1897 | 26.2 | 1923 | 5.8 | 1949 | 134.7 | 1975 | 15.5 |
| 1872 | 101.7 | 1898 | 26.7 | 1924 | 16.7 | 1950 | 83.9 | 1976 | 12.6 |
| 1873 | 66.3 | 1899 | 12.1 | 1925 | 44.3 | 1951 | 69.4 | 1977 | 27.5 |
| 1874 | 44.7 | 1900 | 9.5 | 1926 | 63.9 | 1952 | 31.5 | 1978 | 92.5 |
| 1875 | 17.1 | 1901 | 2.7 | 1927 | 69.0 | 1953 | 13.9 | 1979 | 155.4 |
| 1876 | 11.3 | 1902 | 5.0 | 1928 | 77.8 | 1954 | 4.4 | 1980 | 154.6 |

2

### 3. Butterfly diagram

The numbers and latitudes of sunspots have varied over 8 complete cycles between 1874 and 1976. The latitude of each spot is plotted against the dates on which it was visible. The diagram shows in dramatic form how the first spots of a new cycle appear in small numbers at relatively high latitudes. Numbers increase as the maximum is approached, and at maximum the mean latitude of sunspots is about ± 15°. At the end of the cycle a few spots are seen close to the equator. The distribution of spots on the diagram resembles the wings of a butterfly.

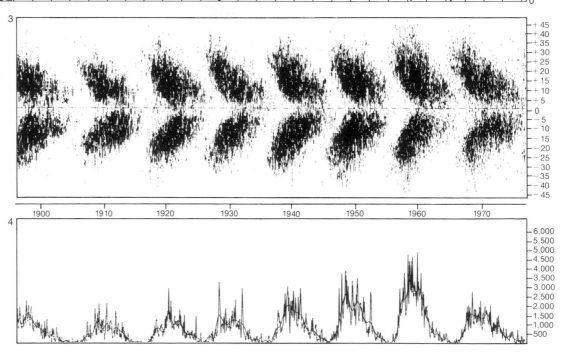

### 4. Sunspot areas

This graph plots the area of the visible disc covered by sunspots, expressed in millionths of the visible hemisphere. The smooth curve through the "spikes" shows the variation in mean annual sunspot areas. The variation in sunspot areas follows the same basic pattern as the variation in sunspot numbers shown in the butterfly diagram.

# Active Regions

Associated with sunspots are various other indications of solar activity, phenomena such as prominences and flares (*see* pages 40–41 and 42–5), faculae (*see* page 27), plages and filaments, which are all controlled by the intense localized magnetic fields that characterize "active regions".

### Plages

Named after the French word for "beaches", plages are areas of intensified brightness visible in various monochromatic lines; they look, in a way, like bright, sandy beaches against the background of the chromosphere. Plages coincide approximately in position with the photospheric faculae, and represent regions of enhanced density and temperature which are, in turn, overlaid by hotter regions of the corona; they delineate the area of enhanced magnetic field, which constitutes a complete active region. Like photospheric faculae, they appear before the emergence of sunspots and persist long after the spots themselves have faded away. Indeed, the term "chromospheric faculae" is sometimes used to describe them.

The brightness of plages is believed to be due to the increased flow of energy into the solar atmosphere which is brought about by the action of the concentrated magnetic fields. The deposition of energy transported by mechanical means is believed to be the main source of heating in the solar atmosphere (*see* pages 30–31). Thus the process responsible for heating the plages is basically the same as that which produces the bright regions that outline the chromospheric network. Although the very strong fields of 2,000 to 4,000 G found in sunspot umbrae clearly limit the flow of heat to those regions of the photosphere, the weaker overall fields in plages (which average 100 to 200 G) enhance the deposition of energy in those features and in the corona above.

Plages generally appear brighter in shorter wavelengths, which indicates that they are brighter at higher altitudes. Radio plages seen at much longer, decimetric, wavelengths are relatively high-density regions of the corona which overlay the visible plages and

have brightness temperatures of about $10^6$ K. They are the main source of the S-component of solar radio emission (*see* page 24).

### Filaments

$H_\alpha$ spectroheliograms reveal the presence of dark filaments of absorbing material which may be up to, or even in excess of, 100,000 km in length. They form along the neutral line (the line dividing regions of opposite magnetic polarity in an active region—*see* page 34) and may persist for several months (6 to 10 months is not uncommon); sometimes they fade from view and then re-form in more or less the same locations.

Filaments are tubes and loops of relatively concentrated matter at chromospheric temperatures lying above the sunspot groups, in the upper chromosphere and low corona. Seen against the bright background of the solar disc they show up because of their absorption of light, but seen at the limb they are about 100 times brighter than the background corona, and show up clearly as emission features; they are then known as prominences.

### Structure and evolution of an active region

Most active regions are bipolar magnetic regions (BMRs), which have approximately equal amounts of flux of the opposite magnetic polarities. Although in sunspots the average flux in the p-spot tends to be higher than that in the f-spot, the difference is ironed out in the overall field of the active region. Unipolar spot groups usually turn out to be old BMRs from which the f-spots have vanished (the p-spot, being stronger, tends to last longer); the distribution of faculae and plages in the vicinity usually has the form typical of a bipolar region, with a magnetic region of opposite polarity to the visible spot usually found in the location of the "missing" f-spot. The neutral line is picked out along parts of its length by filaments and by plage "corridors", which consist of darker lanes that lie between bright plages.

The formation of an active region is generally a quite rapid

1A

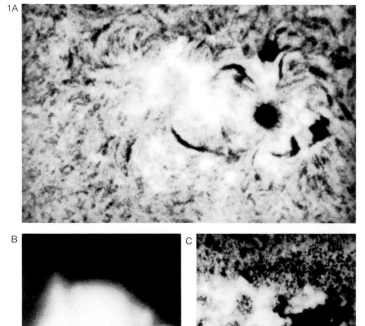

B

C

**1. Active region**
A group of sunspots, together with a plage and several dark filaments are visible in the Hα image of a typical active region (**A**). The coronal loops above this region are shown up at X-ray wavelengths (**B**), while the magnetogram (**C**) reveals the distribution of magnetic polarities (black and white representing opposite magnetic polarity). The loops in the corona can be seen arching across and joining up opposite polarities in the chromosphere below.

2A

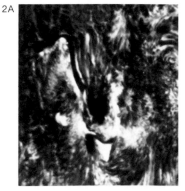

30 July        17.35

16.18

B

**2. Development of an active region**
This sequence of two sets of images was taken at the Big Bear Solar Observatory, California, in 1972. The top set shows an active region in Hα, best for revealing plages, filaments and the magnetic pattern of the fibrils, while the lower set is in Hα–0.1 nm, providing a clearer image of the sunspots. As the region develops, the neutral line dividing the preceding spots from the following spots rapidly becomes more twisted and the spots grow in size.

process, a typical region taking about 10 days to build up to its maximum extent. As the local magnetic field begins to increase with the emergence of magnetic flux from below the photosphere, features such as fibrils begin to line up along the field lines. A small plage will often make its appearance and thereafter, although not always, spots will begin to form. Prominence and flare activity escalate rapidly during the first couple of weeks.

Usually after about two to four weeks (unless it is a particularly large group) almost all of the sunspots will have disappeared, but the BMR itself can persist for several months, declining slowly by spreading out and breaking up as its field is eroded. The size of an active region usually determines its lifetime. While flares are normally associated with active sunspot groups, they can also occur in spotless BMRs.

## Other aspects of solar activity

X-ray pictures reveal that the face of the Sun is speckled with bright points that correspond to small short-lived active regions where sunspots are absent; they endure for between a few minutes and a few days, but typically survive for only a few hours. Although much smaller than sunspot groups they display the characteristics of bipolar magnetic regions with intense localized fields, which appear to follow the same polarity rules with regard to p- and f-regions. Bright points tend to be found at the boundaries of supergranular cells where the solar field is enhanced but, although they show a slight preference for middle latitudes, they are distributed widely over the entire solar globe, including the polar regions. At any one time there are typically about 100 bright points on the visible disc.

There is some debate as to just how much magnetic flux is contained in these ephemeral regions, but it does seem to be a significant fraction of the total that emerges from the Sun.

There is some evidence to suggest that their numbers increase as the frequency of other aspects of solar activity—sunspots, active regions, flares—declines. This may imply that the level of solar

activity evens out over the cycle such that the sunspot cycle alone may not give a full account of the degree to which magnetic activity fluctuates. A tendency to increase in numbers towards the minimum of the solar cycle is also displayed by polar faculae, the numbers of which reach a minimum around solar maximum and increase thereafter. Polar faculae are small, round, bright points seen in optical wavelengths; they are typically about 2,000 km in diameter and last, on average, for only some 30 min. Recent results indicate that during the minimum of 1973–75 the total magnetic flux above latitudes $\pm 40°$ increased by about 250 percent. The increase in numbers of polar faculae—which are, after all, magnetic phenomena—may be linked to this change.

Although the most obvious aspect of the overall magnetic field of the Sun is the existence of regions of opposite polarity at opposite poles, there are other large-scale aspects. As the decaying fields from BMRs spread out and are dissipated, with a net drift of f-spot polarity to the polar region, the flux also shows a predilection to form large-scale regions, or sectors, of alternating net magnetic polarity; these zones can span a few tens of degrees to more than 100 degrees of heliographic longitude and may endure for as long as three years. The individual elements of flux change continually while the overall pattern remains essentially constant. Although of weak overall field strength, these large unipolar magnetic regions appear to have a significant effect on the behavior of the solar wind (*see* pages 68–9).

The wealth of new data from sources such as Skylab and the Solar Maximum Mission has shown clearly that all forms of solar activity, as well as the structures of the chromosphere and the corona (*see* pages 46–8), are controlled by solar magnetic fields which exist as a patchwork of isolated elements and twisted flux tubes. If these emerge singly, they give rise to phenomena such as spicules, while if they burst through in concentrated bundles, they create the large-scale bipolar regions, sunspots and the multitude of varied forms of activity which keep the Sun in a state of turmoil.

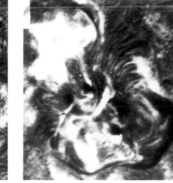

1 August    20.50        3 August    19.59        5 August    17.20        7 August    14.59

16.46        20.20        17.20        20.50

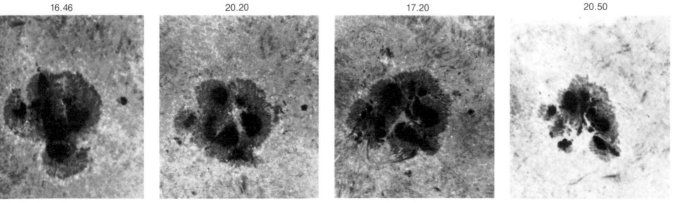

# Prominences and Filaments

Prominences are among the most beautiful and spectacular of solar phenomena. They appear at the limb of the Sun as flame-like clouds in the upper chromosphere and inner corona, and consist of clouds of material at lower temperatures but higher densities than their surroundings. Typical temperatures at their cores fall in the range of 10,000 K to 30,000 K (about one-hundredth of the coronal temperatures), while typical densities are about 100 times the value of the ambient corona. As a result, the gas pressure inside a prominence is roughly the same as that of its surroundings, although the mass of a given volume of prominence material will be about 100 times greater than that of an equivalent volume of coronal material at the same altitude.

Normally, prominences may be seen only in the light of certain Fraunhofer lines, such as $H_\alpha$, but during a total eclipse they may be seen directly in white light. In recent years space-borne instruments have made it possible for prominences to be studied in ultraviolet and even soft X-ray wavelengths. At X-ray and extreme ultraviolet wavelengths the prominences are often seen as dark shadows against the corona, since at these wavelengths the level of coronal emission is much higher than that of the relatively cool prominence material.

Filaments, which were formerly known as "dark flocculi", are the same type of phenomenon as prominences; the only difference between the two is the mode in which they are rendered visible, whether as emission or absorption features. Prominences are visible in emission at the limb because their relatively low temperature and high density ensure that compared with the background there is a concentration of non-ionized hydrogen atoms, which are responsible for $H_\alpha$ emission. At such wavelengths, therefore, the emission from the prominence far exceeds the background brightness of the corona. However, when seen against the bright background of the disc the absorption produced by these clouds predominates, and they are seen as dark filaments against the disc.

Prominences give the impression of being comprised of material surging upwards from the chromosphere, but in fact the great majority consist of material condensing from the corona and flowing downwards into the chromosphere, although prominences of the former kind do exist.

## Quiescent prominences

Among the most stable and long-lived of solar phenomena, quiescent prominences may retain their overall shape and structure for periods ranging from a few months to as much as a year before breaking up, suddenly disappearing, or, occasionally, erupting violently into space and dispersing. Their typical dimensions are: length, about 200,000 km; height, 40,000 km; thickness 5,000 km to 8,000 km. They exist as long, thin, vertical sheets of material often taking on the appearance of a viaduct with numerous arches. There is considerable vertical structure and, although the overall shape remains remarkably constant, a considerable flow of material (usually downwards) takes place within the prominence structure.

Quiescent prominences occur along the neutral line separating regions of opposite polarity in a sunspot group, in active regions, or between larger-scale regions of opposite polarity. Along these lines the field lines run parallel to the solar surface, their slope changing from "up" to "down" as they flow from one side of the dipole to the other. Rather like elastic strings, they stretch under the weight of the dense prominence material, and the stretching can be regarded as increasing the tension, thus allowing the field to support the greater weight of the prominence, which is typically 100 times the weight of an equal volume of ambient coronal material.

Closely related to quiescent prominences are the active region filaments, which form in the same kind of locations and also last for long periods; these, however, are characterized by the flow of material along their lengths, whereas quiescent prominences exhibit less movement of material, and what there is takes place mainly in a vertical direction. The filaments tend to form a few weeks after the formation of a spot group and then far outlive the spots themselves. Initially making an angle of about 40° with the local meridian, they become stretched out in length and make greater and greater angles with the meridian, approaching closer to an east–west direction with successive rotations because of the

**1. Prominence/filament**
As the Sun rotates, a prominence is carried over the limb and is observed as a filament on the disc. This sequence was recorded between 24 and 26 July 1980.

**2. Quiet filaments**
A crown of quiet filaments is often observed at times of minimum solar activity. This Hα photograph was taken on 4 October 1966.

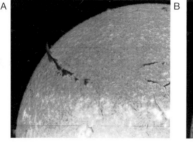

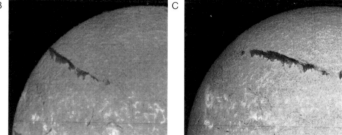

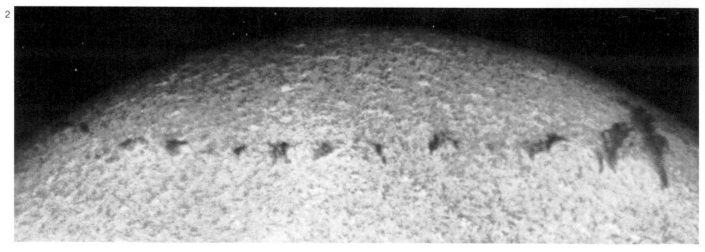

effect of differential rotation. On average a filament will increase in length by about 100,000 km per solar rotation, and they can stretch to lengths as great as 1,000,000 km.

### Active prominences

Shorter lived and usually smaller than quiescent prominences, active prominences may show dramatic changes in form within minutes. They have a mean length of about 60,000 km. They are normally associated with sunspot groups and have two basic characteristic forms: arches (or loops) on the one hand, and condensation and knots on the other. In loop prominences and "coronal rain", material from the corona descends, some of this material having previously been ejected from the chromosphere. In developing active regions, arch prominences appear at a fairly early stage as a result of the emergence of magnetic flux from below the surface. In arch prominences of this type, material descends also from above, flowing down through both feet of the arch. Arch prominences connect regions of opposite polarity in developing sunspot groups, running across the neutral line, unlike quiescent prominences which run along it. The lifetime and content of these short-lived features are controlled by the rate of condensation of material from the corona and the rate of flow along the magnetic field lines into the chromosphere. Depletion of the source of material and changes in the local magnetic field can terminate it.

In contrast, surges, puffs and sprays consist of material being hurled upwards from below. These are the most violently active of prominence phenomena and are mostly related to flares: in surges which occur in active regions, chromospheric material is shot into the corona at speeds of around 100 to 200 $km\,s^{-1}$. Thousands of millions of tonnes of matter can soar to heights of several hundred thousand kilometers before falling back, usually along the same track as the ascent. The falling material may break up and be lost from sight before it reaches the surface. The typical lifetimes of surges are about 10 to 20 min, and they frequently recur. Puffs often occur just prior to surges, as sudden expansions of material.

Sprays are even more violent impulsive events, in which chromospheric material is ejected in excess of the solar escape velocity (618 $km\,s^{-1}$) and is scattered widely. They are essentially more energetic versions of surges, and are invariably associated with the occurrence of flares.

Magnetohydrodynamic waves (*see* page 31) emitted from flares may also set prominences into vertical oscillation and may trigger their disappearance, although in many cases another prominence of similar shape will re-form in the same region. Prominences may erupt, following a flare outburst, usually as a growing arch whose center expands rapidly and disappears while the ends remain visible and rooted in the chromosphere below.

3

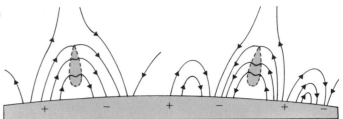

**3. Prominence support**
The relatively dense material of a prominence appears to be supported by the local magnetic field. Two possible field configurations are shown in the diagram. The support results from a force that is produced by a current flowing perpendicular to the direction of the magnetic field lines. This current in turn produces a magnetic field that distorts the shape of the original field lines.

**4. Arch prominences**
This series of arches seen in Hα light rises to a height of over 65,000 km above the level of the photosphere.

**Classification of prominences**

A. Prominences originating from the corona
   S. Spot prominences
     a. Rain
     f. Funnels
     l. Loops
   N. Nonspot prominences
     a. Coronal rain
     b. Tree trunk
     c. Tree
     d. Hedgerow
     f. Suspended mound
     m. Mound
B. Prominences originating in the chromosphere
   S. Spot prominences
     s. Surges
     p. Puffs
   N. Nonspot prominences
     s. Spicules

4

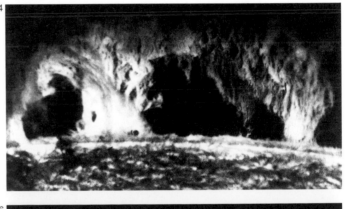

6

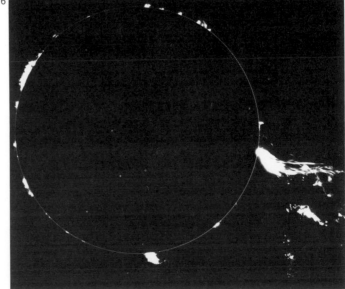

5

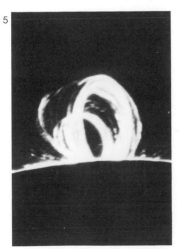

**5. Loop prominence**
The loop structure of this prominence outlines the magnetic field above a sunspot group. In spite of its appearance, which suggests material flowing upwards, this prominence, like those in the previous two illustrations, in fact consists of material condensing out of the corona and flowing downwards. It is seen here in the light of the green line due to FeXIV.

**6. "Westward Ho"**
This eruption, nicknamed the "Westward Ho" event, occurred on 2 March 1969. A large cloud of material was observed moving away from the Sun at a rate of 270 $km\,s^{-1}$ over a period of about 2 hr 15 min.

# Flares I

Flares are sudden violent releases of energy that occur in the vicinity of active regions. They eject atomic particles and emit radiation across the entire range of the electromagnetic spectrum from $\gamma$-rays and "hard" X-rays (wavelengths less than $10^{-10}$ m) to radio wavelengths of several kilometers. The first flare to be recorded was seen by the English astronomer Richard Carrington, in 1859; it was visible in white light, but it is only relatively rarely that a "white light flare" can be seen against the brilliance of the photosphere. Flares are normally seen by Earth-based observers in monochromatic light, particularly $H_\alpha$ and K-calcium.

A flare consists of an $H_\alpha$ region emitting at temperatures of the order of 10,000 K, embedded within which are smaller bright regions or "kernels" some 1,000 to 10,000 km in diameter. High-temperature flare emission is seen in extreme ultraviolet wavelengths (1 to 100 nm). The individual sources of this radiation usually take the form of small arch-shaped regions, typically less than 10,000 km in size.

Flares usually occur close to the neutral line in complex active regions. In $H_\alpha$ a flare is generally seen to commence in a number of compact bright points of light within a plage, and the region of enhanced emission rapidly spreads—within a few minutes—to cover an area typically of some $10^9$ km$^2$. Where two or more bright spots are seen, they are found to lie on opposite sides of the neutral line and, in larger flares, the individual bright patches merge into two rows giving rise to a "two ribbon" flare. After the disturbance

caused by the flare, the magnetic field sometimes appears to restore itself approximately to its original configuration, since after a few hours a similar flaring phenomenon may recur in the same site.

The $H_\alpha$ emission results from a temperature increase of several thousand Kelvin in a thin region at upper chromospheric or low coronal altitudes; observations made at the solar limb indicate that flares typically reach heights of 5,000 to 15,000 km above the photosphere. The flare reaches maximum brightness within a few minutes and declines more slowly thereafter, the total duration of such events ranging from about 10 min to several hours, with 20 min being typical for smaller flares. In $H_\alpha$, flares are classified in order of importance according to the area they cover and the maximum brilliance attained (*see* Table 1). Events smaller than 2 square degrees—angles measured about the Sun's center—on the solar surface (less than $3 \times 10^8$ km$^2$ or 100 millionths of the area of the visible hemisphere) are known as subflares. By an alternative system, flares are classified according to their peak brilliance in soft X-rays (0.1 to 0.8 nm), giving a better guide to the energy output and resulting terrestrial effects (*see* Table 2).

The numbers of flares occurring is closely related to the solar cycle. An approximate guide to the daily frequency of flares of importance class 1 or greater is given by R ÷ 25, where R is the Wolf Relative Sunspot Number (*see* pages 36–7). For example, at a high maximum with R equal to 150, there would on average be $150/25 = 6$ flares per day, while at minimum several days could pass without

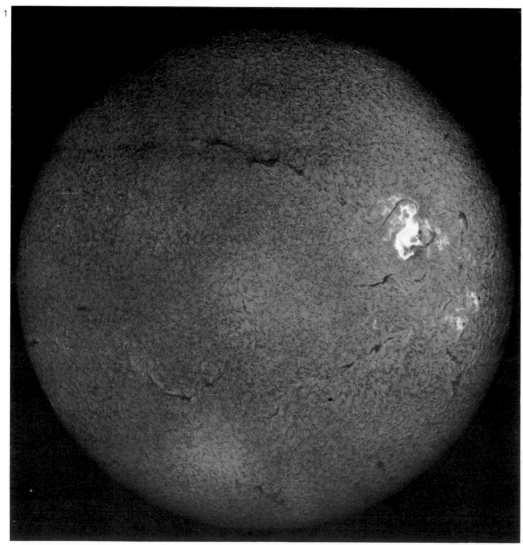

**1. Flare**
The great flare of 7 August 1972, shown here at its peak, was one of the largest ever recorded. Increased auroral activity could be observed following the flare as well as shortwave radio blackouts. The major part of the flaring activity is concentrated into two ribbons on either side of the neutral line.

**Table 1. Classification of flares – $H\alpha$**

| Area (millionths of hemisphere) | (square °) | Class |
|---|---|---|
| < 100 | < 2.0 | s |
| 100–250 | 2.0–5.1 | 1 |
| 250–600 | 5.2–12.4 | 2 |
| 600–1,200 | 12.5–24.7 | 3 |
| > 1,200 | > 24.7 | 4 |

The maximum brightness of a flare in $H\alpha$ is added to the importance class: F = faint, N = normal, and B = bright. The least important flare is 1F and the most important 4B.

**Table 2. Classification of flares – soft X-rays**

Peak flux, $\phi$, in the range 0.1 to 0.8 nm

| | Class |
|---|---|
| $\phi < 10^{-5}$ W m$^{-2}$ | C |
| $10^{-5} \leqslant \phi < 10^{-4}$ | M |
| $\phi > 10^{-4}$ | X |

A one-digit number from 1 to 9 is added as a multiplier. Thus a C5 event has a flux of $5 \times 10^{-6}$ W m$^{-2}$ and X7 corresponds to $7 \times 10^{-4}$ W m$^{-2}$. The class C0 is used for events below $10^{-6}$ W m$^{-2}$. This classification dates from 1 January 1969.

any flare activity. The average frequency of white light flares is less than one per year.

Prior to the onset of a flare, adjustments take place in the magnetic field of the region in which it is about to occur. For example, if a quiescent filament is present close to the location it will be triggered into activity. The first phase of some flares, and certainly of the more energetic ones, is a sharp burst of hard X-rays followed by a more gradual rise and fall of soft X-rays. Simultaneous flashes may be seen in optical and extreme ultraviolet wavelengths; SMM results seem to show that the hard X-ray bursts coincide in time with the sudden intensification of ultraviolet emission near the tops of small loop structures within the flare region while the hard X-rays themselves come from the feet of the loop. The temperature in flare kernels may be as high as $10^9$ K.

The hard X-rays are produced by streams of electrons that are rapidly accelerated to speeds of up to half the velocity of light; the radiation is produced as a result of close collisional encounters between these "relativistic" particles and the surrounding particles in the solar atmosphere. Radiation of this type, emitted by electrons that are sharply slowed down, is known as "bremsstrahlung" radiation. Strongly polarized microwave radiation is also emitted by the "synchrotron" process (*see* page 25) as the streams of electrons move out from the flare site. The microwave bursts span a wide range of frequencies from above $10^{10}$ Hz down to about $10^8$ Hz (wavelengths from below 0.01 m to about 1 m).

3

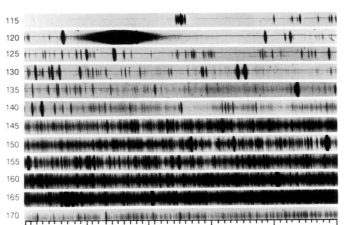

**2. Development of a flare**
These 12 filtergrams, taken on 8 August 1968, show the development of a Class 2B flare. The center of the Hα line (**A**) gives the clearest view of the flare. Off-center wavelengths (**B** and **C**) allow the sunspot group at the center of the flare region to be observed in greater detail.

**3. Ultraviolet flare spectrum**
Part of the ultraviolet spectrum of a solar flare was photographed by Skylab on 15 June 1973. Thousands of emission lines (black in this negative print) appear between 115 and 175 nm. The broad band centered on a wavelength of 121.6 nm is the Lyman-α line of hydrogen.

18ʰ13.1ᵐ   18ʰ26.1ᵐ   18ʰ44.4ᵐ   19ʰ2.3ᵐ

2A

Hα

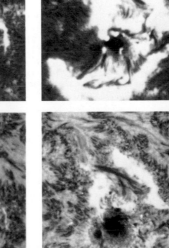
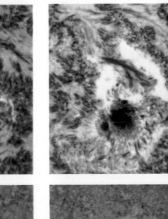
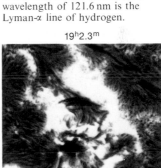
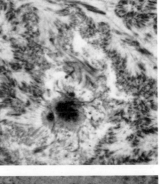
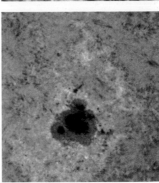

B

− .06 nm

C

− .09 nm

# Flares II

High-speed streams of electrons and sometimes protons may be ejected from flares, together with bulk masses of material. The particles travel outward through interplanetary space and have important effects on the terrestrial magnetosphere and atmosphere (*see* pages 140–41). Streams of low-speed (non-relativistic) electrons are also emitted; a typical large flare event expels up to $10^{39}$ of these electrons. The high-speed electrons are emitted in smaller numbers (say $10^{36}$ in a large flare) and a similar number of protons is ejected in a major flare.

As the fast electrons accelerated by the flare event pass out through progressively more rarefied layers of the solar atmosphere, they set the coronal plasma oscillating in such a way as to produce radio emissions. Solar radio emissions at wavelengths of meters or tens of meters are classified into five major types of "noise" bursts.

Type I noise storms—which are not specifically associated with flares—consist of long series of bursts, each individual event appearing as a spike of 0.1 to 2 sec duration superimposed on a general increase in solar radiation at these frequencies. These noise storms may last from a few hours up to several days. Type II bursts, in the 300 to 10 MHz range to 1 to 30 m wavelength, usually occur 10 to 30 min after the onset of a flare and are due to the effects of a shockwave ascending through the corona at speeds in the region of 800 to 2,000 km s$^{-1}$. Type III bursts, in the 800 MHz to 10 MHz range occur typically 2 to 4 min after the onset of a flare. They consist of short-lived spikes that drift in a few seconds from high to low frequencies, and are therefore referred to as "fast drift bursts". The source of these bursts are plasma oscillations generated by the high-speed electrons moving at up to half the speed of light.

Type IV bursts are complex in form, spanning a wide range of frequencies. The radiation arises by the synchrotron process, and events of this type usually occur after the Type II bursts in large flares and proton flares (flares that emit energetic protons and other heavy particles—solar cosmic rays). Part of the burst moves out through the corona with speeds in the range 100 to 1,400 km s$^{-1}$; these "moving type IV bursts" are emitted from a blob of hot plasma moving outward and carrying with it its own magnetic field. Such bursts have been shown to follow in the wake of loop-shaped coronal transients (*see* pages 66–7). Type V bursts are similar to Type III, occurring a little later, and might be caused by the trapping or delaying of some of the high-speed electrons.

The sequence of events in a flare would seem to be as follows. Over a period of hours or days prior to the event large amounts of energy are built up and stored in the magnetic field over a complex active region. Somehow, after activation of some trigger mechanism, the energy is released very rapidly indeed and as a result streams of electrons are accelerated to near-relativistic speeds; the interactions of these electrons with the magnetic field and the surrounding plasma give rise to a wide variety of emissions from X-rays to radio waves. The sudden heating of local areas of the chromosphere and corona results in the expulsion of considerable masses of material, up to several thousand million tonnes, which may move out into interplanetary space. Shock waves spread out through the photosphere, chromosphere and corona at speeds of 1,000 to 2,000 km s$^{-1}$, setting up violent oscillations in prominences up to hundreds of thousands of kilometers distant.

## Theories of flares

Despite decades of intensive research there is still no clear accepted theory that accounts satisfactorily for all, or even any, of the aspects of the flare mechanism. There is no doubt, however, that the primary energy source is stored magnetic energy. No other source can provide the necessary energy in a concentrated volume over the

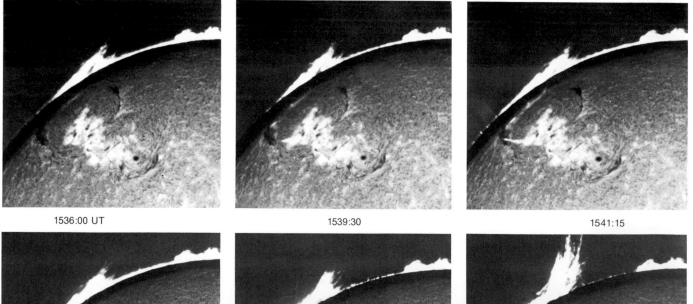

| | | |
|---|---|---|
| 1536:00 UT | 1539:30 | 1541:15 |
| 1543:15 | 1548:30 | 1612:15 |

short timescale involved. The rate at which energy is released from a typical flare is about $10^{21}\,\mathrm{J\,s^{-1}}$ (about $10^{-5}$ of the solar luminosity). Exceptionally, flares can emit at up to $10^{-3}$ of the entire solar output. If the duration of the flare is assumed to be about 1,000 sec, then its total energy output would be some $10^{24}\,\mathrm{J}$. Certainly, there is plenty of magnetic energy available; it has been shown that the $10^{25}\,\mathrm{J}$ required for a large flare could be attained by depleting the field strength of a 3,000 G sunspot group by about 20 percent over a volume of $10^{12}\,\mathrm{km^3}$.

Flares occur much more frequently in magnetically complex active regions than in simple bipolar groups, and there is no doubt that the complexity of the magnetic field and the presence of warped, twisted, neutral lines is an important factor in flare production. Studies have shown that the number of flares seen in $H_\alpha$ in an active region is closely correlated with the number of kinks in the neutral line. The emergence of new flux from below the photosphere in an existing active region further enhances the complexity of the region and induces flare activity. The largest single X-ray emission spike so far detected (by the SMM on 29 March 1980) was inferred to originate from the site of emerging magnetic flux of opposite polarity to the existing field.

There have been many theories as to the form of magnetic field that leads to the sudden energy release, but it is generally agreed that field "reconnection" or "annihilation" is the process best able to release the requisite quantities of energy (*see* diagram 2).

According to one model originally proposed by P.A. Sweet and by E.N. Parker, above a bipolar region and between the regions of opposite polarity there can exist a "neutral sheet" separating lines of force with opposite directions. A slight instability is sufficient to bring together such lines so that they join up, or reconnect, with the sudden release of energy and the expulsion of particles and plasma.

According to an alternative point of view, advocated primarily by D.S. Spicer, a twisted arch (or loop) of magnetic flux joining regions of opposite polarity provides the source and site of the energy release. Instabilities in the tube lead to the breaking up of the tube into many magnetic "islands", and the resultant reconnection of field lines converts magnetic energy into the requisite form for flare phenomena. This process is known as the "tearing mode instability". However, in neither picture is it as yet possible to release energy fast enough to account for observations.

The magnetic arch picture tends to be favored by extreme ultraviolet and X-ray observations, which show that flares occur either in single arches or in a series of arches (an "arcade"). SMM observations indicate that these arches are the sites of X-ray, ultraviolet and $H_\alpha$ bright points at the onset of flares.

Even with an appreciation of the basic mechanisms of energy release there are still formidable difficulties in the way of a full understanding of such problems as how the energy is channelled into the appropriate forms and how the particle acceleration takes place. The whole subject is immensely complicated, and the difficulties experienced in trying to account for such phenomena on the Sun indicates how important it is to exercise great caution when making statements about the physics of distant stars.

The study of flares is of the utmost interest for a wide variety of reasons: the X-ray, ultraviolet and particle emissions have direct effects on the Earth, and large flares pose hazards for astronauts in space; an understanding of the flare mechanism would provide vital clues in the search for energy from controlled thermonuclear fusion by magnetic confinement on Earth, with important implications for the global energy crisis. The same kind of process is thought to be responsible for the behavior of flare stars—faint red stars which, from time to time, flare up to several times their normal brightness.

**1. Flare sequence**
Two superimposed images show the development of a small flare with an associated spray on 21 May 1967.

**2. Flare mechanism**
According to one model (**A**) a neutral sheet exists between magnetic field lines with opposite directions above a bipolar magnetic region. Reconnection of field lines results in a sudden release of magnetic energy (**B**) with the expulsion of streams of particles. A large blob of plasma may ascend through the corona (**C**) carrying magnetic flux with it. An alternative possibility (**D**) for the site of the primary release of flare energy is the neutral point between an established loop of magnetic flux (the larger loop) and a newly emerging loop. A third possibility (**E**) is that flaring occurs in magnetic arches connecting regions of opposite magnetic polarity. Theory indicates that the field lines in such an arch will follow helical paths and that flaring can occur at many points along the arch.

**3. Radio noise storms**
The frequencies of the various emissions have been plotted against time elapsed since the onset of a flare. The diagram is schematic, and not all flares are necessarily followed by all types of noise storm.

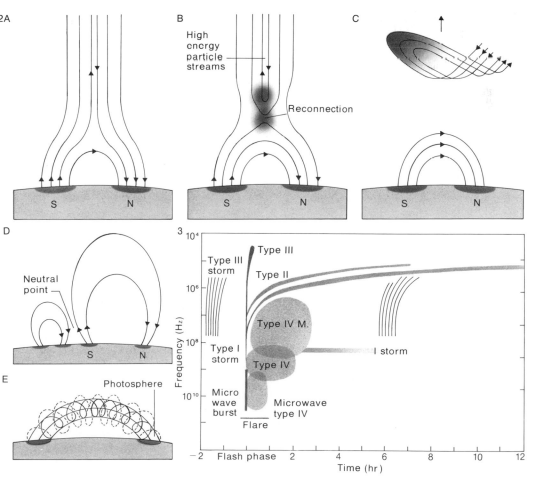

# Corona I

The corona is the outer atmosphere of the Sun. It emits on average slightly less visible light than the full Moon but, because of the level of scattered sunlight close to the Sun in the sky, it cannot be seen with the unaided eye except during a total solar eclipse. In the 1930s the invention of the coronagraph (*see* page 21) allowed the inner corona to be studied at other times from high-altitude sites when conditions were favorable, but it is only since the early 1970s, when it has been possible to place coronagraphs in orbit, that the corona could be studied continuously under ideal conditions. Another major development has been the growth of extreme ultraviolet and X-ray techniques, which have allowed the corona to be studied all over the solar disc rather than merely at the limb since the background (cool) photosphere does not swamp the (hot) corona at these wavelengths.

The appearance of the corona can vary markedly, since its overall shape changes with the solar cycle. At sunspot minimum the corona typically has a fairly symmetrical structure with long streamers extending outward from the equatorial regions, while polar plumes emerge from the polar regions rather like the pattern adopted by iron filings in proximity with the poles of a bar magnet. At maximum the corona is more conspicuous and usually rather more evenly spread over the whole disc.

Long extensions of the corona ("streamers") are found over active regions of the photosphere, and may take the form of "fans" (broad extensions usually found over quiescent prominences) or of narrower "rays". Both types of feature clearly follow the geometry of the local magnetic fields. In the low corona, arch-shaped coronal condensations are found, following closed arches or loops of magnetic field between regions of opposite polarity within active regions, or larger arches spanning regions of opposite polarity in separated regions. High above such arches extensive streamers may be found following the open pattern of field lines out into interplanetary space.

The general distribution of fans, rays and other coronal features follows the latitude of solar active regions. Thus, at solar minimum the streamers tend to be concentrated in equatorial regions, while at maximum they are more widely spread in latitude. The greatest extensions occur above sunspots, where the extensions at minimum have been observed out to colossal distances, over 20 $R_S$ on occasion.

### X-ray and extreme ultraviolet observations
Because of its immense temperature, the hot coronal gas is a strong emitter of X-rays, and by examining these wavelengths, concentrations and structures in the corona are rendered visible over the entire solar disc. When observed in this way the corona is seen to consist of three principal types of regions:
1. Active regions and coronal bright points, which are regions of bright X-ray emission associated with strong, closed loops of magnetic field.
2. Quiet regions, of much fainter X-ray emission, with weak coronal magnetic fields which nevertheless appear to be closed on a large scale; in general they are larger and fainter than active regions.
3. Coronal holes (*see* pages 66–7).

### Magnetic influence
Magnetic loop structures seem to be the basic building blocks of the

entire corona. Material is concentrated to higher densities and
temperatures within these structures, while open lines spread out
over the tops of the closed regions and then stream out into space.
The underlying loop, overlaid by the open field lines, gives coronal
streamers their characteristic bulb shape above active regions. The
basic magnetic connections between active regions, shown up as
loops at X-ray wavelengths, seem to persist for the duration of the
lifetime of those active regions, but the individual loops survive for
much shorter times, usually less than a day.

The optical corona is divided into three main components: the
K-corona, which exhibits a continuous spectrum in which the
Fraunhofer lines are invisible; the F-corona, which also exhibits a
continuous spectrum of reflected photospheric light but in which
the Fraunhofer lines are visible; and the E-corona, consisting of a
series of faint emission lines. Overall, the spectrum comprises a
faint continuum on which are superimposed weak emission lines.

The K-corona is the inner part of the corona. Its light is
photospheric light scattered by very fast-moving electrons, and the
absence of Fraunhofer lines is caused by the very high Doppler
shifts (*see* page 27) associated with the random velocities of the
particles: the Fraunhofer lines of the photospheric spectrum are
"smeared" out by these shifts, and are therefore not discernible
against the general background light.

The E-corona is composed of light emitted by relatively slow-
moving ions, giving rise to an emission line spectrum. The existence
of emission lines due to such highly ionized species provided the
first clear evidence that the corona must be at a very high
temperature, in excess of $10^6$ K.

**4. Coronal magnetic field**
In this image of the corona, a
number of theoretically
extrapolated field lines have been
plotted over an Hα image
superimposed on an eclipse
photograph (November 1966).

**5. Coronal loops**
A number of well-formed loops in
the inner corona, which were
found to correspond with bright
prominences seen in Hα light, can
be seen in this photograph taken
in the light of FeXIV.

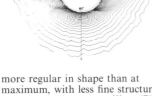

**1. Corona**
The eclipse of 12 November 1966
occurred during a year of
relatively high solar activity (1969
was a solar maximum). Coronal
streamers can be seen stretching
out to a distance of about 4.5 $R_S$;
typical "Helmut" streamers lie
over the prominences visible on
the limb; arch structures are
apparent directly above the
prominences.

**2. Corona at minimum**
A photograph taken on 30 June
1954 (**A**) shows that the corona is

more regular in shape than at
maximum, with less fine structure
to be seen. The contour lines (**B**)
represent equal intensities spaced
at intervals of 0.2 magnitude.

**3. Ultraviolet solar images**
The UV images (**B**, **C** and **D**)
were taken at wavelengths
representing progressively higher
temperatures ($10^4$ K, $3.25 \times 10^5$ K,
and $2.25 \times 10^6$ K respectively).
They thus depict the Sun at
successively greater altitudes
above the chromosphere, as seen
in the Hα image (**A**).

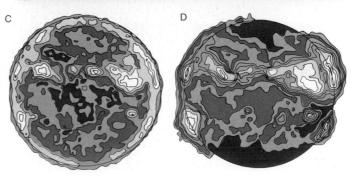

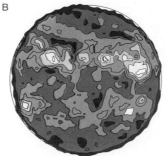

# Corona II

The F-corona, or outer corona, is responsible for most of the light beyond a distance of about 2 R$_S$. It comprises relatively slow-moving particles of interplanetary dust, and is really not so much a part of the solar atmosphere as that part of the interplanetary medium which happens to be close to the Sun. Photospheric light is scattered from these dust particles, and because of their relatively slow motions Doppler broadening is insufficient to blur out the Fraunhofer lines. The "white light" corona visible from the Earth at eclipses and from space with coronagraphs comprises the combined contributions of the K- and F-coronas.

## Temperature of the corona

The temperature of the corona involves a similar problem to that of the chromospheric temperature (*see* pages 30–31): according to the second law of thermodynamics, heat cannot be transferred from a cooler to a hotter body, and it therefore appears impossible at first sight for the corona to be a higher temperature than the 6,000 K photosphere. For a time astronomers remained puzzled by the anomalous figures, but the very high temperature of the corona has been confirmed by a wide variety of observational techniques, including radio observations, the already mentioned absence of Fraunhofer lines in the K-corona, and the presence of highly ionized metals in the E-corona. More detailed measurements of the widths of forbidden lines broadened by the Doppler effect and of the relative intensities of various lines confirm the view that the particles in the corona have a high average level of energy—in other words, the gas has a high temperature. A nominal value for the temperature of the lower K-corona is $2 \times 10^6$ K, although the temperature in different regions of the corona can range from about 1 to $5 \times 10^6$ K. The density of material in the corona, like its temperature, decreases with increasing distance from the Sun: the

mean density in the inner corona is (in terms of electron densities) $10^{15}$ m$^{-3}$, as compared with a value some 10,000 times lower at an altitude of about 2 R$_S$. Thus although the coronal temperature is very high, and although the energy of each coronal particle is correspondingly high, the particle density is so low that the total energy contained in and emitted by the corona is relatively small.

Recently the widely held view of the basic process by which the corona is heated to such high temperatures has been questioned; although the detailed mechanisms were a matter of dispute it was felt that shock waves resulting from the bubbling, noisy convective activity in the photosphere were propagated through the chromosphere and deposited energy in the corona. However, the highest levels of energy radiation are emitted from the regions of strongest closed magnetic structures, and some astronomers have suggested that magnetic effects, rather than mechanical ones, may be the basic source of coronal heating. Two mechanisms are suggested: magnetic field annihilation and MHD waves.

Electrons flowing under the action of coronal magnetic fields experience resistance and so release energy in the form of radiation. In complex twisted magnetic structures—such as may occur when a magnetic flux tube is emerging from the photosphere to form an active region (*see* pages 38–9)—concentrated tubes are readily prone to kinking: lines of force with opposing directions may meet and join up, releasing energy rapidly in the process (as in flares). This mechanism may be responsible for at least some of the energy dissipation in the corona. At the very least, even if the conventional mechanical energy theory is correct, the magnetic field plays the dominant role in channelling the flow of matter and energy through the corona. For the moment, however, it is not certain whether acoustic wave energy or magnetic energy is the primary source of the coronal heating; most probably both processes are at work.

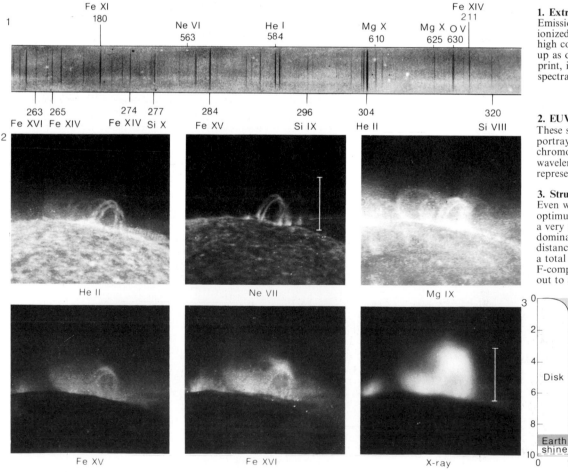

**1. Extreme ultraviolet spectrum**
Emission lines from several highly ionized atoms resulting from the high coronal temperatures show up as dark lines in this negative print, in which three different spectral orders can be seen.

**2. EUV and X-ray images**
These simultaneous images portray the Sun in six different chromospheric and coronal wavelengths. The vertical line represents a height of 240,000 km.

**3. Structure of the corona**
Even when viewing conditions are optimum, it is only possible to see a very small portion of the dominant K-corona, out to a distance of about 1.2 R$_S$. During a total eclipse, the E- and F-components may also be seen, out to a distance of 4 R$_S$ or more.

# Plates

Solar corona
Voi, Kenya, East Africa
16 February 1980

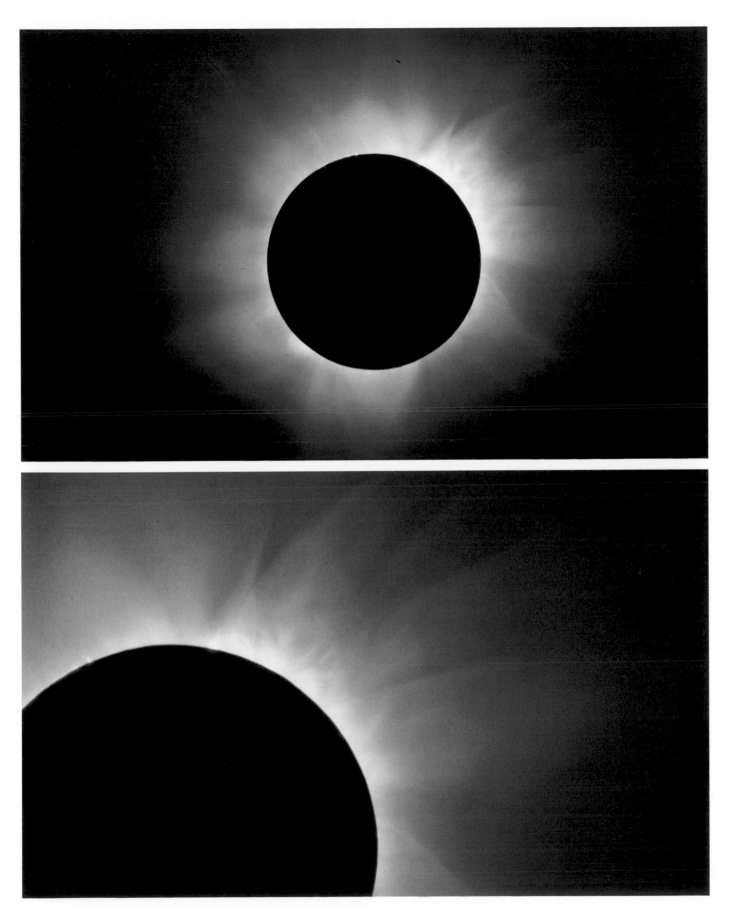

49

Visible solar spectrum
Sacramento Peak Observatory

Flash spectrum
Pacific Ocean
12 October 1977

Visible solar spectrum
Sacramento Peak Observatory

Flash spectrum
Pacific Ocean

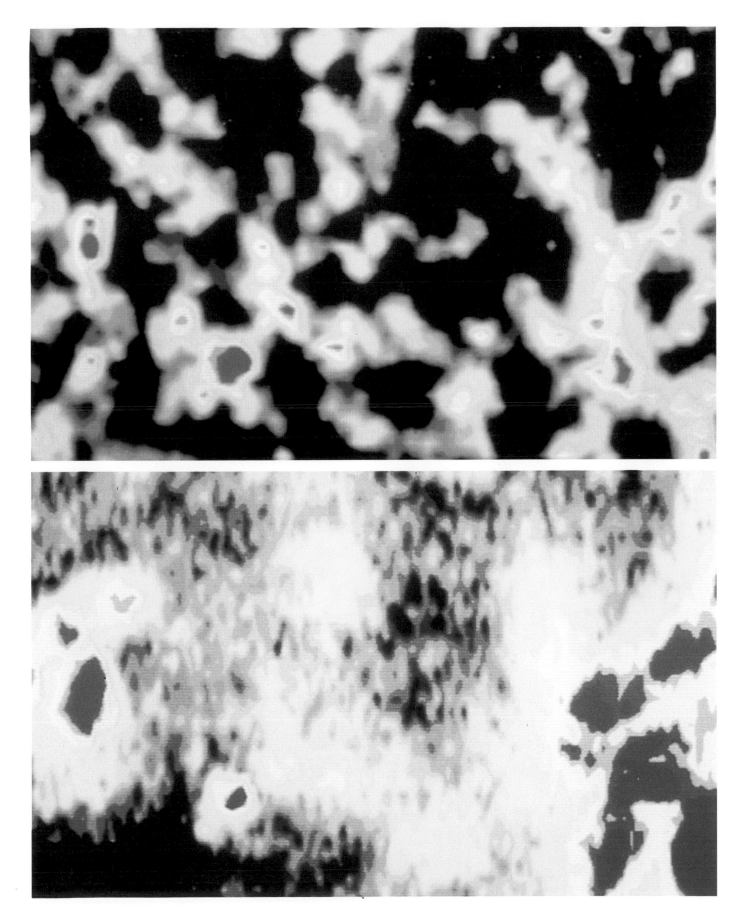

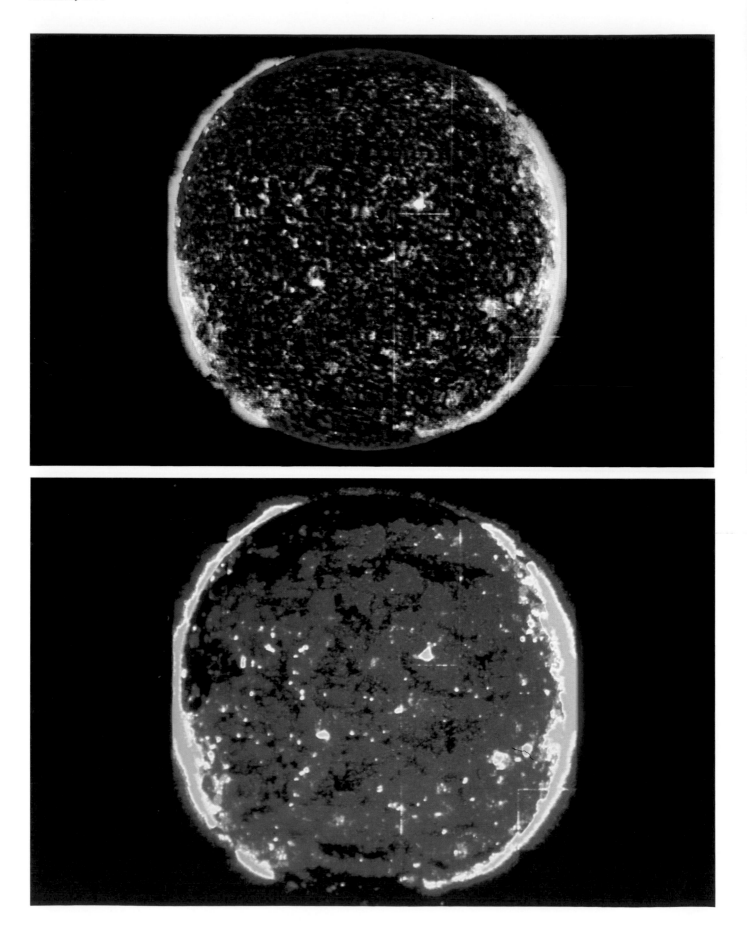

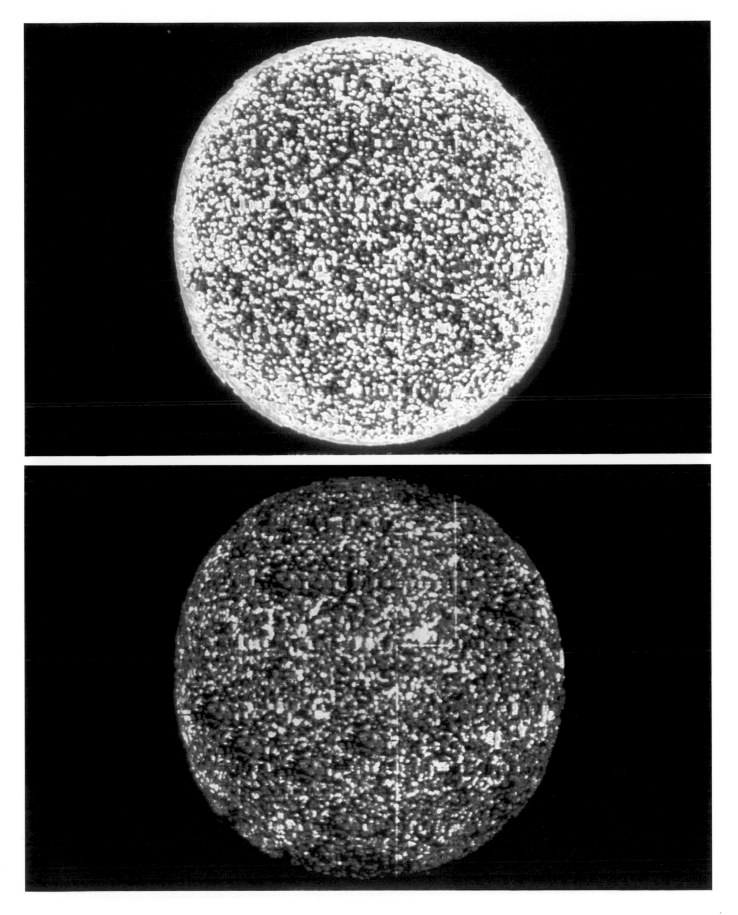

Loops of active region
Skylab
4 December 1973

Magnetic loop
Solar Maximum Mission
27 March 1980

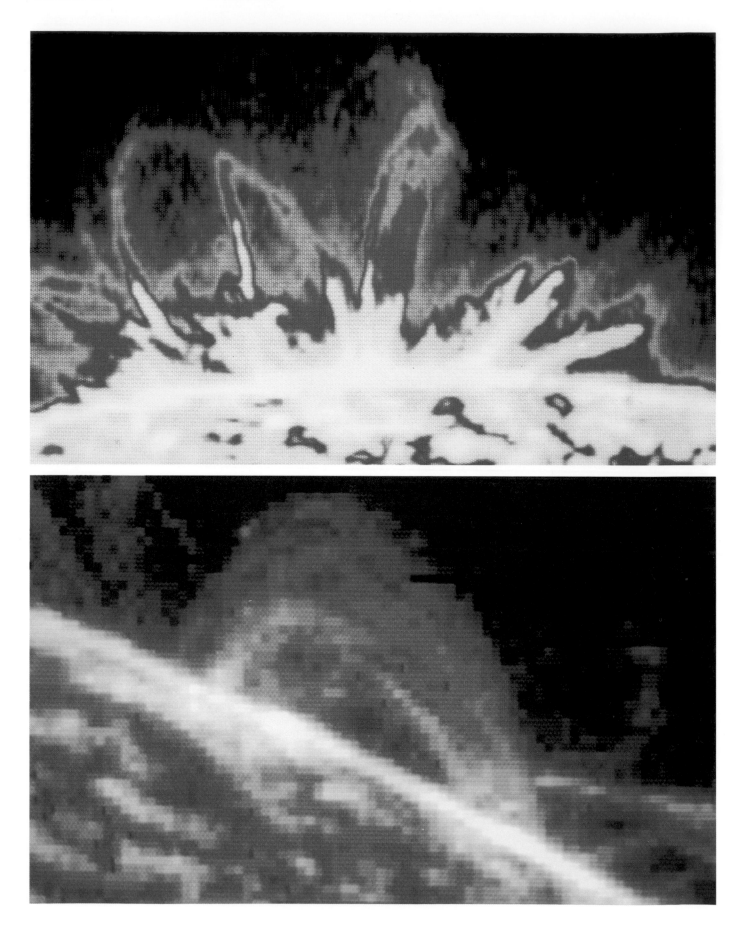

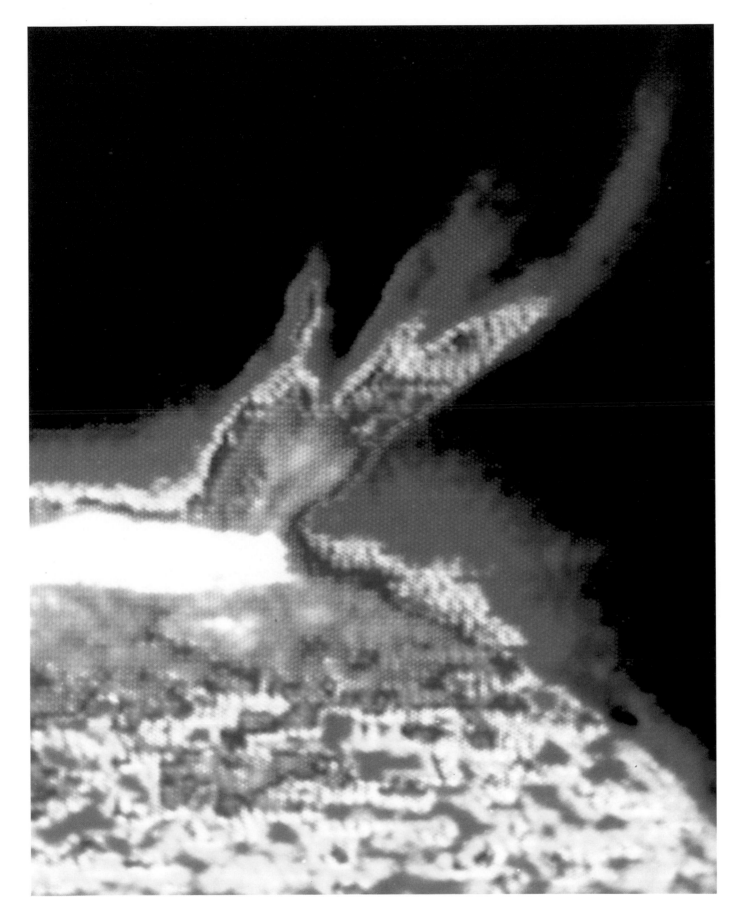

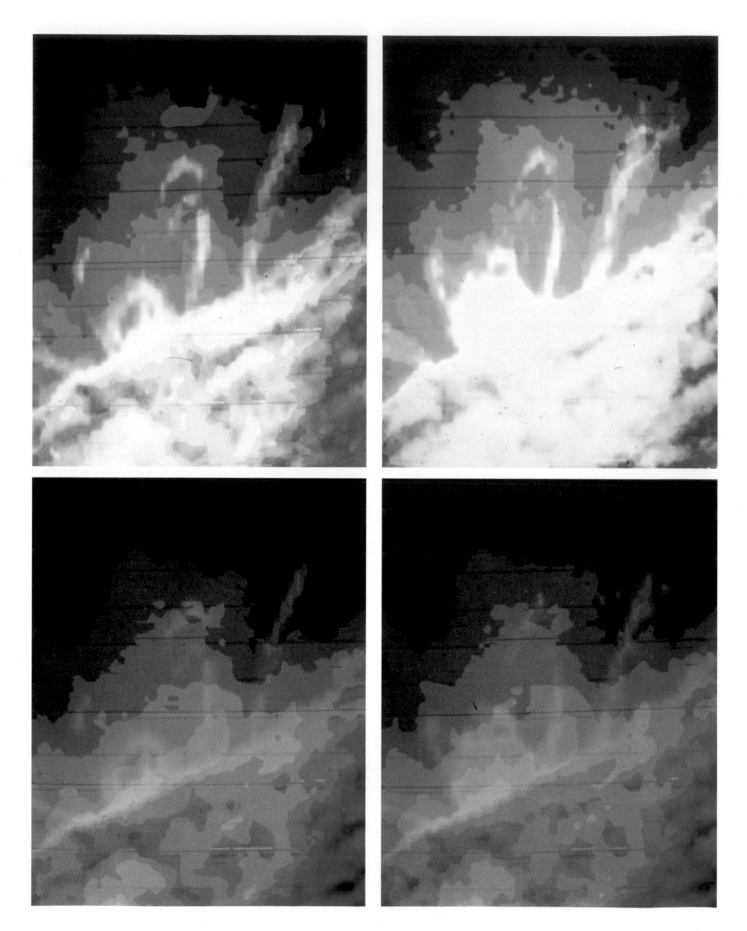

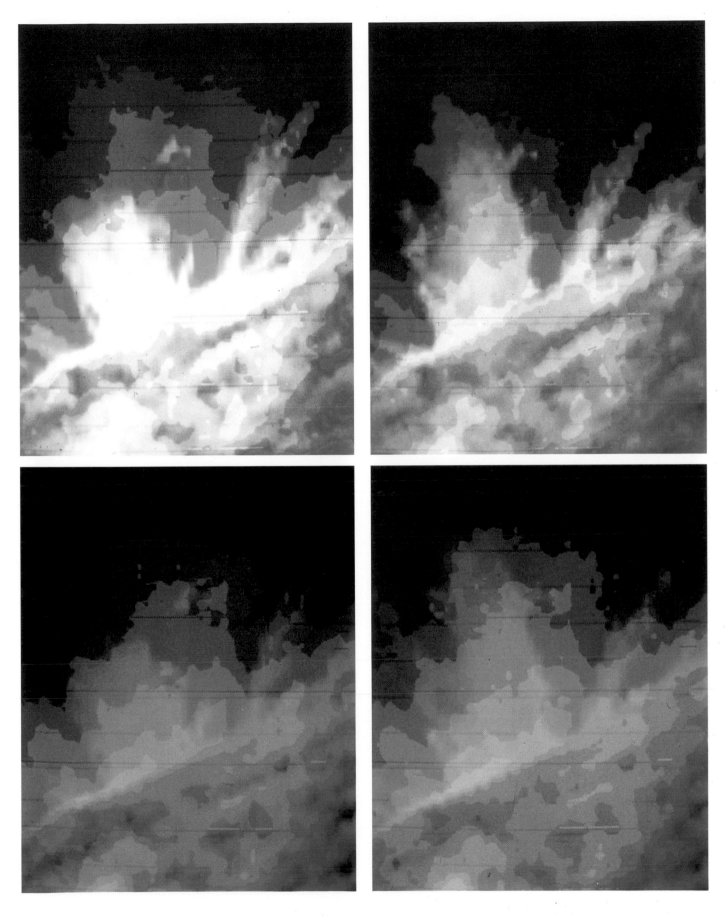

Solar limb
Skylab
5 January 1974

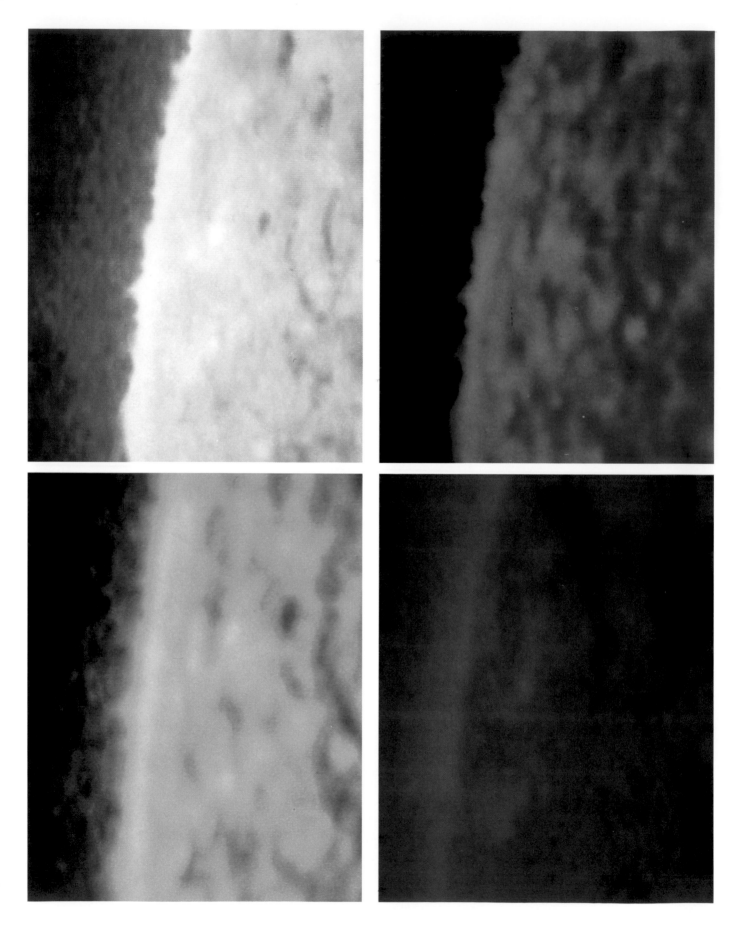

Active region
Skylab
14 August 1973

X-ray Sun
Skylab
16 June 1973

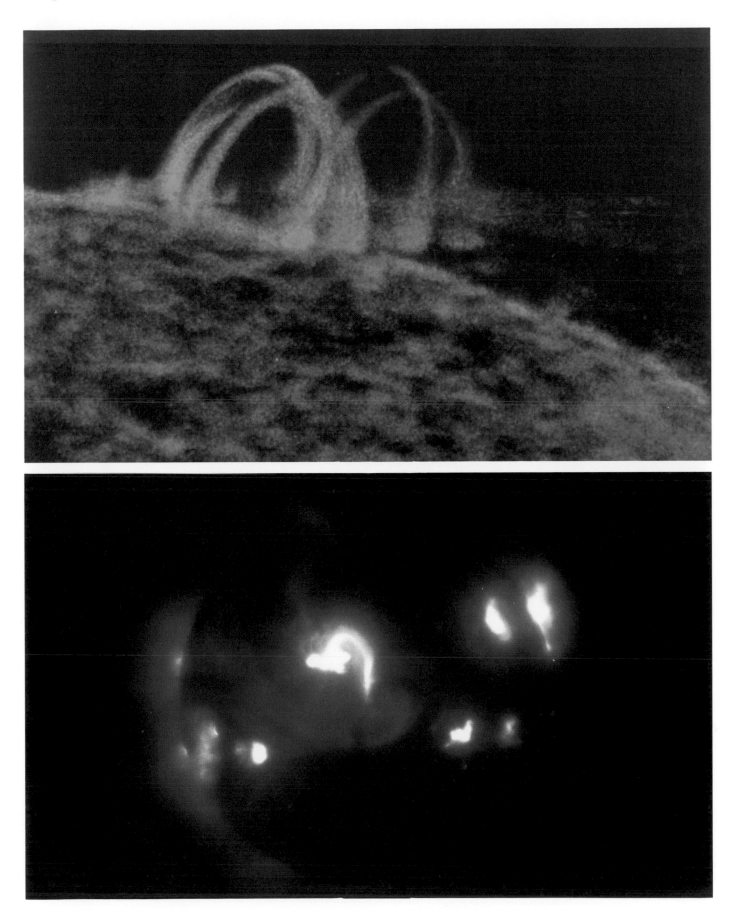

Active region
Skylab
14 August 1973

X-ray Sun
Skylab
16 June 1973

X-ray image (false color)
Skylab
1 June 1973

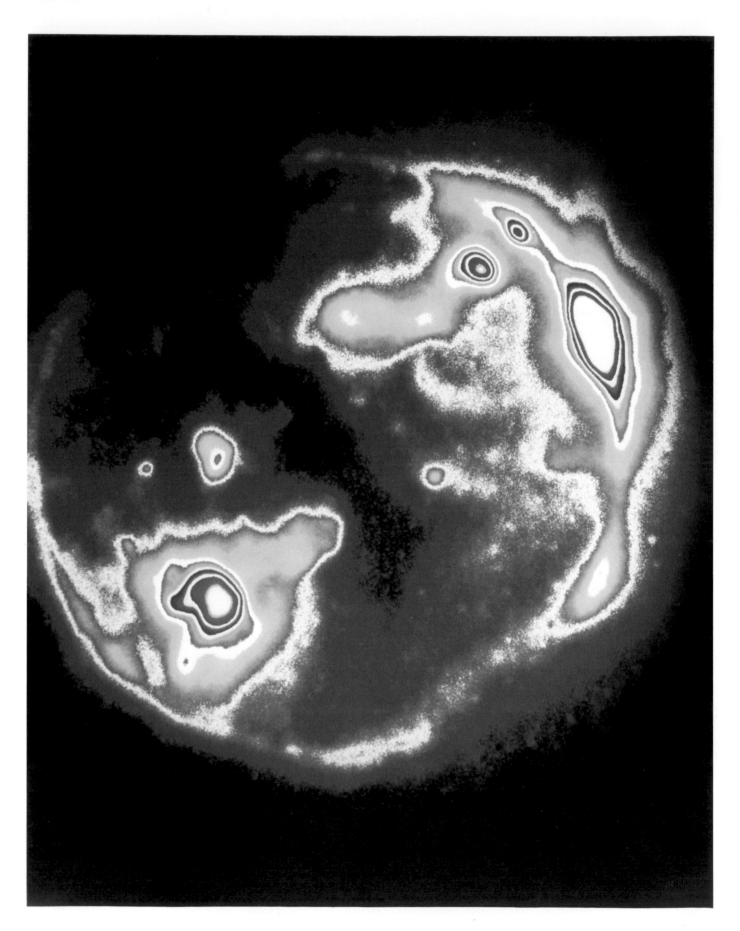

Solar corona (false color)
High Altitude Observatory
10 June 1973

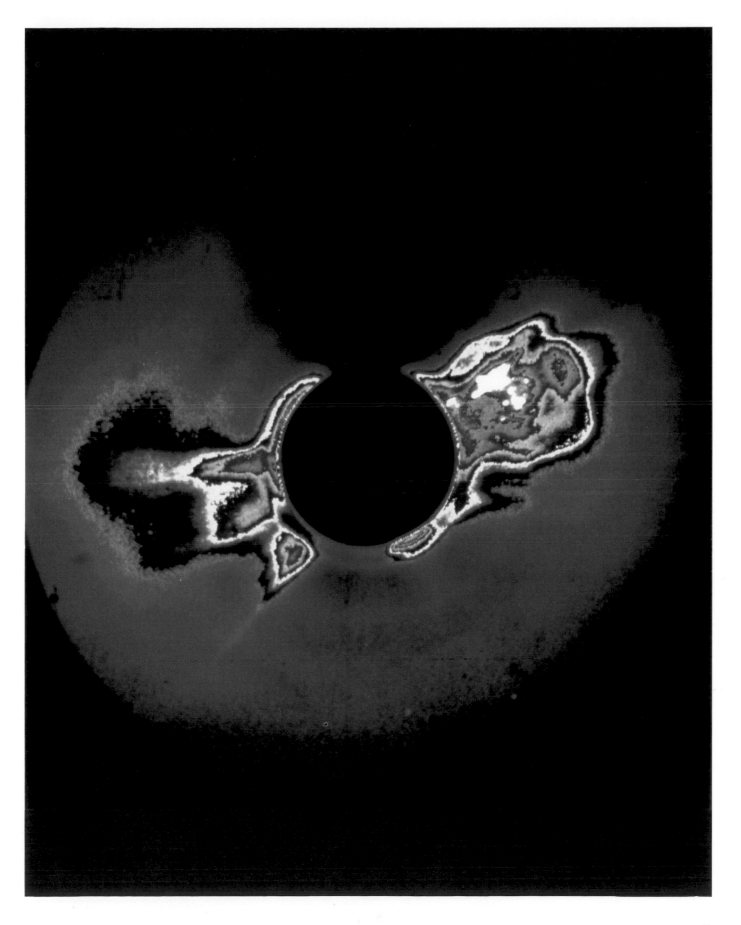

X-ray image
Skylab
21 August 1973

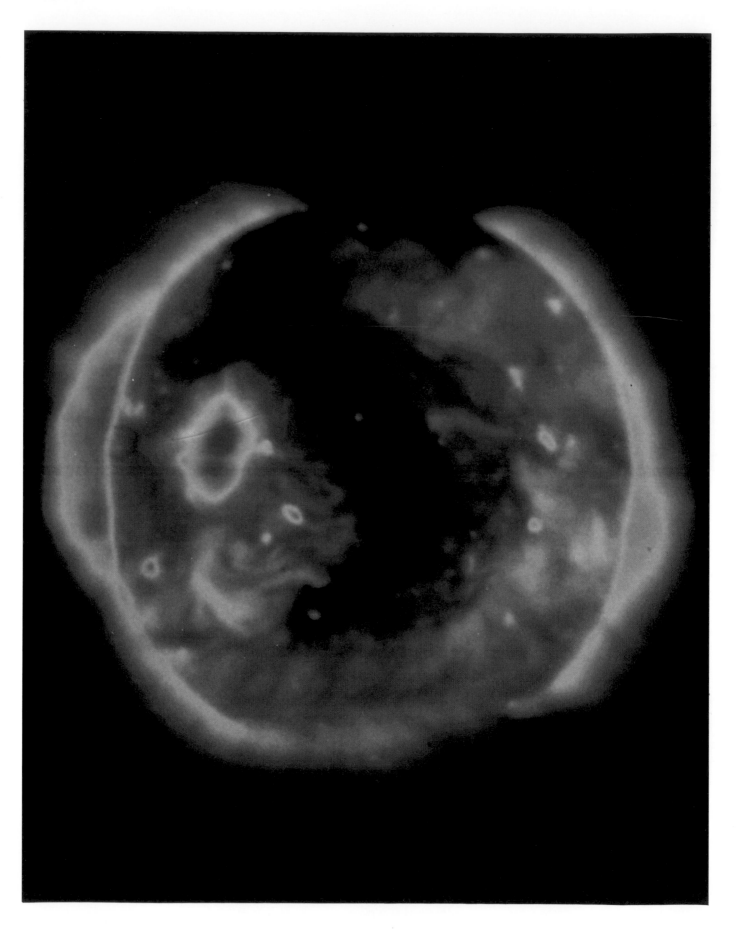

Extreme ultraviolet image (false color)
Skylab
20 August 1973

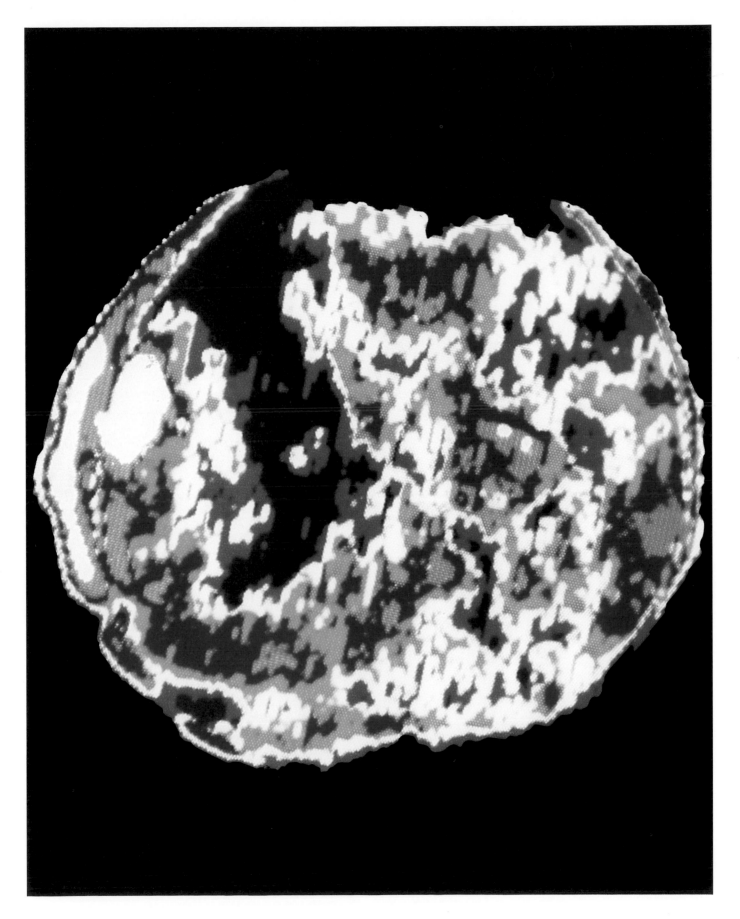

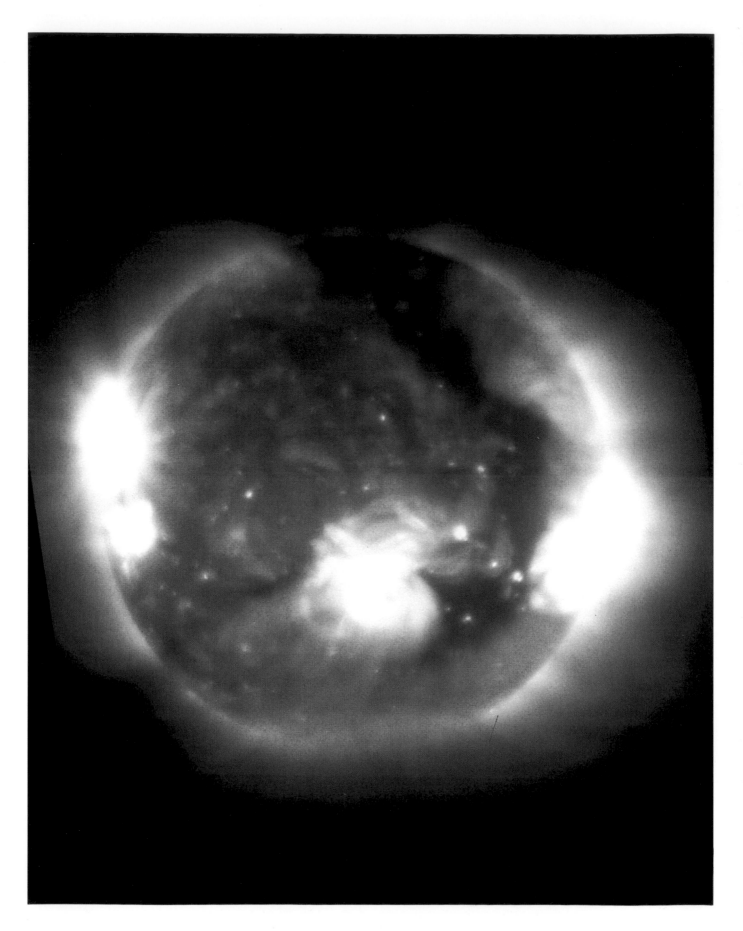

**Notes to color plates**

**Page 49.** Much fine structure is visible in these magnificent photographs of the corona. Several prominences can also be seen.

**Page 50.** (Top) The visible portion of the Sun's spectrum reveals hundreds of dark absorption lines, each produced by a particular element at a particular level of ionization. For example, the two very broad lines near the center of the top row (in the violet region) are characteristic of singly ionized calcium (Fraunhofer's H and K lines). (Bottom) A number of chromospheric emission lines can be seen in this spectrum photographed at second contact: $H\gamma$ in the dark violet (434 nm), $H_\beta$ in the blue (486.1 nm), the strong $D_3$ line due to neutral helium in the red, and $H\alpha$ (656.3 nm). The bright horizontal line is the result of light from a Baily's bead.

**Page 51.** These two ultraviolet images illustrate the way in which the chromospheric network becomes indistinct at high altitudes (*see* page 29). At a wavelength of 97.7 nm (top), representing chromospheric temperatures of around 20,000 K, the network is clearly defined; in the coronal light of 62.5 nm (bottom)—representing a temperature of about $1.4 \times 10^6$ K—the pattern has virtually disappeared.

**Page 52 and 53.** The first image (top left) is a composite of the other three, each of which was taken at a different ultraviolet wavelength: 121.6 nm (bottom left), 97.7 nm (top right), and 62.5 nm (bottom right). 121.6 nm is the Lyman-alpha line of hydrogen; this image shows the chromospheric network in the middle chromosphere. The next image (a wavelength due to CIII) represents a level higher in the chromosphere; the spicules are seen end-on, and appear as bright speckles. The final image is in the light of MgX, and shows the corona, with coronal holes at the polar regions (*see* pages 66–7).

**Page 54** (Top). This false-color image shows ionized material following the magnetic loop patterns above an active region. The wavelength is 103.2 nm, due to the light of OVI. (Bottom). A similar loop appears in this SMM image taken at a wavelength of 154.8 nm (CIV).

**Page 55.** An eruptive prominence (*see* pages 40–41) is shown here in a computer-enhanced image taken in the light of the HeII (30.4 nm).

**Pages 56 and 57.** Two sequences (reading from left to right across the page) show the rapid development and eruption of chromospheric material breaking loose of the Sun's magnetic loops. Both sequences are computer-enhanced ultraviolet composites.

**Page 58.** An ultraviolet composite (top left) with three component images: 97.7 nm (CIII), 103.2 nm (OVI), and 62.5 nm (MgX). The high chromosphere (top right) has been colored blue, and shows sharply defined spicules at the limb. The transition region between the chromosphere and corona has been colored green (bottom left), while the corona itself (bottom right) is shown in red.

**Page 59** (Top). A weaker secondary image appears next to the predominant NeVII image at a wavelength of 46.5 nm. (Bottom). High-temperature X-ray emissions from the corona, reaching a maximum of over $5 \times 10^6$ K, appear concentrated along a band near the equator. The bright curve near the center represents a huge arch.

**Page 60.** In this X-ray image, false colors have been used to represent different temperatures, with white showing the highest temperature. A large coronal hole (black) extends downward from the north polar region.

**Page 61.** The temperature structure of the corona is revealed in this false-color image of a white light coronagram.

**Page 62.** An X-ray image of the solar disc shows a large "S-shaped" coronal hole extending from the Sun's north polar region towards the equator.

**Page 63.** A false-color enhancement of an extreme ultraviolet image showing the same coronal feature as that seen in the X-ray image on page 62.

**Page 64.** This X-ray image shows the corona as it appeared a month before the images on pages 62 and 63. The developing coronal hole is already apparent.

# Coronal Holes and Transients

Coronal holes delineate regions of weak magnetic field in which the field lines are "open"—in other words, they do not curve around to form closed loops connecting regions of opposite polarity. Instead they spread out into interplanetary space, diverging rapidly so that the angular size of a coronal hole increases greatly with increasing distance from the Sun: for example, at a distance of $3\,R_S$ the angular area spanned by a hole can be about seven times greater than it is at a level near the base of the corona. Holes show up as dark patches in ultraviolet and X-ray photographs and can cover a large proportion of the visible disc; they are normally present in the polar caps, and it is from these polar coronal holes that the polar plumes emerge, having their bases in bright points.

With temperatures of about $1 \times 10^6$ K they are two to five times cooler than coronal active regions, and cooler, too, than the less bright quiet regions; densities are also correspondingly lower, at about 30 percent of the neighboring bright regions.

Out of coronal holes flow streams of particles, following the field lines and making up the "solar wind" (*see* pages 68–9). These particles emerge from the coronal holes and are accelerated to their maximum velocities within 2 to $3\,R_S$ of the surface.

Since the discovery of coronal holes their evolution has been observed in some detail. In outline, the process begins with the appearance of a small isolated hole near the equator. This feature grows in size, rotating with the Sun as if it were attached to a solid surface, and eventually links up with the polar region of the same magnetic polarity. The hole then begins to shrink in area (*see* diagram 1). The entire sequence of events lasts for up to seven or eight months.

## Coronal transients

First studied in detail by the coronagraphs aboard Skylab, coronal transients take the form of giant loops of coronal material hurled out through the corona into interplanetary space. Usually triggered by flares involving mass ejection or, more frequently, by eruptive prominences, these loops or arcs travel outward at speeds of between 200 and 900 km s$^{-1}$ (according to Skylab and SMM measurements). Observations made in 1980 close to solar maximum showed that these events spanned a broader range of latitudes (the first 22 events studied spanned 68°N to 54°S) than the Skylab events, which had been observed at a time relatively closer to sunspot minimum (less than 50° latitude). They were also less strongly clustered around the equator; in this they mirror the latitude behavior of phenomena such as coronal extensions, active regions and prominences.

Throughout the duration of the SMM, commencing in 1980, space-borne and ground-based observations were correlated; in particular the SMM coronagraph was used in conjunction with the radioheliograph at Culgoora, Australia, to yield spectacular observations of coronal transients. The loops themselves are relatively thick and structureless and are followed by cavities of less than average coronal density as they move outward. Typical transients of the Skylab era had masses of about $5 \times 10^{12}$ kg and moved at speeds of 470 km s$^{-1}$; the energy ultimately contained in this material was about 10 times greater than the radiative energy observed in the initiating flares, and the magnetic energy involved in the restructuring of the field may have been even greater. Thus the total active region energy release associated with flares over

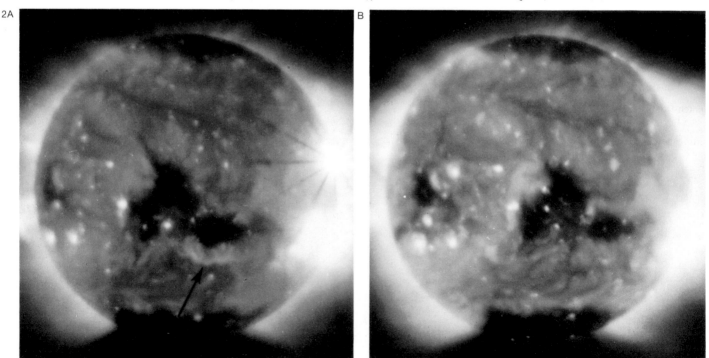

**1. Evolution of a coronal hole**
Small, isolated coronal holes (**A**) appear to link up with one of the polar caps (**B**). Later the hole contracts to the polar region (**C**).

**2. Coronal hole**
These X-ray images showing a coronal hole were taken on 9 and 10 October 1973. A transient X-ray brightening is indicated by an arrow (**A**).

time scales much larger than the flare itself goes mainly into magnetic and mass motion forms. However, during the flare proper (triggered by the slow evolution of the active region structure) power is released more or less equally into radiation and mass motion. Transients seem to be responsible for major rapid restructuring of the corona. For example, in the case of the event of 7 April 1980, before the transient there was only one streamer visible in the quadrant over which it occurred, but afterward there were five.

The frequency, scale and violence of these events has led to a significant modification of the view of the corona and its structure. Coronal transients are (geometrically) the largest scale impulsive events so far observed and, since their effects propagate through the solar wind to the Earth, they can be noticed here.

The corona, then, is seen to be entirely different from the earlier conception of a relatively smooth, quiet, homogeneous solar atmosphere. Instead its very existence as well as its detailed and overall structure has been shown to be in part determined by the existence of closed magnetic loops that both channel the energy flow and, possibly, provide a major source of heating. It is entirely non-homogeneous, having a clumpy structure comprising hot, bright, relatively dense active regions, cooler, less bright, quiet regions, and even cooler, faint, rarefied holes from which flows the solar wind. Finally it is seen to be wracked by powerful transient events by means of which material is propagated through the corona at speeds of up to 1,000 km s$^{-1}$ leading to major restructuring. The fact that these events have a high frequency even near solar minimum implies that they are not tied to the most violent flare events: rather they tend to be associated with prominence activity.

3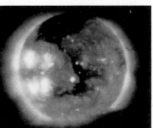

**3. Coronal transient**
This disruption of the corona occurred on 10 June 1973, triggered by an eruptive prominence. The disturbance travelled rapidly through the corona, taking less than half an hour to cover a distance of nearly a million kilometers

**4. X-ray images of coronal hole**
This sequence of images shows the development of a coronal hole during six solar rotations. Unlike photospheric and chromospheric features, coronal holes appear to rotate as if they were attached to a solid object.

**5. Polar hole**
In this negative image the polar hole shows up as a bright "bald" patch from which polar plumes may be seen emanating. The image was made in the light of MgIX (at a wavelength of 36.8 nm) in the ultraviolet.

4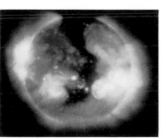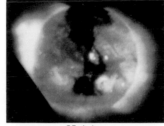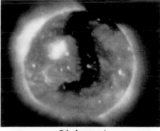

27 June     25 July     21 August

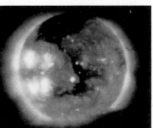 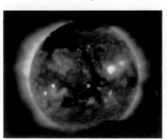 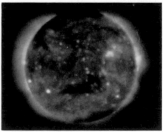

16 September     14 October     10 November

5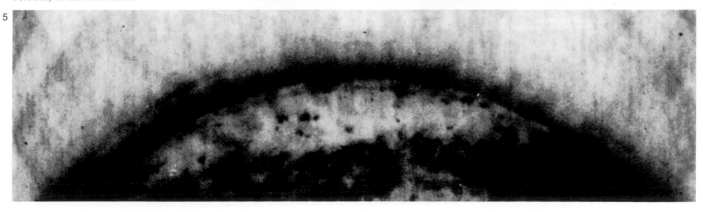

# Solar Wind

It was suggested in the nineteenth century that the Sun might be the source of transient clouds of plasma that travelled through interplanetary space, impinging on the Earth's atmosphere and giving rise to phenomena such as the aurorae and magnetic storms (disturbances in the Earth's magnetic field). The idea was the subject of a major paper by Sir Oliver Lodge, published in 1900, and in 1932 J. Bartels pointed out that moderate magnetic storms, presumably not caused by violent transient solar flares, had a tendency to recur at 27-day intervals. This interval is the synodic rotation period of the Sun, and Bartels postulated the existence of regions of the Sun that were responsible for producing magnetic disturbances. Such regions came to be known as "M-regions".

Evidence for a more widespread and continuous emission of particles from the Sun came from studies of the tails of comets. By 1958 E. N. Parker had developed a model of the corona that showed that it must be in a constant state of expansion, losing material into the interplanetary medium in all directions in a phenomenon that Parker called the "solar wind". Parker showed that because of the very high temperature of the corona the pressure exerted would ensure an outward flow of material: even at very great distances this pressure could not be contained by any known external pressure. Just as a sand pile will always flow out if its slope exceeds a certain value, so the corona would expand and lose material, which would be replaced from below. Spicules and macrospicules may be major sources of this material from the chromosphere, but the exact mechanism for the supply of matter is not yet clearly understood.

Direct confirmation of the existence of a continuous solar wind came from Soviet and American interplanetary space probes (in particular, Mariner 2) and the solar wind has subsequently been studied *in situ* by a wide variety of spacecraft.

## Properties of the solar wind
The solar wind, although gusty and variable, blows continuously with velocities that can range from as low as 200 km s$^{-1}$ to as high as nearly 900 km s$^{-1}$. The typical value is between 400 and 500 km s$^{-1}$, and the particles normally take about four or five days to reach the Earth. The wind consists of a roughly equal number of electrons and protons, together with a small proportion of heavier ions and nuclei, of which the most prominent are alpha particles (helium nuclei), which comprise, typically, about 4 to 5 percent by number of the total. At the Earth's distance from the Sun, the Earth's particle density is on average some $5 \times 10^{-6}$ m$^{-3}$, but it can vary by a factor of over 100. (For comparison the particle density at sea level in the Earth's atmosphere is $2.5 \times 10^{25}$ m$^{-3}$.) The temperature of the plasma is defined by the "velocity dispersion" of the particles—the magnitude of their random velocities relative to the mean flow of the wind. Near the Earth this temperature is on average about $10^5$ K, and the Earth is thus enshrouded by high temperature (but extremely rarefied) plasma. These parameters indicate that the Sun is losing about $10^9$ kg s$^{-1}$ to the solar wind. At this rate, however, the entire solar mass would take some $6 \times 10^{13}$ years to blow away completely, which is $10^4$ times longer than the Sun will continue in its present form.

## The interplanetary field
Close to the Sun the entire structure is controlled by closed magnetic loops that contain most of the material (*see* page 47), but at larger distances, open magnetic field lines spread out into interplanetary space, somewhat like the spokes of a wheel. Within a few solar radii this magnetic field causes the corona and solar wind plasma to co-rotate with the globe of the Sun. However, beyond a certain distance, the energy of the plasma exceeds that of the field. Although individually the solar wind particles move out more or less radially from the Sun, collectively they form a spiral pattern, in rather the same way as water droplets from a garden sprinkler. Because the solar wind is ionized, it is electrically conductive, and the field is frozen into the plasma. Thus the field lines must also take up the spiral pattern of the particle stream. By about the Earth's distance from the Sun, the field lines of the interplanetary field

**1. Idealized magnetic field**
A simple "dipole" field structure (shown as areas of tone) has been imposed on a simplified model of the corona: the resulting pattern of field lines reveals a "neutral sheet" extending from the equator from a distance of about 2 R$_\text{S}$.

**2. Solar wind**
As the Sun rotates, the particles of which the solar wind is composed fan out rather like droplets of water from a garden sprinkler: close to the Sun, however (within the circumscribed region in the diagram), the Sun's magnetic field is sufficiently strong to cause the particles to co-rotate with the Sun as though they were rigidly attached.

**3. Model of magnetic structure**
In these schematic diagrams, the Sun's magnetic structure at the base of the corona is represented as consisting of two components: a pattern of alternating positive and negative polarities near the equator (**A**), and tilted dipole field at the poles (**B**). The two rotate at different rates, and their sum depends on their relative phase.

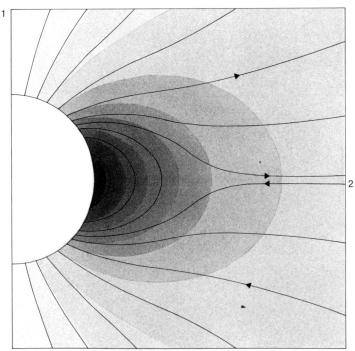

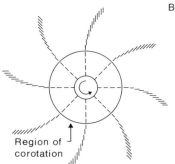

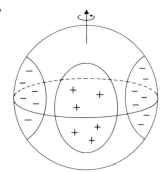

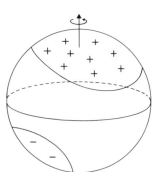

Region of corotation

make an angle of some 45° with the radial direction from the Sun, while the solar wind particles themselves move almost radially (*see* diagram 2). Observations show that although on the small scale the field lines wave around in a chaotic way, when averaged over a longer period they follow the expected spiral pattern.

Near the solar equator the field is usually found to be divided into two or four sectors of alternate polarity. These sectors are rooted in the large-scale regions of the photosphere, which have a net single polarity (*see* diagram 3). Since the field lines carried by the wind are rooted in the photosphere it is hardly surprising that the interplanetary field should mirror the sector polarities evident in the photosphere.

A consistent view of the nature of these sectors and the source of the high-speed streams is now beginning to emerge. Around sunspot minimum, active regions are concentrated in the equatorial regions of the Sun, and the net polar fields are well established. If the solar field were a simple dipole field it could be represented by closed magnetic structures in the inner corona at the equator, and open field lines emerging from higher latitudes. The open lines would bend over the closed equatorial loops and eventually extend out radially above and below the solar magnetic equator. A "neutral sheet" would be formed across which the direction of the field lines is reversed and this sheet would spread out through the interplanetary medium. However, this simple pattern is disturbed by the sectors of net single polarity in the photosphere, which warp the neutral sheet. At the solar surface the neutral line, instead of following a straight line along the equator, waves to and fro across the equator in a roughly sinusoidal form between heliographic latitudes of around ±40° (*see* diagram 4). The pattern extends outward, to produce a warped neutral sheet in interplanetary space, and as the Sun rotates the warped boundary passes the Earth at regular intervals, giving an Earth-based observer the impression of sectors of opposite magnetic polarity (depending on whether the neutral sheet passes above or below the Earth). The duration, and hence angular diameter of the sectors, varies in a seasonal fashion

as the Earth passes above or below the solar equator.

The magnitude of the seasonal effect indicates that the vertical extent of the distortions is only about ±0.25 A.U. at the distance of 1 A.U. Thus the spread in latitude of the wobbles may be only about 15° at 1 A.U. compared to 40° at the solar surface. This assertion has been confirmed by space-probes that have moved some distance from the ecliptic. For example, Pioneer 11 (which flew past Jupiter in 1974 and Saturn in 1979) reached 16° above the ecliptic in February 1976 and experienced an interplanetary field with a constant direction for several months. In effect it was for a time above the warped neutral sheet and was experiencing essentially one polarity of an overall dipole field.

The solar wind speed is low over any neutral line or sector boundary and attains high speed away from such a boundary. If the warps in the neutral sheet are sufficiently great, a high-speed stream will be observed when each resulting "sector" passes by. These sectors are often, but not always, filled by a coronal hole. Since large coronal holes are almost always present at each pole, it may be that the major part of the solar wind flows from these holes. Very high-speed streams (750 km s⁻¹) have been observed when the polar coronal holes have extensions towards the solar equator, and it would appear that high-speed streams emanate from polar regions, or from any coronal hole.

The overall effective dipole field—particularly when well established, as it is around sunspot minimum—controls the general form of the interplanetary field and the flow of solar wind particles in such a way that it is likely that the major part of the wind emanates from the polar regions. The high speed of the streams emanating from the polar hole extensions is probably typical of the wind flow over those major parts of the hemispheres away from the vicinity of the warped neutral sheet, above and below 20° or so of the solar equatorial plane. The Out-Of-The-Ecliptic/Solar Polar mission, scheduled for the early 1980s, will travel via Jupiter before doubling back over one of the Sun's poles, and should greatly clarify a phenomenon which is so far unexplored.

**4. Coronal brightness maps**
In an idealized situation (**A**), the Sun's tilted dipole field would give rise to a sinusoidal wave pattern in the neutral line. (**B**) is an actual example showing the coronal brightness as measured for Carrington rotation 1602.

**5. Three-dimensional model**
The Sun's magnetic and density structure for Carrington rotation 1602 (the same as diagram 4B) is illustrated as it might be imagined in three dimensions. The neutral sheet is distorted in accordance with the inferred pattern.

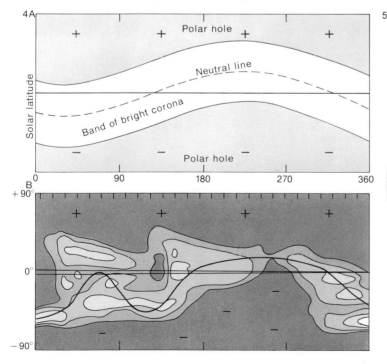

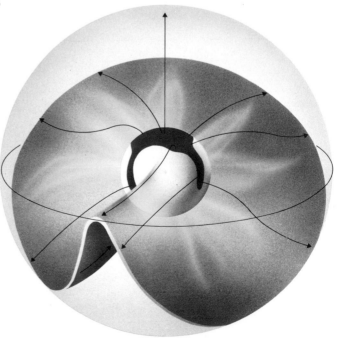

# Solar Energy

Over the years solar physicists have built up a convincing model of the interior of the Sun and the mechanism by which it produces its energy. According to this "standard model", the density, temperature and pressure increase sharply towards the center of the Sun, where the temperature is estimated to be about 15 million K; the density is estimated at $1.58 \times 10^5$ kg m$^{-3}$ (about 12 times the density of lead) and the pressure $2.5 \times 10^{11}$ Earth atmospheres. The core is believed to extend out to about $0.25\,R_S$ and, although this comprises only one sixty-fourth of the Sun's volume, it is believed to contain about 50 percent of the total mass; within this zone about 99 percent of all solar energy is generated. However, most (50 percent) of this energy is generated in only 5 or 10 percent of the solar mass. Despite the very high pressure, the intensely high temperature ensures that the central matter is kept in a fully ionized gaseous state, comprised mainly of protons and electrons moving about at very high speeds. A state of "hydrostatic" equilibrium exists, in which the Sun's tendency to collapse under its own gravitational self-attraction is balanced by the outward pressure of the hot gas inside.

It is assumed that originally the whole of the Sun had the same chemical composition as is now evident in its outer layers, with proportions (by mass) of about 78 percent hydrogen, 20 percent helium and 2 percent other elements (principally carbon, nitrogen and oxygen—*see* pages 14–15), and that the interior has been modified by thermonuclear reactions so that now the fraction of hydrogen in the core is 36 percent, about half the value at the surface. Beyond about $0.25\,R_S$ the composition is believed to be essentially uniform and equal to the surface value.

### The proton-proton chain
The reaction believed to be principally responsible for the production of energy in the solar core is the "proton-proton" reaction or "p-p reaction" (*see* diagram 1). In a series of stages, four hydrogen nuclei (protons) are fused together to form one helium nucleus, comprising two protons and two neutrons. The helium-4, once formed, is stable because the temperature of the core is too low for the next stage of thermonuclear reaction, involving carbon, to take place. The mass of the end product is 0.7 percent less than the

components that went to assemble it: this small percentage loss of mass is converted to energy. The well-known relationship between mass (m) and energy (E), derived from the Special Theory of Relativity, is $E = mc^2$, where c is the speed of light. Since c is a large number, a very large amount of energy can be released by the destruction of small quantities of matter. In order to sustain the present luminosity just over $4 \times 10^6$ tonnes of matter must be converted into energy every second by this process.

Several different variations can occur in the reaction, but 95 percent of the helium nuclei are produced by the basic p-p chain.

### The solar neutrino problem
Although there is no real doubt about the basic process that sustains the Sun's output—the fusion of hydrogen to form helium—serious doubts have recently been raised about the precise chain involved. Two of the principal strands of observational evidence responsible for these doubts concern the observed flux of neutrinos, and the solar oscillations (*see* pages 84–85).

The first step in the proton-proton reaction releases neutrinos. Neutrinos are a common type of "particle" (symbol $v$) with zero electrical charge and (it is usually assumed) zero rest mass. Neutrinos travel at the speed of light and very rarely interact with matter, passing through solid objects almost as if they were transparent. Thus neutrinos created at the Sun's core are able to emerge from the surface in a fraction of a second. The neutrinos produced in the p-p reaction turn out to be of rather low energy, usually less than 0.42 MeV. In the comparatively rare alternative routes of the reaction chain higher energy neutrinos are released, particularly the very rare route in which beryllium is converted into boron before breaking down to form helium-4. The beryllium–boron chain occurs only in about 0.01 percent of reactions.

The Sun, then, should be emitting neutrinos, most of rather low energy, with a small proportion of high-energy neutrinos. The rate of their production depends sensitively on the internal temperature of the solar core, so that, if it were possible to measure the flux of neutrinos emerging from the Sun, this would provide a measure of the internal temperature. However, since neutrinos can penetrate the Sun so easily, it is evident that they are extremely difficult to

### 1. Proton-proton chain
The main source of the Sun's energy is believed to be a chain of nuclear reactions in which hydrogen nuclei (protons) are converted to give helium (an atom containing two neutrons and two protons). The simplest route involves only three steps (**A-B-C**). Two protons combine to give a deuterium nucleus (a proton and a neutron), which emits a positron and a neutrino (**A**); the deuterium then combines with a proton, yielding a nucleus of helium-3,

which emits a photon (**B**); finally the helium-3 combines with another helium-3 nucleus (formed in the same way), producing a nucleus of ordinary helium-4 plus two extra protons (**C**). However, several variations in this chain are possible (**A-B-D-E-F**, or **A-B-D-G-H-I**). In these rarer alternative routes, involving beryllium-7, lithium-7 and boron-8, neutrinos with different energies are emitted. The neutrinos given off by boron-8 (**H**) should be detectable by the Brookhaven instrument.

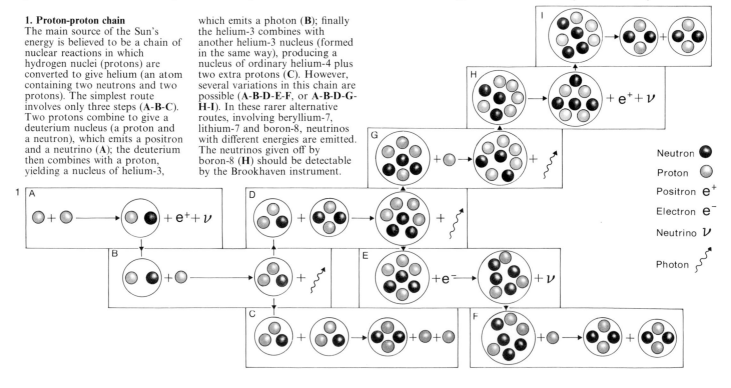

Neutron ● 
Proton ○ 
Positron $e^+$ 
Electron $e^-$ 
Neutrino $v$ 
Photon ∿

capture and measure. Nevertheless, in 1964 a neutrino "telescope" was established by the Brookhaven National Laboratory capable of detecting solar neutrinos. The instrument consisted of some 400,000 liters of perchloroethylene ($C_2Cl_4$)—dry-cleaning fluid—containing large amounts of chlorine in a tank located at the bottom of a mine in South Dakota, at a depth of some 1,500 m. The isotope chlorine-37, which comprises about 25 percent of the chlorine content of the dry-cleaning fluid, is effective at capturing the energetic neutrinos. In the process it is converted into radioactive argon (and an electron is given off). The radioactive argon is flushed from the tank by means of helium, and collected in a charcoal trap cooled to the temperature of liquid nitrogen (77 K). The decay of the argon is registered on particle detectors and the number of argon atoms is measured. Thus the flux of neutrinos entering the detector can be determined.

The chance of an individual chlorine-37 atom capturing a neutrino is very tiny indeed: if standard theory is correct, a given chlorine-37 atom in the detector should capture a neutrino once in about $10^{28}$ years—far in excess of the estimated age of the universe. However, the container holds a very large number of atoms. Neutrino astronomers have invented a unit of measurement known as the solar neutrino unit (SNU), where $1 \text{ SNU} = 10^{-36}$ neutrino captures per second per target atom. According to the standard model, the detector should measure between 6 and 7 SNU, equivalent to about 1 neutrino capture a day. In the 15 or so years that the experiment has been operating it has failed to detect anything like the expected number of neutrinos. The measured flux is 2.2 SNU, about one third of the predicted value.

If the temperature of the Sun's core were about 10 percent lower than implied by the standard model it would explain the low flux of neutrinos; on the other hand, a lower core temperature would reduce the Sun's luminosity to well below the observed value—unless the Sun is also less opaque than is supposed. But a lower opacity would in turn imply a lower proportion of heavy elements in the outer region, and this would be in conflict with the estimated age of the Sun. It has been suggested that the outer layers of the Sun were accreted from interstellar materials, which might account for a difference in composition between the interior and the exterior.

Another explanation is that the output of energy from the core fluctuates over long periods of time, and that the present level of photospheric radiation characterizes an earlier era when the core was more energetic (photons taking $10^7$ years to escape) while the neutrino flux represents the present level of core activity. More erratically it has been conjectured that a tiny "black hole" resides in the solar interior, digesting material and contributing about half of the radiation output.

A different approach to the problem is to argue that it is the nuclear theory that is at fault. On the one hand, there is a degree of uncertainty about the rate of reactions involving boron; on the other, it may be that more needs to be known about the neutrinos themselves. It has been argued that if the neutrino had a tiny but finite mass it would oscillate between two or more states: there are known to be two types of neutrino ("electron family" neutrinos and "muon family" neutrinos), and there is good reason to suppose that a third type ("tau-neutrinos") also exists. If the neutrinos produced by the decay of boron are distributed fairly evenly between the three types, then the expected rate of detection would be reduced by one third, since chlorine-37 is only capable of detecting electron-family neutrinos. More refined calculations indicate that this effect would only reduce the expected flux by half.

What is needed is a form of neutrino detector capable of picking up the much more common low-energy neutrinos produced in the first stage of the p-p chain. The element gallium appears to offer one of the best possibilities. Plans are afoot to build a gallium detector of some 50 tonnes in the Homestead mine, but funding is a problem: 50 tonnes of gallium—well in excess of annual world production—would cost some 50 million dollars. A 20-tonne gallium detector is under development. Meanwhile the Soviet Union has established a neutrino observatory deep under the Caucasus mountains. A neutrino detector filled with 330 tonnes of white spirit is already complete and should be capable of improving sensitivity over the Brookhaven detector by a factor of about 5. Indium-115 provides another viable detection agent; about 4 tonnes of indium would suffice to achieve a detection of about 1 neutrino per day over the whole energy range. The cost would be between 10 and 20 million dollars.

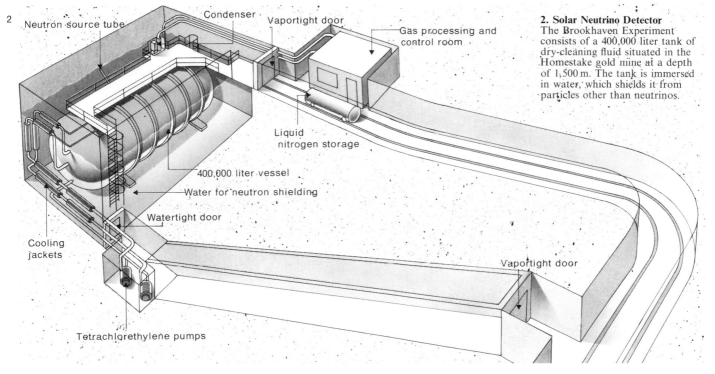

2

Neutron source tube

Condenser

Vaportight door

Gas processing and control room

Liquid nitrogen storage

400,000 liter vessel

Water for neutron shielding

Watertight door

Cooling jackets

Vaportight door

Tetrachlorethylene pumps

**2. Solar Neutrino Detector**
The Brookhaven Experiment consists of a 400,000 liter tank of dry-cleaning fluid situated in the Homestake gold mine at a depth of 1,500 m. The tank is immersed in water, which shields it from particles other than neutrinos.

# Solar Oscillations

In 1973, in the process of making very precise measurements of the ratio of the polar to equatorial diameters of the Sun, R.H. Dicke discovered that the Sun is oscillating, vibrating like a jelly, in such a way that the equator bulges as the poles are flattened, and the poles are elongated as the equator contracts (*see* diagram 1). Since that time it has been established that the Sun oscillates with a variety of periods from about 5 min to just over 1 hr. The difficulty in making such observations can hardly be overstressed: the maximum amplitude (vertical motion) of the solar surface is only about 5 km and maximum velocity about $10 \, \text{m s}^{-1}$. There are many stars, including conspicuously variable stars such as Cepheids and Long-period variables, that oscillate in a much more pronounced fashion, and this phenomenon was well established in these examples long before the discovery of solar oscillations.

It is not very surprising that the Sun should oscillate in this way. Just as a bell has a natural frequency of oscillation at which it will tend to vibrate when it is struck, so the Sun has natural frequencies at which it will vibrate given a suitable "hammer". A natural oscillation like a bell's vibration dies away due to internal damping unless it is struck periodically in order to maintain its vibration. Likewise, within the Sun there must be a mechanism that keeps the oscillations going: it is assumed that fluctuating nuclear processes in the core are in some way responsible.

Global oscillations of the Sun offer astronomers a means of studying the structure of the solar interior, just as seismic waves from phenomena such as earthquakes have made it possible to determine the interior structure of the Earth. By analyzing the frequencies of solar oscillations attempts are being made to achieve an equivalent understanding of the Sun's interior. Although the range of oscillation periods is broadly consistent with the standard model, there appear to be discrepancies; for example it appears that the range of frequencies is not consistent with the value for the (assumed) depth of the convective zone.

The 5 min oscillation had been discovered as early as 1960, and is generally believed to be a surface ripple affecting only the outermost 10,000 km or so of the solar globe. The other frequencies that have been measured are believed to represent the fundamental radial oscillation of the entire Sun and its harmonics. If the standard model of the interior is correct, then a fundamental period of about 1 hr would be expected and, on the face of it, the observed results seem roughly consistent with the way the Sun's interior is believed to be constructed. However, it has been pointed out that if the oscillations arise in the deep interior, then—because of damping mechanisms—the oscillation seen at the surface should be weaker than those in the interior. Attempts to calculate the oscillation magnitudes towards the core required to match the observations appear to indicate that they would become of such great amplitude that they would disrupt the solar interior.

In 1976 a Soviet research team and a group from Birmingham, England, independently announced the discovery of a long-period global oscillation of 2 hr 40 min. These measurements were not direct measurements of the motion of the solar limb but spectroscopic measurements of selected spectral lines. The Doppler shifts in these lines were measured by comparison with laboratory sources, and periodic variations were interpreted as oscillations of the solar surface as it alternately approached and receded from the observer. If this long-period oscillation were a true fundamental period, then the standard model could not be correct: to account for such a period, the Sun would have to be more homogeneous throughout, and the density could not increase so sharply towards the center as proposed by the standard model. Moreover, the core would be too cool to account for the Sun's luminosity by the proton-proton reaction. Indeed, it appears that if the Sun was shining by means of the p-p reaction *and* if the 2 hr 40 min period was the fundamental oscillation, then the central temperature would be less than half the conventional value, and the luminosity would be less than one ten-thousandth of the observed value.

It was suggested that the 2 hr 40 min period was not an oscillation at all but an apparent effect induced by supergranulation. As a result of the rotation of the Sun, supergranular cells of 15 to 30 thousand km diameter would cross the field of view in periods of between 2 and 4 hr. Since these cells contain rising gas, which

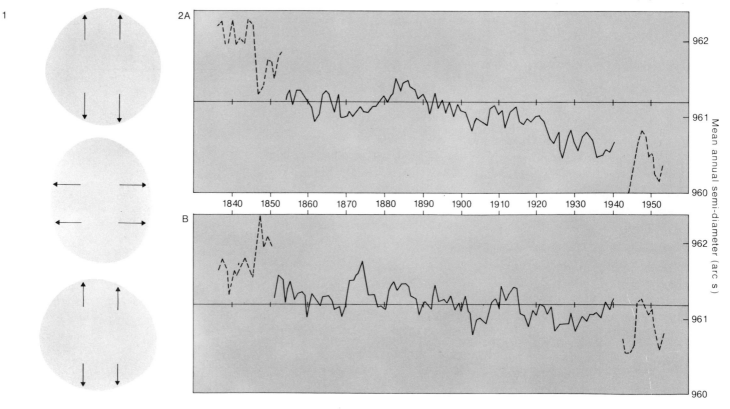

spreads out and falls back to the interior, a significant number of cells moving across the strip of Sun studied by the Soviet instrument would introduce an apparent periodic Doppler shift. However, it does not appear to be possible to explain away the long-period oscillation in this way, and it has been confirmed by other groups since then; there is even suggested evidence for a small periodic variation in the geomagnetic field consistent with the propagation of the surface disturbances by means of the solar wind. Another objection raised was that the observed variations might result from the fact that the distance travelled through the Earth's atmosphere by light from the Sun changes with the Sun's varying altitude in the sky. This possibility seems to be eliminated by the work of a French group at the South Pole, where the solar altitude is near constant. Further checks will soon be made by simultaneous observations from remote sites. It seems certain there is some periodic effect to explain, but whether the oscillation is a true global oscillation or a surface effect, or a "gravity wave" like waves in the ocean, remains a matter of debate.

More recent analyses of the 2 hr 40 min oscillation suggest that it could be accounted for by reducing the core temperature by about 10 percent. This would leave a major problem in accounting for the observed solar luminosity, but it is interesting to note that that same degree of reduction of core temperature would probably suffice to account for the low level of neutrino flux.

## Is the Sun shrinking?
In 1979 evidence was put forward indicating that the Sun was shrinking. From an analysis of solar diameter measurements made at the Royal Observatory, Greenwich, over a period of about 120 years from 1836 to 1954, it was suggested that the diameter was decreasing by about 0.1 percent per century. If this figure was correct, and represented a uniform rate of decrease, then the Sun would have been twice its present size about 100,000 years ago, and would shrink to a point in the next 100,000 years. Such a conclusion is patently absurd, but it was suggested that the Sun expands and contracts periodically, possibly over hundreds of years. If the rate of

energy generation at the core (the fundamental driving force behind these variations) was to fluctuate, then, it was argued, the gravitational energy released in the shrinkage would supplement the energy released from the core. Given the present rate of shrinkage, the extra energy would account for a lower central temperature and would fit in with the low observed neutrino flux.

The validity of the results has been challenged by a number of investigators. During the period spanned by the Greenwich observations both the telescope used and the chronometer were changed. Furthermore, increasing atmospheric pollution at Greenwich over the period would affect the brilliancy of the solar limb and further contribute to an apparent shrinkage.

Different types of evidence were also adduced. For example, the eclipse of 1567 seen from Rome was not fully total, the Moon's disc being surrounded by a ring of light: it was argued that the Sun must have been larger then than at present to account for this. However, detailed analysis of the orientation of the Moon on that date shows that the phenomenon of "Baily's beads" (due to light shining through valleys at the lunar limb) would have ensured that the eclipse in question could not have appeared total. Similarly analysis of the duration of transits of Mercury over the past 250 years and of the duration of eclipses over a similar period appear to show quite clearly that the solar diameter has not altered significantly over that time. However, although it would appear that the Sun is not shrinking to the extent that was first suggested, the matter is still not settled, and the possibility of small-scale periodic changes in solar diameter—linked perhaps to small changes in the solar constant—certainly cannot be ruled out.

Understanding of the solar interior is in a state of flux. No-one seriously doubts that the Sun shines by means of thermonuclear reactions converting hydrogen to helium, but the precise mechanism is open to doubt. The Sun has been shown to be variable, although only on a relatively minor scale, and its variability is sufficient to exert climatological effects on Earth. The comfortable view of a steady and unchanging Sun has been replaced by a slightly confused picture of a somewhat inconstant Sun.

**1. Solar oscillation**
The distortions of the Sun's shape have been greatly exaggerated in this illustration. The oscillations consist of alternate bulges at the poles and at the equator.

**2. Changes in the Sun's diameter**
These graphs show the mean annual semi-diameter of the Sun in the horizontal (A) and vertical planes (B). The measurements made at Greenwich were obtained by measuring the time taken by the Sun to cross the meridian (for the horizontal value), and by measuring the angles from the zenith to the upper and lower

limbs (for the vertical value). The most reliable results are indicated by solid lines; the vertical divisions mark changes in the instruments or methods used.

**3. Transits of Mercury**
Timing the transits of Mercury provides a more accurate measure of the solar diameter than that used for diagram 2, but the results are still not conclusive. Readings obtained in this way are shown here as black dots, with vertical lines representing the uncertainties of each value. The open circles give values deduced from total eclipses (see diagram 4).

**4. Baily's beads**
An exaggerated profile of the Moon has been superimposed on a print made from a frame of cine film, recording the eclipse of 20

May 1966. The brightness of the Baily's beads corresponds with the depth of the lunar valleys. The film has been used to measure the solar diameter.

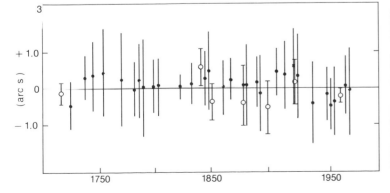
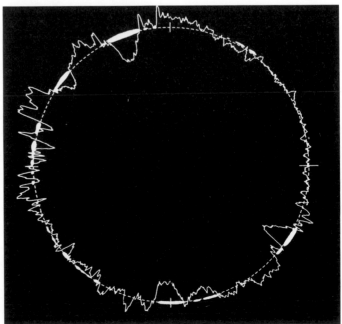

# Life Cycle of the Sun

The problem of understanding the life cycle of the Sun involves the more general problem of the evolution of stars. Since stars live for up to several thousand million years the evolution of a single star clearly cannot be observed from birth through to death. However, it is possible to examine a large cross-section of stars at different stages in their individual life cycles and to try to build up a picture of the stages through which they pass. The first step is to arrange stars into groups: the most widely used system of classification is the "MKK" system (after Morgan, Keenan and Kellman), also known as the Yerkes system. This system divides stars into 10 principal classes, or "spectral types", according to the relative strength or weakness of certain absorption-lines in their spectra (*see* diagram 1). The various spectral types are designated by a letter (O, B, A, F, G, K, M, R, N, S), corresponding to a decreasing level of effective temperature from O (hottest) to S (coolest); each class is further subdivided into 10 subclasses numbered from 0 to 9. According to this system, the Sun's spectral class is G2.

## Hertzsprung-Russell diagram

If the spectral types (or temperatures) of a large number of stars are plotted against their luminosities (or against their absolute magnitudes) several interesting features emerge in the resulting "Hertzsprung-Russell diagram" (*see* diagram 2). Most stars lie in a narrow band sloping from the upper left of the graph to the lower right: in other words, the hotter they are the brighter they are. The Sun belongs to this group, known as the "main sequence". There is also a sizeable group of stars lying above and to the right of the main sequence, representing cool stars very much larger than main sequence stars of the same spectral class (in other words, with the same temperature), and therefore much more luminous: these are known as "red giants". Another important group lies well below and to the left of the main sequence. These are "white dwarfs"— very hot stars whose sizes are comparable with that of the Earth. Their luminosities are therefore very low.

It turns out that these three "types" of star represent different stages through which a star like the Sun will pass during the course of its life. A star is believed to form when an interstellar gas cloud collapses under its own weight. The temperature of the cloud increases as the collapse proceeds and a protostar is formed. The luminosity of the newly forming star is provided by the release of gravitational energy, but as the contraction proceeds the density at the center becomes large and the central temperature eventually reaches about $10^7$ K, at which point nuclear fusion reactions can begin, converting hydrogen to helium. These reactions are capable of supplying sufficient energy to halt the shrinkage of the star as the thermal gas pressure balances the gravitational pull. A state of equilibrium is reached and the star becomes a stable main sequence star with a temperature and luminosity determined by its initial mass: the more massive the star the hotter and brighter it will be. It remains in this state with little change of surface properties for as long as there is sufficient hydrogen fuel in the core to sustain its output. For a star like the Sun, the main sequence stage is thought to last for about $10^{10}$ years. Since the estimated age of the Sun is about $4.6 \times 10^9$ years, it should remain on the main sequence for a further 5 or $6 \times 10^9$ years.

Eventually, the core of the star will become exhausted of hydrogen and clogged with helium "ash". The core, shrinking under the weight of the overlying layers of material, will heat up hydrogen in a shell surrounding the core to a level where nuclear fusion is possible. Indeed, it appears that the rate of energy production will increase substantially, causing the star to expand until internal pressure and gravitational attraction are once again in a state of balance. This process causes the star to expand into a red giant, and its luminosity increases perhaps 1,000 times. When the Sun reaches this stage, the planet Mercury will be swallowed up into the solar globe and life on Earth will become impossible.

When the core temperature of a red giant reaches about $10^8$ K a new core reaction commences, converting carbon to helium; this process ushers in another relatively stable state of existence. A star cannot remain a red giant for long, however, since the enhanced luminosity implies that it is using up its reserves of nuclear fuel at a greatly increased rate. After at most a few hundred million years as a red giant, the shrinking dead core encompasses essentially the whole star and the star, its nuclear energy spent, collapses to

**1. Spectral classes O to M**
The peak of continuum emission shifts towards longer wavelengths as the temperature decreases from O type stars (hottest) to M type (coldest). The main absorption-line features are:
O: ionized and neutral helium, ionized metals, weak hydrogen
B: neutral helium, ionized metals, hydrogen stronger
A: hydrogen dominant, ionized metals
F: hydrogen weaker, neutral and singly ionized metals (particularly calcium)
G: singly ionized calcium prominent, hydrogen weaker, neutral metals
K: strong metallic lines, some molecular bands
M: neutral metals, strong titanium oxide bands.

**2. Hertzsprung-Russell diagram**
Stars pass through various positions on the diagram in the course of their lifetimes, spending the longest period as part of the "main sequence". The Sun's evolutionary track up to the present is shown as a bold line; its predicted development (dashed line) is shown schematically.

become a white dwarf. The shrinkage is halted not by thermal pressure but by the pressure of the high-speed electrons flying around in the interior as a result of the high density. When this stage is reached, the mass is not substantially less than before, but the density is greater by a factor of about a million. A teaspoonful of white dwarf material, if brought back to Earth, would weigh several tonnes. As a white dwarf, the Sun will slowly radiate away internal energy, cooling down over thousands of millions of years eventually to become a cold dark body—a black dwarf.

This is the sequence of events for the Sun sketched out by current theories of stellar evolution. However, the questions that are now being raised about the mechanisms operating in the solar interior may yet prove to have far-reaching consequences for the whole question of stellar evolution.

**Origin of the solar system**
The account of stellar evolution that has emerged involves a number of problems. Clearly nature has solved the problems, for stars exist; and there is good observational evidence for new stars being located within clouds of interstellar material, but it is difficult to see how stars can form from gas clouds: unless the initial cloud is tens or even hundreds of thousands of solar masses, the internal pressure due to temperature, magnetic fields and rotation would be too great for it to collapse under its own weight. Some kind of trigger would appear to be required to cause clouds to collapse.

One possible trigger involves the spiral arms of the galaxy, in which regions of enhanced density, comprising gas and stars, are separated by less dense regions between the arms. As a cloud of gas rotates about the galactic center it passes in and out of these spiral arms. When it collides with a denser region the cloud will be compressed, and this may trigger its collapse. The collapse is thought to lead to fragmentation of the cloud, and to the formation of individual massive stars. Such stars go through their life cycles very rapidly indeed (about 10 million years) and many are believed to blow themselves violently apart in cataclysmic eruptions known as "supernovae". A supernova scatters into the interstellar regions the heavy elements built up in its interior by nuclear reactions,

contaminating the original material of the galaxy (believed to be hydrogen and helium). Later generations of stars such as the Sun are constructed out of this contaminated mix. The supernova eruption, sending shock waves through the surrounding interstellar medium, provides one of a number of possible mechanisms for compressing smaller clouds and leading to the formation of individual stars of solar mass.

Useful evidence is provided by meteorites, presumed to be the oldest solid material in the Solar System. They contain the decay products of short-lived radioactive materials, and the relative abundance of these elements suggests that a supernova occurred within a million years of the solidification of the meteorite materials. This supernova may well have been the trigger that led to the formation of the Sun and the rest of the Solar System.

Although there is considerable dispute as to the mode of formation of the Solar System, the consensus view is that as the proto-solar cloud collapsed it formed a spinning flattened nebula with a hot center which eventually became the Sun. The angular momentum of the Sun was transferred to the surrounding disc of gas by the action of the solar magnetic field and by friction, slowing the Sun to its present modest rate of revolution and accelerating the surrounding material. The Sun then contracted to become a main sequence star, adjusting its diameter to suit the rate of energy generation and decreasing in luminosity as it did so. As the outer cloud cooled, material began to solidify and eventually accreted to form the planets; the outer giant planets, moving in the cooler regions of the solar system, retained vast envelopes of hydrogen and helium which the inner planets, having higher temperatures and lower gravities, were unable to do. Young stars are known to have very strong stellar winds, and the strong solar wind of this early phase is believed to have swept away most of the residual gas from the system, leaving it essentially as it is today. According to this view the Sun and planets originated at the same time and from the same initial cloud of interstellar material. There are, however, many uncertain links in the chain and no-one can be confident that the mystery has been solved. Revolutionary changes of opinion may lie only just around the corner.

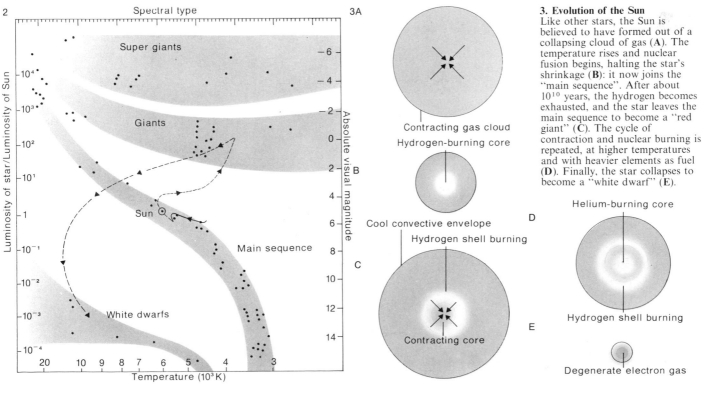

**3. Evolution of the Sun**
Like other stars, the Sun is believed to have formed out of a collapsing cloud of gas (**A**). The temperature rises and nuclear fusion begins, halting the star's shrinkage (**B**): it now joins the "main sequence". After about $10^{10}$ years, the hydrogen becomes exhausted, and the star leaves the main sequence to become a "red giant" (**C**). The cycle of contraction and nuclear burning is repeated, at higher temperatures and with heavier elements as fuel (**D**). Finally, the star collapses to become a "white dwarf" (**E**).

# MERCURY

# Characteristics

Five of the planets in the Solar System were known in ancient times. Four of these—Venus, Mars, Jupiter and Saturn—are so bright that they are easy to recognize. The fifth, Mercury, is much less prominent because it always keeps relatively close to the Sun in the sky: the angular separation between the two is never more than 27°45′. Mercury is best seen, therefore, either low in the west after sunset, or low in the east before dawn. It was certainly known to the Egyptians, and the Greeks named it after Hermes (Mercury), the fleet-footed messenger of the gods. But few details were known until the first flyby of Mariner 10 in March 1974, since when the planet has been recognized as a rather Moon-like body.

Mercury orbits the Sun at a mean distance of 57,910,000 km, but this ranges from 45,900,000 km at perihelion (when Mercury is closest to the Sun) to 69,700,000 km at aphelion (when Mercury is furthest from the Sun). The orbit is more eccentric (0.2056) than any other planet except Pluto, a fact that produces a marked effect upon surface temperatures. Mercury has an orbital period of 87.969 Earth days. The belief in a synchronous axial rotation was not disproved until 1962, when W. E. Howard and his colleagues in the United States found that the dark side of Mercury was warmer than it should have been if it received no sunlight at all. Radar established the axial rotation period at 58.65 Earth days, or two-thirds of the orbital period. The smallest of the four inner planets, Mercury's equatorial diameter is 4,878 km. Its mass is 0.055 and the volume 0.056 that of the Earth.

At night the temperature on the surface of Mercury is as low as 90 K, whereas the maximum day temperature may reach more than 600 K. This enormous temperature difference between the night and day sides of the planet is the largest of any known object in the Solar System. Such variations result from Mercury's close position to the Sun and its slow rotation period. Solar radiation is intense and the day is long.

A small telescope will reveal the phases of Mercury. When the planet is between the Sun and the Earth (inferior conjunction) it is new and can only be seen in transit. The planet displays phases like the Moon: crescent, half (dichotomy), gibbous and full, when it is virtually behind the Sun (superior conjunction) and barely observable. The mean synodic period (from one inferior conjunction to the next) is 115.9 Earth days. Mercury is brightest when gibbous. Its small size and the fact that it is never seen against a dark background make observations of surface markings very difficult. The temperature gradient between Mercury's light and dark sides suggests that the surface is similar to the Moon, which consists of an insulating blanket of dust pulverized by meteoritic impact.

## Atmosphere
The escape velocity on Mercury is a mere 4.25 km/sec, and the planet essentially does not possess an atmosphere. The observa-

**1. Size of Mercury**
With an equatorial diameter of 4,878 km, Mercury compares with the Moon (3,476) and Mars (6,794).

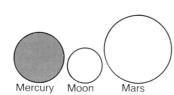

Mercury    Moon    Mars

**Physical data**

| | Mercury |
|---|---|
| Equatorial diameter | 4,878 km |
| Ellipticity | 0.0 |
| Mass | $3.3022 \times 10^{23}$ kg |
| Volume (Earth = 1) | 0.056 |
| Density (water = 1) | 5.43 |
| Surface gravity (Earth = 1) | 0.377 |
| Escape velocity | 4.25 km s$^{-1}$ |
| Equatorial rotation | 58.65 days |
| Axial inclination | 0° |
| Albedo | 0.06 |

**2. Spin-orbit coupling**
It is now known that the axial rotation period of Mercury is not synchronous but is in fact two-thirds of the orbital period. In the diagram a marker indicates the direction of one of Mercury's two "hot poles", that is the points on Mercury's equator that face the Sun directly when the planet is at perihelion. The numbers indicate the position of the planet in its orbit during two revolutions around the Sun. It can thus be seen that the planet has rotated three times during two orbits. Each position is separated by an interval of 22 Earth days. Both spin and the orbital period are anticlockwise. This spin-orbit coupling was confirmed by Mariner 10.

**3. Interior of Mercury**
The most anomalous property of Mercury is its high mean density, which almost equals that of the Earth, even though Mercury is only one-third the size. This would suggest a composition of 65–70 percent by weight of a heavy material, probably iron, leaving 30 percent silicates, and making Mercury twice as rich in iron as any other planet in the Solar System. Study of the magnetic field and evidence of volcanic activity indicates that Mercury is chemically differentiated. If most of the iron is concentrated in a core the radius of the core would be 70–80 percent of the planet's radius. The outer region would then be made of silicate rock like Earth's mantle.

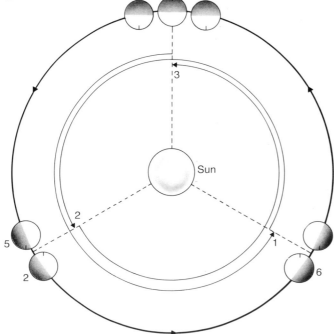

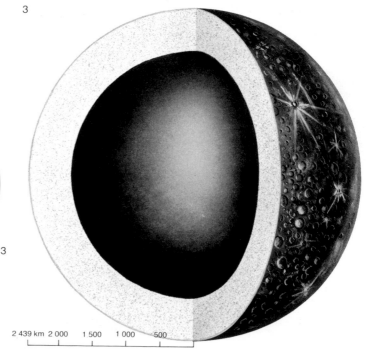

2 439 km   2 000   1 500   1 000   500

tions made by Mariner 10 indicate that the surface pressure is only $10^{-12}$ mb, which is only 1,000-billionth of the value for the Earth. The most abundant gas in this tenuous atmosphere is helium, which is more abundant than on the Moon. The helium possibly originates from the radioactive decay of uranium and thorium, or alternatively it may be trapped in the vicinity of Mercury by the magnetic field having been blown out from the Sun in the solar wind.

## Magnetic properties

The discovery by instruments mounted on the Mariner 10 spacecraft that Mercury has a magnetic field was a surprising result. It is, however, rather weak, with a strength of only one-hundredth of that of the Earth. Mercury's field, like the Earth's, is closely aligned with the planet's axis of rotation. It is generally thought that the Earth's magnetic field is generated by a dynamo-like action in the core of the planet resulting from slow-moving currents of molten metal induced by the planet's rotation.

Mercury has a density of $5.4\,g\,cm^{-3}$. It is almost certain to possess a large iron core. However, the planet rotates slowly, which is why the presence of a magnetic field was not completely unexpected. An alternative explanation to the dynamo action would be that Mercury possesses a fossil field acquired early in its history and which has never been subsequently lost, or that the magnetic field has been induced by interactions between the planet and the Sun's magnetic field. Neither of these proposals can fully account for the magnetic field. The high temperatures that are required for the generation of a magnetic field destroy induced magnetism once the dynamo field-generating process stops. The visible appearance of the planet does not provide evidence of internally generated deformation or volcanism for at least three billion years. Consequently, not even the dynamo explanation is completely consistent with the primitive state of the planet's crustal development to the present.

Although the magnetic field of Mercury is weak, it still provides a definite interaction with the solar wind particles streaming out from the Sun. A well-defined bow shock can be observed, despite the absence of an atmosphere and ionosphere, together with transient charged particles and other plasma phenomena. But the field is too weak to permit the development of a radiation belt of trapped particles like the Van Allen belts around the Earth. The effect of the solar wind pressure does, however, alter the geometry of the magnetosphere, as in the case of other planets in the Solar System. On the sunside, the field is compressed towards the planet and then stretched out into a long tail on the side away from the Sun. There is little doubt that the closest planet to the Sun still holds some mysteries associated with its evolution and internal characteristics.

## Nomenclature

The Sixteenth International Astronomical Union Commission defined the zero meridian of Mercury as the subsolar point at the first perihelion passage in 1950, that is, the point on the equator that lay directly under the Sun when Mercury was at its closest. After the Mariner 10 flight, the reference point was taken as a tiny crater, Hun Kal, and it was assigned longitude $20\,°W$. Although Hun Kal is now the Mercurian "Greenwich", it could not be located on the zero meridian, because this was in the dark hemisphere during the spacecraft's encounters.

Mariner 10 made three passes of Mercury before its power finally failed. Unfortunately the same regions were in sunlight during each pass, so that not all the surface could be mapped. Only about half the Caloris Basin could be examined, for example. However, there is no reason to assume that the unexplored regions are essentially different from those imaged from Mariner 10.

The old nomenclature was abandoned, and the I.A.U. Commission listed six major classes of topographical features (*see* pages 82–3): craters, scarps (dorsa), mountains (montes), plains or basins (planitia), ridges (rupes) and valleys (valles). Valleys are generally named after radar installations (e.g. Arecibo Vallis); scarps after famous ships of discovery (such as Endeavour Rupes) and craters after great contributors to the arts and humanities such as Homer and Mozart, though there are notable exceptions such as Kepler and Hun Kal. Plains and basins take their names after the word for Mercury in various languages; thus Suisei Planitia is the Japanese form. The main exceptions are the north polar Borealis Planitia and also Caloris Planitia (the Caloris Basin), literally "the hot basin" because it lies near one of Mercury's two hottest regions, where the Sun is overhead at perihelion and the temperature is thus higher than anywhere else on Mercury.

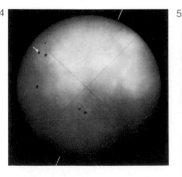

**4. Transit of Mercury**
Mercury cannot be seen in transit with the naked eye, but through a telescope is much darker than any sunspots on the Sun's disc.

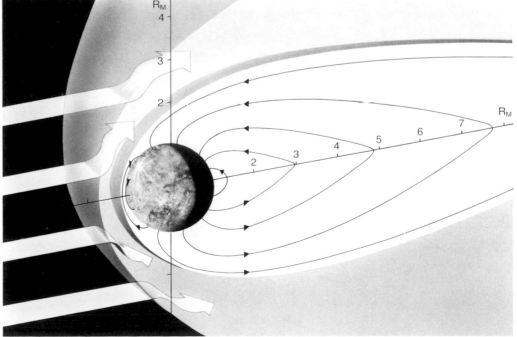

**5. Magnetosphere**
One of the major discoveries by Mariner 10 was that Mercury did possess a significant magnetic field. Observations of the solar wind showed the magnetosphere to be similar to a scaled-down version of the Earth's, but it is far too weak for radiation belts to form.

# Observational Background

### Telescopic Observations

Telescopes were first used in astronomy in the seventeenth century, but they were not powerful enough to show surface details on Mercury. The first observer to detect the phases is believed to have been Johannes Hevelius. In the following century Mercury was closely studied by Johann Hieronymus Schröter, the first great observer of the Moon, but even Schröter could see virtually nothing. He did publish some drawings which showed features that he believed to be mountains up to 20 km in height, and dark ill-defined patches.

A new series of observations was begun in 1881 by G. V. Schiaparelli, using 22 cm and 49 cm refractors at Milan. Schiaparelli adopted the best method of observing Mercury, that is when it is high in the sky. This means that the Sun is also high, but the results are better than those obtained by observing Mercury when it is close to the horizon.

Schiaparelli's map showed streaky features, which he described as brownish in color. He stressed that the features were very hard to define, and that his map was necessarily of dubious accuracy. From his observations he concluded that Mercury had what is termed a captured or synchronous rotation, so that its axial rotation period was the same as its revolution period round the Sun—88 Earth days. If this were the case Mercury would keep the same hemisphere turned towards the Sun all the time. There would be a region of permanent day and a region of everlasting night, though between these two extremes there would be a "twilight zone" in which the Sun would rise and set, but always keep fairly close to the horizon.

In the meantime Percival Lowell made many drawings of Mercury and recorded linear features similar to the "canali" on Mars. Unlike the Martian network, however, Lowell believed the features on Mercury to be natural in origin. An English observer, W. F. Denning, compared patches on Mercury with features on Mars from which he derived a rotation period of 25 hr.

### Antoniadi's map of Mercury

Before the flight of Mariner 10 in 1974, the accepted chart of Mercury was that compiled by E. M. Antoniadi between 1924 and 1929. Antoniadi, a Greek-born astronomer who spent most of his life in France, had the advantage of being able to use one of the world's best telescopes, the 83 cm refractor at the Observatory of Meudon outside Paris. He published his findings in *La Planète Mercure* in 1934. Like Schiaparelli, he made all his observations in broad daylight, and wrote: "On my drawings I have recorded only the patches which were seen with certainty. In general the patches were very pale and difficult to distinguish, but they were genuine, and their color seemed greyish, similar to that of the lunar seas."

The names he gave to the various patches were derived from Greek and Egyptian mythology; for instance, he called one shading the "Solitudo Hermæ Trismegisti", or Wilderness of Hermes the Thrice Greatest. He considered it likely that these patches were of the same basic nature as the uplands and maria of the Moon.

His descriptions of such features were precise. Thus his Solitudo Iovis was "a rounded patch, very dark; dimensions very nearly equal to those of France. . . . It was nearly always easy to see in the great refractor, and very dark when it was not enfeebled by local veils." The reference to "local veils" is very significant. Antoniadi knew that the Mercurian atmosphere must be very tenuous, but he nevertheless believed that it was able to support fine dust in suspension, and that obscuration of surface features was as frequent as on Mars. Also like Sciaparelli, he was confident that the axial rotation period was equal to the orbital period.

Antoniadi also observed transits of Mercury and Venus and compared the two. When Venus enters transit it appears to be surrounded by a luminous ring, an effect caused by the planet's extensive atmosphere. This is not present on Mercury and Antoniadi recognized that Mercury's atmosphere is too tenuous to

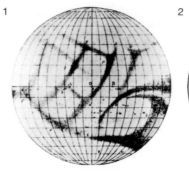
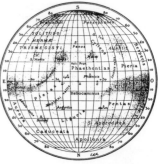

**1. Schiaparelli's map**
Based on observations between 1881 and 1889 at the Brera Observatory, Schiaparelli's map showed brown streaks against a rosy background. He found the patches difficult to distinguish against the background and they later proved to be illusions.

**2. Antoniadi's chart**
This was based on observations between 1924 and 1929 and only included features that Antoniadi had seen on at least one occasion. Observations of the dark patches and lighter areas led him to the idea of a synchronous rotation.

**3. Antoniadi's theory of rotation**
The Greek astronomer believed that only one hemisphere of the planet ever faced the Sun, that is, that the rotation was synchronous. This explained the apparently fixed surface markings.

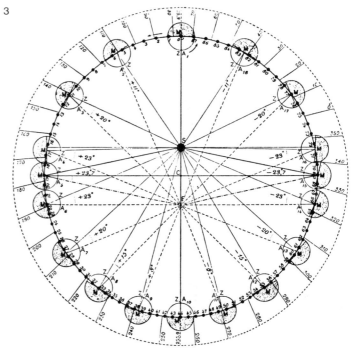

produce the luminous ring effect. He also observed the "black drop" effect on Mercury: when the planet passes in front of the Sun's disc it appears to draw a strip of blackness after it. Antoniadi correctly attributed this to disturbances in the Earth's atmosphere rather than the effects produced by an extensive atmosphere on Mercury itself.

### Later maps of Mercury

Various other charts of the surface features were produced in the 1950s and 1960s. One, from A. Dollfus and H. Camichel, was in reasonable agreement with that of Antoniadi. And in 1953 Dollfus detected a very tenuous atmosphere. Subsequently C. Chapman in the United States compiled a map from a selection of drawings and photographs collected from all over the world. He was able to

**4. Map by Camichel and Dollfus**
This map of Mercury was produced during the 1950s and included all the features that were regarded as being well defined. Antoniadi's nomenclature was used, though it was modified and extended. There is some similarity between this map, Antoniadi's, and the later map of C. Chapman.

**5. Map by C. Chapman**
C. Chapman in the United States produced the definitive map of Mercury before the Space Age. He compiled the map from a selection of 130 drawings and photographs made by observers all over the world, which was considered to be the best available evidence. The results, however, bear little resemblance to the reality that Mariner presented, and illustrate the great difficulty in observing Mercury from Earth.

**6. Mariner 10 approach sequence**
On its journey of exploration of the inner Solar System, the Mariner 10 spacecraft travelled more than 1,600 million km. As it approached Mercury the planet appeared to be a Moon-like body, heavily cratered and with large flat basins. The bright spot that was the first feature to be seen was a bright-rayed crater 25 km in diameter, later named Kuiper.

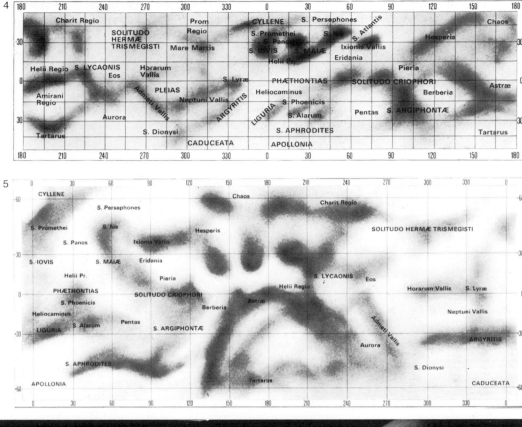

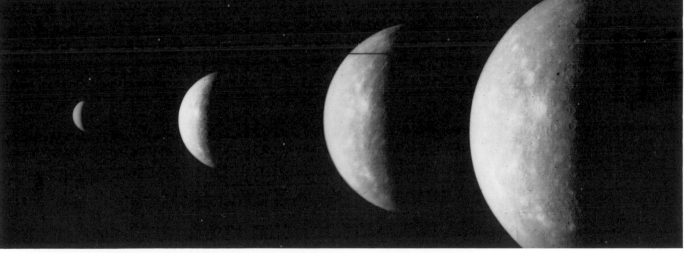

retain Antoniadi's nomenclature, but he dismissed any idea of a definite atmosphere surrounding the planet.

Although it was recognized that no Earth-based map of Mercury could be reliable, most astronomers believed that the main features shown by Antoniadi and others had at least a basis of reality. However, the results from Mariner 10 showed otherwise. The actual surface features bear no resemblance to those on the earlier maps, and Antoniadi's nomenclature has had to be discarded.

Moreover, Antoniadi and others were misled through no fault of their own. Mercury's rotation period is not synchronous; it is only 58.6 Earth days, so that there is no permanent day zone, night zone or twilight zone. But every time Mercury is best placed for observation, the same region of the surface is turned towards the Earth, so that the features on view are the same.

**The results from Mariner 10**
It is in fact Mariner 10 that has provided almost all our knowledge of the Mercurian surface: it is the only probe that has so far bypassed the planet. It was also the first two-planet probe, since on its way to Mercury it encountered Venus (*see* pages 428–9).

As soon as Mariner 10 came within useful range of Mercury, it became clear that the surface is cratered. The first crater to be positively identified—a ray-center—was named Kuiper, in honor of the late Gerard P. Kuiper, who played such a major role in the early days of practical planetary research. As the range narrowed, and the images became more detailed, it was found that the surface is very complex. There are few of the great plains which characterize the Moon, but the Caloris Basin has been compared with the lunar Mare Orientale.

# Surface Features

As Mariner 10 approached the innermost planet, the landscape could easily have been mistaken for that of the Moon. There are, however, important differences: the surface of Mercury is more complex and geologists have in fact identified at least 12 different types of terrain. There are five main distinctions and these relate approximately to the evolution of the planet: heavily cratered terrain, intercrater plains, the Caloris Basin, smooth plains, and young craters.

Craters (*see* pages 84–5) are the main feature on Mercury and the heavily cratered terrain is the oldest part of the surface, as on the Moon. Unlike on the Moon, however, there are ancient intercrater plains. The chronology of these two types of terrain is not easily determined for the plains do not clearly overlie the craters. The plains are peppered with what are thought to be secondary craters produced by debris from the formation of the heavily cratered terrain. Both areas could have been formed by overall surface melting, or the plains might have been produced by volcanism that coincided with meteoritic bombardment.

The Caloris Basin (*see* pages 86–7) is the largest structural feature on Mercury and its formation must have been a key event in the evolution of the landscape. The crater is 1,300 km in diameter and was probably created by an enormous impact. The scale of such an event resulted in a variety of associated topographical features. A series of mountain ranges appears to have been thrown up as a result of the impact, and also associated with the Caloris Basin is what is termed weird terrain. At the antipodes of the Basin, this type of hummocky landscape is believed to have been produced by seismic activity as a result of the impact. Waves may have been set up that reverberated right around the planet until they met, producing a shock that produced the mountain blocks of the weird terrain. Between the mountain ranges by the Caloris Basin are smooth plains that are thought to have formed at a later stage in Mercury's geological evolution and probably as a result of volcanic activity. These plains resemble the lunar maria and examples are found on other basin floors on Mercury.

Finally, there are younger craters: many are rayed and a surrounding ejecta blanket testifies to the craters' relative youth. There are other topographic features, however, that are worthy of mention. Wending their way across the planet's surface and over different types of terrain are linear features that take the form of scarps or troughs. They are evidently the result of some sort of tectonic activity and may be several hundred kilometers in length. The orientation of these features (NW–SE and NE–SW) suggests that the tectonic activity happened when a slowing of Mercury's rotation produced a flexing of the crust and hence surface disruptions. Alternatively, crustal buckling might have resulted from an expansion or contraction of the crust reflecting internal changes. In places the linear features transect other types of terrain and therefore postdate them, but in other cases the lineaments are obliterated by a crater or another feature and must have been formed previously.

## Mercury's geological evolution

Mercury is similar to the Moon in that it is considered to be relatively unevolved. The surface features suggest that for the most part there has been little activity apart from meteoritic impact and a limited amount of volcanism. It seems from Mariner 10 evidence that Mercury underwent a period of heavy bombardment very early in its history—4 to 4.5 thousand million years ago. Bombardment was followed by volcanism that infilled the basins and craters and produced the plains. The planetary-scale ridges do suggest a subsequent phase of compression or an event that was linked in some way to the internal evolution of the planet. On the whole, however, the sequence of events that formed the surface on Mercury is remarkably similar to those events associated with the Moon (*see* pages 154–5).

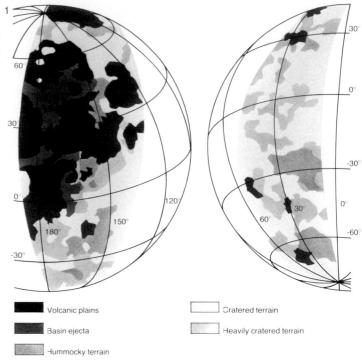

Volcanic plains

Basin ejecta

Hummocky terrain

Cratered terrain

Heavily cratered terrain

**1. Geological map of Mercury**
Only 35 percent of the surface of Mercury was mapped by the Mariner 10 spacecraft, but the resulting photomosaics (*see* pages 88–9) enabled a fairly representative map of surface features to be plotted. This shows the liberal distribution of plains and craters and demonstrates the similarity between Mercury and the Moon in this respect. The 1,300 km Caloris Basin straddles the terminator and is centered at 30°N, 190°W. Each map covers topographical features within approximately 60° of the terminator in each hemisphere. Future unmanned missions will return to Mercury since more than half the planet remains to be seen.

**2. Craters and plains**
This photograph of part of the northeast quadrant of Mercury as viewed by Mariner shows an area bordering Planitia Borealis and demonstrates the two distinct types of Mercurian landscape: craters and plains. The two are juxtaposed where there is a gradation of the surface from a very rough area surrounding the pole to the relatively smooth plain. The largest craters to be seen here are about 75 km across and may have superimposed secondary craters. The large crater in the immediate foreground is named Jokai and the nearest one of the group of three craters to the right of center is named Mansart. The plain is thought to be volcanic in origin.

3

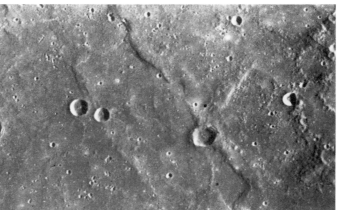

4

### 3. Cratered terrain
This heavily cratered terrain near the south pole of Mercury has several ray systems, and a number of scarps are also very prominent. Smooth intercrater areas are evident in places.

### 4. Ridged plain
The ridge crossing this image is typical of several features in the region of Tir Planitia. The ridge apparently predates the crater that straddles it. Linear features such as this are not yet adequately explained, but they may result from a contraction of the planet's core.

### 5. Discovery scarp
This ridge or scarp runs in an NE–SW direction and extends over a distance of more than 400 km. It is the most striking example of its kind seen so far on Mercury. The scarp cuts through the crater Rameau and a smaller crater, and would seem to have been formed by crustal uplift occurring as a result of compression of the crust.

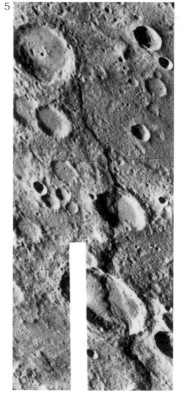

5

6

7

### 6. Weird terrain
The only known example of its kind on Mercury, this hilly region is antipodal to the Caloris Basin and is believed to have formed as a result of seismic waves generated by the colossal impact that created the Caloris Basin.

### 7. Santa Maria Rupes
A low, rounded ridge about 3 km high, Santa Maria Rupes crosses intercrater plains and old craters.

# Craters

The surface of Mercury presents a massively cratered face which in many ways resembles the far side of the Moon. This was first revealed by Mariner 10, which during its three encounters collected more than 2,500 useful images with a best resolution of 100 m, but covering less than half of the total surface. The craters range in size from small pits about 100 m across to massive rings upwards of 1,000 km in diameter, there being a range of types similar to those of the Moon. There are certain aspects of crater form and distribution, however, which make the Mercurian craters distinct from their lunar counterparts. These differences are almost certainly related to the greater gravitational acceleration prevailing at the surface of Mercury (about 2.2 times that of the Moon).

Large regions of the planet are covered with a plethora of craters giving rise to high crater densities, similar to those encountered in the lunar Fra Mauro formation and considered to have an age of about 4,000 million years. Between these heavily cratered areas are regions with a lower density of large craters but a high density of craters within the size range 5–10 km. These regions are called intercrater plains. Such terrain is probably the most widespread on the surface of Mercury, but the precise relationship of its age to that of the heavily cratered terrain is in some doubt.

The youngest craters, particularly those imaged under extreme low-lighting conditions near the limb, appear to have well-developed ray systems that can extend for hundreds of kilometers from their focus. An analysis of such craters and other fresh young craters below a diameter of about 30 km suggests they were formed prior to the great Caloris Basin.

It is perhaps the arrangement of material external to crater rims that most clearly distinguishes Mercury's craters from those of the Moon. As is usual with craters produced by ballistic means, hummocky rim materials grade outward into radially patterned debris and this in turn leads into a zone of secondary craters and more discontinuous ejecta. On Mercury, however, the passage into the secondary crater zone takes place over a much shorter distance. Furthermore, the density of secondaries in this zone is far higher than it is on the Moon. Another significant aspect is that in the zone of secondary craters, the craters are often overlapping and form linear chains or grooves that may extend towards the crater's rim.

The consensus view is that Mercury's craters were formed by meteoritic impact, as were the Moon's, although some hold the view that there is evidence of an endogenic origin for many of the craters, the presence of aligned crater rows and elongated secondaries being cited in favor of this.

## 1. Crater types
Crater morphology is similar to that found on the Moon. Craters are generally circular, have ejecta rim deposits, fields of secondary craters, terraced inner walls and central peaks or even concentric inner rings. The smallest craters are bowl shaped; with increasing size there may be a central peak, then inner terracing of the walls. A further increase in crater dimensions sees more frequent central peaks or clusters of peaks, and in the very largest structures complete or partial concentric inner rings may develop. There are similar morphological tendencies on the Moon, but the change from one type to the next occurs at lower diameters on Mercury than on the Moon. About 80 percent of Mercurian craters between 10 and 20 km in diameter are terraced and there is complete terracing at more than 20 km, whereas on the Moon only 12 percent of craters in the 10–20 km class are terraced and complete terracing occurs only in craters over 40 km in diameter.

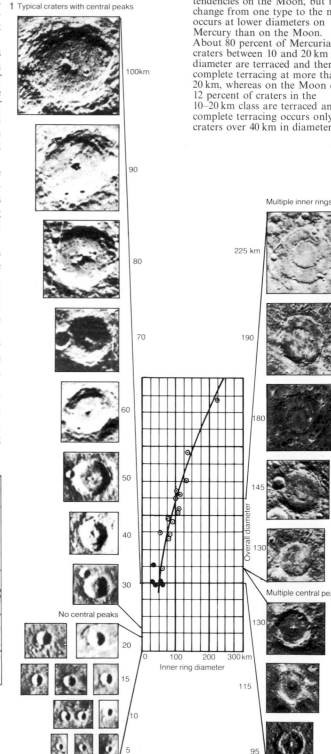

## 2. Secondary craters
A Mariner 10 image (**A**) of the crater Verdi reveals an extensive zone of secondary craters. An enlarged image enabled them to be mapped (**B**). For any given crater diameter, the extent of ejected material is less for Mercury than it is on the Moon. This is because the gravitational acceleration is twice as great on Mercury as on the Moon.

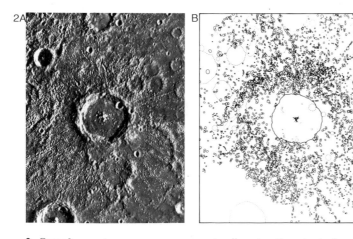

### 3. Degas
Degas is a bright ray-crater. Craters such as this are thought to be relatively young since rays emanating from them cross all other formations. These high-albedo, wispy filaments consist of fine particles of ejecta.

### 4. Concentric rings
Both Strindberg, which is about 175 km in diameter, and Ahmad Baba, 115 km in diameter, have been partially filled by lava flows, but parts of the concentric mountain rings produced by the impact can still be seen.

### 5. Schubert and Chekhov
The crater Schubert has been flooded with smooth plains material so the inner ring of mountains has been obliterated whereas it remains in nearby Chekhov. Smaller, younger craters overlap the old.

### 6. Hun Kal
Hun Kal is typical of the small, bowl-shaped craters on Mercury. It is only 1.5 km in diameter.

Situated 0°.58 south of the equator, Hun Kal (20 in the language of the Maya) is centered on the twentieth meridian.

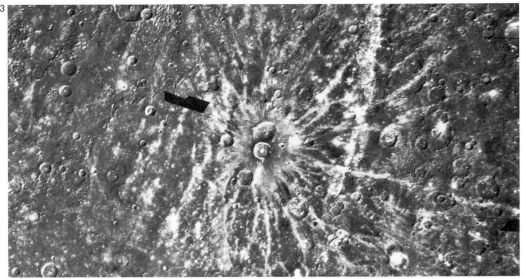

3

4

5

6

### 7. Brahms
Brahms is a large crater north of the Caloris Basin. It has a central peak complex, terraced walls and displays a radial flow of ejecta deposits all around.

### 8. Teddy Bear
The configuration of these overlapping craters situated just outside the eastern rim of the Caloris Basin earned this nickname for obvious reasons.

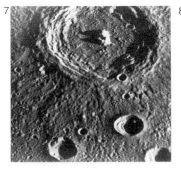

7

8

# Caloris Basin

The Caloris Basin was the largest structure imaged by Mariner 10. It has a diameter of 1,300 km, but during each of the three flybys only half of the complete structure was visible. Caloris resembles the lunar Imbrium basin in many respects, in particular the size and overall morphology, but there are differences among the details. Surrounding the basin, and comparable to the lunar Apennines, is a rim of mountains, Caloris Montes, which rise to between 2 and 3 km above the floor. A distinct gap or breach in this chain exists south of the large crater March, on the southeastern rim.

The floor of the basin differs from anything found on either the Moon or Mars. Evidently it has been infilled with materials since its formation and these materials have been broken up by a large number of fractures. Two sets of fractures can be distinguished: one has a radial orientation, the other is concentric. The greatest concentration is found towards the center of the basin. In addition to the fractures there are large numbers of ridges which display a similar arrangement but which are more numerous towards the outer regions of the floor. Both the ridges and the cracks appear related to a period of structural "settling".

Surrounding the main ring of mountains is a rather ill-defined outer scarp which sits some 100–150 km beyond to the northeast. Between this outer feature and the main rim is an area of relatively smooth hills and domes, some of which may be of volcanic origin. Also extending out from the basin is an extensive radial system of valleys and ridges which closely resembles that of the lunar Imbrium basin and, like it, extends for about one basin diameter. Within this zone the Mercurian surface is distinctly hummocky in some areas and relatively smooth in others. Some geologists have interpreted the latter to be representative of lava flows, but the hummocky terrain is more likely to be in the nature of ejecta from the basin-forming process. The consensus view is that Caloris is the result of a major impact event, the last in the planet's long history.

## 1. Caloris Basin

The composite mosaic (opposite) of the Caloris Basin made up from Mariner 10 images was in fact constructed from photographs with very different resolution limits so that the degree of detail is not the same everywhere. The two diagrams should be of assistance in the identification of features in the photograph. The first diagram (A) shows fractures and other features associated with the Basin. The second diagram (B) shows the ridges and associated features. The basin floor with its central fractures and outer region of sinuous ridges can be clearly seen on the photograph. Note the large number of small craters on the outer eastern floor and a number of sizeable fresh-looking craters around the edge of the fractured zone. These bear witness to post-Caloris crater-forming events. The large crater in the extreme northeastern corner of the image (Van Eyck) predates Caloris and was filled with material ejected during the basin's formation. Radial ridges NE and SE of the basin are believed to be ejectamenta. The smooth surface between the ridges and the surface of the floor of the basin may be of a comparable age, both being post-Caloris and produced by melting resulting from the event that produced the basin. The ridges on the basin floor are probably similar in origin to the lunar "wrinkle ridges", resulting from the compression of lava.

## 2. Fractures

A detail of the floor of the Basin shows the characteristic ridges and fractures. Fractures transect the ridges at various angles and the length and the width of the fractures increase towards the center of the Basin.

## 3. Domical terrain

A close-up from the large image shows what is termed a domical terrain in the northeastern quadrant of the Caloris Basin between the rim of the Basin and the outer scarp.

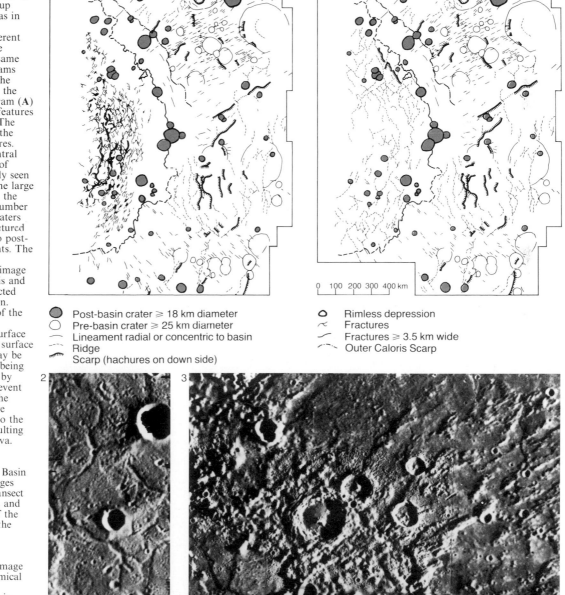

Post-basin crater ⩾ 18 km diameter
Pre-basin crater ⩾ 25 km diameter
Lineament radial or concentric to basin
Ridge
Scarp (hachures on down side)

Rimless depression
Fractures
Fractures ⩾ 3.5 km wide
Outer Caloris Scarp

0  100  200  300  400 km

# Photomosaics

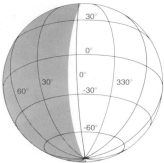

**Incoming**
As Mariner 10 drew close to Mercury a series of photographs was taken and combined into a photomosaic. 18 images were taken at an interval of 42 sec and each one was computer-enhanced to produce this overall image of the planet. Mariner was at a distance of 200,000 km when the photographs were taken which was, in effect, six hours away from Mercury, and the date was 29 March 1974. About two-thirds of the planet seen in this image are in the southern hemisphere.

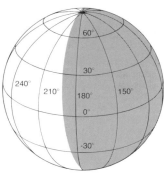

**Outgoing**
As Mariner 10 left Mercury, having completed its mission, more photographs were taken and again combined to make up a photomosaic. This time the mosaic shows rather more of the illuminated surface of the planet: the north pole is at the top of the image and the equator runs horizontally across the image roughly two-thirds of the way down from the top. Again, the photographs, 18 of them in all in this mosaic, were taken six hours after Mariner's closest approach.

# Northwest Quadrant

The general topography of the whole of this quadrant is dominated by the Caloris Basin, part of which is shown. Unfortunately, the center lies beyond the terminator, at latitude 30°N, longitude 190°W, and was not photographed. The Caloris Basin is estimated at 1,350 km in diameter, and is therefore comparable in size with the lunar Mare Imbrium.

Outside the mountain ring lies a complex terrain of ancient craters and basins which have been strongly sculptured by shallow valleys radial to the Caloris Basin.

Many areas have also been flooded by the smooth plains material, but others are covered with rougher material, making up a terrain of hummocky plains. This area extends over 1,000 km around Caloris, and it contains five of the six other plains so far named—Tir, Budh, Odin, Sobkou and Suisei. The effects of the Caloris event die away with increasing distance, and there is a steady gradation back into the intercrater plain kind of terrain which is so characteristic of the other quadrants.

There are four prominent ray-craters in this quadrant, counting the close pair in Sobkou Planitia as a single center. The map shows that the peculiar curved ray seen to the south and east of these in the more distant views breaks down into straight segments in the map projection. The very prominent component running north-north-east is clearly a northward extension of Heemskerck Rupes. The large crater Mozart is of special interest. It was obviously formed after the Caloris Basin, and must be one of the youngest of the main craters.

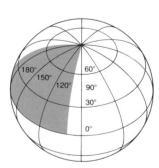

**Using the quadrant scale**
To use the quadrant scale (right) measure the distance from the subsolar point to the center of the feature concerned with a centimeter ruler. Locate this measurement on the horizontal centimeter scale. Measure the feature concerned. Place the ruler vertically at the point previously located on the centimeter scale. On the ruler read up from the base line the number of centimeters that the particular feature measures against the kilometer scale.

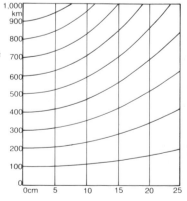

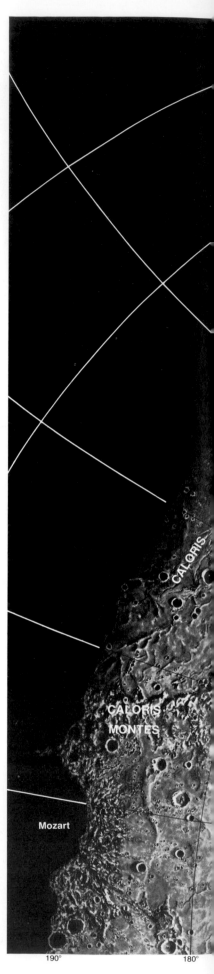

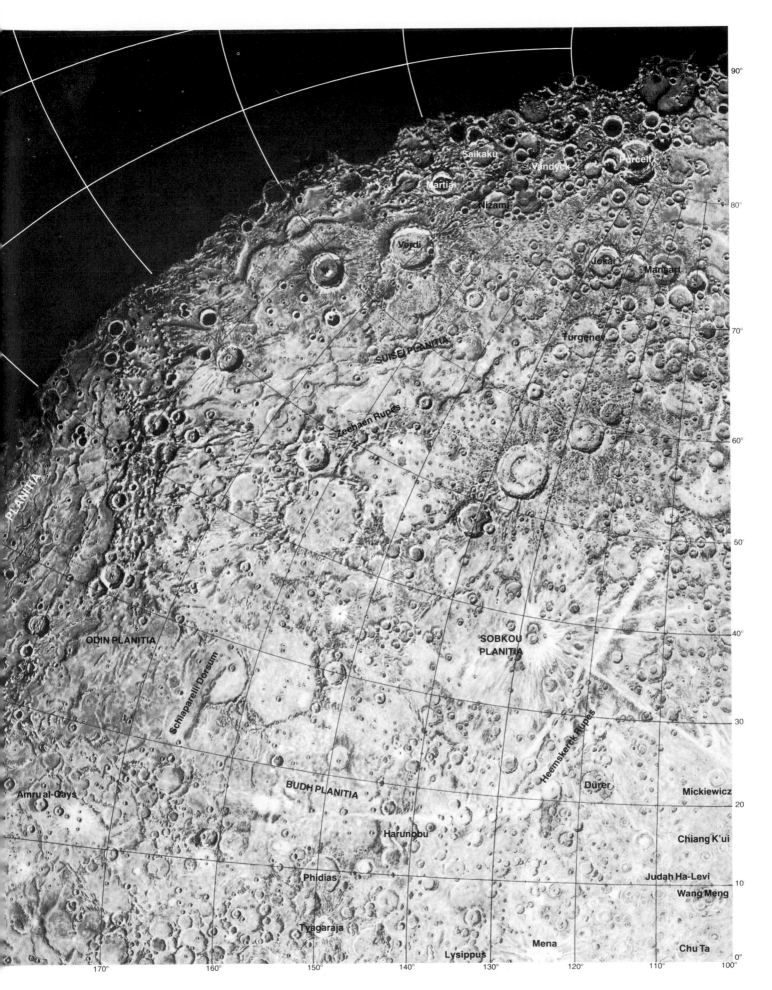

SUISEI PLANITIA

Saikaku

Vandyck

Purcell

Martial

Nizami

Jokai

Mansart

Verdi

Turgenev

Zeehaen Rupes

PLANITIA

SOBKOU
PLANITIA

ODIN PLANITIA

Schiaparelli Dorsum

Heemskerck Rupes

Dürer

Mickiewicz

Amru al-Qays

BUDH PLANITIA

Harunobu

Chiang K'ui

Judah Ha-Levi

Phidias

Wang Meng

Tyagaraja

Mena

Chu Ta

Lysippus

90°

80°

70°

60°

50°

40°

30°

20°

10°

0°

170°   160°   150°   140°   130°   120°   110°   100°

# Northeast Quadrant

| | | | | | | |
|---|---|---|---|---|---|---|
| Abu Nuwas | 17.5°N,21°W | Molière | 16°N,17.5°W | Antoniadi Dorsum | 28°N,30°W | |
| Al Hamadhani | 39°N,89.5°W | Monet | 44°N,9.5°W | | | |
| Al-Jāhiz | 1.5°N,22°W | Monteverdi | 64°N,77°W | Borealis Planitia | 75°N,85°W | |
| Aristoxenes | 82°N,11°W | Myron | 71°N,79.5°W | | | |
| Asvagosha | 11°N,21°W | Polygnotus | 0°,68.5°W | Endeavour Rupes | 38°N,31°W | |
| Chaikovskij | 8°N,50.5°W | Praxiteles | 27°N,60°W | Santa Maria | | |
| Deprez | 81°N,92°W | Proust | 20°N,47°W | Rupes | 6°N,20°W | |
| Derzhavin | 44.5°N,35.5°W | Rajnis | 5°N,96.5°W | Victoria Rupes | 50°N,32°W | |
| Donne | 3°N,14°W | Rodin | 22°N,18.5°W | | | |
| Gauguin | 66.5°N,97°W | Rubens | 59.5°N,73.5°W | Haystack Vallis | 4°N,46°W | |
| Giotto | 12.5°N,56°W | Sholem Aleichem | 51°N,76.5°W | | | |
| Gluck | 37.5°N,18.5°W | Sinan | 16°N,30°W | | | |
| Goethe | 79.5°N,44°W | Sor Juana | 49°N,24°W | | | |
| Handel | 4°N,34°W | Stravinsky | 50.5°N,73°W | | | |
| Holbein | 35.5°N,29°W | Tansen | 4.5°N,72°W | | | |
| Homer | 1°S,36.5°W | Ts'ai Wen-Chi | 23.5°N,22.5°W | | | |
| Hugo | 39°N,47.5°W | Tung Yuan | 73.5°N,55°W | | | |
| Kuan Han-ch'ing | 29°N,53°W | Velazquez | 37°N,54°W | | | |
| Lermontov | 15.5°N,48.5°W | Vivaldi | 14.5°N,86°W | | | |
| Li Po | 17.5°N,35°W | Vyāsa | 48.5°N,80°W | | | |
| Lu Hsun | 0.5°N,23.5°W | Wren | 24.5°N,36°W | | | |
| Melville | 22°N,9.5°W | Yeats | 9.5°N,35°W | | | |
| Mistral | 5°N,54°W | | | | | |

Because of the flight path of Mariner 10, both the strip of surface on this quadrant and the whole of the reverse hemisphere were inaccessible. Unfortunately, the regions to either side of the missing strip appeared highly foreshortened in the only views that Mariner 10 was able to obtain, so that analysis of them is far from easy. There is some smooth plains material (Borealis Planitia) and a large lava-flooded basin, 310 km in diameter, lying between the Planitia and the Pole. This lunar-type region is completely separated from the much more extensive plains near the Caloris Basin by a belt of heavily cratered terrain.

There are seven prominent ray-craters in this quadrant. These appear to be concentrated towards the subsolar region, but this may be misleading, because rays appear only under high illumination—and ray-craters on Mercury are thus easier to detect when the Sun is almost above them. Of the more interesting features, one is Rodin, a particularly large and well developed double ring basin. Its associated field of ejecta and secondary cratering is very well displayed. Around it are other major craters such as Molière and Abu Nuwas. Of the four named valleys, one (Haystack) is a crater-chain. Also in the quadrant is one of the two named ridges, Antoniadi Dorsum, which seems to be a southern extension of Endeavour Rupes. It would appear that the scarp has turned into a ridge simply because the level on the upthrust side of the fault has declined more abruptly. Equally it has been suggested that Goethe is geologically related to the Borealis Planitia.

**Using the quadrant scale**
To use the quadrant scale (right) measure the distance from the subsolar point to the center of the feature concerned with a centimeter ruler. Locate this measurement on the horizontal centimeter scale. Measure the feature concerned. Place the ruler vertically at the point previously located on the centimeter scale. On the ruler read up from the base line the number of centimeters that the particular feature measures against the kilometer scale.

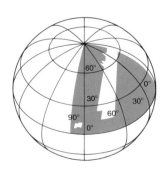

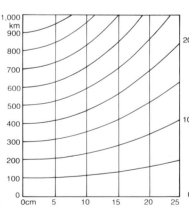

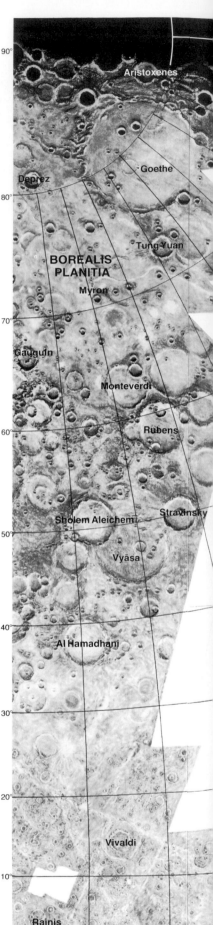

Monet

Sor Juana

Victoria Rupes

Gluck

Derzhavin

Holbein

Endeavour Rupes

Hugo

Antoniadi Dorsum

Melville

Velázquez

Ts'ai Wen-Chi

Rodin

Kuan Han-ch'ing

Wren

Praxiteles

Abu Nuwas

Molière

Proust

Li Po

Sinan

Lermontov

Giotto

Asvaghosha

Yeats

Santa Maria Rupes

Chaikovskij

Tansen

Mistral

Handel

Donne

Polygnotus

Homer

Al-Jahiz

Lu Hsun

70°    60°    50°    40°    30°    20°    10°

# Southwest Quadrant

| | | | | | |
|---|---|---|---|---|---|
| Bach | 69°S,103°W | Mena | 0.2°S,125°W | Gjoa Rupes | 65°S,163°W |
| Beethoven | 20°S,124°W | Milton | 25.5°S,175°W | Hero Rupes | 57°S,173°W |
| Bello | 18.5°S,120.5°W | Philoxenus | 8°S,112°W | Pourquoi Pas | |
| Bernini | 79.5°S,136°W | Po Chü-I | 6.5°S,165.5°W | Rupes | 58°S,156°W |
| Cervantes | 75°S,122°W | Rublev | 14.5°S,157.5°W | | |
| Chao-Meng Fu | 87.5°S,132°W | Schoenberg | 15.5°S,136°W | | |
| Chopin | 64.5°S,124°W | Scopas | 81°S,185°W | | |
| Dickens | 73°S,157°W | Sophocles | 6.5°S,146.5°W | | |
| Eitoku | 21.5°S,157.5°W | Theophanes | 4°S,143°W | | |
| Goya | 6.5°S,152.5°W | Tolstoj | 15°S,165°W | | |
| Ictinus | 79°S,175°W | Ts'ao Chan | 13°S,142°W | | |
| Kālidāsā | 17.5°S,178°W | Vālmiki | 23.5°S,141.5°W | | |
| Keats | 70.5°S,156°W | Van Gogh | 76°S,139°W | | |
| Leopardi | 73°S,185°W | Wagner | 67.5°S,114°W | | |
| Lysippus | 1.5°N,133°W | Yun Sön-Do | 72.5°S,110°W | | |
| Mark Twain | 10.5°S,138.5°W | Zeami | 2.5°S,148°W | | |
| Marti | 75.5°S,169°W | | | | |

Like the southeast quadrant, much of this area is made up of intercrater plains. There are, however, some significant differences. Fewer scarps are included, and it also seems that there are various younger craters and basins, so that the surface in this quadrant as a whole is somewhat less ancient than that of the southeast. Some of the younger craters have more perfect walls, together with ejecta sheets and fields of secondary craterlets.

There is one region of smooth plains material situated at the top left. This is a southern extension of Planitia Tir, most of which lies in the northeast quadrant. The other apparently smooth patch, at the top right, is the subsolar region. The area appears smoother than it really is, because the Sun is vertically above it and casts no shadows.

The three largest craters have diameters greater than those of any comparable features in the southeast quadrant. The senior in size is Beethoven, 625 km wide, which has a floor flooded with smooth plains material, though it includes several features and one named crater, Bello. Another

huge flooded crater is Tolstoj. The third really vast formation lies at latitude 45°S, longitude 177°W, and has a diameter of 400 km. These three basins are presumably of comparable age with the large formations in the southeast quadrant. Apart from the Caloris Basin, Beethoven is the largest known enclosure.

The south polar area is very rough, and contains various prominent craters, of which Cervantes is typical. The polar point lies just inside the rim of Chao Meng-Fu, the floor of which is in permanent shadow.

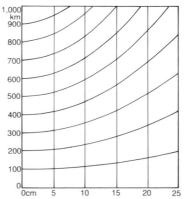

**Using the quadrant scale**
To use the quadrant scale (right) measure the distance from the subsolar point to the center of the feature concerned with a centimeter ruler. Locate this measurement on the horizontal centimeter scale. Measure the feature concerned. Place the ruler vertically at the point previously located on the centimeter scale. On the ruler read up from the base line the number of centimeters that the particular feature measures against the kilometer scale.

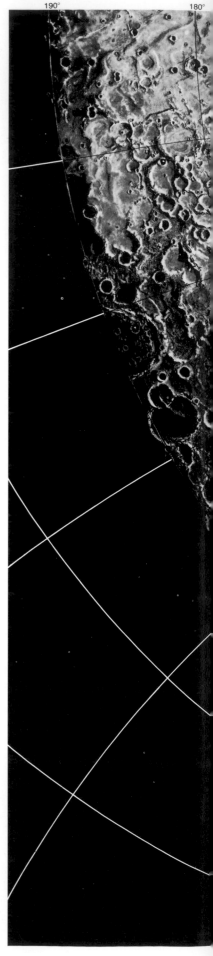

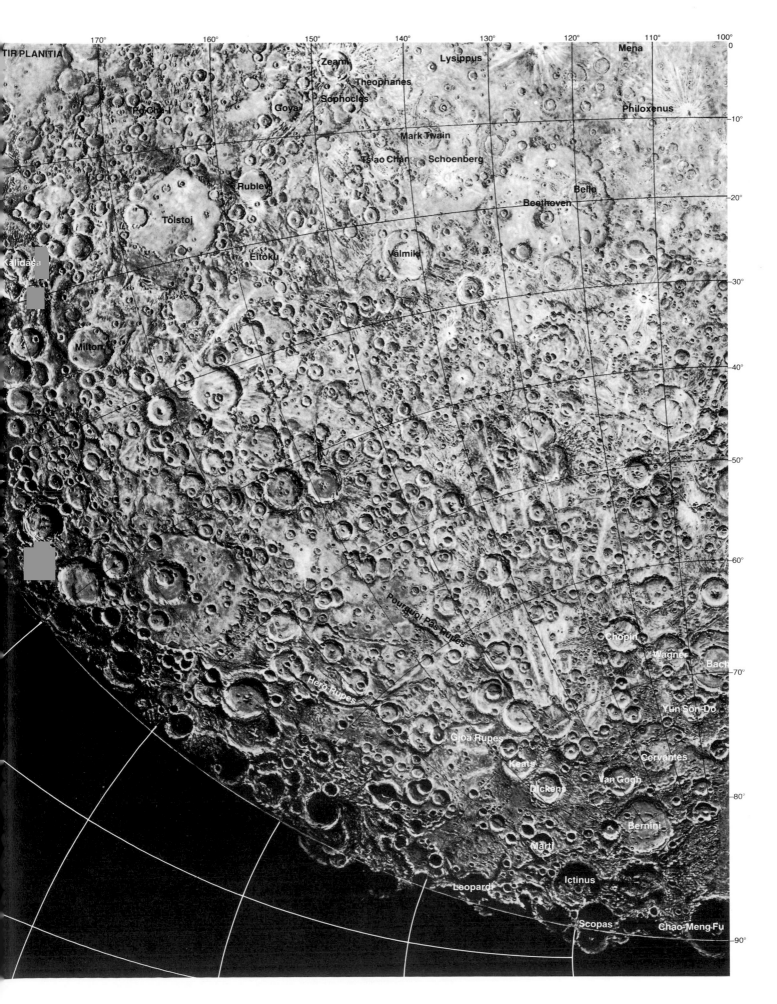

TIR PLANITIA

Po Chü-i
Zeami
Lysippus
Mena
Theophanes
Sophocles
Philoxenus
Goya
Mark Twain
Ts'ao Chan
Schoenberg
Rublev
Bello
Beethoven
Tolstoj
Eitoku
Valmiki
Kalidasa
Milton
Pourquoi Pas Rupes
Chopin
Wagner
Bach
Hero Rupes
Yun Son-Do
Gjoa Rupes
Cervantes
Keats
Van Gogh
Dickens
Bernini
Marti
Ictinus
Leopardi
Scopas
Chao-Meng-Fu

95

# Southeast Quadrant

| | | | | | |
|---|---|---|---|---|---|
| Africanus Horton | 50.5°S,42°W | Machaut | 1.5°S,83°W | Adventure Rupes | 64°S,63°W |
| Andal | 47°S,35.5°W | Mahler | 19°S,19°W | Astrolabe Rupes | 42°S,71°W |
| Balagtas | 22°S,14°W | Matisse | 23.5°S,90°W | Discovery Rupes | 54°S,38°W |
| Boccaccio | 80.5°S,30°W | Mendes Pinto | 61°S,19°W | Fram Rupes | 57°S,94°W |
| Boethius | 0.5°S,74°W | Mofolo | 38°S,29°W | Mirni Rupes | 37°S,40°W |
| Bramante | 46°S,62°W | Murasaki | 12°S,31°W | Resolution Rupes | 62°S,52°W |
| Brunelleschi | 8.5°S,22.5°W | Nampeyo | 39.5°S,50.5°W | Vostock Rupes | 38°S,19°W |
| Byron | 8°S,33°W | Neumann | 37.5°S,34°W | Zarya Rupes | 42°S,22°W |
| Callicrates | 65°S,32°W | Ovid | 69.5°S,22°W | | |
| Camões | 70.5°S,70°W | Petrarch | 30°S,26.5°W | Arecibo Vallis | 27°S,29°W |
| Carducci | 36°S,90°W | Pigalle | 38°S,11.5°W | Goldstone Vallis | 15°S,32°W |
| Carío | 27.5°S,10°W | Polygnotus | 0°,68.5°W | Simeiz Vallis | 13°S,66°W |
| Chekhov | 35.5°S,61.5°W | Po-ya | 45.5°S,19°W | | |
| Coleridge | 54.5°S,66.5°W | Puccini | 64.5°S,46°W | | |
| Copley | 38.5°S,86.5°W | Pushkin | 65.5°S,23°W | | |
| Dvořák | 9.5°S,12.5°W | Rabelais | 59.5°S,62.5°W | | |
| Equiano | 39°S,31°W | Rameau | 54°S,38°W | | |
| Futabatei | 16.5°S,83.5°W | Raphael | 20.5°S,76.5°W | | |
| Ghiberti | 48°S,80°W | Renoir | 18°S,52°W | | |
| Guido d'Arezzo | 38°S,19°W | Repin | 19°S,63°W | | |
| Haydn | 27.5°S,70°W | Rilke | 45.5°S,12.5°W | | |
| Hesiod | 58°S,35.5°W | Rudakī | 3.5°S,51.5°W | | |
| Hiroshige | 13°S,27°W | Sadī | 78.5°S,55°W | | |
| Hitomaro | 16°S,16°W | Schubert | 42°S,54.5°W | | |
| Holberg | 66.5°S,60°W | Sei | 63.5°S,89.5°W | | |
| Homer | 1°S,36.5°W | Shevchenko | 53°S,47°W | | |
| Horace | 68.5°S,50°W | Snorri | 8.5°S,83.5°W | | |
| Ibsen | 24°S,36°W | Sōtatsu | 48°S,18.5°W | | |
| Imhotep | 17.5°S,38.5°W | Spitteler | 68°S,59°W | | |
| Kenkō | 21°S,16.5°W | Sullivan | 17°S,87°W | | |
| Khansa | 58.5°S,52°W | Thākur | 3.5°S,64°W | | |
| Kuiper | 11°S,32°W | Tintoretto | 47.5°S,23°W | | |
| Kurosawa | 52°S,23°W | Titian | 3°S,42.5°W | | |
| Li Ch'ing Chao | 77°S,75°W | Tsurayuki | 62°S,22.5°W | | |
| Lu Hsun | 0.5°S,23.5°W | Unkei | 31°S,62.5°W | | |
| Ma Chih-Yuan | 61°S,77°W | Wergeland | 37°S,56.5°W | | |

Most of this quadrant is heavily cratered. The formations are crowded together, and there are many cases of overlapping. The regions between the main craters, or intercrater plains, are pitted with secondary craters, which most geologists consider to be very old, and it seems that the quadrant has less than its fair share of younger craters and basins. There are, however, three ray craters which are presumably among the most recently formed of the features: these are Copley, Snorri and Kuiper – the first crater to be identified on Mercury.

There is a different kind of terrain in the area between latitudes 20° and 40°S and longitudes 10° and 40°W, which is more hilly and lineated. It is interesting to note that this area is antipodal to the Caloris Basin (Caloris Planitia), though it is smaller. Since this hilly and lineated terrain is found only in this region, it has been suggested that it may have been produced by seismic effects being focused through the globe of Mercury at the time of the formation of the Caloris Basin; however, this is still very speculative.

Everywhere in the quadrant the floors of many of the craters and basins have been flooded with smooth plains material, so that any central mountain peaks and inner rings have been obliterated. A typical example is the large crater Pushkin and its companion Mendes Pinto; others include Raphael, Haydn, Pigalle and Petrarch. Eight of the 16 named scarps are in this quadrant. They are thought to have been formed by crustal compression and provide confirmation of the relatively ancient character of the whole terrain.

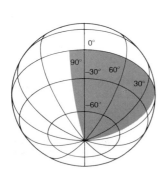

**Using the quadrant scale**
To use the quadrant scale (right) measure the distance from the subsolar point to the center of the feature concerned with a centimeter ruler. Locate this measurement on the horizontal centimeter scale. Measure the feature concerned. Place the ruler vertically at the point previously located on the centimeter scale. On the ruler read up from the base line the number of centimeters that the particular feature measures against the kilometer scale.

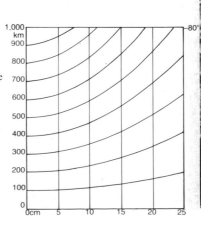

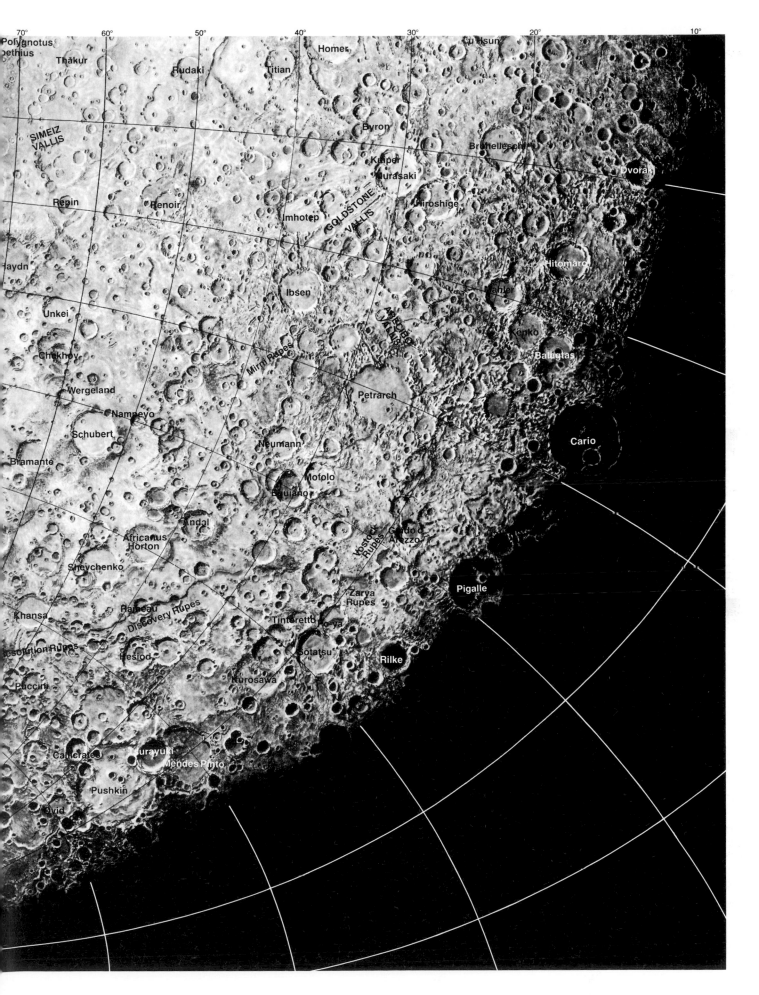

Polygnotus
Aethius
Thákur
Rudaki
Titian
Homer
Cui Hsun
SIMEIZ
VALLIS
Byron
Brunelleschi
Dvorak
Repin
Renoir
Kuiper
Murasaki
Hiroshige
Imhotep
GOLDSTONE VALLIS
Hitomaro
Haydn
Ibsen
Mahler
Unkei
ARECIBO VALLIS
Kenko
Chekhov
Mirni Rupes
Balagtas
Wergeland
Petrarch
Nampeyo
Schubert
Neumann
Cario
Bramante
Mofolo
Equiano
Andal
Africanus
Horton
Guido d'
Arezzo
Shevchenko
Vostok Rupes
Zarya
Rupes
Pigalle
Khansa
Rameau
Discovery Rupes
Tintoretto
Goya
Resolution Rupes
Hesiod
Sôtatsu
Rilke
Puccini
Kurosawa
Callicrates
Tsurayuki
Mendes Pinto
Pushkin
Ovid

# VENUS

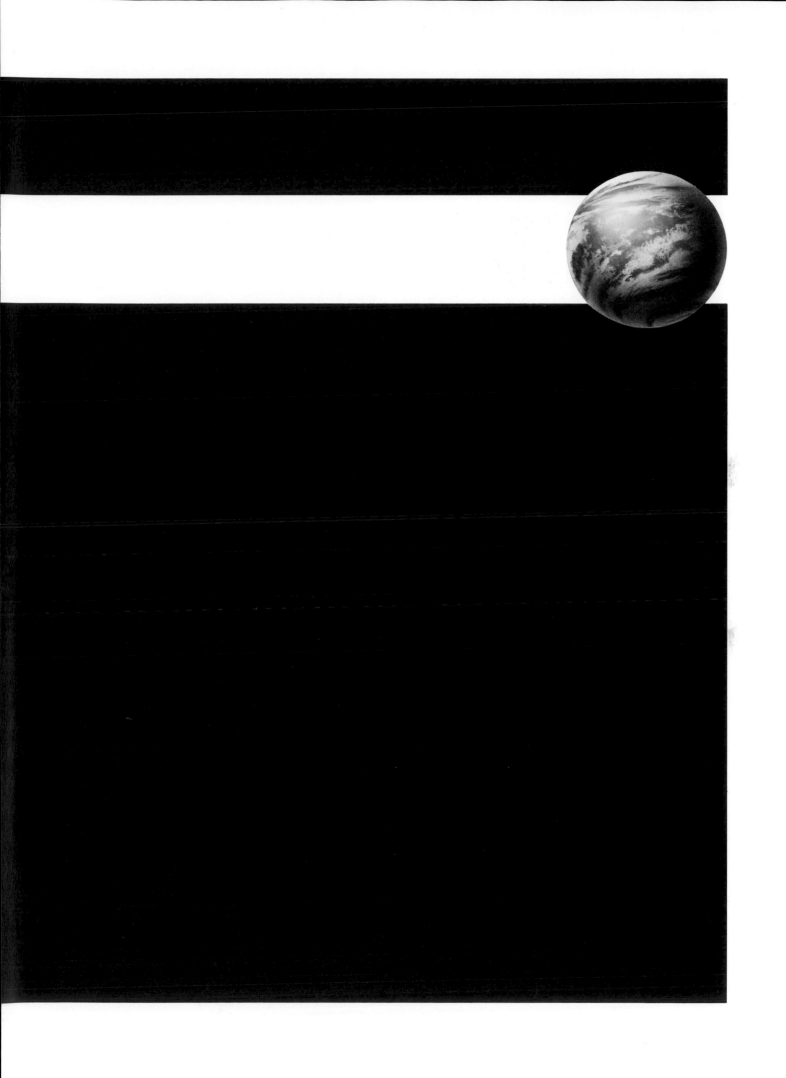

# Characteristics

Venus is the most brilliant of all the planets. At its brightest it is visible with the naked eye even when the Sun is above the horizon, and when seen against a dark background it can sometimes cast shadows on Earth. The most ancient observations of the planet known to us are Babylonian. The poet Homer described Venus as "the most beautiful star set in the sky", and it is hardly surprising that it was named in honor of the goddess of beauty.

Venus orbits the Sun at a mean distance of 108,200,000 km and has the most closely circular orbit of all the planets in the Solar System. At perihelion Venus is at a distance of 107,400,000 km from the Sun, and at aphelion it is 109,000,000 km from the Sun. When at its closest to the Earth it is the nearest object to us in the sky apart from meteoroids and the Moon. The revolution period is 224.7 Earth days. The dense atmosphere (*see* pages 104–109) makes the planet featureless when viewed in visible light, but radar investigations of the surface show that the planet rotates once every 243 Earth days, so the day on Venus is longer than the year. The mean temperature at the cloud tops is − 33°C, but the mean surface temperature is as high as 480°C.

With a diameter (both equatorial and polar) of 12,104 km, Venus is very much the Earth's twin in terms of size. The rotation is retrograde, that is backwards from the rotation of the Sun and nearly all the other planets in the Solar System. Venus has more than 80 percent of the mass of the Earth, a density only two percent less than that of Earth, and a specific gravity of 5.2 (compared with 5.5 for the Earth).

### Brightness

Like Mercury, Venus shows phases. When the planet is new (at inferior conjunction) it is only visible at transit, and when it is full (at superior conjunction) it is on the far side of the Sun. Unlike Mercury, Venus is brightest when at the crescent phase: the apparent diameter shrinks greatly as the phase increases. Its maximum magnitude is − 4.4, but, being closer to the Sun than the Earth, this planet can never be observed throughout the night.

### Interior

At first sight the similarity in size, mass and density of Venus and the Earth would suggest that the two planets have similar internal structures. The Earth has a hot, mobile mantle and a crust that is sufficiently thin for it to break up into large plates (the continents) that move relative to one another. In the geological studies on Earth, the science of plate tectonics (*see* pages 140–141) has become of fundamental importance. The continents represent regions of gravitational equilibrium in relation to their surroundings—a state known as isostatic equilibrium, which requires that the topographical relief be supported by buoyancy arising from a reduced density, rather like an iceberg floats in water. Thus the continents on Earth rise above the ocean basins surrounding them to an extent that depends primarily on the densities and thicknesses of the continents. Venus has continents of a sort, and the degree of isostatic equilibrium has been investigated by studying small perturbations in the motion of the Pioneer Orbiter spacecraft as it moves over the elevated structures. The investigations carried out so far over the two main highland areas, Ishtar and Aphrodite, have not shown any evidence of isostacy.

If the large-scale relief on Venus is in gravitational equilibrium, any substantial differences in the density of the crust must be the result of chemical differentiation. On Earth differentiation has resulted from the large-scale melting of the planet's interior. The dense Venusian atmosphere suggests that a similar melting has taken place in the interior of Venus, followed by out-gassing. Still further evidence for chemical differentiation on Venus is afforded by the significant amount of radioactivity measured on the surface by the Venera 9 and 10 probes.

There is, however, no evidence of widespread tectonic activity on

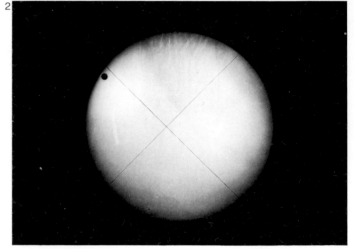

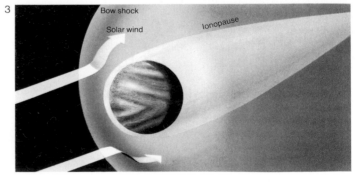

**1. Phases of Venus**
The apparent diameter of Venus changes according to the phase. It is least when Venus is full because the planet is then on the far side of the Sun and at its furthest from Earth. It increases as the phase shrinks.

**2. Transits of Venus**
Transits of Venus are rare but occur in pairs, separated by eight years – after which no more are seen for over a century. The last pair were in 1874 and 1882.

**3. Magnetosphere**
Venus' slow rate of rotation means that internal currents are weak or nonexistent, so the magnetic field is slight. The surface of the planet is protected from the solar wind by the thick atmosphere and by the electrically conductive ionosphere. There is a well-developed bow shock therefore. The magnetosphere displays several plasma effects within it but there are no bands of trapped particles similar to that on the Earth.

## 4. Size of Venus
Venus has a diameter of
12,104 km and is almost the same
size as the Earth (12,756 km).

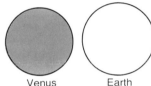

Venus      Earth

## 5. Interior of Venus
Venus has a similar size and mass
to Earth and is thought to have
roughly the same internal
structure. If the two formed in the
same region of the developing
Solar System they are also likely
to share similar elemental
compositions. Venera 8 indicated

### Physical data

|  | Venus |
|---|---:|
| Diameter | 12,104 km |
| Ellipticity | 0.0 |
| Mass | $4.8689 \times 10^{24}$ kg |
| Volume (Earth=1) | 0.86 |
| Density (water=1) | 5.24 |
| Surface gravity (Earth=1) | 0.902 |
| Escape velocity | 11.18 km/s$^{-1}$ |
| Rotation period | 243.01 days |
| Axial inclination | 178° |
| Albedo | 0.76 |

that the radioactivity of some
crustal rocks is similar to that of
terrestrial granites. Venus is
believed to have a liquid core, a
mantle and a rocky crust,
although the dimensions of the
latter two are unknown. The core
is thought to be smaller than that
of the Earth.

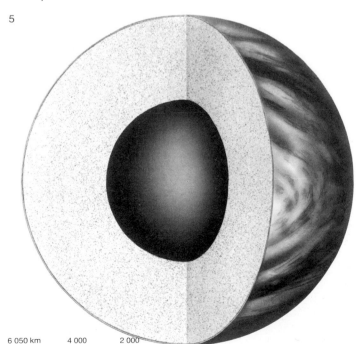

6 050 km      4 000      2 000

Venus, though the development of thin-crusted lowlands and thick-
crusted highlands suggests that Venus must have experienced a
period of widespread convection in its mantle. On the whole the
crust of Venus seems to be thicker than on Earth, which supports
the conclusion that Venus is a one-plate planet. According to
American oceanographers, the depth of crust beneath Aphrodite
Terra may be as much as 160 km. Ishtar and Aphrodite may be
regarded as "continents" that have been uplifted in the past and
have since remained almost static. But the main evidence comes
from the fact that though there is every reason to believe that
volcanism exists on Venus at the present moment, it is restricted to
only a few areas instead of being widespread as on Earth.

## Evidence from lightning
It has been suggested that lightning strikes are almost continuous at
the surface of Venus. "Whistlers" (electromagnetic impulses) are
most frequent over Beta Regio and the eastern end of Aphrodite,
which has been nicknamed the Scorpion's Tail. It is not thought
that these signals come from clouds over the two sites because the
clouds there are too insubstantial. It is more likely that the lightning
is produced immediately above ground level, within violently
disturbed dust and ash shot out from volcanic vents that represent
weak points in the crust.

Soil samples obtained by the Venera 14 probe in the Phoebe
Regio and Beta Regio regions are similar in composition to
terrestrial ocean deposits, though they do not appear to be as
volcanic as their terrestrial counterparts. The composition of the
soils at the two sites is also slightly different, the lowlands of Beta
Regio being younger than the remnants of the ancient crust which
were sampled in the rolling uplands of the Phoebe Regio site. The
observed tectonic activity at both sites was localized. It is possible
that the development of full-scale plate tectonics was stopped by
the evaporation of water and the slow rotation of the planet. Even
the small difference in density between Venus and the Earth may
indicate small but significant differences in the evolutionary history
of the two planets.

In the early days of the Solar System Venus and Earth probably
began to develop along similar lines. Although the surface of Venus
is extremely hot and dry now, the climate was probably much more
hospitable at one time, and scientists believe it may well have been
covered by an ocean of water. This is based on the fact that the ratio
of deuterium to hydrogen in the atmosphere of Venus is 100 times
greater than in seawater on Earth. This would suggest that there
was once at least 100 times as much water in liquid form as there is
today in the form of water vapor. It has been calculated that this
would be enough to cover the entire surface to a depth of 9 m. At
this stage in the evolution of the Solar System, the Sun was
probably cooler and therefore less luminous. As the Sun became
hotter, however, oceans on Venus boiled and evaporated and, also
as a result of the increased temperatures, carbon dioxide was driven
out of limestone rocks on the surface into the atmosphere. There
was a runaway "greenhouse effect" (*see* pages 104–105).

## Magnetism
While Venus is thought to possess an iron-rich core only slightly
smaller than that of the Earth, it does not appear to have an
intrinsic magnetic field. On such a slowly rotating planet the
dynamo-like mechanism such as that in the Earth's core would
create only an insignificant field. The maximum possible strength of
the Venus magnetic field is $5 \times 10^{-5}$ that of the Earth. Nevertheless
the solar wind is prevented from reaching the surface of the planet
by the dense atmosphere, more specifically by magnetic eddy
currents induced in the ionosphere. Thus Venus has a well-
developed bow shock region, which exhibits plasma effects but
contains no trapped particles. Magnetism on Venus is therefore
intermediate between the Earth and Moon.

# Observational Background

In 1610, Galileo used his primitive "optick tube" to look at Venus, and he observed the planet's phases. This was of great significance since according to Ptolemaic theory, which stated that the Sun and planets moved around the Earth, Venus could never show a full cycle of phases. Galileo could see no details on the surface of the planet, which was only to be expected in view of the fact that Venus is permanently hidden by its dense, cloud-laden atmosphere, but his observation of the phases of Venus contributed greatly to the downfall of the Ptolemaic theory of the universe.

A map of Venus was produced in 1645 by F. Fontana, an amateur astronomer who used a small-aperture, long-focus refractor that was home-made and probably not much better than Galileo's telescope. Fontana believed that he had recorded seas and continents, but there is no doubt that they were purely illusory.

In 1667 G. D. Cassini observed Venus from Bologna and recorded a number of bright, dusky patches. From a series of observations he estimated the rotation period of Venus to be 23 hr 21 min. Later, in Paris, Cassini was unable to recover the markings he had seen and his son, J. J. Cassini, was equally unsuccessful.

### Discovery of the atmosphere

On rare occasions Venus passes in transit across the face of the Sun. These transits were formerly regarded as important, since they provided a method of measuring the length of the Earth–Sun distance or astronomical unit. There were transits in 1761, 1769, 1874 and 1882; the next will occur in 2004. The method of measurement has become totally obsolete, but at the 1761 transit the Russian astronomer M. V. Lomonosov observed a blurred appearance around the disc of the planet when seen against the Sun, and correctly attributed it to the presence of an atmosphere at least as extensive as that of the Earth. Lomonosov's findings made little impact at the time, but J. H. Schröter confirmed the existence of an atmosphere around Venus later in the eighteenth century.

### Schöter and Herschel

The two great planetary observers of the late eighteenth century were Johann Hieronymus Schröter, an amateur who set up his observatory at Lilienthal, near Bremen, and William Herschel, who discovered the planet Uranus and who came to England from his native Hanover while still a young man and constructed the best telescopes of the time. Schröter recorded vague, dusky patches on the brilliant disc of Venus, and also believed that he had detected lofty mountains. Herschel was no believer in Schröter's mountains, but he too saw some dusky features, and both observers agreed that the visible surface was nothing more than a layer of cloud. Schröter also recorded the "ashen light", or faint luminosity of the night side of Venus when the planet is in its crescent stage. The precise nature of the ashen light is still not known, but it seems to be a real phenomenon, and may be due to electrical effects in the upper atmosphere of the planet.

Since the markings on Venus were so elusive, it was extremely difficult to measure the rotation period. Schröter calculated the value to be 23 hr 21 min 19 sec in 1789 and then he modified it to 23 hr 21 min 7.9 sec in 1792.

During the seventeenth and eighteenth centuries several observers described what they believed to be a satellite of Venus. However, there can be no doubt that no satellite exists. Venus is so brilliant that it tends to produce ghost images when seen through a telescope, and no doubt the "satellite" was the result of this effect.

### Rotation period

The rotation period of Venus was not known until modern times. Estimates ranged from 22 hours up to 224.7 days. If the latter value had been correct, then the rotation would have been synchronous, the same hemisphere being turned permanently towards the Sun. The value was difficult to determine, because of the lack of well

**1. Goddess of beauty**
Canova's sculpture of Venus in Florence epitomizes classical ideas of the goddess of beauty. Since the planet Venus is more brilliant than any other star or planet in the sky, it has always been considered a sight to behold. The Chinese called Venus Tai-pe, meaning "beautiful white one"; the Babylonians worshipped "Ishtar" as "the bright torch of heaven"; but the Greeks and Romans named the planet after the goddess of beauty, and they too worshipped in temples dedicated to Venus.

**2. Drawings by Bianchini 1726**
Francesco Bianchini was one of a number of 17th- and 18th-century astronomers who thought they saw features on the surface of Venus. Bianchini viewed the planet with a 6.4 cm refractor and claimed to have seen continents and seas. He drew maps based on features that he had seen, but we have to dismiss all of these observations as optical illusions prevalent at the time.

**3. Inhabitants of Venus**
Imaginative speculation about possible inhabitants of Venus abounded in the 19th century. Many such illustrations appeared in popular journals (**A**) or even the more reputable scientific publications (**B**).

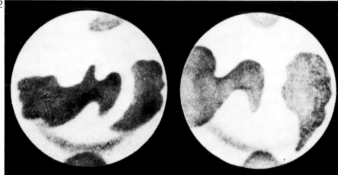

marked features on the surface. Neither was photography of much help. The first really good photographs were taken in 1923 by F. E. Ross at Mount Wilson; at ultraviolet wavelengths some distinct features were shown, but all were badly defined and could be dismissed as cloud phenomena. Until the development of radar, it was generally believed that the rotation period must be of the order of three to four weeks.

Better photographic results were obtained in the 1950s by G. P. Kuiper at the McDonald Observatory in Texas, N. A. Kozyrev in the Soviet Union, A. Dollfus at the Pic du Midi in France and R. S. Richardson at Mount Wilson. Kuiper's photographs and Richardson's revealed a vaguely banded appearance to the planet. Other French astronomers, Boyer and Guèrin, observed the dark Y-shaped feature and from it they calculated a four-day, retrograde rotation. This is in fact true of the upper clouds.

Before the Space Age, Venus was still described as a planet of mystery. There were two theories concerning the surface. The American astronomers F. L. Whipple and D. H. Menzel believed the planet to be covered mainly with water, while others regarded it as a fiercely hot desert without a trace of moisture. The one piece of positive information was that the atmosphere was rich in carbon dioxide. Since this gas in effect shuts in the Sun's heat, it was logical to assume that the surface temperature must be high. Earth-based radar methods provided new information; apparently the rotation was very slow indeed, and the temperature was estimated to be several hundred degrees centigrade.

**Missions to Venus**
The first Venus probe was Venera 1, launched by the Soviet Union in February 1961. It was not successful, as contact was lost within a few weeks of launch, but Mariner 2, the US probe despatched on 26 August 1962, met with better fortune. In December it bypassed

Venus at only 35,000 km, and sent back information that was fatal to the oceanic theory. The temperature was far too high for liquid water to exist; the modern value is 480°C. The solid surface rotation period was confirmed as being slow, and it is now known to be 243 days, which is longer than Venus' "year"—a case unique in the Solar System. Moreover, Venus rotates from east to west, not west to east, and the upper atmosphere rotates much faster than the levels below. The reason for this curious state of affairs is still not known. Finally, Mariner 2 and later probes confirmed that Venus has no detectable magnetic field.

The next American probes were Mariner 5 (1967) which sent back images of the cloud tops, and Mariner 10 (1973) which by-passed Venus on its way to Mercury. Meanwhile the Russians had paid great attention to the planet with their Venera probes, and in 1975 Veneras 9 and 10 sent back pictures direct from the surface. Then, in 1978, came a veritable armada—the United States Pioneer mission, involving an orbiter and a series of small entry probes, and two more Veneras, 11 and 12. The Pioneer Orbiter was able to send back detailed radar maps of much of the surface of Venus, and further radar maps were provided by Veneras 15 and 16, which reached the neighborhood of the planet in 1983. In 1985 the Soviet Vega probes, *en route* to Halley's Comet, dropped balloons into Venus' atmosphere, obtaining information from various levels, and in May 1989 the American spacecraft Magellan was launched on a mission which will, it is hoped, improve the radar coverage of Venus; it is scheduled to reach the neighborhood of the planet in August 1990.

Obviously Venus, with its dense atmosphere and its hostile environment, is a difficult target for spacecraft, but by now a tremendous amount of information has been obtained—conditions there are akin to the conventional picture of hell!

**4. Schröter's drawings 1796**
Schröter observed the planets from 1178 until 1814. He did not find Venus an easy object and between 1779 and 1888 he did not detect any markings. Then he observed a "filmy streak", which he drew in his book (1796) *Aphroditographische Fragmente*.

**5. Lowell's map 1897**
Percival Lowell saw features on a number of planets, and in particular on Mars and Venus. He produced a map of Venus with named features, and rejected the idea of an obscuring atmosphere.

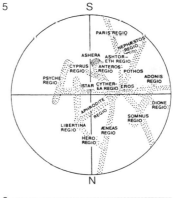

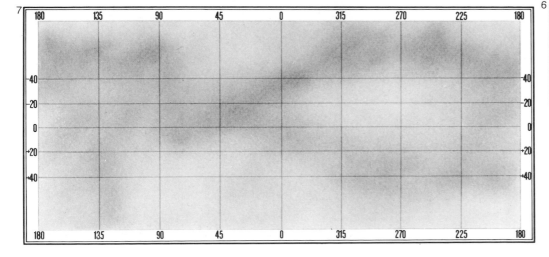

**6. The phantom satellite**
Many observers thought they saw a satellite accompanying Venus. Montaigne of Limoges drew its position in 1761 relative to Venus.

**7. Boyer's map**
Based on photographs taken at the Pic du Midi Observatory, C. Boyer drew a map of Venus during the 1950s. It included the equatorial Y shape.

# Atmosphere I

**Composition of the atmosphere**
Carbon dioxide is the major constituent of Venus' atmosphere, with nitrogen, oxygen, sulphur dioxide and water vapor amounting to a few percent. The major interest for scientists, however, is in the inert gases that are present, since they provide information about the evolutionary history of Venus' atmosphere. Measurements of the quantities of these gases have now been obtained from the probes on the Pioneer Venus mission.

The atmosphere on Venus seems to contain greater amounts of both the isotopes of argon (Ar–38, Ar–36) than the Earth's atmosphere, but a smaller amount of Ar–40, which is produced by the radioactive decay of potassium. On Earth about 10 percent of the argon produced by radioactive decay processes is released into the atmosphere as a result of the weathering of rocks. On the Moon, where no weathering occurs, the present release of Ar–40 amounts to about six percent of the total production. This implies that other mechanisms of release of trapped gases are as important as weathering. It would be expected, therefore, that the amount of Ar–40 in the atmosphere of Venus would be quite large. The fact that this is not the case suggests that Venus has less potassium than the Earth, or that the release mechanism is less efficient.

Venus also seems to possess more krypton than might be expected. The value is less than that of the Sun but more than found on the Earth, which provides further constraints on theories of the origin of the planets. It is possible that Venus received a large input of various gases from the Sun in the early history of the Solar System during a period when the solar wind was much denser than it is today. These gases would have impacted the mass of material condensing to form the planet. The proto-Venus would have

blocked off the enriched solar wind, preventing it from blowing out to the developing Earth and Mars. This may help to account for today's Earth and Mars being relatively deficient in the two noble gases krypton and xenon, as well as in argon.

Undoubtedly Venus has three times as much krypton as the Earth and 700 times more primordial argon than krypton. This compares with an argon to krypton ratio on Earth and Mars of only 30 to 1. The Sun on the other hand has 2,000 times more argon than krypton. The Pioneer Venus results certainly suggest that these noble gases may indeed have come from the Sun and formed a veneer on the dust grains as they condensed out of the solar nebula.

There is a large amount of carbon dioxide in Venus' atmosphere compared with that of the Earth, and the amount of atmospheric nitrogen on Venus is similarly large, in fact three times the terrestrial value. However, since the amount of nitrogen fixed in the Earth's crust is two or three atmospheres, the total amount of nitrogen on the two planets is about the same. On Mars, the abundance of nitrogen is far less. The other important constituents in Venus' atmosphere are closely associated with the complex chemistry and the processes of cloud formation.

The distributions of hydrogen sulphide ($H_2S$), carbonyl sulphide (COS), sulphur dioxide ($SO_2$) and water vapor ($H_2O$) are probably associated with the sulphur cycle and the production of sulphur and sulphuric acid droplets in the various cloud layers. It is equivalent to only about $10^{-5}$ times the amount in the terrestrial oceans.

This may not always have been the case, however. Light hydrogen and heavy hydrogen (deuterium) have different escape rates, and over geological time scales the variation in the deuterium to hydrogen ratio provides information about the history of water

### Composition of Earth's atmosphere

| | % volume |
|---|---|
| Nitrogen | 76.084 |
| Oxygen | 20.946 |
| Argon | 0.934 |
| Carbon dioxide | 0.031 |
| Neon | $1.82 \times 10^{-3}$ |
| Helium | $5.24 \times 10^{-4}$ |
| Methane | $1.5 \times 10^{-4}$ |
| Krypton | $1.14 \times 10^{-4}$ |
| Hydrogen | $5 \times 10^{-5}$ |
| Nitrous oxide | $3 \times 10^{-5}$ |
| Carbon monoxide | $10^{-5}$ |
| Xenon | $8.7 \times 10^{-5}$ |
| Ozone | up to $10^{-5}$ |
| Water (average) | up to 1 |

### Composition of Venus' atmosphere

| | % volume |
|---|---|
| Carbon dioxide | $\approx 97$ |
| Water vapor | $\approx 1 \times 10^{-1}$ |
| Carbon monoxide | $5 \times 10^{-3}$ |
| Hydrogen chloride | $6 \times 10^{-5}$ |
| Hydrogen flouride | $5 \times 10^{-7}$ |

**1. Temperature and pressure**
Temperature and pressure in the atmosphere of Venus were measured by both the Mariner and Venera spacecraft, enabling profiles to be compiled. At the surface atmospheric pressure is 90 times what it is on the surface of Earth. At a height of 50 km pressure reaches 1 atmosphere (that is, the mean sea-level pressure on Earth). The temperature and pressure gradients on Venus are compared in **A**. A temperature profile of Venus (**B**) shows two subdivisions of the atmosphere. The lower one

(troposphere) is heated largely by the greenhouse effect, and the temperature decreases with altitude. The upper part, or thermosphere, is heated by the absorption of solar radiation, and the temperature increases with altitude. Night and day temperatures are compared in the diagram and temperatures on Venus are also compared with those on Earth. The atmosphere on Earth is more complex, and it reveals a middle subdivision where the temperature is raised locally by the absorption of ultraviolet radiation by ozone.

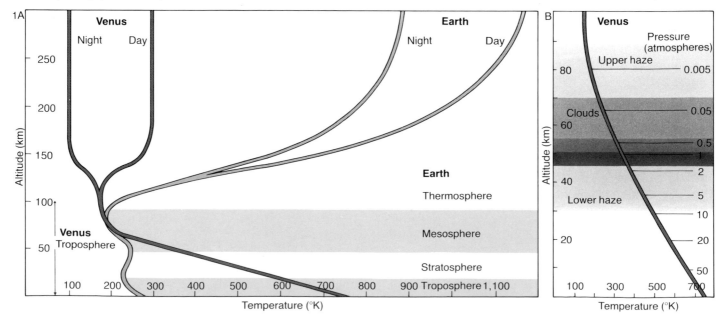

on the planet. Currently the ratio on Venus is about 1 to 160. Originally, the ratio is thought to have been about 1 to 15,000, which is found now on Earth. This suggests that Venus has lost what was probably once an ocean through the outflow of hydrogen to space. In this earlier epoch the Sun would have been cooler too, so that Venus could have possessed an ocean similar to those now found on Earth. It is possible that somewhere on Venus' surface there may be fluvial channels that relate to this earlier period in the climatic history of Venus.

The subsequent increase in the radiant heat of the Sun would have initiated the runaway greenhouse effect, which ultimately raised the surface temperature to a level where the ocean of water evaporated and escaped from the planet. It is possible that Venus has lost an amount of water equivalent to one-third of that currently contained in the Earth's oceans.

### Structure of the atmosphere

Observations have now been made throughout the Venusian atmosphere from the surface to an altitude of 200 km, and a detailed comparison of the structure may be made with that of the Earth's atmosphere.

On the day side of Venus there is a terrestrial-type thermosphere with temperatures increasing from about 180 K at 100 km to about 300 K in the exosphere. The thermosphere does not exist on the night side of the planet, where the temperature falls from about 180 K at 100 km to as low as 100 K at 150 km. Transition from day- to night-side temperatures across the terminator is very abrupt.

Between an altitude of 100 km and the cloud tops at about 70 km the atmospheric temperature is distinctly variable. Diurnal fluctu-

ations of as much as 25 K have been observed at the 95 km level. About 90 percent of the volume of the entire atmosphere lies between the surface and height of 28 km, and at this level the atmosphere resembles a massive ocean; it is dense and very sluggish in response to solar heating, which naturally is feeble at these depths.

### The greenhouse effect

The surface temperature of Venus is almost uniformly 737 K and the massive atmosphere is virtually all carbon dioxide. Why is this so different from the Earth? The basic makeup of the Earth and Venus is very similar, although there is as much carbon dioxide in the atmosphere of Venus as we find in rocks such as limestone on the surface of the Earth. The orbit of Venus is nearer to the Sun and the greater heat rapidly warms the surface of the planet, producing the so-called greenhouse effect (*see* diagram 3).

The carbon dioxide in the atmosphere of Venus is not sufficient to maintain the greenhouse-like situation above. It is responsible for about 55 percent of the trapped heat; a further 25 percent is due to traces of water vapor, while sulphur dioxide, which constitutes only 0.02 percent of the atmosphere, traps 5 percent of the remaining infrared radiation. Clouds and hazes are responsible for the remaining 15 percent of the greenhouse effect.

The catastrophic change in the Venusian environment should serve as an example to those of us on Earth, where the carbon dioxide content is increasing as a result of the greater use of fossil fuels, while the natural sink, the equatorial rain forests, is being reduced. Weather and climate are stabilized by the oceans, which cover two-thirds of the Earth's surface, but the danger from the aforementioned trends, though as yet slight, is not insignificant.

**2. Heat budget**
A comparison of absorbed solar energy and emitted infrared radiation for both the Earth (**A**) and Venus (**B**) are plotted here. The values have been averaged out over all seasons, times of the day and longitude.

**3. Greenhouse effect**
In the early days of the Solar System the Sun radiated much less energy than it does now. There was much more water on the surface of Venus and a state of equilibrium was maintained as the amount of heat absorbed by

the surface was equal to that emitted from it (**A**). As increased solar heating warmed the surface more quickly, more carbon dioxide was liberated, increasing the opacity of the atmosphere. Solar radiation was still able to penetrate the increasingly carbon dioxide atmosphere, so continuing to warm the surface, but the atmosphere was now opaque to radiation emitted by the surface. Surface temperature therefore continued to rise (**B**). By the time surface and atmosphere were in thermal and chemical equilibrium, no surface water remained (**C**).

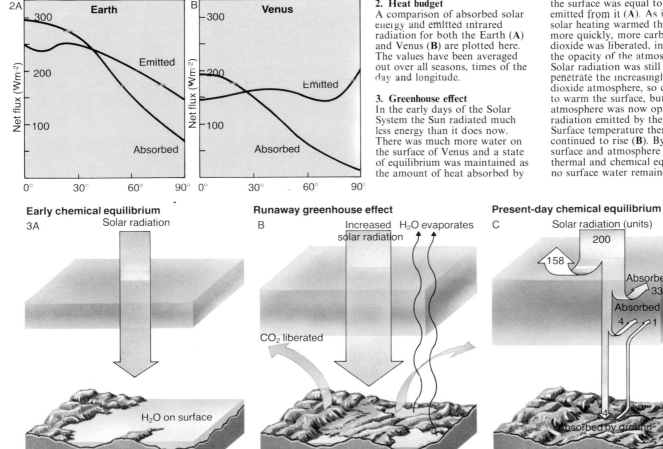

2A **Earth**

Net flux (Wm⁻²)

Emitted

Absorbed

0°   30°   60°   90°

B **Venus**

Net flux (Wm⁻²)

Emitted

Absorbed

0°   30°   60°   90°

**Early chemical equilibrium**
3A   Solar radiation

H₂O on surface

**Runaway greenhouse effect**
B   Increased solar radiation   H₂O evaporates

CO₂ liberated

**Present-day chemical equilibrium**
C   Solar radiation (units)

200

158

Absorbed by clouds
33

Absorbed by CO₂
4   1

Absorbed by ground

# Atmosphere II

### Circulation within the atmosphere

Without doubt one of the most surprising properties of Venus' atmosphere is the rapid rotation at cloud top level. The rotation period is only approximately four days, which is very fast when compared with the solid body rotation of 243 days. Indeed, this is the largest super rotation in the Solar System, with a 60:1 rotation of atmosphere to planetary period compared with about 7:1 for Titan. Furthermore, the mechanisms responsible for generating Venus' meteorology and the strong winds cause atmospheric features to move in a direction opposite to, and 20 times faster than, the overhead motion of the Sun relative to a fixed point on the surface of Venus.

Almost all solar energy is absorbed in the neighborhood of the cloud tops, so providing the heating that drives the observed atmospheric motions. The effects of the heating are conveyed to higher levels by transport processes. In the stable upper atmosphere of Venus, internal gravity waves play a major role in the vertical transport of heat. These mechanisms do not act simultaneously, thus producing a tilt in the vertical pattern of convection which results in a net motion in a direction opposite to that of the Sun.

High-resolution images of Venus show many examples of small-scale structures in the clouds. These whirls and swirls are generally referred to as eddies, which typically cover a wide range of spatial scales. They include convection cells, small-scale gravity waves and planetary-scale waves, covering a range from several km up to that of the planetary radius. Eddies are very important for the transfer of energy and momentum in planetary atmospheres, and this is certainly the case with Venus.

The dominance of the largest-scale eddies is evident in those images which are so often characterized by the single dark "Y" feature which encircles the planet. This feature probably results from the superposition of two large-scale waves whose individual parts drift slowly in and out of phase before dissipating. This situation may well explain the change in the shape of the Y to a C and the occasional absence of the feature.

In other regions the picture is more complicated. Velocities increase to either side of the equator, and at a latitude of approximately 50° the rotational period may be as short as two days. Streaks in the clouds show a flow towards the polar vortex, where the kinetic energy of rotation is eventually dissipated. In order to maintain the planetary angular momentum balance the return flow of the atmosphere must occur at deeper levels. Hadley cells (*see* diagram 3) are essential characteristics of the flow, providing the latitudinal redistribution of heat in the atmosphere. Winds decrease rapidly nearer the surface: at the cloud tops they have velocities of about 100 ms$^{-1}$; at an altitude of 50 km velocities are about 50 ms$^{-1}$; and they decrease to only a few meters per second near the surface. The overall situation can be reduced to a zonal retrograde rotation dominating the circulation between two scale heights above the clouds. Superimposed on this atmospheric rotation is a cloud level pattern of Hadley cells. Although the circulation shows unexpected variability over time, large zonal wind motions as well as small poleward meridional motions at the cloud tops are persistent features. It would seem that the Hadley cells or eddies are essential for the distribution of heat in the deep atmosphere although it is difficult to say just how many of these features are present.

The stratosphere of Venus is not well defined. On Earth this region is created through the absorption of solar energy by ozone, but there does not seem to be an equivalent process taking place in Venus' atmosphere.

The upper atmospheric motions are particularly interesting on Venus. The zonal winds decrease from 100 ms$^{-1}$ at the cloud tops to less than 500 ms$^{-1}$ at an altitude of 100 km. In the stratosphere and mesosphere vertical and horizontal propagation of planetary-scale waves transport heat towards the pole in a manner similar to that of the Earth's atmosphere. A significant difference, however, between the two atmospheres is that on Venus the thermal structure of the stratosphere is affected by the morphology and radiative properties of the ubiquitous clouds. Rather surprisingly, the upper atmosphere of Venus seems to be rather insensitive to solar activity, suggesting that this region is not controlled by the usual heating mechanisims that are associated with planetary thermospheres.

**1. Circulation of clouds**
A series of ultraviolet images taken by the Pioneer Venus Orbiter on consecutive days shows the rotation of cloud markings in only four Earth days. The apex of the Y is clearly seen in image 1 and then appears again in image 5, having moved across the disc from east to west.

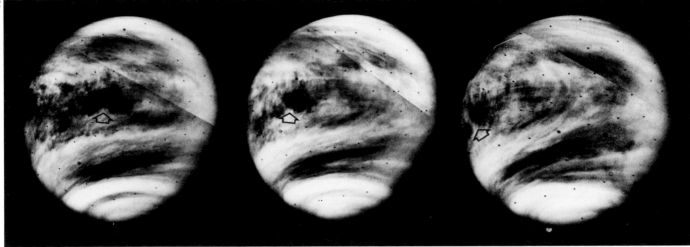

## 2. First optical evidence
Mariner 10 provided the first proof of the four-day rotation of the upper clouds of Venus with three photographs taken at seven-hour intervals. Contrast has been enhanced in the photographs and an arrow marks the same region on each one. The images were taken in the ultraviolet in order to enhance the amount of detail of cloud formations across the disc.

## 5. North polar phenomenon
A crescent-shaped collar of cloud surrounds the north pole of Venus at about 70°N. It rises to approximately 15 km above the cloud tops. In the center of this dipolar structure are two rotating clearings in the cloud straddling the pole.

## 6. Dipole
The dipolar structure represents a deviation from the circulatory patterns and may result from planetary-scale waves of unknown origin. There are anomalous and variable temperatures throughout the structure which shows up on this thermal image of the pole.

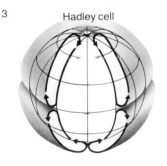

Hadley cell

Westward winds

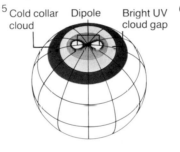

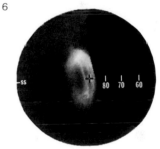

5 Cold collar cloud   Dipole   Bright UV cloud gap

6

## 3. Hadley cells
A Hadley circulation pattern is the response of a planetary atmosphere to differential latitudinal heating. Air rises in warm regions, flows to cool zones, sinks and returns to warm latitudes. On Venus there are probably several cells.

## 4. Prevailing winds
Venus has a slow retrograde rotation, and winds in the lower and middle atmosphere blow from east to west. They blow in the direction of the planet's rotation at all latitudes, varying from 1 m/sec at ground level to 100 m/sec at the cloud tops.

## 7. Wind velocity profiles
The graph compares zonal wind velocity profiles obtained from the Pioneer Venus and Venera probes. Differences between the profiles may reflect variations in the wind field over long periods of time on the planet. The diagram includes several results that were obtained from the Pioneer probe.

## 8. Zonal wind velocities
The graph shows the relative zonal velocities on Venus. They are measured relative to the planet's rotation and are shown alongside an image of the bands. The graph compares results from Mariner 10 in 1977, Mariner 10 in 1978 and Pioneer Venus in 1980. The results are somewhat speculative due to little data.

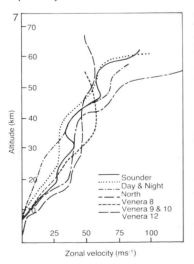

7

Altitude (km)

— Sounder
······ Day & Night
─ ─ North
─·─ Venera 8
─··─ Venera 9 & 10
── Venera 12

Zonal velocity (ms⁻¹)

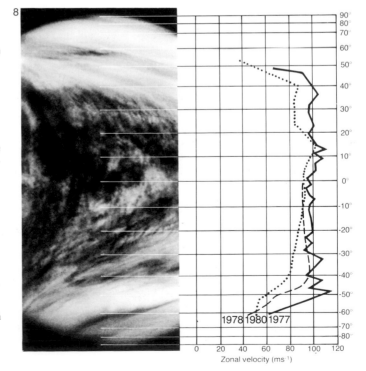

8

1978 1980 1977

Zonal velocity (ms⁻¹)

# Atmosphere III

## Cloud types

The yellow color of the clouds on Venus has long been cited as evidence that, unlike the white clouds of the Earth, they do not consist of water vapor. They extend to an altitude of more than 70 km with a haze layer overlying the main cloud deck. The haze is composed of sulphuric acid droplets of radius between 1 and 2 $\mu$m. It may not be a permanent layer, however, and observations indicate that it may vanish over periods of several years.

The middle cloud region at an altitude of between 52 and 56 km contains liquid droplets and solid particles of radius 10 to 20 $\mu$m. The densest cloud layer lies between 49 and 52 km, where the visibility is reduced to less than 1 km, and this is where the largest particles are found. Just below this opaque layer is a distinct thin region of reduced opacity.

Between 32 and 48 km lies a thin haze of particles almost 1 $\mu$m in size and with concentration of 1 to 20 cm$^{-3}$. The upper part of this layer from 45 to 47 km is the most dense. Below the lower boundary the atmosphere seems to be free of particles all the way down to the planet's surface. This does not mean, however, that the visibility is unlimited. The atmosphere on Venus is so dense that even in the absence of clouds the scattering of light by gas molecules alone probably restricts the limit of vision.

The various spacecraft measurements indicate that the concentration of cloud particles is consistent with a source at the top and a sink at the bottom of the atmosphere, with downward flow by mixing and by gravitational settling. This suggests that there is a source of sulphur and sulphuric acid at the top of the middle cloud region at an altitude of about 57 km. The source is probably photochemical in nature requiring ultraviolet radiation that does not penetrate far into the sulphuric acid cloud. The sulphuric acid droplets lose water on the way down, but the increased amounts of water ($H_2O$), and perhaps sulphur dioxide ($SO_2$) and sulphur trioxide ($SO_3$) as they fall under the influence of gravity, form a layer of dilute acid, which is concentrated in the lower cloud region. It is possible that the large size of particles, greater than 10 $\mu$m, encourages precipitation on Venus, either from the dilute sulphuric acid in the lower cloud region or initiated by sulphur from the middle cloud region which could pick up dilute sulphuric acid as it falls below 5 km.

Besides water, there are several substances that are almost nonexistent above the clouds but well represented below them. The abundance of sulphur dioxide varies in this way by a factor of more than 200 and oxygen by a factor of 60. These higher abundances arise when sulphuric acid and sulphur rain out of the clouds, vaporize and are even dissociated in the heat below, so that they chemically attack the gases of the lower atmosphere.

The presence of sulphur dioxide and not carbonyl sulphide (COS) as the major sulphur-bearing compound in the lower atmosphere is a major surprise. Previously it had been thought that the abundance of COS indicated that the carbon monoxide (CO) abundant in the atmosphere might be affected by certain minerals on the surface, and that its value was therefore representative of the amount existing beneath the clouds; but new measurements suggest that CO is generated almost entirely by photochemical processes, and no significant amounts of COS should be expected.

Perhaps one of the most startling findings is the suggestion that lightning occurs at some level beneath the clouds. The Venera 11 and 12 spacecraft have detected 13 minutes of electromagnetic

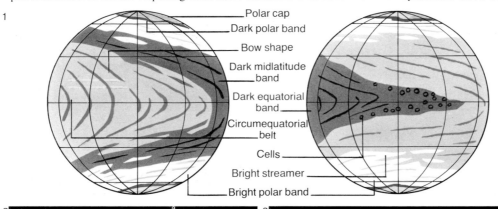

Polar cap
Dark polar band
Bow shape
Dark midlatitude band
Dark equatorial band
Circumequatorial belt
Cells
Bright streamer
Bright polar band

**1. Cloud belts**
With more and more photographs of details of features within the Venusian atmosphere, scientists were able to identify broad bands according to latitude and certain cloud features within the bands. Cloud movements were revealed from time-lapse motion pictures made up from individual frames. The equatorial band has fine streams of cloud, and Y- and C-shaped formations that spread out in the direction of rotation. Each polar region has a hood of cloud with spiral patterns that connect with equatorial bands.

**2. Venus from Mariner 10**
A mosaic of photographs taken by Mariner 10 in 1974 reveals the now familiar dark and light bands across the planet. Many similar images of Venus have been taken in ultraviolet light. This is in order to bring out the details of the upper levels of the atmosphere which would appear very bland at visible wavelengths. In ultraviolet light a wealth of detail emerges, indicating major features of the circulation of Venus' upper atmosphere.

**3. Venus from Pioneer 1979**
The clouds of Venus were again photographed in ultraviolet by the Pioneer Venus Orbiter in 1979. More detail is apparent than from Mariner, especially of the equatorial Y-shaped feature. This enabled scientists to get a better idea of the very complex wave motions that are responsible for this phenomenon.

signals similar to terrestrial lightning storms. They began at an altitude of 32 km and continued down to only 2 km above the surface. Thunder and lightning may be very much in evidence on Venus, and there is also strong evidence of active vulcanism.

The two main volcanic regions are Beta Regio, where there are two massive shield volcanoes (Rhea Mons and Theia Mons) and Atla Regio, the so-called "Scorpion's Tail". From studies of over a thousand Pioneer orbits, American astronomers have found evidence for clusters of lightning bolts similar to those generated in terrestrial volcanic plumes, so that apparently Beta Regio and Atla sit over powerful upflowing convective plumes deep in Venus' interior magma. The amount of atmospheric sulphur dioxide showed a sharp increase in 1978, and this may have been due to a violent volcanic outburst in one of the two main active regions. If so, the sulphur dioxide may have formed into small aerosol particles of sulphuric acid, and it was suggested that the sulphur may have been blown up as high as 60 km with a force ten times greater than any Earth eruption over the past century. We should be able to obtain much more information from the Magellan spacecraft once it enters Venus' orbit in August 1990.

## Cloud motions
Although optically Venus is a featureless disc, observations at ultraviolet wavelengths are extremely rewarding and show a wide range of cloud structures. At infrared wavelengths a dipole structure is evident in the north polar region. It consists of two clearings in the clouds, at locations straddling the pole and rotating around it in about 2.7 days. The clearings are thought to be evidence of subsidence of the atmosphere at the center of the polar

vortex. The absence of descending motion elsewhere suggests that a single large circulation cell may fill the northern hemisphere at levels near the cloud tops. A crescent-shaped collar region consisting of anomalous and variable temperatures and cloud structures surrounds the pole at about 70°N and rises to perhaps 15 km above the mean height of the cloud tops. It sometimes shows spiral breaks. This feature and the double vortex eye are large, persistent deviations from the mean circulation due to the planetary-scale waves of unknown origin.

The polar regions were found to be appreciably brighter during the first phase of the Pioneer Venus mission than they had been during the brief Mariner 10 flyby. The planetary scale dark Y feature, previously identified both by Earth-based telescopic observers and in the Mariner 10 images, rotates around the planet with a period of from four to five days, but often changes its precise characteristics from one rotation to the next. Sometimes it has been known to disappear completely. Bow-shaped features and cellular features occur at all longitudes but are more easily observed near to the subsolar point. They seem to be composed of both diffuse bright cells 500 to 1,000 km in diameter with dark rims, and dark cells with bright rims. All the cloud features show a globally co-ordinated oscillation of orientation relative to latitude circles. The zonal winds near the equator have velocities of about 100 ms$^{-1}$ with a retrograde motion that is in the same direction as the planetary rotation. While the basic structure of the zonal rotation has been seen throughout the Earth-based and spacecraft observations, some variability has been found in the structure and strength of the jets at 35°S, 15°N and 45°N between the Mariner 10 and Pioneer Venus studies. This relates to instabilities in weather systems.

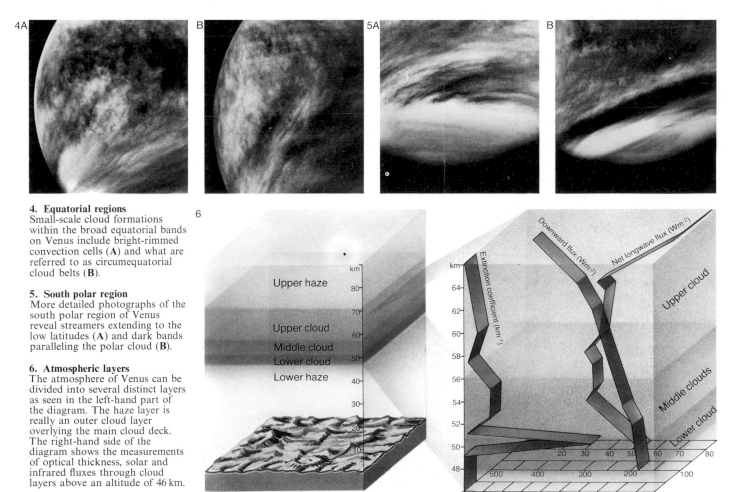

**4. Equatorial regions**
Small-scale cloud formations within the broad equatorial bands on Venus include bright-rimmed convection cells (**A**) and what are referred to as circumequatorial cloud belts (**B**).

**5. South polar region**
More detailed photographs of the south polar region of Venus reveal streamers extending to the low latitudes (**A**) and dark bands paralleling the polar cloud (**B**).

**6. Atmospheric layers**
The atmosphere of Venus can be divided into several distinct layers as seen in the left-hand part of the diagram. The haze layer is really an outer cloud layer overlying the main cloud deck. The right-hand side of the diagram shows the measurements of optical thickness, solar and infrared fluxes through cloud layers above an altitude of 46 km.

# Surface Features I

Before 1962 our knowledge of the surface of Venus was negligible and the ideas of F. L. Whipple and D. H. Menzel, that the clouds were composed of water and that the surface of Venus was covered with oceans, were still widely believed. It has long been clear to scientists that, since direct photography of the surface of Venus is impossible, the only method is to use radar. Earth-based experiments had provided some results, but almost all of the present detailed knowledge has come from one probe, the orbiting Pioneer which was launched by the United States in 1978. It reached its target at the end of that year and was still operating and sending back information in 1982.

Radar echoes from Venus are dominated by the effect known as specular reflection: this is rather like the rays of the Sun glinting off small waves on the surface of the sea. The total power of the reflected signal depends upon the nature of the surface, and it is possible to measure altitudes and depressions with considerable accuracy. The fact that so much of the surface has now been mapped (80 percent) is because the Pioneer orbiter has been so long lived and successful a mission.

### Types of terrain

The first important point is that a huge, rolling plain covers about 70 percent of the surface (or, rather, 70 percent of the areas of the surface that have so far been mapped). Depressions cover another 20 percent, while highlands (10 percent) are concentrated mainly in only two areas, one well to the north of the equator and the other slightly to the south.

The overall impression that has been gained is that Venus has a relatively static crust that may have evolved to become a single plate rather than several moving plates as on Earth (*see* pages 134–5). Much of the surface is comparatively smooth, lying within 500 m to either side of the mean level. There is no sea level for reference, as on Earth, since Venus is devoid of oceans, and the best reference is the mean radius, which is 6,051.2 km—a satisfactory compromise since Venus, unlike the Earth, appears to be almost perfectly spherical. It may well be that the intense heat on Venus has contributed to its smooth surface. Over a sufficient period of time the rocks of crater walls will deform plastically back to the local level. Earlier radar experiments indicated that there are craters on Venus, but in general they are relatively large and shallow. Most meteorites plunging through the atmosphere of Venus would almost certainly be broken up before impact, and it is therefore reasonable to assume that at least some of the craters on Venus are of volcanic origin.

### Highlands

The uplands occupy about 10 percent of the total mapped surface, but they are of fundamental importance to any understanding of Venus' geological history. There are two main areas: Ishtar Terra in the north, and Aphrodite Terra, which crosses the equator although most of it lies in the southern hemisphere.

Ishtar Terra is about the size of Australia; the average height above the mean radius is 3 km and it is a lofty plateau, carrying several mountainous areas. These mountains are made up of three sections: Maxwell Montes (Maxwell Mountains), Lakshmi Planum and Freyja Montes. Lakshmi Planum is a smooth plateau bounded on the west and north by the Akna and Freyja Montes respectively. This is a volcanic region, and volcanic features seem to be more in evidence than in the other highland area, Aphrodite. The central area has few defined features, and may be composed of young lavas. The huge escarpments round the edge of the plateau are fairly steep.

The Maxwell Mountains, the highest on Venus, lie at the eastern end of Ishtar. They rise to almost 11 km above the mean radius and 8.2 km above the adjoining plateau, so that they are lofty by any standards. The highest areas run in a north-west/south-east direction, with lower projections spreading out to the east and west. This is the roughest area so far detected on Venus, and the transition to the smooth lava plain to the west is abrupt.

The other important highland area, Aphrodite Terra, measures 9,700 by 3,200 km, which is about the same size as the African continent. It lies almost parallel with the equator, and is made up of mountain areas to the east and west, separated by a lower region. The mountains are rough and broken; those to the west attain 7 km above the mean radius, while the eastern peaks have an altitude of about 3 km. Adjacent to Aphrodite is Atla Regio, sometimes nicknamed the Scorpion's Tail, which is apparently one of the main volcanic regions of Venus.

Beta Regio, in the northern hemisphere, contains the two large shield volcanoes Rhea Mons and Theia Mons, which rise to some 4 km above the plains and lie on a fault line running roughly from north to south. In structure they seem to be very like terrestrial shield volcanoes such as Mauna Loa and Mauna Kea in Hawaii, but are much more massive—because plate tectonics of Earth type seem not to apply to Venus. Several smaller volcanoes are nearby. Finally there is the smaller Alpha Regio, south of a line joining Phoebe Regio to Aphrodite. It is 2 km high, and rough in nature.

High features on Earth are supported by isostatic processes, but some of those on Venus are probably the result of convection. Others indicate more recent volcanic activity. The fact that some highland areas on Venus are as high as they are suggests either that the outermost layers of the planet are not very plastic or that the topography is continuously being renewed. In the former case, the rocks of the surface would have to be very dry. On Earth such high mountains could not be supported at Venus' temperatures. All this indicates the complex interelations between atmosphere and surface processes, and how the different evolutionary paths of Venus and the Earth have produced a range of very different surface features on each planet today.

### Basins

Basins are comparatively rare on Venus, but they do exist. The most extensive seems to be Atalanta Planitia in the northern hemisphere. Its depth is 1.4 km below the mean radius of the planet, and it is about as large as the Gulf of Mexico on Earth. It seems to be smooth and has been likened to a mare basin on the Moon, although the analogy should not be taken too far in view of the fact that Venus and the Moon are different in nature. No large craters have been found in Atalanta, or in other basins such as Niobe, Leda, Sedna and Guinevere, and it may well be that lowland areas on Venus are relatively young.

### Valleys

The most important rift valley on Venus is Diana Chasma, in the Aphrodite region, which sinks to more than 2 km below the mean radius and 4 km below the adjacent ridges. Its maximum width is 280 km, so that it is much larger than any comparable feature on Earth, and can be compared with the Valles Marineris on Mars. It is probably of tectonic origin; close to it is the smaller Dali Chasma, which is presumably connected with the same system.

### Sappho

One of the smaller but very interesting features on Venus is Sappho, which is more than 200 km in diameter and lies in a region that is slightly higher than its surroundings. Linear structures radiate from it, and these are thought to be volcanic flows. Some of the bright rings surrounding small darkish areas on the plains may also be made up of volcanic ejecta.

**1. Topographic map of Venus**
A topographic map of the surface of Venus based on a 1 km contour interval does not reveal many types of features like those on Earth. There is no evidence of plate tectonics—slowly moving slabs of crust that give rise on Earth to features such as deep oceanic trenches, mid-ocean ridges or mountain chains. Detailed radar images of features marked can be seen below (2A–E).

**2. Radar images**
In these radar images from the Arecibo Observatory light areas are generally rough terrain and dark regions are smooth. (A) shows Ishtar Terra, which is made up of the dark Lakshmi Planum and the bright Maxwell Montes, which can be seen in close-up in (B). This region is 750 km from north to south and is the highest region of Venus. Other features shown are Rhea Mons (C) and Theia Mons (D), two shield volcanoes, and what are thought to be crater-like features (E).

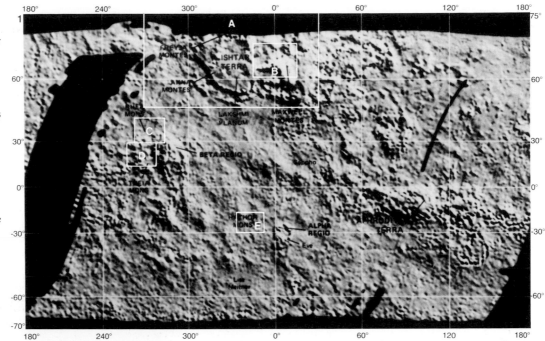

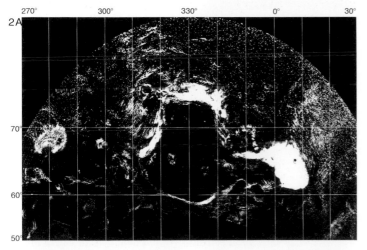

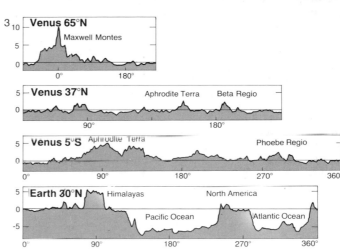

**3. Topographic cross sections**
In the three topographic cross sections of Venus and one of Earth, it is interesting to compare, for example, Ishtar Terra on Venus and the Earth's Himalayas. Both rise steeply above the surrounding terrain.

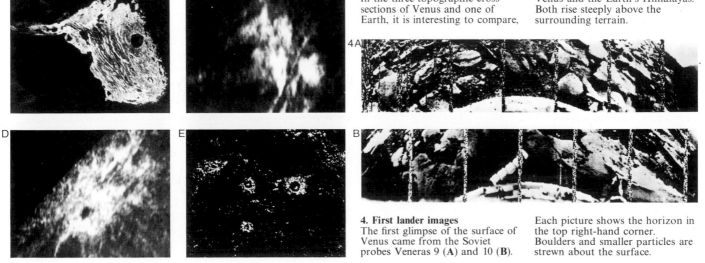

**4. First lander images**
The first glimpse of the surface of Venus came from the Soviet probes Veneras 9 (A) and 10 (B).

Each picture shows the horizon in the top right-hand corner. Boulders and smaller particles are strewn about the surface.

# Surface Features II

Our present knowledge of Venus is derived almost entirely from the Soviet soft-landers so far as the surface is concerned. Veneras 9 and 10 were the first to send back direct pictures, in 1976; each transmitted for roughly an hour before being put permanently out of action.

The two sites—almost 2000 km apart—were not identical. That of Venera 9 was compared with scree on a hill slope, while that of Venera 10 appeared to be older, with a smoother surface. The light-level was rather higher than had been expected (it was compared with that in Moscow at noon on a cloudy winter day) and the wind velocity was low. The rocks were similar to erupted basalts, apparently richer in potassium, thorium and uranium than comparable rocks on Earth.

The next two Venera landers, 11 and 12 (1978) also transmitted for an hour each. In 1982 Veneras 13 and 14 achieved startling

success, and even sent back color pictures; the general hue of the rock was reddish-brown, and the sky was orange. Venera 13 landed southeast of Phoebe Regio on a rock-strewn plain, with obvious outcrops and patches of dark, fine-grained material. There was evidence of chemical erosion. Venera 14 landed some 900 km southeast of Venera 13; this time the scene was rather different, with less fine-grained material but reasonably regular layers of rocky, broken plates, differing in color and perhaps also in composition. Soil analysis results from Venera 13 indicated a basalt with a high percentage of potassium, while the Venera 14 basalts were similar to those found on the seabed on Earth.

The Pioneer Venus 2 mission of 1978 involved an entry probe and also a "bus" which dispatched several small landers which sent back data during their descent. They were not intended to survive impact though one of them did so (briefly).

# Plates

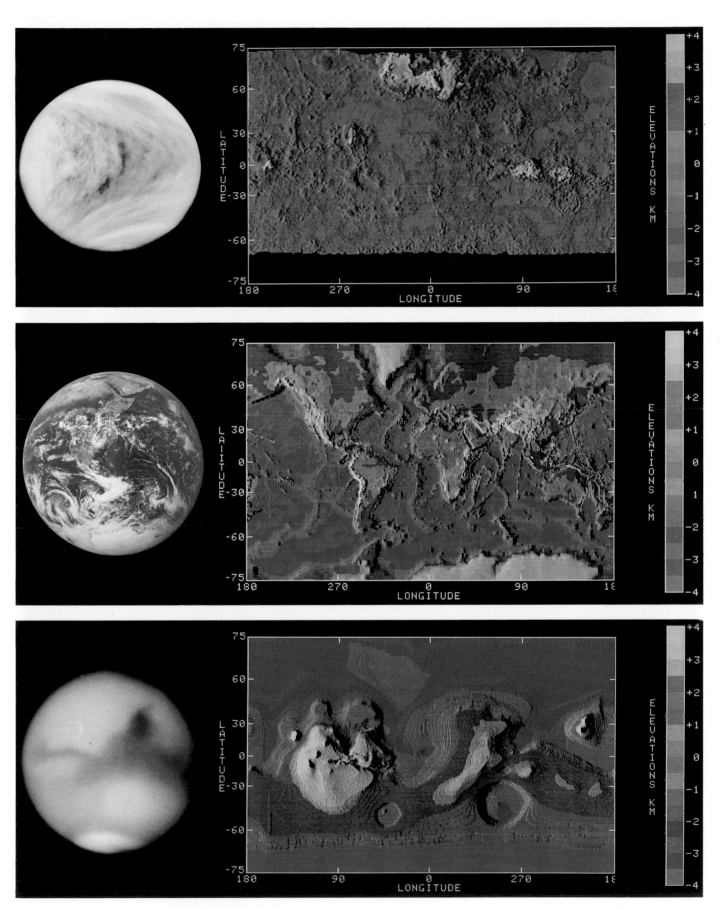

Topographic map of Venus
Pioneer Venus

Topographic globes of Venus
Pioneer Venus

Venus surface
Venera 14

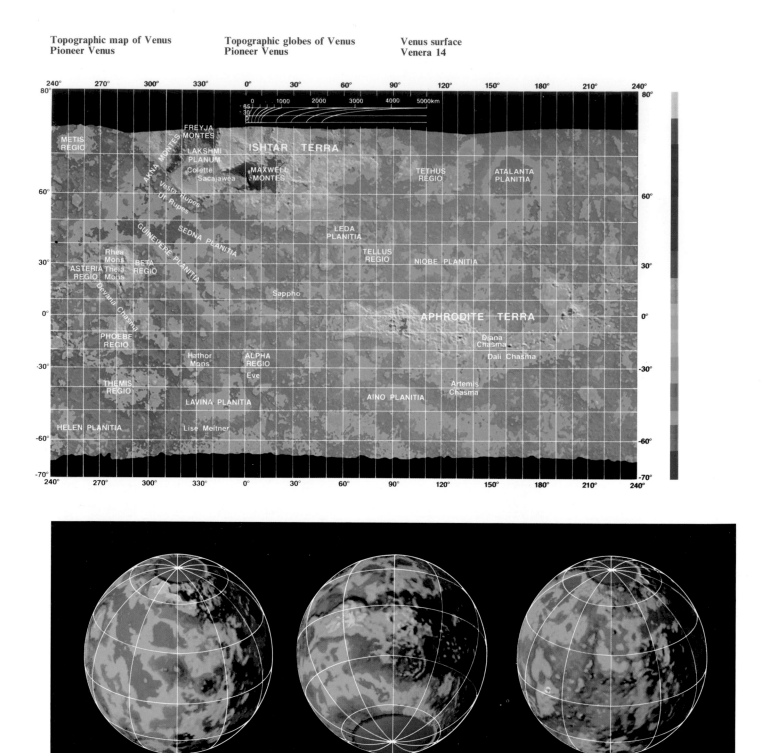

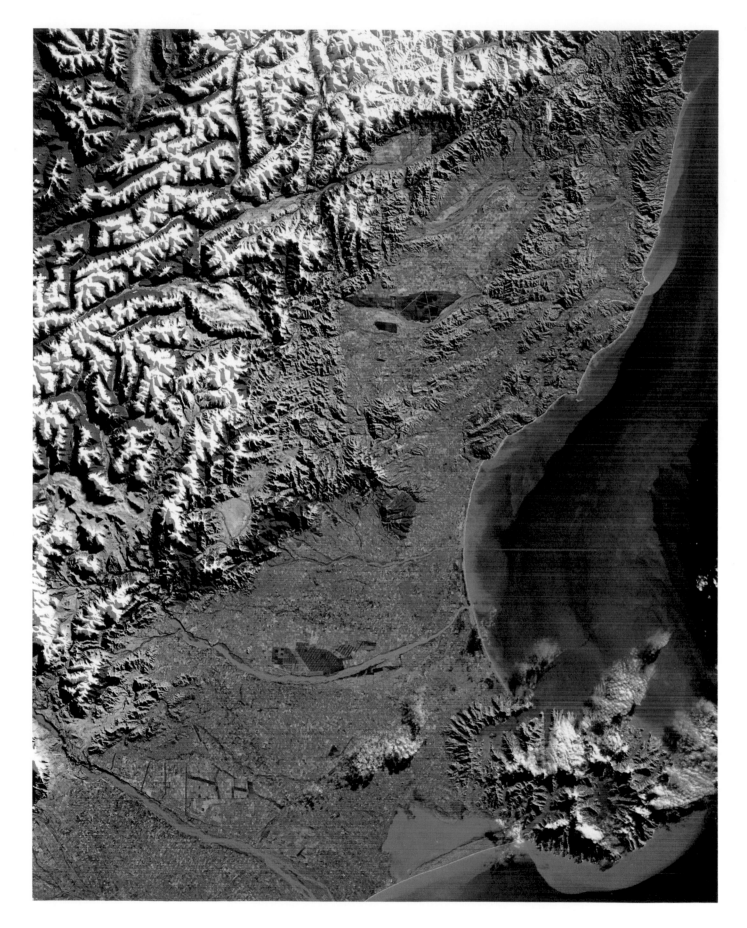

Baluchistan desert, Pakistan
Landsat
25 November 1972

**Sea ice off Newfoundland**
**Landsat**
**13 February 1974**

Mars from Earth
1971
Lowell Observatory

South pole, Mars
Viking

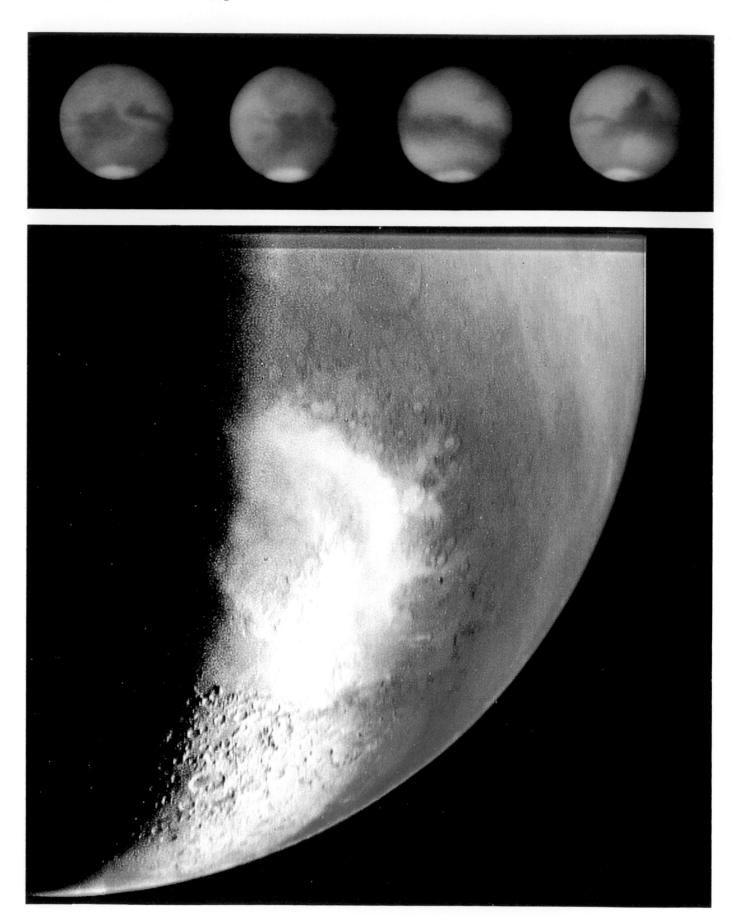

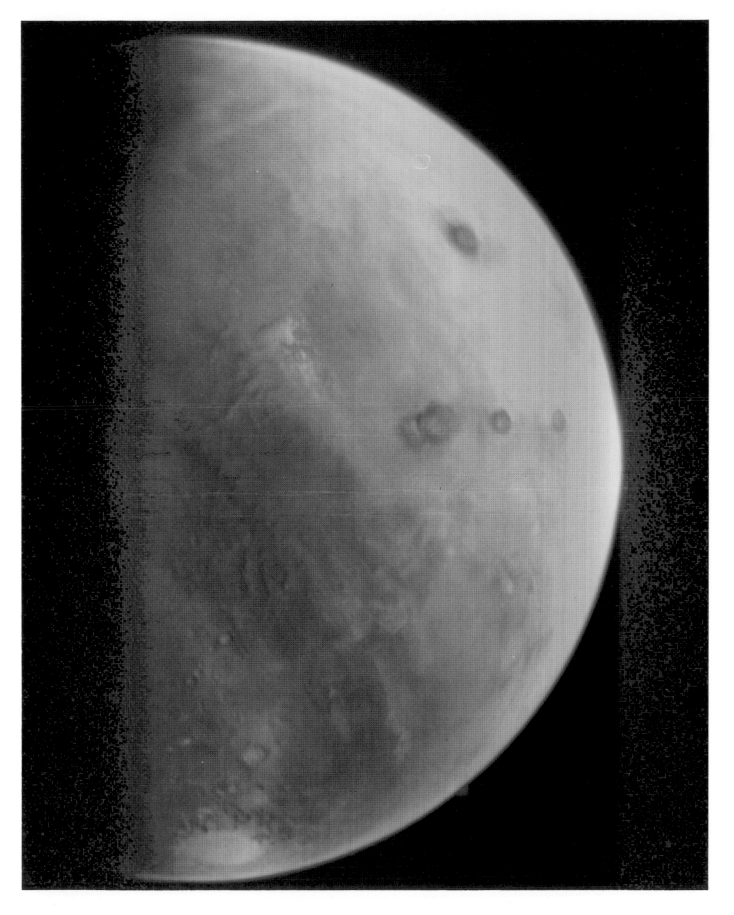

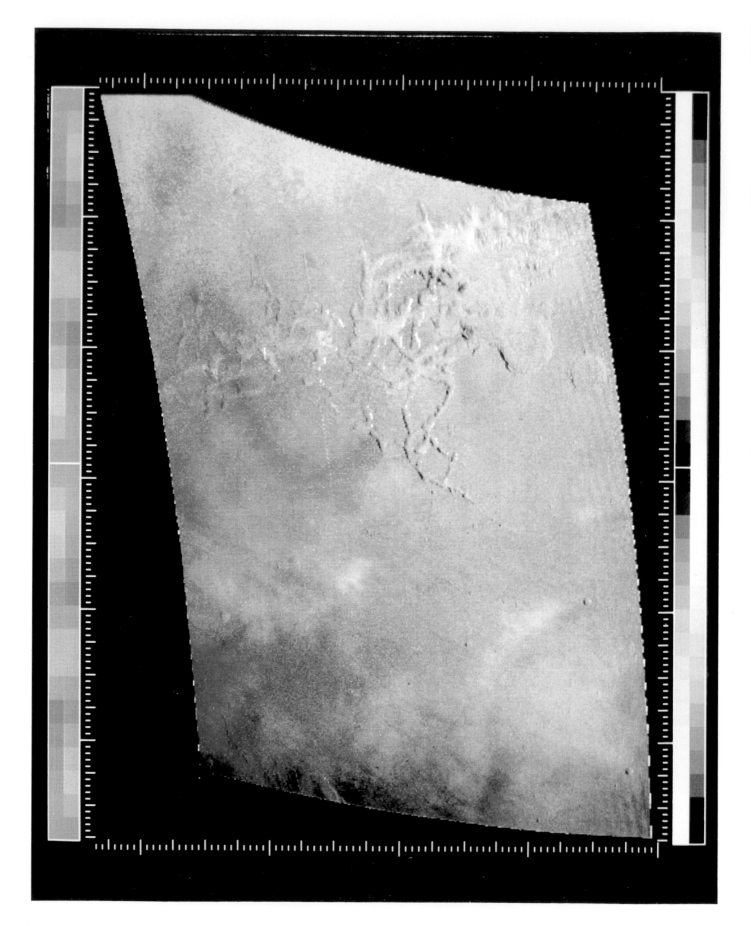

Noctis Labyrinthus, Mars
Viking 1
12 October 1976

Valles Marineris, Mars
Viking

Argyre Basin, Mars
Viking 2
22 February 1977

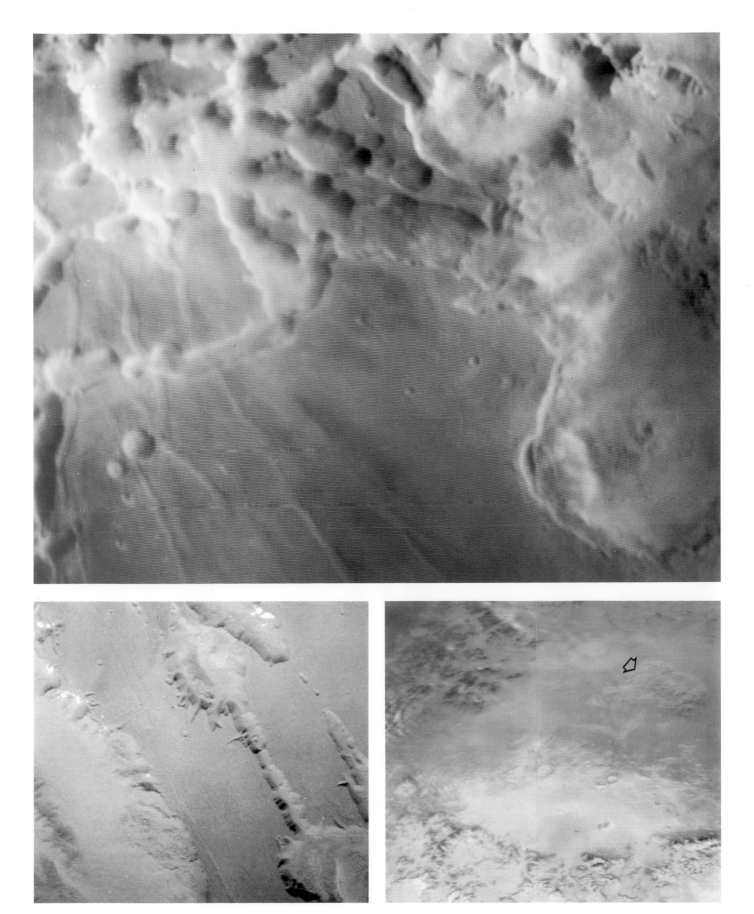

Noctis Labyrinthus, Mars
Viking 1
12 October 1976

Valles Marineris, Mars
Viking

Argyre Basin, Mars
Viking 2
22 February 1977

**North polar region, Mars**
**Viking**

**South polar cap, Mars**
**Viking**

**Lampland region, Mars**
**Viking**

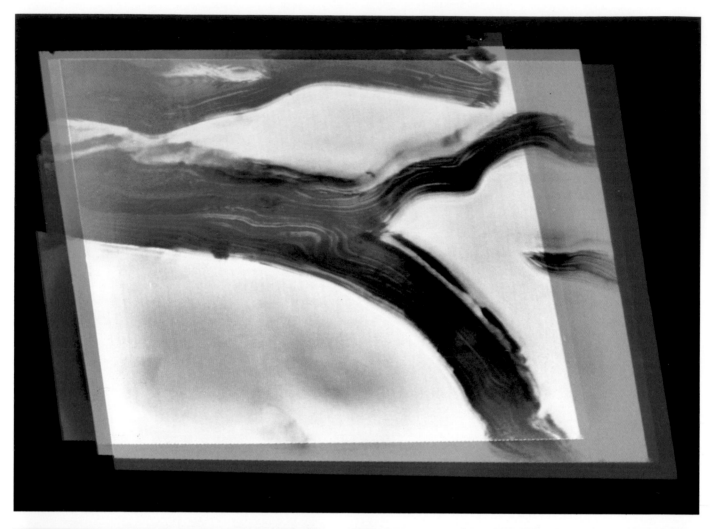

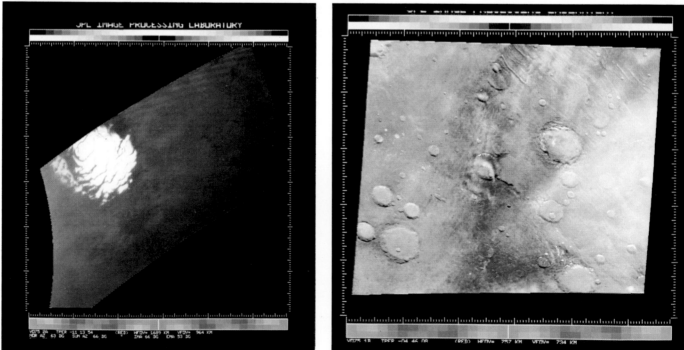

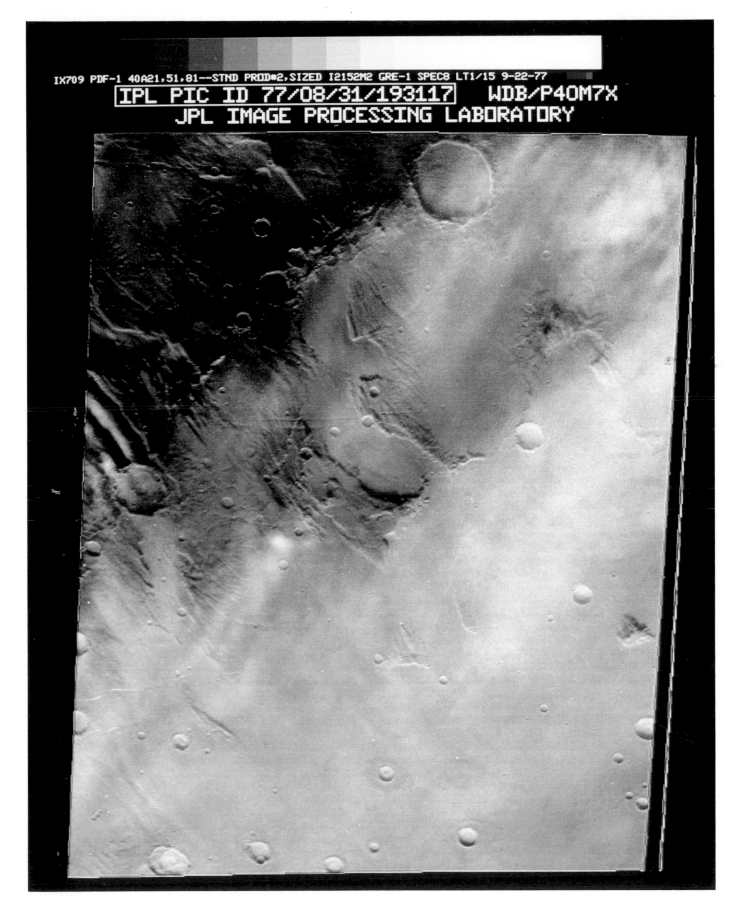

IX709 PDF-1 40A21,51,81--STND PROD#2,SIZED I2152M2 GRE-1 SPEC8 LT1/15 9-22-77

IPL PIC ID 77/08/31/193117    WDB/P40M7X

JPL IMAGE PROCESSING LABORATORY

## Notes to Color plates

**Page 113.** Computer-generated radar maps of Venus, Earth and Mars. (Top). Topographic map of Venus in Mercator projection. Each color corresponds to an elevation interval of 500 m, and the resolution at the equator is about 100 km. Data for this map was obtained by the Radar Mapper Experiment aboard the Pioneer Venus Orbiter. (Center). Topographic map of the Earth in Mercator projection with the same elevation interval as the Venus map. Resolution at the equator is about 111 km. Data for this map was compiled by the Rand Corporation, Scripps Institute of Oceanography and the US Defense Mapping Agency. (Bottom). Topographic map of Mars in Mercator projection with the same elevation interval as the Venus map. Resolution at the equator is about 60 km. Data for this map was compiled by the US Geological Survey from various spacecraft missions and ground-based radar experiments.

**Page 114** (Top). Topographic map of Venus in Mercator projection. Surface elevations are indicated by color contours, showing violet in the lowest regions and pink in the highest. Data for this map was obtained by the radar altimeter aboard the Pioneer Venus Orbiter. (Center). Representations of Venus plotted on imaginary globes by the computer graphics laboratory at JPL. Surface features are shown more nearly as their true relative sizes, although heights are exaggerated and artificial sunlight has been added for clarity. (Bottom). The surface of Venus as seen from Venera 14. Variations in color of the layers of rocky, broken plates is evident. Part of the spacecraft is seen in the foreground (*see* page 112).

**Page 115.** The Earth as seen from Meteosat 1 in geostationary orbit above the Gulf of Guinea. Clearly identifiable cloud patterns occur over a wide area (*see* page 138).

**Page 116.** South Island, New Zealand, is clearly dominated by the Southern Alps, a chain of complexly folded, faulted and metamorphosed rocks. Two groups of rocks form a pair of crustal blocks and are separated (top) by the Alpine Fault. Two large, now extinct volcanoes are seen (bottom right). Altitudes are in excess of 3,000 m.

**Page 117.** The high plain of the Ganges gives way to the Churia Range in Nepal, at the base of the Himalayas—a huge mass of faulted rocks compressed and raised when the Indian subcontinent collided with Asia (*see* page 134). The northern crest region contains many peaks in excess of 8,000 m, including Mount Everest (top right edge) at 8,848 m.

**Page 118.** The Grand Canyon of the Colorado River is 1.6 km deep and approximately 19 km from rim to rim. Incised in horizontal Paleozoic sedimentary rocks, its relatively undisturbed cross-section is unequaled in the world. The Kaibab Plateau is seen under snow; altitudes are in excess of 2,700 m. The snow-covered San Francisco Peaks surrounded by lava flow and cinder cone volcanoes are visible (bottom right).

**Page 119.** The Mirjawa Range (bottom center) exposes folded and faulted sedimentary rocks of Cretaceous and Tertiary age. To the north is a series of longitudinal sand dunes.

**Page 120.** The swirling mass of brash and drift ice is seen off the east coast of Newfoundland. The prevailing wind is westerly. Scattered, low cumulus clouds occur over the main ice mass.

**Page 121.** False-color photography discloses the internal temperature structure of the hurricane Camille. Cool air near the eye shows as grey and blue; increasing temperatures show from yellow to red. The mature hurricane may be 400 km in diameter and tower upward more than 16 km. The dense walls of cumulus-type clouds, which enclose the calm, low-pressure eye, mark the region of highest wind speeds, up to 175 km/hr near the center, and violent precipitation.

**Page 122** (Top). In this sequence of Mars as seen from Earth in 1971, distinctive features, such as Syrtis Major Planitia, can be identified despite the global dust storm which delayed the start of Mariner 9's mapping mission. (Bottom). In this view of the south polar region frost is clearly evident. The smooth, circular area is the large impact basin of Argyre, approximately 1,300 km in diameter (*see* page 234).

**Page 123.** In this view from Viking 1 of the Tharsis mountains, the row of three huge volcanoes standing about 20 km above the surrounding plain are clearly seen (center). Olympus Mons, Mars' largest volcano, dominates the area (top); the Argyre basin is also clearly visible (bottom). The bright area around Argyre is probably the result of discontinuous thin carbon dioxide ice (*see* pages 234–5).

**Page 124.** One of the most conspicuous features seen from Viking is the complex fracture pattern of Noctis Labyrinthus. Each graben is approximately 20 km across, comparable to the Grand Canyon. This area marks the head of the great rift valleys, Ius Chasma and Tithonius Chasma, to the east (*see* page 235).

**Page 125** (Top). In this closer view of Noctis Labyrinthus bright clouds of water ice can be observed in and around the tributary canyons. The distribution of the clouds against the rust colored background of the Martian desert is clear. The area covered is about 10,000 km². (Bottom left). In this detail of Valles Marineris, Candor Chasma continues where Tithonius Chasma finishes (from bottom right to top left). The much longer Ius Chasma is the depressed area (bottom left) that is approximately 100 km across at this point. (Bottom right). From Viking 2 a turbulent, bright dust cloud (arrowed) more than 300 km across can be seen inside Argyre Planitia. It is apparently moving eastward under the influence of strong winds that also create condensate lee-wave clouds to the west of the basin. Large depressions like Argyre and Hellas seem to be favored locations for the formation of dust storms.

**Page 126** (Top). In this view of the north polar region, the white areas are frost and the orange areas are valleys with stratified walls showing through the ice cap. The stratification may result from interbedded ice layers with dust layers on top. (Bottom left). Unlike its counterpart, the south polar cap varies considerably in size, becoming much reduced in the height of summer. At its smallest the remnant of the cap is centered on a point 400 km from the pole. (Bottom right). To the northwest of Argyre Planitia is the region of Thaumasia Fossae; in this detail the largest crater (center right) is Lampland (*see* pages 234–5).

**Page 127.** This region of Thaumasia Fossae borders on the southern edge of the smooth, yellow Solis Planum. Telescopically, Solis Planum is one of the most variable areas on Mars; observers since 1877 have noted pronounced changes of shape and intensity of the dark patch.

**Page 128** (Top). The Martian surface, taken from Viking Lander 2, is strewn with rocks out to the horizon ranging up to several meters in size. Some may have come from the nearby impact crater Mie, which is about 1 km across. (Bottom). In this view the 8-degree tilt of the spacecraft has not been electronically rectified. Here the pebbled surface between the rocks is covered in places by small drifts of very fine material, similar to the drifts seen at the Lander 1 site in Chryse. The reddish cast of the sky in both images is probably due to reflection and absorption of sunlight by reddish sediment in the lower atmosphere (*see* pages 232–3).

# THE EARTH

# Characteristics

The Earth is the third planet in order of distance from the Sun. It is more accurately considered as a binary planetary system with the Moon: nowhere else in the Solar System as we know it is there such a closely matched pair of associated bodies (apart from Pluto and its satellite Charon), although, on a closer inspection, the two are rather dissimilar in nature (*see* pages 143–209). The Moon is about one quarter of the size of Earth and approximately one-eighteenth of its mass. Planets ordinarily have satellites that are about one-thousandth of their parent planet's mass.

The Earth orbits the Sun at a mean distance of 149,597,900 km (one astronomical unit). Its orbital eccentricity is slight (0.017) and the Earth–Sun distance varies from 147,000,000 km at perihelion to 152,000,000 km at aphelion. The revolution period is 365.256 d and the rotation period 23 hr 56 min 4 sec.

The largest of the inner group of planets, the Earth has an equatorial diameter of 12,756 km, but it is less than one-tenth the size of Jupiter. Its oblateness amounts to 0.003, which is slight but not inappreciable. Despite the highest mountains and the deepest ocean trenches, the Earth is remarkably smooth. If the planet was a globe 1 m in diameter, these features would reach only 1 mm either side of the mean sea level. It is the densest of the planets, with a specific gravity of 5.517.

The Earth viewed from space appears blue, brown and green with an ever-changing pattern of white. The surface is unique as far as we know in that it comprises two different states of matter—solid and liquid. The boundary between ocean and land is the only place known in the universe where solid, liquid and gaseous matter co-exist stably. The oceans cover two-thirds of the Earth's surface. The third that rises above the water level is rock, dust and soil, which are rich in silicate minerals. There are two polar caps, one in the north and one in the south, the size of which varies according to season.

## Terrestrial terrains

A global map of the Earth's terrain reveals it to be very different from the other terrestrial planets. Some features, such as craters, are much less widespread, while others, volcanic plains for instance, are more common. Still others, like folded mountain chains and platform deposits, are relatively new to the Earth and not seen elsewhere in the inner Solar System. The Earth is divided up into a series of rigid, lithospheric plates, and the formation and interaction of these plates are largely responsible for the large-scale topographical and structural features on Earth (*see* pages 134–5). The basaltic plains of the ocean floors are the most widespread feature of the terrestrial terrain, although they did not form in the same way as on other planets (*see* pages 12–13). There are meteoritic craters: Precambrian shields (covering about 10 percent of the Earth's surface) are probably the nearest equivalent to the ancient "cratered terrains" of other planets but they are much younger and the density of craters is much less.

The face of the Earth is continually being modified by weathering and erosion—arising from the particular characteristics of the Earth's atmosphere (*see* pages 136–9) and surface and the interaction of the two. Chemical reactions and frost shattering weather rocks, producing debris which is worn away and transported elsewhere by erosion. Water, running as a liquid, alternatively freezing and thawing, or frozen as ice and snow, is the principal agent of erosion, but wind also erodes. Weathering and erosion together sculpt the landscape and eventually lower it. Also, loose material falls, slides or creeps downhill under the force of gravity. It has been estimated that the average rate of erosion of the land is 8.6 cm per 1,000 years. The rate does vary from place to place, however, depending on the climate. Erosion is rapid in steep areas that have heavy rainfall and also in semi-arid regions where there is

**1. Seasons**
The Earth's seasons result from the planet's axial inclination of 23.5° to the perpendicular to the orbital plane. When the northern hemisphere is tilted towards the Sun it has its summer, while the southern hemisphere is in winter. Twice a year (the equinoxes) the Sun is overhead at the equator. at noon.

**3. The Earth's surface**
If the Earth were devoid of its water and vegetation then the characteristics of its geology could be plainly seen. These include the continental plates, mountain chains, subduction trenches (where one plate goes beneath another) and oceanic ridges (where newly formed crust is emerging onto the surface).

**2. Size comparison**
The Earth is roughly four times bigger than its satellite (3,476 km) and a twin of Venus (12,104 km).

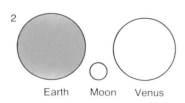

#### Physical data

| | Earth |
|---|---|
| Diameter | 12,756 km |
| Ellipticity | 0.0034 |
| Mass | $5{,}9742 \times 10^{24}$ kg |
| Volume | $1.084 \times 10^{12}$ km³ |
| Density (water=1) | 5.52 |
| Surface gravity | $9.78 \, \text{m s}^{-2}$ |
| Escape velocity | $11.18 \, \text{km s}^{-1}$ |
| Equatorial rotation | 23.9345 hr |
| Axial inclination | 23.44° |
| Albedo | 0.29 |

only patchy vegetation cover, but the rate is slow in deserts and cold lowland areas. Material removed by erosion is deposited at the bottom of rivers or oceans. These sediments are eventually compacted by pressure into a hardened state. They are then known as sedimentary rocks. Igneous rocks are formed by the solidification of magma, molten material produced by volcanic processes. Finally, metamorphic rocks are formed when other rocks are transformed by heat (but without melting) and pressure. Characteristically, terrestrial rocks result from a succession of deposits.

### Cycles on Earth

The dynamic activity of the atmosphere and crust on Earth is a unique feature of this planet, and both have been substantially affected by the most strikingly unique feature of the Earth—the evolution of living organisms. A number of cycles are the foundation of interrelationships between organisms and environment.

The hydrological cycle involves the evaporation of water from the oceans, the transportation of water vapor in the atmosphere, precipitation as rain or snow, and runoff via rivers back into the oceans. In addition, plants take up groundwater during photosynthesis and in the nutrients they absorb from the soil. Water vapor is returned to the atmosphere during transpiration. Carbon dioxide is also absorbed during photosynthesis: the carbon is used for food and oxygen is returned to the atmosphere. Carbon dioxide is released in transpiration and when organisms die and decay.

Nitrogen, sulphur and phosphorus are similarly transferred between land, sea and air. All three are involved in plant growth, grazing, predation and decay. Nitrogen in the atmosphere and the soil is absorbed by plants by means of bacteria in root nodules that can fix nitrogen gas and convert it into a usable form. Sulphur is reduced on land by anaerobic bacteria in swamps and bogs and

collects in the sea from the decomposition of calcareous deposits. Phosphorus is recycled by means of phosphatizing bacteria on land and algae and diatoms in the sea.

All these cycles involve the absorption, storage and distribution of solar energy, which is used in various forms to sustain life on Earth. The biosphere—that part of the Earth and its atmosphere in which living things exist, and extending some 16 km above and below sea level—receives 99.8 percent of all its energy from the Sun. Twenty percent of this is used to drive the hydrological cycle and 0.02 percent is used in photosynthesis by plants. The rest of the solar radiation reaching Earth is either reflected back into space (about 30 percent), absorbed and eventually released as waste heat (47 percent), or absorbed by the ozone layer of the atmosphere (1–3 percent). A small fraction is used to drive winds and waves.

### Origins of life

The traditional view of the origins of life on Earth is that an organic "soup" accumulated in the early oceans and it was in this that life was spawned. However, experiments have shown that amino acid "building blocks" could also have been formed by a process of freezing and thawing. It is generally believed that the luminosity of the Sun has slowly increased over geological time, gradually warming up the Earth's surface, which in early days would have been characterized by frozen accumulations of volatiles. On the other hand, reduced gases such as ammonia, hydrogen and methane were probably present before oxygen accumulated in the early atmosphere, and these may have absorbed radiant heat from the surface, making the Earth warmer than it is even now, which is consistent with the soup theory. Still other scientists argue that life need not necessarily have formed on Earth at all but been transported here by meteorites.

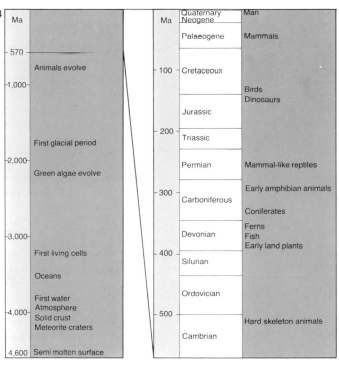

**4. The Earth's history**
The timechart plots the history of the planet from its initial formation about 4,600 million years ago. It shows the major evolutionary landmarks both in terms of the Earth's geology and the origins and development of life forms. The fossil record dates back to the Cambrian era and shows that abundant life on Earth dates only from about 570 million years ago. The evolution of mammals on a timescale such as this occurs a relatively short period of time before the present.

# Geology

Spacecraft images of the planet Earth reveal the familiar pattern of land masses and oceans, and the changing cloud cover that is our weather. Visible also are high mountain ranges, fractures, deserts, huge river systems and polar ice caps. Present-day knowledge of the Earth is very different from that of a century ago. New technology has enabled geophysicists to elucidate where the Victorians could only speculate. Today, for instance, we know of the existence of vast submarine mountains, then totally unknown and unexpected, and how the Earth's "heat engine" powers this dynamic planet.

Oceanic regions cover about 70 percent of the total area of the globe. Although a thin veneer of sedimentary rocks covers most of the ocean floors, the substructure is composed of high-density volcanic rocks called basalts, which are widespread also on Mars, the Moon and probably on Venus and Mercury. Radiometric dating reveals that most oceanic crust is less than 200 million years old. Land areas are much less extensive and display deep fractures and immense fold mountain ranges. The rocks that form the continental crust are relatively low-density igneous and metamorphic rocks. Both continental and oceanic materials comprise the Earth's crust, which, together with the upper part of the next layer down (the mantle), make up the lithosphere, a relatively mobile but rigid layer 70–100 km thick, which is moving slowly over a semi-molten layer of the upper mantle (the asthenosphere).

## Sea-floor spreading
Rising from the floor of the oceans are extensive linear mountain ranges, the best example of which, the Mid-Atlantic Ridge, divides the Atlantic from north to south. Iceland sits astride the northern end and, as is characteristic of such ridges, is the site of active volcanoes and seismic activity. Extrusion of basaltic lavas from the crests of oceanic ridges adds materially to the oceanic crust almost constantly, this material rising up from the mantle beneath. The newly produced oceanic crust then moves slowly away on either side of the ridge crests at a rate of between 1 and 10 cm each year. This process is called sea-floor spreading.

The new crust being created at the spreading ridges moves laterally away from its source. Since the Earth's surface area is not increasing significantly, there must be a compensatory loss of crust somewhere else. Around every major ocean, especially the Pacific, oceanic crust is either pushed or drawn back down towards the mantle, the consumption process occurring in vast submarine trenches along which adjacent slabs of lithosphere converge. The rigid lithospheric slabs are called plates and the Earth's crust is composed of 15 of them.

## Subduction zones
The steeply inclined zones of underthrusting along which one plate is deflected beneath another are called subduction zones. Usually the more mobile plate is deflected and the thin veneer of sedimentary rocks attached to this are mostly scraped off and deformed during the process, while the rest is eventually melted or reabsorbed at depths of between 100 and 300 km. Since it is much less dense than mantle material, the melted crust may then rise towards the ocean floor to be erupted as lava. This lava, together with the scraped-off sedimentary materials, may form an island arc on the edge of the overriding plate. The Aleutians represent such an arc.

Running down the edges of some continents are chains of high fold mountains such as the Rockies and the Andes. The incidence of

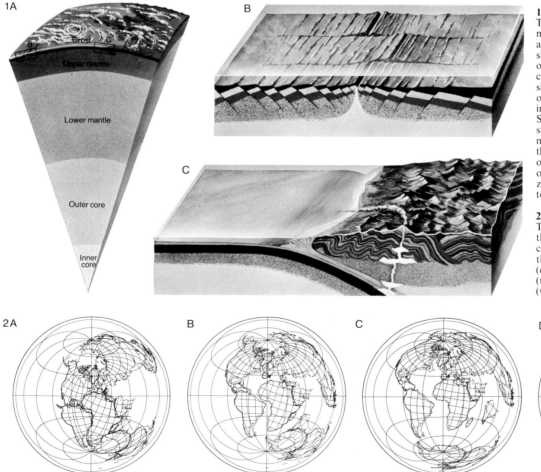

**1. Interior and crust**
The Earth is made up of three main layers—the crust, mantle and core (**A**). The crust can be subdivided into continental and oceanic material. Continental crust is largely granite, rich in silicon and aluminium (sial): oceanic crust is mostly basalt, rich in silicon and magnesium (sima). Sima also underlies continental sial. The mantle is composed of magnesium and iron silicates, and the core of molten nickel and iron oxides. New crust is made at ocean ridges (**B**), and subduction zones (**C**) are where it returns to the mantle,

**2. Continental drift**
The continents are derived from the break-up of an ancient super-continent, Pangaea, and evolved thus: (**A**) 220 million years ago (early Triassic); (**B**) 100 million (mid-Cretaceous); (**C**) 60 million (Cenozoic); (**D**) present day.

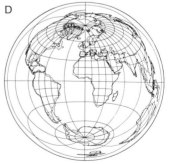

volcanoes and seismic activity is again high in such regions. Here, two lithospheric plates are converging but in this case the associated subduction zone borders a continental plate, the oceanic material being consumed beneath the continental lithosphere. In this situation the remelted material from the descending slab rises to form a chain of major volcanoes which pierce the high Andes and Rockies, while large masses of granite have invaded the mountain roots. The frictional effects of plate movements along subduction zones cause earthquakes. Earthquakes are also common along the spreading ridges and also along linear fractures which have the effect of pulling the ridges apart. Many such fractures transect the Mid-Atlantic Ridge.

## Plate tectonics

The concept that explains lithospheric plate activity is called plate tectonics and this is central to our understanding of the Earth's crustal dynamics. All plate boundaries are formed by combinations of spreading ridges and linear fractures, subduction zones and collision zones (where continental plates are colliding).

Fold mountain ranges are thought to form as a result of the activity of moving plates in regions where abnormally thick piles of sedimentary strata (20 km thick in places) and associated volcanic rocks have accumulated over hundreds of millions of years along the length of a mobile zone. The process of accumulation, burial, uplift, deformation and subsequent erosion is known as an orogenic cycle. The Alps and Himalayas are typical examples: they have been intensively deformed during the last 120 million years and then rapidly uplifted during the last 25 million. Smaller ranges like the Grampians are between 250 and 500 million years old but

have been uplifted many times, the last occasion being during the last 50 million years. Many lesser ranges, the western Appalachians being a case in point, were formed in rather a special way. Here flat-lying sedimentary rocks resting on a rigid crystalline basement, were rucked up rather like a rug on a polished wooden floor.

By means of isotopic age-dating it is possible to see how the continents have evolved. They appear to consist of ancient crystalline cores, many of which existed as long as 2,500 million years ago. These cores were once welded together into a super-continent called Pangaea, which eventually split apart. Our present continents are the relics of this splitting operation, but they have been added to over millennia by the welding on of orogenic belts. Continental growth is essentially a slow accretionary process. The movements of the continental masses over the last 200 million years are now quite well understood. Some areas that are now close together, like north Africa and southern Europe, were once widely separated, while Britain and North America were once united. The idea of continental drift, as this is called, was proposed many years ago but has only recently found firm support thanks to the technology now available to the modern geophysicist.

Finally, rocks on the surface of Earth are continually being sculptured. Mechanical and chemical disintegration, the further breakdown of debris by the action of wind, ice and water, and its redistribution from the higher to the lower parts of the surface, all bear witness to the dynamic atmosphere. The very widespread veneer of sedimentary rocks found on Earth contains evidence, not only of past geography and climate, but also of the remains of fauna and flora that once flourished. This fossil record, testifying to such a diverse array of events, is as yet unique in the Solar System.

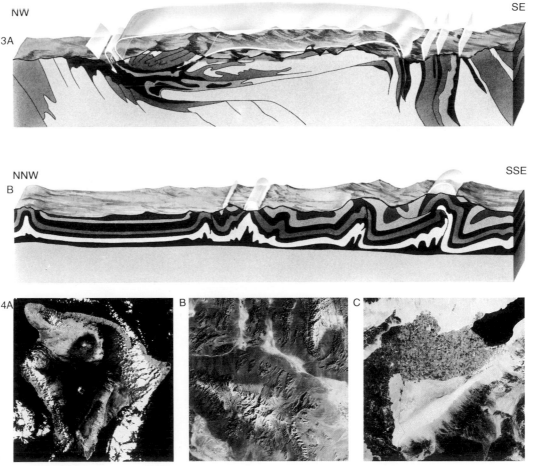

**3. Shaping of the surface**
Continental regions consist of stable shields separated by mobile belts where material is pushed up into mountain chains by collisions between crustal plates. The shapes of mountains are determined by the way in which they are formed. Molten rock is intruded between and below strata, thus bending them. Pressure also causes strata to bend. Where rocks do not bend easily they are faulted upwards in large blocks. The highly folded strata of the European Alps (A) have not only been folded and crumpled but moved northwards as nappes. Where pressure and heat were great, sedimentary rocks have been metamorphosed into schists and gneisses. On a smaller scale, a section through the Juras (B) in central Europe shows a folded and faulted belt. One theory suggests that the upper strata were perhaps deformed independently of the basement.

**4. The many surfaces of Earth**
Geological processes on Earth have produced volcanoes such as Mauna Loa on Hawaii (A). This is one of the youngest volcanic shields and was formed by the submarine outpouring of basaltic lavas on to the Pacific floor. Basin and range topography in Nevada (B) exhibits intricately folded and faulted mountain ranges separated by weathered debris and alluvial basins. Man's impact on the landscape can be seen in the chequered land use in Coachella Valley, California (C).

# Atmosphere

## Composition of the atmosphere

Nitrogen accounts for approximately 80 percent of the volume of the terrestrial atmosphere, with oxygen constituting most of the remainder. There are traces of water vapor, carbon dioxide and ozone, all of which are important because of their ability to absorb infrared radiation and therefore affect atmospheric temperatures. Many of the other gases of the atmosphere, such as argon, krypton, neon and xenon, are inert and they therefore do not play a significant role in atmospheric processes.

The chemical composition of the atmosphere is remarkably uniform, except at high altitudes, where there is some variability. The water vapor content is extremely variable on a local scale and depends on temperature and the proximity of the oceans. If the average surface temperature were only a few degrees higher than it is now, the oceans would evaporate, drastically altering the composition of the atmosphere. This may well have happened on Venus some time in the planet's history.

## Radiation budget

The atmosphere of the Earth receives almost all its energy from the Sun. Not all this radiation is absorbed: albedo values for the Earth vary over a wide range according to the type of surface. It may be as much as 80 percent for fresh snow, 35 percent for desert regions, 20 percent for vegetation, 10 percent for dark soil and only 6 percent for oceans. Albedo values for clouds vary from 10–80 percent depending on their thickness. The average albedo for the Earth therefore varies according to the amount of cloud and the extent of snow cover in different seasons and at different latitudes.

Clouds play an important role in maintaining the delicate balance between solar and thermal radiation levels, screening solar radiation from the surface and preventing longwave radiation from escaping into space. Approximately 50 percent of the Earth's surface is covered by cloud at any one time. On a local scale, changes of 100 percent in cloud cover may occur during any one day, but globally there is only about one percent variation throughout the year. This is because of the continual exchange between atmosphere and oceans in the water cycle.

Energy loss at the surface of the Earth is not by longwave heat only. Heat loss also results from the evaporation of water and by heat convection. A considerable proportion of the energy emitted into space originates in the atmosphere, with the rest coming directly from the surface. As measured from space the effective temperature of the Earth is 253 K, but the average surface temperature is 289 K, the difference being due to radiation being emitted from the atmosphere.

The extent to which the temperature of the surface and lower regions of the atmosphere differs from the effective temperature depends on the mass of the atmosphere and also on the constituents. Water vapor, carbon dioxide and ozone all are instrumental in the transfer of solar and thermal infrared radiation within the atmosphere. The atmosphere of Venus illustrates well what happens if the amount of carbon dioxide in an atmosphere is increased, thus producing the greenhouse effect (*see* pages 104–5).

Small changes in the amount of cloud, surface properties and the composition of the atmosphere all therefore affect the distribution of radiative energy within the atmosphere, and it is changes between these factors that contribute to climatic change. The distribution of solar heating is also closely linked to the variations

### Composition of the atmosphere

|  | % volume |
|---|---|
| Nitrogen $N_2$ | 76.084 |
| Oxygen $O_2$ | 20.946 |
| Argon A | 0.934 |
| Carbon dioxide $CO_2$ | 0.031 |
| Neon Ne | $1.82 \times 10^{-3}$ |
| Helium He | $5.24 \times 10^{-4}$ |
| Methane $CH_4$ | $1.5 \times 10^{-4}$ |
| Krypton Kr | $1.14 \times 10^{-4}$ |
| Hydrogen $H_2$ | $5 \times 10^{-5}$ |
| Nitrous oxide $N_2O$ | $3 \times 10^{-5}$ |
| Carbon monoxide CO | $10^{-5}$ |
| Xenon Xe | $8.7 \times 10^{-5}$ |
| Ozone $O_3$ | up to $10^{-5}$ |
| Water (average) | up to 1 |

**1. Structure of the atmosphere**
The variations in atmospheric temperature and pressure result from the distribution of solar heating. There are three maxima in the temperature profile: at and near ground level (troposphere), where visible and infrared radiation is absorbed; in the stratosphere, where ultraviolet radiation is absorbed by ozone; and in the thermosphere, where ultraviolet radiation is absorbed in a process of photo-ionization. This is the source of what is known as the ionosphere, which extends upwards of about 50 km. The troposphere is the region of the atmosphere that contains all the recognizable weather. The tropopause is the boundary between this region and the stratosphere. Above the next boundary—the stratopause—are the mesosphere, thermosphere and the rarefied exosphere.

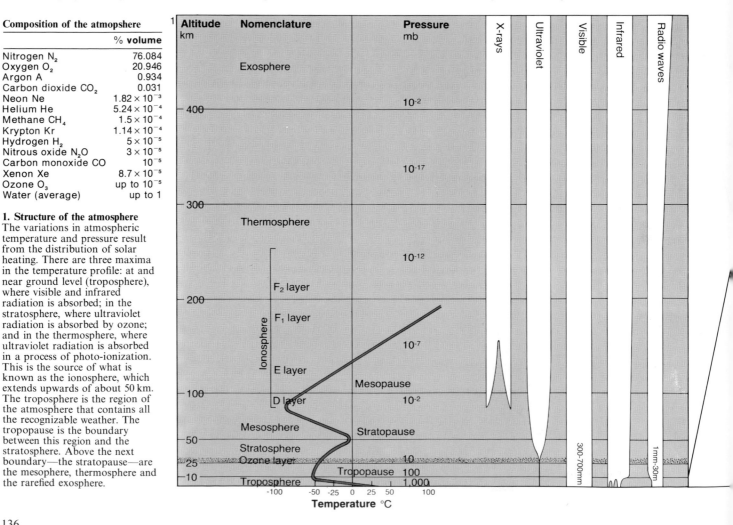

of temperature and pressure that determine the recognizable layers within the terrestrial atmosphere (*see* diagram 1), although chemical changes do play a role.

## Clouds

Water exists on Earth in three forms: as a gas, liquid or solid. As a gas it is largely opaque to planetary radiation but transparent to solar radiation; as water or ice (or snow) it either absorbs solar radiation or reflects it strongly. Alternatively, water exists as droplets in clouds. Clouds play a major role in the distribution of water vapor within the water cycle and they are crucial to weather processes within the atmosphere. They are continually created by atmospheric motion, and the variable motions of the atmosphere produce variable cloud types, depending on whether air masses are rising or descending and whether they are large- or small-scale atmospheric features (*see* diagram 2).

Large-scale cloud systems reflect the movements of air masses and winds. In equatorial regions there is a band of cloud that stretches around the globe and this marks the position of the inter-tropical convergence zone. Over tropical oceans cumulonimbus clouds develop where the winds from two hemispheres meet. These clouds extend to high levels of the atmosphere, sometimes into the stratosphere. They have a small horizontal extent since they are the product of the rapid vertical motion of air.

These cloud systems move in an easterly direction and their intensity changes during the course of a day. They can develop into intense cyclonic disturbances (hurricanes in the Atlantic, typhoons in the Pacific), which are powered by energy drawn from the warm ocean below. Air converges towards low pressure at the center of the system: it spins faster as it does so and spirals inwards and upwards. Condensation of water vapor evaporated from the ocean produces latent heat that maintains the instability of the system. Torrential rainfall and thunderstorms result. Winds of up to 240 km/hr can be produced when the hurricane is at its most intense. Some depressions affecting higher latitudes originate as hurricanes: the cyclonic circulation increases in size gradually as the storm moves eastwards. Other higher-latitude clouds are associated with jet streams higher in the atmosphere. These are revealed as large wave structures on images of the Earth.

Along the western boundaries of continents coastal stratus clouds develop and these are associated with cold offshore ocean currents. Other cloud features are associated with surface features. Islands sometimes deflect air streams so that perturbations in the clouds look like a ship's wake. Mountain systems greatly affect weather systems, producing lee-wave clouds. These appear like the seemingly stationary ripples behind a submerged object in a shallow stream. They can be accompanied by lenticular clouds when there is sufficient moisture for condensation to occur during the air's ascent to a wave crest with evaporation during its descent to a wave trough.

Other orographic clouds include banner clouds, which form when winds blow past an isolated, cone-shaped mountain, and rotor clouds, which are saucer-shaped clouds that sometimes form under lee waves as a result of turbulent eddies. Wave clouds usually form at 4–5 km (altocumulus), but they can be as low as 1–2 km (stratocumulus) or as high as cirrus levels (6 km to the tropopause). Clouds also include fogs, formed by condensation associated with radiative cooling of the Earth's surface.

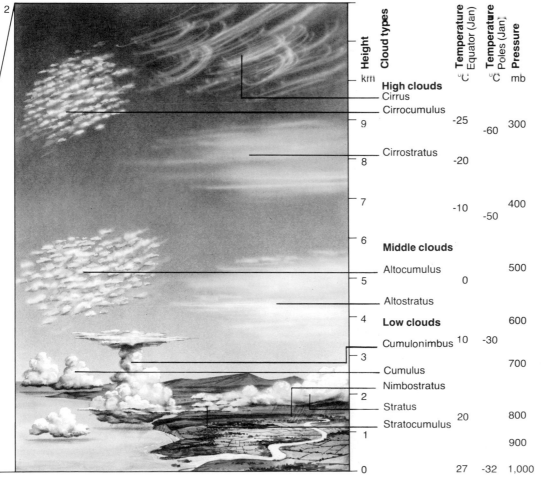

| Height km | Cloud types | Temperature °C Equator (Jan) | Temperature °C Poles (Jan) | Pressure mb |
|---|---|---|---|---|
| | **High clouds** | | | |
| | Cirrus | | | |
| | Cirrocumulus | | | |
| 9 | | -25 | -60 | 300 |
| 8 | Cirrostratus | -20 | | |
| 7 | | -10 | -50 | 400 |
| 6 | | | | |
| | **Middle clouds** | | | |
| | Altocumulus | | | 500 |
| 5 | | 0 | | |
| | Altostratus | | | |
| 4 | **Low clouds** | | | 600 |
| | Cumulonimbus | 10 | -30 | |
| 3 | | | | 700 |
| | Cumulus | | | |
| | Nimbostratus | | | |
| 2 | | | | |
| | Stratus | | | |
| | Stratocumulus | 20 | | 800 |
| 1 | | | | 900 |
| 0 | | 27 | -32 | 1,000 |

**2. The troposphere**
This bottom layer of the Earth's atmosphere contains 80 percent (by mass) of all the atmosphere. Temperature falls with height according to the heating from below. The layer as a whole is characterized by wind speeds that increase with height, a great deal of moisture at lower levels, and a significant air movement vertically. The height of the troposphere varies according to latitude and the season of the year. It may be only 8 km deep at the poles, but 16–19 km over the equator. The coldest tropopause temperatures are therefore not found over the winter pole but at the equator, where the greater height cancels out the effects of high surface temperatures. Atmospheric motions produce irregular and varying patterns of rising and descending air, which therefore ultimately produce a correspondingly irregular and varying distribution of clouds. Clouds are divided into three groups according to height. A cloud is also distinguished by its shape: layered or stratified clouds are produced when air rises slowly over a large area, and heaped clouds form when air rises rapidly in relatively narrow-based convection currents.

# Climate

**Atmospheric circulation**

Variations in solar radiation with latitude and season are determined by the geometry of the Earth's orbit. Different intensities of this radiation produce atmospheric temperature differences and, therefore, pressure gradients. The global circulation patterns that are established by pressure fields at different levels have many regular features, especially when observations are averaged out over a long period of time. Examples of this regularity are the Icelandic low-pressure system and the Siberian high-pressure system. The complexities of atmospheric circulation on a global scale result partly from the sheer size of the system and partly from the many interactions between the various physical processes in the atmosphere and on the surface of the Earth.

Winds are generated by differences in pressure. They redistribute warm and cold air and water vapor and these effects eventually feed back on the winds again. Air movements are strongly influenced by

topography. The presence of mountains, for example, causes water vapor to condense and this produces a release of energy. The distribution of continents and mountains is very important in controlling weather patterns. In addition, the atmosphere continually exchanges energy with the oceans and this also has a great influence on meteorology and climatology on Earth. The oceans in fact have a heat capacity that is 1,000 times greater than that of the atmosphere itself.

A pressure map plotted at 500 mb reveals a series of waves encircling the planet. These are not evident on a surface pressure map: they are directly influenced by the development of cyclones and anticyclones, and the precise structure of the waves is determined by surface meteorology and, therefore, the distribution of land and sea or changes in snow cover, for example. The effects of these pressure waves on meteorology are significant too, and indeed this relates to the inability of the atmosphere to repeat the

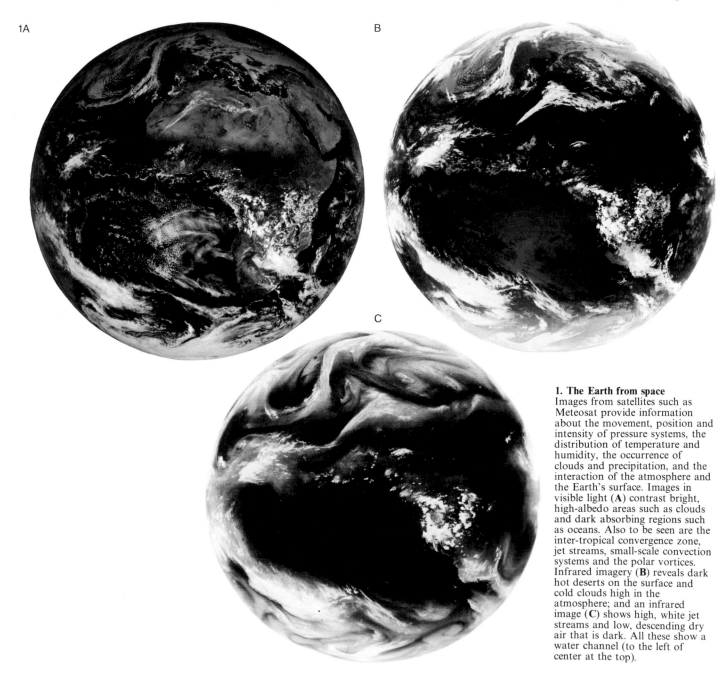

1A

B

C

**1. The Earth from space**
Images from satellites such as Meteosat provide information about the movement, position and intensity of pressure systems, the distribution of temperature and humidity, the occurrence of clouds and precipitation, and the interaction of the atmosphere and the Earth's surface. Images in visible light (**A**) contrast bright, high-albedo areas such as clouds and dark absorbing regions such as oceans. Also to be seen are the inter-tropical convergence zone, jet streams, small-scale convection systems and the polar vortices. Infrared imagery (**B**) reveals dark hot deserts on the surface and cold clouds high in the atmosphere; and an infrared image (**C**) shows high, white jet streams and low, descending dry air that is dark. All these show a water channel (to the left of center at the top).

same weather pattern every year. We therefore observe differences in the weather from one summer to the next. This tends to hinder accurate weather forecasting, but satellites are continually monitoring weather patterns. Polar-orbiting satellites observe global atmospheric patterns, while geostationary satellites monitor diurnal changes over one region.

## Climatic change

We know of great climatic changes in the Earth's past, but even on a less dramatic scale than the last Ice Age, which ended about 10,000 years ago, there have been interesting fluctuations right up to the present day. There was the post-glacial climatic period 7,000–5,000 years ago Before Present (BP); the colder climatic period of the Iron Age (2,900–2,300 BP); a warmer interval in the Early Middle Ages about 1,000–800 years BP; and the Little Ice Age between 500 and 125 years BP, which was particularly severe in the late seventeenth century. The beginning of the present century was unusually warm.

Despite the complicated interaction of factors determining the characteristics of the atmosphere and meteorological processes, it is possible to decipher cyclic variations in climatic patterns over a long period. Over many millions of years tectonic activity in the Earth's crust has moved the continents in such a way (see pages 134–5) as to allow the formation of polar caps, but there are also astronomical cycles over tens of thousands of years that assist or hinder the development of surface ice. These are known as the Milanković cycles.

The eccentricity of the Earth's orbit varies from nearly circular to more elliptical and back again over a period of some 90,000–100,000 years. The longitude of perihelion, which determines the season of closest approach to the Sun, varies over 21,000 years. The Earth is currently closest to the Sun during the southern hemisphere summer and furthest from the Sun during the northern hemisphere summer, but this situation will be reversed in about 10,000 years. Also, the obliquity of the Earth relative to the orbital plane changes during a period of 40,000 years, which affects the seasonal differences between winter and summer. The axial inclination varies between 21.8° and 24.4°. Such cycles can affect the amount of heat received by polar latitudes by one percent. Mars experiences similar cyclical changes (see page 216).

Changes in the composition of the atmosphere also affect the climate. There is presently concern about the carbon dioxide content being increased as a result of man's use of fossil fuels and depletion of the Earth's forests. Carbon dioxide is absorbed by

plants during photosynthesis and exhaled by living organisms in respiration. It also enters the atmosphere from the lithosphere, particularly during volcanic eruptions. A doubling of the carbon dioxide content, estimated by some scientists to be likely by the year 2000, could increase global temperatures by about 2 K by means of the greenhouse effect. However, the oceans absorb $CO_2$, although the solubility of the gas does depend on temperature, and it is unlikely that a runaway greenhouse effect could occur on Earth as it did on Venus in the past.

Man's activities may also affect the amount of ozone in the atmosphere. It is thought that it may have been depleted even more in the past by natural means. When the Earth's magnetic field undergoes reversals, the absence of a magnetosphere may allow cosmic rays to enter the atmosphere unchecked, thus destroying the ozone. This could even happen during intense solar activity such as a large solar flare. High-energy particles in the stratosphere result in ionization and the production of large amounts of nitrogen oxides, which destroy ozone. Many living species became extinct at the end of the Cretaceous period and this coincided with a resumption of polarity reversals. It is, however, generally believed that the demise of the dinosaurs came about after a catastrophic event, such as a collision with an asteroid, significantly altered the atmosphere, removing its protective shield.

Fluctuations in the solar constant could profoundly affect the climate on Earth. It has been estimated that the variability in solar output of approximately one percent would alter the global temperature by about 1 K. We do not know whether changes of this magnitude have ever occurred: between February 1980 and August 1981 there was a decrease of 0.1 percent, which may be affecting the weather now. During the Little Ice Age few sunspots were observed; this is known as the Maunder Minimum and other such periods have occurred in the past, traceable in the records of ancient Chinese astronomers.

Volcanic eruptions release large amounts of material into the atmosphere. This can reduce the transmitted sunlight, thus lowering the temperature of surface layers on Earth. There is also concern about the spread of deserts, since the resultant changes in albedo will affect the atmosphere and ultimately reduce precipitation. This will in turn affect cloud distribution and finally the general circulation patterns. The finely balanced interrelationships between the Earth's atmosphere and its surface are therefore also inextricably linked to man's activities within the environment and the longterm effects thereof.

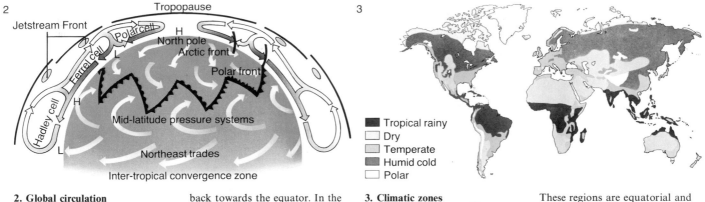

**2. Global circulation**
On Earth the principal driving force behind the circulation is the differential heating between the equator and the poles. Near the equator there is a Hadley circulation: warm air rises and is transported polewards while cooler air sinks and is carried back towards the equator. In the mid-latitudes circulation is complicated, however, by eddies. Tropical and polar air meet at the polar front. Where warm air slides above colder air there is a frontal zone and depressions result. The diagram shows an idealized circulatory pattern.

**3. Climatic zones**
The Earth exhibits a wide range of climates across its surface and there have been many systems of classification. Wladimir Köppen's system has been the basis of many others: it has five regions defined on the basis of temperature and rainfall, and hence vegetation. These regions are equatorial and tropical rainy climates, dry climates, temperate climates, humid cold climates and polar climates. All the Earth's climatic zones reflect the interaction of features of atmospheric circulation with different aspects of local geography.

# Solar Interaction with the Earth

Apart from providing light and heat the Sun interacts with the Earth in a wide variety of more subtle ways. Its influence on the Earth's magnetic field, for example, is considerable. Like the solar field, the geomagnetic field may be represented (to a first approximation) as a simple dipole, with a mean strength at the poles of about 0.6 G; the position of the Earth's magnetic poles alters slowly with time. The region in space within which the Earth's field is the predominant magnetic influence is called the "magnetosphere", and if it were not for the effect of the solar wind, this region would extend out to a distance of about 100 Earth radii ($R_E$), affecting the motion of charged particles within this region. However, the pressure of the solar wind compresses the sunward side of the magnetosphere to within 8 to 10 $R_E$, while in the opposite direction, interaction with the wind draws out the field lines into a "magnetotail", stretching well beyond the orbit of the Moon. The magnetotail has been detected as far as 1,000 $R_E$ from the planet. It contains a "neutral sheet", on either side of which the field lines have opposite direction.

The boundary of the magnetosphere, across which the solar wind cannot readily flow, is called the "magnetopause". The solar wind flows much faster (about eight times faster on average) than the speed at which sound waves can pass through it, and the magnetosphere can thus be regarded as effectively "moving" supersonically through the wind. Just as a supersonic aircraft will pile up material in front of it forming a shock wave, so the magnetosphere produces a shock wave in the solar wind called the "bow shock". On the sunward side this bow wave precedes the magnetopause by about 3 or 4 $R_E$. At the shock front, the speeds of solar wind particles are abruptly slowed from about 400 km s$^{-1}$ to about 250 km s$^{-1}$, and the kinetic energy of motion lost in this process is transformed into heat, raising the temperature of the plasma to several million K, 5 to 10 times higher than normal for the wind. Between the bow shock and the magnetopause is a region

known as the "magnetosheath", characterized by jumbled and chaotic magnetic field lines.

## Magnetic storms

Variations in the solar wind produce changes in the magnetosphere that are reflected in the terrestrial magnetic field at ground level in phenomena known as magnetic storms. The bulk of solar flare particles reach the Earth after about 2 days from the occurrence of the flare; the shock wave preceding the cloud of plasma compresses the magnetosphere, rapidly intensifying the geomagnetic field at ground level. This phase, which takes place over a timescale of a few minutes, is called the "sudden storm commencement" (SSC). It is followed by the "initial phase" (IP), lasting from about 30 min to a few hours, during which time the flare plasma flows past the Earth while the field strength remains higher than it was before SSC. During the next phase, the "main phase" (MP), the particle population is enhanced and particles are accelerated by the release of magnetic energy, particularly in the magnetotail where field lines reconnect. These effects cause a flow of current in the magnetosphere, which itself generates a field opposed to that of the Earth; as a result there is a drop in the geomagnetic field strength lasting between a few hours and about a day. As the current decays, the field strength returns to normal, possibly over a few days. Magnetic storms of this kind may recur at 27-day intervals where the source is a persistent active region.

## Van Allen belts

There are certain restricted regions in the Earth's magnetosphere within which highly energetic charged particles are trapped by the magnetic field. These regions, which were discovered in 1958 by the space-probes Explorers 1 and 3, are known as the "Van Allen belts". Conventionally two zones are shown, at about 1.5 and 5 $R_E$ from the center of the Earth, but there is no clear-cut division;

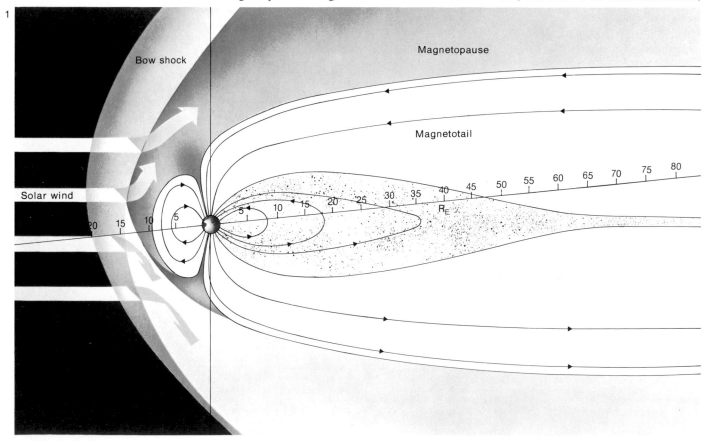

however, the inner belt has a higher concentration of high-energy protons and electrons than the outer belt. The charged particles in these belts are bound to the magnetic field and spiral to and fro along the north–south lines, since they are reflected back when they meet the converging lines near the poles; the period of one to-and-fro oscillation is of the order of 1 sec. The population of particles in these belts is maintained by the solar wind, by cosmic rays and by particles sprayed from the upper atmosphere by cosmic ray impacts, combined with acceleration of these particles in the Earth's field by processes similar to those occurring in solar flares.

## The ionosphere
Solar X-ray and ultraviolet radiation photoionizes atoms and "photodissociates" molecules in the Earth's upper atmosphere resulting in the presence of free electrons stripped from their atoms and molecules. These have a tendency to recombine, but new ions are created at an equal rate, and at any one time the concentration of ions is sufficient to cause variety of phenomena. A number of regions of ionization in the atmosphere are recognized, each owing its level of ionization to different wavelengths of radiation. Of particular interest is the D-region, extending up to 90 km above ground level, where ionization is due to ultraviolet radiation (notably Ly$_\alpha$), extreme ultraviolet, X-rays and cosmic rays. This layer reflects long-wave radio waves in the wavelength range of 10 to 100 m allowing radio signals to be used for communication around the globe, but preventing ground-based long-wave radio astronomy. While the ionosphere reflects long wavelengths, it also absorbs very short wavelengths, and so prevents the penetration to ground level of a range of frequencies originating in space. This absorption also gives rise to enhanced levels of ionization which occur after solar flares and related events because of an increased incidence of X-ray radiation. As a result, some radio frequencies are sometimes absorbed and the signal may fade.

Much more drastic ionospheric disturbances are caused by violent flares called "solar proton events", in which copious quantities of higher energy protons and other nuclei are ejected with energies of tens or hundreds of mega electron volts (MeV). These particles penetrate the barrier normally provided by the magnetosphere and enter the Earth's atmosphere by spiralling along the field lines at high geomagnetic latitudes. They can produce sufficient ionization to black out radio signals and severely attenuate cosmic radio waves for periods lasting a few days. The frequency of these Polar Cap Absorptions (PCA) correlates well with the solar cycle: 10 or more events may occur in a year near maximum, while near solar minimum there may be none at all.

## The ozone layer
Solar ultraviolet radiation impinges on the upper atmosphere and breaks down molecular oxygen ($O_2$) into separate oxygen atoms. In the atmosphere between 15 and 50 km altitude, ozone ($O_3$) is formed by a combination of O and $O_2$. Ozone is destroyed by ultraviolet radiation, particularly in the 210 to 310 nm range. In this process it heavily absorbs the radiation, which is harmful to living tissue. Equilibrium is normally maintained between creation and destruction of ozone. The peak concentration of ozone tends to occur at an altitude of about 20 km, a region known as the "ozone layer", but even there its concentration is only about 10 parts per million. Nevertheless this thin concentration is essential for the preservation of human life on this planet.

The ozone layer is subject to fluctuation. Recently it has been found that there are "holes" in the layer over the poles, and this has given rise to speculation that human activity may be responsible; certain chemicals reaching the upper atmosphere might well have damaging effects. As yet there is no final proof that this is the case, but there is every reason for anxiety. And there is no doubt that the situation will have to be very carefully watched.

**1. Earth's magnetosphere**
On its sunward side the Earth's magnetosphere is compressed by the highly energetic particles of the solar wind, which collide with the Earth's magnetic field. On the opposite side, the "magnetotail" stretches far out into space.

**2. Van Allen belts**
The Earth is encircled by zones of radiation known as "Van Allen belts", in which charged particles spiral to and fro, trapped by the Earth's magnetic field; the belts are therefore inclined at an angle to the Earth's rotational axis.

**3. Ionosphere and ozone layers**
X-ray and ultraviolet radiation from the Sun ionizes atoms in the upper regions of the Earth's atmosphere, producing a layer known as the "ionosphere". This layer reflects certain long-wave radio signals. Variations in solar activity alter the concentration of ions in this region, and may possibly influence the frequency or intensity of thunderstorms. At a lower altitude (about 20 to 30 km) lies the "ozone layer", which is also the product of solar ultraviolet radiation.

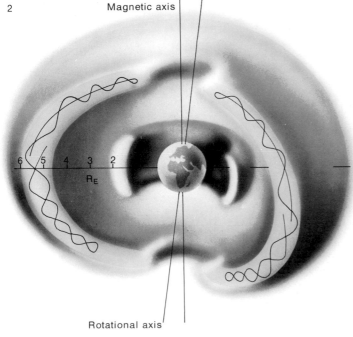

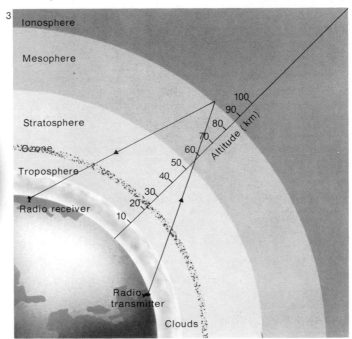

# THE MOON

# Characteristics

Although relatively insignificant in terms of the Solar System, the Earth's Moon is outstandingly large and massive in relation to its primary. It has a diameter of 3,476 km, over a quarter the size of the Earth's diameter, and a mass of $7.3483 \times 10^{22}$ kg, as compared with a figure roughly 81 times larger for the Earth. There are five satellites that are larger than the Moon—three in Jupiter's system, one in Saturn's and one in Neptune's—but all these move around giant planets which are several hundred times more massive than their respective satellites. The Earth–Moon system has sometimes been described as a "double planet" since the mass ratio between the two bodies is distinctly less unequal than for other planets and their satellites.

## Atmosphere and surface conditions

The Moon is a hostile world for one main reason: its lack of atmosphere. Because of its small size and low mass, the Moon's gravitational field is relatively weak; in fact the acceleration due to gravity on its surface is only one-sixth that on the surface of the Earth. Consequently, the Moon's "escape velocity" is low—2.37 km s$^{-1}$—which is not high enough for it to retain an appreciable atmosphere. This has been confirmed both by Earth-based astronomers and by the Apollo astronauts. The density of the Moon's atmosphere must be $10^{-14}$ times less than that of the Earth, making it virtually indistinguishable from a hard vacuum. However, an experiment set up in December 1972 by the astronauts

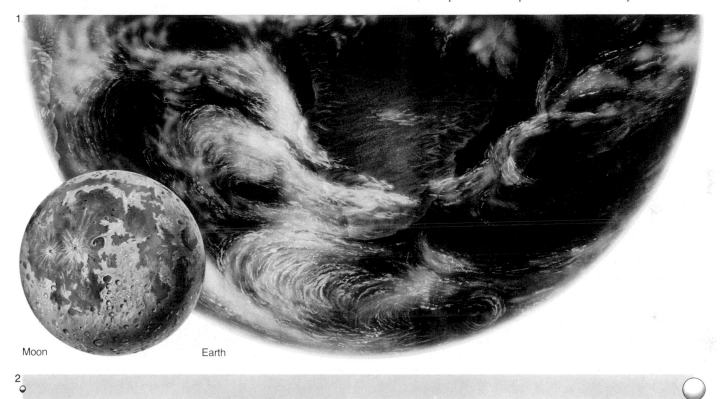

Moon          Earth

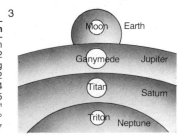

### 1. Scale of the Moon
The Moon's diameter (3,476 km) is more than a quarter that of the Earth; the mass ratio is 1:81.

### 2. Earth–Moon distance
The Moon orbits the Earth at a mean distance of 384,392 km, less than 10 times the circumference of the Earth. At closest approach it is 406,679 km; at farthest approach it is 406,679 km from the Earth (measured from center to center).

### Physical data

|  | Moon |
| --- | --- |
| Equatorial diameter | 3,476 km |
| Ellipticity | 0.002 |
| Mass | $7.3483 \times 10^{22}$ kg |
| Volume (Earth=1) | 0.02 |
| Density (water=1) | 3.34 |
| Surface gravity (Earth=1) | 0.165 |
| Escape velocity | 2.37 km s$^{-1}$ |
| Axial inclination | 1.53° |
| Albedo | 0.07 |

### 3. Satellites and their primaries
Relative to its primary (Earth), the Moon is unusually large and massive. By comparison Triton has only about one-eighth the diameter of Neptune, with a mass ratio of 1:750; for the largest satellites of Saturn (Titan) and Jupiter (Ganymede) the ratio is even smaller. This unique mass ratio between the Earth and the Moon has led to a number of theories as to the Moon's origin (*see* page 154).

3

| Moon | Earth |
| --- | --- |
| Ganymede | Jupiter |
| Titan | Saturn |
| Triton | Neptune |

of Apollo 17 revealed the presence of an excessively tenuous atmosphere on the Moon's surface; gases detected included hydrogen, helium, neon and argon, but the density of these gases is so low that the constituent molecules would practically never collide with one another.

Thus, whereas the surface of the Earth has been subject to a wide variety of climatic conditions which have influenced its geological history, the Moon's surface has been unaffected by these weathering processes. Similarly, the absence of free water on the Moon means that another of the principal causes of erosion on the Earth is lacking, and it would therefore be expected that the Moon's surface should differ substantially from that of the Earth.

Other forces, on the other hand, play a considerably greater part in shaping the surface of the Moon. There is, for example, a steady bombardment by particles and fragments of rock, ranging in size from large boulders (which are fairly rare) to particles of dust. The Moon is also bombarded by the "solar wind", which is emitted by the Sun and which consists mainly of protons and electrons; since these particles are electrically charged, they tend to be deflected by the Earth's magnetic field, whereas the Moon's magnetic field (*see* page 151) is too weak to provide the same protective influence.

# Observational Background

The main dark areas of the Moon are easily visible with the naked eye, and the first of all lunar maps seems to have been that of the British scientist W. Gilbert in or about 1600; it was published posthumously in 1651, and some of the details are plainly recognizable. The first telescopic map was the work of Thomas Harriott, one-time tutor to Sir Walter Raleigh, in July 1609, but the earliest serious work was carried out from 1610 by Galileo, who made drawings of special features and also attempted to estimate the heights of the lunar mountains by measuring the lengths of their shadows. His results gave altitudes which were rather too high, but they were at least of the right order, and showed the Moon's mountains are higher, comparatively, than those of the Earth.

In Wales Sir William Lower is said to have used a telescope to look at the Moon around 1611. His drawings have not been preserved, but he made the picturesque comment that the full moon looked rather like "a tart which his cook had made".

## Nomenclature

The pioneer attempts to draw up a useful nomenclature were made by Langrenus in 1645 and by Hevelius in 1647. Hevelius, a city councillor of Danzig (now Gdansk), used geographical analogies; for instance, the crater now called Plato was called "the Greater Black Lake". His nomenclature never became popular, and it is said that after his death the copper engraving of his lunar map was melted down and made into a teapot.

Hevelius's nomenclature was superseded by that of the Italian Jesuit Giovanni Riccioli, who published a map in 1651 based on observations by his pupil, Grimaldi. Riccioli named the mountains after Earth ranges such as the Alps and Apennines, and the craters after famous persons, usually (though not always) astronomers. His system has survived and has since been extended to cover minor features and also those on the Moon's far side. It is reasonably satisfactory, and will certainly never be altered now, but inevitably later scientists have had to be allotted less prominent features. Newton's name, for example, is attached to a formation not far from the south pole. Moreover, Riccioli was not impartial in his choices: he did not believe in the Copernican System (according to which the Earth moves around the Sun) and therefore "flung Copernicus into the Ocean of Storms"; Galileo was similarly allocated a very small insignificant crater. Grimaldi and Riccioli himself, meanwhile, have large, prominent walled plains near the western limb, identifiable under any conditions of solar illumination because of their dark floors. The maria were given "romantic" names like "Serenitatis" and "Imbrium" to distinguish them from the craters. At a much later date, two small maria near the limb were named after astronomers (Mare Humboldtianum and Mare Smythii). There are also some unexpected names; Julius Caesar has a crater of his own, presumably because of his association with calendar reform. One crater is named Hell. This does not, however, indicate any exceptional depth but honors a Hungarian astronomer, Father Maximilian Hell.

## Development of selenography

In 1775 the German astronomer Tobias Mayer produced a small but fairly accurate lunar map and also introduced a system of coordinates; but the real "father of selenography" was Johann Hieronymus Schröter, chief magistrate of the little town of Lilienthal, near Bremen. Schröter built a private observatory and studied the Moon and planets from 1778 until his observatory was destroyed by the invading French soldiers in 1813. He made hundreds of drawings of lunar formations and was the first to make detailed observations of the crack-like features known as clefts or rills. Despite his admittedly clumsy draughtsmanship his work was of the utmost value, and his name is commemorated in both a crater and valley. The next progress was due to a Berlin banker, Wilhelm Beer, and his colleague, Johann Mädler. Using the small refractor

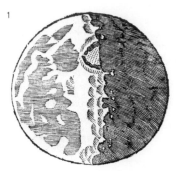

**1. Galileo's observations, 1609**
Ptolemaeus (lower center), Copernicus (white patch, left center) and Mare Imbrium (upper left) are recognizable in these drawings from *Sidereus Nuncius*.

**2. Lunar map by Hevelius, 1647**
Hevelius produced a map showing many recognizable features, but his system of nomenclature, based on geographical analogies, has not survived.

**3. Lunar map by Riccioli, 1651**
Riccioli's map (*Almagestum Novum*) was better than that of Hevelius. Riccioli introduced the system of naming features after personalities. Many of the features shown on the map are recognizable; even the ray-systems from the craters Tycho, Copernicus and Kepler are shown. Riccioli also drew the Moon at different phases.

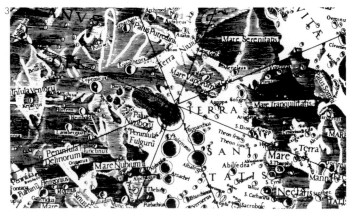

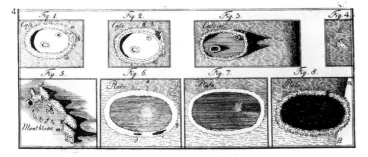

**4. Lunar details by Schröter, 1791**
Schröter never produced a complete map, but he made thousands of drawings. These sketches show Mont Blanc and the craters Cassini and Plato.

**5. Map by Beer and Mädler, 1837**
This map was a masterpiece of careful observation. It remained the standard work for over 40 years after being compiled, and in general is remarkably accurate.

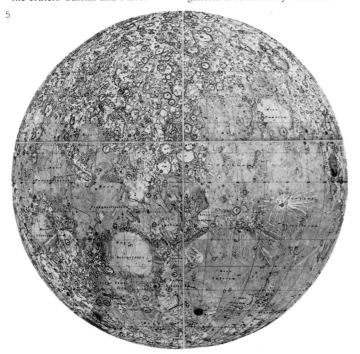

**6. Detail by J. Schmidt, 1878**
Schmidt's map, based on preliminary work by Lohrmann much earlier, was extremely detailed and superseded that of Beer and Mädler.

**7. Far side preliminary chart, 1959**
This early chart was drawn from the photographs obtained from Luna 3. Some features are recognizable, such as Tsiolkovsky; others were misinterpreted.

in Beer's observatory, they produced in 1837 a lunar map which was a masterpiece of careful, accurate observation, and followed it up with a book, *Der Mond*, containing a detailed description of every named feature. Unlike Schröter, they correctly believed the Moon to be dead and changeless. Interest was greatly stimulated in 1866, when Julius Schmidt, a German who had emigrated to Greece and become Director of the Athens Observatory, announced that the crater Linné, recorded by Beer and Mädler on the grey plain of the Mare Serenitatis, had disappeared to be replaced by a white patch. Even though it now seems unlikely that any real change occurred in Linné, the repercussions of Schmidt's announcement were far-reaching and the report was probably directly responsible for a number of new maps. Schmidt himself published an elaborate version in 1878, based on a chart begun much earlier by Wilhelm Lohrmann, a Dresden land surveyor; in England Edmund Neison produced a map and a book, and the British Association planned an even larger chart, although little of it was ever actually completed.

**Modern selenography**
The first photograph of the Moon was taken in 1840 by J. W. Draper, and from the 1870s photography became an essential basis of all lunar mapping. Photographic atlases were produced—notably by Loewy and Puiseux, at the Paris Observatory, in the 1890s and by W. H. Pickering in 1904. Pickering, one of the few professional astronomers to take an interest in the Moon at that time, showed each part of the Moon under five different conditions of illumination, though it is true that the pictures are of poor quality by modern standards. The members of the Lunar Section of the British Astronomical Association were both enthusiastic and skilful, and the outline chart drawn by its first Director, Thomas Gwyn Elger, is still useful. It was compiled mainly from direct visual observations, but with a photographic basis.

In 1840, when photography was just beginning, the French astronomer François Arago stated that by this new technique it would be possible to complete lunar mapping in a matter of weeks. He was wrong; it took more than a century and a quarter, but today the Moon is better mapped than some regions of the Earth.

Before the Space Age, study of the Moon was regarded as essentially an amateur province. Further maps were produced, notably by Walter Goodacre in 1930 and by H. P. Wilkins in 1946, and were probably better than the official chart published in 1935 by the International Astronomical Union, based on photographic micrometrical measurements by W. H. Wesley and M. A. Blagg. As flight to the Moon became a real possibility, however, professional observatories took over the task of lunar cartography, although they owed much to the older amateur work.

Particular attention had always been paid to the libration regions of the Moon (*see* page 147), which are highly foreshortened and therefore difficult to chart. It was possible for amateurs to make interesting discoveries: for instance, the great ringed structure known as the Mare Orientale was first seen (and named) by Wilkins and the present writer just after the war.

During the 1950s elaborate photographic atlases were compiled, mainly by American astronomers at the Lunar and Planetary Laboratory at Tucson, Arizona, and these naturally made the older charts virtually obsolete. The system of nomenclature was refined, and new names were added where appropriate. By 1959, when the first unmanned lunar probes were launched, maps of the Moon's Earth-turned hemisphere were reasonably good except in the libration regions, but obviously they could not show the very small features. The final phase began with the American Orbiters of 1966–68. These were put into closed orbits around the Moon and sent back thousands of highly detailed photographs, covering almost the whole of the surface. Without them, it is not likely that the manned Apollo missions could have been attempted.

# Orbit and Phases

As observed from the Earth, the Moon appears to move in an elliptical orbit with the Earth at one focus, just as the planets move in ellipses around the Sun. In fact, it would be slightly more accurate to say that the Earth and Moon revolve together around the "barycenter", the center of gravity of the system, but since the Earth is much more massive than the Moon the barycenter lies within the terrestrial globe, 1,630 km below the surface. The eccentricity of the Moon's orbit is 0.0549, and its distance from the Earth (measured from center to center) ranges from 356,410 km at its closest approach or "perigee" to a maximum value of 406,679 km at "apogee"; the mean distance is 384,392 km. At apogee the apparent diameter of the Moon is only nine-tenths of that at perigee; the difference is inappreciable to the naked eye, but easy to measure. The line joining the perigee and the apogee is called the "line of apsides".

The plane of the lunar orbit is inclined to the plane of the ecliptic (defined by the Earth's orbit around the Sun) by an angle of 5°9'. The Moon therefore crosses the ecliptic at two points, once as it moves from south to north (the "ascending node") and once as it moves from north back to south (the "descending node").

However, the Moon's movement around the Earth is complicated in several ways by the gravitational pull of the Sun, which causes perturbations in the orbit. It is perhaps surprising to learn that the Sun's pull on the Moon is more than twice as strong as that of the Earth; viewed from outside the Solar System, the Moon would appear to revolve around the Sun, with an orbit that is always concave to the Sun. (There is no chance, however, that the Earth and the Moon will part company, since the Sun attracts the two bodies almost equally.)

The perturbing influence of the Sun can be analyzed into six main effects. One is to cause periodic variations in the Moon's eccentricity, which oscillates between 0.044 and 0.067; this variation is called "evection". Similarly, the inclination of the orbit varies between 4°58'and 5°19'. Another effect is that the Moon's perigee advances in the same direction as the Earth's rotation, taking a period of 8.85 years to complete one revolution. A further consequence of the Sun's gravitation is that the line joining the Moon's nodes has retrograde motion along the ecliptic (in other words, in the opposite direction to the movement of the perigee); the period, known as the "nutation period", is 18.61 years. Finally, calculations of the position of the Moon must take into consideration the fact that the gravitational pull of the Sun on the Moon is weaker when the Moon is on the far side of the Earth than when it is situated between the Earth and the Sun (an effect known as "variation"), as well as allowing for the changes in the Earth's distance from the Sun during a year (the "annual equation").

The problem of giving a complete mathematical account of the Moon's orbit is one of the most difficult in astronomy. A more detailed description than has been given so far would have to take into account not only the effects of the Sun, but also such minor influences as the planets and the exact shapes of the Earth and the Moon; but for the purposes of giving a brief outline of the lunar orbit, these terms are sufficiently small to be ignored.

## Phases of the Moon

Having no light of its own, the Moon shines only because of the light it reflects from the Sun: at any given moment the hemisphere turned away from the Sun will be dark. The apparent shape of the Moon in the sky (its "phase") therefore depends on what position it has reached in its orbit around the Earth. When it lies exactly between the Earth and the Sun, so that its sunlit hemisphere cannot be seen from Earth, the Moon is said to be "new"; shortly afterward a crescent will be seen. As the Moon moves further along its orbit, the size of the crescent increases as the Moon "waxes", and eventually half the surface turned towards the Earth is illuminated; this point is termed the "first quarter", the Moon having completed

1
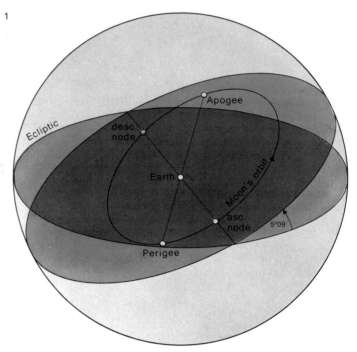

**1. Orbit of the Moon**
The Moon's orbit is shown projected against a celestial sphere centered on the Earth. The plane of the orbit is inclined to the plane of the ecliptic by an angle of 5° 9'. The Moon's nodes, defined by the two points at which the orbit crosses the ecliptic, move backward along the ecliptic at a rate of just over 19° per year. The line of apsides (joining the perigee and apogee) moves in the opposite direction with a period of 8.85 years.

**2. Phases of the Moon**
The phase of the Moon depends on the angle between the Sun, the Moon and the Earth. One cycle of lunar phases is completed in the interval between two successive similar alignments of the three bodies, as represented by the first and last positions shown here (**A**). The appearance of the Moon from Earth (**B**) is shown in relation to the corresponding position of the Moon in its path. The Moon takes slightly less time to repeat its initial position against the stellar background; in the penultimate position in diagram **A** (bold line) the Moon is shown one sidereal month after the start of the sequence.

2A
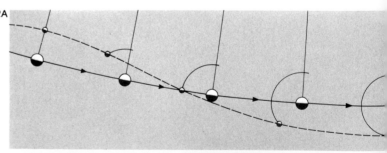

B
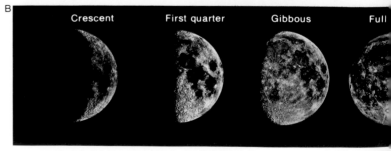

Crescent    First quarter    Gibbous    Full

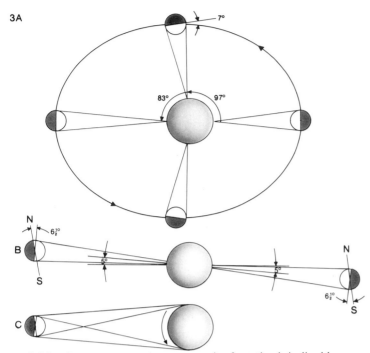

3A

a quarter of its orbit. Next it becomes "gibbous", when more than half the sunlit hemisphere is visible, and eventually "full". The maximum magnitude of the full Moon is − 12.5 (*see* Glossary). Continuing its orbit, the Moon begins to "wane", once again becoming gibbous until it reaches the "last quarter", and then dwindling to a crescent. With the following new Moon, the cycle begins again. The "age" of the Moon is the number of days that has elapsed since the previous new Moon.

During the crescent phase, the unilluminated part of the Moon's disc is often clearly visible. This phenomenon is due to the reflection of light by the Earth, as was explained by Leonardo da Vinci.

The boundary between the bright and dark hemispheres of the Moon is known as the "terminator". Because the lunar surface is rough, the terminator is jagged in appearance, rather than smooth, and it often happens that a mountain, catching the first rays of the Sun over the horizon, appears as an isolated point of light slightly beyond the dark side of the terminator. The points or horns of the Moon when it is between new and full are known as the "cusps".

### The Moon's period

It takes 27.32166 days for the Moon to complete one revolution, measured against the background of fixed stars. This period is known as a "sidereal month". There are, however, several other kinds of month. For example, because the Earth is itself moving around the Sun, the interval between one new Moon and the next is slightly longer than a sidereal month (*see* diagram 2) and is known as a "synodic month": it is equal to 29.53059 days. This period is also sometimes referred to as a "lunar month" or "lunation". There is also the "draconic" or "nodal" month (measured between successive passages through the same node), the "anomalistic month" (perigee to perigee) and the "tropical month" (successive conjunctions with the vernal equinox—*see* Glossary).

The Moon moves eastward against the stellar background by about 13° per day and therefore rises later on successive days; the time-difference between one moonrise and the next is known as the "retardation". Its average value is 50 min, but because the Sun and Moon do not move at uniform speeds, and because their orbits are inclined to the equator, this value varies throughout the year. In the northern hemisphere the retardation is greatest around March, when the angle of the Moon's path to the horizon is at its steepest; it is least around September, when the angle is shallowest.

The full Moon that occurs nearest the time of the autumnal equinox (21 September) is called the "Harvest Moon" and at this lunation the retardation may be reduced to as little as a quarter of an hour. The full Moon following Harvest Moon is known as "Hunter's Moon", which also rises early in the evening; again the retardation is less than average.

### Librations

The Moon's orbit around the Earth is "captured" or "synchonous": its rotational period is the same as its period of revolution and it therefore keeps the same hemisphere pointing towards the Earth all the time. However, because the lunar orbit is not circular, the velocity of the Moon varies, being greatest at perigee and least at apogee. At the same time, the rate of axial spin is constant, so that in the course of one lunar orbit it is possible to see a little more than half of the globe; this apparent "wobble" is termed the "libration in longitude". There is also a "libration in latitude" because of the inclination of the Moon's orbit, so that an Earth-based observer can see a short distance around alternate poles. Finally, there is a daily or "diurnal" libration, due to the axial rotation of the Earth itself: at moonrise an observer can see a little way beyond the eastern limb and at moonset a little beyond the western limb. The combined effect of these three librations is to allow slightly more than 59 percent of the Moon's surface to be examined by any Earth-based observations.

**3. Librations**
The Moon's libration in longitude (**A**) is a consequence of the non-uniformity of its speed in its elliptical orbit around the Earth (exaggerated in this diagram). The Moon revolves at a constant rate, so that after one quarter of the rotational period it has rotated through 90°; at the same time, however, it has moved through 97° of its orbit, so that from Earth the Moon's central meridian would appear to be displaced by about 7°. The libration in latitude (**B**) results from the combined effect of the 5° inclination of the Moon's orbit to the Earth's equator, and the $1\frac{1}{2}$° tilt between the Moon's axis and its orbit: as a result the Moon's

axis of rotation is inclined by about $6\frac{1}{2}$° relative to the Earth's axis, so that the north and south polar regions are alternately tilted slightly towards the Earth. Finally, the diurnal libration (**C**) allows an Earth-based observer to view the Moon from a slightly different angle in the morning as compared with the evening, because his viewpoint changes with the Earth's daily rotation.

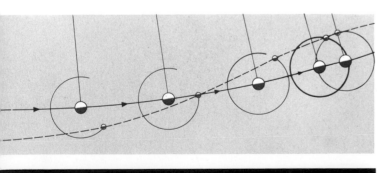

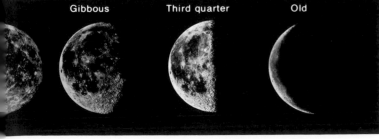

Gibbous    Third quarter    Old

# Eclipses and Tides

It has been known since ancient times that the twice-daily rise and fall of the oceans was in some way related to the position of the Moon, but it was not until the problem was investigated by Sir Isaac Newton that an adequate explanation of the tides was given. Essentially, the tides occur as a result of the Moon's gravitational attraction. Water is more easily "pulled" than land; therefore, the water in the Earth's oceans tends to heap up under the Moon, producing a high tide, while there is a corresponding high tide on the opposite side of the Earth. As the Earth spins, the water-bulges do not spin with it; they tend to remain under the Moon, so that each region experiences two high tides per day.

It may not be immediately clear why there is a high tide on the far side of the Earth with respect to the Moon. To show why this happens, imagine that the Earth is surrounded by a uniform shell of water, and that it and the Moon are isolated in space, so that they tend to draw together by the influence of gravity (*see* diagram 4). The water on the Moon-facing side of the Earth will be accelerated most strongly, and will heap up. The water on the side of the Earth directly away from the Moon will be least accelerated, so that it will be "left behind", and will also tend to heap up. This is why there are two high tides and two low tides per day (there are also land tides, but these are very slight compared with the tides in the oceans).

In practice the situation is complicated by several other factors. Local conditions influence the exact time of the tides because of irregularities in the depth and form of the oceans. The Sun also influences the tides, either reinforcing or opposing the pull of the Moon. When the Moon is either full or new ("syzygy") its gravitational pull is combined with that of the Sun to produce particularly strong tides known as "spring tides"; when the Moon is in its first or third quarter ("quadrature") the pull of the Sun acts at right angles to that of the Moon so that the two partly cancel one another, giving weak or "neap" tides.

Moreover, frictional forces in the oceans produce an appreciable lag in the time taken by the waters to heap up, while the rotation of the Earth causes the bulge to be carried slightly ahead of the actual position of the Moon rather than directly beneath it. One important consequence of this frictional effect is that the Moon's gravity acts as a brake on the Earth's rotation, so that the Earth's period is gradually increasing. Studies of the growth lines in fossil corals suggest that 350 million years ago the Earth day was approximately 3 hr shorter than it is today, so that there were about 400 days in a year. Other studies, based on ancient eclipse records, indicate that the rate of increase of the Earth's period is about 0.0016 sec per century.

As the Earth's rate of spin decreases, energy is transferred to the Moon's orbit. The Moon is gradually accelerating, and as it does so its distance from the Earth increases. It may therefore be deduced that in the remote past the Earth and the Moon were closer together. (This is one of the reasons that first led G. H. Darwin to propose the "fission hypothesis"—*see* page 154.) In a similar way tidal forces acting on the slightly ellipsoidal shape of the Moon have slowed down the Moon's rotation over the ages, thus producing the present situation in which its period of axial rotation is exactly equal to its period of revolution. In fact, all major planetary satellites have similarly captured rotations.

## Eclipses

Both the Earth and the Moon cast long conical shadows out into space. When the Earth passes through the Moon's shadow-cone, a "solar eclipse" occurs and can be seen on Earth by an observer situated in the area covered by the shadow; when in turn the Moon passes through the Earth's shadow there is a "lunar eclipse". Unlike an eclipse of the Sun, a lunar eclipse is visible from an entire hemisphere of the Earth.

During a lunar eclipse the Moon does not usually vanish completely, because some of the Sun's rays are bent or "refracted"

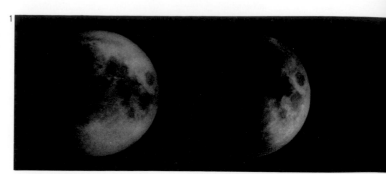

**1. Lunar eclipse**
This sequence of photographs shows the start of the total eclipse of the Moon that occurred on 24 June 1964; the Moon is shown moving into the shadow cone of the Earth. In 450 BC, Anaxagoras of Clazomenae reasoned that because the Earth's shadow on the Moon was curved, the Earth itself must therefore be spherical.

**2. Conditions of an eclipse**
The Moon's orbit is inclined to the plane of the ecliptic; an eclipse, either lunar or solar, can only occur if the points at which the Moon crosses the ecliptic lie on or near the line joining the Earth and the Sun (**A**). Otherwise the shadow cone of each body will pass above or below the surface of the other (**B**).

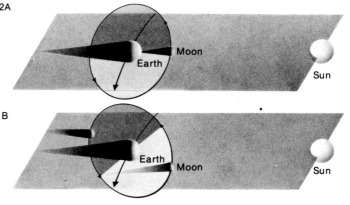

**3. Types of eclipse**
In a lunar eclipse (**A**), the Moon passes first into the "penumbra" of the Earth's shadow before reaching the darker "umbra". An eclipse of the Sun (**B**) only appears total from the limited region of the Earth's surface that is covered by the umbral shadow cone; from inside the penumbra the eclipse appears partial, and the solar corona is not seen. An annular eclipse (**C**) occurs when the Moon is near apogee, and its shadow cone does not reach the Earth's surface.

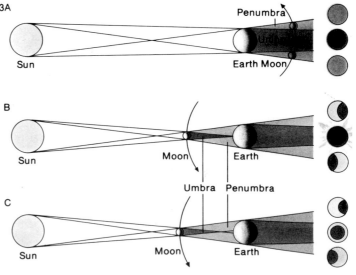

onto the lunar surface by the atmosphere surrounding the Earth. The Moon's appearance during an eclipse, therefore, depends on the conditions in the Earth's atmosphere, which can give the Moon a reddish or coppery color; on other occasions the Moon may be so dark as to be invisible to the naked eye.

An eclipse need not be "total"; if the disc of the Sun or Moon is not entirely obscured by shadow because the bodies do not line up exactly, then the eclipse is said to be "partial". Thus a partial eclipse of the Sun would be seen by an observer standing on Earth in the area covered by the Moon's "penumbra", the region of partial shadow that surrounds the main cone of shadow or "umbra" (*see* diagram 3). In the case of solar eclipses, there is a third type of eclipse, called "annular", in which the Moon is directly in line with the Sun, but its disc appears too small to produce a total eclipse, so that the Sun's rim can be seen forming a bright ring around the Moon. Annular eclipses occur because the distance of the Moon from the Earth is sometimes greater than the length of the Moon's umbra, so that the conical shadow fails to reach any point on the Earth's surface.

A solar eclipse can only occur when the Moon is new, a lunar eclipse only when it is full. However, because the Moon's orbit is inclined to the plane of the ecliptic, eclipses do not occur at every full or new Moon. In order for there to be an eclipse, the Moon must be at or near one of its nodes. If the angle between the line of the nodes and the Sun or the Moon is greater than 18° 31′ there can be no solar eclipse, while if this angle is greater than 12° 15′ there can be no total lunar eclipse. The maximum number of eclipses of both kinds that can occur in a year is seven, the minimum, two. Eclipses tend to come in twos or threes, a lunar eclipse always being preceded or followed by a solar eclipse. The duration of eclipses is variable, depending on the exact geometry of the Earth, the Sun and the Moon. For a lunar eclipse totality can last for as long as 1 hr 44 min; for a solar eclipse the maximum is 7 min 40 sec. Annular eclipses can last longer.

The position of the Moon's nodes in relation to the Sun is clearly of particular importance in determining whether or not an eclipse will occur. Because of the retrograde motion of the nodes (*see* page 146), the ascending node will come into conjunction with the Sun in slightly less time than it takes the Sun to complete one orbit: in other words, the synodic period of the ascending node is less than a year. This period, known as the "eclipse year", is equal to 346.62003 days. The Sun will be in line with one of the nodes twice during this interval.

**The Saros**

It was discovered by the ancient Babylonians that eclipses with very similar circumstances tend to recur after an interval of about 6,585 days, equal to 18 years 10 or 11 days (depending on the number of leap years). This period is called the "Saros" and can be used as a reasonable guide to predicting eclipses. The Greek philosopher Thales of Miletus is reported to have used the Saros to predict the eclipse of 28 May 585 BC.

The method works because of a mathematical coincidence. Nineteen eclipse years (6,585.78 days) happens to be almost exactly equal both to 223 synodic months (6,585.32 days) and to 239 anomalistic months (6,585.54 days). In other words, the same configuration of the Sun, the Moon and the Earth is almost exactly repeated after one Saros interval. For example, on 29 January 1953 there was a total lunar eclipse, visible from England. One Saros later, on 10 February 1971, there was another total eclipse, although it was not fully visible from England because the Moon set before the eclipse was over.

Because the three periods are not exactly equal, each eclipse in a given series differs slightly from the previous one, and eventually the cycle comes to an end. At any given moment several Saros series will be in progress, overlapping one another.

**4. Cause of the tides**
The phenomenon of the tides may be simplified by ignoring frictional forces and imagining a non-rotating Earth that is covered with water. Under such circumstances the water on the side of the Earth closest to the Moon would experience the strongest gravitational attraction, while those on the far side would be weakly attracted; thus two tidal bulges would be formed (**A**). These would follow the Moon as it orbited the Earth. Since, however, the Earth is rotating (**B**), its surface is swept past the bulges twice daily. At the same time frictional forces cause the bulges to be carried round slightly ahead of the Moon, rather than directly below it.

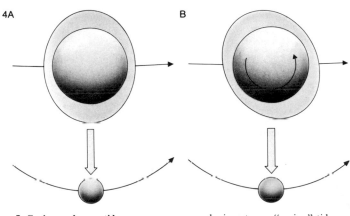

**5. Spring and neap tides**
The Sun as well as the Moon influences the Earth's tides, although to a lesser extent. As a result the height of the tides varies with the lunar cycle: at new or full Moon (**A**) the effect of the Sun reinforces that of the Moon, producing strong "spring" tides; at the first and third quarters (**B**) the Sun's attraction partially cancels that of the Moon, giving weak "neap" tides. The height of the water level at a given location fluctuates as shown (**C**) during half a lunation.

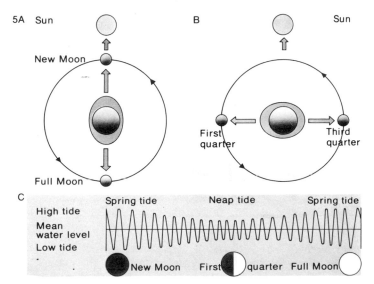

# Interior

The mean density of the lunar globe is 3.34 times that of water, a value which is much lower than those of the Earth (5.5), Mercury (5.4), Venus (5.2) or even Mars (3.9). It is therefore reasonable to assume that there is no large, heavy, iron-rich core comparable with that of Earth, although magnetics investigators argue that there must be a small core to be accounted for.

It is important to establish first of all whether the Moon is hot inside, or whether it is a cold globe throughout. Experiments designed to measure the heat outflow were among the most important of those carried out by the later Apollo missions, but unfortunately they were not entirely successful. Instruments were carried on board Apollos 15, 16 and 17, but that of Apollo 16 was broken while being set up, and attempts to carry out makeshift repairs proved to be abortive. However, the results from the other two experiments were conclusive. Apollo 15 showed that the outflow is about half that of the Earth—the temperature rose by 1 degree K at a depth of 2 m—and the experiments from Apollo 17 gave results of the same order. The "cold Moon" theory was at once abandoned, and it became clear that the interior is at a high temperature. The lunar heat flow may be accounted for by the presence of the radioactive elements potassium, uranium and thorium (*see* pages 158–9).

### Moonquakes

In the terrestrial globe earthquakes produce waves of various types, some of which can travel through molten or liquid material while others cannot; research into the behavior of such waves has provided most of the information about the Earth's interior, and it was hoped that the same would be true of the Moon. Each Apollo ALSEP, therefore, included a seismometer or "moonquake recorder". It was found that there are three distinct kinds of moonquake. The first kind are those produced by the impacts of meteorites, the second are those produced artificially, either by exploding small charges on the surface during Apollo missions or (more importantly) by crashing the IVB ascent stage of the Saturn rocket, or the Lunar Modules after the astronauts have rejoined the orbiting section of Apollo, and the third are due to internal movements in the lunar globe.

Of the meteoritic impacts, the most important so far detected occurred in July 1972; the object is thought to have weighed about 1,000 kg. Meanwhile, the Module impacts produced unexpected results. It was found that the vibrations set up did not die abruptly away; they lasted for over an hour, so that the materials below the crust did not immediately dampen the shock.

The third group, natural moonquakes, has proved to be the most valuable. Natural moonquakes may be produced in a variety of ways: for example, deep moonquakes, which originate at depths of 600 to 950 km, are thought to be of tidal origin, while shallow moonquakes may be caused by the expansion and contraction of the surface rock under the influence of solar heating. Moonquakes had been predicted long before the Apollo period, mainly because observers had detected minor glows and local obscurations on the surface—known today as TLP or Transient Lunar Phenomena—which were attributed to minor disturbances; it was found that they were commonest in certain specific areas, notably that of Aristarchus, and that they were most common near perigee, when the lunar crust was under maximum gravitational strain. Suggestions that TLP would be associated with moonquakes proved to be well founded ideas.

There were found to be less than 3,000 moonquakes per year on average, and they never exceed a value of 2 on the Richter scale, which means that by terrestrial standards they are very mild; the total yearly energy released is only about $2 \times 10^6$ J, against $10^{17}$ to $10^{18}$ J for Earth (they will present no hazards to future lunar bases). But they are evident enough, and they have led on to a picture of the Moon's interior which is probably reasonably accurate.

1A

### 1. Interior
The structure of the Moon (**A**) differs markedly from that of the Earth (**B**). Both crust and mantle are thicker on the Moon, no doubt because the internal temperatures are much lower. The upper regolith, made up of shattered bedrock, is relatively shallow; below comes the crust, extending down to 60 km. The upper crust (to 20 km) is more solid rock; the remainder is made up of materials with properties similar to those of the anorthosites and feldspar-rich gabbros of the highlands. Below the crust lies the mantle, and below this again is the asthenosphere, assumed to be a region of partial melting. Finally there is the relatively dense core.

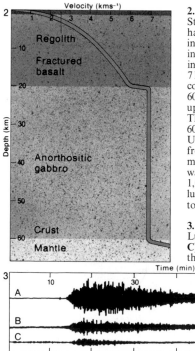

### 2. Moonquake waves
Studies of natural moonquakes have yielded considerable information about the lunar interior. At 20 km there is a sharp increase in the wave velocity to 7 km s⁻¹, which remains fairly constant down to a depth of 60 km, indicating that most of the upper cracks have been filled in. There is an abrupt change at 60 km, the bottom of the crust. Useful information also came from the impact of a 1,000 kg meteorite on 17 July 1972; the waves set up indicated that below 1,000 to 1,200 km in depth the lunar rocks are hot enough to be molten.

### 3. Seismographic traces
Lunar seismic signals (**A**, **B** and **C**) take much longer to die away than those of the Earth (**D**).

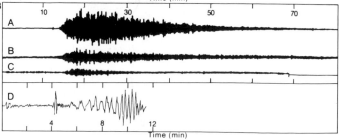

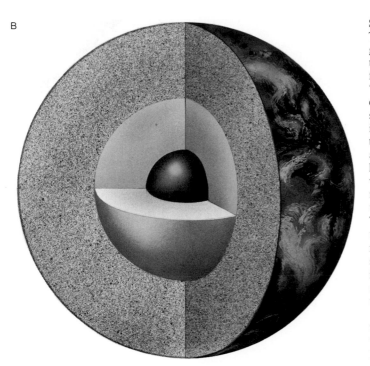

**4. Mascons, epicenters and TLP**
Mascons were discovered from
the orbital movements of probes
circling the Moon. Mascons are
associated with circular maria
such as Serenitatis and Imbrium,
and smaller formations of the
same basic type such as Grimaldi.
They indicate differences in the
thickness or composition of the
crust in these regions as compared
with their surroundings. Also
shown are the main epicenters of
moonquake regions and regions in
which red TLP have been reliably
reported. It is clear that the
distribution of these three types of
features is not entirely random.

4 🔲 Mascons

★ Deep seismic epicentres

● TLP sites

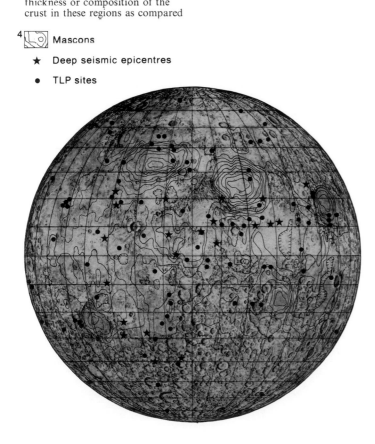

## Structure

The velocities of moonquake waves (both natural and artificial) give useful information about the Moon's layered structure. Down to 1 km in depth the compressional waves have velocities increasing from 100 m s⁻¹ to 900 m s⁻¹. This indicates a change from the "regolith", the outermost layer of the Moon, to firmer and more consolidated material. By a depth of 20 km the velocities reach 6 km s⁻¹, indicating fractured basaltic material (*see* pages 152–3), which increases in density with depth. At 20 km there is a velocity increase to 7 km s⁻¹, and this remains constant down to 60 km depth, which agrees well with the idea of anorthositic gabbro composition, which has properties similar to the surface highlands. Below 60 km the velocity goes up to 9 km s⁻¹, but the data are not yet reliable. This marks the limit of the Moon's crust. The lower new rocks, rich in the substances known as pyroxene and olivine, are relatively dense. They go down to 150 km. Beneath this is a region extending down to 1,000 km, termed the lithosphere; it is solid and rigid, with the main moonquakes occurring near its base. Below 1,000 km the shear waves are weakened; this is the asthenosphere, which may indicate the presence of a liquid or at least partially melted core with a diameter of from 1,200 to 1,800 km (as was shown from studies of the meteoritic impact of July 1972).

Finally there is the true core, the composition of which is uncertain. If it were pure iron, it could not have a diameter greater than 1,000 km, because it would have a detectable effect on the Moon's motion, and because of the high melting point of iron—1,808 K—it would be hard for such a core to form and remain molten. However, if the core contained some iron sulphide (FeS), the melting point would be only about 1,300 K and the entire core might be up to 1,400 km in diameter. It must be admitted, however, that as yet knowledge of the Moon's interior is very incomplete; the best hope of obtaining more reliable information will be if by chance a sufficiently large meteorite falls in the right place while a seismometer network is operating on the Moon.

It is interesting to note that the crustal thickness is greater on the Moon's far side than on the visible hemisphere and may go down to over 70 km. This again must be due to the fact that the rotation has been synchronous for so long a period.

Of great importance are the mascons, which appear to be regions of denser material lying not far below the Moon's surface. They were first detected by studies of the movements of the US probe Orbiter 5. If a probe passes over a region where the lunar material is unusually dense, it will speed up slightly (the effect is known as a "positive gravity anomaly"); if it passes over an area where the material is less dense, it will slow down ("negative gravity anomaly"). This was the case with Orbiter 5. Positive anomalies were found with some of the regular maria—Imbrium, Serenitatis, Crisium, Nectaris, Humorum, Humboldtianum, Orientale, Smythii, Aestuum and the walled plain Grimaldi, which is very dark floored and presumably of the same nature. One explanation is that they were produced by the transformation of lunar basalts to denser rock at the edges of the circular formations. On the other hand, craters of the Copernicus type tend to show negative gravity anomalies, and so do the large unfilled basins, most of which lie on the far side of the Moon.

Finally, what of the Moon's magnetic field? Today the overall field is negligible; this was established as long ago as 1959 by the second Luna probe. Yet there are some regions which show traces of magnetic fields here and there—for instance, at the small crater named North Ray Crater, studied by the astronauts of Apollo 16. There is also a local field near the formation Van de Graaff, on the far side, which is a depression 4 km deep and is slightly richer in radioactive materials than its surroundings. The evidence from paleomagnetic studies of rocks indicates that there used to be an appreciable overall magnetic field around 3,000 million years ago, which has now disappeared.

# Regolith

In contrast with most of the planets, whose colors are striking, the Moon's appearance gives the impression of an almost complete absence of local color. Mars has its red deserts and its darker areas, Saturn is yellow, while Uranus is decidedly green, and the overall yellow hue of Jupiter is modified by features such as the Great Red Spot; but on the Moon there is nothing to relieve the greyness of the plains except the blackness of the shadows and brighter areas of the lunar highlands.

The descriptions given by the Apollo astronauts fully confirm this impression. Charles Conrad, commander of Apollo 12—which came down in the Mare Nubium—commented that "the Moon is just sort of very light concrete color. In fact, if I wanted to look at something I thought was the same color as the Moon, I'd go out and look at my driveway."

The surface of the Moon is covered by what is termed the "regolith", which may be defined as a debris blanket made up of loose material of rock fragments and "soil", overlying solid bedrock. Conventionally, soils are defined as particles less than 1 cm in diameter, while larger particles are termed rocks. The thickness of the regolith varies considerably. Direct measurements have been made only from the sites of the successful Apollo missions (11, 12 and 14 to 17 inclusive)—a mere half-dozen in all—but they were obtained from diverse areas and provide reliable clues. It seems that over the maria the average depth of the regolith is from 4 to 5 m, while over the highlands it may go down to 10 m and even deeper in places—perhaps as much as 30 m, although this may be rather exceptional. On the other hand, the Apollo 16 astronauts found that at North Ray Crater the regolith went down to a depth of only a few centimeters.

Because the regolith is loose, it retains impressions. Pictures of the prints left by the astronauts have been widely published, and the prints left by the footpads of space vehicles that have landed provide similar evidence of the fine, powdery quality of the lunar surface. The prints themselves will remain visible for a very long time before they are eventually obliterated. Yet despite the lack of atmosphere, which means that there is no wind and indeed no "weather" on the Moon, the regolith does not stay undisturbed indefinitely. In particular, there is a constant bombardment from space. Meteorites can reach the lunar surface without being destroyed (as happens over the Earth—over 2,000 meteorites have now been located, but only a few have been observed actually during their descent) and so can the much smaller and more numerous micrometeorites. In fact the regolith is churned up, although the effects are very slow and would be quite undetectable over a period of many human lifetimes. The regolith has itself been produced by this kind of external action, and there has been plenty of time for this to happen, since the lunar surface solidified so long ago. The age of the highland regolith is of the order of 4,000 million years.

## Composition

A few terms must be introduced here. "Magma" is molten subsurface rock, which becomes "igneous rock" when it solidifies and "lava" when it comes out onto the surface. "Basalts" are fine-grained, dark, igneous rocks. They are composed chiefly of plagioclase feldspar and pyroxene with or without olivine. "Anorthosite" is an igneous rock made up chiefly of plagioclase; some anorthosites may indeed be almost pure plagioclase. They are relatively rare on Earth and are found mainly in Precambrian rocks such as those of Quebec and Labrador, but they are common on the Moon, although not identical in composition. "Pyroxene" is a calcium–magnesium–iron silicate and is the most common mineral in lunar mare rocks, making up about half of most specimens; it forms yellowish-brown crystals. "Olivine", a magnesium–iron silicate, is also found and is characterized by pale green crystals. "Gabbro" is a coarse-grained igneous rock, dark in color; it may be considered to be a coarse-grained equivalent of

**2. Lunar Rock Exteriors**
This montage shows a selection of lunar rock exteriors. (**A**) The fine-grained crystalline texture is typical of a mare basalt. Both dark and light minerals are distinguishable. (**B**) Mare basalts can also have this highly "vesicular" appearance—the vesicles were formed by the release of gas as the rock cooled. The nature of the gas is, however, unknown. (**C**) Highland basalts are almost entirely plagioclase and therefore are almost completely white in color. The dark patch on this sample is splashed impact-produced glass. Zap pits produced by micrometeorite impacts are also visible on the lower central area of the rock (*see* 5A). (**D**) A typical "polymict" breccia, consisting of larger fragments set in a fine-grained matrix; this is

not a primary rock, its constituents were formerly components of other materials.

**3. Lunar soils**
In addition to single mineral grains, which are in fact comparatively infrequent, igneous rock fragments, breccias and glass spherules (**A**) abound in lunar soils. These objects, which can be either round or dumbell shaped, are mainly the result of impact-produced glass that solidified in flight. When the glass has landed back on the surface before solidification, "glassy agglutinates" are formed (**B**). These particles, which can account for up to 60 percent of a lunar soil, consist of the whole spectrum of soil components held together by a fragile web or matrix of splashed glass.

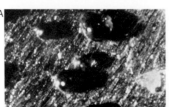

### 1. Lunar surface
The fine texture of the regolith can be seen clearly in the impression made by the footpad of Surveyor III. The vehicle bounced slightly on touchdown, leaving two footprints before it finally came to rest.

### 4. Lunar rock interiors
The minerology of rocks can best be studied after samples have been cut into thin slices and polished. A typical mare basalt (**A**) is revealed as being mainly pyroxene (grey), plagioclase (white), and ilmenite (black) by microscopic study. (**B**) When viewed in polarized light plagioclase appears black and white striped. This rock, which is called "Genesis Rock", is an anorthosite made up almost entirely of plagioclase, with small amounts of pyroxene. (**C**) Microscopic study also reveals the different textures of the rocks. Compare the gabbro shown here (pyroxene, dark) with the basalt in (**4A**). The crystals' shapes are clearly different and thus are said to have different "habits". (**D**) This photomicrograph exhibits a breccia texture— larger fragments are imbedded in a finer-grained matrix. The rock is not a typical breccia, because only one mineral (olivine) is present, but is called a "dunite". A sample from this specimen yielded a radiometric age of 4,600 million years, nearly the age of the Moon.

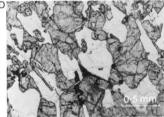

### 5. Lunar rocks and soil grains
The scanning electron microscope reveals a crater on a glassy spherule like one of those in (**3A**), showing that micrometeorites extend the range of impacts on the Moon over a scale of many orders of magnitude. (**B**) In this transmission electron micrograph, electron diffraction demonstrates that the interior of a particle remains crystalline while the exterior has been rendered amorphous by intense solar wind bombardment. (**C**) Cosmic ray particles also leave evidence or "tracks" of their presence in the lunar soil; these pits were produced by etchng material that has been damaged by radiation. (**D**) Solar wind bombardment also causes microscopic chemical changes—these dark spherules are iron formed in this way.

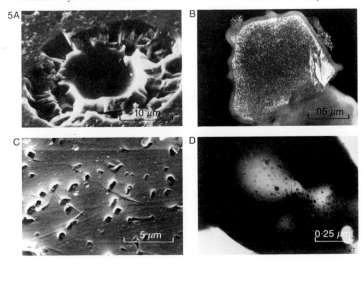

basalt. Lunar basalts and gabbros may also contain ilmenite, an opaque iron–titanium oxide. "Breccias" are fragments made up of shattered, crushed and sometimes melted pieces of rocks; these pieces may be igneous rocks of various types, glasses, or bits of other breccias. It is worth noting that breccias have been collected from every landing site on the Moon. There are, for instance, widespread layers of breccias at the sites of Apollo 14 (Fra Mauro) and Apollo 16 (the so-called Cayley Formation, in the region of Descartes). One mineral first found on the Moon has been named "Armalcolite" (after the three astronauts of Apollo 11— *Arm*strong, *Al*drin and *Col*lins): it is an opaque oxide of iron, titanium and magnesium. It is reasonably common, particularly in the rocks returned from the Taurus–Littrow area, and has been recognized in terrestrial rocks since it was identified in the Apollo 11 samples.

No entirely new substances have been found on the Moon. All lunar material is made up of the elements familiar on Earth, though there are modifications in composition due to the fact that the materials have developed under conditions which have been completely different.

The regolith is not the same over the maria as over the highlands. On the maria the rock fragments are chiefly basalt, with pyroxene, plagioclase and other minerals. On the highlands the fragments are chiefly plagioclase-rich rocks and broken plagioclase crystals. As yet there is no direct information about the regolith on the Moon's far side (at least, there are no samples of it), but it is not likely to be substantially different from that on the Earth-turned hemisphere.

Glasses are very common in the regolith. Much of the soil is composed of tiny rock fragments held in a matrix of glass—such particles are termed agglutinates. Occasionally discrete glass beads of various colors are found, such as, for example, emerald-green glasses found in the Apollo 15 soils which are rich in calcium. These beads are formed when glass which has been melted—usually as the result of impacts—re-solidifies in flight. The resulting particles can be either spherical or dumbell shaped. One startling discovery was that of the "orange soil" at Shorty Crater in the Apollo 17 site. It was found to be almost entirely tiny colored glass particles, from 0.1 to 0.2 mm in diameter, and to be no younger than the surrounding material; the age has been given as 3,800 million years. One ingenious attempt to explain the abundance and uniformity of composition is that the orange glass was formed by an impact into a lava lake, and was subsequently excavated by Shorty Crater. Samples of soil brought back as cores up to 2.5 m in length show distinct layering, and demonstate how the regolith has developed as a result of a sequence of impacts.

Many samples from the regolith have been brought back for analysis, and all have interesting features of their own. For example, there is the so-called "Genesis rock" from Apollo 15, which is an anorthosite 4,000 million years old, but still not the oldest lunar rock; and from Apollo 12 comes Sample 12013, which is lemon sized and made up of a dark grey breccia, a light grey breccia and a vein of solidified lava. Its dark portions contain much more than the usual quantity of potassium, as well as having an increased proportion of uranium and thorium, so that it is the most radioactive rock known from the Moon.

There is a certain amount of meteorite material (about 1 to 2 percent) in the regolith, although certainly not nearly as much as some authorities had expected. Finally, it had been suggested that tektites might come from the Moon—tektites are aerodynamically shaped objects found in restricted fields on the Earth, notably in Australia, and their origin is a completely mystery. They are usually button sized and are unlike any other objects found. However, their chemical composition is so unlike that of known lunar materials that whatever they may be, they did not come from the Moon, either shot out from lunar volcanoes or hurled earthward by meteoritic impacts.

# Lunar Chronology

### Origin of the Moon

There have been many different theories proposed to explain the origin of the Moon. One important fact is that for a satellite, the Moon is exceptionally large and massive in relation to its primary; the Earth is a mere 81 times as massive as the Moon, and there are sound reasons for regarding the Earth-Moon system as a double planet rather than as a planet and a satellite.

Broadly speaking, the theories fall into three classes: fission, capture or binary accretion (*see* diagram 1).

The fission hypothesis was proposed by G. H. Darwin (son of Charles Darwin) in 1878. On this picture, the Earth and the Moon were originally one body which was fluid and in quick rotation. The speed of rotation made the body first pear-shaped and then dumbbell-shaped; eventually the "neck" of the dumbbell broke, and the Moon moved away as an independent body. The theory sounded attractive, but it was found that there were serious mathematical objections to it; for instance, the rate of spin would have had to have been improbably great. The theory has now been generally abandoned. Neither has a modification, involving the breaking-away of Mars as well as the Moon, been better received.

A more popular idea is that the Moon was formerly an independent planet, which came close to the Earth and was moving in such a manner that it was unable to break free again. This is certainly not out of the question though it does involve some very

special assumptions which have led to some authorities to regard it with a considerable amount of reserve.

Thirdly, the Earth and the Moon may have been formed in the same way, in the same part of the Solar System and at about the same time, so that they have always remained gravitationally linked. A modification of this general picture involves the production of a ring of material around the newly-forming Earth, and that this material collected together to form the Moon.

As yet it is impossible to decide between these two rival theories, but it is at least significant that the Earth and the Moon are made up of the same kinds of materials.

### Surface evolution: The mare basins

It had been tacitly assumed that the Apollo results would clear up all the outstanding problems presented by the Moon, but this is far from being the case. It has even been said that every lunar expert sees in the Apollo results, a confirmation of his own particular theories. The long-standing controversy concerning the origin of the walled formations, in particular, has still not been settled. Many authorities consider that the mare basins and the craters are essentially of impact origin, produced by the falls of meteorites; others maintain that the craters are endogenic (produced by internal processes). Of course, there can be no doubt that both processes have operated, just as they have done on the Earth and Mars. The argument centers on which of these processes was the dominant factor. It is probably true to say that the impact hypothesis is much the more popular in the United States, while European authorities are more divided.

In any case, there can be no doubt that the first major features to be produced on the cooling crust of the Moon were the deep basins which we still see as the maria—Imbrium, Serenitatis, Tranquillitatis, Orientale and the rest. In all probability some of the earliest basins have been obliterated by later activity. The maria are not all of the same age. It seems, for instance, that some, such as Tranquillitatis and Fecunditatis, date back about 4,500 million years; Serenitatis perhaps 4,000 million years, and Orientale about 3,850 million years. The generally accepted order of basin formation is now taken to be: oldest—Tranquillitatis, Fecunditatis, Nubium; intermediate—Serenitatis, Nectaris and Humorum; and youngest—Crisium, Imbrium and Orientale.

When the main basins had been formed, there followed a period of tremendous volcanic activity. Magma flowed out from below the crust, and the basins were filled, so that they were indeed seas—of lava, not of water (and by now there is not the slightest chance that the maria were ever aqueous; all the lunar rocks, maria or highland, are devoid of hydrated materials). Radioactive heating may have played a major role, and it is also significant that the mare lavas are not all identical. For instance, basalts rich in aluminium are somewhat older than those which are relatively rich in titanium.

For almost 700 million years magma poured out from the Moon's interior. Many of the basins were joined by lava flows; thus the boundary between the Mare Serenitatis and the Mare Tranquillitatis has been breached, while the Mare Imbrium, the most prominent of all the regular maria, is connected with the Oceanus Procellarum and the Mare Nubium. There is also a break in the boundary between the Apennine and the Caucasus ranges leading into the Mare Serenitatis. However, most of the basins on the Moon's far side remained unfilled.

The lava-flows ended about 3,200 million years ago, probably rather abruptly. The Moon was starting to take on its present appearance, and even at that remote period the outlines of the great plains such as the Mare Imbrium would have been recognizable. The key to this understanding has been the measurement of the relative abundance of certain long-lived radioactive isotopes. In fact, surprisingly, we know rather more about the appearance of the Moon at that time than we do about that of the Earth.

### 1. Origin of the Moon

According to the "fission theory" (**A**) the Moon once formed part of the Earth, and broke away as a result of the Earth's rapid rotation. The "precipitation hypothesis" (**B**) suggests that the Moon was built up by accretion of particles spun off the Earth soon after the Earth itself was

formed. Another possibility is that the Moon was formed together with the Earth and in the same manner. Finally, the "capture hypothesis" (**C**) proposes that the Moon began as an independent body, formed elsewhere in the Solar System, and that it was captured by the gravitational field of the Earth during a close approach.

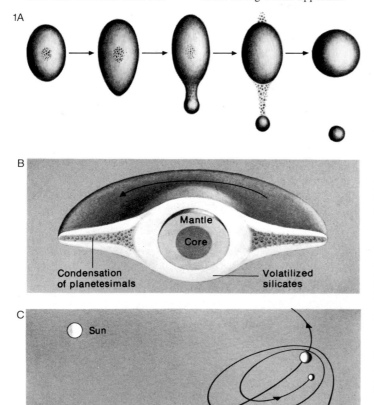

1A

B

Mantle

Core

Condensation of planetesimals

Volatilized silicates

C

Sun

Earth

Moon

**2. 4,600 million years**
Pre-Imbrian period: formation of the Moon; melting and separation of the crust by the process of "magnetic differentiation", which produced the lunar highlands (*see* pages 158–9). As the molten material cools, it begins to crystallize, and the lower-density crystals tend to rise to the surface. Some material as old as 4,600 million years has been found, but most of the original crust has been destroyed by cratering.

**7. Reconstruction of the Moon**
This sequence shows the Moon as it is believed to have appeared (**A**) in the middle of the Imbrian period after the formation of the last of the mare basins, before they had been flooded; (**B**) at the end of the Imbrian period, soon after the formation of the mare material approximately 3,300 million years ago; and (**C**) as it appears at the present time (according to telescopic photographs made in 1966).

**3. 4,500 to 3,850 million years**
Formation of the great mare basins (*see* pages 156–7) during a period of cataclysmic activity and bombardment. The Tranquillitatis basin may be as old as 4,500 million years; Orientale, only 3,850 million years.

**4. 3,900 to 3,200 million years**
Imbrian period: the mare basins were flooded with lava, which poured out of the interior of the Moon in successive stages. These formed layers of mare material and left the maria looking much as they do today.

**5. 3,200 to 1,000 million years**
Eratosthenian period: following the rather abrupt end of the flooding period, crater formation was restricted to the effects of minor volcanism and impacting meteorites. Copernicus may be about 1,000 million years old.

**6. 1,000 million years to the present day**
Copernican period: the Moon's recent history is characterized by almost complete quiescence, both volcanic and meteoritic, although a few of the ray-craters may date back less than 1,000 million years.

# Maria

The grey plains or "maria" occupy about 15 percent of the Moon's surface. They are the most obvious of all the lunar features and are easily visible with the naked eye. The maria are of two distinct types, regular and irregular; the regular maria being basically circular depressions with mountainous borders that are frequently incomplete. The maria on the Earth-turned face of the Moon lie between 2 and 5 km below the mean radius of the globe.

The most prominent of all is the Mare Imbrium, whose formation had profound effects over much of the Moon's surface. It is surrounded by mountains, although in places the border has been obliterated, and there are indications of an inner ring whose southern boundary is traced by the craters Archimedes, Timocharis, Lambert and C. Herschel. Studies of the overall arrangement are most significant. It seems that lunar "mountain ranges" are not of the same type as those found on Earth, but are in fact the "walls" of the regular maria or the uplift caused by cratering; the Mare Imbrium, for example, is surrounded by the Apennines, Alps and Caucasus, while the adjoining Mare Serenitatis is bounded in part by the Caucasus and Haemus ranges.

Most of the main maria are connected, forming one huge system, although the separate basins are of different ages and have different depths. The chief exceptions are the Mare Crisium, in the northeast quadrant of the visible hemisphere, and the Mare Orientale, which lies very close to the western limb and actually extends onto the far side; other separate seas are the Mare Humboldtianum and the Mare Smythii.

Other maria are less regular in outline and are certainly older than the Mare Imbrium. Examples include the Mare Nubium, Mare Tranquillitatis, Mare Fecunditatis and Oceanus Procellarum, which are independent basins but have been less well preserved. The genuinely irregular maria are quite distinct, and some appear to be mere coatings of dark material. The best example is the Mare Frigoris, which has no definite shape at all, while the Mare Australe, in the south-east quadrant close to the limb, and the Mare Undarum, close to the Mare Crisium, do not appear to be true basins.

It is difficult to draw a distinction between a "mare" and a large "walled plain". Grimaldi, for example, is a vast low-walled enclosure, more or less circular, with a diameter of almost 200 km. It has a dark floor of marial type, and if it had been placed closer to the center of the disc, there is every possibility that it would have been classed as a "mare" rather than a "crater"; its diameter is not much less than half that of the Mare Crisium. Thus Grimaldi and the Mare Crisium may well be of the same fundamental type, a conclusion reinforced by the fact that both are associated with mascons (*see* pages 150–51); there are other walled plains with mare-type interiors, such as Plato, for instance, with a diameter of nearly 100 km. Many authorities doubt whether there is any essential distinction between a large, dark-floored walled plain and a regular mare, except in size. It is also worth noting that Grimaldi is the darkest patch anywhere on the visible side of the Moon, and the interior of Plato is comparable.

Then there are the "lakes" and "marshes", such as the curiously coloured Palus Somnii, which adjoins the Mare Tranquillitatis and is bounded by two of the rays issuing from the brilliant crater Proclus. Most spectacular of all is the Sinus Iridum or "Bay of Rainbows", which leads out of the Mare Imbrium. It is bordered on its limbward side by moderately high mountains, and under suitable conditions of illumination it seems to project beyond the terminator; it has been nicknamed the "Jewelled Handle". There are traces of an old border between it and the Mare Imbrium, and since it is circular in form (although foreshortened into an ellipse as seen from Earth) it is presumably a separate basin.

Some of the maria are surrounded by external mountain arcs. One such case is the Mare Nectaris. The feature once known as the Altai Range, and now more appropriately known as the Altai

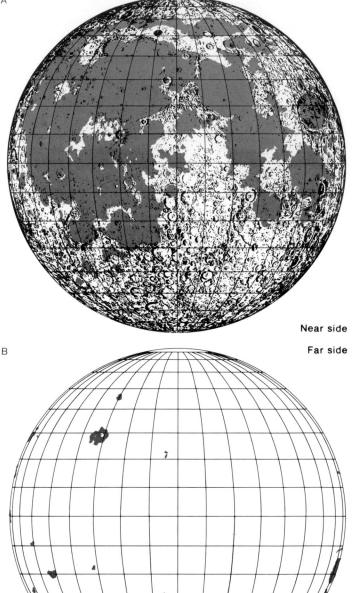

Near side

Far side

**1. Distribution of lunar maria**
A large proportion of the Moon's near side (**A**) is occupied by maria, while the far side (**B**) is notably lacking. The reason for this difference is thought to be associated with the Earth's gravitational effect on the Moon, but the mechanism involved is not yet understood. The diagram shows how many of the near-side maria flow into one another, forming the familiar pattern clearly visible from Earth. The regularly shaped maria are surrounded by mountainous walls.

**Chemical composition of basalt samples (weight%)**

|  | Earth Ocean-floor | Moon Maria | |
| --- | --- | --- | --- |
| $SiO_2$ | 49.2 | 37.0 | – 49.0 |
| $TiO_2$ | 1.4 | 0.3 | – 13.0 |
| $Al_2O_3$ | 15.8 | 7.0 | – 14.0 |
| $FeO$ | 9.4 | 18.0 | – 23.0 |
| $MnO$ | 0.16 | 0.21 | – 0.29 |
| $MgO$ | 8.5 | 6.0 | – 17.0 |
| $CaO$ | 11.1 | 8.0 | – 12.0 |
| $Na_2O$ | 2.7 | 0.1 | – 0.5 |
| $K_2O$ | 0.26 | 0.02 | – 0.3 |
| $P_2O_5$ | 0.15 | 0.03 | – 0.18 |
| $Cr_2O_3$ |  | 0.12 | – 0.70 |

156

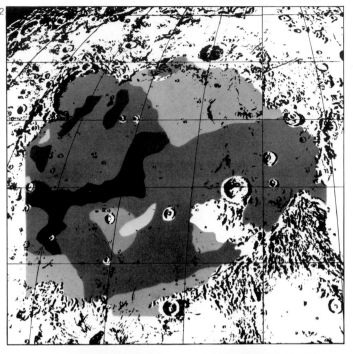

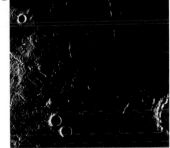

**2. Imbrium lava flows**
Differences of color and crater density in the Mare Imbrium have revealed a difference in ages of the lavas. In the diagram the darkest shading represents the most recent lava flows.

**3. Mare Imbrium**
Under different lighting conditions, the boundaries between geological units appear to be different, as in these contrasting pictures taken when the angle of the sun was 17° (**A**) and when it was 2° (**B**).

**4. Typical mare features**
The maria (**2**) differ from the highland regions (**1**) in being smoother and generally darker in color. They are marked by a variety of features, such as sinuous rills (**3**), wrinkled ridges (**4**) and dome-like features with summit craters (**5**). Sinuous rills are interpreted as lava channels, often originating from craters in relatively high regions of maria and flowing into lower lying areas. Wrinkle ridges are believed to result from folding in the surface layers of the maria.

Scarp, is concentric with the mare border and was certainly produced by the same event. The Mare Orientale, meanwhile, has several rings, some of which are visible from Earth and have been named. The circular outer scarp—the Cordillera Mountains—is over 900 km in diameter and is extremely massive; within this ring lie the Rook Mountains, making up another circular scarp almost 650 km in diameter.

On the Moon's far side the whole situation is different. Apart from a section of Orientale, there are two features that are known as maria—Moscoviense and Ingenii—but neither really qualifies. Mare Moscoviense is dark floored and was recognizable on the very first pictures of the far side sent back by Luna 3 in 1959, but it is smaller than some of the other far-side basins such as Apollo and Hertzsprung, while the so-called Mare Ingenii is merely an irregular darkish area near the interesting walled formation Van de Graaff.

Crater frequency on the maria is, of course, noticeably less than in the highlands, and it is reasonable to assume that most of the marial craters are fairly young by lunar standards. Copernicus is a striking example. Not far to the east is the famous "ghost", Stadius, which is 70 km in diameter, with walls that are barely traceable; it is pitted with small craterlets. Ghost-craters are found on most of the maria, and in many cases the so-called "wrinkle-ridges" are in fact parts of the "walls" of such features, though other wrinkle-ridges are probably compressional features. There are prominent wrinkle-ridges on the Mare Serenitatis, but very few craters; the largest of them, Bessel, is only 19 km in diameter.

There are craters belonging to the basin-forming era whose "seaward" walls have been destroyed by lava-flows, so that they have become bays. The Sinus Iridum comes into this category, and another excellent example is Fracastorius, in the south of the Mare Nectaris, where the old wall can just be traced, but is of negligible height. Doppelmayer and Hippalus, leading off the Mare Humorum, have been similarly damaged, though each has the remains of a central peak, and Hippalus is associated with a major system of rills.

**Composition**
So far, specimens have been obtained from only a few lunar sites, but it has been found that these specimens are not identical—there are local variations, which is by no means unexpected. The rocks found in the maria are of basaltic origin—as are all the lunar rocks. They differ from the highland basalts in that they may be obtained as distinct samples rather than merely as components of complicated breccias. Their nearest terrestrial counterparts are the lavas that are continuously erupting from the spreading ocean ridges, but there are very marked chemical differences (*see* Table). The lunar mare basalts contain much less silica ($SiO_2$) and aluminium but far greater amounts of iron, magnesium and titanium. The local differences between rocks are reflected in the relative proportions of these elements. Apollos 11 and 17 sampled areas where titanium content was very high—near Hadley Rill the rocks were rich in magnesium and silica. Volatile elements such as potassium and sodium are also depleted on the Moon. Likewise neither mare nor highland rocks contain water or even traces of hydrated minerals. They must have been formed under extremely dry conditions. Any grandiose idea of extracting water for the benefit of future colonists has had to be abandoned. Moreover, since aqueous geologic processes are frequently involved in the concentration on Earth of economic deposits of important metals such as copper, there now seems little likelihood of the Moon becoming a future source of raw materials.

One of the consequences of the unusual chemistry of lunar mare rocks is a very low viscosity, hence the scale of lunar lava flows. Rocks of this fluidity were not known before the lunar missions and this ignorance contributed to the difficulty of accepting that the maria were not some kind of sedimentary formation.

# Highlands

The highlands or "terrae" are the most ancient parts of the lunar surface. They were the first to solidify, and were not penetrated by the tremendous volcanic flows that flooded the mare basins. The highlands were produced by the process known as "magmatic differentiation". Materials in an igneous melt separate chiefly according to their densities: as the lighter materials crystallize they float to the top, forming lower-density regions. Measurements carried out largely by the orbiting sections and the sub-satellites of the Apollo vehicles have confirmed that the density in the highlands is indeed less: it averages 2.75 to 3 times that of water, as against a value of 3.3 to 3.4 for the maria. The highland crust appears to be in a state of "isostatic equilibrium" (that is to say, in elevated regions the material tends to be less dense than in depressions, so that the total weight pressing down is much the same everywhere).

But although the highland surface is ancient even by lunar standards, it has not remained completely undisturbed since it became solid. The cataclysmic activity and bombardment before 3,900 years ago has severely brecciated almost all the highland rocks so that remnants of the primitive crust may only be found disseminated in such samples. A veil has been drawn over the early history of the highlands of the Moon; it will only be disclosed by painstaking effort. There are also indications of tectonic activity: for instance there are many faults, or pairs of faults close together, bordering grabens. Rills are less well-marked than in the marial regions, and most of the large so-called valleys, such as those of Rheita and Reichenbach, turn out to be chains of confluent craters; there are also small "strings of beads", such as the 225 km chain near the crater Abulfeda, which can only be volcanic. On a smaller scale, too, there are visible tracks of boulders that have rolled down slopes and left their impressions, probably because they have been jolted by sudden crustal movements. There was one excellent example of this in the area of North Massif, at the Apollo 17 landing site, and there is another inside the crater Vitello on the edge of the Mare Humorum, so that such events are not confined to the terrae.

Before the Apollo and Luna missions the highlands were less well explored than the maria, because the features are so crowded; there are craters in profusion, and the amount of detail to be seen is daunting to even the most enthusiastic telescopic observer. It was evident that landing in a rough region presented more problems than coming down in one of the smoother plains, and only one of the manned missions was aimed at the highlands. This was Apollo 16, which touched down in the so-called Cayley Formation in the region of Descartes. It was indeed unfortunate that the heat-flow experiment failed and that the data in this field of research remain incomplete. However, it has been established that magnetic fields found locally in the highlands are stronger and less uniform than in the maria.

## Composition

The highland rocks, although still basalts, are not identical in composition with those of the maria, and are certainly very different from the terrae or continental rocks on Earth, of which granite is an example (*see* Table).

On the basis of tiny chips found in the Apollo 11 lunar soils obtained from the Mare Tranquillitatis, it was postulated that the lunar highlands would be anorthosite, a material made up chiefly of plagioclase and far richer in aluminium and calcium than the mare basalts. This prediction was almost correct. The major rock type found when Apollo 16 visited the Cayley highland formation was anorthositic gabbro, which contains more pyroxene than anorthosite and possibly was formed under conditions of slower cooling leading to larger and more evenly shaped crystals. All the mare regions of the Moon contain a proportion of highland materials which have been thrown there by major impacts. Consequently a second highland rock type known as "KREEP", rich in potassium

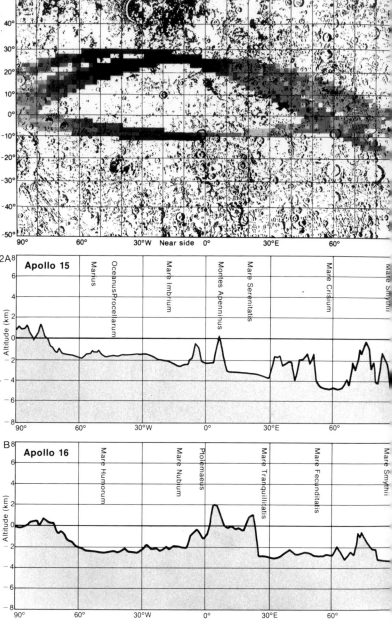

**1. Radioactive regions**
The orbiting command modules of Apollo 15 and 16 carried γ-ray spectrometers which measured the level of natural radioactivity in the lunar soil. The results indicate the concentration of thorium, uranium and potassium, since it is the decay of these elements that is the main source of the radioactivity. In this geochemical map the darkest shading represents the highest level of radioactivity, revealing the distribution of KREEP—which is rich in the elements concerned—on the lunar surface.

**2. Altitude profiles**
These graphs show the altitudes of features relative to a sphere of radius 1.738 km.

**3. KREEP sample**
KREEP is one of the two types of rock characteristic of the lunar highlands. This sample was brought back by Apollo 15.

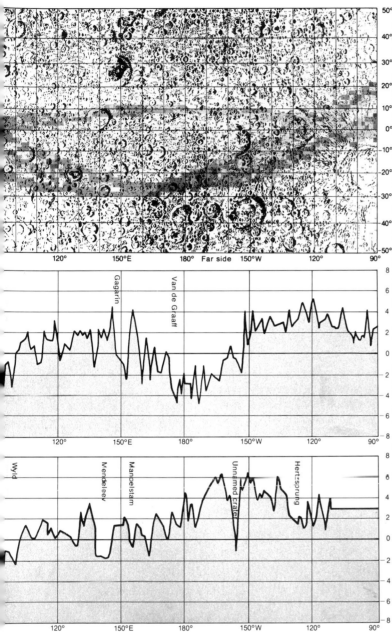

120°  150°E  180° Far side 150°W  120°  90°

120°  150°E  180°  150°W  120°  90°

120°  150°E  180°  150°W  120°  90°

**4. Highland terrain**
The densely cratered region to the north-east of Tsiolkovsky on the far side of the Moon provides a good example of the appearance of the lunar terrae. The materials in this kind of region are thought to be the oldest on the Moon, but cratering has altered the rocks from their original forms.

⁵⁰°(K), rare earth elements (REE) and phosphorus (P), was soon recognized among the Apollo 12 rocks and soils. KREEP is concentrated among other places at Fra Mauro, the Apollo 14 site, and may represent material excavated from Mare Imbrium. The importance of KREEP is that it must have been formed by partial melting of the primitive crust—partial melting being a process that concentrates elements such as potassium with large atomic diameters. KREEP is also rich in radioactive elements uranium and thorium, therefore its distribution on the lunar surface may easily be mapped by orbiting geochemical instruments (*see* diagram 1).

**Distribution**

Altogether the highlands cover 85 percent of the total surface of the Moon, but they are not evenly distributed. The roughest part of the Earth-turned hemisphere is the south-east quadrant, where there are few walled formations with dark interiors (Stöfler, 145 km in diameter, is an exception). In the south-west highlands there are large, light-floored enclosures such as Bailly, almost 300 km across—more than half the diameter of the Mare Crisium—which has been aptly described as "a field of ruins". There are also highland areas around both the poles. The north polar crater has been appropriately named Peary, in honor of Robert Edwin Peary, the famous Arctic explorer.

The far side of the Moon is made up almost exclusively of terrae, with an average elevation of 5 km above the mean radius of the Moon's globe. Instead of maria, there are huge, circular depressions such as Apollo, Hertzsprung and Korolev, which are of marial size in many cases but are not filled with basalt. The regions furthest away from the regular ringed maria (Imbrium and Orientale in particular) have presumably been last disturbed by the events that produced the near-side basins and showered ejecta in all directions. Unfortunately there is as yet no chance of obtaining samples from them, since landing on the Moon's far side would be too hazardous to attempt in the present stage of technological development, although doubtless it will be done eventually. The far side of the Moon would be an ideal site for a radio astronomy observatory, since it would be shielded from the interference of all Earth transmissions.

The general aspect of the far-side uplands differs markedly from that of the terrae on the familiar hemisphere, as has been stressed by all the astronauts who have had direct views of it (the crews of Apollo 8, and 10 to 17 inclusive). It is not only the absence of maria that distinguishes it; there are differences in tone and in crater arrangement, and apparently the grid system is very much less marked. Of the crater-valleys, the most striking is that associated with the giant walled plain Schrödinger, at latitude 75°S.

One characteristic of all the uplands, on both the near and the far sides of the Moon, is the absence of mountain ranges of terrestrial type. On the Moon there is a clear distinction between highlands and mountains. The main lunar ranges form the boundaries of the circular maria and were produced in association with these phenomena. The highlands are the remnants of the lunar crust.

**Chemical composition of rock samples (weight%)**

| | Lunar Highland Rocks | | Average Earth Continental Granite |
| | Anorthositic gabbro | KREEP basalt | |
|---|---|---|---|
| SiO₂ | 44.5 | 48.0 | 72.3 |
| TiO₂ | 0.39 | 2.1 | 0.3 |
| Al₂O₃ | 26.0 | 17.6 | 14.0 |
| FeO | 5.77 | 10.9 | 2.4 |
| MnO | 0.07 | 0.14 | 0.05 |
| MgO | 8.05 | 8.70 | 0.5 |
| CaO | 14.9 | 10.70 | 1.4 |
| Na₂O | 0.25 | 0.70 | 3.1 |
| K₂O | 0.08 | 0.54 | 5.1 |
| Cr₂O₃ | 0.06 | 0.18 | – |

# Craters

The entire surface of the Moon is dominated by walled circular enclosures known as "craters", although in many cases "walled plain" would be a better term. Some of them are truly immense. For example Bailly, close to the Moon's southwestern limb as seen from the Earth (and therefore very foreshortened) is almost 300 km in diameter. Craters more than 100 km across are common, and many of them have either central peaks or groups of central elevations. The basic form is circular, but they frequently break into each other, and in many cases there are formations distorted almost beyond recognition. The distribution is not random; when one formation intrudes into another, it is almost always the smaller crater which breaks into the larger. A typical case is that of Thebit, 60 km in diameter, which is broken by a smaller crater (Thebit A) which is in turn interrupted by a still smaller crater (Thebit F). Lines of vast enclosures also occur, such as that running down the Moon's eastern limb as seen from Earth (Furnerius, Vendelinus, Petavius, Langrenus and through to Mare Crisium and Cleomedes). These lines appear to be related to the central meridian of the near hemisphere, which can hardly be coincidence; it has been attributed to the effects of the Earth's pull, and certainly there is every reason to assume that the lunar rotation has been synchronous since very early in the Moon's evolution.

On the maria there are numerous "ghost craters", such as Stadius, with ramparts rising to only a few meters or tens of meters above the outer surface. Some craters, notably Tycho and Copernicus, are the focal points of systems of bright rays, best seen under conditions of high solar illumination and which are at their most conspicuous near the time of full Moon. Plateaux are less common, but there is one splendid example, Wargentin, 89 km in diameter, adjoining the huge walled plain Schickard in the south-west part of the Moon. Wargentin is filled with lava almost to the rim.

Lunar craters have sunken floors, with rims rising to only modest elevations above the outer surface, so that in shape they are much more like shallow saucers than steep-sided mine-shafts. For instance, the walls of Bailly rise to only 3.6 km above the deepest part of the floor, a ratio of 82:1. Many other characteristics are shown by the craters. Some, such as Grimaldi and Plato, have dark, mare-type floors; there are very brilliant craters, such as Aristarchus, which is often visible when illuminated only by Earthshine and has sometimes been mistaken for a volcano in the process of erupting.

It is also worth noting that some of the central mountains of craters have symmetrical summit pits; Alpetragius is an excellent example of this. Other craters, such as Vitello and Taruntius, are "concentric", with complete inner rings. It is rare to find a crater with both an inner peak ring and a central elevation, though a few examples are known, notably Antoniadi (140 km in diameter) and Compton (175 km), both on the Moon's far side.

Isolated peaks are common—Pico and Piton, on the Mare Imbrium, are excellent examples—and there are high chains of mountains, although in general these chains form the borders of the regular maria (for instance, the Apennines and the Alps make up part of the border of the huge Mare Imbrium). Some of these peaks are extremely lofty and may exceed the heights of terrestrial ranges such as the Rockies; since the Moon is a much smaller world than the Earth, the relative heights of the lunar peaks are much greater. Originally their heights were measured by their shadows, but today more refined methods are available. Since there is no sea-level on the Moon, the altitudes have to be related to the mean radius of the globe.

Valleys are of various kinds. The celebrated Alpine Valley slices through the mountains, while the so-called Rheita Valley, in the southeastern highlands, is made up of craters which have run together. Then there are crack-like features or "rills", known also as rilles or "rimae". Some of them, such as those associated with Hyginus and Ariadaeus, are visible with very small telescopes under

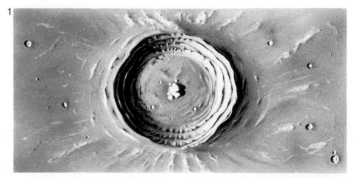

**1. Terraced crater**
The inner walls of this type of crater show well-defined terraces. The floor is sunken and there is a prominent central peak. Such craters are often surrounded by ejected material. Examples: Theophilus and Alpetragius.

**2. Concentric crater**
Craters with multiple rings are found in all sizes on the Moon. The nested form of the smaller examples of such craters has been compared to nested volcanoes on Earth. Examples include Taruntius, Hesiodus, Marth.

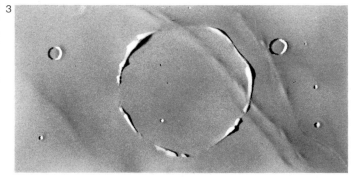

**3. Ghost crater**
Some craters appear to have been flooded by mare material so that their walls are barely traceable.

**4. Ray-crater**
Certain craters are distinguished by systems of bright rays which may extend for great distances.

5A

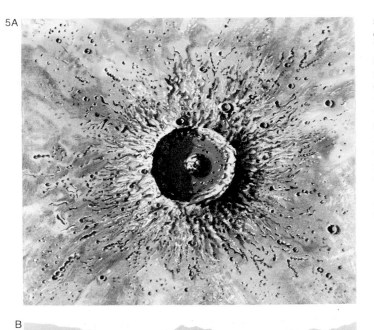

B

### 5. Crater morphology
Walled formations on the Moon are of various types, although in a typical example the basic form is circular (**A**). The main rampart may be very massive and terraced; the floor is sunken, and may contain a central peak or group of peaks. The surrounding area is generally covered by a blanket of ejecta, and pitted with secondary craters. Seen in profile (**B**) a crater is shallow, although smaller craters are comparatively deep relative to their diameters.

### 6. Terrestrial craters
Various craters on Earth have been identified as being of impact origin: two proven examples, where remnants of the meteorite that struck have been found on the site, are the Meteor Crater in Arizona (**A**), and Wolf Creek in Australia (**B**). Meteor Crater is 1,265 m in diameter and 174 m deep; its estimated age is 22,000 years. Wolf Creek is 853 m in diameter and 46 m deep. Volcanic craters (apart from those on volcanic cones) are also found on Earth; a good example is Hverfjall in Iceland (**C**). Craters have also been produced artificially by the use of explosives (**D**): this example, in western Canada, bears a strong likeness to the far-side crater Tsiolkovsky (*see* page 189), although it is very much smaller in scale (about 100 m in diameter and 6.5 m deep).

6A

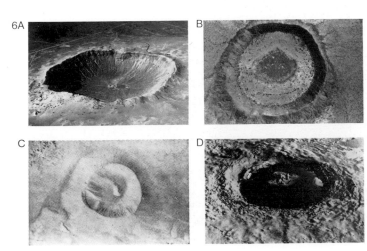

suitable conditions of illumination, and in many cases there are complicated systems, such as those of Triesnecker, in the Mare Vaporum–Sinus Medii area, and Hippalus, on the border of the small comparatively regular Mare Humorum. However, not all rills are regular and crack-like. Many—notably those of Hyginus—are composed of chains of small craterlets that have run together, often with the destruction of their common walls; they have been likened to "strings of beads". Finally there are "faults", the most striking example of which is the Straight Wall in the Mare Nubium (which, incidentally, is not straight, and is not a wall). The surface drops sharply to the west, so that the feature appears as a dark line before full Moon (on account of its shadow) and as a bright line afterward (because its inclined face is being illuminated by the Sun).

### Grid system
The Moon is criss-crossed with faults and ridges that form an important pattern known as the "grid system". Sometimes the faults are due to almost vertical slippages, while in other cases the movements have been more nearly horizontal. In many cases the ridges form parts of the walls of craters, though often enough the craters themselves are so broken and distorted that they are hard to identify. It has been pointed out by various authorities that there are two principal "families" of faults and ridges, running in specific directions and almost at right angles to each other. This makes up the main grid system, although it is complicated by minor families such as those of the Alps and Caucasus regions. The presence of the grid system indicates that the Moon's crust was subjected to strain over a long period, and it has also been found that small craters in some regions are non-circular, with their longest axes lying in the direction of one or the other of the families of the main grid.

### Far side of the Moon
The far side, totally unknown before the flight of Luna 3 in 1959, is distinctly different from the familiar near side. The most striking difference is the absence of large maria. Of all the main near-side maria, only the Mare Orientale extends far onto the averted regions, and there is nothing on the far side even remotely comparable with, say, the Mare Imbrium. Even the so-called Mare Moscoviense with a diameter of about 420 km is relatively small. There are craters everywhere, and even basins of mare size that are light-floored rather than filled with dark mare material. There seems no reasonable doubt that this difference is due to the fact that the lunar rotation has been synchronous since a fairly early stage in the story of the Earth–Moon system.

### Hot spots
During lunar eclipses (*see* pages 148–9) there is an abrupt cooling of the Moon's surface, and this fact has proved useful to astronomers investigating the Moon's geology from Earth. Since the Moon is without atmosphere and the surface materials are very poor at retaining heat, the sudden fall in temperature is quite dramatic. Yet not all areas cool down at the same rate. Infrared techniques have revealed that there are some regions which cool much less rapidly than their surroundings, and these have become known as "hot spots"; the best example is the great ray-crater Tycho, in the southern uplands of the Moon. These so-called hot spots are also warmer than their environs during the long lunar nights. There is no suggestion that internal heat is responsible; the effect is due solely to a difference in the nature of the surface materials.

From time to time there have been suggestions that the sudden cold may produce visible effects upon certain special features of the surface, notably the craterlet Linné on the grey plain of the Mare Serenitatis, which is surrounded by a white nimbus. However, these results are extremely dubious, and it is not likely that any real effect occurs. Nor have major structural changes occurred on the Moon for many millions of years.

# **Near Side** Orthographic

**North-west**

This section is dominated by the Mare Imbrium, the largest and most important of the Moon's regular "seas". Its borders include the Apennines, the Alps, the Jura Mountains and the Carpathians; the Apennines are the most impressive of all the lunar ranges. Leading out of the Mare Imbrium is the beautiful Sinus Iridum. Other mare areas include the Sinus Roris, part of the Mare Frigoris and a small part of the Oceanus Procellarum. Walled formations include the dark-floored Plato, and Archimedes on the Mare Imbrium (the other formations of the Archimedes group, Aristillus and Autolycus, are just in the north-east section of the map). Of special importance is Aristarchus, the brightest point on the entire Moon; nearby is the darker-floored Herodotus, together with the magnificent Schröter's Valley.

**Center west**

Much of the huge Oceanus Procellarum is to be found in this section, with part of the Mare Nubium and smaller marial areas such as the Sinus Aestuum. The whole region is dominated by Copernicus, the magnificent ray-crater with its high, terraced walls and central mountains; Kepler is another ray-crater, and there is yet another, Olbers, closer to the limb. Other very important walled formations are: Ptolemaeus, Alphonsus and Arzachel, which form a splendid chain; Grimaldi, Riccioli and Gassendi on the northern border of the Mare Humorum. Part of the Apennines range extends into the section, ending near the deep, well-marked formation Eratosthenes. The Riphaean Mountains are quite prominent, as are the Carpathian Mountains forming part of the border of the Mare Imbrium.

**South-west**

The main "sea" areas here are those of the Mare Humorum and part of the Mare Nubium; on the Mare Nubium is the celebrated Rupes Recta or "Straight Wall", the most famous fault on the Moon. There are few craters on the Mare Humorum (Gassendi, at the northern edge, is just in the center section of the map), but there are some well-marked "bays", such as Doppelmayer and Hippalus. High peaks, associated with the Mare Orientale, lie along the limb. The southern area is mainly highland, with some huge walled plains, such as Schickard, Clavius and the irregular Schiller, as well as the great chain composed of Purbach, Regiomontanus and Walter; but the most prominent feature in this region is the ray-crater Tycho. Seen under high light, the Tycho rays dominate the entire area, and indeed much of the Moon's disc.

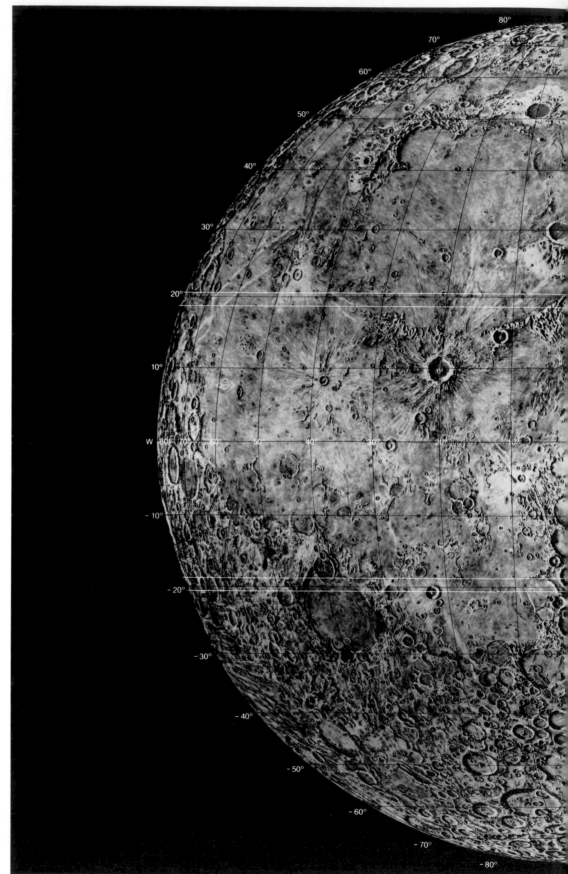

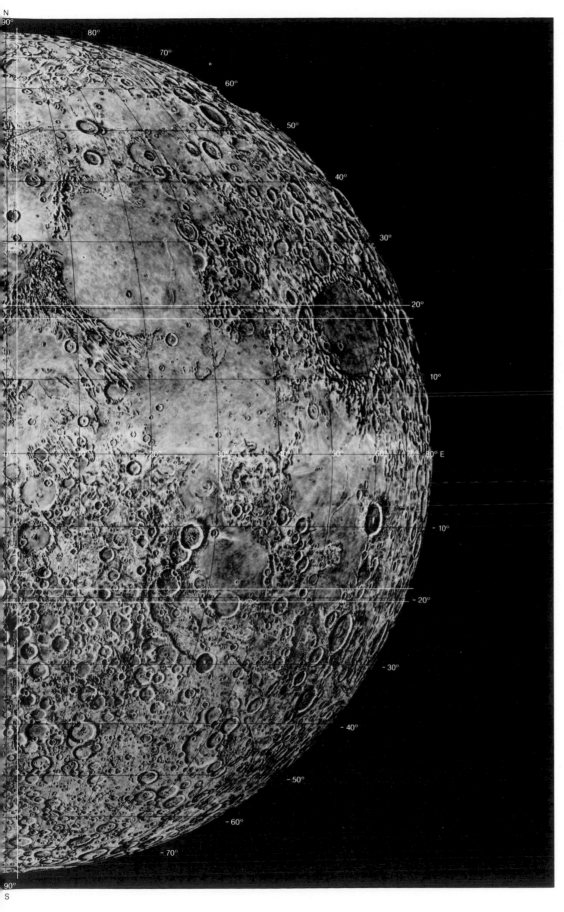

## North-east

The main "sea" area in this section is the Mare Serenitatis, almost all of which is included. There are few major formations on its floor; the most prominent is the small but easily recognized Bessel, and here too is Linné, the minor feature once suspected of having shown definite change in the nineteenth century, although this is now discounted. A small part of the Mare Crisium is included, together with a large section of the irregular Mare Frigoris; on the limb is the well-marked Mare Humboldtianum. Of the walled formations, Posidonius on the edge of the Mare Serenitatis is prominent; so are the Atlas–Hercules, Aristoteles–Eudoxus, and Aristillus–Autolycus pairs. The Alpine Valley comes into this section, and there is an important system of rills associated with the well-known crater Bürg.

## Center east

There are wide marial areas here. Almost all of the well-marked Mare Crisium is included, and the whole of the Mare Tranquillitatis, together with much of the Mare Fecunditatis and the Mare Nectaris, and the smaller Mare Vaporum. Objects of special interest include the Messier twins on the Mare Fecunditatis, the ray-crater Proclus near the distinctive Palus Somnii, and the prominent rills associated with Hyginus and Ariadaeus. The great walled formations Langrenus and Vendelinus lie on the boundary of the Mare Fecunditatis. Close by the border of the Mare Nectaris are the prominent craters Theophilus, Cyrillus and Catharina, making up one of the most magnificent "chains" on the Moon. "Tranquillity Base", the landing site of Apollo 11, falls in this section.

## South-east

This is the main upland region of the visible hemisphere of the Moon; there are almost no "sea" areas apart from the patchy, irregular Mare Australe near the limb. The Altai Scarp, associated with the Mare Nectaris system, runs in from the center section of the map, and there is also the crater-valley of Rheita, with the almost equally prominent valley associated with Reichenbach. Large walled formations include Petavius, with its magnificent internal rill; Furnerius and the huge, ruined Janssen. Right on the limb is Wilhelm Humboldt, always hard to study from Earth because of its extreme foreshortening, but clearly shown on Orbiter and Apollo pictures; the floor contains many rills. Other major formations are Stöfler and Maurolycus. The whole area is crowded with walled formations of all types.

# **Northwest Region** Orthographic

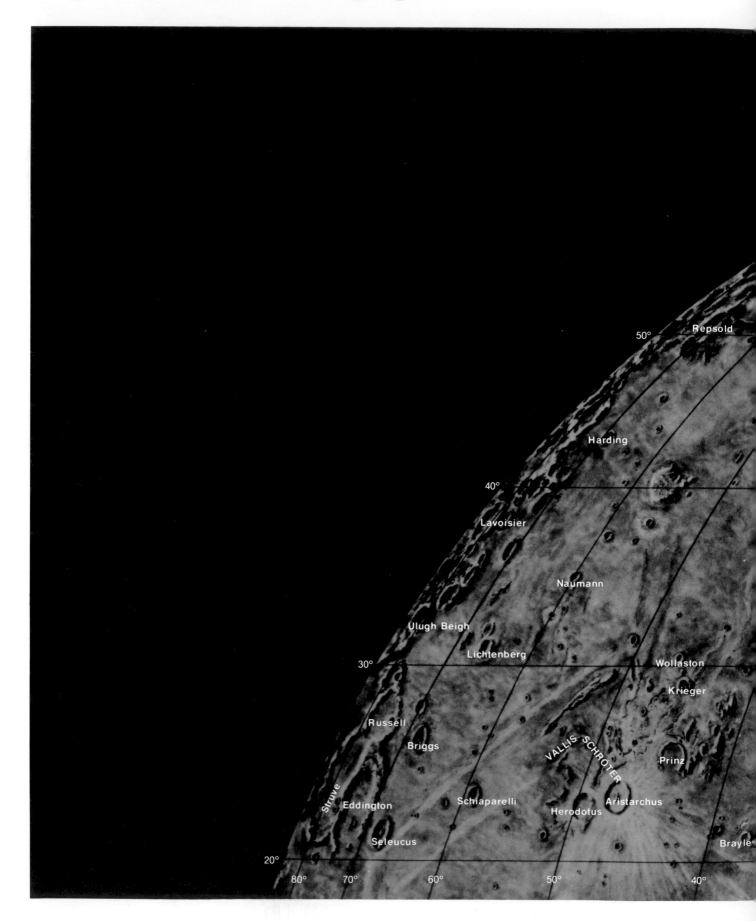

Repsold

50°

Harding

40°

Lavoisier

Naumann

Ulugh Beigh

Lichtenberg

30°

Wollaston

Krieger

Russell

Briggs

VALLIS SCHROTER

Prinz

Struve

Eddington

Schiaparelli

Aristarchus

Herodotus

Seleucus

Brayle

20°

80°    70°    60°    50°    40°

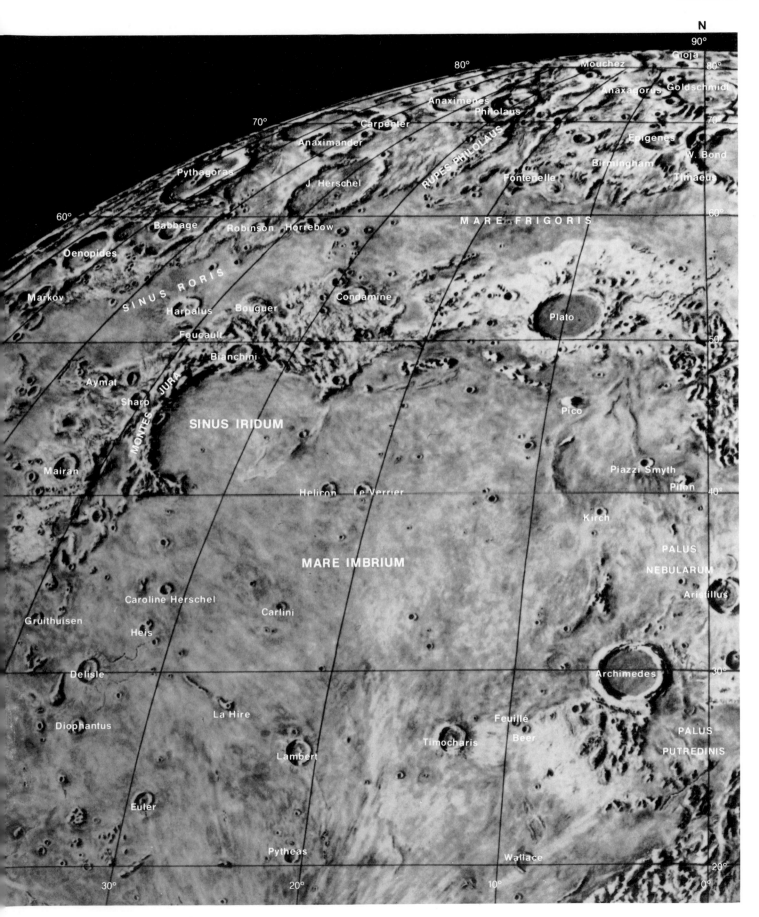

N
90°
80°                                                                                  Gioja
                                                                    Mouchez                    80°
80°                                                          Anaxagorus  Goldschmidt
                                        Anaximenes              Philolaus
70°                        Carpenter          Philolaus                          Epigenes
              Anaximander                                                                    70°
                                                                Birmingham
        Pythagoras              J. Herschel                              W. Bond
                                                  Fontenelle                    Timaeus
60°                                                M A R E   F R I G O R I S              60°
        Babbage      Robinson  Horrebow
Oenopides
                                                                                  Plato
Markov                                          Condamine
                        Harpalus      Bouguer
              Foucault
                        Bianchini
                                                                                        50°
        Aymat                                                              Pico
              Sharp
                        SINUS IRIDUM                                      Piazzi Smyth
Mairan                                                                          Piton      40°
                        Helicon    Le Verrier                              Kirch
                                                                              PALUS
                        MARE IMBRIUM                                        NEBULARUM
                                                                              Aristillus
              Caroline Herschel
Gruithuisen                    Carlini
        Heis
                                                                              Archimedes    30°
Delisle
                        La Hire
Diophantus                                                  Feuillé           PALUS
                                                    Timocharis  Beer         PUTREDINIS
                        Lambert
Euler
                        Pytheas                          Wallace                  20°
30°              20°                          10°                          0°

MONTES JURA
SINUS RORIS
RUPES PHILOLAUS

165

# Northeast Region Orthographic

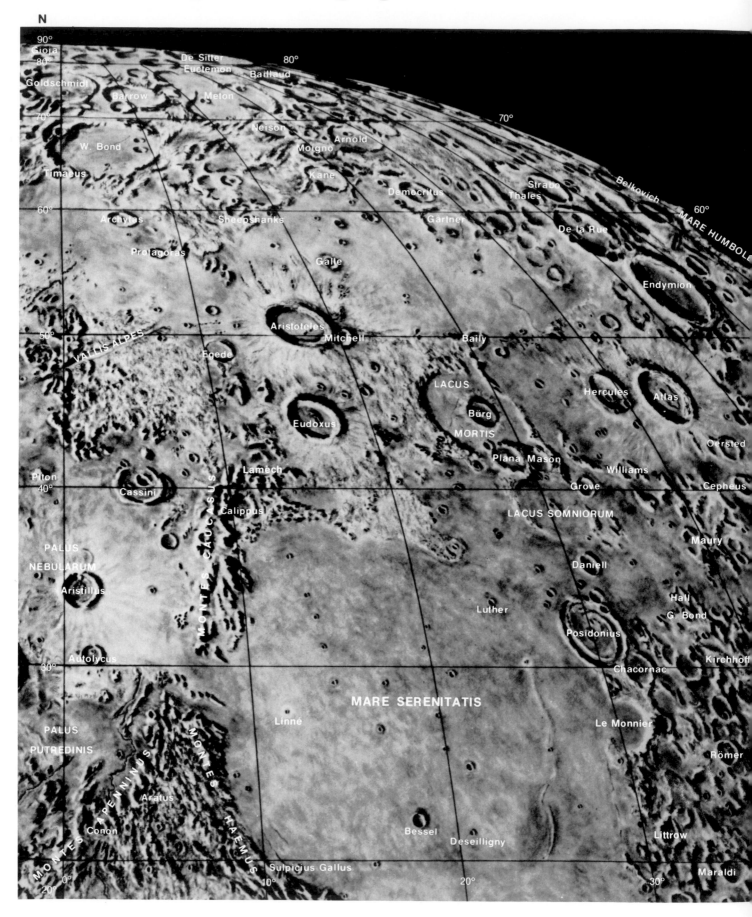

N

90°
Gioja
80°
De Sitter
Euctemon
Baillaud
80°
Goldschmidt
Barrow
Melon
70°
70°
Neison
Arnold
W. Bond
Moigno
Kane
Democritus
Strabo
Thales
Belkovich
60°
MARE HUMBOLD
Timaeus
60°
Archytas
Sheepshanks
Gärtner
De la Rue
Protagoras
Galle
Endymion
Aristoteles
50°
Mitchell
Baily
VALLIS ALPES
Egede
LACUS
Hercules
Atlas
Bürg
Eudoxus
MORTIS
Oersted
Plana Mason
Lamech
Williams
Cepheus
Piton
40°
Cassini
Grove
LACUS SOMNIORUM
Calippus
Maury
PALUS
Daniell
Hall
NEBULARUM
G. Bond
Aristillus
Luther
Posidonius
Kirchhoff
Autolycus
30°
Chacornac
MARE SERENITATIS
Le Monnier
PALUS
Linné
Römer
PUTREDINIS
MONTES APENNINUS
MONTES HAEMUS
MONTES CAUCASUS
Aratus
Bessel
Conon
Deseilligny
Littrow
Sulpicius Gallus
Maraldi
0°
10°
20°
30°
20°

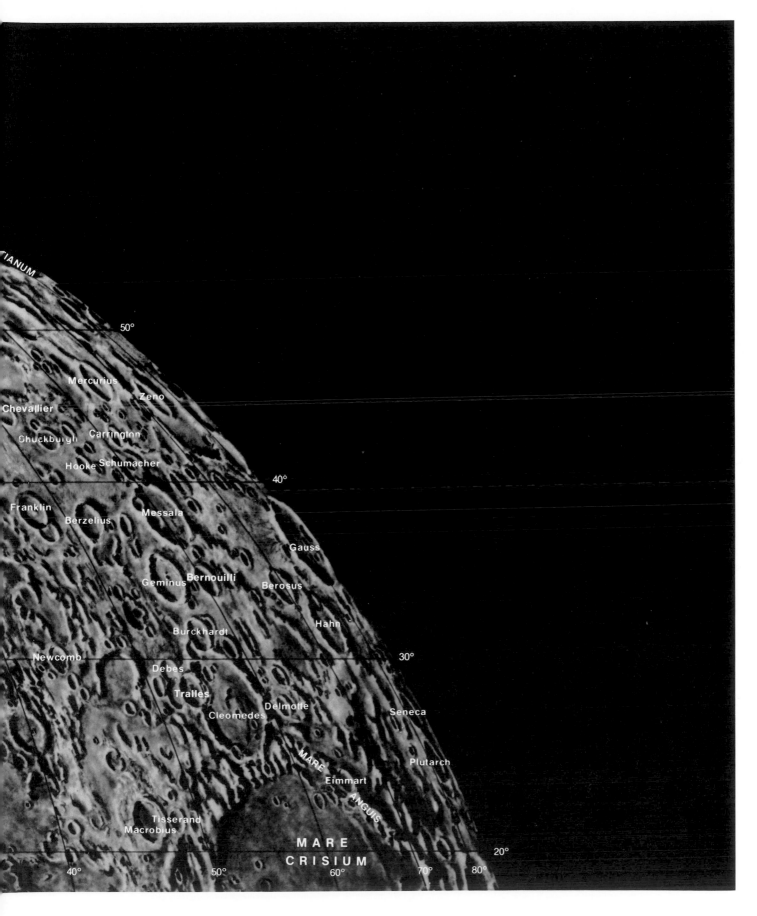

TIANUM

50°

Mercurius

Zeno

Chevallier

Chuckburgh

Carrington

Hooke Schumacher

40°

Franklin

Berzelius

Messala

Gauss

Bernouilli

Geminus

Berosus

Hahn

Burckhardt

Newcomb

30°

Debes

Tralles

Delmotte

Seneca

Cleomedes

Plutarch

MARE

Eimmart

ANGUIS

Tisserand

Macrobius

MARE

20°

CRISIUM

40°

50°

60°

70°

80°

# West Central Region Orthographic

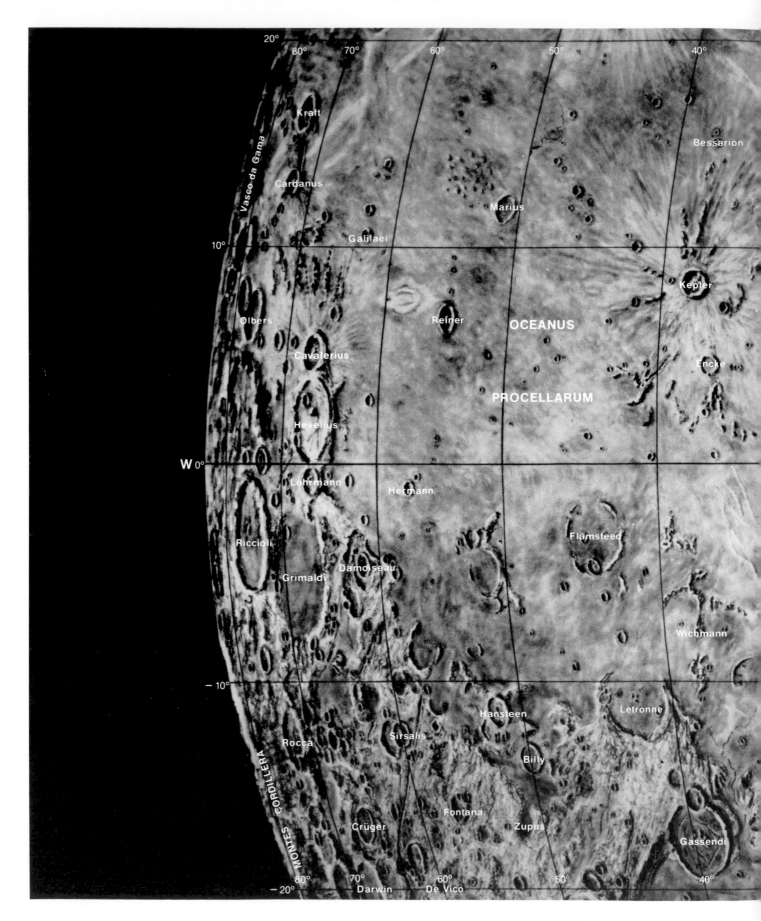

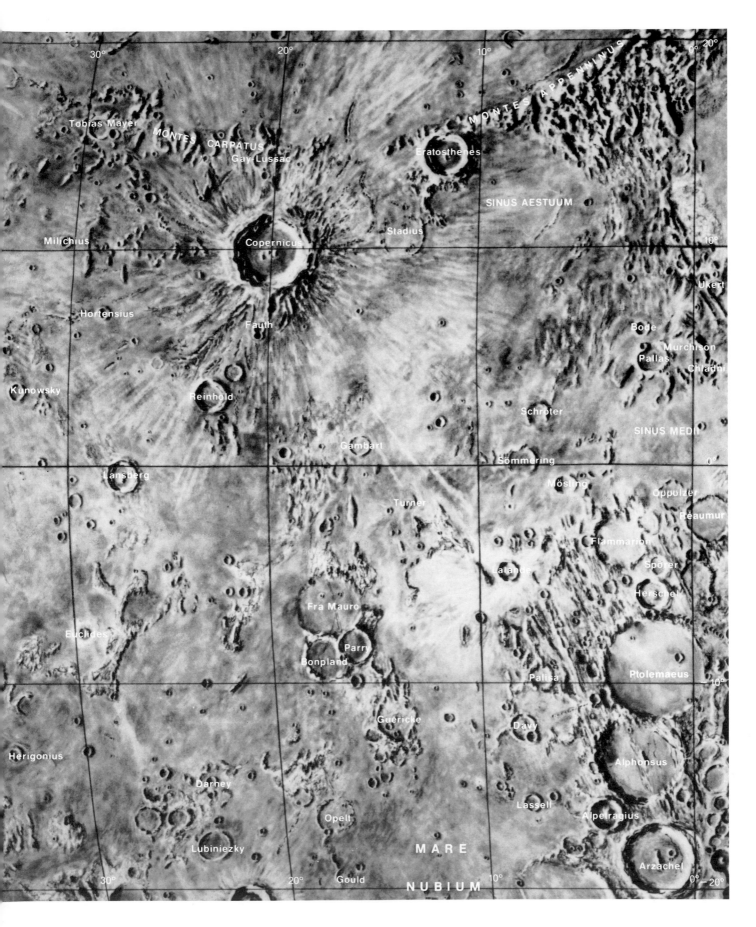

# East Central Region Orthographic

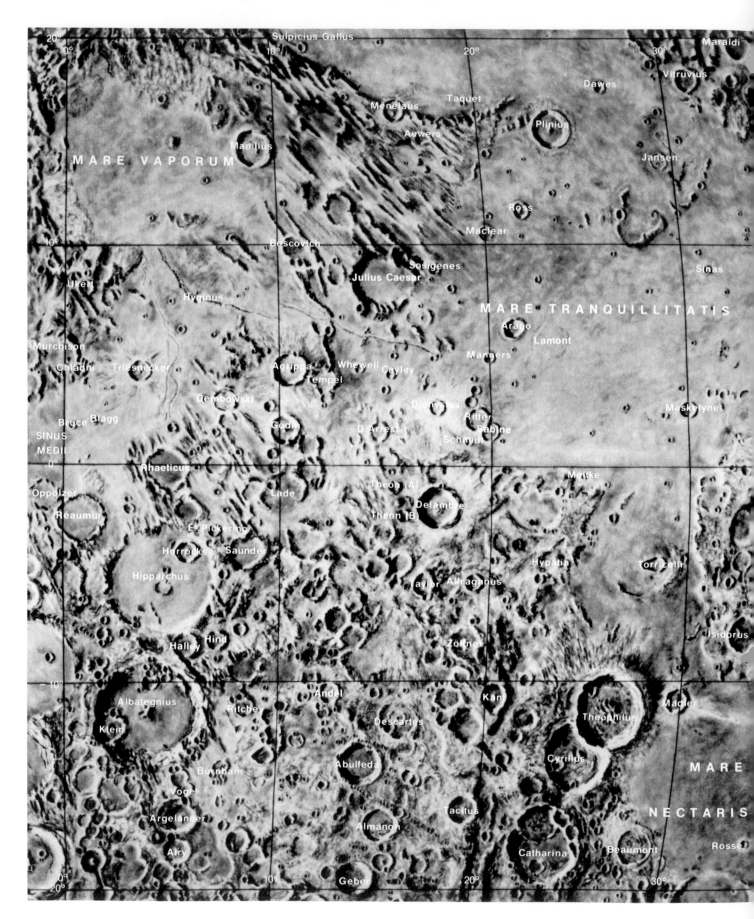

MARE VAPORUM

MARE TRANQUILLITATIS

SINUS MEDII

MARE NECTARIS

Sulpicius Gallus
Maraldi
Vitruvius
Dawes
Menelaus
Taquet
Auwers
Plinius
Manilius
Jansen
Ross
Maclear
Boscovich
Sosigenes
Julius Caesar
Sinas
Ukert
Hyginus
Arago
Lamont
Murchison
Manners
Chladni
Triesnecker
Agrippa
Whewell
Cayley
Tempel
Maskelyne
Dembowski
Dionysius
Bruce
Blagg
Godin
D'Arrest
Ritter
Sabine
Schmidt
Rhaeticus
Moltke
Oppolzer
Lade
Theon (A)
Delambre
Réaumur
Theon (B)
E. Pickering
Hypatia
Horrocks
Saunder
Torricelli
Hipparchus
Taylor
Alfraganus
Halley
Hind
Zöllner
Isidorus
Albategnius
Andel
Kant
Mädler
Ritchey
Theophilus
Klein
Descartes
Cyrillus
MARE
Burnham
Abulfeda
Vogel
Tacitus
NECTARIS
Argelander
Almanon
Catharina
Beaumont
Rosse
Airy
Geber

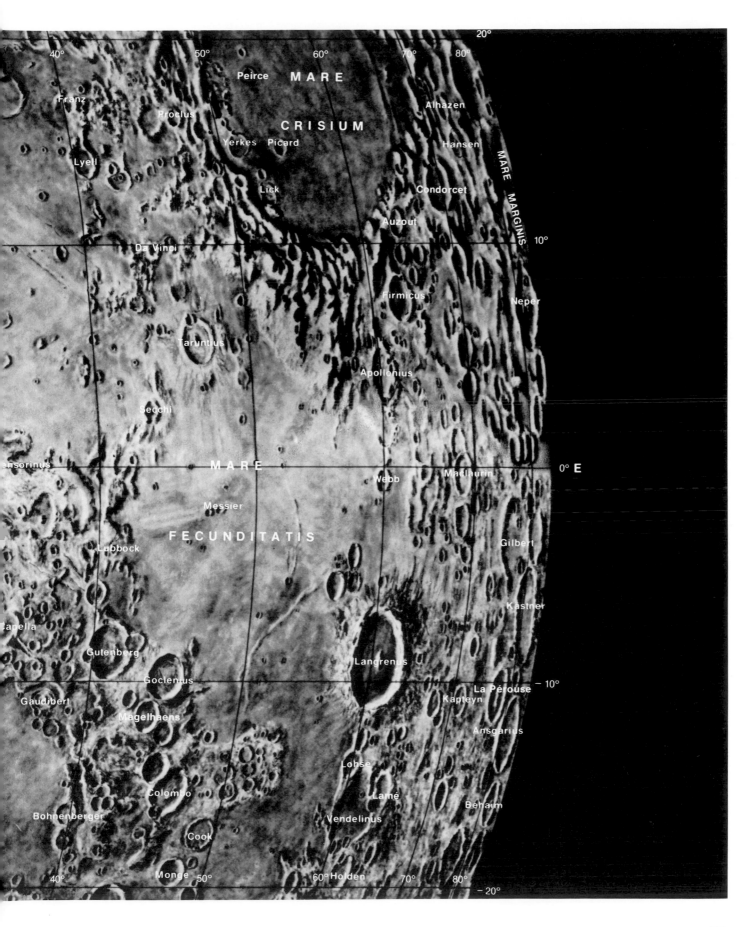

# Southwest Region Orthographic

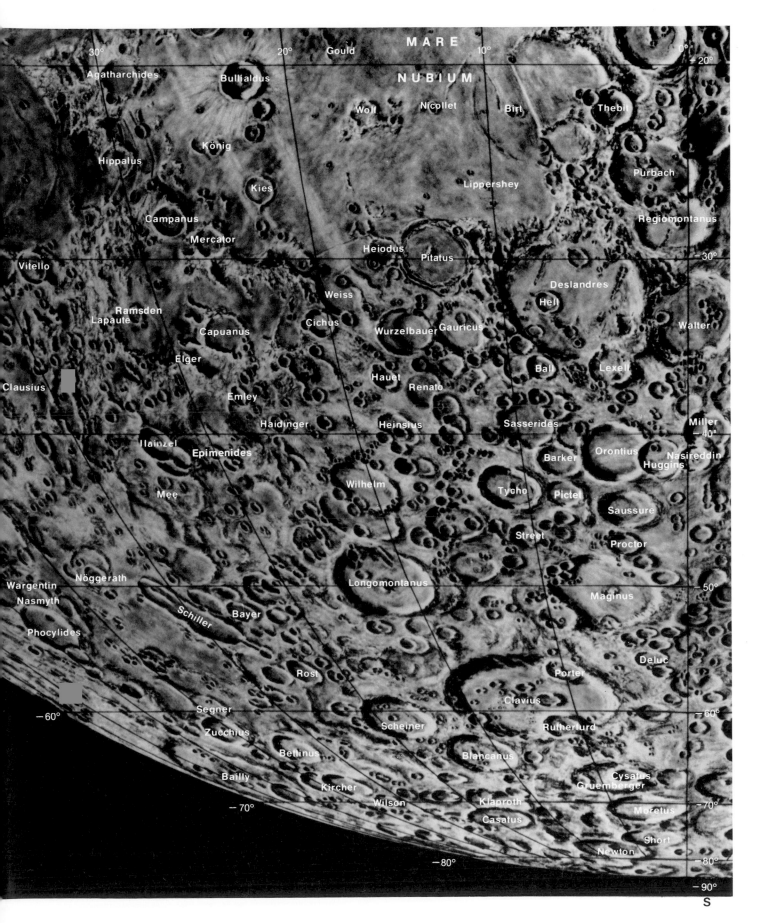

MARE

NUBIUM

30°    20°    Gould    10°    0°
-20°
Agatharchides    Bullialdus    Wolf    Nicollet    Birt    Thebit
König    Lippershey    Purbach
Hippalus
Kies    Regiomontanus
Campanus
Mercator    Heiodus    -30°
Vitello    Pitatus
Weiss    Deslandres
Ramsden    Hell
Lapaute    Cichus    Walter
Capuanus    Wurzelbauer  Gauricus
Elger    Ball    Lexell
Hauet
Clausius    Emley    Renato
Haidinger    Heinsius    Sasserides    Miller
-40°
Hainzel    Orontius    Nasireddin
Epimenides    Barker    Huggins
Wilhelm    Tycho    Pictet
Meę    Saussure
Street    Proctor
Nöggerath    Longomontanus    Maginus
Wargentin    -50°
Nasmyth    Schiller    Bayer
Phocylides    Deluc
Rost    Porter
Segner    Clavius    -60°
Zucchius    Scheiner    Rutherfurd
Bettinus    Blancanus
Bailly    Cysatus
Kircher    Gruemberger
Wilson    Klaproth    Moretus    -70°
Casatus    Short
Newton
-80°
-90°
S

# Southeast Region Orthographic

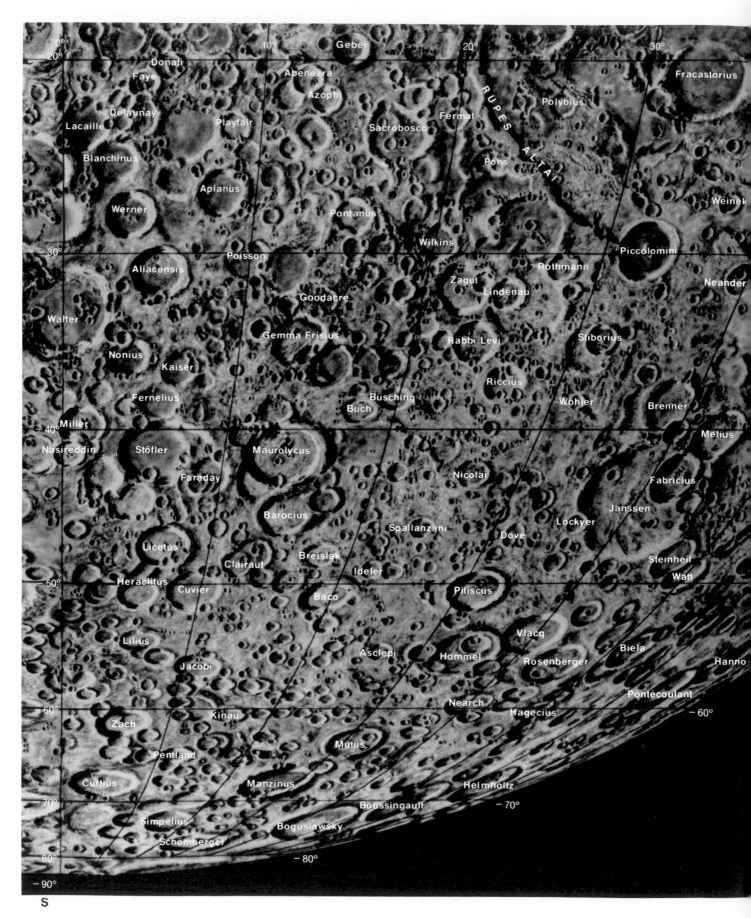

S

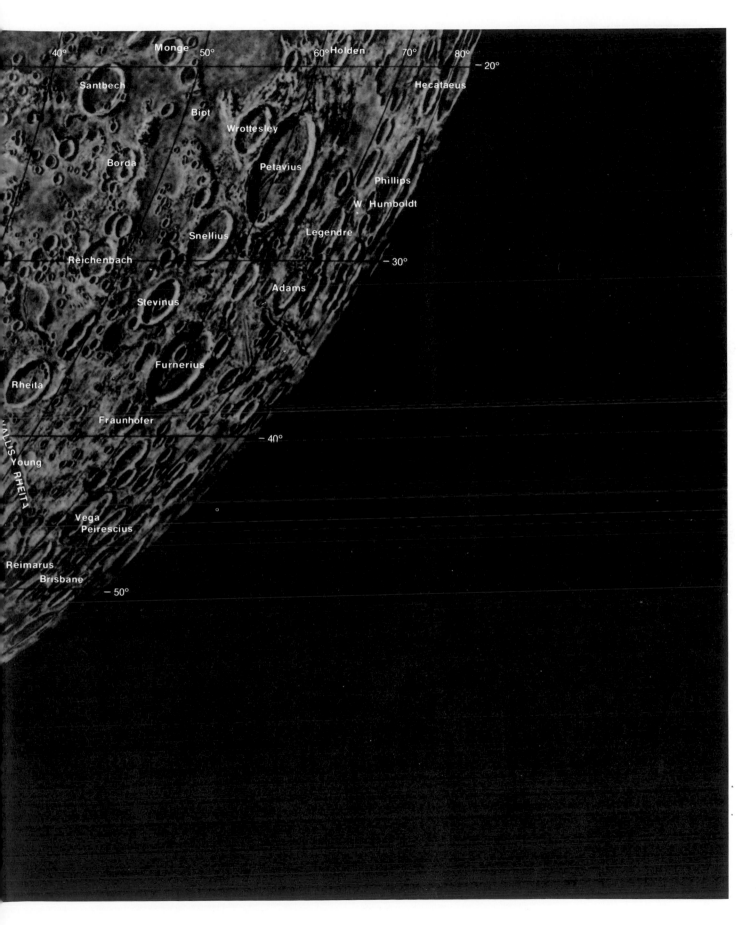

# Near Side Mercator

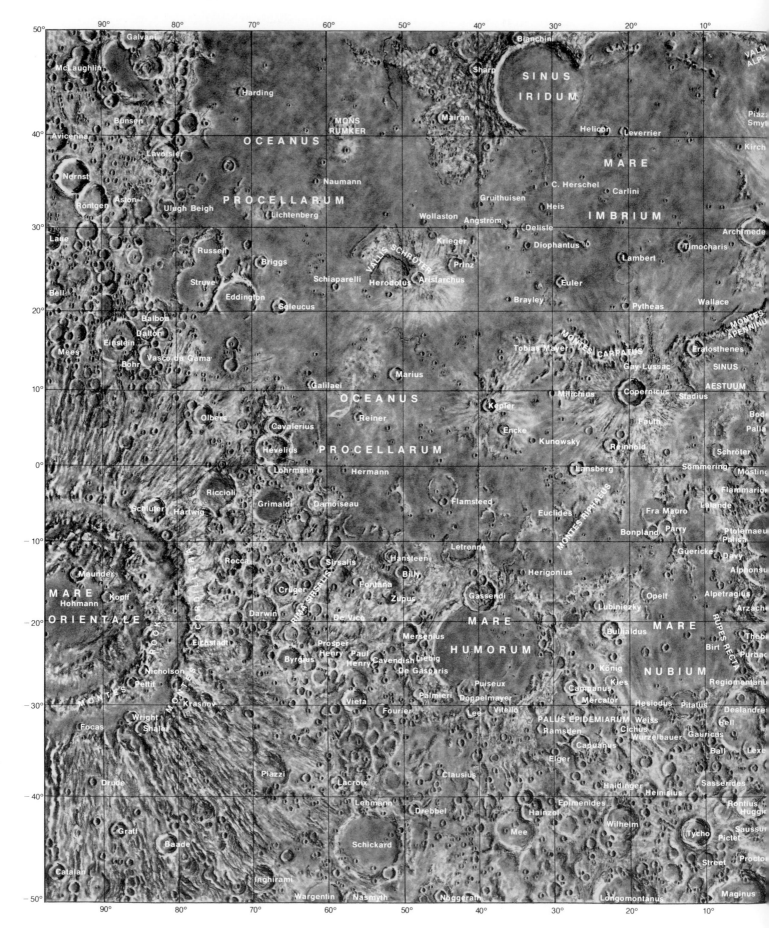

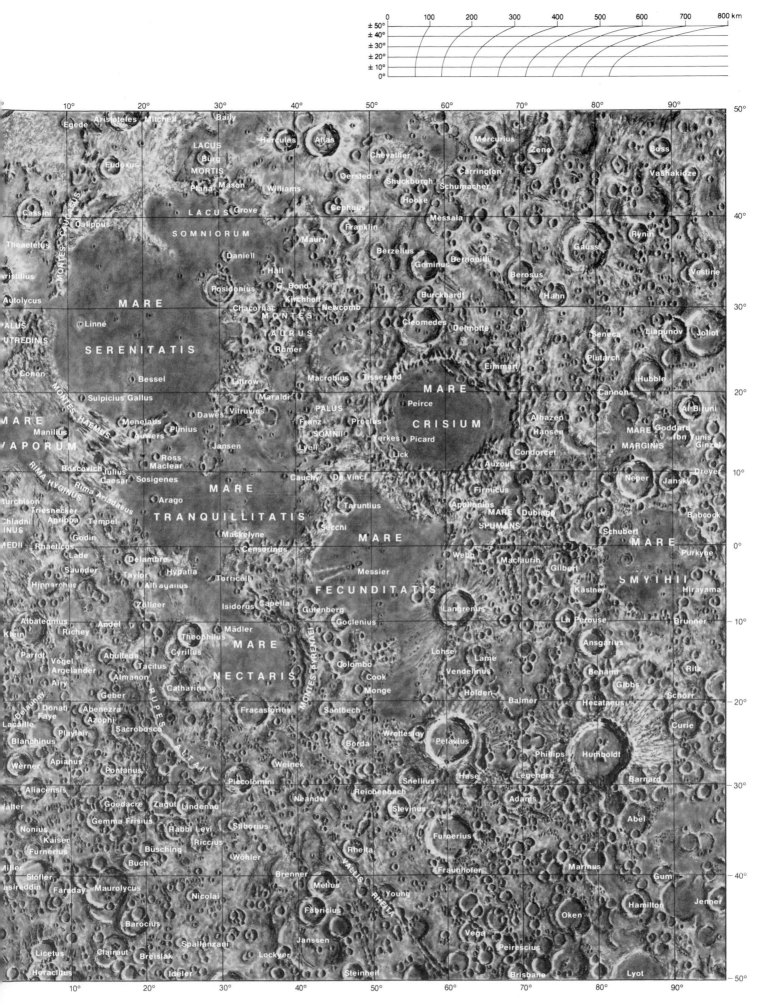

# Far Side Mercator

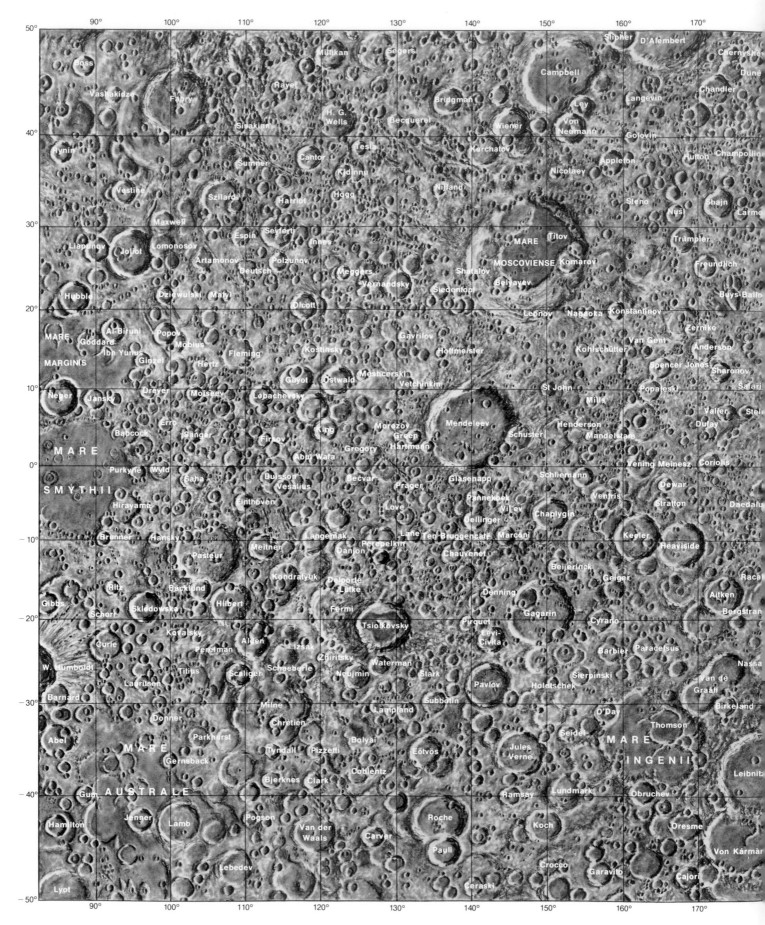

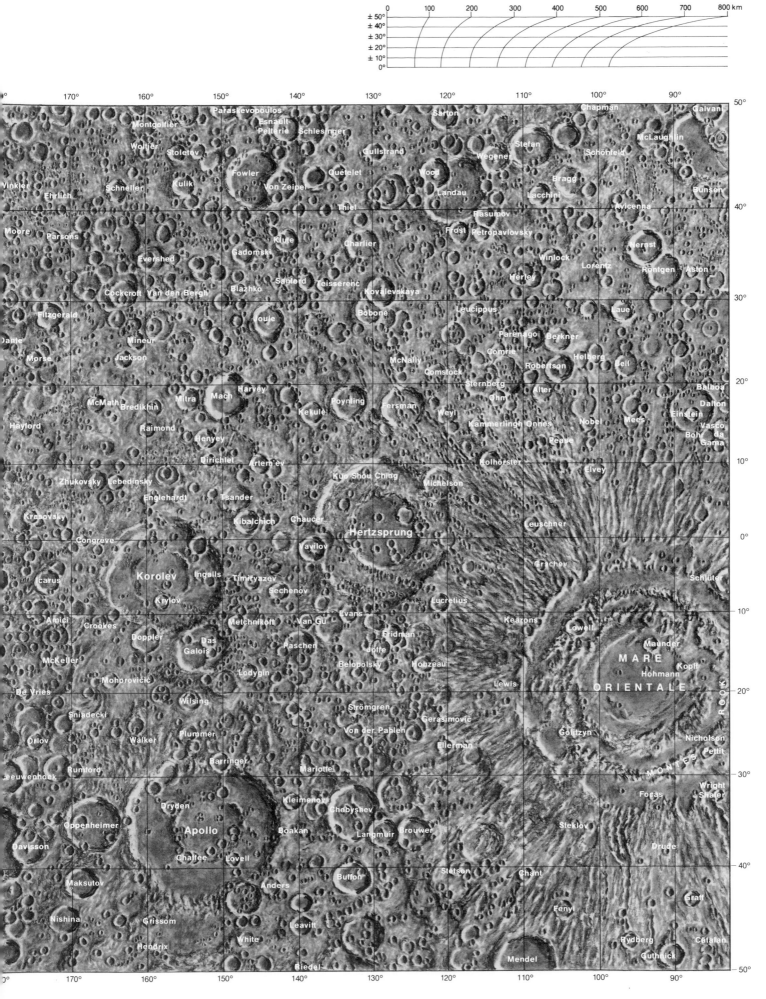

±50°
±40°
±30°
±20°
±10°
0°

0    100    200    300    400    500    600    700    800 km

50°
170°    160°    150°    140°    130°    120°    110°    100°    90°

Vinkler
Montgolfier
Wolijer
Stoletov
Schneller
Kulik
Paraskevopoulos
Esnault-Pelterie
Schlesinger
Gullstrand
Quételet
Von Zeipel
Thiel
Sarton
Chapman
Galvani
Stefan
Schönfeld
McLaughlin
Wegener
Wood
Bragg
Lacchini
Bunsen
Landau
Avicenna
Rasumov
40°

Moore
Parsons
Evershed
Cockcroft
Van den Bergh
Klute
Gadomski
Blazhko
Sanford
Teisserenc
Charlier
Kovalevskaya
Frost
Petropavlovsky
Winlock
Lorentz
Herzberg
Röntgen
Aston
Merley
30°

Fitzgerald
Jarte
Morse
Mineur
Jackson
Joule
Bobone
Parenago
Berkner
Comrie
Helberg
Robertson
Bell
Laue
McNally
Comstock
Sternberg
Ohm
Alter
Nobel
Mees
Balboa
Dalton
Einstein
Vasco da Gama
Bohr
20°

McMath
Bredikhin
Mitra
Mach
Harvey
Poynting
Fersman
Weyl
Kammerlingh Onnes
Pease
Haylord
Kekule
Raimond
Henyey
Dirichlet
Artem'ev
Kolhörster
Elvey
10°

Zhukovsky
Lebedinsky
Englehardt
Tsander
Kibalchich
Chaucer
Kuo Shou Ching
Michelson
Grachov
Schlüter
Krasovsky
Vavilov
Congreve
Hertzsprung
Leuschner
0°

Icarus
Korolev
Ingalls
Timiryazev
Sechenov
Krylov
Lucretius
Kearons
Lowell
10°

Amici
Crookes
Doppler
Das
Galois
Metchnikoff
Van Gu
Fridman
Joffe
Belopolsky
Houzeau
Maunder
Kopff
Hohmann
McKeller
Paschen
Lodygin
Lewis
MARE
ORIENTALE
Golitzyn
20°

De Vries
Mohorovicic
Wilsing
Strömgren
Gerasimovic
Von der Pahlen
Ellerman
Nicholson
Pettit
Sniadecki
30°
Orlov
Walker
Plummer
Barringer
Mariotte
ROOK
Wright
Shaler
Rumford
Focas
eeuwenhoek
Kleimenov
Chebyshev
Drude
Dryden
Boakan
Langmuir
Brouwer
MONTES
Oppenheimer
Apollo
Steklov
40°
Davisson
Chaffee
Lovell
Stetson
Chant
Nishina
Buffon
Grissom
Anders
Fenyi
Graff
Maksutov
Leavitt
Rydberg
Catalan
Hendrix
White
Mendel
Guthnick
50°
0°    170°    160°    150°    140°    130°    120°    110°    100°    90°

179

# North and South Poles Stereographic

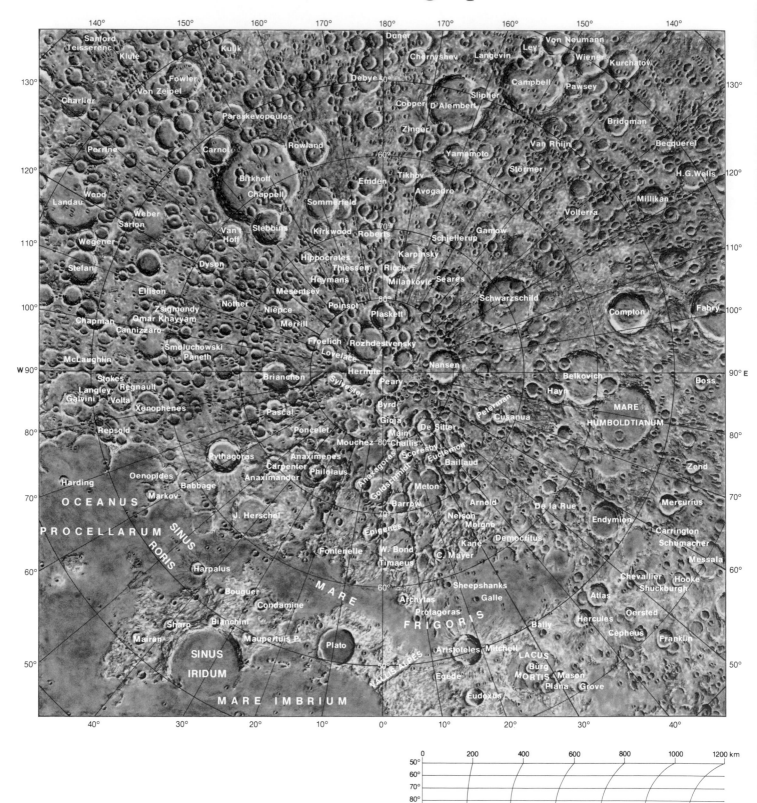

North Polar Region

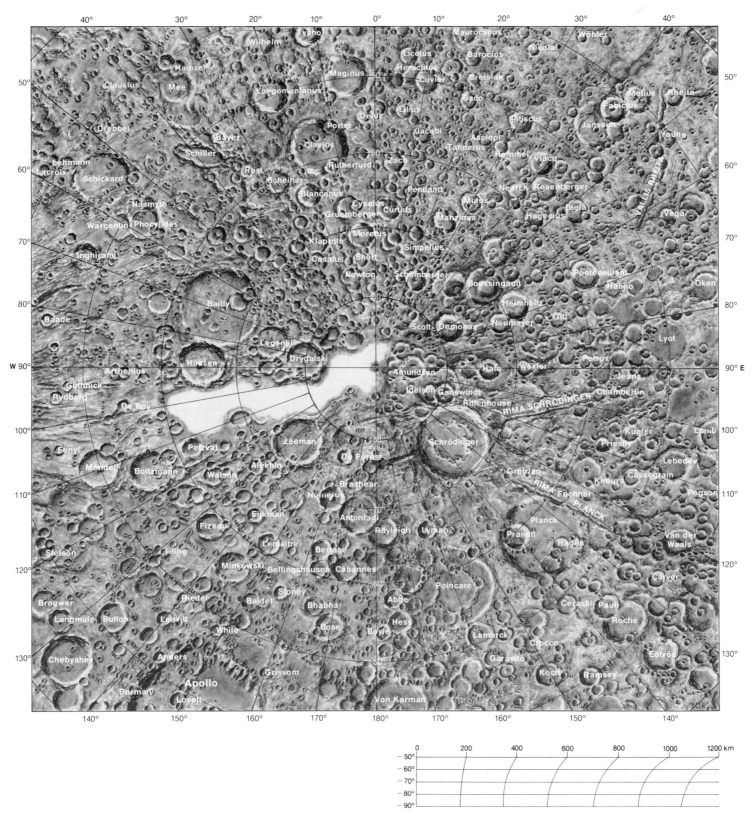

40°　30°　20°　10°　0°　10°　20°　30°　40°

Tycho
Wilhelm
Maurocycus
Wöhler
Hainzel
Licetus
Nicolai
Maginus
Heraclius
Barocius
Clausius
Mee
Cuvier
Breislak
Mefius
Rheita
Longomontanus
Baco
Fabricius
50°　50°
Drebbel
Delut
Lilius
Pitiscus
Janssen
Bayer
Porter
Jacobi
Asclepi
Young
Schiller
Clavius
Zach
Tannerus
Hommel
60°　60°
Lacroix
Rost
Rutherfurd
Pentland
Nearch Rosenberger
Lehmann
Scheiner
Vlacq
Schickard
Blancanus
Mutus
Biela
Vega
Nasmyth
Cysatus
Curtius
Manzinus
Hagecius
Wargentin Phocylides
Gruemberger
70°　70°
Klaproth
Moretus
Simpelius
Inghirami
Casatus
Short
Pontécoulant
Newton
Schömberger
Hanno
Oken
Bailly
Boussingault
80°　80°
Baade
Helmholtz
Gill
Legentil
Neumayer
Drygalski
Scott Demonax
Lyot
W 90°
Hausen
Amundsen
Hale
Wexler
90° E
Arrhenius
Idelson Ganswindt
Petrov
Gutnnick
Rittenhouse
Jeans
Rydberg
RIMA SCHRÖDINGER
Chamberlin
De Roy
Kugler
Lamb 100°
100°
Zeeman
Schrödinger
Priestly
Fenyi
Petzval
De Forest
Mendel
Lebedev
Boltzmann
Watson
Grotrian
Kimura Cassegrain
Alekhin
RIMA
Brashear
Fechner
Pogson
110°　110°
Numerov
PLANCK
Eijkman
Antoniadi
Planck
Van der
Fizeau
Rayleigh Lyman
Prandtl
Waals
Lemaitre
Hagen
Stetson
Tiling
Bertage
Minkowski
Carver
120°　120°
Bellingshausen Cabannes
Poincaré
Riedel
Stoney
Ceraski Pauli
Brouwer
Baldet
Bhabha
Abbe
Roche
Langmuir Buffon
Leavitt
White
Bose
Hess
Lamarck
Chebyshev
Boyle
Crocco
Eötvös
130°　130°
Anders
Garavito
Koch
Ramsey
Grissom
Apollo
Von Karman
Borman
Lovell

140°　150°　160°　170°　180°　170°　160°　150°　140°

0　200　400　600　800　1000　1200 km
-50°
-60°
-70°
-80°
-90°

**South Polar Region**

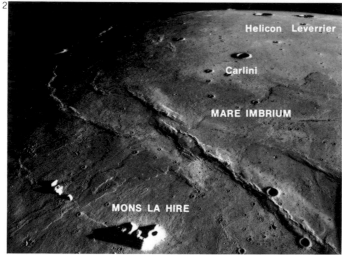

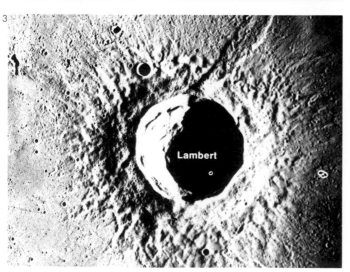

Lambert

Helicon Leverrier

Carlini

MARE IMBRIUM

MONS LA HIRE

Aristarchus

Prinz

MONTES HARBINGER

Aristarchus

Copernicus

**North-west region (left)**

1. Key features: Aristarchus (diameter 37 km), Prinz (50 km); Montes Harbinger. This is one of the most fascinating regions of the Moon. Prinz is an incomplete crater lacking its southern (upper) wall. The Harbinger Mountains are really clumps of hills rather than a true range, and the whole region is rich in rills. (Apollo 15)

2. Key features: Mare Imbrium area; Peak La Hire; Carlini (diameter 8 km), Helicon (29 km); Leverrier (25 km). The mountain mass La Hire is 1,500 m high with a summit craterlet. Beyond the ridge is the well-marked bright crater Carlini, the twin craters Helicon (left) and Leverrier (right), all of which are well sighted when observed from the Earth. (Apollo 15)

3. Lambert (diameter 29 km). Situated in the Mare Imbrium, this is a well-formed crater with walls which are massive and terraced, though not particularly bright. When this picture was taken, much of the interior of Lambert was in shadow, concealing the central formation—which is a crater instead of a mountain mass. Adjoining Lambert, below it on the picture, is a large "ghost ring". (Apollo 15)

4. Aristarchus (diameter 37 km). The central peak, to the upper left in the picture, casts a pronounced shadow; the floor of the crater is rough, and the wall is of great complexity. (Lunar Orbiter 5)

5. Copernicus (diameter 97 km). This great ray-crater has massive, terraced walls and a regular outline. The wall is well-defined, with a sharp crest. The floor of the crater is pitted with craterlets, and is evidently sunken; the wall rises only to a modest height above the surface outside the formation. (Lunar Orbiter 5)

**South-west region (right)**

6. Key features: Mare Nubium; Fra Mauro (diameter 81 km), Bonpland 58 km), Parry (42 km). Note the "rolling rock" in Fra Mauro. This was the intended landing-area of the ill-fated Apollo 13; Apollo 14 did in fact land in the region, taking astronauts Shepard and Mitchell onto the surface. (Apollo 16)

7. Key features: part of Ptolemaeus (diameter 148 km), Alphonsus (129 km), Arzachel (97 km). The floor of Ptolemaeus is rich in craterlets, but there is no central peak. (Apollo 16)

8A and B. Tycho (diameter 84 km). This great ray-crater is one of the most perfect of the walled formations; the rays are seen only under high illumination, and so are not shown here. The walls are high and terraced, and there is a central mountain complex. The area enclosed in the rectangle is shown in more detail on the right-hand picture, and the roughness of the floor leading up to the base of the inner wall is very evident. (Lunar Orbiter 5)

9. Gassendi (diameter 89 km). Like many other areas rich in rills, Gassendi has been the site of a number of reported TLP (see page 150). (Lunar Orbiter 5)

10. Key features: Rima Sirsalis; Sirsalis (diameter 32 km), Crüger A (22 km), De Vico A (30 km), De Vico T (35 km), De Vico (24 km). Sirsalis intrudes into its "twin", Sirsalis A. The rill runs past Crüger A and cuts the irregular De Vico A. (Lunar Orbiter 4)

11. Key features: Hippalus area; Hippalus (diameter 61 km), Agatharchides A (48 km), Campanus (38 km), Mercator (38 km). Hippalus has a central peak and is associated with a fine system of curved, parallel rills, (Lunar Orbiter 4)

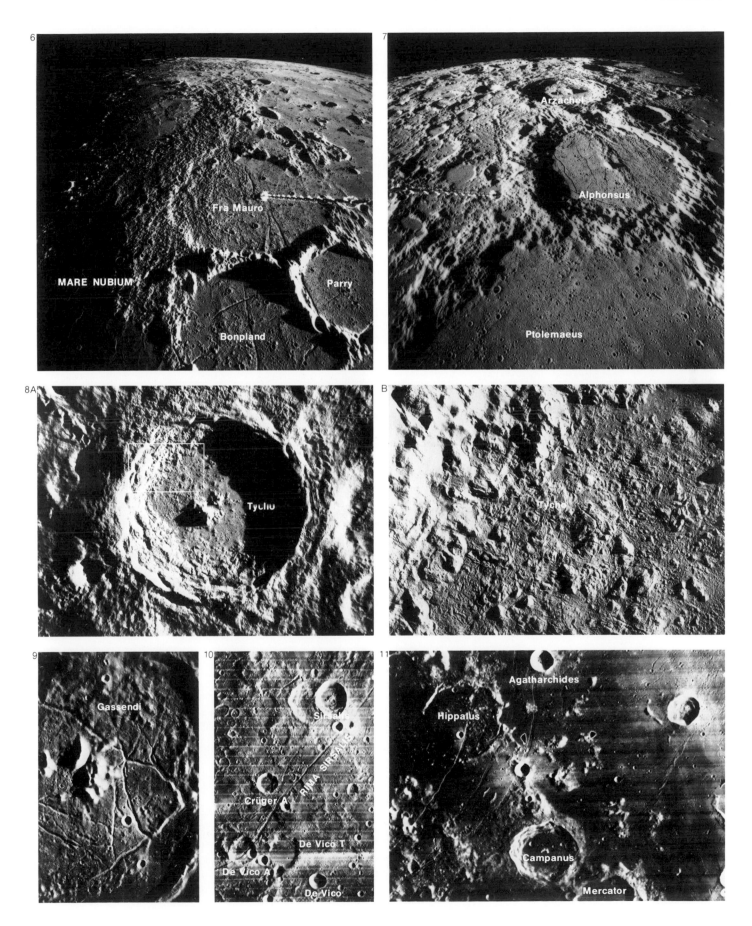

# Near Side East Region

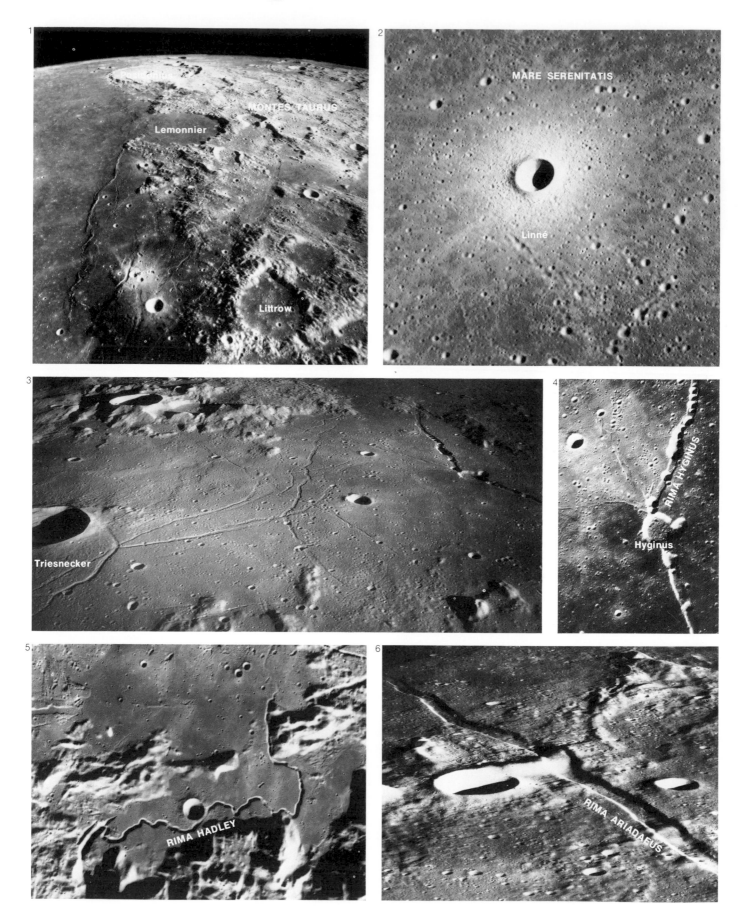

**North-east region (left)**

1. Key features: Posidonius (diameter 96 km), Lemonnier (55 km), Littrow (35 km); Taurus Mountains. The photograph shows part of the Mare Serenitatis. It was in this general area that the Soviet Lunokhod 2 crawled around in 1973, and where it remains. The so-called Taurus Mountains do not make up a connected range. Apollo 15 landed in the Taurus–Littrow area in December 1972. (Apollo 17)

2. Linné (diameter 4 km). This is a good example of a fresh, sharp-edged crater of probable impact origin. It is located on the western region of the Mare Serenitatis. (Apollo 15)

3. Triesnecker (diameter 23 km). Triesnecker is regular and notable mainly because of its association with a system of rills, which are not crater-chains—unlike the so-called Hyginus Rill, which appears to the upper right. This region lies between the Mare Vaporum and the Sinus Medii. (Apollo 10)

4. Key features: Rima Hyginus; Hyginus (diameter 6 km). The Hyginus Rill is essentially a crater-chain. The regular crater above Hyginus on the picture is Hyginus B. (Lunar Orbiter 5)

5. Rima Hadley. This photograph shows a region near the edge of the Mare Imbrium; it was near here that astronauts Scott and Irwin landed, and drove in their Lunar Roving Vehicle almost to the edge of the rill. (Apollo 15)

6. Rima Ariadaeus. Unlike the Hyginus formation, Ariadaeus is a genuine rill and not a crater-chain. It is over 240 km long. (Apollo 10)

**South-east region (right)**

7. Rupes Altai. This feature, also known as the "Altai Scarp", forms part of the ring system associated with the Mare Nectaris. It rises to a greater height above the surrounding region on the east than it does above the region to the west. The scarp has a relatively low crater density. (Lunar Orbiter 4)

8. The Messier twins (diameters approximately 13 km). This is a strange pair of craterlets in the Mare Fecunditatis. Messier A was formerly known as W. H. Pickering. From it extends the unique "comet" ray towards the western edge of the Mare. The craters show marked apparent changes at different conditions of illumination, but suggestions of real structural changes in historic times may safely be dismissed. (Apollo 15)

9. Key features: Stöfler (diameter 145 km), Faraday (64 km). Part of Stöfler has been destroyed by the intrusion of Faraday. (Lunar Orbiter 4)

10. Langrenus (diameter 137 km). Langrenus is a vast walled plain with high terraced walls rising to 2,700 m above the sunken floor. (Apollo 15)

11. Theophilus (diameter 101 km). The terraced inner ramparts rise to over 5,000 m above a floor which contains a massive central group of elevations. (Apollo 16)

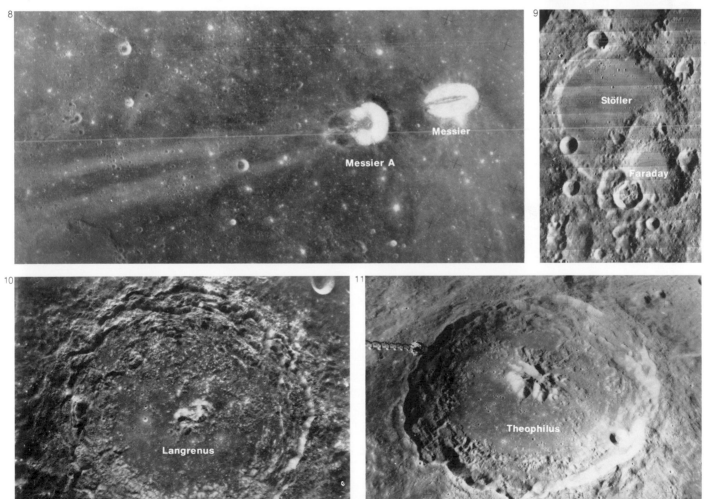

# Far Side West Region

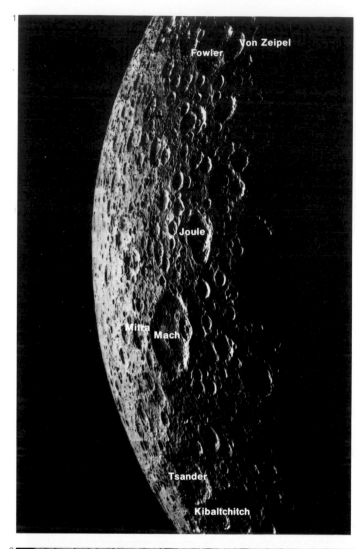

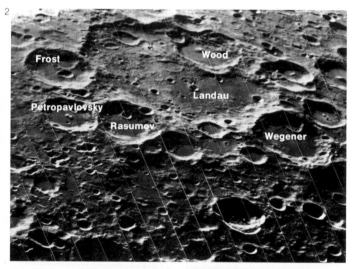

**North-west region**
1. Key features: Fowler (diameter 140 km), Von Zeipel (70 km), Joule (120 km), Mitra (80 km), Mach (180 km), Tsander (140 km), Kibaltchitch (100 km). Fowler has a darkish floor containing some small craterlets; its western wall is broken by a small formation, Von Zeipel. Tsander and Kibaltchitch make up a pair; they lie between two huge enclosures, Hertzsprung to the west on the unilluminated part of the Moon and Korolev to the east (not shown in this photograph). (Lunar Orbiter 5)

2. Key features: Landau (diameter 240 km), Wood (110 km), Wegener (100 km), Frost (80 km), Petropavlovsky (80 km), Rasumov (80 km). Landau is considerably deformed by intruding craters of which the most prominent is Wood. Wegener has a somewhat asymmetrical elevation on its floor; it touches Landau to one side, and on the other it abuts on Stefan (not shown in this photograph). (Lunar Orbiter 5)

3. Key features: Hertzsprung (diameter 440 km), Michelson (130 km). Hertzsprung is one of the major features of the Moon's far side. It has an inner section which is darker than the outer part of the interior; the outer ring is broken by Michelson to the north-west and Vavilov the south-east. (Lunar Orbiter 5)

4. Woltjer (diameter 130 km). This and its companion Montgolfier (not shown in this photograph) lie to the east of Fowler, which is shown in photograph 1 on this page. (Lunar Orbiter 5)

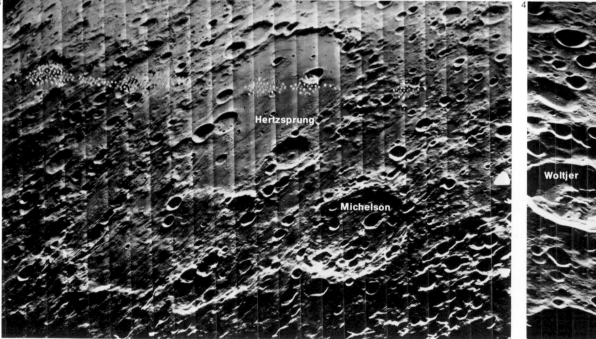

186

**South-west region**

5. Key features: Mare Orientale (diameter 900 km); Schlüter (80 km), Einstein (150 km). The Mare Orientale is one of the most important features of the Moon. Its basin is comparable in size with that of the Mare Imbrium, but contains little dark lava except near its center. Its outer ring is made up of the Rook Mountains, which are observable from Earth. This photograph also shows Grimaldi and Riccioli, which belong to the near side of the Moon. Schlüter is well-marked with a central peak. Einstein (named "Caramuel" on some of the older maps) has a central crater, high walls and a long ridge on its floor; from the Earth it is visible on the limb only under ideal conditions. (Lunar Orbiter 4)

6. Key features: Korolev (diameter 360 km), Krylov (50 km), Doppler (90 km). Korolev is the second of the trio of huge ringed formations, the others being Hertzsprung and Apollo. Korolev has two named interior craters, Krylov and Ingalls. (Lunar Orbiter 1)

7. Krylov (diameter 50 km). A regular, well-marked crater with a central peak, inside Korolev. A crater-row runs west from it to the wall of Korolev. (Lunar Orbiter 1)

8. Key features: Apollo (diameter 520 km), Barringer (60 km), Dryden (50 km), Chaffee (60 km), Lovell (30 km). Chaffee is one of a curved line of formations lying on the inner ring of Apollo. (Lunar Orbiter 5)

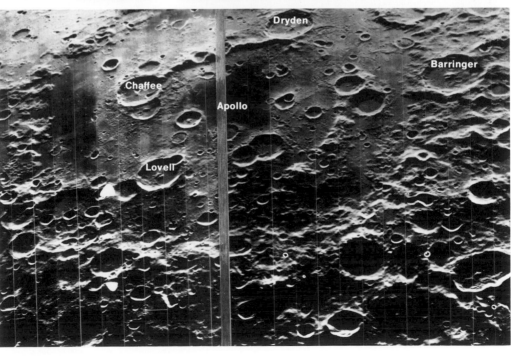

# Far Side East Region

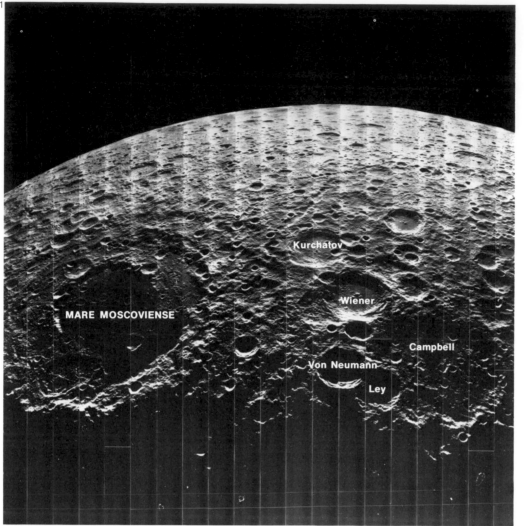

**North-east Region**

1. Key features: Mare Moscoviense; Campbell (diameter 260 km), Ley (80 km), Von Neumann (110 km), Wiener (140 km), Kurchatov (110 km). The Mare Moscoviense was one of the first features to be identified on the Moon's far side; it was shown as a dark patch on the Luna 3 pictures of 1959. It is certainly not a major mare, but it does have a dark floor, and there is an outer ring. This whole area is very rough. (Lunar Orbiter 5)

2. Key features: Compton (diameter 175 km), Fabry (210 km), Szilard (140 km), Seyfert (110 km), Cantor (100 km), H. G. Wells (160 km), Millikan (140 km). This is another very rough highland area, lying on the far side of the Moon towards the north pole, beyond the Mare Humboldtianum. Compton is of special interest in as much as it has both a central elevation and an inner ring of peaks—a very rare combination unknown on the near side. (Lunar Orbiter 5)

3. Key features: Lobachevsky (diameter 80 km), Guyot (110 km) Ostwald (120 km), Fleming (130 km), King (90 km), Olcott (70 km). This region lies on the far hemisphere beyond Mare Smythii. It contains some well-formed craters; for instance Lobachevsky has fairly regular walls (slightly polygonal in outline) and a group of central elevations. King also is well-formed, with terraced walls; a low-walled crater makes up a chain with King and Ostwald. Terracing is also noticeable in the crater Olcott. (Apollo 16)

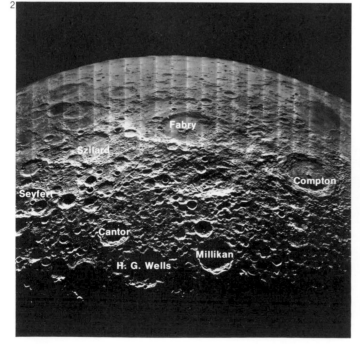

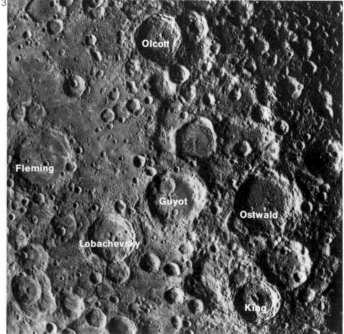

**South-east region**

4. Key features: Gagarin (diameter 240 km), Pavlov (130 km), Jules Verne (130 km), Mare Ingenii (350 km), O'Day (80 km), Thomson (100 km), Van de Graaff (210 km), Aitken (130 km), Heaviside (140 km), Keeler (140 km). This rough region includes the so-called Mare Ingenii, an irregular formation with a darkish floor; it has no real claim to the title of "mare". Much the more interesting feature is the crater Van de Graaff. It is in many ways unique; it is a depression about 4 km deep, and is associated with a strong magnetic anomaly, possibly indicating buried volcanic rock. Van de Graaff also shows greater radioactivity than its surroundings. (Lunar Orbiter 2)

5. Key features: Tsiolkovsky (diameter 198 km), Fermi (210 km), Alden (90 km), Milne (250 km), Waterman (70 km), Neujmin (110 km), Roche (190 km), Pauli (100 km), Pavlov (130 km), Delporte (50 km), Lutke (50 km), Chrétien (90 km), Scalinger (100 km). This region adjoins the area shown in photograph 4 (Pavlov appears on both). (Lunar Orbiter 3)

6. O'Day; this is a fairly prominent crater adjoining Mare Ingenii. (Lunar Orbiter 2)

7. Key feature: Tsiolkovsky (diameter 190 km). This formation, one of the most remarkable on the Moon, has about half the diameter of the Mare Crisium. Part of the floor is very dark, and obviously consists of mare material. (Apollo 15)

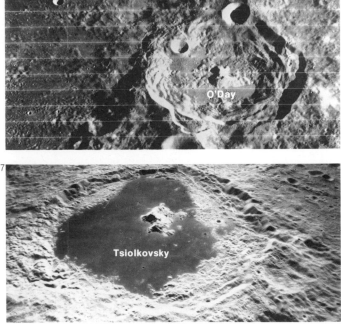

Crater Index

The names of lunar features appearing in the maps are listed with their latitude and longitude. Features other than craters are indicated by SMALL CAPITALS.

| Name | Lat. | Long. |
|---|---|---|
| Abbe | 58°S | 175°E |
| Abel | 36°S | 85°E |
| Abenezra | 21°S | 12°E |
| Abulfeda | 14°S | 14°E |
| Abul Wafa | 2°N | 117°E |
| Adams | 32°S | 69°E |
| AESTUUM, SINUS | 12°N | 9°W |
| Agatharchides | 20°S | 31°W |
| Agrippa | 4°N | 11°E |
| Airy | 18°S | 6°E |
| Aitken | 17°S | 173°E |
| Albategnius | 12°S | 4°E |
| Al Biruni | 18°N | 93°E |
| Alden | 24°S | 111°E |
| Alekhin | 68°S | 130°E |
| Alfraganus | 6°S | 19°E |
| Alhazen | 18°N | 70°E |
| Aliacensis | 31°S | 5°E |
| Almanon | 17°S | 15°E |
| ALPES, VALLIS | 49°N | 2°E |
| Alpetragius | 16°S | 4°W |
| Alphonsus | 13°S | 3°W |
| ALTAI, RUPES | 24°S | 22°E |
| Alter | 19°N | 108°W |
| Amici | 10°S | 172°W |
| Amundsen | 83°S | 103°W |
| Anaxagoras | 75°N | 10°W |
| Anaximander | 66°N | 48°W |
| Anaximenes | 75°N | 45°W |
| Andĕl | 10°S | 13°E |
| Anders | 42°S | 144°W |
| Anderson | 16°N | 171°E |
| Ångström | 30°N | 42°W |
| ANGUIS, MARE | 23°N | 69°E |
| Ansgarius | 14°S | 82°E |
| Antoniadi | 69°S | 173°W |
| APENNINUS, MONTES | 20°N | 2°W |
| Apianus | 27°S | 8°E |
| Apollo | 37°S | 153°W |
| Apollonius | 5°N | 61°E |
| Appleton | 37°N | 158°E |
| Arago | 6°N | 21°E |
| Aratus | 24°N | 5°E |
| Archimedes | 30°N | 4°W |
| Archytas | 59°N | 5°E |
| Argelander | 17°S | 6°E |
| ARIADAEUS, RIMA | 7°N | 13°E |
| Aristarchus | 24°N | 48°W |
| Aristillus | 34°N | 1°E |
| Aristoteles | 50°N | 18°E |
| Arnold | 67°N | 38°E |
| Arrhenius | 55°S | 91°W |
| Artamonov | 26°N | 104°E |
| Artem'ev | 10°N | 145°W |
| Arzachel | 18°S | 2°W |
| Asclepi | 55°S | 26°E |
| Aston | 32°N | 88°W |
| Atlas | 47°N | 44°E |
| AUSTRALE, MARE | 50°S | 80°E |
| Autolycus | 31°N | 1°E |
| Auwers | 15°N | 17°E |
| Auzout | 10°N | 64°E |
| Avicenna | 40°N | 97°W |
| Avogadro | 64°N | 165°E |
| Aymat | 57°N | 55°W |
| Azophi | 22°S | 13°E |
| Baade | 47°S | 83°W |
| Babbage | 58°N | 52°W |
| Babcock | 4°N | 94°E |
| Backlund | 16°S | 103°E |
| Baco | 51°S | 19°E |
| Baillaud | 75°N | 40°E |
| Bailly | 66°S | 65°E |
| Baily | 50°N | 31°E |
| Balboa | 20°N | 84°W |
| Baldet | 54°S | 151°W |
| Ball | 36°S | 8°W |
| Balmer | 20°S | 70°E |
| Barbier | 24°S | 158°E |
| Barker | 42°S | 8°W |
| Barnard | 32°S | 86°E |
| Barocius | 45°S | 17°E |
| Barringer | 29°S | 151°W |
| Barrow | 73°N | 10°E |
| Bayer | 51°S | 35°W |
| Beaumont | 18°S | 29°E |
| Becquerel | 41°N | 129°E |
| Bečvar | 2°S | 125°E |
| Beer | 27°N | 9°W |
| Behaim | 16°S | 71°E |
| Beijerinck | 13°S | 152°E |
| Belkovich | 60°S | 92°E |
| Bell | 22°N | 96°W |
| Bellingshausen | 61°S | 164°W |
| Belopolsky | 18°S | 128°W |
| Belyayev | 23°N | 143°E |
| Bergstrand | 19°S | 176°E |
| Berkner | 25°N | 105°W |
| Berlage | 64°S | 164°W |
| Bernouilli | 34°N | 60°E |
| Berosus | 33°N | 70°E |
| Berzelius | 37°N | 51°E |
| Bessarion | 15°N | 37°E |
| Bessel | 22°N | 18°E |
| Bettinus | 63°S | 45°W |
| Bhabha | 56°S | 165°W |
| Bianchini | 49°N | 34°W |
| Biela | 55°S | 52°E |
| Billy | 14°S | 50°W |
| Biot | 23°S | 51°E |
| Birkeland | 30°S | 174°E |
| Birkhoff | 59°N | 148°W |
| Birmingham | 64°N | 10°W |
| Birt | 22°S | 9°W |
| Bjerknes | 38°S | 113°E |
| Blagg | 1°N | 2°E |
| Blancanus | 64°S | 21°W |
| Blanchinus | 25°S | 3°E |
| Blazhko | 31°N | 148°W |
| Bobone | 29°N | 131°W |
| Bode | 7°N | 2°W |
| Boguslawsky | 75°S | 45°E |
| Bohnenberger | 16°S | 40°E |
| Bohr | 13°N | 86°W |
| Boltzmann | 78°S | 99°E |
| Bolyai | 34°S | 125°E |
| Bond, G. | 32°N | 36°E |
| Bond, W. | 64°N | 3°E |
| Bonpland | 8°S | 17°W |
| Borda | 25°S | 47°E |
| Borman | 37°S | 142°W |
| Boscovich | 10°N | 11°E |
| Bose | 54°S | 170°W |
| Boss | 46°N | 90°E |
| Bouguer | 52°N | 36°W |
| Boussingault | 70°S | 50°E |
| Boyle | 54°S | 178°E |
| Bragg | 42°N | 103°W |
| Brashear | 74°S | 172°W |
| Brayley | 21°N | 37°W |
| Bredikhin | 17°N | 158°W |
| Breislak | 48°S | 18°E |
| Brenner | 39°S | 39°E |
| Brianchon | 77°N | 90°W |
| Bridgman | 44°N | 137°E |
| Briggs | 26°N | 69°W |
| Brisbane | 50°S | 65°E |
| Brouwer | 36°S | 125°W |
| Bruce | 1°N | 0° |
| Brunner | 10°S | 91°E |
| Buch | 39°S | 18°E |
| Buffon | 41°S | 134°W |
| Buisson | 1°S | 113°E |
| Bullialdus | 21°S | 22°W |
| Bunsen | 41°N | 85°W |
| Burckhardt | 31°N | 57°E |
| Bürg | 45°N | 28°E |
| Burnham | 14°S | 7°E |
| Büsching | 38°S | 20°E |
| Buys-Ballot | 21°N | 175°E |
| Byrd | 83°N | 0° |
| Byrgius | 25°S | 65°W |
| Cabannes | 61°S | 171°W |
| Caesar, Julius | 9°N | 15°E |
| Cajori | 48°S | 168°E |
| Calippus | 39°N | 11°E |
| Campanus | 28°S | 28°W |
| Campbell | 45°N | 152°E |
| Cannizzaro | 55°N | 100°W |
| Cannon | 20°N | 80°E |
| Cantor | 38°N | 118°E |
| Capella | 8°S | 36°E |
| Capuanus | 34°S | 26°W |
| Cardanus | 13°N | 73°W |
| Carlini | 34°N | 24°W |
| Carnot | 52°N | 144°W |
| CARPATUS, MONTES | 15°N | 24°W |
| Carpenter | 70°N | 50°W |
| Carrington | 44°N | 62°E |
| Carver | 43°S | 127°E |
| Casatus | 75°S | 35°W |
| Cassegrain | 52°S | 113°E |
| Cassini | 40°N | 5°E |
| Catalan | 46°S | 87°W |
| Catharina | 18°S | 24°E |
| CAUCASUS, MONTES | 36°N | 8°E |
| Cauchy | 10°N | 39°E |
| Cavalerius | 5°N | 67°W |
| Cavendish | 25°S | 54°W |
| Cayley | 4°N | 15°E |
| Censorinus | 0° | 32°E |
| Cepheus | 41°N | 46°E |
| Ceraski | 49°S | 141°E |
| Chacornac | 30°N | 32°E |
| Chaffee | 39°S | 155°W |
| Challis | 78°N | 9°E |
| Chamberlin | 59°S | 96°E |
| Champollion | 37°N | 175°E |
| Chandler | 44°N | 171°E |
| Chant | 41°S | 110°W |
| Chaplygin | 6°S | 150°E |
| Chapman | 50°N | 101°W |
| Chappell | 62°N | 150°W |
| Charlier | 36°N | 132°W |
| Chaucer | 3°N | 140°W |
| Chauvenet | 11°S | 137°E |
| Chebyshev | 34°S | 133°W |
| Chernyshev | 47°N | 174°E |
| Chevallier | 45°N | 52°E |
| Chladni | 4°N | 1°E |
| Chrétien | 33°S | 113°E |
| Cichus | 33°S | 21°W |
| Clairaut | 48°S | 14°E |
| Clark | 38°S | 119°E |
| Clausius | 37°S | 44°W |
| Clavius | 58°S | 14°W |
| Cleomedes | 27°N | 55°E |
| Coblentz | 38°S | 126°E |
| Cockcroft | 30°N | 164°W |
| Colombo | 15°S | 46°E |
| Compton | 55°N | 104°E |
| Comrie | 23°N | 113°W |
| Comstock | 21°N | 122°W |
| Condorcet | 12°N | 70°E |
| Congreve | 0° | 168°W |
| Conon | 22°N | 2°E |
| Cook | 18°S | 49°E |
| Cooper | 53°N | 176°E |
| Copernicus | 10°N | 20°W |
| CORDILLERA, MONTES | 27°S | 85°W |
| Coriolis | 0° | 172°E |
| CRISIUM, MARE | 18°N | 58°E |
| Crocco | 47°S | 150°E |
| Crookes | 11°S | 165°W |
| Crüger | 17°S | 67°W |
| Curie | 23°S | 92°E |
| Curtius | 77°S | 5°E |
| Cusanus | 72°N | 70°E |
| Cuvier | 50°S | 10°E |
| Cyrano | 20°S | 157°E |
| Cyrillus | 13°S | 24°E |
| Cysatus | 66°S | 7°W |
| D'Alembert | 52°N | 164°E |
| Dalton | 18°N | 86°W |
| Damoiseau | 5°S | 61°W |
| Daniell | 35°N | 31°E |
| Danjon | 11°S | 123°E |
| Dante | 25°N | 180° |
| Darney | 15°S | 24°W |
| D'Arrest | 2°N | 15°E |
| Darwin | 20°S | 69°W |
| Das | 14°S | 152°W |
| Da Vinci | 9°N | 45°E |
| Davisson | 38°S | 175°W |
| Davy | 12°S | 8°W |
| Dawes | 17°N | 26°E |
| Debes | 29°N | 52°E |
| Debye | 50°N | 177°W |
| De Forest | 76°S | 162°W |
| De Gasparis | 26°S | 50°W |
| Delambre | 2°S | 18°E |
| Dellinger | 7°S | 140°E |
| De la Rue | 67°N | 55°E |
| Delaunay | 22°S | 3°E |
| Delisle | 30°N | 35°W |
| Delmotte | 27°N | 60°E |
| Delporte | 16°S | 121°E |
| Deluc | 55°S | 3°W |
| Dembowski | 3°N | 7°E |
| Democritus | 62°N | 35°E |
| Demonax | 85°S | 35°E |
| Denning | 16°S | 143°E |
| De Roy | 55°S | 99°W |
| Descartes | 12°S | 16°E |
| Deseilligny | 21°N | 21°E |
| De Sitter | 80°N | 40°E |
| Deslandres | 32°S | 6°W |
| Deutsch | 24°N | 111°E |
| De Vico | 20°S | 60°W |
| De Vries | 20°S | 177°W |
| Dewar | 3°S | 166°E |
| Dionysius | 3°N | 17°E |
| Diophantus | 28°N | 34°W |
| Dirichlet | 10°N | 151°W |
| Donati | 21°S | 5°E |
| Donner | 31°S | 98°E |
| Doppelmayer | 28°S | 41°W |
| Doppler | 13°S | 160°W |
| Dove | 47°S | 32°E |
| Drebbel | 41°S | 49°W |
| Dreyer | 10°N | 97°E |
| Drude | 39°S | 91°W |
| Dryden | 33°S | 157°W |
| Drygalski | 80°S | 55°W |
| Dubiago | 4°N | 70°E |
| Dufay | 5°N | 170°E |
| Dunér | 45°N | 179°E |
| Dyson | 61°N | 121°W |
| Dziewulski | 21°N | 99°E |
| Eddington | 22°N | 72°W |
| Egede | 49°N | 11°E |
| Ehrlich | 41°N | 172°W |
| Eichstadt | 23°S | 80°W |
| Eijkman | 62°S | 141°W |
| Eimmart | 24°N | 65°E |
| Einstein | 18°N | 86°W |
| Einthoven | 5°S | 110°E |
| Elger | 35°S | 30°W |
| Ellerman | 26°S | 121°W |
| Ellison | 55°N | 108°W |
| Elvey | 9°N | 101°W |
| Emden | 63°N | 176°W |
| Emley | 38°S | 28°W |
| Encke | 5°N | 37°W |
| Endymion | 55°N | 55°E |
| Engelhardt | 5°N | 159°W |
| Eötvös | 36°S | 134°E |
| EPIDEMIARUM, PALUS | 31°S | 26°W |
| Epigenes | 73°N | 4°W |
| Epimenides | 41°S | 30°W |
| Eratosthenes | 15°N | 11°W |
| Erro | 6°N | 98°E |
| Esnault-Pelterie | 47°N | 142°W |
| Espin | 28°N | 109°E |
| Euclides | 7°S | 29°W |
| Euctemon | 80°N | 40°E |
| Eudoxus | 44°N | 16°E |
| Euler | 23°N | 29°W |
| Evans | 10°S | 134°W |
| Evershed | 36°N | 160°W |
| Fabricius | 43°S | 42°E |
| Fabry | 43°N | 100°E |
| Faraday | 42°S | 8°E |
| Fauth | 6°N | 20°W |
| Faye | 21°S | 4°E |
| Fechner | 59°S | 125°E |
| FECUNDITATIS, MARE | 4°S | 51°E |
| Fenyi | 45°S | 105°W |
| Fermat | 23°S | 20°E |
| Fermi | 20°S | 123°E |
| Fernelius | 38°S | 5°E |
| Fersman | 18°N | 126°W |
| Feuillée | 27°N | 10°W |
| Firmicus | 7°N | 64°E |
| Firsov | 4°N | 112°E |
| Fitzgerald | 27°N | 172°W |
| Fizeau | 58°S | 133°W |
| Flammarion | 3°S | 4°W |
| Flamsteed | 5°S | 44°W |
| Fleming | 15°N | 109°E |
| Focas | 34°S | 94°W |
| Fontana | 16°S | 57°W |
| Fontenelle | 63°N | 19°W |
| Foucault | 50°N | 40°W |
| Fourier | 30°S | 53°W |
| Fowler | 43°N | 145°W |
| Fracastorius | 21°S | 33°E |
| Fra Mauro | 6°S | 17°W |
| Franklin | 39°N | 48°E |
| Franz | 16°N | 40°E |
| Fraunhofer | 39°S | 59°E |
| Freundlich | 25°N | 171°E |
| Fridman | 13°S | 127°W |
| FRIGORIS, MARE | 55°N | 0° |
| Froelich | 80°N | 110°W |
| Frost | 37°N | 119°W |
| Furnerius | 36°S | 60°E |
| Gadomski | 36°N | 147°W |
| Gagarin | 20°S | 150°E |
| Galilaei | 10°N | 63°W |
| Galle | 56°N | 22°E |
| Galois | 16°S | 153°W |
| Galvani | 49°N | 85°W |
| Gambart | 1°N | 15°W |
| Gamow | 65°N | 143°E |
| Ganswindt | 79°S | 110°E |
| Garavito | 48°S | 157°E |
| Gärtner | 60°N | 34°E |
| Gassendi | 18°S | 40°W |
| Gaudibert | 11°S | 37°E |
| Gauricus | 34°S | 12°W |
| Gauss | 36°N | 80°E |
| Gavrilov | 17°N | 131°E |
| Gay-Lussac | 14°N | 21°W |
| Geber | 20°S | 14°E |
| Geiger | 14°S | 158°E |
| Geminus | 35°N | 57°E |
| Gemma Frisius | 34°S | 14°E |
| Gerasimovic | 23°S | 124°W |

| Name | Lat | Long |
|---|---|---|
| Gernsback | 36°S | 99°E |
| Gibbs | 19°S | 83°E |
| Gilbert | 4°S | 75°E |
| Gill | 64°S | 75°E |
| Ginzel | 14°N | 97°E |
| Gioja | 89°N | 9°W |
| Glasenap | 2°S | 138°E |
| Goclenius | 10°S | 45°E |
| Goddard | 16°N | 90°E |
| Godin | 2°N | 10°E |
| Goldschmidt | 75°N | 0° |
| Golitzyn | 25°S | 105°W |
| Golovin | 40°N | 161°E |
| Goodacre | 33°S | 14°E |
| Gould | 19°S | 17°W |
| Grachev | 3°S | 108°W |
| Graff | 43°S | 88°W |
| Green | 4°N | 133°E |
| Gregory | 2°N | 127°E |
| Grimaldi | 6°S | 68°W |
| Grissom | 45°S | 160°W |
| Grotrian | 66°S | 128°E |
| Grove | 40°N | 33°E |
| Gruemberger | 68°S | 10°W |
| Gruithuisen | 33°N | 40°W |
| Guericke | 12°S | 14°W |
| Gullstrand | 45°N | 130°W |
| Gum | 40°S | 89°E |
| Gutenberg | 8°S | 41°E |
| Guthnick | 48°S | 94°W |
| Guyot | 11°N | 117°E |
| HAEMUS, MONTES | 16°N | 14°E |
| Hagecius | 60°S | 46°E |
| Hagen | 56°S | 135°E |
| Hahn | 31°N | 74°E |
| Haidinger | 39°S | 25°W |
| Hainzel | 41°S | 34°W |
| Hale | 74°S | 90°E |
| Hall | 34°N | 37°E |
| Halley | 8°S | 6°E |
| Hamilton | 44°S | 83°E |
| Hanno | 57°S | 73°E |
| Hansen | 44°S | 83°E |
| Hansky | 10°S | 97°E |
| Hansteen | 12°S | 52°W |
| Harding | 43°N | 70°W |
| Harpalus | 53°N | 43°W |
| Harriot | 33°N | 114°E |
| Hartmann | 3°N | 135°E |
| Hartwig | 7°S | 82°W |
| Harvey | 19°N | 147°W |
| Hase | 30°S | 63°E |
| Hauet | 37°S | 18°W |
| Hausen | 65°S | 90°W |
| Hayford | 13°N | 176°W |
| Hayn | 63°N | 85°E |
| Healy | 32°N | 111°W |
| Heaviside | 10°S | 167°E |
| Hecateus | 23°S | 84°E |
| Heinsius | 39°S | 18°W |
| Heis | 32°N | 32°W |
| Helberg | 22°N | 102°W |
| Helicon | 40°N | 23°W |
| Hell | 32°S | 8°W |
| Helmholtz | 72°N | 78°E |
| Henderson | 5°N | 152°E |
| Hendrix | 48°S | 161°W |
| Henry, Paul | 24°S | 57°W |
| Henry, Prosper | 24°S | 59°W |
| Henyey | 13°N | 152°W |
| Heraclitus | 49°S | 6°E |
| Hercules | 46°N | 39°E |
| Herigonius | 13°S | 34°W |
| Hermann | 1°S | 57°W |
| Hermite | 85°N | 90°W |
| Herodotus | 23°N | 50°W |
| Herschel | 6°S | 2°W |
| Herschel, C. | 34°N | 31°W |
| Herschel, J. | 62°N | 41°W |
| Hertz | 14°N | 104°E |
| Hertzsprung | 0° | 130°W |
| Hesiodus | 29°S | 16°W |
| Hess | 54°S | 174°E |
| Hevelius | 2°N | 67°W |
| Heymans | 75°N | 145°W |
| Hilbert | 18°S | 108°E |
| Hind | 8°S | 7°E |
| Hippalus | 25°S | 30°W |
| Hipparchus | 6°S | 5°E |
| Hippocrates | 71°N | 146°W |
| Hirayama | 6°S | 93°E |
| Hoffmeister | 15°N | 137°E |
| Hogg | 34°N | 122°E |
| Hohmann | 18°S | 94°W |
| Holden | 19°S | 63°E |
| Holetschek | 28°S | 151°E |
| Hommel | 54°S | 33°E |
| Hooke | 41°N | 55°E |
| Horrebow | 59°N | 41°W |
| Horrocks | 4°S | 6°E |
| Hortensius | 6°N | 28°W |
| Houzeau | 18°S | 124°W |
| Hubble | 22°N | 87°E |
| Huggins | 41°S | 2°W |
| Humboldt, W. | 27°S | 81°E |
| HUMBOLDTIANUM, MARE | 55°N | 75°E |
| HUMORUM, MARE | 23°S | 38°W |
| Hutton | 37°N | 169°E |
| Hyginus | 8°N | 6°E |
| HYGINUS, RIMA | 8°N | 6°E |
| Hypatia | 4°S | 23°E |
| Ibn Yunus | 14°N | 91°E |
| Icarus | 6°S | 173°W |
| Ideler | 49°S | 22°E |
| Idelson | 81°S | 114°E |
| IMBRIUM, MARE | 36°N | 16°W |
| Ingalls | 4°S | 153°W |
| INGENII, MARE | 35°S | 164°E |
| Inghirami | 48°S | 70°W |
| Innes | 28°N | 119°E |
| IRIDUM, SINUS | 45°N | 32°W |
| Isidorus | 8°S | 33°E |
| Izsak | 23°S | 117°E |
| Jacobi | 57°S | 12°E |
| Jackson | 22°N | 163°W |
| Jansen | 14°N | 29°E |
| Jansky | 8°N | 87°E |
| Janssen | 46°S | 40°E |
| Jeans | 53°S | 91°W |
| Jenner | 42°S | 96°E |
| Joffe | 15°S | 129°W |
| Joliot | 27°N | 93°E |
| Joule | 27°N | 144°W |
| JURA, MONTES | 46°N | 38°W |
| Kaiser | 36°S | 6°E |
| Kamerlingh Onnes | 15°N | 116°W |
| Kane | 63°N | 26°E |
| Kant | 11°S | 20°E |
| Kapteyn | 13°S | 70°E |
| Karpinsky | 73°N | 166°E |
| Kästner | 6°S | 80°E |
| Kearons | 12°S | 113°W |
| Keeler | 10°S | 162°E |
| Kékulé | 16°N | 138°W |
| Kepler | 8°N | 38°W |
| Kibalchich | 2°N | 147°W |
| Kidinnu | 36°N | 123°E |
| Kies | 26°S | 23°W |
| Kimura | 57°S | 118°E |
| Kinau | 60°S | 15°E |
| King | 5°N | 120°E |
| Kirch | 39°N | 6°W |
| Kircher | 67°S | 45°W |
| Kirchhoff | 30°N | 39°E |
| Kirkwood | 69°N | 157°W |
| Klaproth | 68°S | 22°W |
| Kleimenov | 33°S | 141°W |
| Klein | 12°S | 3°E |
| Klute | 37°N | 142°W |
| Koch | 43°S | 150°E |
| Kohlschütter | 15°N | 154°E |
| Kolhörster | 10°N | 115°W |
| Komarov | 25°N | 153°E |
| Kondratyuk | 15°S | 115°E |
| König | 24°S | 25°W |
| Konstantinov | 20°N | 159°E |
| Kopff | 17°S | 90°W |
| Korolev | 5°S | 157°W |
| Kostinsky | 14°N | 118°E |
| Kovalevskaya | 31°N | 129°W |
| Kovalsky | 22°S | 101°E |
| Kraft | 13°N | 73°W |
| Krasnov | 31°S | 89°W |
| Krasovsky | 4°N | 176°W |
| Krieger | 29°N | 46°W |
| Krylov | 9°S | 157°W |
| Kugler | 53°S | 104°E |
| Kulik | 42°N | 155°W |
| Kunowsky | 3°N | 32°W |
| Kuo Shou Ching | 8°N | 134°W |
| Kurchatov | 39°N | 142°E |
| Lacaille | 24°S | 1°E |
| Lacchini | 41°N | 107°W |
| La Condamine | 53°N | 28°W |
| Lacroix | 38°S | 59°W |
| Lade | 1°S | 10°E |
| LA HIRE, MONS | 28°N | 25°W |
| Lalande | 4°S | 8°W |
| Lamarck | 23°S | 69°W |
| Lamb | 43°S | 101°E |
| Lambert | 26°N | 21°W |
| Lamé | 15°S | 65°E |
| Lamèch | 43°N | 13°E |
| Lamont | 5°N | 23°E |
| Lampland | 31°S | 131°E |
| Landau | 42°N | 119°W |
| Lane | 9°S | 132°E |
| Langemak | 10°S | 119°E |
| Langevin | 44°N | 162°E |
| Langley | 52°N | 87°W |
| Langmuir | 36°S | 129°W |
| Langrenus | 9°S | 61°E |
| Lansberg | 0° | 26°W |
| La Pérouse | 10°S | 78°E |
| Larmor | 32°N | 180° |
| Lassell | 16°S | 8°W |
| Laue | 29°N | 96°W |
| Lauritsen | 27°S | 96°E |
| Lavoisier | 36°N | 70°W |
| Leavitt | 46°S | 140°W |
| Lebedev | 48°S | 108°E |
| Lebedinsky | 8°N | 165°W |
| Lee | 31°S | 41°W |
| Leeuwenhoek | 30°S | 179°W |
| Legendre | 29°S | 70°E |
| Legentil | 73°S | 80°E |
| Lehmann | 40°S | 56°W |
| Leibnitz | 38°S | 178°E |
| Lemaître | 62°S | 150°W |
| Lemonnier | 26°N | 31°E |
| Leonov | 19°N | 148°E |
| Lepaute | 33°S | 34°W |
| Letronne | 10°S | 43°W |
| Leucippus | 29°N | 116°W |
| Leuschner | 1°N | 109°W |
| Leverrier | 40°N | 20°W |
| Levi-Civita | 24°S | 143°E |
| Lewis | 19°S | 114°W |
| Lexell | 36°S | 4°W |
| Ley | 43°N | 154°E |
| Liapunov | 27°N | 88°E |
| Licetus | 47°S | 6°E |
| Lichtenberg | 32°N | 68°W |
| Lick | 12°N | 53°E |
| Liebig | 24°S | 48°W |
| Lilius | 54°S | 6°E |
| Lindenau | 32°S | 25°E |
| Linné | 28°N | 12°E |
| Lippershey | 26°S | 10°W |
| Littrow | 22°N | 31°E |
| Lobachevsky | 9°N | 112°E |
| Lockyer | 46°S | 37°E |
| Lodygin | 18°S | 147°W |
| Lohrmann | 1°S | 67°W |
| Lohse | 14°S | 60°E |
| Lomonosov | 28°N | 98°E |
| Longomontanus | 50°S | 21°W |
| Lorentz | 34°N | 100°W |
| Love | 6°S | 129°E |
| Lovelace | 82°N | 107°W |
| Lovell | 39°S | 149°W |
| Lowell | 13°S | 103°W |
| Lubbock | 4°S | 42°E |
| Lubiniezky | 18°S | 24°W |
| Lucretius | 9°S | 121°W |
| Lundmark | 39°S | 152°E |
| Luther | 33°N | 24°E |
| Lütke | 17°S | 123°E |
| Lyell | 14°N | 41°E |
| Lyman | 65°S | 162°E |
| Lyot | 48°S | 88°W |
| Mach | 18°N | 149°W |
| McKeller | 16°S | 171°W |
| McLaughlin | 47°N | 93°W |
| Maclaurin | 2°S | 68°E |
| Maclear | 11°N | 20°E |
| McMath | 15°N | 167°W |
| McNally | 22°N | 127°W |
| Macrobius | 21°N | 46°E |
| Mädler | 11°S | 30°E |
| Magelhaens | 12°S | 44°E |
| Maginus | 50°S | 6°W |
| Main | 82°N | 3°E |
| Mairan | 42°N | 43°W |
| Maksutov | 41°S | 169°W |
| Malyi | 22°N | 105°E |
| Mandelstam | 4°N | 156°E |
| Manilius | 15°N | 9°E |
| Manners | 5°N | 20°E |
| Manzinus | 68°S | 25°E |
| Maraldi | 19°N | 35°E |
| Marconi | 9°S | 145°E |
| MARGINIS, MARE | 13°N | 87°E |
| Marinus | 50°S | 75°E |
| Mariotte | 29°S | 140°W |
| Marius | 12°N | 51°W |
| Maskelyne | 2°N | 30°E |
| Mason | 43°N | 30°E |
| Maunder | 14°S | 94°W |
| Maurolycus | 42°S | 14°E |
| Maury | 37°N | 40°E |
| Maxwell | 30°N | 99°E |
| Mayer, C. | 63°N | 17°E |
| Mayer, Tobias | 16°N | 29°W |
| MEDII, SINUS | 0° | 0° |
| Mee | 44°S | 35°W |
| Mees | 14°N | 96°W |
| Meggers | 24°N | 123°E |
| Meitner | 11°S | 113°E |
| Mendel | 49°S | 110°W |
| Mendeleev | 5°N | 140°E |
| Menelaus | 16°N | 16°E |
| Mercator | 29°S | 26°W |
| Mercurius | 56°N | 65°E |
| Merrill | 75°N | 116°W |
| Mersenius | 21°S | 49°W |
| Mesentsev | 72°N | 129°W |
| Meshcerski | 12°N | 125°E |
| Messala | 39°N | 60°E |
| Messier | 2°S | 48°E |
| Metchnikoff | 11°S | 149°W |
| Metius | 40°S | 44°E |
| Meton | 74°N | 25°E |
| Michelson | 6°N | 121°W |
| Milanković | 77°N | 170°E |
| Milichius | 10°N | 30°W |
| Miller | 39°S | 1°E |
| Millikan | 47°N | 121°E |
| Mills | 9°N | 156°E |
| Milne | 31°S | 113°E |
| Mineur | 25°N | 162°W |
| Minkowski | 56°S | 145°W |
| Mitchell | 50°N | 20°E |
| Mitra | 18°N | 155°W |
| Möbius | 16°N | 101°E |
| Mohorovičić | 19°S | 165°W |
| Moigno | 66°N | 28°E |
| Moiseev | 9°N | 103°E |
| Moltke | 1°S | 24°E |
| Monge | 19°S | 48°E |
| Montgolfier | 47°N | 160°W |
| Moore | 37°N | 178°W |
| Moretus | 70°S | 8°W |
| Morozov | 5°N | 127°E |
| Morse | 22°N | 175°W |
| MORTIS, LACUS | 44°N | 27°E |
| MOSCOVIENSE, MARE | 25°N | 147°E |
| Mösting | 1°S | 6°W |
| Mouchez | 86°N | 35°W |
| Murchison | 5°N | 0° |
| Mutus | 63°S | 30°E |
| Nagaoka | 20°N | 154°E |
| Nansen | 81°N | 91°E |
| Nasireddin | 41°S | 0° |
| Nasmyth | 52°S | 53°W |
| Nassau | 25°S | 177°E |
| Naumann | 35°N | 62°W |
| Neander | 31°S | 40°E |
| Nearch | 58°S | 39°E |
| NEBULARUM, PALUS | 38°N | 1°E |
| NECTARIS, MARE | 14°S | 34°E |
| Neison | 68°N | 28°E |
| Neper | 7°N | 83°E |
| Nernst | 36°N | 95°W |
| Neujmin | 27°S | 125°E |
| Neumayer | 71°S | 70°E |
| Newcomb | 30°N | 44°E |
| Newton | 78°S | 20°W |
| Nicholson | 26°S | 85°W |
| Nicolaev | 35°N | 151°E |
| Nicolai | 42°S | 26°E |
| Nicollet | 22°S | 12°W |
| Niépce | 72°N | 120°W |
| Nijland | 33°N | 134°E |
| Nishina | 45°S | 171°W |
| Nobel | 15°N | 101°W |
| Nöggerath | 49°S | 45°W |
| Nonius | 35°S | 4°E |
| Nöther | 66°N | 114°W |
| NUBIUM, MARE | 19°S | 14°W |
| Numerov | 71°S | 161°W |
| Nušl | 32°N | 167°W |
| Obruchev | 39°S | 162°E |
| O'Day | 31°S | 157°E |
| Oenopides | 57°N | 65°W |
| Oersted | 43°N | 47°E |
| Ohm | 18°N | 114°W |
| Oken | 44°S | 78°E |
| Olbers | 7°N | 78°W |
| Olcott | 20°N | 117°E |
| Omar Khayyam | 58°N | 102°W |
| Opelt | 16°S | 18°W |
| Oppenheimer | 35°S | 166°W |
| Oppolzer | 2°S | 1°W |
| Oresme | 43°S | 169°E |
| ORIENTALE, MARE | 19°S | 95°W |
| Orlov | 26°S | 175°W |
| Orontius | 40°S | 4°W |
| Ostwald | 11°N | 122°E |
| Palisa | 9°S | 7°W |
| Pallas | 5°N | 2°W |
| Palmieri | 29°S | 48°W |
| Paneth | 63°N | 95°W |
| Pannekoek | 4°S | 140°E |
| Paracelsus | 23°S | 163°E |
| Paraskevopoulos | 50°N | 150°W |

# Crater Index

# Plates

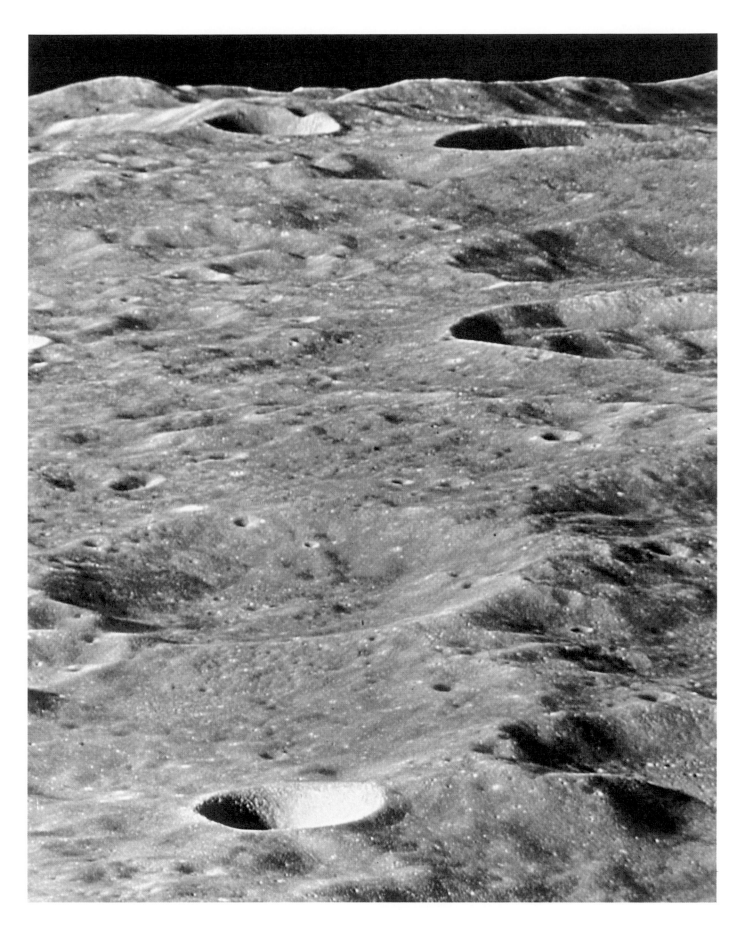

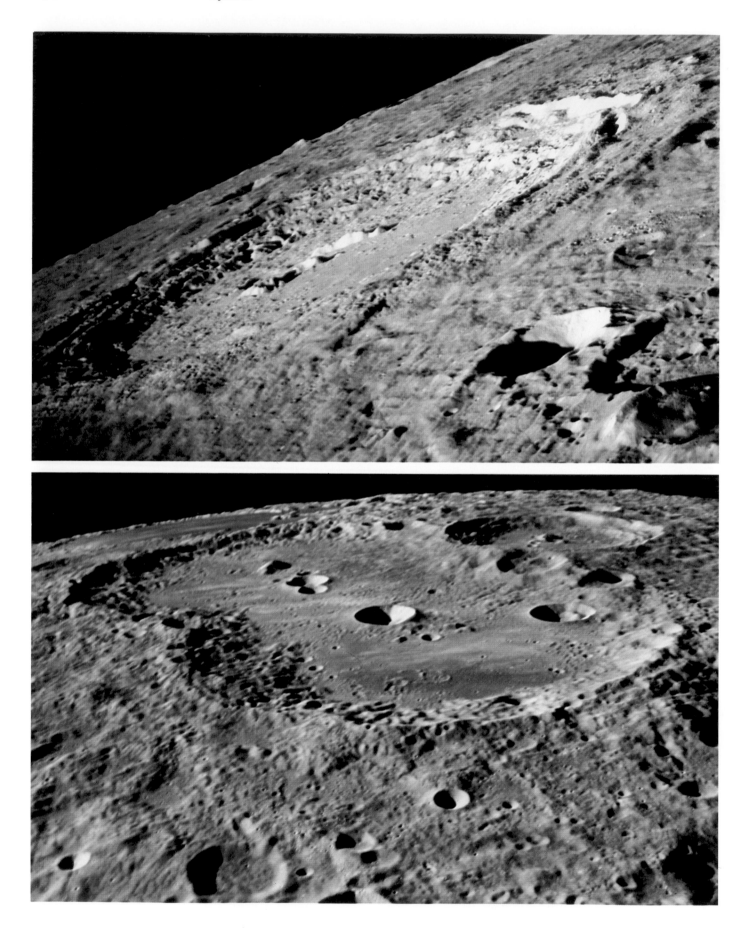

**Eratosthenes (top)**
Apollo 17

**Doppler (bottom)**
Apollo 17

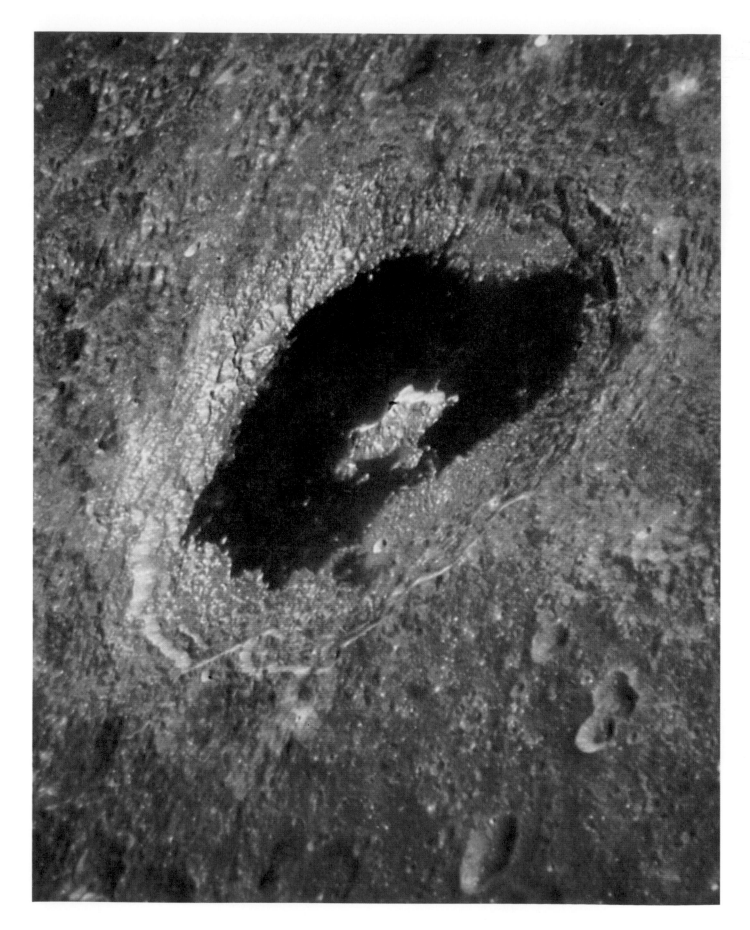

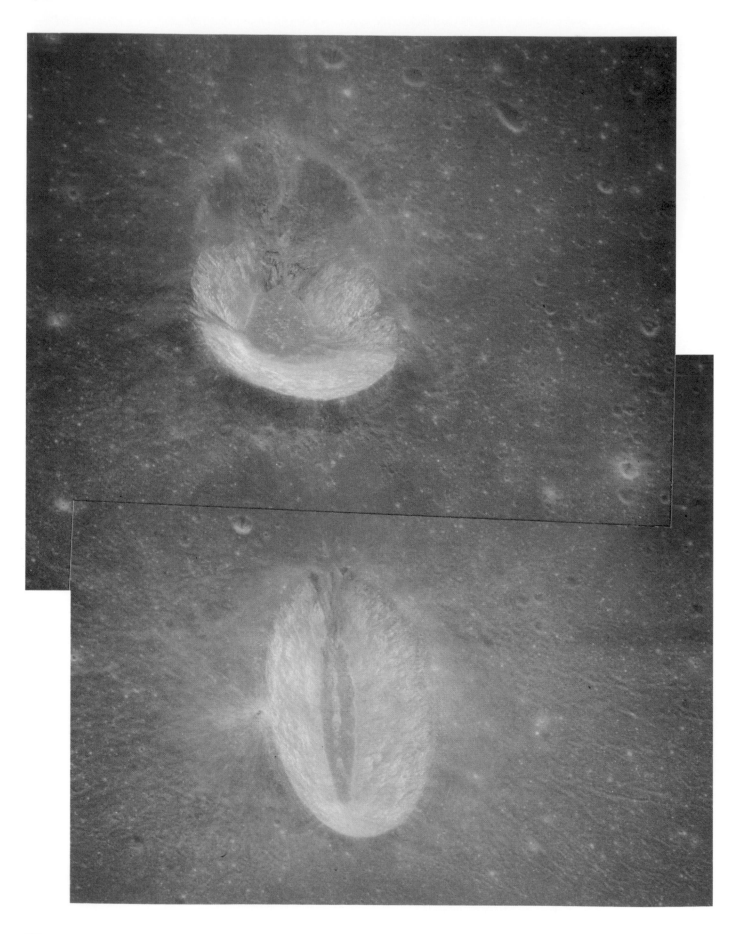

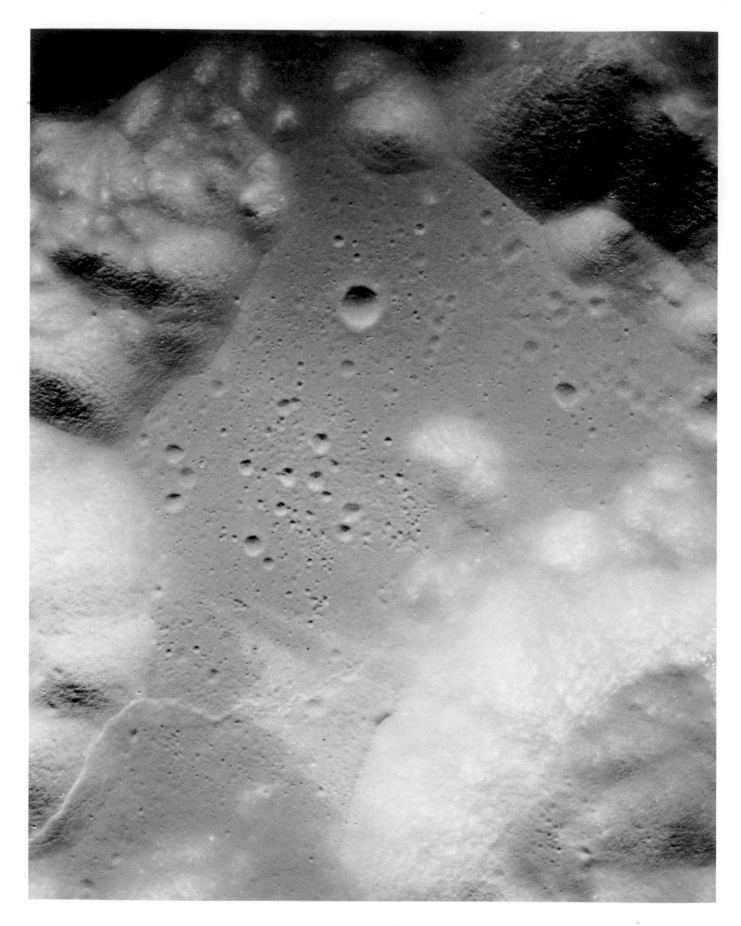

Lunar basalt
Apollo 11
Reflected light (top), polarized light (bottom)

## Notes to Color Plates

**Page 193.** Taken from the Lunar Module of Apollo 11 during the return journey at a distance of 16,000 km. The disc is centered at about 7°E.

**Page 194.** Detail of north-east section of the near side, showing two prominent maria, Crisium and Fecunditatis.

**Page 195.** Mare region to the north-west of Aristarchus, photographed from the Command Module of Apollo 15.

**Page 196.** Vertical view of highland terrain on the far side of the Moon (3°S, 155°E) photographed by Apollo 8.

**Page 197.** Oblique view of highland terrain; the region shown is close to that in the previous photograph (10°S, 155°E).

**Page 198.** (Top). Oblique view of the crater Copernicus (10°N, 20°W) showing central peaks and terraced walls. The small crater to the right of the foreground is Gay-Lussac A, while the slightly larger crater in the bottom right-hand corner is Gay-Lussac. (Bottom). The large, irregularly shaped feature is Van de Graaff (27°S, 172°E); several smaller craters intrude, the largest (towards the top right-hand corner in the photograph) is Birkeland (30°S, 174°E).

**Page 199.** (Top). Eratosthenes (15°N, 11°W). The large crater on the limb in this photograph is Copernicus. (Bottom). The main feature, Doppler (13°S, 160°W), adjoins a much larger crater, Korolev, which can be seen stretching away towards the limb. The walls of Korolev are comparatively worn down, and both Doppler and Korolev are pitted with many smaller craters.

**Page 200.** Tsiolkovsky (21°S, 128°E) is one of the most prominent and unusual features of the far side of the Moon.

**Page 201.** (Top). Humboldt (27°S, 81°E) has a remarkable system of radial and concentric fractures on its floor. (Bottom). Detail of Humboldt, showing the fractures.

**Page 202.** The twin craters Messier (above) and Messier A (lower) are situated in the Mare Fecunditatis (at 2°S, 48°E). Messier A has a distinctive pattern of rays.

**Page 203.** (Top). Parry (8°S, 16°W) is the large walled plain in the foreground of the photograph; Fra Mauro is to the right, and Bonpland can be seen in the background. A graben-like rill, Rima Parry, runs across the picture, hitting a small, well-formed crater on the common wall of Parry and Bonpland. (Bottom). This unnamed crater on the far side of the Moon was photographed from the Lunar Module of Apollo 10 during its descent to within 15 km of the lunar surface.

**Pages 204/205.** Schröter's Valley. The crater at one end of Schröter's Valley is known as the Cobra Head; such irregular depressions are characteristic of sinuous rills. Schröter's Valley is one of the largest sinuous rills on the Moon.

**Page 206.** Vertical view of Taurus-Littrow (20°N, 31°E), the site of the Apollo 17 landing (indicated by a white arrow). The mountain mass to the bottom right of the photograph is called South Massif.

**Page 207.** View from the lunar surface of the valley of Taurus-Littrow, looking north-east. In the foreground is a large boulder field.

**Page 208.** (Top). Mare basalt from Apollo 11 mission, taken in reflected light (sample number 10003,152). The following minerals are easily identifiable: plagioclase (dark grey, elongate); pyroxene (light grey, surrounding plagioclase); ilmenite (light blue); cristobalite ($SiO_2$) (very dark grey-brown, with many cracks—one grain may be seen near the center, several others in the lower left-hand corner); troilite (FeS) (bright pinkish mineral, uncommon in this sample—one small grain near the center). Other minerals are discussed in the main text—*see* pages 152–3. (Bottom). Same sample of mare basalt as in the previous photograph, this time viewed under plane polarized light. Plagioclase shows up as white, pyroxene as multicolored, ilmenite as black, cristobalite as white with cracks, and troilite as black.

# MARS

# Characteristics

Mars, the first planet in the Solar System beyond the orbit of the Earth, is a comparatively small world, but before the advent of the Space Age it was considered by many to be the planetary twin of the Earth. The atmosphere was believed to be similar to that on Earth, the polar caps were thought to contain water-ice, temperatures were comparable with (though much lower than) the Earth's, and there was much speculation about possible life forms. At its brightest Mars outshines all the planets apart from Venus and its red color, clearly visible from the Earth, led observers in ancient times to name it in honor of the God of War.

Mars has a revolution period of 687 Earth days and an orbit that is considerably more eccentric than that of the Earth. The distance from the Sun ranges between 249,000,000 km and as little as 206,000,000 km. This means that not all oppositions are equally favorable. The closest approaches occur when Mars is near perihelion and opposition at the same time: this will happen next in September 1988, when the apparent diameter of the disc will be almost 24 arcseconds, and Mars' distance from the Earth will be only 58,400,000 km. Oppositions occur at mean intervals of 780 days, and since the disc is comparatively small the opportunities for useful observations from Earth are somewhat restricted.

Mars' axis of rotation is at a 24° angle to the orbital plane (compared with 23.4° on the Earth), so there are four seasons during the year on Mars just as on Earth, except that the Martian year is the equivalent of 23 Earth months. Observers on Earth have long watched the progression of the seasons on Mars. During spring in the northern hemisphere, for example, the north polar cap shrinks and material in more temperate latitudes darkens in color and appears to increase in extent. Solar heating is about 40 percent greater when Mars is at perihelion than at aphelion. The rotation period on Mars is 24 hr 37 min 22.6 sec—only 41 minutes longer than the terrestrial day.

The surface temperature range on Mars is between 148 and 310 K. On the whole Mars is very cold. Temperatures at the equator range from a summer high of about 26°C in the early afternoon to −111°C just before sunrise. In polar regions temperatures rarely rise above −123°C all year round.

The diameter of Mars is 6,794 km at the equator and 6,759 km at the poles. The mass of Mars is only 0.11 that of the Earth, and the planet's escape velocity—5 km/sec—is sufficient to retain only a very thin atmosphere (*see* pages 216–19). Mars is not only smaller than the Earth but less dense: it has in fact the lowest density (3.93 gm/cm³) of any of the terrestrial planets, being only slightly denser than the Moon. Our knowledge of the internal structure of Mars, however, is limited and our understanding based rather on theoretical models. Direct observations have not confirmed whether or not the planet has a core. The first probes to fly past Mars revealed a very weak magnetic field. Mariner 4 encountered a bow shock but was unable to determine whether or not the field was intrinsic to Mars or induced by the solar wind. The assumption, therefore, that Mars has an iron-rich core that is smaller than that of the Earth is, at best, a tentative suggestion. The lack of an appreciable magnetic field means that, in this respect, Mars resembles the Moon rather than Mercury. If the internal structure does resemble that of the Earth—with a core, mantle and relatively thin outer crust—the core temperature must be considerably lower than in the case of the Earth.

In shape Mars is by no means a perfect sphere as there is a pronounced bulge, centered at 101°W 14°S in the region called Tharsis, which rises to 10 km above mean datum (taken as 6.1 mb atmospheric pressure in the absence of oceans). The planet is characterized by a wide diversity in physiography, revealed first by Mariner 9 and subsequently by the Viking Orbiters. The surface can be divided into two rather unequal hemispheres, the dividing line running somewhat obliquely to the equator. In general terms the southern is the older of the two, this being indicated by the very

great density of craters present. Counts reveal a similar number per unit area to that in the lunar highlands. This older terrain is also between 1 and 3 km above datum. The northern hemisphere is at a much lower elevation and is less densely cratered. Much of the surface reveals the effects of volcanic processes, and several extremely massive shield volcanoes are particularly striking. These rise from the broad Tharsis "bulge" and from a smaller but similar raised area in Elysium, to the west.

Extending eastward from Tharsis is an immense canyon system, Valles Marineris, while radiating outward from it are large arrays of tensional faults (graben); both are presumably related to the formation of Tharsis itself. North of the canyons are numerous fascinating outflow channels, which appear to have been produced during a period of catastrophic flooding, between 3,500 and 3,000 million years ago. Along the boundary between the ancient cratered terrain and the younger volcanic plains is a swath of country where the two interlace. The ancient surface has apparently wasted away and in places has collapsed into "chaotic terrain", across which there has been considerable mass movement of the eroded debris.

The youngest surfaces exposed on Mars appear to be those around the poles. Here well-laminated deposits appear, these being

**1. Size of Mars**
Mars (diameter 6,794 km) is half as big as the Earth (12,756) and twice as big as the Moon (3,476).

1

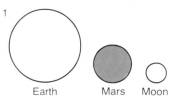

Earth　　Mars　Moon

**Physical data**

|  | Mars |
| --- | --- |
| Equatorial diameter | 6,794 km |
| Ellipticity | 0.0059 |
| Mass | 6.4191 × 10²³ kg |
| Volume (Earth = 1) | 0.15 |
| Density (water = 1) | 3.93 |
| Surface gravity (Earth = 1) | 0.379 |
| Escape velocity | 5.02 km s⁻¹ |
| Equatorial rotation | 24.6229 hr |
| Axial inclination | 23.98° |
| Albedo | 0.16 |

**2. Interior of Mars**
Experiments to investigate the gravity of Mars and hence the internal distribution of mass have indicated that the planet behaves more like a fluid than a rigid body. This conflicts with the fairly widespread belief that Mars has a

2

core. Some scientists studying the Tharsis shield volcanoes believe that a thick lithosphere would be required to supply the volcanoes. On the other hand the mechanics of the volcanic activity could be the result of the upwelling motion of mantle convection currents.

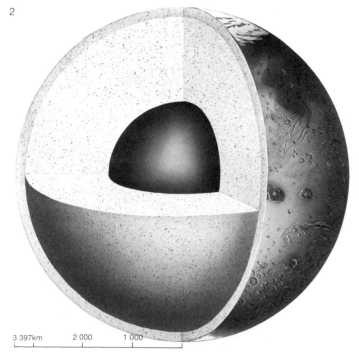

3 397km　　2 000　　1 000

212

cut by deep valleys which expose the layering beautifully. Very few impact craters are found and although they must once have been present, it is evident that they have long since been either removed or covered up by sedimentary debris.

## Geological history

While Mars' early history is not well understood, it is likely that the resurfacing of the northern hemisphere took place very early on, perhaps as long as 4,000 million years ago. Many scientists believed this was in some way connected with the formation of Mars' inner core region.

Subsequent to this resurfacing there was considerable volcanic activity and it is probable that this was directly connected with the rise of Tharsis. Certainly the extensive fracturing must have been related to this, and probably also the opening of Valles Marineris. Much speculation still exists relating to how long the volcanic activity went on; some scientists believe its demise to have been relatively recent, others hold the view that it ceased several hundreds of millions of years ago.

It is very clear that active erosion and movement of debris still continues on Mars and that it has been ongoing for hundreds of millions of years. It is also clear that many of Mars' volatiles are still locked within, probably as subsurface ground-ice, and that the melting of such ice during volcanic activity may have played a significant part in the formation of the amazing outflow channels.

## Nomenclature

In late 1971, Mariner 9 began to send back high-quality pictures, and showed that the surface features on Mars are very varied. Further pictures were obtained from the Viking orbiter from 1975, so that there are now reliable maps of the entire surface.

Antoniadi's nomenclature, which dates from the 1930s, has been retained as far as possible. The classes of features are as follows: catena (a crater-chain or line of craters); chasma (a canyon or steep-sided depression); dorsum (a ridge or an irregular, elongated elevation); fossa (a ditch or long, narrow, shallow depression); labyrinthus (valley complex); mensa (mesa, a flat-topped prominence with steep edges); mons (mountain or volcano); patera (irregular crater, or complex crater with scalloped edges); planitia (smooth, low plain or basin); planum (plateau); tholus (a hill or isolated peak); vallis (a valley or sinuous channel, often with tributaries); vastitas (extensive plain).

3

**Cratered terrain**

Basin ejecta

Cratered

Heavily cratered

**Volcanic regions**

Volcanic constructc

Volcanic plains

**Others**

Channel deposits

Polar deposits

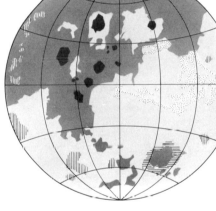

**3. Geological features**
A geological map of Mars reveals a complex variety of erosional, volcanic and tectonic features that distinguish the planet from the more simple surface features of Mercury or the Moon. The map also reveals the dichotomy of the northern hemisphere plains and the southern hemisphere cratered uplands, the cause of which as yet remains a mystery in the overall geological evolution of the planet Mars.

**4. Cratered terrain**
This region is about 300 km east of Hellas Planitia. There are ancient craters, barely visible, and younger craters (A). Erosion is represented by sediment around large remnants (B).

**5. Smooth plains**
These smooth plains in western Chryse Planitia represent lava flows that reach the surface along fault lines and have then been eroded by wind or flowing water. Impact craters are common.

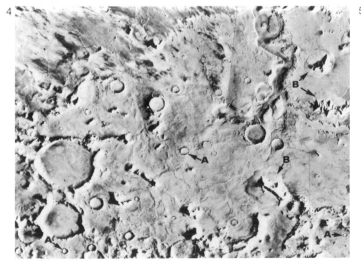

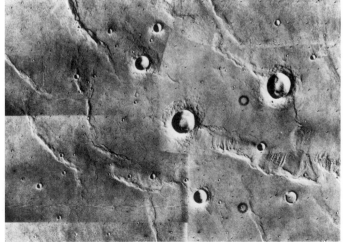

# Observational Background

### Early telescopic work

The first telescopic observations of Mars were made by Galileo in 1610. He was unable to see any surface features, but he did detect the phase; when well away from opposition, Mars may appear of the same phase as the Moon when it is a few days from full. The first drawing of Mars of real value was made by Christiaan Huygens in 1659. It shows a triangular patch which, though exaggerated in size, certainly corresponds to the dark feature now known as Syrtis Major. From his observations, Huygens concluded that the rotation period of Mars must be of the order of 24 hr; it is in fact 24 hr 37 min 22.6 sec. A Martian day is known as a "sol".

G. D. Cassini, an Italian astronomer who later became Director of the Paris Observatory, observed Mars in 1666 and observed the white caps covering the planet's poles. Further studies of the polar caps were made by G. Maraldi in 1719; he found that the caps were not centered exactly at the poles of rotation.

### Nature of the surface features

It was natural to assume that the white caps were made of ice and snow; this was the view of William Herschel, who made observations of Mars between 1777 and 1783 and gave a value for the rotation period which was very close to the true figure. Herschel also observed Mars as it passed close to a star, and from the fact that the star's light was not affected he concluded that the Martian atmosphere must be very tenuous. Herschel also determined the axial inclination of Mars (23.98°), which is only slightly greater than that of the Earth (23.44°). The present north polar star of Mars is Deneb (Alpha Cygni).

Most of the early observers assumed that the dark patches on the Martian surface were seas, while the bright orange areas were continental. However, it became apparent that the Martian atmosphere was too thin and too dry for widespread oceans to exist, and in 1860 E. Liais, a French astronomer living in Brazil, put forward the alternative theory that the dark patches were old seabeds filled with primitive vegetation. This view persisted until less than 20 years ago. It was at least certain that the surface features were permanent, and not merely cloud effects as in the case of Venus, and maps were drawn up. Clouds were also observed in the planet's atmosphere, though many of them were believed to be more in the nature of dust storms than Earth-type clouds.

### The canals

1877 marks the modern phase of telescopic observation of Mars. In that year G. V. Schiaparelli at Milan drew a much improved new chart of Mars, and reported that crossing the deserts there were a number of fine, straight, artificial-looking lines. He called them *canali*; the word was inevitably translated into English as "canals", and the suggestion was made that they were genuinely artificial, built by the inhabitants of the planet as part of a global irrigation system. Mars was known to be short of water, and it was believed that the canal system was designed to take water from the polar caps to the warmer regions, which were centers of population. In 1879 Schiaparelli reported that some of the canals became double over short periods, which again indicated a non-natural origin. The canals were confirmed in 1886 by Perrotin and Thollon, using the large refractor at the Nice Observatory, and subsequently the canals were drawn by many observers. Percival Lowell, who founded his observatory at Flagstaff in Arizona primarily to study Mars and equipped it with a fine 61 cm refractor, wrote several books in the firm belief that the canal system could not possibly be anything but artificial.

Other observers, however, using equipment just as powerful as Lowell's, either failed to see the canals at all or else drew them as broad, diffuse streaks. The problem was not finally resolved until the first close-range pictures were obtained from the Mariner probes. It is now clear that the canal system does not exist in any

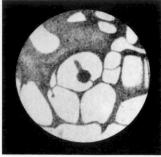

**1. Huygens, 1659**
Huygens' sketch was the first of Mars to show a recognizable feature. It is undoubtedly the most conspicuous dark area on the planet—Syrtis Major—though its size is exaggerated.

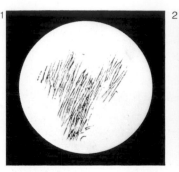

**2. Schiaparelli, 1877**
Schiaparelli first observed his "canali" in 1877 with a 20 cm refractor. He drew the network in detail, but the features were not described by other observers until at least seven years later.

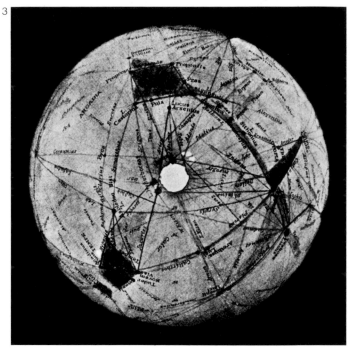

**3. Lowell, 1895**
From 1895 onwards Lowell described a highly developed, elaborate system of irrigation channels from the north pole constructed by Martians.

**4. Antoniadi's map**
Antoniadi believed the apparently artificial nature of the Martian features to be the result of an optical illusion. His observations were remarkably accurate.

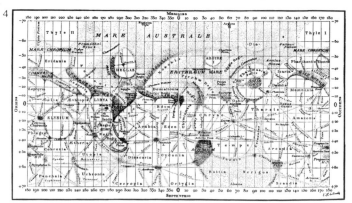

form, and that the features shown by Lowell and others were due to nothing more than tricks of the eye or telescopic aberration.

## The dark areas

Variations in some of the dark areas were noted periodically, but the idea of depressions covered with some kind of organic matter still seemed reasonable up to the time of the flight of the first successful Mars probe, Mariner 4, in 1965. The probe passed Mars at a distance of only 10,000 km, and provided information that revolutionized our ideas about Mars. The atmosphere was very tenuous with a ground pressure of no more than 10 mb, and was composed chiefly of carbon dioxide. The dark areas were merely low-albedo features, and not all of them were depressions; indeed, the most prominent of them, Syrtis Major, proved to be a lofty plateau. Instead of being smooth and undulating the surface was found to be cratered, so that superficially at least Mars resembled the Moon more than the Earth.

## Later probes

Following Mariner 4, Mariners 6 and 7 (1969) sent back improved pictures of the surface. In 1971 Mariner 9 was put into an orbit round Mars, and much of the surface was mapped in detail. Since then there have been only two American Mars probes, the soft-landing Vikings, but the Russians have sent several, none of which has been successful; some have missed the planet, while others have either lost contact or have gone out of touch very soon after landing. In 1988 two Soviet probes were launched with the main object of studying Phobos, the inner Martian satellite, but these too failed, and few data were returned.

## The Vikings

The Viking probes, launched in 1975, were identical craft. Each was made up of an orbiter, which remained in a closed path around Mars carrying out a program extending that of Mariner 9, as well as acting as a relay, and a lander, which could be separated at a suitable moment and brought down to a gentle landing, partly by parachute and partly by rocket braking. The Vikings reached the vicinity of Mars in June and August 1976, and careful surveys of the intended landing sites were carried out. It was essential to bring the lander down on to an even surface; if the vehicle ended up at a sharp angle it would have been unable to communicate.

Both landings were successful. Viking 1 came down on 20 July at latitude 22°.4N, longitude 47°.5W, in the ochre plain of Chryse; Viking 2 followed on 3 September at latitude 48°N, longitude 226°W, in another plain, Utopia. Luck as well as skill was needed. The first lander (Viking Lander 1, or VL1) came down only 7.5 m from a boulder that was large enough to have caused extensive damage had the landing been made on top of it. The only failure of the entire mission was that of the VL1 seismometer. The VL2 seismometer performed well, however, and showed that "Mars-quakes" are very slight and rare occurrences.

It had been anticipated that there would be marked differences between Chryse and Utopia, since Utopia is much further north, but in fact the two sites looked similar. Each was rock strewn; the fine-grained material so much in evidence was rusty red, while most of the rocks were similar, indicating that a red oxide had formed a thin veneer over darker bedrock. There was evidence of old streambeds, and the rocks themselves were clearly vesicular. The sky proved to be pink: the thin Martian atmosphere is not capable of scattering the shorter wavelength radiations from the Sun that make the sky blue from Earth, and the pinkness results from the thin dust which pervades the atmosphere. There were, however, minor differences between the two sites. Utopia looked rather "cleaner" than Chryse, and there were fewer large boulders; Chryse was compared to "a forest of rocks", and there were dunes and windblown dust.

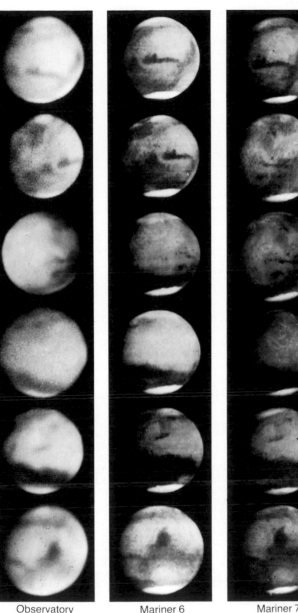
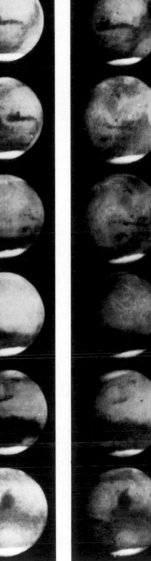

| Observatory | Mariner 6 | Mariner 7 |

**5. Light and dark markings**
Light and dark surface features on Mars are compared in images taken on Earth (left) at the New Mexico State University Observatory, from Mariner 6 (center) and Mariner 7 (right) in July and August 1969. Both Mariners flew by Mars and photographed the surface.

**6. Atlantis on Mars**
Mariner 4 was the first craft to fly past Mars—in July 1965. This picture covers about 270 by 240 km in the Atlantis region.

**7. The "Giant's Footprint"**
An oblique view from Mariner 7 of two adjacent craters looked uncannily like a large footprint.

# Atmosphere I

The dynamics of the Martian atmosphere are in many ways less complicated than those of the atmosphere of Earth. In the first place there are no oceans on the surface of the planet to affect the transport of heat and to provide moisture. Also, the Martian atmosphere is much thinner than that of the Earth, so it responds to heat and heat loss more rapidly, and it has a smaller heat capacity. On the other hand, the topography varies in height much more than on Earth and this affects the winds on Mars.

The atmospheric pressure on Mars is less than one-hundredth of what it is on Earth. In effect, this means the air is as rarefied on the Martian surface as it is at a height of 30,000 m on Earth. At pressures such as this and in Martian temperatures, liquid water becomes unstable and freezes on the surface. Another effect of such a rarefied atmosphere is that the transportation of material along the surface (saltation) and the raising of fine material to form dust clouds are a conspicuous characteristic of the Martian landscape.

## Composition of the atmosphere

Ninety-five percent of the atmosphere of Mars is made up of carbon dioxide, with about two percent nitrogen and between one and two percent argon. There are traces of water vapor, carbon monoxide, oxygen, ozone, krypton and xenon. The abundance of water, oxygen and ozone is variable according to season and geographical location. As on Earth, the Martian atmosphere is composed of gases released from the interior by way of volcanism, in this case mainly through the giant volcanoes of the Tharsis shield. The amount of water vapor might at first indicate that the atmosphere is dry but, considering Martian temperatures, it is relatively moist.

There are important sources and sinks that maintain the observed distribution of water vapor on Mars. In summer there is a residual cap at each pole. At the north pole the cap is known to be composed of water ice, but the cap at the south pole is composed of dry carbon dioxide ice. The reason for this is as yet a mystery but it may have something to do with the frequent Martian dust storms. These are generated in the southern hemisphere and they may transport water vapor northwards. Sublimation of water ice at the north pole provides an abundant supply of water vapor.

In the absence of a protective ozone layer, the surface of Mars is exposed to ultraviolet radiation from the Sun. Water vapor as well as carbon dioxide is photodisassociated as a result of this radiation. Such chemical cycles within the Martian atmosphere result in surface oxidation.

## Evolution of the atmosphere

Studies of the present constituents of the atmosphere of Mars suggest that it was much thicker in the past. Noble gases, such as krypton and argon, and nitrogen are good indicators of the evolution of an atmosphere. Noble gases are chemically inert and are not removed from the atmosphere by chemical reactions. Studies of the carbon dioxide content suggest that degassing from the interior of the planet is not as complete as it is on Earth. The nature of the nitrogen component indicates that there was once an atmospheric pressure as great as on Earth. If all volatiles had been released at once, water may have covered the planet to a depth of 200 m. There are what appear to be old stream beds and other water-sculptured terrains on parts of the surface (*see* pages 224–5).

The interaction between the polar caps and the atmosphere on Mars produces some of the unique characteristics of Martian

### Composition of Mars' atmosphere

|  | % volume |
| --- | --- |
| Carbon dioxide | 95.32 |
| Nitrogen | 2.7 |
| Argon | 1.6 |
| Oxygen | 0.13 |
| Carbon monoxide | 0.07 |
| Water vapor | 0.03 |
| Neon | 2.5 ppm |
| Krypton | 0.3 ppm |
| Xenon | 0.08 ppm |
| Ozone | 0.03 ppm |

### 1. Atmospheric layers

By earthly standards, Mars' atmosphere is very thin, and it has only a little influence on surface temperatures. It can be thought of as having the same vertical layering, but there are differences arising from its composition. There is much less ozone than on Earth so there is no reversal of the temperature gradient that distinguishes the stratosphere on Earth. Lower temperatures are found in the thermosphere, too, because of the cooling effect of carbon dioxide.

### 2. Zonal winds

The diagram shows the averaged latitudinal profiles of zonal wind velocities at upper and lower levels in the Martian atmosphere. In mid-latitudes of the northern hemisphere west winds prevail, with weaker east winds in the southern hemisphere.

### 3. Climatic change

Mars takes nearly twice as long to orbit the Sun as the Earth, but day length and obliquity are close to terrestrial values. Mars therefore exhibits a seasonal pattern similar to the Earth's. The eccentricity of Mars' orbit, however, gives the planet a variable orbital speed and the seasons therefore have different lengths. The north pole is tilted away from the Sun at perihelion so the northern hemisphere autumn lasts for only 142 Martian days and winter lasts for 156 days. (Spring lasts for 194 days and summer 177.) The southern summer is as long as the northern winter and other seasons are similarly reversed. Martian obliquity varies by about 10° over some one million years, which is a huge variation compared with the Earth, and solar radiation at polar regions may vary by more than 100 percent as a result. Additional radiation striking the Martian surface may release more carbon dioxide into the atmosphere, making it more massive and ultimately warmer. Changes in obliquity will also affect the redistribution of water vapor and the generation of dust storms. There may well be wetter periods in the future and a more hospitable climate. Mars also exhibits a precession of the longitude of perihelion.

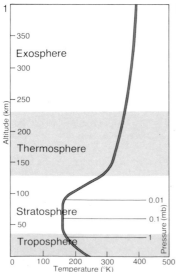

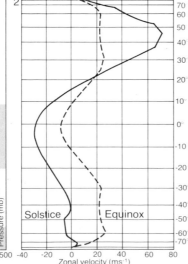

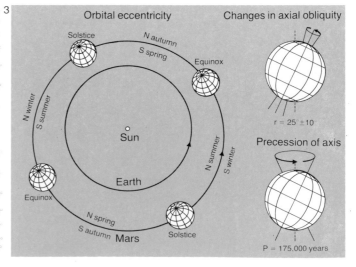

weather. Approximately 20 percent of the carbon dioxide in the atmosphere is recycled between cap and atmosphere during each of the Martian seasons. This produces variations in pressure.

Atmospheric carbon dioxide freezes to form a frost at about $-123°C$ on Mars. This occurs during the nights of late autumn, winter and early spring, and its accumulation is only halted by solar heating during the days. During these seasons more frost accumulates during the nights than sublimes during the days and the result is frost caps of carbon dioxide alternately at the poles as winter comes first to one and then the other pole.

### Circulation of the atmosphere

Since there are no oceans on Mars, the whole surface of the planet responds relatively quickly to solar heating. In the summer hemisphere the hottest place is not at the equator but at tropical or subtropical latitudes where the Sun is directly overhead at midday. As a result there is a single Hadley cell that spans both sides of the equator, rather than a cell in each hemisphere as there is on Earth. On Mars warm air rises in the summer hemisphere and sinks in the winter hemisphere. A unique feature of Martian circulation is the result of the condensation of carbon dioxide at high latitudes in the winter hemisphere. This produces a lower pressure than elsewhere and this pressure gradient produces a strong planetary-scale circulation towards the region where a polar cap is forming. This is known as a condensation flow, and it dominates wind flow at all latitudes on Mars.

The fact that the atmosphere is thin and composed largely of carbon dioxide, which is a good transmitter of infrared radiation, means that there is a great diurnal temperature range in the lower levels of the atmosphere. This produces very strong thermal tides which are propagated in the direction of the movement of the Sun across the Martian sky.

These characteristics of circulation mean that in the summer hemisphere winds tend to be dominated by topography since there is only a slight temperature variation between equator and pole to drive the general circulation. In the winter hemisphere there is a large temperature gradient between equator and pole in addition to the condensation flow.

4

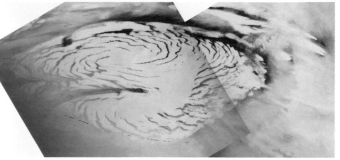

5

**4. The south pole**
This photomosaic of the south polar region of Mars reveals a remnant cap of carbon dioxide ice during the summer of the southern hemisphere. To the right of the cap is a basin with a diameter of about 800 km. This is thought to be the remains of a huge impact crater. The polar terrain exhibits strange patterns and textures. Glacial or aeolian deposits extend from beneath the cap and partially fill the craters surrounding the pole.

**5. The north pole**
The north polar cap is (unlike the south polar cap) made up of water ice. The cap has a striking spiral pattern that is probably the result of radial wind erosion and the sublimation of water and carbon dioxide from the surface. Layered deposits of rock and ice have been eroded, re-covered with ice and eroded again. The rock is probably composed of silt, clay and silicate particles that make up a dust which makes the ice appear to be rather dirty.

# Atmosphere II

## Martian weather

The thin Martian atmosphere responds rapidly to radiative and convective processes, and large-scale atmospheric motions are largely controlled by solar heating. The tenuous atmosphere also means that there is an inefficient transport of heat by means of winds and there are therefore large temperature contrasts from region to region. Another characteristic of Martian meteorology is that latent heat is released when carbon dioxide and water vapor condense to form the polar caps, and this further complicates weather processes on the planet. Finally, dust storms that occur near perihelion affect the thermal balance and stability in the atmosphere, as do the large topographic features, such as volcanoes and trenches, on the Martian surface.

Weather varies greatly on Mars both with season and time of day. In winter there is a large temperature difference between the equator and the poles. This results in westerly winds and low-pressure systems similar to those on Earth. In summer light easterly winds prevail. These are tidal, resulting from the planet's quick response to solar heating, and there is little weather as such.

## Clouds

The Martian atmosphere is close to water vapor saturation and clouds are to be expected. There is nothing resembling terrestrial cumulus clouds, but there are four general cloud types: convective clouds, wave clouds, orographic clouds, and fog. Early morning fog occurs in low-lying areas when ground frost is vaporized by the Sun. There are also clouds not of water vapor but carbon dioxide. At high altitudes and in polar regions during the Martian winter temperature can fall low enough for carbon dioxide to condense, forming clouds of dry ice. It may be that dry ice snowstorms contribute to the polar caps, but it is more likely that they are formed when carbon dioxide condenses on contact with the cold Martian surface.

## Dust storms

The creation of dust storms, some of which engulf the whole planet, is a peculiarly Martian phenomenon. Viking observed more than 35 during 1977 and two of these developed into global systems. The eccentricity of the Martian orbit means that insolation (exposure to solar radiation) is 40 percent greater when the planet is nearest the Sun than when it is furthest away, but it is doubtful whether even this would produce wind velocities great enough to raise the dust storms that have been observed.

Viking has revealed two mechanisms involved in the generation of dust storms. They coincide with the retreat of the south polar cap, when a large temperature gradient exists between the newly exposed surface and the remnant cap. It is thought that in the vicinity of large topographic features this gradient would be large enough to induce winds strong enough to lift surface dust.

The second mechanism is winds that are the result of the dominance of tidal circulation over planetary scale waves. The degree of strength of winds depends on the amount of atmospheric

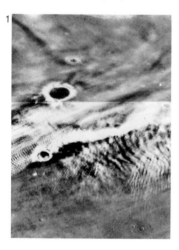

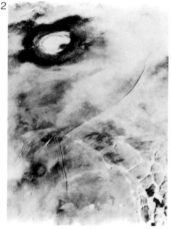

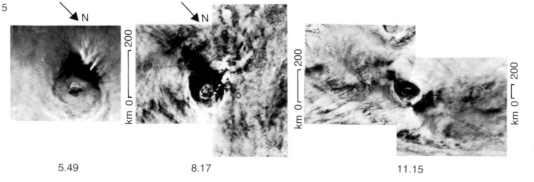

**1. Lee-wave clouds**
Gravity lee-wave systems have long been known in the Earth's atmosphere and they have also been observed on Mars. In this image a lee-wave system extends for at least 800 km to the east of the crater Kunowsky, which is about 90 km in diameter.

**2. Bore-wave clouds and fog**
Many Viking images were studied in order to determine diurnal air flow. Katabatic winds (caused by air flowing downwards) in the Tharsis region produced this thin bore-wave system near Arsia Mons. 1,000 km across, it moved northeastwards in the mornings.

**3. Cirrus clouds**
Cirrus cloud patterns have also been observed on Mars. These bright cloud formations were photographed during the Martian winter. Studies of a number of Viking plates show that these clouds recurred at the same location over several days.

**4. Cyclone**
Many features of Martian meteorology bear a resemblance to conditions on Earth. This image shows a definite cyclonic motion to the airflow generating these clouds. Such instability may result as on Earth from the mixing of different air masses.

5.49　　　　　8.17　　　　　11.15

**5. Diurnal weather patterns**
This sequence of images shows the behavior of airflow in the vicinity of Ascraeus Mons. The first image shows a plume on the western flank of the volcano similar to lee-wave clouds. The cloud gradually changes, becoming more turbulent, and the Tharsis Ridge is covered with a low-level thin layer of cloud. Finally, convective heating produces cloud on the flanks of the volcano and there is no longer any cloud produced by high-level easterly wind systems.

heating and this is influenced by the amount of dust in the air. If the temperature is heavily dust-laden, the winds will increase in intensity until they are strong enough to raise dust. This regenerative process starts on a local scale when tidal winds and disturbances arising from the local topography cause small-scale dust storms or dust devils. These are thought to be fairly widespread on Mars, and the fact that Viking took photographs of a pink sky suggests that the atmosphere may be constantly dust-laden.

Since dust in the atmosphere raises the temperature of the air, dust storms affect the condensation of water and carbon dioxide at the poles. This could have been instrumental in the formation of the layered deposits that have been observed in the polar landscapes.

### Climatic change

There is ample evidence on Mars of the work of a fluid agent. There are numerous examples of what appear to be ancient stream beds and other water-sculptured terrain. There are also channels that emerge at the head of canyons below areas of collapsed terrain, and it is thought that these may have been formed by underground water or the melting of subsurface ice.

This water may have resulted from a cataclysmic event such as a meteoritic impact or a volcanic eruption. If a large amount of water was suddenly added to the atmosphere, it might stay there long enough to produce precipitation and subsequent erosion. The thicker atmosphere of an earlier epoch may well have produced flowing water on the surface, and cyclic changes in the planet's orbit (*see* page 216) could certainly produce climatic changes on a grand scale. Astronomical perturbations to the Earth's rotation are now believed to have influenced the advance and retreat of the terrestrial ice ages. On Mars water ice is locked up in polar regions and carbon dioxide may be contained within surface layers. Such volatiles could be released into the atmosphere, thus affecting the climate, by any global effects that would alter the amount and distribution of solar radiation received by the surface.

**6. Dust storm 1971**
A Martian dust storm as viewed from Earth can be seen in five images taken from the Lowell Observatory in Arizona of the Yellow Storm in 1971. From left to right they show the surface before the storm, at the beginning, with the surface partly obscured, then completely obscured and partly cleared.

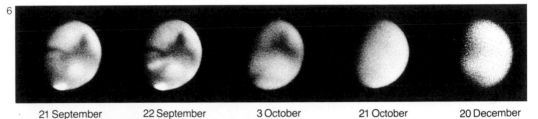

| 21 September | 22 September | 3 October | 21 October | 20 December |

**7. Global dust storm**
Viking Orbiter 2 photographed Mars' southern hemisphere when it was almost completely obscured by a developing global-scale dust storm. This was just after the summer solstice, when other dust storms have been observed. Dark areas are clear regions where the surface is visible.

**8. Before and after**
The appearance of Mars can be totally transformed by a dust storm. The Viking Orbiter took a photograph of the Sinai Planum region of Mars to the south of Valles Marineris both before (**A**) and after (**B**) such a dust storm. The form of the dust cloud can clearly be seen.

# Surface Features

**Surface features**

The two hemispheres of Mars are not identical. In the south craters are numerous, and there are several large basins, the most important of which are Hellas (1,800 km in diameter) and Argyre (800 km). Hellas is in fact the deepest basin on Mars, at 3 km below the "standard level" where the atmospheric pressure is 6.1 mb. Before the Mariner flights Hellas was believed to be a high snow-covered plateau. Smoother plains lie between the thickly cratered areas of the southern hemisphere, and there are also small gullies and elaborate systems of channels. There are some densely cratered regions in the northern hemisphere, but there are also large plains where the number of craters is much less. The main dark areas of the northern part of the planet, Acidalia Planitia (Antoniadi's "Mare Acidalium"), is a smooth plain.

Both volcanic and impact craters (*see* page 222) occur on Mars. From its superficial appearance, the southern hemisphere would seem to be generally older than the northern, and the larger craters, however they were formed, may date back 4,000 million years.

Mars has the largest volcanoes (*see* pages 228–9) known in the Solar System. The main volcanic region is Tharsis, which crosses the equator, although most of it lies in the northern hemisphere. There is a definite "bulge" in the planet where Olympus Mons towers to 25 km. Its volume is between 50 and 100 times that of Mauna Loa, the largest shield volcano on Earth.

The longest valley system on Mars is known as the Valles Marineris (*see* pages 226–7) and is very complex. It can be traced for a total length of over 4,000 km. Noctis Labyrinthus is a huge complex of canyons which covers at least 120,000 km².

In many areas of the Martian surface are features that look like dry riverbeds (*see* pages 224–5). They seem in some cases to be associated with drainage systems from the giant volcanoes, and it is difficult to avoid the conclusion that they were cut by running water. Liquid water cannot now exist on the Martian surface, because the atmospheric pressure is too low; therefore, if this interpretation of the channels is correct, Mars must formerly have had a much denser atmosphere than it has now.

**Composition of the polar caps**

The polar caps (*see* pages 218–19) wax and wane according to the Martian seasons; they are at their most extensive in winter and least in summer. The southern summers are shorter but hotter than those of the northern hemisphere, while the winters are longer and colder because Mars, like the Earth, comes to perihelion during the southern summer. This means that the climate of the southern part of Mars' hemisphere is more extreme than that of the north, and the cap shows a greater range in size.

The seasonal caps are due largely to solid carbon dioxide, which condenses out of the atmosphere at the onset of the cold season. There are, however, residual caps, which never disappear, and it is possible that these differ from each other. The northern residual cap is almost certainly water-ice, while the southern is more probably a mixture of water-ice and carbon dioxide ice. Layered deposits are found at both poles at latitudes higher than about 80°N and S; the southern area is cratered, the northern covers a region of plains.

Samples of surface materials were collected by the scoops of the two Viking landers, and the results of analysis were transmitted satisfactorily back to Earth. The rocks are purely volcanic and do not seem in general to be very different from those of the lunar maria. Iron accounts for about 18 percent (by weight), silicon for up to 45 percent; other elements detected include magnesium, sulphur and aluminium. Potassium seems to be less common than in terrestrial rocks of the same general type. It may be assumed that permafrost extends for some depth, and there is likely to be underground ice. Certainly there is a striking difference between Mars and the Moon in this respect, since the lunar samples show no trace of any hydrated materials.

**1. Viking Lander 1 panorama**
This spectacular 100° panorama shows the desert-like Chryse Planitia. The drifts of fine-grained sediments are layered, presumably the result of changing wind directions. At this time the drifts were being eroded and internal stratification shows just to the right of the lander's meteorology boom (center).

**2. Martian surface details**
High-resolution photographs of the surface of Utopia Planitia near the Viking Lander 2 show in (**A**) rocks that have vesicles (small holes). This might indicate volcanic processes or meteoritic impact. (**B**) shows a thin coating of water-ice over the rocks and soil. This remains after the evaporation of carbon dioxide.

**The search for life**
One of the most important aspects of the Viking exercise was to see whether any traces of Martian life could be found. It was recognized that any organisms would be primitive, but it is probably true to say that many scientists believed that indications of life would be found.

The first step was to secure samples. There was an initial alarm with the grab of VL1 because a latch-pin jammed and prevented the collecting maneuver from being completed. The pin was released, however, and on 28 July (the eighth sol after the VL1 descent) a sample was secured. There were three main experiments.

In the pyrolitic release experiment the sample was heated in the test chamber for five days. The atmosphere in the chamber was similar to that of Mars, but with the addition of carbon-14, which is radioactive. The plan was to treat the Martian samples with this substance, and see whether they assimilated any of it. After an incubation period of 11 days, the chamber was heated sufficiently to

3

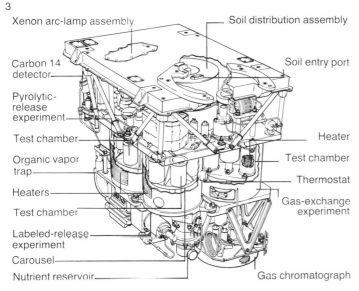

Xenon arc-lamp assembly — Soil distribution assembly

Carbon 14 detector

Pyrolytic-release experiment

Test chamber

Organic vapor trap

Heaters

Test chamber

Labeled-release experiment

Carousel

Nutrient reservoir

Soil entry port

Heater

Test chamber

Thermostat

Gas-exchange experiment

Gas chromatograph

**3. Biological laboratory**
One of the main objectives of the Viking missions to Mars was to investigate the chemistry of soil samples in order to detect possible signs of life. Each Lander was equipped with this apparatus for the gas-exchange experiment, the labelled-release experiment and the pyrolytic-release experiment, each of which had several test chambers on a carousel. This enabled about six soil samples from each landing site to be investigated.

**4. Argyre Planitia**
Two kinds of Martian landscape can be seen in this oblique view from a distance of 19,000 km. At center left is the smooth plain Argyre Planitia, believed to be a large, ancient impact basin. The rim deposits form an arcuate mountain range that may have been uplifted during impact. Argyre Planitia is surrounded by heavily cratered terrain. Craters can be seen almost to the horizon, above which are haze layers of carbon dioxide.

break up any organic compounds (this procedure is termed pyrolysis). The chamber was then flushed out with helium, and the vaporized pyrolysis products put into a detector capable of identifying the carbon-14 taken up by the Martian organisms—if there were any. The initial results were somewhat contradictory, but on the whole they could be explained by chemical rather than biological activity.

The labelled release experiment also involved carbon-14. It was assumed that the addition of water to a Martian sample would trigger off biological processes if any life were present. The sample was moistened with a nutrient that contained carbon-14 as well as various other substances that are absorbed by terrestrial organisms, and the atmosphere above the sample was monitored for any traces of radioactivity; organisms would be expected to give off gas containing carbon, and the carbon-14 would therefore betray it. Any radioactivity in the chamber atmosphere would indicate life. Again the results were not as clear-cut as had been hoped, and they

too could be accounted for by chemical processes. The third and final Viking test was the gas exchange experiment. It was assumed that any biological activity on Mars would involve the presence of water, and the idea was to see whether providing a sample with suitable nutrients in water solution would persuade any organisms to release gas, thereby altering the composition of the artificial atmosphere inside the test chamber. Once more there was no proof of the existence of life.

Trenches were dug round the spacecraft so that as many samples as was practicable could be examined, but Martian life obstinately refused to show itself. Whether Mars has always been sterile is quite another matter. The presence of running water in the past indicates that the climate used to be more favorable than it is now, and certainly the orbital eccentricity of the planet, plus the fact that the axial inclination ranges between 14° and as much as 35° over a long period, may mean that the climate varies widely over tens of thousands of years.

# Craters

Both impact and volcanic craters (*see* pages 228–9) are found on the surface of Mars. Impact features range in size from small pits to large ring structures 200 km or more in diameter. As with lunar and terrestrial impact craters, those on Mars are approximately circular in outline, have raised rims and depressed floors and a surrounding blanket of ejectamenta. There are, however, marked differences between lunar and Martian impact craters.

Most of the fresher craters are of what is termed the rampart variety, that is, they have one or more sheets of ejecta, each of which has an outer ridge or rampart. Craters in the 5–10 km range typically have but a single ejecta layer, which extends out for about one crater radius. Larger craters usually have more layers, while the outer margins of the blanket frequently have complex lobed outlines. Outside of the blanket, rows of small pits, hummocks and hollows are common, often crossed by faint radiating striations.

These features strongly suggest that whereas lunar impact debris is deposited simply under the effects of a ballistic trajectory, Mars' impact craters have a more complicated origin. The morphology indicates that after ballistic ejection the debris moved across the ground as a surface-hugging and largely coherent lubricated flow.

Pedestal craters are rather different. These are particularly numerous in the northern hemisphere, between 30°N and 70°N. In this case the crater sits at the center of a raised pedestal of what is apparently ejected debris. Some pedestals show the striations and other features described above for rampart craters, but others lack these and presumably are degraded versions of the original ejecta blankets. In extreme cases the central crater may lack even a raised rim and is situated in the middle of a flat-topped plateau raised above the general level of the surrounding terrain. Wind action is probably responsible for many of the features seen.

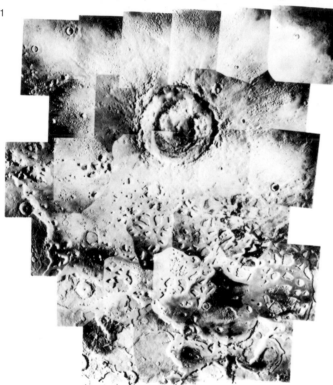

**1. Lyot**
A large impact crater 176 km in diameter, Lyot has an inner rim of impact debris. There are several secondary craters to the north of Lyot but, strangely enough, none to the south, an area that seems to be older than Lyot. Erosional processes may have removed secondary craters.

**2. Arandas**
Arandas, 28 km in diameter, has a complex lobed outline, which suggests that the ejecta blanket was produced by the flow of material. This may have taken the form of a fluidized mixture of fragments of ejecta and water from ground-ice that was melted as a result of the impact.

**3. Rampart crater**
A typical Martian rampart crater shows a sheet of ejecta material about one crater radius in extent with an outer ridge or rampart.

**4. Young crater**
A fresh crater—about 30 km in diameter—is seen by a dry river channel in the Lunae Planum.

**5. Pedestal craters**
Craters may be situated on well-developed platforms standing above the surrounding plains.

**6. Flooded craters**
These are ancient impact craters and they have been flooded by lava from the volcano Arsia Mons, which is 1,500 km distant.

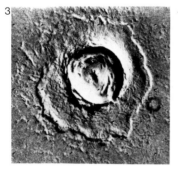

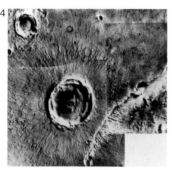

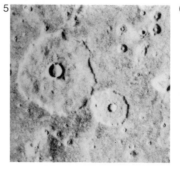

# Wind Features

Landforms arising from the work of the wind may be either constructional or result from abrasion by the wind (and those particles held in suspension by it) of preexisting rocks. On Mars depositional wind features, strangely, are nearly all located in the northern hemisphere and between latitudes 85°N and 75°N, in a belt some 500 km wide. The asymmetry itself is odd, but there are other differences between the Earth and Mars in this respect, since on our planet most dune fields are in the low and mid-latitudes.

The dunes surrounding the north pole form a nearly continuous belt within which there is a fairly consistent pattern. Typically the dunes are only slightly sinuous, between 0.5 and 2 km apart and up to 100 km long, and set transverse to the prevailing winds. Towards the edges of the zone or close to upstanding features such as craters and hillocks, the continuity breaks up and barchans (crescent-shaped dunes) may form. In places the dunes clearly overlie the laminated polar terrain. In Borealis Chasma, carved from this terrain, curving dunes partially fill the chasm, the dune material apparently having been driven south from the polar regions.

Elsewhere the most common wind features are streaks or blotches associated with the smaller craters. Bright streaks are thought to represent accumulated fine debris, while dark streaks presumably occur where bedrock has been exposed. In wind-tunnel experiments, winds with a vortical motion produce similar features.

In southern Amazonis, south of Olympus Mons, there are particularly striking examples of linear ridges and valleys termed yardangs. These streamlined features are aligned parallel to prevailing winds and are accompanied by pits and hollows. On average they are between 0.5 and 1 km wide and up to 50 km long. On Earth such features are formed only in the drier deserts, by the stripping action of the wind.

### 1. Dune fields
This dune field—in Borealis Chasma—extends down from the north polar cap of Mars (seen at the top of the picture). The ridges of the dunes curve round through an angle of 45° from the northern to the southern edge of this triangular field.

### 2. Dark streaks
These dark streaks were probably formed when wind blowing around and over the craters produced a turbulence that removed a thin layer of bright dust, exposing the darker subsurface. The light patches may be deposits from dust storms.

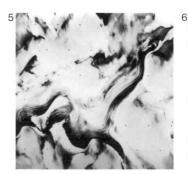

### 3. Sand dunes
Many dune fields in the south polar region, such as this on the floor of a 60 km crater, are similar to sand dunes on Earth.

### 4. Transition zone
This north polar dune field shows a transition from dunes with a transverse ridge structure to isolated linear dunes.

### 5. Layered deposits
Layered sedimentary terrain at the poles (seen partially covered with frost) was deposited by wind.

### 6. Yardangs
Yardangs are wind-eroded linear features that are flat topped. Similar features occur on Earth.

# Channels

Perhaps the most controversial of all Martian surface features are the channels. Did they form as a result of the action of running water or not? Under the currently prevailing surface conditions on Mars liquid water is unstable, so the idea of erosion by running water implies a significantly different atmosphere and climate.

It is possible to recognize three kinds of channel: run-off channels, outflow channels and fretted channels. Run-off channels typically have a V-shaped cross-section, start small and increase in size downstream, and have well-developed tributary networks. The large branching channels Nirgal and Ma'adim Vallis can be assigned to this group, but more typical are the large numbers of smaller networks that incise the ancient cratered terrain. The fact that this kind of channel is confined to the older terrain suggests that the channel-forming epoch was an early development.

The second group of channels, called outflow channels, emerge fully fledged from areas of chaotic terrain. Unlike the first group, outflow channels do not have tributaries and are widest and deepest near to their source. By terrestrial standards they are extremely large. On Earth the only channels of comparable dimensions, in the eastern part of Washington State, USA, are thought to have been formed by catastrophic flooding. A similar origin has been proposed for those on Mars.

Fretted channels are similar to networks located in regions of fretted terrain but themselves are not located in such areas. They have tributaries, are rather flat floored and broaden downstream. Characteristically, features resulting from mass wasting (the mass movement of material caused by gravity) are found on their floors.

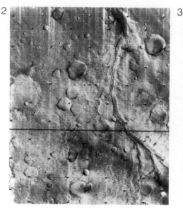

**1. Chaotic terrain**
The rugged landscape on Mars termed chaotic terrain may have been the source of a great flood in the past. The jumbled blocks may be the remains of the crust, which caved in after subterranean melting of ice, thus allowing water to issue forth. The channel would suggest water flowing at high speed. Named Tiu Vallis, it extends from eastern Vallis Marineris to Chryse Planitia.

**2. Run-off channel**
Ma'adim Vallis is a run-off channel in the southern cratered terrain. About 600 km in length, it shows tributaries in its upper course, and widens downstream.

**3. Fretted channel**
This fretted channel—about 190 km in length—has a flat floor, steep walls and angular stretches along its course. It is in the Ismenius Lacus region of Mars.

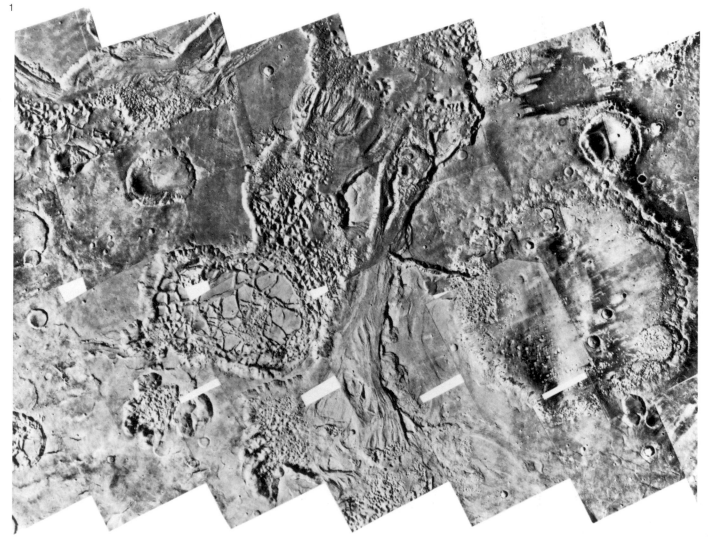

## 4. Outflow channel

Capri Chasma represents another outflow channel from a headward region of chaotic terrain. This oblique shot, covering an area approximately 300 by 300 km, shows the source region enclosed by cliffs and the characteristic hummocks of the Martian chaotic terrain. The rapid release of water that could have produced such a flood channel may have resulted from melting of ground-ice by volcanic activity beneath the Martian surface, or it may have issued from previously buried aquifers, although why this should happen has not yet been adequately explained.

## 5. Drainage networks

Dense, dendritic networks of channels such as this are a common feature of the southern highlands of Mars. Dendritic patterns do suggest a fluvial origin of the channels, although it should be added that faulting, the erosional work of the wind, and erosion by lava have all been called upon to explain these phenomena. The largest crater here is about 35 km wide.

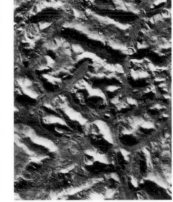

## 6. Fretted terrain

This landscape is transitional between an ancient cratered area and a sparsely cratered area. The main features of this region are similar to features on Earth where surface materials flow slowly aided by the alternate freezing and thawing of ice.

## 7. Flow patterns

Teardrop-shaped "islands" at the mouth of Ares Vallis were formed when flow was diverted around obstacles such as these craters. Similar patterns are formed on Earth by flood waters.

## 8. River channels

Sinuous riverbeds wind their way across this area of lava flows, which are broken by faults that form ridges. It is believed that in places water might have collected behind the ridges, eventually breaking through and forming "water gaps".

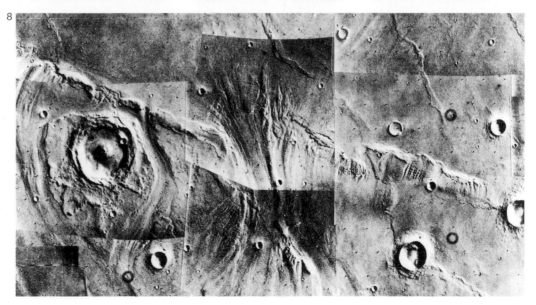

# Canyons

Visible even on long-distance images of Mars is the great canyon system, which straddles the globe just south of the equator between longitudes 30° and 110°W. Called Valles Marineris, this 4,000 km-long network begins on the east side of the Tharsis Bulge and ends in an immense region of chaotic terrain between Chryse Planitia and Margaritifer Sinus. At its deepest it is some 7 km deep and individual canyons are up to 200 km in width. In the impressive central section, where there are three roughly parallel, interconnecting rifts, the total width is 700 km.

Within the canyon system there is little direct evidence for the action of running water, although laminated deposits do occur on the floor in places. In the east, where the system ends amongst chaotic terrain, there is ample evidence of fluvial activity in the form of channels. Faulting seems to have played a major part in shaping the overall form and trend of individual canyons, and where the effects of subsequent erosion are apparent, particularly on the canyon walls, later side-canyons and indentations follow a distinctly linear pattern, typical of fault control.

The canyon can be considered in three sections: in the west there is a maze of relatively short interconnected canyons, called Noctis Labyrinthus; there is a main central section of W–E, well-defined rifts; and lastly there is a zone of irregular depressions, which merge eastwards into the chaotic terrain. The canyons of Noctis Labyrinthus are incised into the Tharsis Bulge and originate close to the focus of its radial fracture system. To the north, these fractures trend NNE–SSW, to the east they run E–W, while to the south they have a NW–SE orientation.

The main 2,400 km-long central canyon section is multiple for its entire length. Parallel to the ESE–WNW-trending canyons are down-faulted blocks and crater chains. In the central part of this section are three huge troughs 200 km wide, and the remnants of an eroded plateau separate them from one another.

East of 52°W there is a change in character, so that in this eastern part canyon walls are not linear and the floors become very hummocky. The two principal canyons here (Ganges Chasma and Eos/Capri Chasma) pass eastward into chaotic ground, where the influence of mass wasting and fluvial activity is more apparent.

### 1. Valles Marineris

Valles Marineris was one of the notorious canali seen by several observers. We now know it to be a large fracture running out from the Tharsis Shield. This photomosaic taken by the Viking Orbiter 1 very clearly shows the three massive shield volcanoes that sit astride the Tharsis Ridge—Arsia Mons, Pavonis Mons and Ascraeus Mons—as well as the vast canyon system that is Valles Marineris. Similar features occur on Earth (East African Rift Valley), the Moon and Venus. Valles Marineris may have formed as magma was withdrawn from beneath it to supply the volcanic uplift region.

### 2. Noctis Labyrinthus

Between the Tharsis volcanoes and Valles Marineris is a complex system of fractures resulting from an extension of the crust. The area has the highest elevation of the region and is believed to be the apex of uplift. The smoother Noctis Lacus area has wind-sculptured features, to the west of which is a complex system of criss-crossing fractures.

### 3. Scale of Valles Marineris

A comparison of, for example, the Tithonius Lacus region of Valles Marineris and the Grand Canyon shows the former to be four times deeper, six times wider and ten times longer.

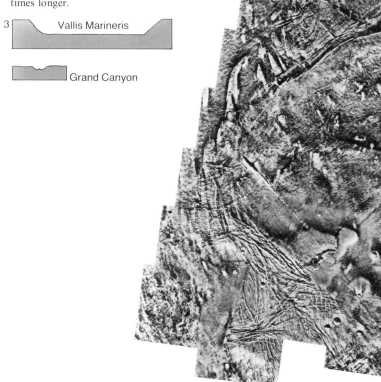

Vallis Marineris

Grand Canyon

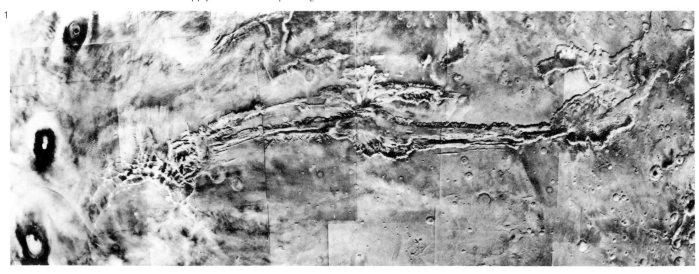

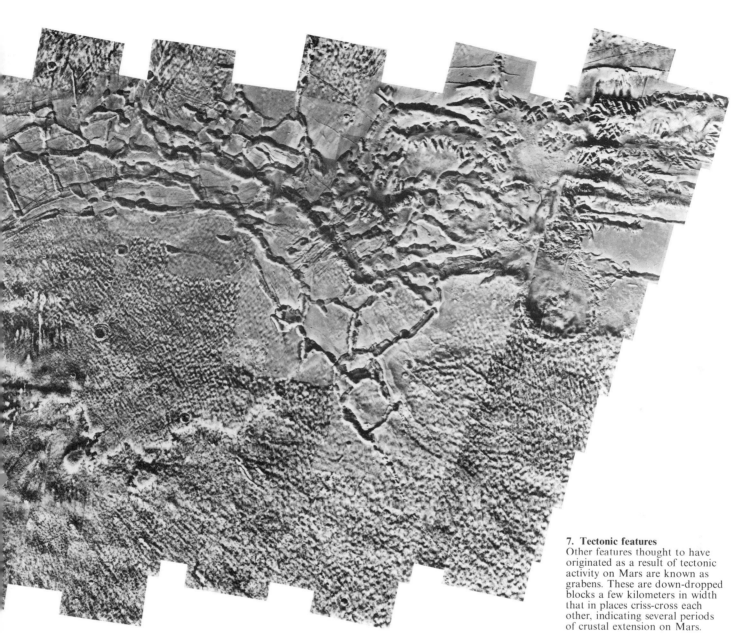

### 7. Tectonic features
Other features thought to have originated as a result of tectonic activity on Mars are known as grabens. These are down-dropped blocks a few kilometers in width that in places criss-cross each other, indicating several periods of crustal extension on Mars.

### 4. Tithonius Lacus
Part of Tithonius Lacus shows numerous tributary canyons and a central, heavily eroded ridge the height of the surrounding plain. Two original parallel canyons may have merged into one.

### 5. Landslides
Landslides have in places given the rim of the Valles Marineris

canyons a scalloped edge. In one place half of a crater remains on the rim of the canyon while half has crashed on to the valley floor below the canyon walls.

### 6. Deposition by water
Dendritic patterns south of Tithonius Chasma are thought to have been formed from material deposited by running water.

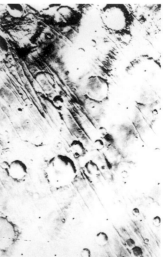

# Volcanoes

The great shields of Tharsis and Elysium sit astride the distinct bulges in the Martian crust and they have associated with them fluidal lava flows, pits, channels and small lava domes. Slopes on the flanks of these volcanoes are generally low, less than 6°. Each of the Tharsis shields—Arsia Mons, Pavonis Mons and Ascraeus Mons and the nearby Olympus Mons—share certain characteristics but differ in detail. All have summit calderas (collapse craters), which are usually surrounded by fractures. The calderas are often nested, indicating subsidence in several stages. Elysium Mons itself is significantly different in form from the other Martian shields: it has a particularly distinct asymmetric main shield, which extends to a broad ridge about 200 km across and 2 km in height which trends northward from the central edifice. On this are lava flows and conical hills resembling terrestrial cinder cones. A ring of fractures surrounds the whole structure.

To the north of Tharsis sits the immense Alba Patera, a fractured low shield with a partly buried central caldera and surrounding ring fractures that have a diameter of some 600 km. The entire structure is probably about 1,600 km in diameter and is the source of many lava flows, some sheet-like, some tabular and others fed by long lava tubes. The flanks are much gullied and channels appear to have been cut not by water but by fluid hot lava.

Much more ancient volcanic structures are located near to the margins of the Hellas Basin. These resemble the Alba structure in general form but are clearly degraded so that they now have little relief: Hadriaca Patera and Tyrrhena Patera are two of these.

That there is some genetic relationship between the shields and the tectonic activity associated with Tharsis and Elysium cannot be doubted. Vast arrays of fractures, mainly graben, characterize the Tharsis region and it would seem reasonable to suppose that the rise of the bulge was directly connected with both the fracturing and the various phases of volcanicity.

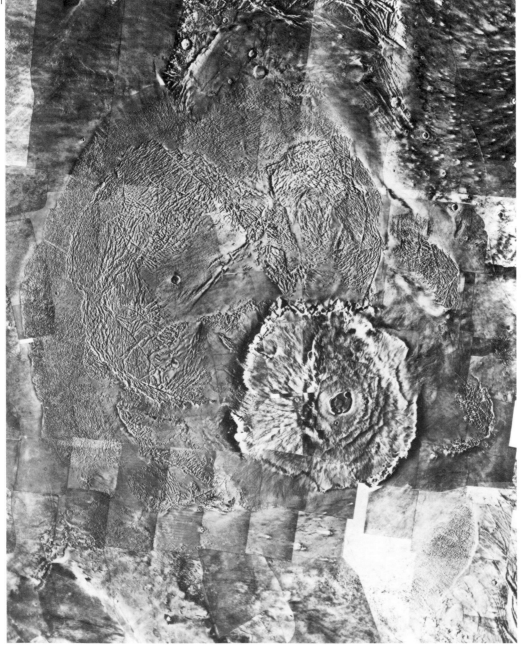

**1. Olympus Mons**
Rising 22 km above surrounding plains and at least 550 km in diameter, Olympus Mons is a giant among volcanic shields. It is made up of thousands of individual flows which can be traced for hundreds of km on the plains beyond the volcano. Unlike Hawaiian volcanoes on Earth, Olympus Mons is more or less circular like the shields of Iceland. Neither does it have large rift zones, although there are some fissures on the lower slopes. However, the form in general is similar to that of terrestrial shields: the gentle slopes were formed from fluid lava and show flow channels. The summit caldera is made up of multiple collapse craters.

**2. Caldera**
The summit of Olympus Mons has an enormous caldera which is at least 80 km in diameter. There may be concentric fault systems. Large calderas on volcanoes on Earth usually form as a result of collapse and Olympus Mons' probably formed in the same way.

**3. Lava flows**
The youngest lava flows seem to have occurred after the formation of the Olympus Mons scarp and here they make up the plains' surface at the foot of the scarp.

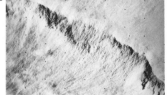

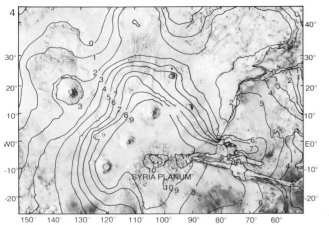

### 4. Tharsis region
This region includes the most prominent volcanoes on Mars. The contours outline the Syria Rise, an enormous bulge in the Martian crust 5,000 km across and 7 km high. This bulge is the site of Ascraeus Mons, Pavonis Mons and Arsia Mons.

### 5. Tharsis Ridge
This VO1 mosaic shows an area of intense crustal faulting. A cluster of volcanoes is seen to the right. It is thought that these fractures occurred at the same time as the bulge in Mars' surface was formed.

### 6. Arsia Mons
A mosaic of the Arsia Mons area shows the central caldera with concentric graben and embayments on the northeastern flank that have given rise to lava flows.

### 7. Tyrrhena Patera
Concentric graben and radial structures surround this ancient volcano. The large number of superimposed craters testify to its relative senility.

### 8. Types of eruption
The diagram distinguishes the features resulting from various kinds of basaltic eruption. A high-volume eruption from a fissure produces virtually no surface features (**A**); lower-volume eruptions from central vents produce shield volcanoes (**B**); and moderate eruptions from rift zones produce plains (**C**).

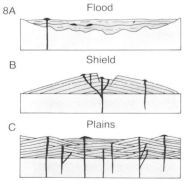

8A      Flood

B      Shield

C      Plains

# Northwest Quadrant

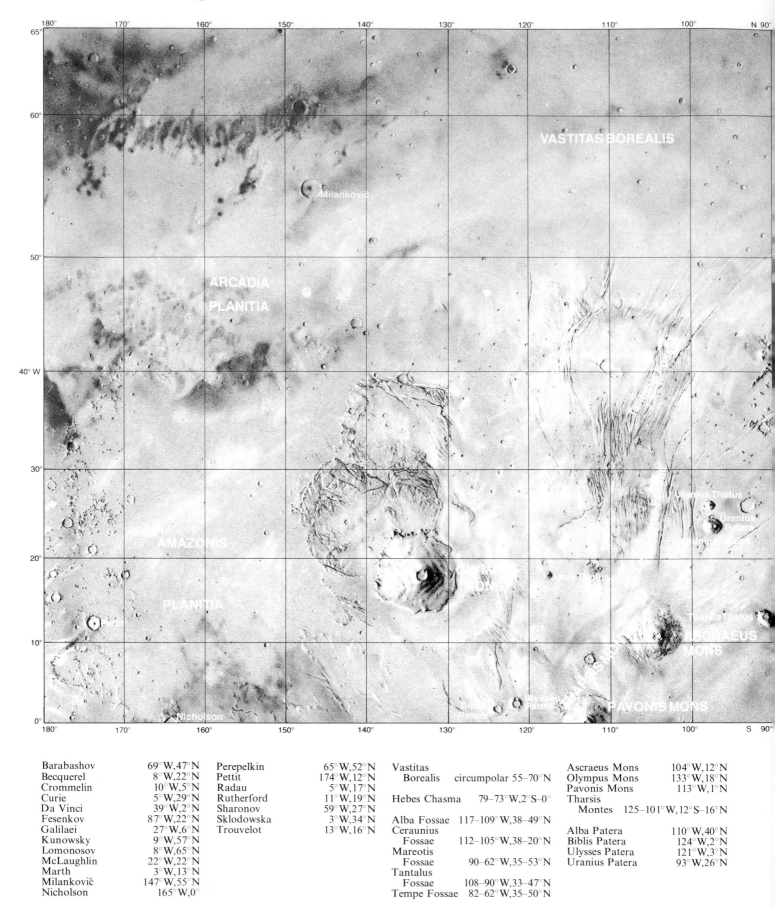

| | | | | | | | |
|---|---|---|---|---|---|---|---|
| Barabashov | 69°W,47°N | Perepelkin | 65°W,52°N | Vastitas | | Ascraeus Mons | 104°W,12°N |
| Becquerel | 8°W,22°N | Pettit | 174°W,12°N | Borealis circumpolar 55–70°N | | Olympus Mons | 133°W,18°N |
| Crommelin | 10°W,5°N | Radau | 5°W,17°N | | | Pavonis Mons | 113°W,1°N |
| Curie | 5°W,29°N | Rutherford | 11°W,19°N | Hebes Chasma 79–73°W,2°S–0° | | Tharsis | |
| Da Vinci | 39°W,2°N | Sharonov | 59°W,27°N | | | Montes 125–101°W,12°S–16°N | |
| Fesenkov | 87°W,22°N | Sklodowska | 3°W,34°N | Alba Fossae 117–109°W,38–49°N | | | |
| Galilaei | 27°W,6°N | Trouvelot | 13°W,16°N | Ceraunius | | Alba Patera | 110°W,40°N |
| Kunowsky | 9°W,57°N | | | Fossae 112–105°W,38–20°N | | Biblis Patera | 124°W,2°N |
| Lomonosov | 8°W,65°N | | | Mareotis | | Ulysses Patera | 121°W,3°N |
| McLaughlin | 22°W,22°N | | | Fossae 90–62°W,35–53°N | | Uranius Patera | 93°W,26°N |
| Marth | 3°W,13°N | | | Tantalus | | | |
| Milankovič | 147°W,55°N | | | Fossae 108–90°W,33–47°N | | | |
| Nicholson | 165°W,0° | | | Tempe Fossae 82–62°W,35–50°N | | | |

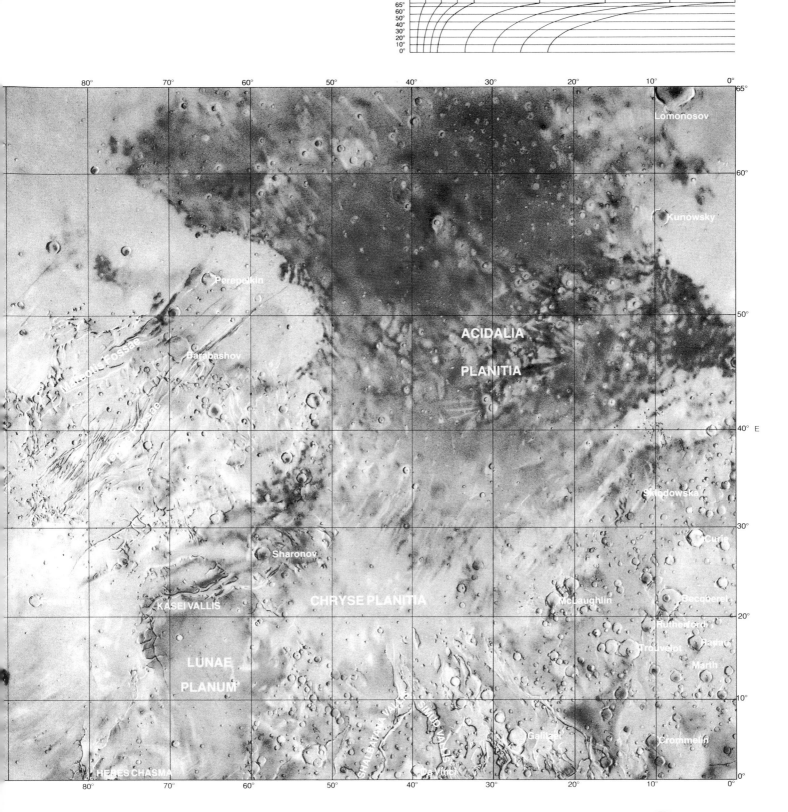

80°    70°    60°    50°    40°    30°    20°    10°    0°

Lomonosov

Kunowsky

Perepelkin

ACIDALIA

PLANITIA

Barabashov

Mareotis Fossae

Tempe Fossae

Skłodowska

Sharonov

Curie

CHRYSE PLANITIA

McLaughlin

Becquerel

KASEI VALLIS

Rutherford

Trouvelot

Rahau

LUNAE

Marth

PLANUM

SHALBATANA VALLIS

SIMUD VALLIS

Galilei

Crommelin

HEBES CHASMA

Da Vinci

80°    70°    60°    50°    40°    30°    20°    10°    0°

65°
60°
50°
40° E
30°
20°
10°
0°

| | | |
|---|---|---|
| Acidalia | | |
|   Planitia | 60° W–0°,55–14° N | |
| Amazonis | | |
|   Planitia | 168–140° W,0°–40° N | |
| Arcadia | | |
|   Planitia | 195–110° N,55–40° N | |
| Chryse | | |
|   Planitia | 51–37° W,19–30° N | |
| Lunae Planum | 75–60° W,5–23° N | |

| | |
|---|---|
| Ceraunius Tholus | 97° W,24° N |
| Jovis Tholus | 117° W,18° N |
| Tharsis Tholus | 91° W,14° N |
| Uranius Tholus | 98° W,26° N |

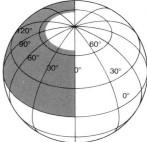

231

# Northeast Quadrant

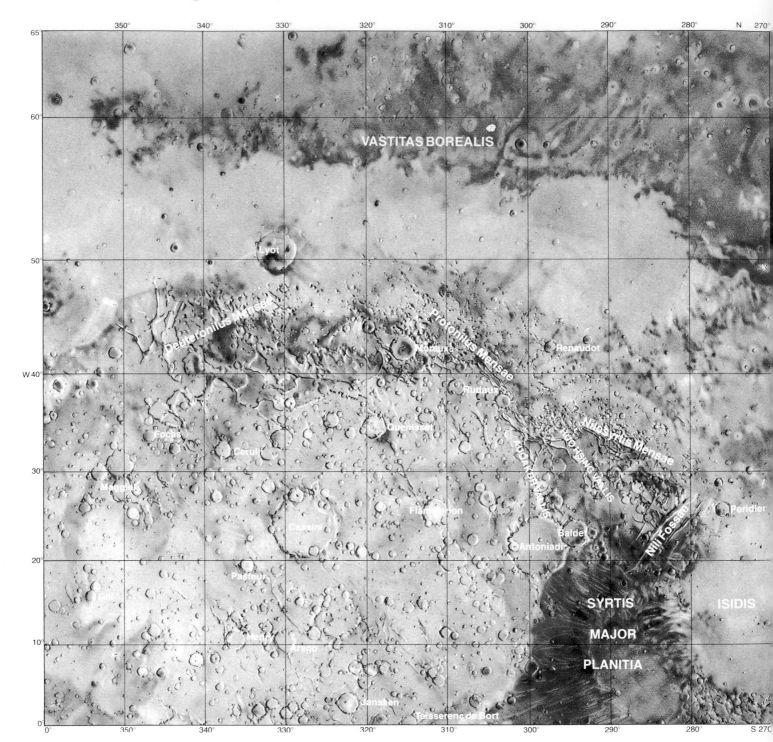

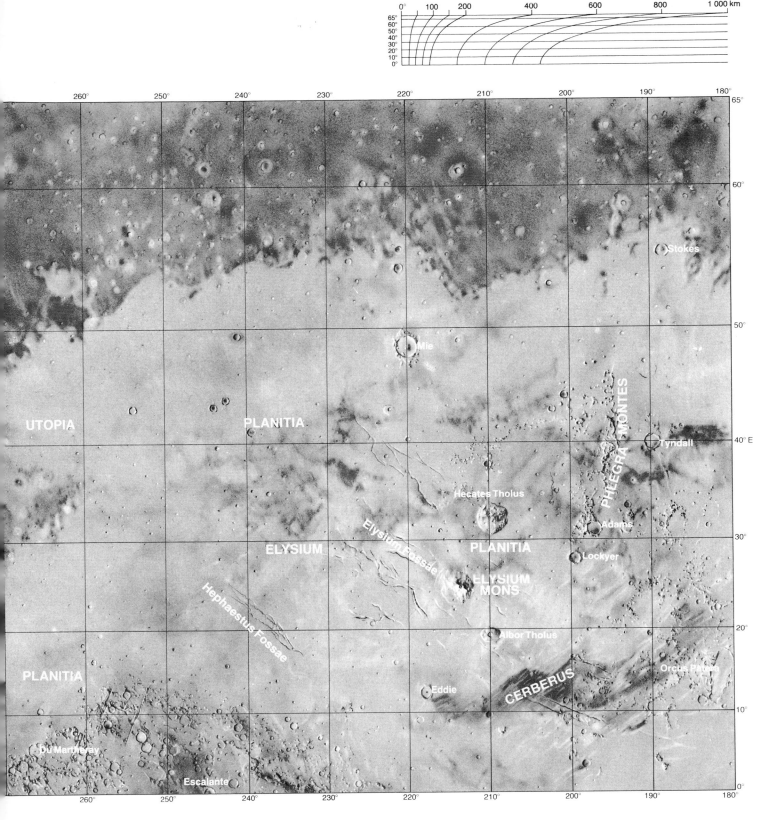

0°   100   200   400   600   800   1 000 km
65°
60°
50°
40°
30°
20°
10°
0°

260°    250°    240°    230°    220°    210°    200°    190°    180°
65°

Stokes

60°

50°

Mie

PHLEGRA MONTES

Tyndall    40° E

UTOPIA    PLANITIA

Hecates Tholus

Elysium Fossae

Adams

30°

ELYSIUM    PLANITIA

Lockyer

Hephaestus Fossae

ELYSIUM MONS

Albor Tholus    20°

PLANITIA    Orcus Patera

CERBERUS    Eddie

10°

Du Martheray

Escalante    0°

260°    250°    240°    230°    220°    210°    200°    190°    180°

Elysium
  Planitia        180–260° W, 10° S–30° N
Isidis
  Planitia        279–255° W, 4–20° N
Syrtis Major
  Planitia        298–283° W, 20° N–1° S
Utopia
  Planitia        310–195° W, 35–50° N

Albor Tholus        210° W, 19° N
Hecates Tholus      210° W, 32° N

Huo Hsing
  Vallis          299–292° W, 34–28° N
Auqakuh
  Vallis          300–297° W, 30–27° N

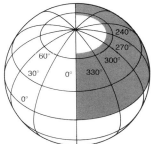

233

# Southwest Quadrant

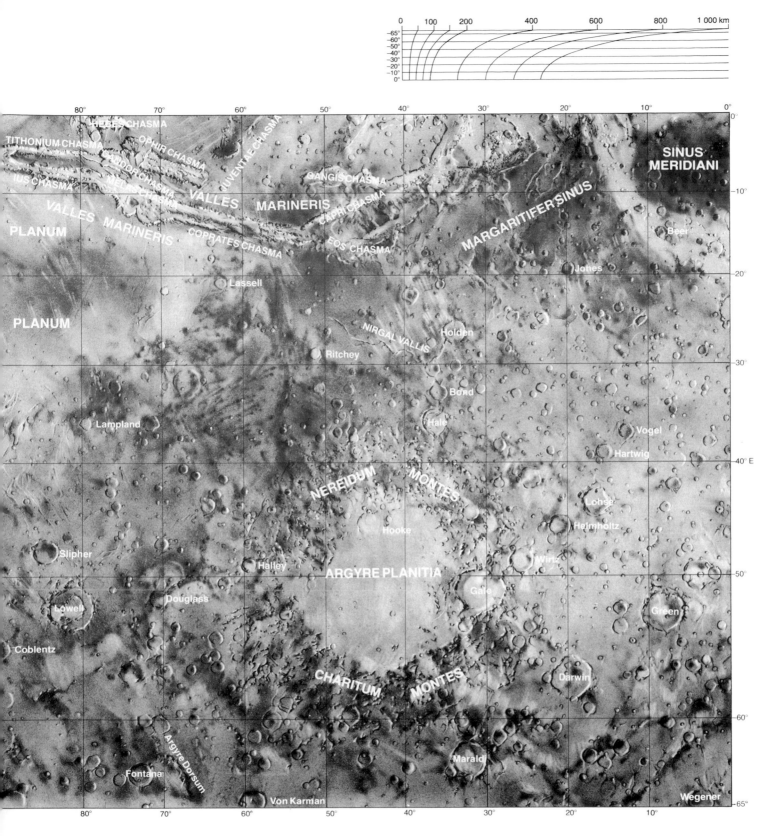

Scale bar: 0 100 200 400 600 800 1 000 km

-65° -60° -50° -40° -30° -20° -10° 0°

80° 70° 60° 50° 40° 30° 20° 10° 0°

HEBES CHASMA
TITHONIUM CHASMA
OPHIR CHASMA
IUS CHASMA
CANDOR CHASMA
MELAS CHASMA
JUVENTAE CHASMA
VALLES MARINERIS
GANGIS CHASMA
CAPRI CHASMA
EOS CHASMA
COPRATES CHASMA
VALLES MARINERIS
PLANUM
PLANUM

SINUS MERIDIANI
MARGARITIFER SINUS
Beer
Jones
Lassell
NIRGAL VALLIS
Holden
Ritchey
Bond
Hale
Vogel
Hartwig
Lampland
NEREIDUM MONTES
Lohse
Helmholtz
Hooke
Slipher
Halley
Wirtz
ARGYRE PLANITIA
Gale
Douglass
Green
Lowell
Coblentz
CHARITUM MONTES
Darwin
Argyre Dorsum
Marald
Fontana
Von Karman
Wegener

80° 70° 60° 50° 40° 30° 20° 10°

-10° -20° -30° -40° E -50° -60° -65°

| Ius Chasma | 92–77°W,10–14°S |
| Juventae Chasma | 54–60°W,2–6°S |
| Melas Chasma | 77–69°W,8–14°S |
| Ophir Chasma | 75–64°W,3–9°S |
| Tithonium Chasma | 92–77°W,3–7°S |
| Claritas Fossae | 108–100°W,12–40°S |
| Medusae Fossae | 160–165°W,0–6°S |
| Memnonia Fossae | 160–147°W,23–18°S |
| Sirenum Fossae | 177–139°W,40–25°S |

| Thaumasia Fossae | 100–85°W,33–45°S |
| Noctis Labyrinthus | 110–95°W,4–14°S |
| Arsia Mons | 121°W,9°S |
| Charitum Montes | 60–27°W,50–59°S |
| Nereidum Montes | 60–30°W,50–38°S |
| Tharsis Montes | 125–101°W,12–16°S |

| Argyre Planitia | 43–51°W,55–40°S |
| Sinai Planum | 83.5°W,14.5°S |
| Solis Planum | 93°W,25°S |
| Syria Planum | 104.5°W,15°S |
| Margaritifer Sinus | 12–45°W,2–27°S |
| Mangala Vallis | 150–152°W,4–9°S |
| Nirgal Vallis | 37–47°W,27–30°S |
| Valles Marineris | 24–113°W,1–18°S |

235

# Southeast Quadrant

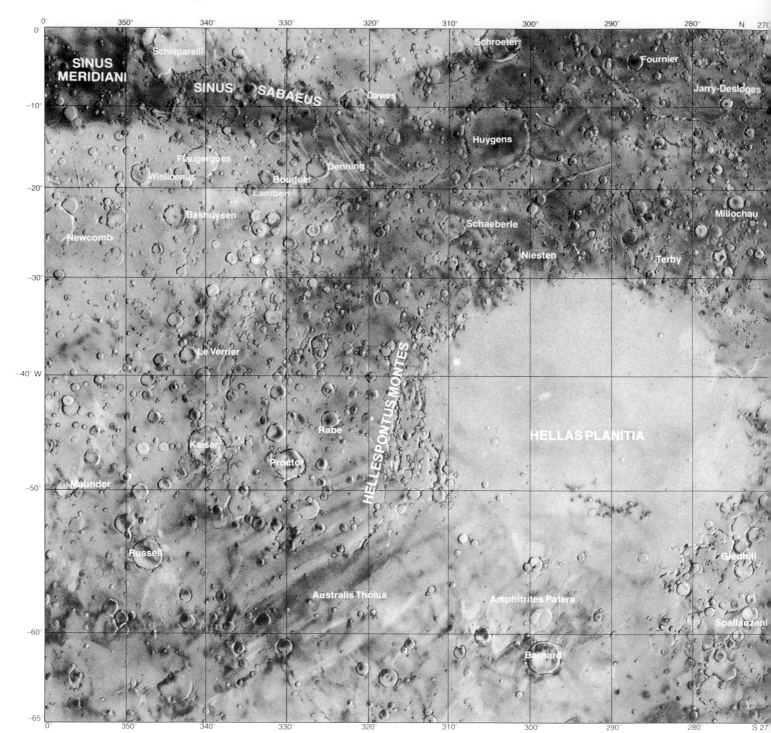

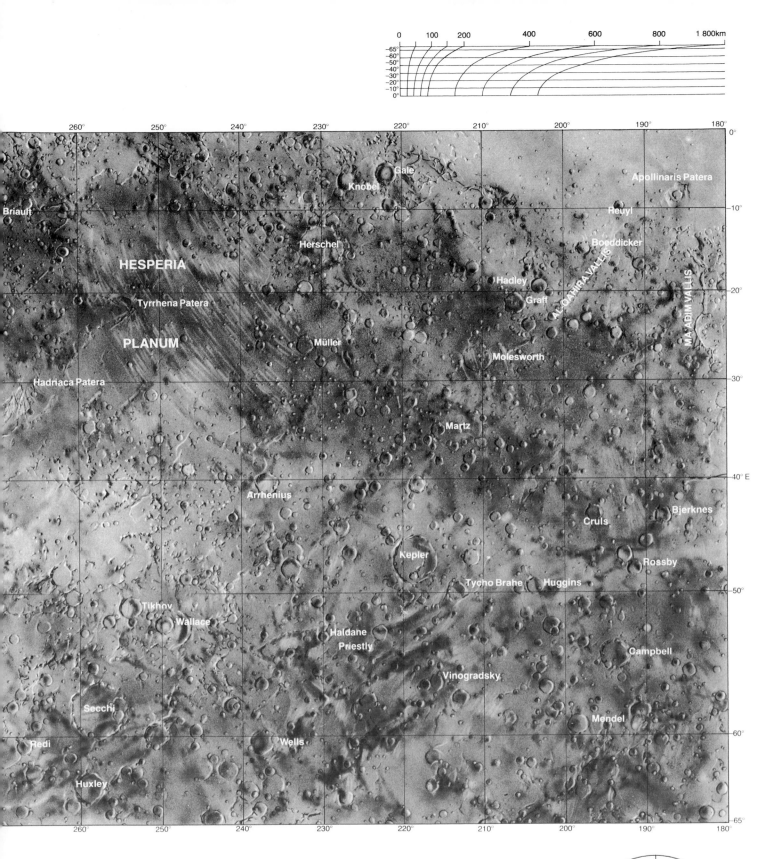

0   100   200   400   600   800   1 800km

-65°
-60°
-50°
-40°
-30°
-20°
-10°
0°

260° 250° 240° 230° 220° 210° 200° 190° 180°

0°
-10°
-20°
-30°
-40° E
-50°
-60°
-65°

Briault

Gale

Knobel

Apollinaris Patera

Reuyl

Herschel

Boeddicker

**HESPERIA**

Hadley

Graff

AL-QAHIRA VALLIS

MA'ADIM VALLIS

Tyrrhena Patera

Molesworth

**PLANUM**

Müller

Hadriaca Patera

Martz

Arrhenius

Bjerknes

Cruls

Rossby

Kepler

Tycho Brahe

Huggins

Tikhov

Wallace

Campbell

Haldane
Priestly

Vinogradsky

Secchi

Mendel

Redi

Wells

Huxley

| | | | | | |
|---|---|---|---|---|---|
| Tycho Brahe | 214°W,50°S | Hellespontus | | Hesperia | |
| Vinogradsky | 217°W,56°S | Montes | 319–310°W,35–50°S | Planum | 258–242°W,10–35°S |
| Wallace | 249°W,53°S | | | | |
| Wells | 238°W,60°S | Amphitrites | | Australis Tholus | 323°W,59°S |
| Wislicenus | 349°W,18°S | Patera | 299°W,61°S | | |
| | | Apollinarsis | | Al-Qahira | |
| | | Patera | 186°W,9°S | Vallis | 200–194°W,19–14°S |
| | | Hadriaca Patera | 267°W,31°S | Ma'adim | |
| | | Tyrrhena Patera | 253°W,22°S | Vallis | 184–181°W,28–16°S |
| | | | | | |
| | | Hellas | | | |
| | | Planitia | 313–272°W,60–30°S | | |

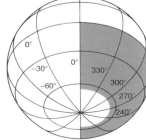

0°
-30°
-60°
0°
330°
300°
270°
240°

# North and South Poles

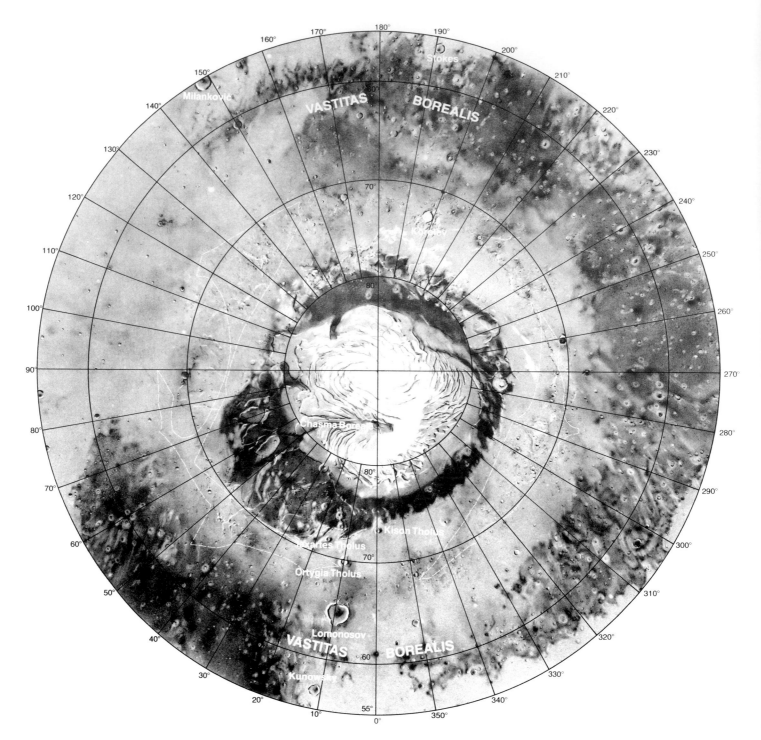

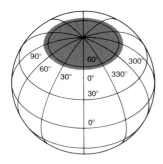

**North polar region**

| | |
|---|---|
| Korolev | 196°W,73°N |
| Kunowsky | 9°W,57°N |
| Lomonosov | 8°W,65°N |
| Milankovič | 147°W,55°N |
| Stokes | 189°W,56°N |
| Chasma Boreale | 50–30°W,85–81°N |
| Vastitas Borealis | 55–70°N |
| Iaxartes Tholus | 15°W,72°N |
| Kison Tholus | 358°W,73°N |
| Ortygia Tholus | 8°W,70°N |

**South polar region**

| | | | |
|---|---|---|---|
| Agassiz | 89°W,70°S | Gilbert | 274°W,68°S |
| Barnard | 298°W,61°S | Heaviside | 95°W,71°S |
| Bianchini | 96°W,64°S | Holmes | 292°W,75°S |
| Burroughs | 243°W,72°S | Hutton | 255°W,72°S |
| Chamberlin | 124°W,66°S | Huxley | 259°W,63°S |
| Charlier | 169°W,69°S | Jeans | 206°W,70°S |
| Clark | 134°W,56°S | Joly | 42°W,75°S |
| Coblentz | 91°W,55°S | Keeler | 152°W,61°S |
| Daly | 22°W,66°S | Kuiper | 157°W,57°S |
| Dana | 32°W,73°S | Lau | 107°W,74°S |
| Darwin | 20°W,57°S | Liais | 253°W,75°S |
| Du Toit | 46°W,72°S | Lyell | 15°W,70°S |
| Fontana | 73°W,63°S | Main | 310°W,77°S |

238

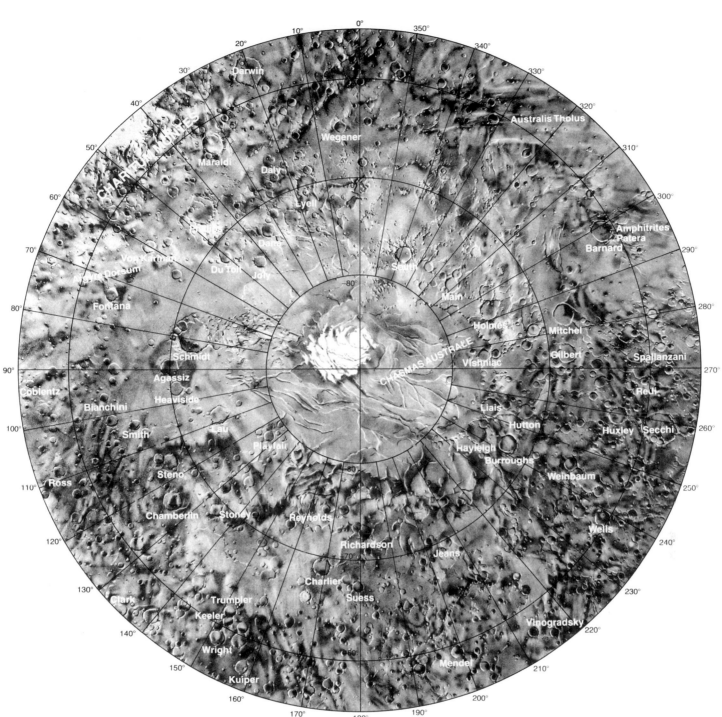

| | | | | | | |
|---|---|---|---|---|---|---|
| Maraldi | 32°W,62°S | South | 339°W,77°S | Chasma | | |
| Mendel | 199°W,59°S | Spallanzani | 273°W,58°S | Australe | 284–257°W,80–89°S | |
| Mitchel | 284°W,68°S | Steno | 115°W,68°S | | | |
| Phillips | 45°W,67°S | Stoney | 138°W,70°S | Charitum | | |
| Playfair | 125°W,78°S | Suess | 179°W,67°S | Montes | 60–27°W,50–59°S | |
| Rayleigh | 240°W,76°S | Trumpler | 151°W,62°S | | | |
| Redi | 267°W,61°S | Vinogradsky | 217°W,56°S | Amphitrites | | |
| Reynolds | 150°W,74°S | Vishniac | 276°W,77°S | Patera | 299°W,61°S | |
| Richardson | 181°W,73°S | Von Karman | 59°W,64°S | | | |
| Ross | 108°W,58°S | Wegener | 4°W,65°S | Australis Tholus | 323°W,59°S | |
| Schmidt | 79°W,72°S | Weinbaum | 245°W,66°S | | | |
| Secchi | 258°W,58°S | Wells | 238°W,60°S | | | |
| Smith | 103°W,66°S | Wright | 151°W,59°S | | | |

239

# Satellites

Mars has two satellites, both of which were discovered by Asaph Hall in 1877 at the US Naval Observatory in Washington. He named them Phobos (fear) and Deimos (terror), after the mythical characters who drove the chariot of the war god Mars. Both satellites are very small, and are quite unlike the Earth's Moon. From Earth they appear only as tiny specks of light, and nothing positive was known about their surfaces until Mariner 9 and the Vikings.

## Dimensions

Both satellites are comparatively close to Mars. Phobos, the inner one, moves in an almost circular orbit 9,270 km from the center of Mars—closer to its parent planet than any other known satellite. The strong gravitational pull of Mars resulting from its close proximity is balanced by Phobos' rapid orbiting. The revolution period is 7 hr 39 min 27 sec, and Phobos therefore completes one journey around Mars in less than one sol. Until the recent discovery of the smaller inner satellites of Jupiter and Saturn, this was a unique case in the Solar System. The orbital velocity of Phobos is increasing very slowly and it has been suggested that Phobos may be spiralling in towards Mars, and that it will crash on to the surface in about 100 million years' time, but this is questionable.

Deimos orbits at a distance of 23,400 km from the center of Mars. The revolution period is 1 day 6 hr 21 min 16 sec and Deimos too has a near circular orbit. Both satellites' orbits lie within 2° of the Martian equatorial plane. This low inclination means that they cross the Martian sky at a fixed zenith angle, which depends on the observer's latitude, throughout the Martian year. The rotation of each satellite is synchronous. When a small, irregular body spins rapidly in the vicinity of a large body, tidal friction will eventually act as a brake on the rotation of the smaller body until it is synchronous with its orbital period about the larger body.

To an observer on Mars neither satellite would be imposing. Phobos, with an apparent magnitude of −3.9, would shed about as much light as Venus does on Earth. It would cross the sky in only 4.5 hr, during which time it would display more than half its cycle of phases, and it would travel backwards across the sky, that is from west to east. The interval between successive risings would be just over 11 hr. Total solar eclipses would never occur, though Phobos would pass across the Sun in transit 1,300 times during the Martian year, taking 19 sec to cross from one limb to the other. The satellite would never rise above the horizon at any Martian latitude higher than 69° N or S, and even when above the horizon it would be frequently eclipsed by the shadow of the parent planet.

Deimos, smaller and further away, would remain above the Martian horizon for 2.5 consecutive sols, and would be visible above the horizon at any latitude lower than 82° N or S. To the naked eye the phases would be almost imperceptible and, with an apparent magnitude of −0.1, Deimos would be less bright from Mars than Sirius is from the Earth. It would transit the Sun 130 times during the Martian year and each transit would take 1 min 48 sec. Like Phobos, it would be often eclipsed by Mars' shadow.

Both satellites are approximately triaxial; the size of Phobos is 20 × 23 × 28 km, and that of Deimos 10 × 12 × 16 km. The Vikings bypassed the satellites at very close range: Viking Orbiter 1 within 88 km of Phobos, and Viking Orbiter 2 within 28 km of Deimos. In 1989 landings were planned on Phobos by the unmanned Soviet spacecraft Phobos 1 and Phobos 2, but both failed, though Phobos 2 did send back a few close-range images before contact with it was lost.

The escape velocities are very low; 15 m/sec for Phobos, 10 m/sec for Deimos. Both satellites are dark objects, with albedoes of between 5 and 7 percent. Their densities are low, approximately 2 gm/cm³. The spectra of Phobos and Deimos resemble those of certain asteroids known as "C-types" (*see* pages 243–5). These are common to the outer regions of the asteroid belt and to the Trojan group of asteroids in the orbit of Jupiter. Phobos and Deimos are

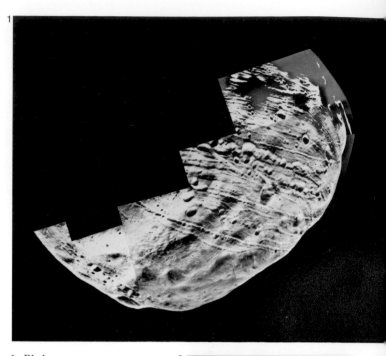

**1. Phobos**
From a distance of 300 km VO1 photographed Phobos and revealed the linear grooves (left). Another image, from 480 km, shows the surface covered with craters and hummocks (right).

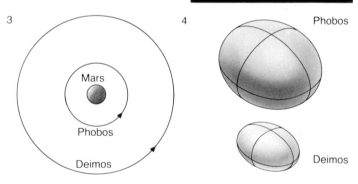

**2. Deimos**
From 500 km, shows a relatively smooth surface with few craters and several large flat areas.

**3. Orbits**
The orbit of Phobos lies just beyond the Roche limit. Deimos lies just beyond the stationary orbit position where orbital and rotation periods are the same.

**4. Shapes and sizes**
The diagram shows the relative sizes of Mars' two moons. They are both ellipsoidal rather than spherical in shape and are thus more like asteroids than moons.

## Physical data

|  | Phobos | Deimos |
| --- | --- | --- |
| Mean distance from center of Mars | 9,270 km | 23,400 km |
| Mean sidereal period | 0.3189 days | 1.2624 days |
| Mean synodic period | 7 hr 39 min 26.6 sec | 1 d 6 hr 21 min 15.7 sec |
| Orbital eccentricity | 0.0210 | 0.0028 |
| Orbital inclination | 1.1° | 1.8° |
| Diameter | 20 × 23 × 28 km | 10 × 12 × 16 km |
| Mass | $9.6 \times 10^{15}$ kg | $20 \times 10^{15}$ kg |
| Mean density | 1.9 g/cm³ | 2.1 g/cm³ |
| Escape velocity | 15 m/sec | 10 m/sec |
| Magnitude at mean opposition | 11.6 | 12.8 |

much darker in appearance than asteroids in the inner part of the belt closer to Mars. The material of which C-types are composed is believed to be similar to carbonaceous chondrite meteorites.

All this suggests that Mars' two satellites were once much nearer to Jupiter, which perturbed them into orbits that allowed their capture by Mars, but their origin is still very much a matter of conjecture. It is certainly very different from that of the Earth's Moon. Most astronomers would agree that Phobos and Deimos have a common origin and that they formed within the asteroid belt (*see* pages 243–5) and were captured by Mars during the final stage of its accretion from the solar nebula (*see* pages 10–11). At this stage the planet would have been still surrounded by an extensive atmosphere, giving the necessary drag to capture the satellites.

### The surface of Phobos

Phobos is cratered and the surface looks superficially like that of the lunar highlands. The largest crater, named Stickney in honor of Asaph Hall's wife (née Stickney), is 10 km in diameter—large in comparison with the diameter of the satellite itself. It has been suggested that the hummocks and blocks around Stickney could be ejecta produced when the crater was formed. This may have been possible, but had the crater been produced by impact, ejecta would have escaped into space because the escape velocity of Phobos is so low. Most craters, however, are not surrounded by ejecta blankets.

Phobos has a series of linear grooves running across parts of the surface. They are about 500 m wide and it is believed they may be the result of fracturing of the surface of the satellite caused by tidal forces exerted on it by the parent planet and by impact with large debris. The collision that may have produced Stickney would probably have fractured Phobos internally. Other features related to tidal forces are the large network of parallel striations that are also found on the surface of Phobos. Prolonged stress may have ruptured faults within the satellites and this is manifested on the surface as the striations, which are on average between 100 and 200 m wide and between 5 and 10 m deep. There is also a network of linear features radiating out from Stickney. Some of them seem to resemble chains of craterlets that have coalesced, and they are undoubtedly associated with Stickney.

The surface of Phobos has now been adequately mapped. The other main craters are Hall and Roche, both about 5 km in diameter, and additional named craters are Todd, Sharpless and d'Arrest. The main ridge associated with the Stickney system is known as Kepler Dorsum.

### The surface of Deimos

Deimos differs from Phobos in certain important respects. There are no craters larger than 3 km in diameter. The two most prominent craters have been named Swift and Voltaire, but all in all, craters are much more subdued on Deimos than on Phobos. Most are flat floored, with distinct breaks of slopes near the base of the walls. It is interesting that neither satellite shows any trace of craters with central peaks or terraced walls. There are no grooves or striations on Deimos and since the satellite orbits at a greater distance from Mars, the rupturing effect of the planet's gravitational forces would be less than in the case of Phobos.

Both Phobos and Deimos are covered with a thin layer of dust. This is known as regolith and is produced by the accumulation of tiny particles that rain down on to the surfaces of satellites. On impact a small amount of dust is produced. In the case of these satellites, much of the dust will have sufficient velocity to escape back into space, but that which remains forms a very thin film that may only be as much as 1 mm thick in places. The regolith on Deimos is thicker than that of Phobos and almost all the craters appear to be partly filled. Why this should be the case cannot be fully explained, but the fact that these small satellites can retain a loose surface covering is an important discovery in itself.

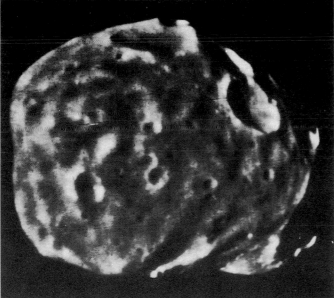

**5. Phobos from Phobos 2**
Phobos, shown by Phobos 2 on 21 February 1989 from a range of 440 km.

**6. Close-up of Deimos**
From only 50 km the surface looks much different: it is pitted with craters and is boulder strewn. Many of the craters are filled or covered with the regolith.

# ASTEROIDS

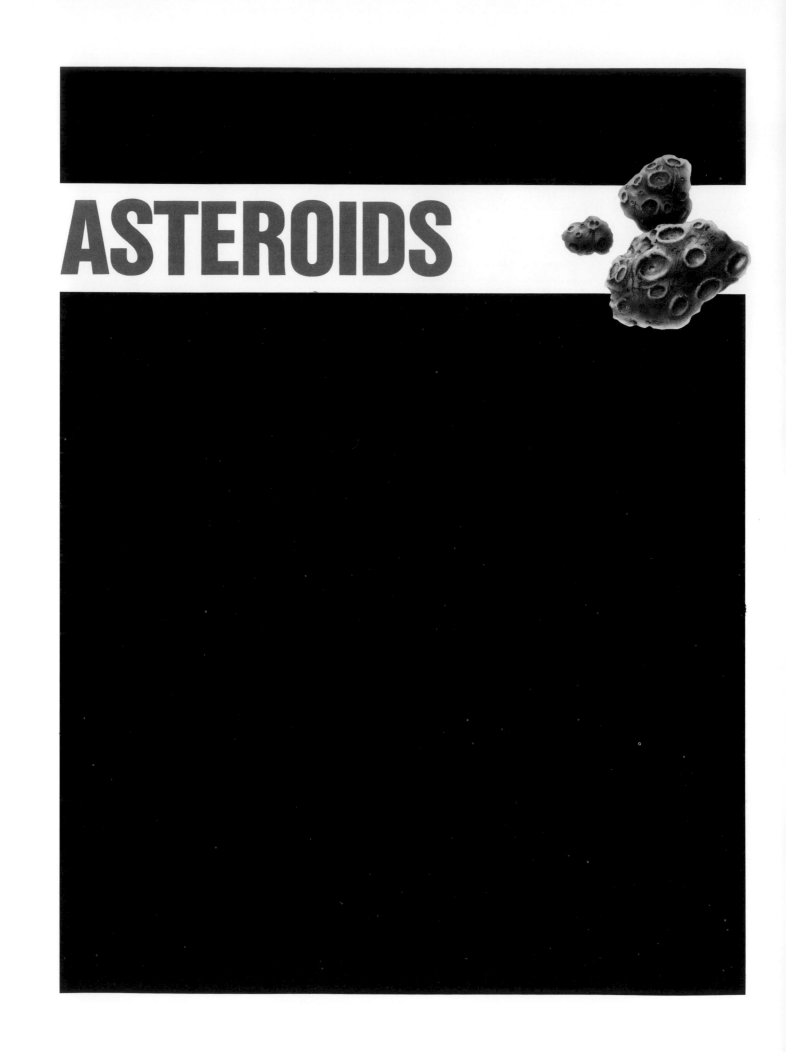

# Characteristics

Most of the minor planets, or asteroids, lie in the region between the orbits of Mars and Jupiter: all are small (only Ceres has a diameter greater than 900 km), and only one (Vesta) is ever visible with the naked eye. By 1990 the number of named asteroids was over 3000.

Asteroid diameters are difficult to measure because of their small discs. Infrared and radar measurements have provided some information and excellent results are obtained from occultations of stars by asteroids, though these events are unfortunately rare. Asteroid shapes range from roughly spherical to elongated and irregular. Some of the latter are now believed to be double or multiple bodies: there are large satellites orbiting 2 Pallas and 12 Victoria, for example. Mass determinations depend upon mutual perturbations. In some cases rotation periods have been determined, ranging from 3 hr to as much as 60 days for 280 Glauke.

The regular asteroids keep strictly to the region between the paths of Mars and Jupiter, but some of them have high orbital inclinations (34°, for example, for Pallas). They tend to collect in "families" due to the dominating gravitational effect of Jupiter; an asteroid with an orbital period that is an exact fraction of that of Jupiter will tend to suffer cumulative perturbations until it has been removed from that particular orbit. Many "field" asteroids are not members of families. A century ago 132 Aethra had the least-known perihelion (1.61 AU) and 153 Hilda the greatest aphelion (3.4 AU), today many are known well outside the main swarm.

| No. | Name | Discoverer | Date | $q$ | $Q$ | $P$ | $e$ | $i$ | $D$ | $A$ |
|---|---|---|---|---|---|---|---|---|---|---|
| 1 | Ceres | Piazzi | 1801 | 2.55 | 2.94 | 4.06 | 0.079 | 10.6 | 940 | 0.054 |
| 2 | Pallas | Olbers | 1802 | 2.11 | 3.42 | 4.60 | 0.237 | 34.9 | 538 | 0.074 |
| 3 | Juno | Harding | 1804 | 1.98 | 3.35 | 4.36 | 0.257 | 13.0 | 285 | 0.151 |
| 4 | Vesta | Olbers | 1807 | 2.15 | 2.57 | 3.63 | 0.089 | 7.1 | 555 | 0.229 |
| 5 | Astraea | Hencke | 1845 | 2.10 | 3.06 | 4.14 | 0.187 | 5.3 | 117 | 0.140 |
| 6 | Hebe | Hencke | 1847 | 1.93 | 2.92 | 3.78 | 0.203 | 14.8 | 195 | 0.164 |
| 7 | Iris | Hind | 1847 | 1.84 | 2.94 | 3.69 | 0.230 | 5.5 | 209 | 0.154 |
| 8 | Flora | Hind | 1847 | 1.86 | 2.55 | 3.27 | 0.156 | 5.9 | 151 | 0.144 |
| 9 | Metis | Graham | 1848 | 2.09 | 2.68 | 3.68 | 0.123 | 5.6 | 151 | 0.139 |
| 10 | Hygeia | De Gasparis | 1849 | 2.84 | 3.46 | 5.59 | 0.100 | 3.8 | 450 | 0.041 |
| 13 | Egeria | De Gasparis | 1850 | 2.36 | 2.80 | 4.14 | 0.085 | 16.5 | 224 | 0.041 |
| 15 | Eunomia | De Gasparis | 1851 | 2.15 | 3.14 | 4.30 | 0.188 | 11.7 | 272 | 0.155 |
| 16 | Psyche | De Gasparis | 1851 | 2.53 | 3.32 | 5.00 | 0.135 | 3.1 | 250 | 0.093 |
| 24 | Themis | De Gasparis | 1853 | 2.76 | 3.52 | 5.56 | 0.121 | 0.8 | 234 | 0.030 |
| 31 | Euphrosyne | Ferguson | 1854 | 2.45 | 3.86 | 5.61 | 0.223 | 26.3 | 370 | 0.030 |
| 44 | Nysa | Goldschmidt | 1857 | 2.05 | 2.79 | 3.77 | 0.151 | 3.7 | 82 | 0.377 |
| 48 | Doris | Goldschmidt | 1857 | 2.93 | 3.30 | 5.50 | 0.060 | 6.6 | 250 | 0.03 |
| 52 | Europa | Goldschmidt | 1858 | 2.75 | 3.43 | 5.45 | 0.111 | 7.5 | 289 | 0.035 |
| 65 | Cybele | Tempel | 1861 | 3.01 | 3.83 | 6.33 | 0.121 | 3.5 | 309 | 0.022 |
| 92 | Undina | Peters | 1867 | 2.97 | 3.43 | 5.72 | 0.072 | 9.9 | 250 | 0.03 |
| 95 | Arethusa | Luther | 1867 | 2.61 | 3.53 | 5.37 | 0.149 | 13.0 | 230 | 0.019 |
| 324 | Bamberga | Palisa | 1892 | 1.70 | 3.59 | 4.40 | 0.330 | 11.2 | 240 | ? |
| 349 | Dembowska | Charlois | 1892 | 2.66 | 3.19 | 5.00 | 0.090 | 8.3 | 144 | ? |
| 433 | Eros | Witt | 1898 | 1.13 | 1.78 | 1.76 | 0.223 | 10.8 | ? | ? |
| 451 | Patienta | Charlois | 1899 | 2.82 | 3.30 | 5.34 | 0.077 | 15.2 | 276 | ? |
| 511 | Davida | Dugan | 1903 | 2.66 | 3.72 | 5.70 | 0.166 | 15.7 | 323 | ? |
| 617 | Patroclus | Kopff | 1906 | 4.48 | 5.94 | 11.88 | 0.140 | 22.0 | 147 | 0.037 |
| 624 | Hector | Kopff | 1907 | 4.99 | 5.25 | 11.59 | 0.024 | 18.3 | 179 | 0.038 |
| 704 | Interamnia | Cerulli | 1910 | 2.58 | 3.53 | 5.35 | 0.155 | 17.3 | 350 | ? |
| 944 | Hidalgo | Baade | 1920 | 2.00 | 9.61 | 14.04 | 0.657 | 42.5 | 15 | ? |
| 1,172 | Aeneas | Reinmuth | 1930 | 4.64 | 5.70 | 11.74 | 0.102 | 16.7 | 130 | 0.044 |
| 1,221 | Amor | Delporte | 1932 | 1.08 | 2.76 | 2.66 | 0.436 | 11.9 | ? | ? |
| 1,566 | Icarus | Baade | 1949 | 0.19 | 1.97 | 1.12 | 0.827 | 22.9 | ? | ? |
| 1,862 | Apollo | Reinmuth | 1932 | 0.65 | 2.29 | 1.78 | 0.566 | 6.4 | ? | ? |
| 2,060 | Chiron | Kowal | 1977 | 8.50 | 18.50 | 50.7 | 0.378 | 6.9 | 110 | ? |
| 2,602 | Moore | Bowell | 1982 | 2.13 | 2.38 | 3.68 | 0.105 | 5.5 | 6 | ? |

**1. Data table**
The 35 asteroids described in the table include the first 10 to be discovered, the 18 largest, the three largest Trojans, and the nine brightest. Information is coded as follows: closest distance to the Sun ($q$) in astronomical units; furthest distance from the Sun ($Q$); revolution period ($P$) in years; orbital eccentricity ($e$) and orbital inclination ($i$); diameter ($D$) in km; and, finally, albedo ($A$).

**2. Sizes and orbits**
A selection of asteroids is shown in the diagram, the choice being based upon interesting physical properties, including the largest and those representing particular families. The bodies are placed at their correct mean distances from the Sun, and part of Mars is shown for a size comparison. The main belt has a volume which exceeds the spherical volume of the space within the orbit of Mars. We now know that asteroids are found throughout the inner Solar System, and the discovery of Chiron beyond the orbit of Saturn may indicate that there are belts of minor planets in the outer Solar System also. They all hold clues as to the origin and subsequent evolution of the Solar System.

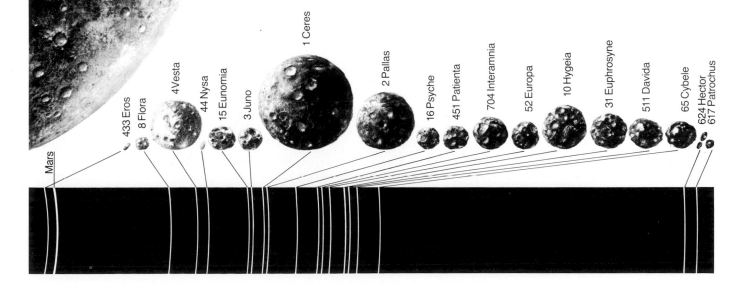

# Classification

In 1772 in Germany, J. E. Bode drew attention to a curious relationship pointed out earlier by Titius of Wittemberg. This relationship, still known as Bode's Law, concerns the distances of the planets from the Sun which seem to follow a regular pattern. The pattern indicated that there should be an extra planet orbiting at a distance intermediate between those of Mars and Jupiter, so in 1800 a systematic telescopic search was begun by a team of astronomers led by J. H. Schröter. The "celestial police", as they called themselves, undertook to search selected regions of the sky close to the ecliptic. In fact the first asteroid, Ceres, was discovered in 1801 by Piazzi, who was not then a member of the "police" (although he joined later). Three more asteroids, Pallas, Juno and Vesta, were discovered before the end of 1807, but the fifth member of the swarm, Astraea, was not found until 1845. Since then the number of asteroids has increased yearly, mainly because asteroid searches are now made photographically.

## Irregular asteroids

Until 1898 it was believed that asteroids were restricted to the zone between the orbits of Mars and Jupiter. Then Carl Witt discovered 433 Eros, which moves in an orbit that brings it closer in than the orbit of Mars so that it may approach to within about 23 million km of the Earth, when it is as bright as magnitude 8.3. These relatively close approaches are rare; the last two were in 1931 and 1975. Eros is a small body, elongated in form, with a longest diameter of about 26 km. During the 1975 approach it was investigated by radar and found to have a decidedly rough surface.

Eros was the first known asteroid to cross the orbit of Mars, but others were soon found; 719 Albert in 1911, 887 Alinda in 1918, 1,036 Ganymede in 1924 and so on; the total number now amounts to several dozen. 1,221 Amor, discovered in 1932, has a perihelion distance of only 162 million km, not far from the Earth's orbit. In 1932 came the discovery of 1,862 Apollo, which actually crosses the

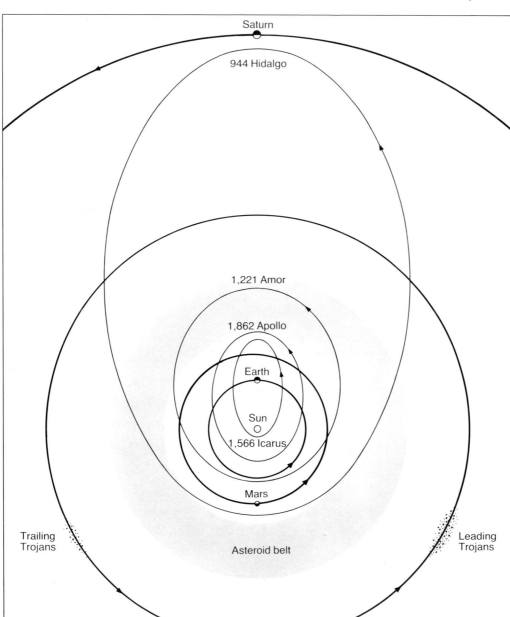

**Trailing Trojans**

**Leading Trojans**

Asteroid belt

### 1. Orbits
The asteroids travel around the Sun in a wide variety of orbits. Most lie in the main belt, but a few of those which do not are shown relative to the orbits of the Earth, Mars, Jupiter and Saturn. The Apollos are named after one of 19 small bodies whose orbits take them within the orbit of Mercury. The family accompanying Jupiter—the Trojans—is also shown. There are about 1,000 of these, but more occur in the forward group than behind. Hidalgo has the most eccentric orbit of all.

### 2. Icarus
During a period of 409 days 1,566 Icarus travels from beyond Mars to within 28 million km of the Sun—closer than Mercury. It is only 2 km in diameter.

### 3. Hidalgo and Chiron
The orbit of 944 Hidalgo ranges between 300 million and 1,440 million km. Its path is more like a comet than an asteroid. Chiron's orbit lies mainly between those of Jupiter and Saturn.

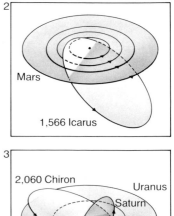

orbit of the Earth. Apollo, "lost" for many years, was re-observed in 1973 and made its last close approach in 1982. Earth-crossing asteroids are now known collectively as Apollos, and are still being discovered. All are very small with diameters of no more than a few km. It has been suggested that Apollo objects could be the nuclei of "dead" comets, but this is rather improbable.

Asteroids whose orbits lie mainly within that of the Earth are termed the Aten group, the best-known members of which are 2,062 Aten and 2,100 Ra-Shalom. Not many Aten asteroids are known but they could be fairly numerous. There is always a chance that the Earth will collide with an Apollo or Aten asteroid, and indeed this must have happened in the past. The great change in terrestrial life which occurred 65 million years ago, when the dinosaurs disappeared, has been attributed to a wide variety of events, in particular an asteroid impact, although the evidence is far from conclusive.

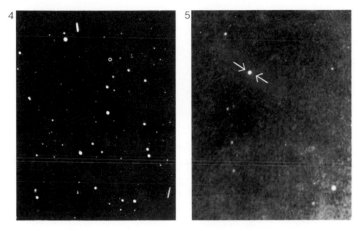

### 4. Vesta
Vesta is one of the four largest and brightest asteroids. Optical measurements of such small bodies are not accurate, however, and infrared techniques are used.

### 5. Asteroid trails
This time-exposure photograph shows three asteroid trails against the stellar background. Max Wolf pioneered this method.

### 6. Gaps in the asteroid belt
If the orbits of asteroids in the main belt are plotted, an unevenness in the distribution is evident between the orbits of Mars and Jupiter. There are seven major gaps between 2.2 and 3.3 AU. These spaces, or Kirkwood Gaps, are where few asteroids are present because of perturbations due to the gravitational field of Jupiter.

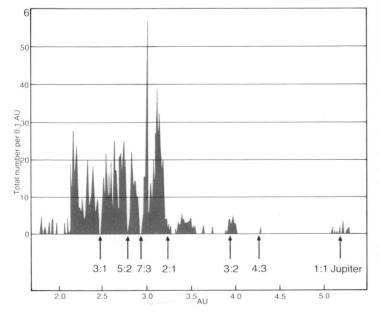

## Chiron
A particularly puzzling object was discovered by Charles Kowal from Palomar in 1977. It spends most of its orbit between those of Saturn and Uranus; it will next come to perihelion in 1996, when its magnitude will be about 15. The orbit may not be stable, and there is a chance that in the future Chiron will be expelled from the Solar System; in 1660 BC it approached Saturn to a distance of 16 million km, so that it was not far beyond the orbit of Phoebe.

Chiron's diameter is uncertain. It has been given an asteroidal number (2060) but in 1988 it brightened unaccountably, and in 1989 astronomers at Kitt Peak reported that it had developed a coma about 5 seconds of arc across. If this is confirmed, then it seems that Chiron may be cometary in nature—and yet it seems far too large to be a comet. Kowal himself summed up the situation neatly when he commented that Chiron was—"well, just Chiron!"

## The Trojans
At a much greater distance from the Sun lie the asteroids of the so-called Trojan group, named after the heroes of the Trojan wars. They move in virtually the same orbit as Jupiter, but are in no danger of being swept up because they keep to stable points either 60° ahead of Jupiter or 60° behind (these are the angles measured at the Sun). By asteroidal standards they are large, and 624 Hector seems to have a greater diameter of about 180 km. Hector shows brightness variations which indicate a very irregular shape.

## Types of asteroids
Asteroids are not identical in composition, as is suggested by their different albedoes. The darkest asteroid known, 95 Arethusa, has an albedo of only 0.019 so it is blacker than a blackboard: the highest albedo, 0.38, is that of 44 Nysa. Attempts have been made to identify asteroidal mineralogies and perhaps relate them to the different kinds of meteorites (*see* pages 406–7). Some asteroids appear to have undergone melting and chemical differentiation like the terrestrial planets, whereas others retain their primordial mineralogies. Why this should be so remains a mystery.

About 75 percent of asteroids are dark and these are C-types or carbonaceous. Some C-types, particularly beyond the main belt and including some Trojans, are reddish in color and are said to have RD (reddish, dark) spectra. About 15 percent of asteroids have moderate albedoes and are red in color. These are S-types and are composed of iron and magnesium silicates. M-types again are moderately bright and have an iron–nickel composition but with no silicates. Some 5–10 percent of asteroids have less common compositions and some are unique. C-types are the predominant kind in the outer belt; both S- and C-types are found equally throughout the inner belt; and M-types and less common types are found throughout.

## Origin of asteroids
Until relatively recently asteroids were merely considered to be the "dregs" of the Solar System, the debris of a former planet, but it has come to be realized that they hold important clues as to the origin and evolution of the planetary system as a whole. It is thought that the asteroids were planetesimals just like any others growing elsewhere in the solar nebula (*see* pages 10–11). Before they could form into planets, however, their orbits were perturbed, becoming tilted and elongated. This resulted in fragmentation and disruption rather than coalescence. They are probably still colliding today but less often. Some scientists believe that Jupiter's gravitational forces were responsible for disrupting the asteroids in the first place and preventing their accretion into a planet. Resonant locations today, such as the Kirkwood Gaps, are fixed, but their effects might have swept through the asteroidal region when primordial nebular gases were driven off in the early days of the Solar System.

# JUPITER

# Characteristics

Jupiter, the largest planet of the Solar System, is fifth in order of distance from the Sun. It is more than 1,330 times greater in volume than the Earth, and 318 times greater in mass. Indeed its mass accounts for more than two-thirds of the total mass of all the planets combined. It has 16 known satellites, four of which are themselves the size of small planets (*see* pages 266–70, 290–310). The ancients aptly named it in honor of the ruler of Olympus.

Jupiter orbits the Sun at a mean distance of 778,360,000 km. The orbit is not, however, circular, and Jupiter's distance from the Sun ranges from 815,700,000 km at aphelion to 740,900,000 km at perihelion. It takes 4,332.59 Earth days for Jupiter to complete one orbit (about 11.86 years). The planet is extremely massive but its density is very low, and this indicates that Jupiter is composed principally of the lighter elements, in particular hydrogen and helium. These are present mainly in the form of gas and liquid. Every visible feature is, in fact, a cloud, and it is not known for certain whether or not Jupiter has a solid core.

Although extremely bright, Jupiter is not the brightest of the planets. Its maximum magnitude is −2.6, which is much less brilliant than Venus at maximum magnitude (−4.4). Even at its faintest, Jupiter far outshines Sirius, the brightest star. Jupiter comes to opposition once in about every 13 months, and because it is so far from the Sun, its opposition magnitude does not vary greatly. The mean synodic period is 398.9 days.

## Nomenclature
Through the telescope Jupiter appears as a flattened yellowish disc crossed by dark streaks which have always been known as the "cloud belts". The bright bands are known as "zones". The surface features are always changing in detail, but several of the belts and zones are relatively stable, changing only slightly in latitude, while others periodically disappear for a while. The more permanent of Jupiter's surface features have been given a standard nomenclature, which is described in detail opposite. North is at the top of the diagram: the leading limb (where features disappear as the planet rotates about its axis) is to the right and the trailing limb (where features first appear as the planet rotates) is to the left.

## Rotation
Like all the other giant planets, Jupiter spins very quickly. This causes an equatorial bulge, producing an ellipsoidal shape which is described as oblate. Thus Jupiter's equatorial diameter is 142,800 km while the polar diameter is only 134,200 km.

Because of its predominantly fluid composition, Jupiter does not spin in the same way as a solid body. Instead, features at different latitudes on the planet rotate at different speeds, with those in the equatorial regions rotating the most rapidly. The movement of the visible surface of Jupiter is extremely complex, and for convenience the planet has been divided into two main regions: a separate rotational period has been adopted for each based on careful observation over a period of many years, and these two rates provide a fairly accurate guide to the relative movement of different features against which longitude can be measured.

The two rotational periods are known as Systems I and II. System I applies to the equatorial regions, bounded by about 9° of latitude on either side of the equator, and has a value of 9 hr 50 min 30.003 sec. This is equivalent to 877.9° rotation in one Earth day. Within this region, however, there is a band of exceptionally rapid rotation known as the Great Equatorial Current. System II applies to everything outside ±9° latitude, and is about 5 min longer than System I, since the polar regions move more slowly. Its exact value is taken as 9 hr 55 min 40.632 sec, or 870.27° per Earth day, although once again this is an average over a range of values.

There is, finally, a third system of rotation, known as System III, which refers to the source of certain "radio" emissions from Jupiter (*see* page 258). The value for this is 9 hr 55 min 29.7 sec.

**North Polar Region (NPR)** Lat. +90° to +55° approx.
Usually dusky in appearance and variable in extent. The whole region is often featureless. The "North Polar Current" has a mean period of 9 hr 55 min 42 sec.

**North North North Temperate Belt (NNNTB)** Mean Lat. +45°
An ephemeral feature often indistinguishable from the NPR.

**North North Temperate Zone (NNTZ)** Mean Lat. +41°
Often hard to distinguish from the overall polar duskiness.

**North North Temperate Belt (NNTB)** Mean Lat. +37°
Occasionally prominent, sometimes fading altogether as in 1924.

**North Temperate Zone (NTZ)** Mean Lat. +33°
Very variable, both in width and brightness.

**North Temperate Belt (NTB)** Mean Lat. +31° to +24°
Almost always visible, with a maximum extent of about 8° latitude. Dark spots at its southern edge are not uncommon.

**North Tropical Zone (NTrZ)** Mean Lat. +24° to +20°
At times very bright. The "North Tropical Current", which overlaps with the North Equatorial Belt, has a period of 9 hr 55 min 20 sec.

**North Equatorial Belt (NEB)** Mean Lat. +20° to +7°
The most prominent of all the Jovian belts. This region is extremely active and shows a large amount of detail, such as dark projections from the southern edge or white spots and rifts in the middle.

**Equatorial Zone (EZ)** Mean Lat. +7 to −7
Covering about one-eighth of the entire surface of Jupiter, the EZ exhibits much visible detail. At the time of the Voyager encounter the northern component was dominated by 13 plume-like features. From Earth the whole zone abounds with white ovals and streaks, and wisps and festoons extending into it from belts on either side.

**Equatorial Band (EB)** Mean Lat. −0.4°
At times the EZ appears divided into two components by a narrow belt, the EB, at or near to the Jovian equator.

**South Equatorial Belt (SEB)** Mean Lat. −7° to −21°
The most variable belt. It is often broader than the NEB and is generally divided into two components by an intermediate zone. The southern component contains the "Red Spot Hollow" (RSH).

**South Tropical Zone (STrZ)** Mean Lat. −21° to 26°
Contains the famous "Great Red Spot". The STrZ was the site of the long-lived "South Tropical Disturbance".

**Great Red Spot (GRS)** Mean Lat. −22°
Although there are other spots on Jupiter's surface, both red and white, the GRS is much the most prominent. It rotates in an anticlockwise direction, and at present measures 26,200 km in length and 13,800 km in width.

**South Temperate Belt (STB)** Mean Lat. −26° to −34°
This belt has never been known to disappear, although it is very variable in width and intensity; at times it appears double.

**South Temperate Zones (STZ)** Mean Lat. −38°
Often wide; may be extremely bright. Spots are common.

**South South Temperate Belt (SSTB)** Mean Lat. −44°
Variable, with occasional small white spots.

**South South South Temperate Zone (SSSTZ)** Mean Lat. −50°

**South South South Temperate Belt (SSSTB)** Mean Lat. −56°

**South Polar Region (SPR)** Lat. −58° to −90° approx.
Like the NPR very variable in extent.

**Size comparison**
Jupiter is the largest planet. Its diameter is 20,000 km larger than Saturn's and far exceeds Earth's.

Jupiter   Saturn   Earth

**Physical data**

|  | Jupiter |
| --- | --- |
| Equatorial diameter | 142,800 km |
| Ellipticity | 0.0637 |
| Mass | $1.899 \times 10^{27}$ kg |
| Volume (Earth=1) | 1,323 |
| Density (water=1) | 1.32 |
| Surface gravity (Earth=1) | 2.69 |
| Escape velocity | 59.6 km s$^{-1}$ |
| Equatorial rotation | 9 hr 50 m 30 s |
| Axial inclination | 3.12° |
| Albedo | 0.34 |

# Observational Background

## The Great Red Spot

The most famous feature on the surface of Jupiter, the Great Red Spot, is known to have persisted for more than three centuries. Although it has disappeared at times, it has always returned, and for the past hundred years its behavior has been monitored almost continuously. A feature which may have been the Great Red Spot was seen by Robert Hooke in 1664, and the *Philosophical Transactions of the Royal Society* (Vol. 1., No. 3. 1664) contains the following account:

" . . . the ingenious Dr. Hooke did some months since intimate to a friend of his that he had, with an excellent 12 feet telescope, observed, some days before he spoke of it (viz. on 1664 May 9) about nine o'clock at night, a spot in the largest of the three observed belts of *Jupiter*; and that, observing it from time to time, he found that, within two hours after, the said spot had moved east to west about half of the diameter. It is situated in the northern part of the southern belt. Its diameter is one-tenth of *Jupiter*; its centre, when nearest, is distant from that of Jupiter about one-third of the semi-diameter of the planet . . ."

The Italian astronomer Giovanni Cassini also reported observations in 1665, and the Spot continued to be observed intermittently until 1713. The next known record is a drawing made by Heinrich Samuel Schwabe, a German apothecary, in 1831. In 1857 it was drawn by William Rutter Dawes, an English clergyman, and in 1870 by Alfred M. Mayer, an astronomer at Lehigh University in Pennsylvania. Drawings made by the fourth Earl of Rosse in the early 1870s also show it unmistakably. It leapt into prominence in 1878, when observers described it as oval, striking and brick-red. Until 1882 it dominated the whole Jovian scene whenever it lay on the hemisphere turned towards Earth. It then faded, and observers began to fear that it would vanish permanently; but it revived in 1891, and since then it has been on view more often than not. There have been times, however, when it has been difficult to locate, such as in 1928–29, 1938 and 1977.

## Theories about the Great Red Spot

Originally the Great Red Spot was believed to be a kind of Jovian volcano: its redness was associated with heat. Before long, however, it became clear that such an explanation was untenable, and an alternative theory (sometimes called the "floating raft theory") was proposed, first by G. W. Hough in 1881, and later by B. M. Peek:

"Consider first the well-known experiment of immersing an egg in a solution of salt and water. If the solution is more concentrated towards the bottom of the containing vessel, as it is likely to be at first, the egg, while remaining completely under water, will float at a level determined by its density. Now replace the solution by Jupiter's atmosphere, of which the density increases rapidly with the depth until it almost certainly approaches that of the liquid state, and let the egg be represented by some solid whose upper surface lies at least some tens of kilometres below the top of the cloud layer . . . Any influence tending to disturb the equilibrium of the density distribution in the atmospheric layers will bring about a change in the level at which the solid will float . . ."

In 1963 a different theory was put forward by R. Hide. He suggested that the Great Red Spot might be the top of a "Taylor column", a relatively stagnant column of fluid which is formed over an obstacle in a rotating fluid. Such columns were first studied in the laboratory by Sir Geoffrey Taylor in 1921. Hide proposed the idea that some topographical feature on the (assumed) surface of Jupiter might be acting as an obstacle to the planet's rotating atmosphere, thus producing a "Taylor column" in the atmosphere.

This hypothesis, however, does not account for the variations in longitude of the Red Spot and, like the "floating raft" theory, it has now been rejected in view of the Pioneer and Voyager discoveries.

1

23 August 1927      27 September 1927

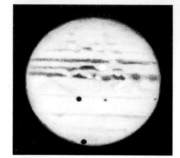
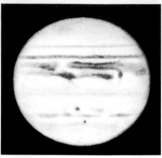

10 October 1927      16 December 1927

11 September 1928      5 November 1928

2

3

4

**1. Drawings by T. E. R. Philips**
This selection of drawings shows Jupiter's changing appearance as recorded by one of its best-known observers. It covers the years from 1927 to 1929.

**2. Theoretical model—1923**
As described by H. Jeffreys, Jupiter's interior consisted of a rocky core of radius 46,000 km, surrounded by an 18,000 km thick layer of ice and solid carbon dioxide, with an atmosphere of negligible density 6,000 km deep.

**3. Theoretical model—1934**
This cut-away drawing illustrates the model proposed by R. Wildt. A very dense, rocky inner core of radius 30,200 km is overlaid by a 27,200 km layer of high-pressure ice, followed by a 12,600 km layer of condensed gases.

**4. Theoretical model—1951**
The model proposed by W. Ramsey and, independently, W. DeMarcus consists of a core of metallic hydrogen with a radius of 61,000 km, surrounded by a layer of liquid hydrogen 8,900 km deep and a shallow atmosphere.

250

August 1891

November 1916

June 1947

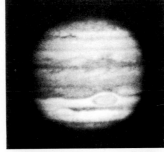

November 1964

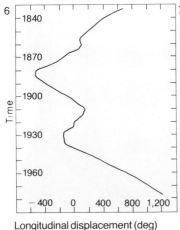

October 1974

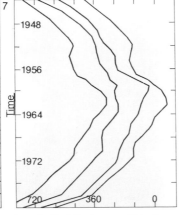

January 1979

**5. Jupiter since 1891**
These photographs reveal the
varying appearance of the planet.
The belts and zones fluctuate in
position and brightness, and the
atmosphere appears to go through
phases of turbulence and quiet. In
the first photograph the GRS is
nearly twice its present length,
while in the following five
photographs it has faded almost
entirely from view.

**6. Historical behavior of the GRS**
The GRS is not a fixed feature.
As this graph shows it has drifted
considerably in longitude since
about 1930.

**7. Historical behavior of the
white ovals**
Like the GRS the white ovals
have exhibited a tendency to drift
in longitude; the average change
amounts to a few degrees per year.

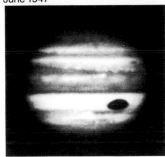

Longitudinal displacement (deg)

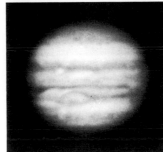

Longitudinal displacement (deg)

## The South Tropical Disturbance

In 1900 some dark humps were observed extending from the South
Equatorial Belts into the South Tropical Zone. They developed
until they formed a shaded part of the whole South Tropical Zone,
and the feature became known as the "South Tropical Disturbance" (STD). Initially its rotational period was 9 hr 55 min 20 sec,
and since the period of the Great Red Spot at the time was 9 hr
55 min 41 sec it was evident that the Disturbance would eventually
catch up with the Spot. The first conjunction occurred in June 1902.
Surprisingly, the Disturbance did not overlap the Spot, but instead
appeared to "leap" past it, reappearing on the far side; the entire
event lasted less than three months. It was repeated regularly until
the Disturbance faded gradually after 1935, and the last definite
observation of it was in 1940. The records of these interactions have
recently been interpreted in terms of the behavior of "solitons" (*see*
page 265).

## The South Tropical Zone Circulating Current

Before the disappearance of the South Tropical Disturbance, some
interactions were observed which gave rise to the name "Circulating Current" given to the prominent current of the South Tropical
Zone. The South Tropical Circulating Current has two components flowing on either edge of the South Tropical Zone: the
northern component flows in an easterly direction at approximately latitude 21°S, while the southern component flows in the
opposite direction at approximately 26°S. Both components have a
zonal velocity of about 50–60 m s⁻¹. Spots moving westward in the
northern component were observed to move suddenly southward
and join the southern component of the current when they came
into contact with the preceding (eastern) edge of the South Tropical
Disturbance. Thus the spots would approach the Disturbance,
cross the South Tropical Zone and then return in the direction from
which they came. Next, they would move towards the Great Red
Spot from the west along the southern component with a relative
velocity of 60 m s⁻¹, but would disappear as they came within
10,000 km of the western edge of the Spot.

Until 1965 no spot in the westerly component of the Circulating
Current had ever been seen to survive an encounter with the Great
Red Spot. In 1965, however, a small double spot was observed
approaching the Great Red Spot along the southern component; it
moved into the narrow channel separating the Great Red Spot
from the South Temperate Belt, attached itself to the perimeter of
the Spot and revolved around it. The direction in which the feature
revolved revealed for the first time the anticlockwise "vorticity" of
the Great Red Spot.

## South Equatorial Belt Disturbances

Since 1919 the South Equatorial Belt has periodically faded and
then dramatically returned to prominence. The revivals always
begin with sudden outbursts of light and dark spots, which are
localized at first, but which soon spread out in turmoil. Thirteen
such disturbances have been observed, and have tended to occur at
intervals of three years or else multiples of three: 1919, 1928, 1943,
1949, 1952, 1955, 1958, 1962, 1964, 1971 and 1975. The years 1943
and 1971 each saw two disturbances.

It has been found that the South Equatorial Belt Disturbances
can be divided into three groups, according to their observed
intensities. Moreover, although the outbreaks appear to be randomly situated, they can all be traced to three sources (corresponding to the intensity grouping) which are stationary relative to one
another. These sources thus define a coordinate system of their
own, whose rotational period is only 0.4 sec longer than that of
System III (*see* page 258), which is thought to refer to the deep
interior of the planet. It therefore seems possible that the South
Equatorial Belt Disturbances have their origins in three "hot spots"
located deep within the Jovian atmosphere.

# Interior

### Planetary magnetism

The first evidence that Jupiter had a magnetic field came in the mid-1950s through the unexpected detection of radio waves from the planet by ground-based radio telescopes. The intense, fluctuating radio emission could only be interpreted in terms of energetic particles interacting with a magnetic field around the planet. The field and belts have now been studied directly by the Pioneer and Voyager spacecraft.

Most of the planets have measurable magnetic fields (though Mars and Venus seem to lack them). However, Jupiter's field is much the most powerful, and it is reasonable to assume that the associated radiation belts are also much the strongest anywhere in the Solar System.

The magnetic field that has been subjected to the most intense study is the Earth's, and it is thought to originate from within the planet's core. The mechanism generating the field may be compared with a dynamo, which is a machine for converting mechanical energy into electrical or magnetic energy. In some ordinary dynamos there is a built-in permanent magnet which supplies the magnetic field, but it is possible to construct them in such a way that they generate their own fields: these are called "self-exciting" dynamos and illustrate the principle that probably underlies all the planetary magnetic fields that have been observed (see diagram 1). The two essential features are the presence of a good electrical conductor and a source of mechanical energy to drive the dynamo.

In the case of a planet, the electrical conductor is thought to be some electrically conducting fluid at or near to the planet's core; but the force driving the system is not known and may involve a combination of motions made up of several factors. The Earth, for example, is believed to have a metallic core composed principally of iron, which may be solid in the center but which is probably dense liquid in the outer regions. Various suggestions have been made to account for movements within the fluid zone sufficient to create the observed magnetic field, including convection brought about by the flow of heat outward from the core and the motions of the Earth's rotation and orbit.

### Interior

It is clear that the origin of the Jovian magnetic field is closely related to the structure of the planet's interior; indeed the existence

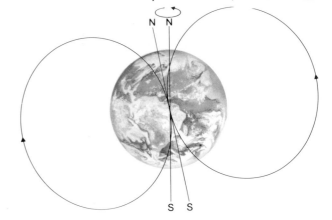

**1. The principle of the dynamo**
A simple model of a disc dynamo consists of a metal disc rotating in a magnetic field between two permanent magnets (**A**). The field produces a force on the free electrons in the disc, pushing them towards the center. As a result there is a difference in electrical potential between the edge and the center of the disc, which would produce a current if the circuit is closed. In a self-exciting dynamo (**B**) this current is used to drive an electromagnetic coil, which replaces the original permanent magnets. The resulting system generates a magnetic field as long as the disc is kept spinning. The model thus demonstrates how mechanical energy may be converted into magnetic energy; some analogous process is thought to be responsible for creating planetary magnetic fields.

**2. The Earth's magnetic field**
At present the Earth's magnetic field is inclined to the axis of rotation by approximately 11°, the same angle as on Jupiter. On Earth the magnetic north is, at present, in the same direction as the geographic north. However, it is known that in the past the Earth's magnetic polarity has reversed itself, and it is conceivable that similar reversals occur on Jupiter.

**3. The Earth's interior**
Most of the information on the Earth's interior is derived from the study of shock waves such as those generated by earthquakes. These have revealed a crust extending to a depth of between 10 and 40 km, which covers a dense shell called the mantle. The mantle is about 3,000 km thick, and below it is a 2,000 km thick outer core surrounding a solid inner core.

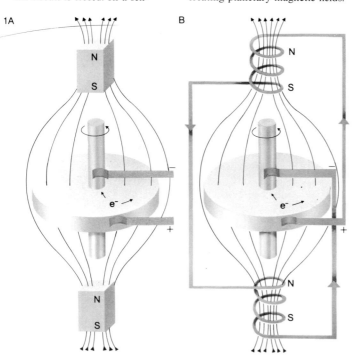

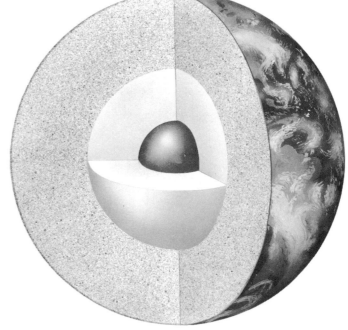

of a magnetic field around Jupiter, and similarly around Saturn, may be interpreted as evidence that these planets contain zones of liquid, although there is no direct evidence in either case. Jupiter's electrically conducting core may be as large as 0.7 or 0.8 Jupiter radii (written as $R_J$).

Jupiter, like Saturn, is known to be composed primarily of the light elements hydrogen and helium. This fact is deduced from a knowledge of Jupiter's density, which is approximately 1.3 times that of water; it is thus apparent that Jupiter must be radically different in composition from the so-called "terrestrial" planets, whose mean densities are in the region of five times the density of water. The giant planets have gravitational fields powerful enough

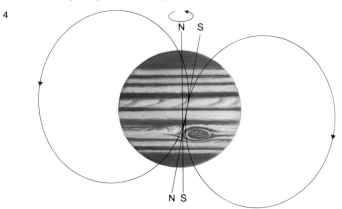

**4. Jupiter's magnetic field**
The magnetic field of Jupiter is inclined to the planet's axis of rotation by an angle of about 11°. Although the field is more complex than that of the Earth, it may be thought of as behaving as though a bar-magnet were embedded at the planet's core. The alignment of north and south magnetic poles is such that a terrestrial compass needle would point south.

**5. Jupiter's interior**
Jupiter is composed almost entirely of gases and liquids, although at present it is believed that there is a small core of rocky materials. The innermost part of the planet around this core is thought to consist of liquid metallic hydrogen, surrounded by a shell of liquid molecular hydrogen. The upper region consists of a deep atmospheric layer of hydrogen and helium.

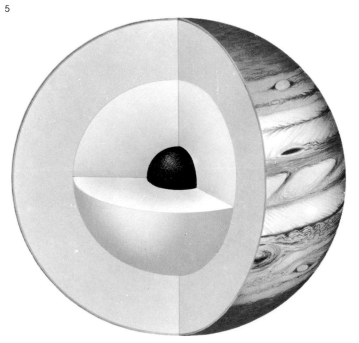

to retain these light elements, and models of their interiors depend heavily on the assumed proportions of hydrogen to helium.

Its interior is now believed to have the following structure: at a depth of about 1,000 km from the visible surface there is a transition zone from gaseous to liquid hydrogen. The temperature at this level is about 2,000 K and the pressure about 5,600 Earth atmospheres (that is to say, 5,600 times the pressure of the Earth's atmosphere at sea-level). Further down, at a level of 3,000 km, the temperature rises to 5,500 K and the pressure to 90,000 atmospheres, and as a result the hydrogen is highly compressed. Twenty-five thousand km below the cloud tops, a distance of more than a third of the radius of the planet, the temperature is over 11,000 K and the pressure is about 3 million atmospheres. In these conditions hydrogen undergoes a dramatic change, rather like the change from a gas to a liquid, altering from its "liquid molecular" form to a state in which it is a good conductor of electricity; in this state it is known as "liquid metallic hydrogen".

From this level onward, the temperature and pressure continue to rise steadily, reaching 30,000 K and 100 million atmospheres at the center. Finally, at the center of the planet, there is thought to be a small rocky core of 10 to 20 Earth masses, composed of iron and silicate materials. The likelihood of the existence of such a core is deduced from what is known about the balance of elements in Jupiter's interior.

**Magnetic field**
The presence of liquid metallic hydrogen in the inner regions of the planet would be of critical importance in explaining the existence of a strong magnetic field because of the material's properties as an electrical conductor. Jupiter's rapid 10 hr rotation might be wholly or partly responsible for setting up a flow of liquid to generate electric ring-currents on the same fundamental principle as the dynamo described earlier. The electric currents create magnetic fields, which in turn influence the strength of the electric currents. In fact, the magnetic fields thus produced will act in such a way as to oppose the force that creates them, limiting the strength of the field and preventing it from increasing indefinitely.

The magnetic field of Jupiter has a strength of 4.2 Gauss measured at the cloud tops. It is therefore more than ten times as strong as the Earth's field of 0.35 Gauss at the surface. The Jovian field is inclined to the axis of rotation by an angle of 10°8, fractionally less than the degree of tilt found on the Earth; and the magnetic axis is displaced from the center of the planet by about $0.1\,R_J$, mainly along the equator.

Beyond a distance of about $3R_J$ (measured from the center of the planet) the major component of Jupiter's field is "dipolar", like that of the Earth: in other words, the field behaves as if there were an immensely powerful bar-magnet embedded inside the planet inclined to the axis of rotation and displaced from the center. At the present time the "polarity" of the field (the alignment of the north and south poles) is opposite to that of the Earth, so that a terrestrial compass taken to Jupiter would point south. However, studies of the residual magnetism in certain types of rock have shown that the polarity of the Earth's field undergoes periodic reversals, at intervals of approximately $2 \times 10^5$ years (on average) for a complete reversal; it is possible that the same phenomenon occurs on Jupiter, though the time-scale of any Jovian field reversal is unknown.

Closer to Jupiter, however, there are some important differences in the structure of its magnetic field as compared with that of the Earth. Although the principal component remains dipolar, the simple model of a bar-magnet is no longer very accurate and must be replaced by a complex system of quadrupole and octopole moments, which act rather like harmonic frequencies superimposed onto a pure musical note. The cause of these high-order magnetic moments may be the complicated circulation patterns that take place in the metallic hydrogen interior.

# Magnetosphere

**Magnetic environment**
Jupiter's magnetic field continues to influence the behavior of charged particles at a distance far above the planet's atmosphere: the region in which its influence is dominant is called the "magnetosphere". Despite its name, the magnetosphere is not in fact spherical, but has a long "magnetotail" streaming away from the direction of the Sun, stretching out almost 750 million km away, beyond the orbit of Saturn. The Earth's magnetosphere resembles that of Jupiter in structure, but is very much smaller. Jupiter's magnetosphere is immense; if it were visible from the Earth, the spherical part would occupy as much sky as the Sun. Its size is, however, variable, extending to a distance of between 50 and 100 $R_J$ in the direction of the Sun. By contrast, it is extremely rare for the Earth's magnetosphere to vary much in size.

Changes in the size of Jupiter's magnetosphere are brought about by its interaction with a continuous stream of particles, called the "solar wind", emanating from the Sun. The particles that make up the solar wind are mainly electrons and protons, and the strength of the wind depends on the level of activity on the Sun, where magnetic storms take place over an 11-year cycle. When the solar wind particles collide with the magnetosphere, they are abruptly slowed down from speeds of approximately 1,500,000 km hr$^{-1}$ to a mere 400,000 km hr$^{-1}$. This rapid deceleration causes a huge rise in the effective temperature, which is increased by as much as a factor of ten. (The term "effective temperature" is used because it depends on the idea that the kinetic energy of particles due to their random movement can be interpreted as heat energy: but in the highly rarefied regions that exist

here, the individual particles are spaced so widely apart that although they may be highly energetic individually, the total kinetic energy in a given volume of space amounts to relatively little heat. Thus spacecraft can safely pass through regions in which the effective temperature is higher than in any part of the Sun.)

The collision between the solar wind and Jupiter's magnetic field forms a shock wave, called the "bow shock", within which there is a turbulent region called the "magnetopause". Just inside the magnetopause the Voyagers measured temperatures of 300–400 million K, the highest found in the Solar System. Finally, the envelope enclosing all the magnetically active regions is known as the "magnetosheath".

Equilibrium is established between the external pressure of the solar wind and the internal pressure of the magnetosphere, which is sensitive to relatively minor changes in the solar wind intensity: the stronger the solar wind, the more compressed the magnetosphere in the direction of the Sun. One of the main reasons why the Earth's magnetosphere is so much smaller than that of Jupiter is that it is much closer to the Sun, and is therefore subjected to a stronger solar wind (in fact, about 25 times stronger). The difference in size

**1. Jupiter's magnetosphere**
A vast region of space surrounding Jupiter is dominated by the planet's magnetic field. This region, the magnetosphere, has a complex structure. The bow shock is situated where the particles of the solar wind collide with the magnetosphere.

The turbulent region immediately inside the bow shock is called the magnetopause, and the whole magnetically active region is enveloped by the magnetosheath. A current sheet of trapped plasma follows close to the magnetic equator; radiation belts dominate the innermost regions.

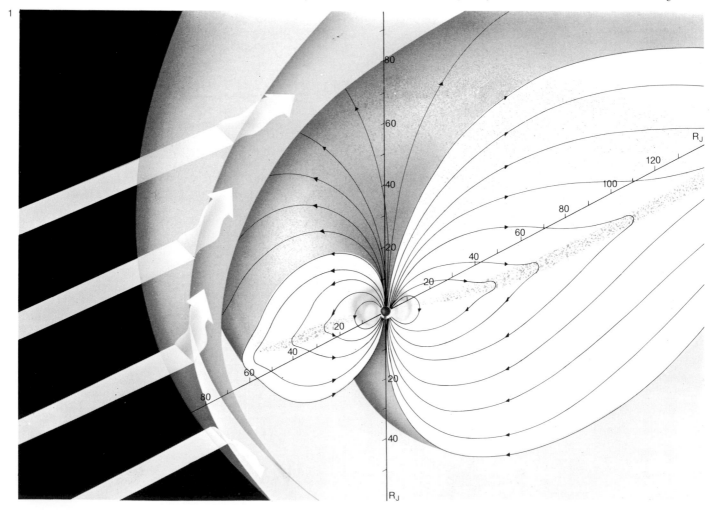

between the two magnetospheres is further explained by the fact that the Earth's field is intrinsically more than ten times weaker than Jupiter's.

In the outer regions of Jupiter's magnetosphere, beyond about 20 R_J, the magnetic field is more or less confined to a single plane which cuts the equator of the planet. Within this disc-like region flow electric currents, carried by low-energy "plasma" trapped in the field. (Plasma is matter which has acquired enough heat not only to break the bonding forces between its molecules, but also to free the electrons from their atoms: it has thus reached a state beyond the condition of existing as a gas and, because it is made up of charged rather than uncharged particles, is able to conduct electricity. The charged atoms are known as "ions".)

The electric currents form a "current sheet" which is warped in opposite directions on either side of the planet, so that it lies above the equatorial plane on one side and below it on the other. It rotates with the planet more or less like a rigid body, producing an up-and-down motion rather like a warped record, which the spacecraft periodically crossed.

The magnetosphere's mixture of trapped particles also changes rapidly. Voyager 2 detected only about a tenth as many high-speed carbon and sulphur ions as Voyager 1, but at the same time it found a higher carbon to oxygen ratio. Ten-hour variations in electron intensity have also been detected inside the magnetosphere, and similar variations have been measured as far as 150 million km from Jupiter: both inside and outside the region the variations are locked together in phase, so that it appears as though Jupiter emits these particles in a rotating beam.

## Radiation belts

High-energy particles are trapped in Jupiter's magnetic field and form belts of intense radiation aligned with the magnetic axis. These belts resemble the Van Allen belts of the Earth, but are 10,000 times more intense at their highest levels. The peak intensities are found within 20 R_J, and Pioneer 10, the first spacecraft to fly through this region, received an integrated dose of 200,000 rads from electrons and 56,000 rads from protons. (For a man, a whole body dose of 500 rads would be fatal.) The radiation seriously jeopardized the spacecraft mission, and the second Pioneer vehicle followed a trajectory which took it over the south pole of the planet, passing rapidly through the most dangerous regions. Nevertheless it was found that there was a greater "flux" or flow of energetic particles at the higher Jovian latitudes than would have been expected from the measurements made by Pioneer 10 alone. Furthermore, it would appear that the flux of energetic particles reaches a maximum to either side of the dipole magnetic equator.

These radiation belts are a serious potential hazard to spacecraft. Intense bombardment by high-energy protons and electrons can saturate the sensitive experimental equipment on board, upset the functioning of the spacecraft's computer systems, and interfere with its communications with the Earth. Although the Pioneer vehicles survived, the Voyagers were carefully constructed to withstand twice the expected dose of radiation. Only minor effects were noted when the first Voyager plunged to within 350,000 km of the planet on 5 March, but Voyager 2, passing Jupiter at the much greater distance of 650,000 km, found the radiation environment three times stronger.

2

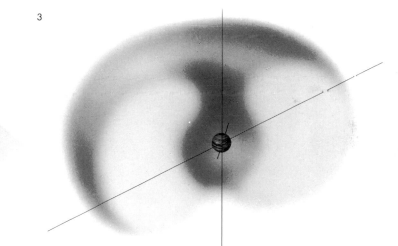

3

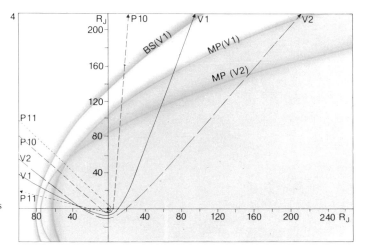

**2. Current sheet**
A disc-like sheet of low-energy plasma is trapped in Jupiter's magnetic field, lying roughly in the plane of the magnetic equator. The charged plasma particles carry electric currents, and the sheet rotates with the planet. Because the magnetic axis is tilted with respect to the axis of rotation, the current sheet appears to wobble up and down as it rotates. Its behavior can be compared to a warped record on a rotating turntable.

**3. Radiation belts**
Within about 20 R_J there exist belts of intense radiation similar to the Van Allen belts of the Earth. These belts are made up of high-energy particles and, like the

current sheet, they are aligned with the magnetic rather than the rotational axis.

**4. Spacecraft trajectories**
The size of the magnetosphere is known to vary considerably. Each of the vehicles encountered the bow shock at different distances from Jupiter. Voyager 2, for example, crossed the bow shock at least 11 times within three days during its approach to Jupiter. Some of the points at which the spacecraft crossed the bow shock are illustrated in the diagram: curves representing Jupiter's magnetopause and bow shock are shown in three different positions, as calculated from Voyager 1 and Voyager 2 data. These curves are based on average values.

# Interaction of Jupiter and Io

## Cosmic rays

Sometimes Jupiter's magnetic field releases bursts of its trapped particles in the form of cosmic rays, the most energetic particles found in nature. The speed of such particles approaches that of light. They are generally atomic nuclei (protons and neutrons) stripped of their electrons, but other types of particles are also found, and the range of types and energy levels indicates that a variety of astronomical sources and processes contribute to the cosmic rays reaching the Earth. Originally it was thought that most cosmic rays came from outside the Solar System, but today it is believed that a substantial proportion are spun off Jupiter. Cosmic rays have been detected as far away as Mercury, a distance of more than 700 million km from Jupiter.

Cosmic rays could pose a hazard to future space travellers. However, although they can cause mutations in living organisms by altering or destroying genes, it is extremely unlikely that life on Earth can be endangered by them since they cannot easily reach the Earth's surface. Nevertheless, it has recently been shown that some computer systems have been affected by extremely random changes within their micro-circuits; the odds that such effects will take place are very small indeed, but it is not impossible for daily life to be indirectly influenced by the presence of cosmic rays.

## Plasma torus and flux tube

Jupiter's seven innermost satellites (*see* pages 266–7) and its system of rings (*see* page 272) all reside inside the hostile environment of Jupiter's magnetosphere, with the furthest of these satellites, Callisto, orbiting near the edge. In this respect, Jupiter differs markedly from the Earth, whose Moon resides outside the main body of the magnetosphere, only passing through the magnetotail as it orbits the planet. Jupiter's satellites are constantly bombarded by the high-energy electrons, protons and other particles that lie in their paths, and the interaction between the satellites and these particles results in a gradual erosion of the surfaces of the satellites. The interaction is particularly vigorous on Io (one of the four major satellites—*see* pages 290–95) and the resulting "sputtering" action is thought to be responsible for the creation of an uncharged or "neutral" atmosphere of sodium, potassium and magnesium that has been observed forming a cloud which stretches out for some distance along Io's orbit around Jupiter. Its vertical extent is about $2R_J$ (*see* diagram 3).

Theoretical ideas suggest that Jupiter and Io are connected by a flux tube of electrons and ions carrying a massive current of 5 million amps at a potential difference of 400,000 volts. The power contained in this torrent of electricity ($2 \times 10^{12}$ watts) is some 70 times the combined generating capacity of all the nations on Earth. This electrical energy may also play a further important role in locally heating the surface of Io, and therefore in assisting the volcanic activity of the satellite (*see* pages 290–91). The material ejected from Io's volcanoes is sent up into the atmosphere as vast

**1. Io's flux tube and torus**
A doughnut-shaped torus surrounds Jupiter along the orbit of Io; it consists of charged plasma particles, and since these are influenced by Jupiter's magnetic field the position of the torus is slightly inclined to Io's orbit. Io is linked to Jupiter by a flux tube carrying an immense current of 5 million amps flowing between Io and Jupiter.

plumes of dust and sulphur dioxide which become ionized, forming a plasma ring or torus around Jupiter in the orbit of Io.

At the time of the Voyager 1 encounter, observations were made of ionized oxygen and singly or doubly ionized sulphur (atoms of sulphur which have lost either one or two of their electrons) with a plasma temperature of $10^5$ K. These observations indicate a considerable change in the Jovian environment since the Pioneer missions four and a half years earlier. A further change was found by the time of the second Voyager encounter: ultraviolet (UV) emissions from the Io plasma torus had doubled in brightness, and the temperature had decreased to $6 \times 10^4$ K.

Since the material in the torus comes from the Ionian volcanoes, changes in the properties of the torus reflect corresponding changes in the volcanic activity of the satellite. Approximately $10^{10}$ ions of sulphur and oxygen $cm^{-2} s^{-1}$ are required to be pumped into the torus to maintain it; the lifetime of the constituent material is estimated to be $10^6$ sec. The torus material is probably transported by diffusion both inward to Jupiter and outward into space.

## Aurorae

Before the Voyager mission it was thought that some solar wind particles would leak through the tail of the magnetosphere and into the polar regions of Jupiter, creating the type of auroral displays familiar on Earth, such as the Aurora Borealis. Auroral displays were indeed found, but they are in fact triggered not by the precipitating solar wind particles, as on Earth, but by the interaction between electrons, streaming in from Io's torus, and the complex Jovian ionosphere (*see* pages 260–61). Auroral activity had been

virtually absent during the Pioneer flybys in 1973 and 1974.

The brightness of aurorae is generally specified in units called "Rayleighs", 1 Rayleigh being equal to $10^6$ photons $cm^{-2} s^{-1}$. On Earth an aurora with a brightness of 1,000 Rayleighs is just visible, while displays of brightness of 40,000 Rayleighs are clearly visible and common enough to appear nightly in some polar latitudes. On Jupiter, on the other hand, the aurorae are even more energetic, with a brightness at UV wavelengths of 60,000 Rayleighs. (The shift towards the ultraviolet marks an overall increase in energy.) The Jovian aurorae observed by Voyager 1 stretched for almost 30,000 km across Jupiter's north pole, making them the largest such phenomena ever seen. At the same time radio emissions of a type commonly associated with the Earth's auroral regions were detected by the Plasma Wave instrument near the Io plasma torus.

### Radio emissions from Jupiter

It has already been mentioned that Jupiter emits waves at radio frequencies. These radio waves are not at all like the signals that carry radio programs, but resemble rather the static or "noise" that causes interference when a radio receiver is playing in the vicinity of flashes of lightning or certain electrical equipment. The radio noise reaching Earth from Jupiter is greater than from any other extraterrestrial source except the Sun and falls into three distinct types: "decametric", "decimetric" and "thermal". Each is characterized by a particular range of wavelengths, although there is some overlap between the decimetric and thermal emissions; but they are quite different in origin and sufficiently distinct in character to enable them to be considered separately.

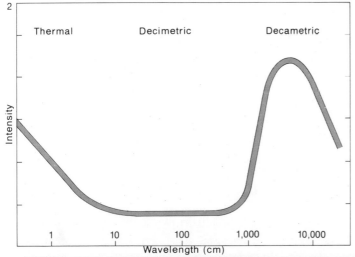

**2. Radiation from Jupiter**
Jupiter emits three distinct types of electromagnetic radiation at radio frequencies. This graph shows wavelength plotted against intensity. The shortest wavelengths, which are produced by thermal emissions, form the sloping part of the curve in the left-hand portion of the graph. The central, straight part of the curve represents emissions known as decimetric, which have a different character and origin from thermal radiation. There is, however, some overlap between their wavelengths. The hump in the right-hand part of the curve represents the third type of emission, decametric. Unlike the shorter wavelength radiation, decametric emissions from Jupiter are irregular.

**3. Io's sodium cloud**
A cloud of neutral gas composed of sodium, potassium and magnesium forms a tenuous atmosphere around Io. This photograph, taken in 1977 from Table Mountain Observatory, shows the cloud stretched out along Io's orbit. A greater proportion of sodium appears to precede the satellite than that trailing behind. This image is in fact a composite, made up of an ordinary photographic image of Jupiter together with an intensified television image of the cloud. The satellite Io and the path of its orbit have subsequently been drawn onto the image.

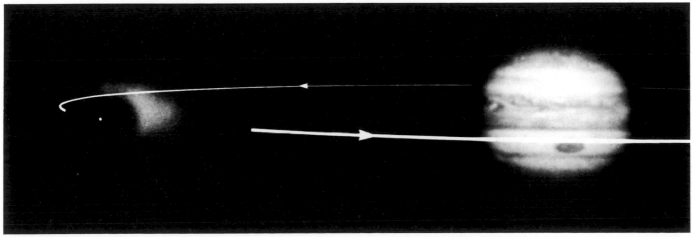

# Radiation

### Decametric emissions

The longest electromagnetic waves emitted by Jupiter have wavelengths ranging from about 7.5 m to 700 m; because the most intense activity occurs at wavelengths measured in tens of meters this type of emission is known as "decametric". Radio astronomers, however, often prefer to identify waves by their frequencies rather than by wavelength. (The two quantities are related by a simple expression: wavelength multiplied by frequency equals the speed of propagation, in this case the speed of light.) Thus Jupiter's decametric waves have frequencies in the approximate range of 40 MHz down to 425 kHz, with the peak of activity occurring between 7 and 8 MHz.

The decametric emissions are not continuous, but are characterized by strong bursts at sporadic intervals. Bursts of intense activity may last for anything between a few minutes and several hours, and are generally separated by long periods of inactivity. It was, in fact, the decametric band of the spectrum that was first observed, and the irregularity of emission and its polarization were two of the main features that led to the idea that the origin of the waves involved a magnetic field (*see* page 252). More conclusive indications came from observations of the decimetric wavelengths, described below.

Although some of the irregular decametric emissions could be created by gigantic electrical discharges such as lightning flashes in the upper atmosphere of Jupiter, they are circularly polarized (*see* Glossary) and exhibit certain regular features which are associated with the relative position of the Galilean satellite Io. It has long been known that the position of Io in its orbit profoundly influences decametric emissions, producing peak effects at certain fixed positions (*see* diagram 1), but the mechanism responsible for this effect is not yet understood.

Observations of the decametric radiation emitted by Jupiter have revealed a third system of rotation, known as System III, comparable with the System I and System II differential rotation of the

visible features described on page 8. System III has a rotational period close to, but not exactly equal to, that of System II, and corresponds to 9 hr 55 min 29.710 sec or 870°.536 per day. System III is, in fact, found to represent the rotation of the magnetosphere; since the magnetic-field lines are joined to the deep interior of the planet where the field is created, System III gives the "true" period of Jupiter itself, in the sense that it behaves as if the planet possessed a solid surface like the Earth. For very accurate work, the velocities of features moving in the Jovian atmosphere are conventionally measured relative to System III, thus overcoming the lack of a fixed reference frame on Jupiter's visible surface.

### Decimetric emissions

At wavelengths below 7.5 m the radiation is continuous in both time and frequency. In this band of the spectrum the shorter wavelengths corresponding to higher frequencies are predominantly produced by thermal radiation, while the lower-frequency, longer wavelengths are non-thermal in origin. These emissions, concentrated in the region of a few meters down to a few centimeters, are referred to as "decimetric".

Decimetric emissions have been observed to emanate from an area larger than Jupiter's disc. They are, moreover, distinguished from Jupiter's thermal emissions by their 30 percent linear polarization and are strongly "beamed" into the plane of the magnetic equator.

The main features of the decimetric emissions can be satisfactorily accounted for as synchrotron radiation, which occurs when electrons are accelerated along a helical path around magnetic field lines at speeds approaching the velocity of light. (Such speeds are often referred to as "relativistic", because the behavior of particles travelling at these very high velocities can only be accounted for in terms of Einstein's Theory of Relativity.) Under these conditions, electrons emit polarized radiation in narrow cones pointing along the direction of the electron's motion.

**1. Decametric emissions**
The irregular decametric emissions from Jupiter are known to be influenced by the position of Io relative to the Earth. The majority of stronger bursts occur when the angle between Io, Jupiter and the Earth corresponds to the values shown below.

**2. Decimetric emissions**
Radiation in the 10 cm band is most strongly emitted from the dark areas of the diagram.

**3. Synchrotron radiation**
When a fast-moving electron spirals around magnetic field lines it emits a narrow beam of strongly polarized radiation. The wavelength is dependent on the strength of the field and the velocity of the electron. Jupiter's decimetric emissions are produced by this so-called synchrotron radiation. By studying the direction of the polarization it is possible to discover the direction of the magnetic field.

1

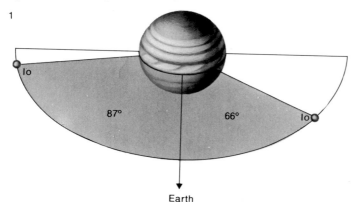

Io   87°   66°   Io

Earth

2

3

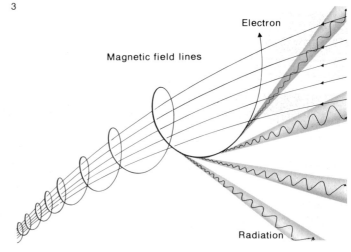

Electron

Magnetic field lines

Radiation

As Jupiter rotates the changing aspect of the magnetic equator as viewed from Earth causes variations in the observed intensity and polarization. The total intensity goes through two maxima and two minima per revolution, the maxima occurring when the magnetic equator is viewed edge on, and the minima when the equatorial plane is viewed from extreme positive and negative latitudes.

In addition to the linear polarization, a small amount of circular polarization has been observed. It is left-handed when Jupiter is viewed from positive magnetic latitudes, indicating that the polarity of the planet's magnetic field is opposite to that of the Earth (see pages 252–3).

**Thermal radiation**
When a body is heated it radiates energy. If a curve is drawn, plotting the intensity of radiation against wavelength, it is found to have a characteristic form which reaches its peak at a certain wavelength depending on the temperature of the body: the hotter the body, the shorter the wavelength at which the peak occurs. This corresponds with common experience: as an object is heated it starts to glow, first dark red, then yellow, and eventually white or white tinged with blue. (Physicists interpret this type of thermal radiation in terms of a theoretical model of a perfect radiator called a "black body", which is assumed to reflect no radiation whatsoever. For such a black body it is possible to predict the exact intensity-wavelength curve for any given temperature; conversely, the temperature may be determined by measuring the radiation.) Thus the thermal emissions of Jupiter at wavelengths of a few centimeters and below carry information about the temperature of the planet, and a particular wavelength is sometimes spoken of as having an "equivalent temperature".

Radiation emitted from inside Jupiter has to pass through the planet's atmosphere before it is detected. As it does so it is selectively absorbed and re-emitted by the atoms and molecules present in the atmosphere. Because the re-radiation is omnidirec-

tional, radiation travelling in a given direction will be depleted; and since every chemical substance is known to absorb at certain characteristic wavelengths, the wavelengths at which depletion occurs provide information about the constituents of the atmosphere (see page 261). More detailed examination of the spectrum, continuing through the thermal radio wavelengths to the infrared and visible wavelengths, can also yield information about the changes of temperature and pressure with altitude in the Jovian atmosphere. If, for example, the temperature of the atmosphere increases instead of decreases with altitude, the gases at higher levels will appear to be emitting rather than absorbing radiation; their characteristic "spectral lines" will show up as stronger rather than weaker emissions at particular wavelengths.

**Thermal maps**
Analysis of the spectrum of Jupiter, particularly of the thermal radiation at infrared wavelengths, has made it possible to build up a picture of the structure of the Jovian atmosphere. Breaks in Jupiter's thick cloud layers allow measurements to be taken over a range of depths down to layers where the pressure is several Earth atmospheres. Infrared radiation penetrates haze more easily than visible radiation, and observations at a wavelength of 5 μm are ideal for studying the deepest Jovian cloud layers, because there is negligible absorption of this wavelength by the main chemical constituents of the atmosphere (ammonia, methane and hydrogen). Other infrared wavelengths are known to be most strongly emitted at certain pressures and can therefore be used to produce "thermal maps" of the varying structure of the Jovian clouds. The atmosphere is too opaque to permit much information to be obtained about layers deep below the clouds, but the available data on the upper regions indicate a structure that may usefully be compared with that of the terrestrial atmosphere. Jupiter's characteristic structure of alternating belts and zones is clearly evident in thermal maps such as the two shown on this page (see diagram 5).

**4. 5 μm maps**
An Earth-based image of Jupiter at a wavelength of 5 μm (A) reveals "hot" regions corresponding with the dark belts seen in the picture made by Voyager 1 (B) about an hour later. These hot 5 μm areas indicate the deeper levels of Jupiter's atmosphere, where there is little or no overlying cloud. At 5 μm the GRS appears cooler than its surroundings.

**5. Thermal maps**
Jupiter is shown at two different infrared wavelengths, 44.2 μm (A) and 16.6 μm (B). Map A thus depicts the temperature structure of the atmosphere at the level of the cloud tops, while map B represents a higher altitude corresponding with the tropopause (the region between the troposphere and the stratosphere). The cool region on map B at about 23°S, 105°W is associated with the GRS; no corresponding feature is apparent on map A.

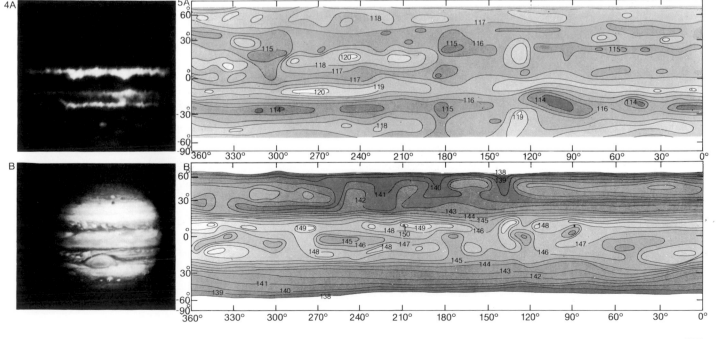

# Atmosphere I

### The structure of the atmosphere
The Earth's uncharged or "neutral" atmosphere is divided into several distinct regions. The innermost layer is called the troposphere; above it lies the stratosphere, followed by the mesosphere and, then, the thermosphere. These layers are primarily distinguished by the way in which temperature varies with altitude as different influences make themselves felt. The thermosphere has an important, electrically charged region within it called the "ionosphere". Finally comes the exosphere, which has no definite boundary, but whose density falls off with altitude until it is no greater than that of the interplanetary medium.

### The troposphere
On Earth the temperature above the planet's surface decreases with increasing altitude until it reaches a minimum at a height of approximately 15 km; this region is the troposphere. One of the main constituents of the troposphere is water vapor, and it is here that the weather processes, in particular the formation of clouds, take place. The dominant mechanism by which heat is transported vertically is convection. Air, warmed by the surface of the Earth, rises and is cooled higher up, while cool air is drawn downwards to replace it and is warmed in turn. The equivalent region on Jupiter is bounded at the upper point by a temperature minimum of about 105 K, which occurs when the pressure is about 0.1 Earth atmospheres. As on Earth, the exact value of the temperature minimum varies with different locations on Jupiter's globe. As observed by Voyagers 1 and 2, the coldest value occurred over the Great Red Spot and in the zones. Below the point of temperature minimum the temperature generally increases with depth, following closely the "adiabatic" gradient of about $2°$ km$^{-1}$. (This is the value for a well-mixed atmosphere, and implies that the region, like the Earth's troposphere, is stirred by convective processes.)

### The stratosphere and mesosphere
By analogy with Earth, the temperature minimum in Jupiter's atmosphere marks the transition between the troposphere and the stratosphere, a region in which the temperature is controlled largely by radiation processes. On Earth the change in temperature gradient (the rate of change of temperature with height) is due to the heating of the upper atmosphere by a layer of ozone, an "allotropic" form of (triatomic) oxygen which absorbs ultraviolet radiation from the Sun. On Jupiter the increase in temperature is due to the heating of the atmosphere by methane, which appears to play a similar role to the Earth's ozone. Dust, possibly from the Jovian ring system (*see* page 272), may also play a part in heating Jupiter's stratosphere.

The temperature gradient of the Earth's atmosphere changes again at an altitude of about 30 km, when temperature once again decreases with height for a further 30 to 40 km; this region is known as the mesosphere. However, not enough is known about Jupiter's atmosphere to describe in any detail the region corresponding to the terrestrial mesosphere.

### The ionosphere
In the upper regions of the atmosphere where the density is very low, the electrical conductivity increases. This part of the atmosphere, within the thermosphere, is another distinct region known as the ionosphere. The name is derived from the high proportion of ionized atoms and molecules contained in this region, and the conductivity is due to the consequent presence of free electrons. Jupiter's ionosphere, extending for more than 3,000 km above the visible surface, is comparable with that of the Earth, which begins at altitudes of around 80 km to 500 km above the surface. The atmosphere here is highly rarefied, and the air molecules are easily ionized by energetic radiation from the Sun with wavelengths of 1,000 Å or less (in other words, ultraviolet radiation). By this process, energy from the Sun in the ultraviolet portion of the spectrum is absorbed by the atoms and molecules in upper regions of the atmosphere, causing them to eject an electron and leave behind a positively charged ion. The Earth has, in fact, several distinct ionized layers, extending upwards as high as 300 km. These layers have properties which make them important in radio communications; they reflect back certain radio frequencies transmitted from the Earth, thus making it possible to send signals around the globe. At the same time, however, they also reflect radio waves back out into space, and therefore obstruct certain types of astronomical research.

The extreme ultraviolet radiation that causes photoionization originates in the upper chromosphere and corona of the Sun. Changes in these regions of the solar atmosphere therefore cause detectable variations in the density of electrons in the ionosphere.

Jupiter's ionosphere is a huge, highly structured region, whose principal constituent is ionized hydrogen (H$^+$ ions) produced by the same process of photoionization as in the Earth's ionosphere. The structure of this region changed appreciably between the Pioneer and Voyager encounters. The ionosphere seen by the Voyager had a much greater vertical extent than had been observed by the Pioneer spacecraft four and a half years earlier, as well as exhibiting greater diurnal variation. There is no doubt that these differences result from increased solar activity (*see* page 254).

With Jupiter the energetic particles in the magnetosphere are likely to provide an important secondary source of ionization.

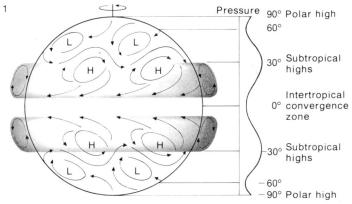

**1. The Earth's circulation**
A large temperature gradient between the equator and the poles dominates the Earth's circulation, producing a longitudinal wave travelling from west to east.

**2. The Earth's atmosphere**
The structure of the Earth's atmosphere is shown in this graph. Changes in the gradient of the curve indicate the important boundaries.

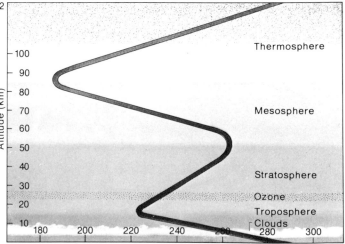

Their contribution to the production of the ionosphere would certainly account for the strong latitudinal variations observed by the Pioneer and Voyager spacecraft, and for the differences in ionosphere profiles.

## Circulation of the atmosphere

It is, then, in the lower levels of the atmosphere that the meteorological processes which produce the visible features of the planet take place. At first glance Jupiter's appearance suggests that its weather systems are quite different from those found on Earth or, indeed, on any of the other terrestrial planets such as Mars or Venus. Jupiter is a completely fluid planet (as least, if it has a solid core, it is only a very small proportion of the total planetary radius) and thus has no solid surface equivalent to that of the Earth affecting the atmospheric motions. There are also fundamental differences in the driving mechanisms of Jupiter's weather system. For example, on Jupiter the temperature at the poles is virtually the same as the temperature at the equator, whereas on Earth the poles are much colder than the equatorial regions, and the transfer of heat outwards from the equator plays an important role in the Earth's meteorology; thus a longitudinal wave is set up moving in a westerly direction, transporting the excess heat to the poles.

Another important difference between the driving mechanisms on Earth and Jupiter affecting their weather is found in their sources of heat. The primary source of energy for the Earth's weather system is solar radiation; by contrast, Jupiter has an internal source of heat energy, but receives only relatively weak incident sunlight because of its distance from the Sun. Finally, the rapid 10 hr rotation of Jupiter plays an important part in shaping the planet's clouds.

Nevertheless, many of the individual features on Jupiter's visible surface may be interpreted through analogies with familiar terrestrial weather-effects, and it may indeed be that the whole of Jupiter's weather system is no more than a multiple form of processes that are relatively well-understood from the Earth's atmosphere. For example, on Earth air will flow outwards from a region of high pressure, producing an anticyclone which spirals in a clockwise direction in the northern hemisphere or anticlockwise in the southern hemisphere; for low-pressure regions, the direction of flow is inwards to produce a cyclone, spiralling in the opposite direction to an anticyclone in either hemisphere. (The direction of flow depends on the Coriolis forces.) In Jupiter's atmosphere the regions of high and low pressure are stretched around the planet by the rapid rotation of the body, producing the characteristic bright zones and dark belts respectively. There is a temperature difference between these regions amounting to 3 K or less, and although this difference is only slight by terrestrial standards, it may, coupled with the rapid rotation, energize the circulation in such a way as to create the observed pattern of alternating easterly and westerly jets. Thus Jupiter's belts and zones may simply be large-scale wave systems, extending to considerable depths throughout the meteorologically active regions, but produced by the same fundamental forces that operate in the Earth's atmosphere.

In recent years considerable progress has been made in studies of the Earth's meteorology with numerical models which represent in detail the physical process and surface conditions over the planet. One such model has been extended to the conditions of the Jovian atmosphere. With this type of approach it is possible to vary different parameters so as to examine the response of the atmosphere. The results, generated by a computer, closely resemble the large-scale features of the Jovian atmosphere, such as the belts, zones and the large-scale spots. It would seem that instabilities in the atmosphere energize the circulation of Jupiter, creating large-scale planetary waves resembling the belts and zones, while the jets result from the turbulent interaction of propagating waves. The absence of any solid surface on Jupiter probably contributes to the distinctive zonal appearance.

3

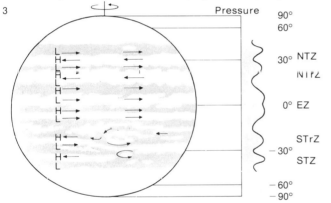

## 3. Jupiter's circulation
On Jupiter there is little flow between latitudes. Instead there are alternating regions of high and low pressure, with high-speed winds between the bands.

## 4. Jupiter's atmosphere
The basic structure of Jupiter's atmosphere resembles that of the Earth, at least at tropospheric levels. However, knowledge of higher levels is still incomplete.

## 5. Atmosphere compared
The principal constituents of a planet's atmosphere can be identified by the way in which they emit and absorb radiation. Although there is no direct correlation between Earth's and Jupiter's constituents, the way the two atmospheric systems behave is comparable. Mathematical models, representing the Earth's atmosphere and programmed with data for Jupiter show flow patterns resembling belts and zones.

4

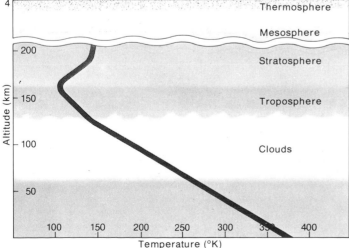

5

**Composition of Jupiter's atmosphere above cloud tops**

|  | % volume |
|---|---|
| Hydrogen | $\approx 90$ |
| HD | $\approx 1.8 \times 10^{-3}$ |
| Helium | $\approx 4.5$ |
| Methane | $\approx 7 \times 10^{-2}$ |
| Deuterated methane | $\approx 3 \times 10^{-5}$ |
| Ammonia | $\approx 2 \times 10^{-2}$ |
| Ethane | $\approx 10^{-2}$ |
| Acetylene | $\approx 10^{-2}$ |
| Water vapor | $\approx 10^{-4}$ |
| Phosphine | $\approx 10^{-6}$ |
| Carbon monoxide | $\approx 10^{-7}$ |
| Germanium Tetrahydride | $\approx 10^{-7}$ |

**Composition of the Earth's atmosphere**

|  | % volume |
|---|---|
| Nitrogen | 76.084 |
| Oxygen | 20.946 |
| Argon | 0.934 |
| Carbon dioxide | 0.031 |
| Neon | $1.82 \times 10^{-3}$ |
| Helium | $5.24 \times 10^{-4}$ |
| Methane | $1.5 \times 10^{-4}$ |
| Krypton | $1.14 \times 10^{-4}$ |
| Hydrogen | $5 \times 10^{-5}$ |
| Nitrous oxide | $3 \times 10^{-5}$ |
| Carbon monoxide | $10^{-5}$ |
| Xenon | $8.7 \times 10^{-5}$ |
| Ozone | up to $10^{-5}$ |
| Water (average) | up to 1 |

# Atmosphere II

### Cloud shapes

The changing shapes of Jupiter's cloud surface have been observed from Earth almost continuously over the past century, and considerable variations have been seen to occur, sometimes over very short periods. In the equatorial region marked changes sometimes take place within a length of time less than the planet's period of rotation. In the four and a half years between the Pioneer and the Voyager flybys, Jupiter's appearance had become considerably more turbulent and dramatic. In the broadest white zone, for example, violent motions can be seen in the Voyager photographs that are not apparent in the earlier photographs. Other features, however, remain stable over long periods; two plume-like clouds trailing behind (to the west of) small bright centers observed by the Pioneer spacecraft have been traced back a whole Jovian year (almost 12 Earth years), while features like the Great Red Spot and the white ovals have lifetimes of hundreds of years. From the Voyager studies it was found that all the easterly jets are unstable, while the corresponding westerlies remain stable.

### The northern hemisphere

Generally speaking Jupiter's northern hemisphere is distinguished from its southern counterpart by its lack of large-scale structures comparable to the Great Red Spot, although it does exhibit smaller features which appear to be comparable in basic morphology. This asymmetry between north and south is one of the planet's most striking characteristics.

At the time of the Voyager 1 encounter, 13 plume-like features were observed between the equator and approximately 10°N, in the Equatorial Zone. Then, a few months later, only 11 plumes were seen (*see* pages 278–9). This sudden change is caused by rapid upward motions in convective cloud systems which have horizontal scales of about 2,000 km and contain individual puffy elements about 100 to 200 km in size. The energy for these streams must be related to the condensation processing occurring in layers of water-cloud beneath the visible clouds. Morphologically, these smaller clouds resemble terrestrial cumulus, and contrast sharply with the diffuse, filamentary features found in many other regions on Jupiter. The flow characteristics of the plumes resemble some of the properties of clouds found in the Earth's tropics, in the Inter-Tropical Convergence Zone (ITCZ), where the crests of a travelling wave trigger convective activity in the atmosphere.

The plumes move in a westerly direction in a strong equatorial jet whose velocity ranges from 100 to 150 m s$^{-1}$. Surprisingly, these plumes are only found along the northern boundary of the Equatorial Zone and have no counterparts in the south. However, the flow that probably triggers the plumes requires a convergence of fluid beneath the visible surface, and in the southern part of the Equatorial Zone such a flow might well be disrupted by the presence

**1. Flow patterns on Jupiter**
In this mosaic, composed of nine separate photographs, details of Jupiter's cloud structure as small as 140 km across can be clearly distinguished. The images were made by Voyager 1 on 26 February 1979 at a distance of 7.8 million km from Jupiter; an orange filter was used. The graph shows the relative zonal velocities of the winds at various latitudes.

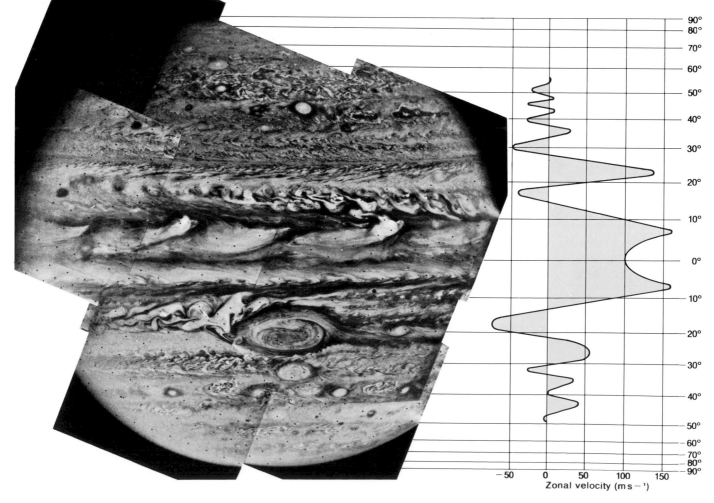

of the giant cloud systems such as the Great Red Spot and the white ovals. Thus these large features may account for the asymmetry between north and south in this region.

Further north, in the North Equatorial Belt (latitudes 9° to 18°N), cloud patterns vary considerably with longitude, with dark brown cloud shapes at irregular intervals. These structures generally have sharp boundaries and sometimes show linear striations in their interiors. The clouds seem to rotate around them cyclonically, indicating that these dark regions represent low-pressure systems.

As the brown clouds moved towards the west with the prevailing current, significant changes occurred in the period between the two Voyager encounters. The most westward feature almost disappeared, and a small but important intrusion of white appeared in the largest of the dark clouds (*see* illustration). It is clear that the different colored clouds are stratified into layers, with the brown clouds situated deeper than the bright layers.

The North Equatorial Belt also marks the highest 5 μm emissions. It is therefore apparent that the visible features in this region are the deepest that can be seen in Jupiter's atmosphere, where the temperatures and pressures are high. (The cloud tops radiate 5 μm waves with an equivalent temperature of about 140 K, as compared with the maximum of 258 K found at the deepest observable levels. Even the highest temperatures, however, are lower than those found on the surface of the Earth on a warm day.)

On the southern edge of the North Temperate Belt (latitudes 20 to 23°N) the clouds take on a tilted linear pattern associated with the opposing currents at 18° to 23°N. These features tilt north-east and form chevron shapes with the rows of puffy cloud tilted in the opposite direction further to the north. The apex of the chevron pattern lies in the strongest part of the westerly jet, which had a

speed of 150 m s⁻¹ during the Voyager encounter. In 1970 these features were found to move at a speed of 163 m s⁻¹, the highest known zonal velocity in the Jovian atmosphere.

The high-speed jet appears as a thin brown line dividing the broad and apparently featureless North Tropical Zone. In the northern boundary of this zone, at 35°N, a row of dark clouds can be seen. The dark areas surrounding the bright centers are associated with high 5 μm radiation. These small circular features measure about 3,000 km in diameter. They are composed of dark rings with white cones, and rotate anticyclonically. On one occasion there was a most extraordinary interaction, when the poleward member of a pair overtook the other and combined with it (*see* illustration). The combined features tumbled for a while and then ejected a streamer to the west, towards the equator. The new, combined spot then continued to proceed to the east.

Between the two Voyager encounters the spacing between the spots in this region altered considerably, indicating strong motions of these spots relative to one another. The mean zonal speed is about 26 m s⁻¹.

Further north the alternating pattern of belts and zones breaks down. In regions north of 35° long recirculating currents are common features, though occasionally there are large cloud systems such as the one at 45°N, 70°W, which appears to be more than 12,000 km in diameter, with a spiral structure in its interior.

**2. Temperature graphs**
In the graph of Jupiter's temperature at the latitude of the plumes, the lower line gives the temperatures at a height of 15 km above the cloud tops, the upper line the higher temperatures at the level of the cloud tops.

**3. Cloud layers**
The bands of colors on Jupiter's visible surface distinguish levels of cloud. The highest clouds, which form the bright zones, consist of ammonia crystals at a pressure of about one atmosphere. About 100 km below these, ammonium hydrosulphide crystals are expected, forming a dark cloud

layer. Below these there is believed to be a layer of bluish water-ice crystals through which a longitudinal wave (arrowed) may travel.

**4. Plumes**
This photograph shows a detail of two of the plume-like features at latitudes 0° to 10°N. Some of these features appear to remain stable over relatively long periods.

**5. Chevron patterns**
A high-speed jet separates the NTrZ from the NTZ. Linear features in the two zones tilt towards one another, forming V-shaped patterns where they meet.

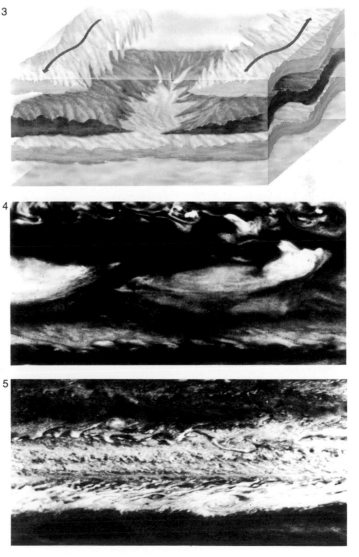

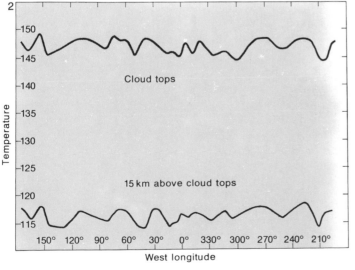

Temperature / West longitude

# Great Red Spot

**The southern hemisphere**

The appearance of Jupiter's southern hemisphere is dominated by the spectacular Great Red Spot and the three smaller, though similar, white ovals. A row of spots can also be seen at 41°S, and these features may also be related in structure to the large spots. The history of observations of the Great Red Spot and early theories about its nature have already been described (*see* pages 250–51), but the information gained from the satellite missions now places major constraints on theories about it.

The Great Red Spot is not a fixed feature. During the period covered by the photograph (*see* illustration) it drifted westward at a rate of approximately 0°.5 per day. In addition it has been found to oscillate relative to the north–south axis, with an average amplitude of 1,800 km. The period of oscillation is 89.85 ±0.1 days. The Great Red Spot is not unique in color; very small red spots have been observed from time to time in the northern hemisphere of the planet. The most prominent was seen during the Pioneer 10 flyby.

Associated with the Great Red Spot is a disturbed region to the west of the feature, in which sudden brightening frequently breaks out. Similar features seem to be characteristic of all the southern hemisphere spots (*see* illustration). In fact, the white ovals and the spots at 41°S resemble the Great Red Spot in several ways. They all rotate in an anticyclonic manner, and many have associated cyclonic circulating currents to the east and cyclonic "wake-like" regions to the west. (The resemblance to a wake is misleading. Movement of features in this region is eastward towards the Great Red Spot rather than westward away from it; such features then appear to be blocked by the Great Red Spot. An exception occurred when some light material crossed the South Equatorial Belt and then centered in the rapid westerly current in the Equatorial Zone.)

At the time of the Voyager 1 encounter, small spots were seen interacting with the Great Red Spot. Cloud vortices, moving in an easterly direction towards the Great Red Spot at a speed of about 55 m s⁻¹, were deflected northward at the eastern cusp, and entered the westerly current flowing on the north side of the Great Red Spot. Some of the cloud spots circulated around the Great Red Spot with a period of six days, approximately half the time taken 20 years ago when similar interactions were first observed.

As the spots rotate around the Great Red Spot, their shapes become distorted by the opposing directions of flow between the regions through which they are moving. In the interval between the two Voyager encounters, a cloud structure started to develop to the east of the Great Red Spot, forming a barrier to the flow (*see* illustration). While cloud vortices continued to approach the Great Red Spot at about 55 m s⁻¹, the presence of this barrier forced them to recirculate in the direction from which they originally came. Considerable shear was set up in this region, causing the cloud vortices to become distorted; as many as four or five such vortices were seen in this recirculating pattern adjacent to the Great Red Spot, which itself continued to rotate just as before.

The appearance of this cloud structure east of the Great Red Spot is highly significant. It may indicate a transition in the appearance of the planet, possibly corresponding to the change from placid to turbulent that took place between the Pioneer and Voyager encounters. This change of appearance may be part of a climatic cycle in the Jovian atmosphere.

**Nature and origin of the Great Red Spot**

Infrared observations have provided some important clues to understanding the Great Red Spot. At 5 μm wavelength, the Spot is much colder than its surroundings, suggesting that it is an elevated high-pressure region; this idea is consistent with its anticyclonic rotation. Temperature maps of the Great Red Spot and of the white

**1. The Great Red Spot**
A time-lapse sequence of photographs showing the GRS after every alternate rotation of Jupiter reveals the flow of material around the GRS. This particular sequence was produced by Voyager 2.

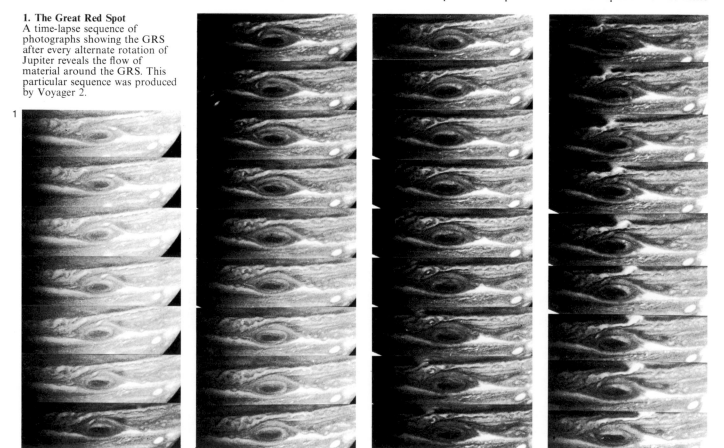

ovals show them to have cold regions above their centers, consistent with a slow upward motion of gas at a rate of a few millimeters per second, cooling "adiabatically" as it rises. ("Adiabatic cooling" takes place when a gas does not undergo a net loss of heat energy, but decreases in temperature only as a result of expansion as its pressure drops; such cooling takes place in a well-mixed atmosphere.) In a typical convective system on the Earth, upward motions are much faster, in the region of a few centimeters per second. If material is flowing upwards in the Great Red Spot and the ovals, there must be a corresponding convergence of material deep beneath the clouds to replace the material flowing up. The new material would slowly spiral up into the circulating system and then flow outwards at the level of the cloud tops.

Perhaps the best clue to understanding the nature and origin of the Great Red Spot comes from the white ovals, which were observed at their formation in 1939. At that time dark features were seen in the South Temperate Zone, stretching all the way around the planet. Gradually these dark regions extended until three large, white oval features formed, each nearly 100,000 km long. In the subsequent 40 years these white ovals have slowly diminished in size, and are now only 11,000 km long. At the time of the Voyager 1 encounter they were positioned at longitudes 5°, 85° and 170°W.

It would seem, therefore, that the white ovals may well disappear in the very near future. They may simply be large-scale gyres formed out of the zone in which they are situated. If they are merely temporary, there would be no need to find a sustained driving mechanism in order to explain them. The Great Red Spot may have originated in a similar fashion, and may also be a transient feature in the Jovian atmosphere, although it appears to be too long-lived for this explanation to be entirely satisfactory. However, like the white ovals, it is known to be shrinking, and is now half the length it used to be a century ago.

There have been, however, several other theories, such as the idea of a "Taylor column" described earlier (see page 250). One interesting recent suggestion is that the Great Red Spot is a kind of "soliton" or "solitary wave", an isolated permanent wave which can flow between two layers of fluid when there is a velocity difference between the layers. The wave takes its energy from the velocity gradient. Solitons have been analyzed from laboratory models, and a simple example is illustrated (see diagram 2). When solitons interact there is a phase shift that looks like an acceleration and the waves then emerge with identical forms as before. Such interactions have been compared with the interactions between the Great Red Spot and the South Tropical Disturbance, which were first observed in 1902 (see pages 250–51), and both the observed flows and the manner in which the features re-established themselves after interacting have a marked resemblance to the behavior of solitary waves. However, although several alternative models are able to reproduce the streamlined patterns, the uniqueness of the feature is difficult to establish. Perhaps the meteorology of Jupiter is in fact hybrid, a combination of several of the suggestions that have so far been advanced. For example, the computer model of the atmosphere (see page 261) would indicate that the Great Red Spot is simply a gyre; but the interaction of the Great Red Spot with the cloud vortices, observed by Voyager, and with the South Tropical Disturbance still needs to be explained. These aspects of the flows are well represented by solitary wave interactions. Perhaps some modified form of solitary wave may occur under the special conditions of Jupiter's environment. A complete explanation of the Great Red Spot would have to demonstrate such an effect within the context of a general mathematical model of the atmosphere as a whole, such as the one used in computer simulations (see page 325). At present the soliton idea seems to be the most promising line of investigation.

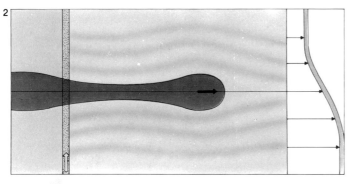

**2. Solitary waves**
The diagram shows a propagating solitary wave in a laboratory demonstration. When the barrier at the left is removed, the lower-density fluid forms a solitary wave which travels straight through the main tank.

**3. Structure of the GRS**
The GRS is believed to be an elevated high-pressure region. It rotates anticyclonically (anticlockwise), presumably drawing material upwards from some level below the visible surface of the cloud layers. This material would then flow back downwards, probably at the edges of the GRS. The exact details of these flows, however, are not known at present. At first sight the GRS and the white ovals appear to resemble terrestrial storm systems such as hurricanes, but this impression is misleading for the reasons given.

**4. Terrestrial hurricane**
On Earth storm clouds are driven by energy derived from a flow of air over a warm surface, such as an ocean. The hurricane is cyclonic at sea-level and weakly anticyclonic at high altitudes, and has a characteristic "eye" in the center. Air is drawn upwards and cools in the upper atmosphere, spreading outwards as it does so. The rotation of the Earth induces the clouds to form into spiral shapes. By contrast, the GRS and white ovals exhibit no central "eye" and no comparable spiral patterns. Moreover, if the Jovian features were storms, it would be necessary to explain why the whole surface of the planet was not covered with similar spots.

Solar UV

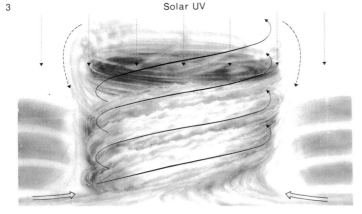

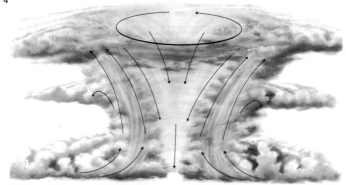

# Satellites I

Jupiter's satellite family, consisting of 16 known satellites, is the most extensive in the Solar System. The four largest satellites (Io, Europa, Ganymede and Callisto) form a group known as the "Galileans", so called because they were observed by Galileo Galilei in the very first days of telescopic research. All the Galileans have synchronous rotation (in other words, each satellite has a period of revolution around Jupiter equal to the period of the rotation of the satellite on its own axis). Three small satellites, Amalthea (discovered in 1892) and three others (Adrastea, Metis and Thebe) found on the Voyager images, move within the orbit of Io. Beyond Callisto, the outermost of the Galileans, there are eight more small satellites which may be asteroidal in nature. Their names, in order of increasing distance from Jupiter, are Leda, Himalia, Lysithea, Elara, Ananke, Carme, Pasiphaë and Sinope. Of these, the outermost four have retrograde motion and may well be captured asteroids. Neither the Pioneer nor the Voyager missions provided any information about the other satellites.

No doubt other small satellites await discovery but will be almost impossible to detect from Earth; we must await the arrival of the Galileo probe in 1995.

Apart from their intrinsic interest, Jupiter's satellites have been of scientific value in that their motions can be used to determine the mass of Jupiter and provide information about the gravitational field of this massive planet.

**The Galileans data**

|  | Io | Europa | Ganymede | Callisto |
|---|---|---|---|---|
| Mass (Jupiter = 1) | $4.696 \times 10^{-5}$ | $2.565 \times 10^{-5}$ | $7.845 \times 10^{-5}$ | $5.603 \times 10^{-5}$ |
| Mass (Moon = 1) | 1.213 | 0.663 | 2.027 | 1.448 |
| Mean density (kg m$^{-2}$) | 3,530 | 3,030 | 1,930 | 1,790 |
| Mean surface gravity (m s$^{-2}$) | 1.80 | 1.46 | 1.43 | 1.14 |
| Escape velocity (km s$^{-1}$) | 2.56 | 2.09 | 2.75 | 2.38 |

## 1. Satellite orbits

Jupiter's satellites fall broadly into three groups. The innermost group, comprising the Galileans together with Amalthea and the three newly discovered satellites, move in very nearly circular orbits in Jupiter's equatorial plane. Another four satellites, Leda, Himalia, Lysithea and Elara, all have a mean orbital distance in the range of 11 million km and make up the second group. Their orbits are more eccentric than those of the inner group and are inclined to them by an angle of slightly less than 30°. The remaining satellites, Ananke, Carme, Pasiphaë and Sinope, form the third group. They move in retrograde orbits, inclined to Jupiter's equatorial plane by an angle of approximately 150° to 160°. All the members of the outermost group orbit Jupiter at a mean distance of over 21 million km, thus the orbits are strongly perturbed by the Sun.

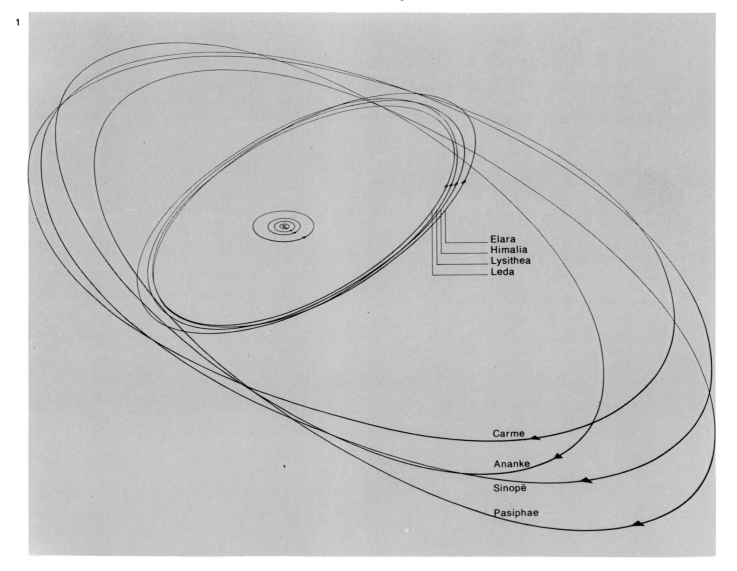

Elara
Himalia
Lysithea
Leda

Carme

Ananke

Sinopë

Pasiphae

**The Jovian Satellites**

| No. | Satellite | Discoverer | Year of discovery | Mean distance from Jupiter km | Diameter km | Magni-tude | Orbital inclination degrees | Orbital eccentricity | Sidereal period days | Mean synodic period d hr min sec |
|---|---|---|---|---|---|---|---|---|---|---|
| XVI | Metis | Synnott | 1980 | 127,600 | 40 | 17.4 | ? | 0 | 0.295 | 0 7 4 48 |
| XIV | Adrastea | Jewitt & Danielson | 1979 | ≈128,980 | 26 × 20 × 16 | <18.9 | ? | 0 | 0.297 | 0 7 9 7 |
| V | Amalthea | Barnard | 1892 | 181,300 | 262 × 146 × 134 | 14.1 | 0.4 | 0.003 | 0.498 | 0 11 57 27.6 |
| XV | Thebe | Synnott | 1980 | 225,000 | 110 × 100 × 90 | 15.5 | ? | 0.013 | 0.678 | 0 16 12 0 |
| I | Io | Galileo, Marius | 1610 | 421,600 | 3,642 | 4.9 | 0.0 | 0.0001 | 1.769 | 1 18 28 35.9 |
| II | Europa | Galileo, Marius | 1610 | 670,900 | 3,130 | 5.3 | 0.5 | 0.0001 | 3.551 | 3 13 17 53.7 |
| III | Ganymede | Galileo, Marius | 1610 | 1,070,000 | 5,268 | 4.6 | 0.2 | 0.0014 | 7.155 | 7 03 59 35.9 |
| IV | Callisto | Galileo, Marius | 1610 | 1,883,000 | 4,806 | 5.6 | 0.2 | 0.0074 | 16.689 | 16 18 05 06.9 |
| XIII | Leda | Kowal | 1974 | 11,100,000 | 10 | 20.2 | 26.7 | 0.1478 | 238.7 | 254 |
| VI | Himalia | Perrine | 1904 | 11,470,000 | 170 | 14.8 | 28 | 0.1580 | 250.6 | 266 |
| X | Lysithea | Nicholson | 1938 | 11,710,000 | 24 | 18.4 | 29 | 0.1074 | 259.2 | 276 |
| VII | Elara | Perrine | 1905 | 11,743,000 | 80 | 16.7 | 28 | 0.2072 | 259.7 | 276 |
| XII | Ananke | Nicholson | 1951 | 20,700,000 | 20 | 18.9 | 147 | 0.169 | 631 | 551 |
| XI | Carme | Nicholson | 1938 | 22,350,000 | 30 | 18.0 | 163 | 0.207 | 692 | 597 |
| VIII | Pasiphaë | Melotte | 1908 | 23,300,000 | 36 | 17.7 | 148 | 0.410 | 744 | 635 |
| IX | Sinope | Nicholson | 1914 | 23,700,000 | 28 | 18.3 | 157 | 0.275 | 758 | 645 |

**2. Scale of the Galileans**
The Galilean satellites are comparable in size with the smaller planets; Ganymede, for example, is slightly larger in diameter than Mercury. With the exception of Europa, all are larger than the Moon.

2

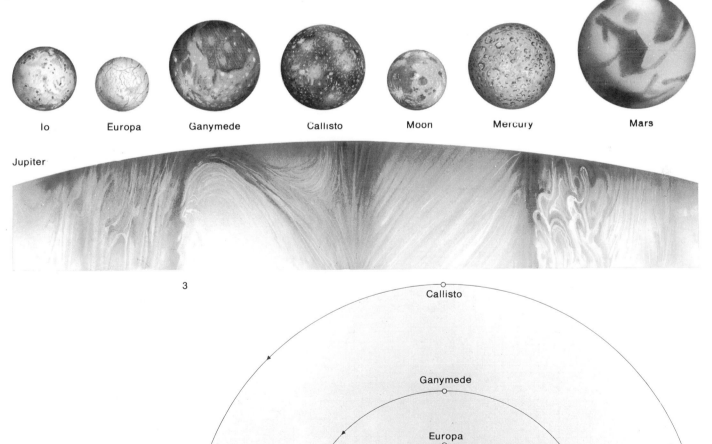

Io  Europa  Ganymede  Callisto  Moon  Mercury  Mars

Jupiter

3

Callisto

Ganymede

Europa

Io

Amalthea

1979 J1
1979 J2
1979 J3

**3. The Galileans**
The orbits of the four largest and most important Jovian satellites lie within 2 million km of the planet. The satellites Adrastea, Metis, Thebe and Amalthea are also shown.

267

# Satellites II

Jupiter's four main satellites were observed by Galileo Galilei for the first time on 7 January 1610. It is possible that they had been seen even earlier by Simon Marius, and it was, in fact, Marius who gave the satellites their familiar names; but until recently these names were not used officially mainly because of the question of priority of discovery, and instead the satellites were referred to as I, II, III and IV. However, Galileo's work was much more reliable and extensive, and he is generally given the principal credit for discovering them. In *Sidereus Nuncius* (1610) he describes the occasion as follows:

" . . . I should disclose and publish to the world the occasion of discovering and observing four Planets, never seen from the beginning of the world up to our own times, their positions, and the observations made during the last two months about their movements and their changes of magnitude; and I summon all astronomers to apply themselves to examine and determine their periodic times, which it has not been permitted me to achieve up to this day . . . On the 7th day of January in the present year, 1610, in the first hour of the following night, when I was viewing the constellations of the heavens through a telescope, the planet Jupiter presented itself to my view, and as I had prepared for myself a very excellent instrument, I noticed a circumstance which I had never been able to notice before, namely that three little stars, small but very bright, were near the planet; and although I believed them to belong to the number of the fixed stars, yet they made me somewhat wonder, because they seemed to be arranged exactly in a straight line, parallel to the ecliptic, and to be brighter than the rest of the stars, equal to them in magnitude . . . When on January 8th, led by some fatality, I turned again to look at the same part of the heavens, I found a very different state of things, for there were three little stars all west of Jupiter, and nearer together than on the previous night."

Following further observations, Galileo wrote:

"I therefore concluded, and decided unhesitatingly, that there are three stars in the heavens moving about Jupiter, as Venus and Mercury round the Sun; which was at length established as clear as daylight by numerous other subsequent observations. These observations also established that there are not only three, but four, erratic sidereal bodies performing their revolutions around Jupiter."

There are reasons for claiming that this was the most important of all Galileo's discoveries. It showed that contrary to the old beliefs there were at least two centers of movement in the universe; in other words, not everything revolved around the Earth. Some church officials were decidedly unenthusiastic, and one opponent refused to look at the satellites through Galileo's telescope—which he believed to be bewitched. Galileo is said to have expressed the hope that the official concerned would have a good view of the Jovian system on his way to Heaven.

## Theoretical questions

Before the advent of the first practical marine chronometer in the eighteenth century, the phenomena of the Galileans (*see pages 270–71*) were regarded as potentially important for the determination of longitude by marine navigators far from land, since the times of the phenomena could be predicted accurately. The method, however, was never actually used with any real success.

More importantly, it was by comparing the times of eclipses of the Galileans that the Danish astronomer Ole Römer established that the velocity of light was finite, and made the first reasonably accurate measurement of it. He realized that the discrepancies between predicted and observed times of satellite phenomena were due to Jupiter's varying distance from Earth, and obtained a value for the velocity of light accurate to within 2 percent.

**1. Galileo Galilei (1564–1642)**
Among Galileo's most significant contributions to science was the discovery of the four Jovian satellites that are now named in his honor. They were originally called by him "Medicean planets", after the Medici family.

**2. Sidereus Nuncius**
Galileo published the first results of his telescopic researches in 1610, in *Sidereus Nuncius*, two pages of which are shown here. Jupiter is represented by a large star, surrounded by the Galileans, illustrated as smaller black stars.

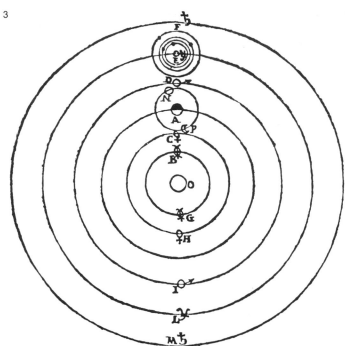

## Maps of the Galileans

In 1961 A. Dollfus and his colleagues at the Pic du Midi Observatory in France produced some preliminary mercator maps of Io, Europa, Ganymede and Callisto. Although inevitably very rough, these maps at least demonstrate that a certain amount of surface detail can be seen from Earth using very large telescopes. The orange color of Io was also discernible. Since the Pioneer and Voyager missions, however, it has been possible to produce far more detailed and accurate maps based on the photographs sent back by the space-probes (*see* pages 294–5, 298–9, 304–5 and 308–9). There are still some gaps in the coverage, and not all the areas were photographed to the same resolution, so in some places the maps are more precise than in others.

**3. The Copernican system**
Galileo's drawing of the Copernican system of the universe, published in *Dialogo* (1632), includes the Jovian satellites that he had discovered. Copernicus believed that the planets moved in circular orbits, with the Sun at the center.

**4. Maps by Dollfus**
Maps of Io (**A**), Europa (**B**), Ganymede (**C**) and Callisto (**D**) were made in 1961.

**5. Ole Römer (1644–1710)**
Römer, inventor of the transit instrument shown here, determined the velocity of light from observations of the times of eclipse of each Galilean.

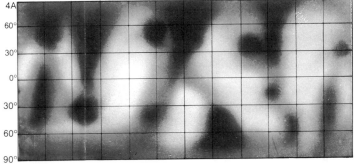
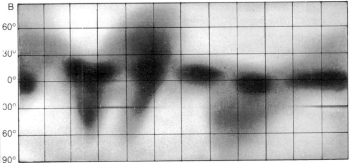
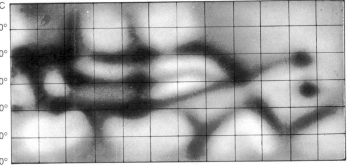

# Satellites III

Any small telescope will show all four Galileans as definite discs. Amateur observers can take great pleasure in watching the satellites' phenomena: eclipses, occultations, transits and shadow transits. Because the orbits all lie practically in the plane of Jupiter's equator, which is itself inclined by only about 3°, the Galileans alternately transit across Jupiter's disc and are occulted behind it during the course of each revolution. Only Callisto is far enough away from Jupiter to pass clear of the disc, which it does when the declination of the Earth as seen from Jupiter is greater than about 2°40'. Generally, several satellites can be seen at any particular moment; it is rare for all four to be out of view simultaneously.

Shadow transits are relatively easy to observe, the shadows appearing as dark spots. As far as actual transits are concerned, the satellites behave differently. Near the start of a transit each satellite will appear as a bright spot, but may soon be lost to view until it reappears as a bright spot once again just before the transit ends. Io may, however, be seen as a dusky spot when a transit is well advanced, particularly when it is in front of a bright zone, while Europa may remain bright for most of the time, although it is clearly visible only when projected against a dark belt; Ganymede and Callisto are visible as grey spots during transit. It is interesting to compare the apparent size of a satellite in transit with its shadow. The shadow will seem the larger due to the effects of penumbra.

Mutual phenomena may also be observed. A partial occultation of Io by Callisto, for example, was seen on 18 February 1932 by two famous observers of Jupiter, W. H. Steavenson and B. M. Peek, while on 8 February 1920, C. S. Saxton, using only a 3 in (7.6 cm) refractor, saw an eclipse of Io by the shadow of Ganymede. Other instances are on record.

## Observations by Pioneer and Voyager

The first attempt to study the Galileans from space-probes was made in December 1973, when Pioneer 10 encountered Jupiter. In December of the following year, images were sent back to Earth by Pioneer 11, but much more detailed information about the satellites was obtained from the two Voyager missions in 1979. The Voyager trajectories were designed to provide the fullest pictorial coverage of the satellite surfaces by taking advantage of their synchronous rotation and rapid motion around Jupiter. Thus Voyager I took pictures of the Jupiter-facing hemispheres of Ganymede and Callisto, passing them at high latitudes to achieve good coverage of their north poles, while Voyager 2 recorded the outward-facing hemispheres and passed further to the south of Ganymede than had Voyager 1; Voyager 2 also obtained images of part of Europa's surface. Both spacecraft were able to achieve detailed coverage of much of the surface of Io, thanks to its short orbital period.

## Metis and Adrastea

Two small inner satellites moving within the orbit of Amalthea were discovered from examination of the Voyager pictures. Metis orbits at less than 57,000 km above the cloud-tops, and has an estimated diameter of about 40 km. Adrastea is even smaller and fainter, and appears to be somewhat egg-shaped, with a longest diameter of 110 km. It seems that both Metis and Adrastea have low albedo (less than 0.05) and may therefore be high-density objects rather than icy in nature, but this remains to be confirmed. It is hoped that images will be obtained from the Galileo probe after 1995.

Metis and Adrastea move in the region of Jupiter's ring, and may therefore play an important role in the stability of the ring itself. Amalthea, the closest-in of the satellites known before the Voyager missions, is well beyond the outer edge of the ring.

## Amalthea

Amalthea was discovered in 1892 by Edward Emerson Barnard, using the 36 in (91 cm) refractor at the Lick Observatory, while he

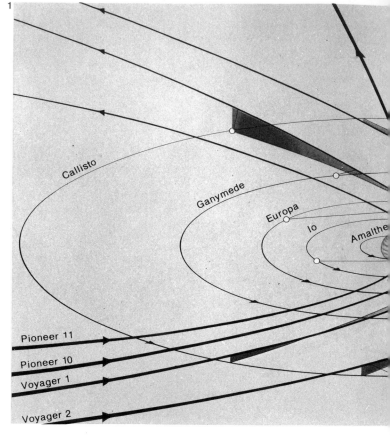

**1. Spacecraft trajectories**
The paths taken by each of the four spacecraft that have visited Jupiter were carefully designed to make optimum use of the time available for observation. Pioneers 10 and 11 concentrated mainly on Jupiter itself, sending back only a few low-definition images of the satellites. The Voyager vehicles, on the other hand, between them provided coverage of between about 50 and 90 percent of the surfaces of the Galileans at high resolution. Europa received the poorest coverage, Io the most extensive. Images of Amalthea were also obtained.

**2. The Galileans**
A small telescope or a good pair of binoculars is sufficient to reveal the Galilean satellites. These two Earth-based photographs were taken at McDonald Observatory on 25 and 30 January 1955.

was carrying out a deliberate search for new satellites. Amalthea was in fact the very last satellite to be discovered by direct visual observation. It is so faint and so close to Jupiter that it is difficult to observe from Earth, but it has been carefully studied, and its orbit was well known even before the Pioneer and Voyager missions. Voyager 1 approached it to a distance of 420,100 km, and Voyager 2 to 558,270 km. Both probes took pictures of it.

Amalthea's orbit is close to the "Roche limit", the limiting distance within which an orbiting body with little gravitational cohesion would be disrupted by the pull of gravity. Amalthea is in fact elongated by the gravitational pull of Jupiter, and has a generally ellipsoidal shape. Its 270 km long axis is pointed towards Jupiter, while its 155 km short axis lies at right angles to the orbital plane. It has an irregular profile, which indicates that it must be fairly dense: a more plastic body would become smooth.

Amalthea has been found to be rather warmer than it would be if it were simply absorbing and re-radiating solar radiation and the radiation it receives from Jupiter. Possibly the additional heating is derived from electrical currents induced by the Jovian magnetic field (*see* pages 252–5). It is red in color, and the surface is probably covered with sulphur, which could only originate from Io.

Four features were shown on the Voyager images; two craters, known as Pan and Gaea, and two mountains known as Mons Ida and Mons Lyctas. Nothing definite is known about them.

As seen from Jupiter, Amalthea would have an apparent diameter of 7'24". As the apparent diameter of the Sun as seen from Jupiter is less than 6', Amalthea would be able to produce a total solar eclipse.

## Thebe

This was the second new satellite to be found on the Voyager pictures; it was discovered while the photographs were being examined to confirm Adrastea. The new body is over 100 km in diameter, and orbits at 151,000 km above the cloud tops, so that it moves outside the orbit of Amalthea but within that of Io. Its revolution period is 16 hr 12 min, but at the moment nothing further is known about it.

**3. Satellite phenomena**
The various possible configurations of the satellites provide a focus of interest for observers. More common examples include shadow transits (A), occultations (B) and eclipses (C). Rarer events, such as mutual occultations and eclipses, may also be observed.

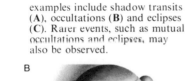

3A B

C 4

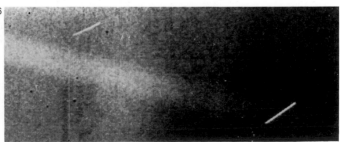

**4. Effect of the penumbra**
The shadow cast by a satellite appears larger than the satellite itself because of the area of partial shadow or "penumbra".

**5. Amalthea**
The irregular-shaped satellite Amalthea was photographed by Voyager 1 on 4 March 1979 from a distance of 425,000 km. Some of the indentations in the photograph may be craters.

**6. Satellite Adrastea**
Jupiter's fourteenth satellite was discovered in this computer-enhanced image made by Voyager 2 on 8 July 1979. The satellite appears as a bright streak at the lower right of the image; the other streak is the track of a star. The grey band is Jupiter's ring.

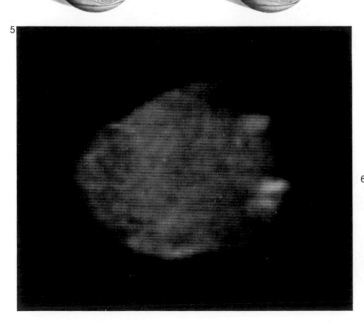

5

6

# Rings

The discovery that Jupiter is surrounded by rings of particles similar to those of Saturn and Uranus was another unexpected result of the Voyager mission. Jupiter's rings were first detected in a narrow-angle frame targeted halfway between Amalthea and the limb of Jupiter at 16 hr 52 min before the closest approach, as Voyager 1 crossed the equatorial plane of the Jovian system. Following this discovery, Voyager 2 was programmed to take additional pictures giving considerably better resolution.

The rings seem to consist of several components. The brightest, rather narrow portion has a radius at the outer edge of $126,380 \pm 140$ km ($1.772 \pm 0.002$ $R_J$). There is also a narrow bright segment $800 \pm 100$ km wide with the inner edge at $125,580 \pm 140$ km ($1.688 \pm 0.002$ $R_J$). The ring particles seem to extend nearly all the way down to the planet itself. This has other important consequences: the upper atmosphere would distribute the tiny particles on a global scale into a haze layer, which would then absorb the incoming solar radiation at visible and ultraviolet wavelengths. It is even possible that the ring material is a source of oxygen to the upper atmosphere that could be related to the carbon monoxide unexpectedly detected in Jupiter's atmosphere.

The ring has a characteristically orange color and seems to be composed of particles of radius about 4 $\mu$m. This is in complete contrast with the Saturnian ring particles, which are estimated to be several centimeters in cross-section.

## Origin and composition

The Jovian rings lie well inside the classical Roche limit for the break-up of a liquid body, which occurs at 2.44 $R_J$. It would therefore seem that the particles must be relatively high-density rocks (dust), rather than the icy material of the Saturnian rings. If, for example, the material is assumed to have roughly the same density as Io ($3.5 \times 10^3$ kg m$^{-3}$), then the edge of the ring would be situated at 1.81 $R_J$, which is not far from the observed position that has been recorded.

It is possible, therefore, that the material that currently resides in the Jovian ring originated from the gravitational break-up of a tiny inner satellite that evolved in the neighborhood of the Roche limit. This, however, would not create much material. Additional sources might include material from Io and other inner satellites, and debris from comets and meteorites.

The newly discovered inner satellites, Metis and Adrastea, may play an important role in relationship with the outer edge of the ring. Adrastea lies precisely at the outer edge of the ring system, and as it moves round Jupiter it sweeps up magnetospheric particles that reside there, giving a sharp edge to the ring system. It is quite possible that there may be other small satellites in the general region of the ring; certainly the discovery of Metis, soon after that of Adrastea, came as no surprise.

**1. Dark side of Jupiter**
The ring showed up particularly brightly when Voyager 2 passed behind Jupiter with respect to the Sun. The planet's shadow obscures part of the ring in the direction of the spacecraft.

**2. Jupiter's ring**
This composite image of Jupiter's ring was made by Voyager 2 on 10 July 1979. The spacecraft was 2° below the plane of the ring, at a distance from Jupiter of 1,550,000 km.

# Plates

North Temperate Zone
Voyager 1 (top)
2 March 1979

Equatorial Zone
Voyager 2 (bottom)
28 June 1979

Great Red Spot
Voyager 2 (top)
3 July 1979

Great Red Spot
Voyager 1 (bottom)
4 May 1979

277

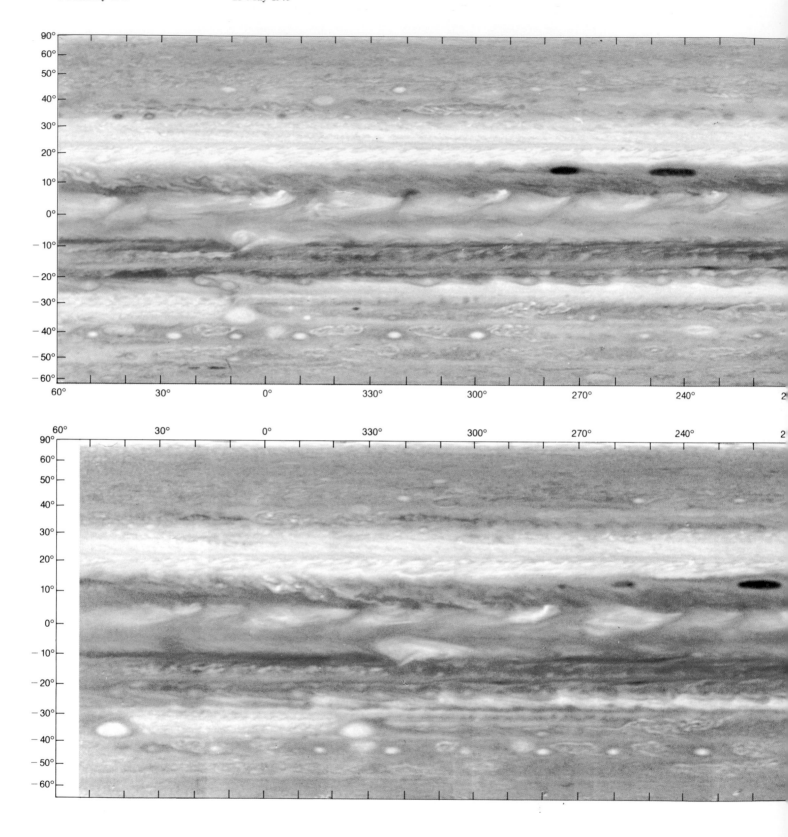

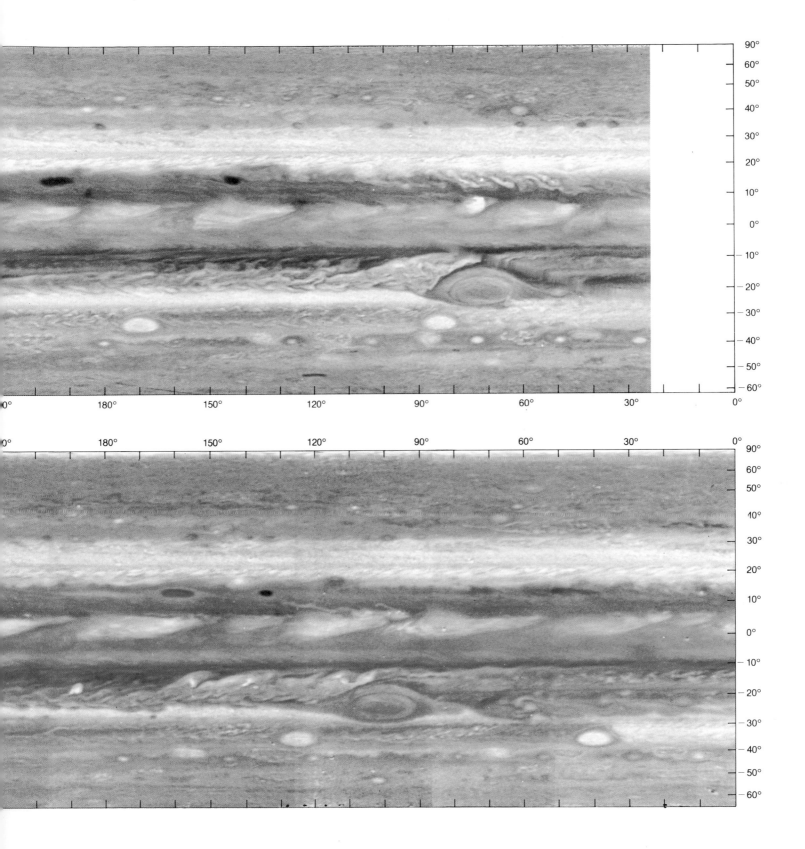

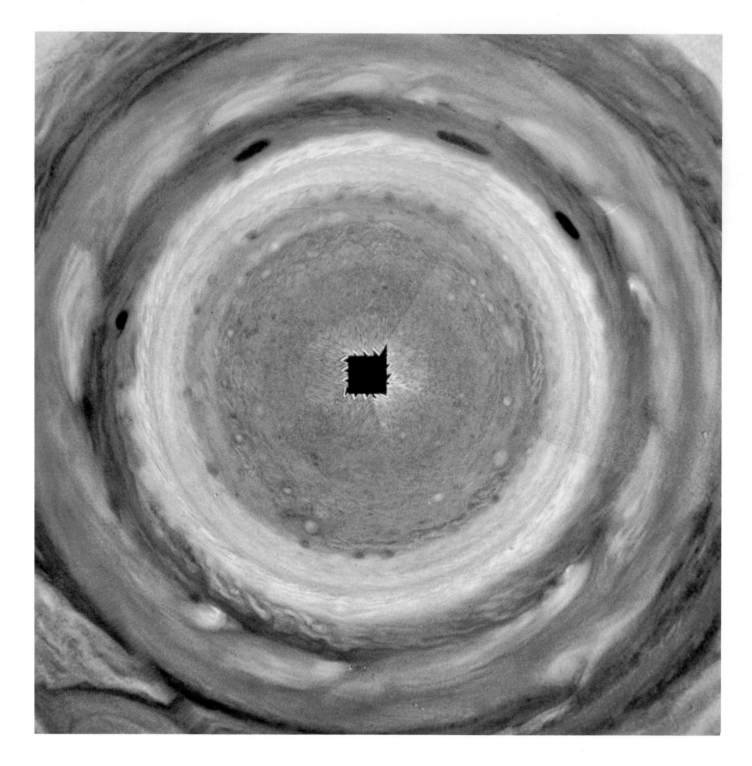

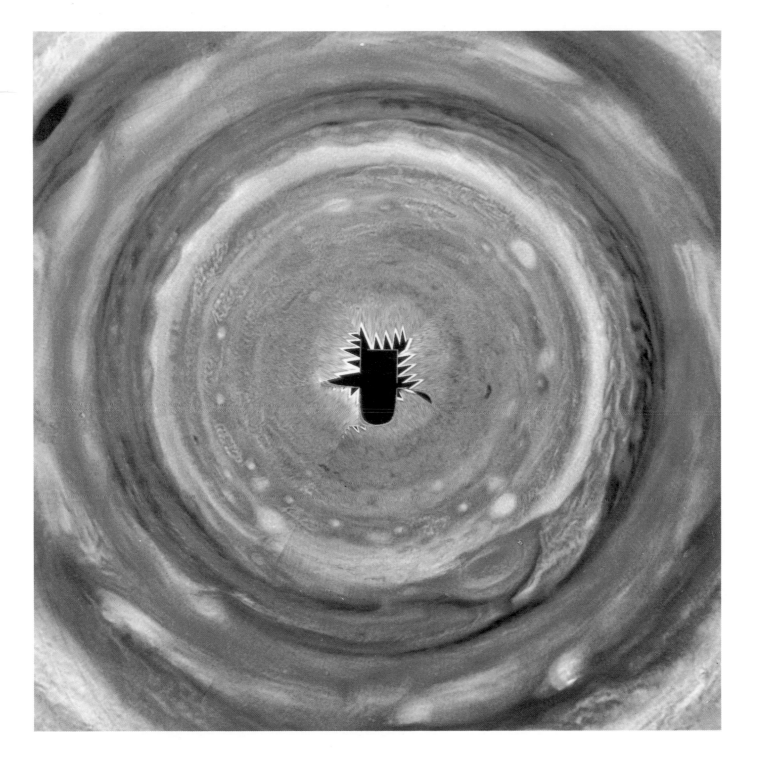

Io
Voyager 1
4 March 1979

282

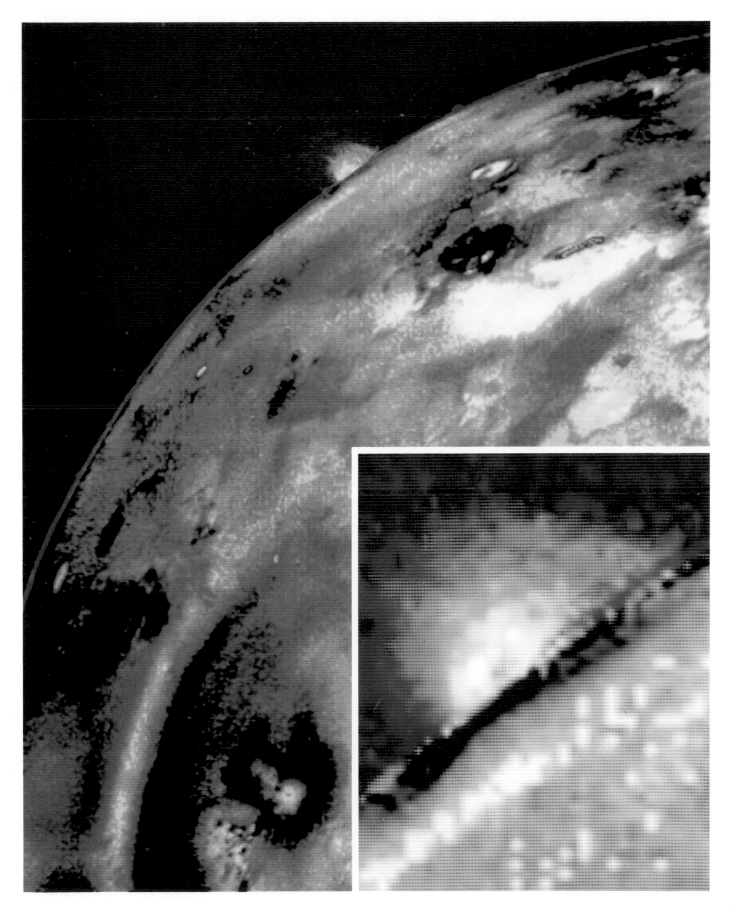

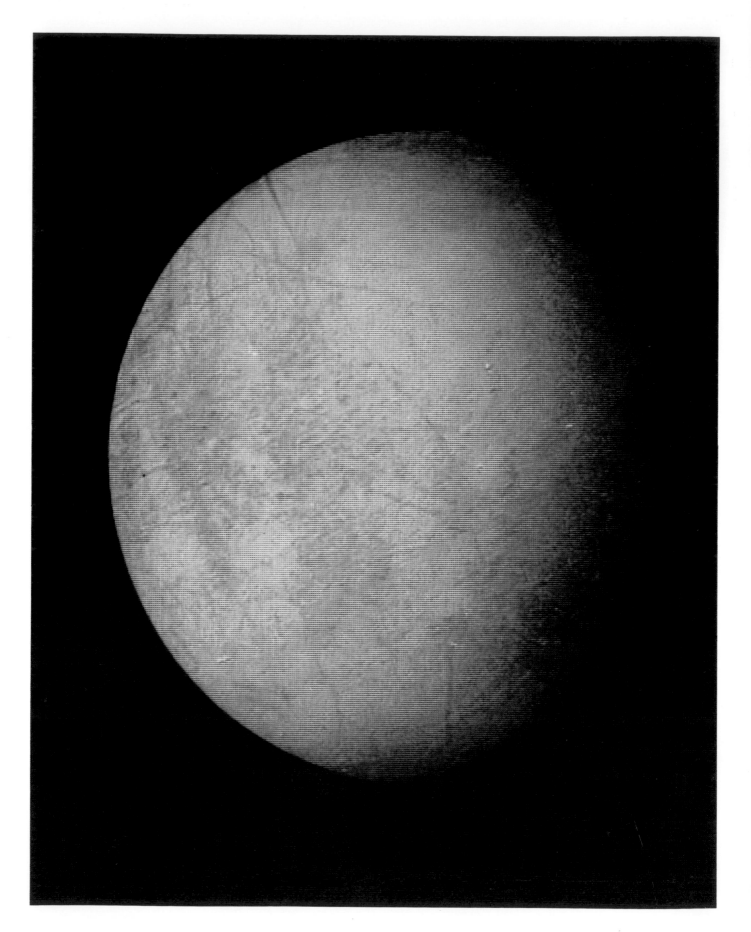

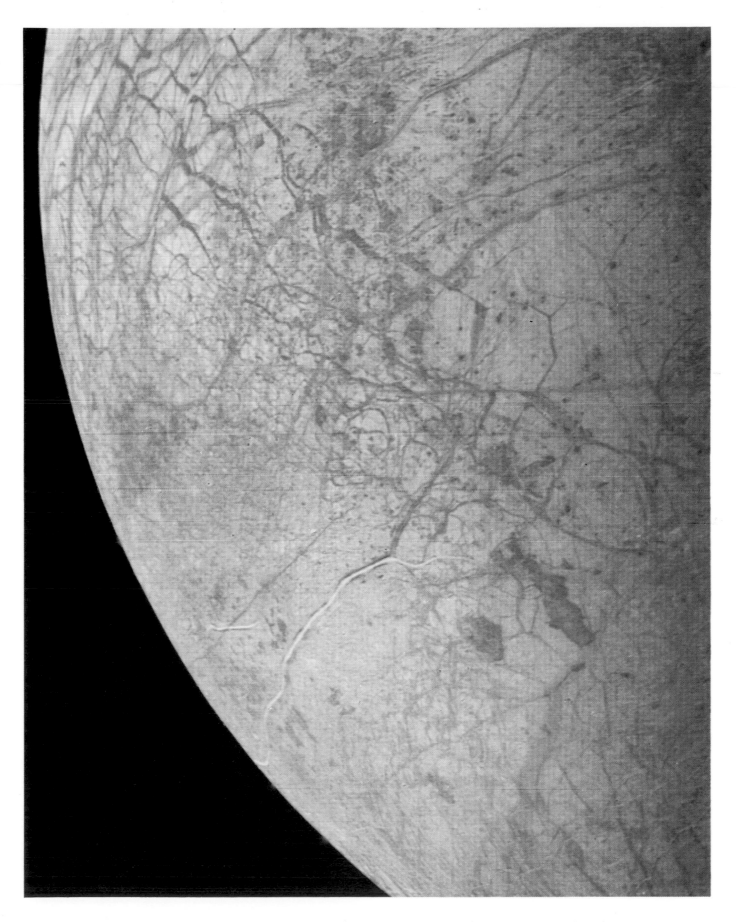

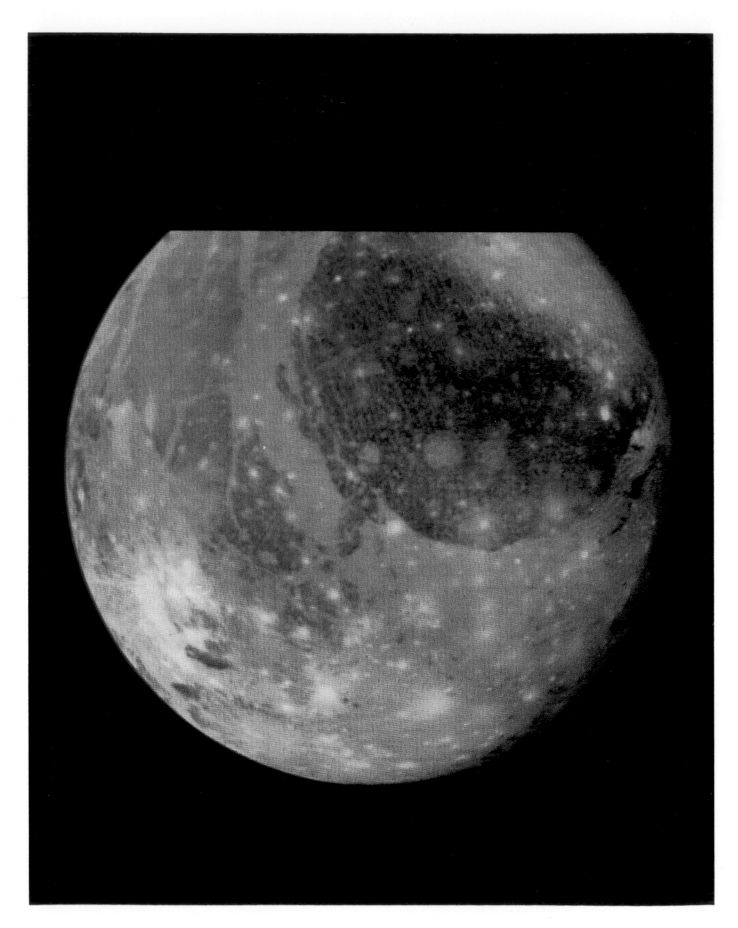

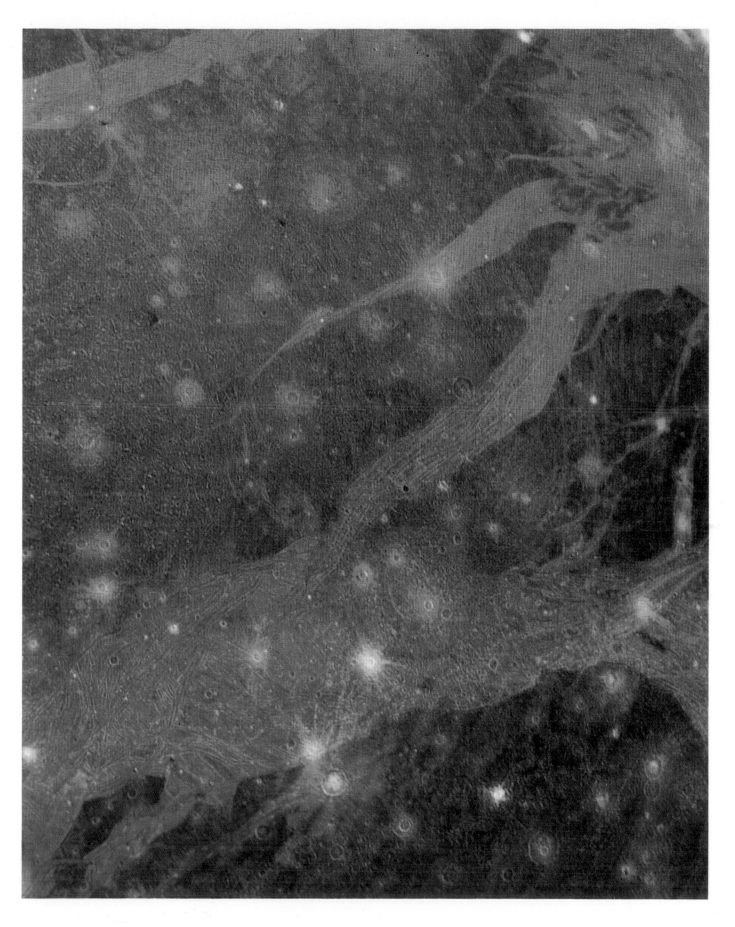

## Notes to Color Plates

**Page 273.** Three of the Galilean satellites can be seen in this image (Callisto is just visible in the lower left-hand corner). The spacecraft was at a distance of 28.4 million km from Jupiter.

**Page 274.** Earth-based image from Mt. Lemmon Observatory, using a 1.54 m telescope. University of Arizona.

**Page 275.** Jupiter's appearance was distinctly calmer at the time of the Pioneer flybys (top) than it appears in the Voyager images (bottom). Significant changes have taken place in the four months that separate these two Voyager images, particularly in the region of the GRS. The large white oval just below the GRS has drifted a considerable distance to the east (*see* pages 250–51).

**Page 276.** (Top). The pale orange line cutting the right-hand corner of the image marks the North Temperate Current, in which wind-speeds reach 120 m s$^{-1}$. A weaker jet further to the north (towards the top of the image) exhibits swirling cloud patterns. (Bottom) Colors have been exaggerated in this image to show up more detail. The NEB with a long-lived dark cloud feature can be seen at the top of the image. Three of the plume-like features in the EZ can be seen immediately below the NEB.

**Page 277.** (Top) The white oval just below the GRS in this image is different from the one seen in a similar position at the time of the Voyager 1 encounter. (Bottom) Blue and red have been deliberately exaggerated in this image to bring out details in the structure of the GRS.

**Pages 278/279.** These cylindrical projections of Jupiter clearly reveal the longitudinal drift of features on Jupiter. The two images are aligned in such a way that the scale of longitude is the same for both of them. The GRS has moved westward by about 30°, while other features have drifted in both directions, some of them at higher speeds.

**Pages 280/281.** This pair of polar stereographic projections shows up features situated at high latitudes. The black shapes at the poles themselves are due to lack of data.

**Page 282.** Io is shown from a distance of about 862,000 km. In the center of the image is the volcano Prometheus.

**Page 283.** Io at a distance of about 490,000 km. The inset shows a computer-enhanced image of the volcanic plume seen on Io's limb.

**Page 284.** Europa at a distance of 1.2 million km. The colors in this image have been slightly enhanced. The bluish areas in the polar regions should in fact appear white.

**Page 285.** Europa at a distance of 241,000 km.

**Page 286.** Ganymede at a distance of 1.2 million km. The large darkish region towards the north-east (top right) is Galileo Regio.

**Page 287.** Ganymede at a distance of 312,000 km. The smallest features visible in this image are about 5 km across.

**Page 288.** Callisto at a distance of 1.2 million km. The large ringed structure known as "Valhalla" can be seen near the limb in the upper left portion of the image.

# Io

The innermost large satellite, Io, has a radius of 1,820 km and an albedo of 0.63. It is the most dense of the Galileans, with a density of $3.5 \times 10^3$ kg m$^{-3}$, about the same as that of the Moon. Before the Voyager missions it had been assumed that, like the Moon, Io would have a surface covered with craters, but in the event this view proved mistaken. Shortly before Voyager 1 made its closest approach to Io, however, it had been suggested that as the satellite was subjected to the gravitational attraction of Jupiter on the one hand and of the other Galileans (particularly Europa) on the other, its surface would be flexed and the resulting friction could produce enough heat to cause a molten interior, with heat being released through the crust in the form of volcanic eruptions.

This prediction was confirmed dramatically on 9 March 1979. At the Jet Propulsion Laboratory a check was being made of the orbital position of Io by measuring its position relative to a faint star, known by its catalogue number of AGK-10021. When the star appeared on the screen of the imaging module, so did a huge umbrella-shaped plume at the edge of Io. This could be nothing but an active volcano, with a plume rising to some 280 km above the Ionian surface.

Pictures from Voyager 1 (which passed Io at a distance of 420,100 km) revealed a surface strikingly different from the lunar appearance that had been expected. The color was red and orange, and the dominant features were violently active volcanoes. Eight eruptions were seen, and the ejection velocities were calculated to be over 1 km s$^{-1}$, more violent than Etna, Vesuvius or even Krakatoa. The explosive plumes were from 70 to 300 km high, with the material sometimes rising to around 100 km and fanning out to form an umbrella-shaped cloud before falling back to the surface. When Voyager 2 passed at a distance of 558,270 km, the general appearance of the scene was much the same, though there had been changes in detail.

According to present ideas of Io's internal structure, there is a sulphur and sulphur dioxide crust overlying a molten silicate interior, while the actual core may well be solid. Various theories have been proposed to explain the volcanism. For instance, it has been suggested that a crust about 20 km deep is made to rise and fall regularly by 100 km or so by tidal forces, and that heat, produced by the friction, then escapes through the surface vents as volcanic eruptions. In the "silicate-volcanism" model, silicon-enriched magma erupts through a silicate crust which is rich in sulphur. The general process could not be unlike that of terrestrial volcanism, allowing for the greater abundance of sulphur on Io; variations in the chemistry and physical conditions of the magma chamber could account for the fact that not all eruptions are of the same type.

Alternatively, there may be a "sea" of sulphur and sulphur dioxide about 4 km deep, with only the uppermost kilometer frozen; this ocean has been forced above the silicate sub-crust because of thousands of millions of years of tidal heating. Violent volcanism occurs in the crust when liquid sulphur dioxide meets molten sulphur and explodes into space as it is decompressed. Sulphur seeping back into the molten silicate interior would then balance the extra sulphur deposited on the surface.

Io's low escape velocity prevents the satellite from retaining an atmosphere of appreciable density. However, even before the Pioneer mission, some astronomers had reported a temporary increase in brightness of about a tenth of a magnitude immediately after Io emerged from an eclipse, and it was suggested that reflective material had been deposited on the surface while the sunlight was cut off, only to evaporate again as soon as the eclipse finished.

Then, in December 1973, Pioneer 10 detected an "ionosphere". Moreover, Earth-based observations also revealed an electrically neutral atmosphere with a surface pressure of below 10$^{-3}$ Pascals (*see* pages 256–7). It was suggested that the surface of Io could be covered with salt, in which case this "sodium cloud" would be produced by the effects of Jovian radiation on the salty covering.

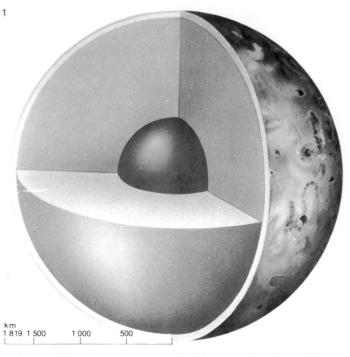

km
1 819  1 500          1 000          500

**1. Interior of Io**
Io's density has been determined more accurately than those of the other Galileans; this knowledge is helpful in constructing models of the interior structure.

Io is known to be rich in sulphur; its crust is probably composed of sulphur and sulphur dioxide, while the interior is thought to consist of molten silicates surrounding a solid core.

**2. Gravitational effects**
Io is subject to the gravitational pull not only of Jupiter but also of the other Galileans, particularly Europa. When Io passes between Jupiter and Europa, the opposing forces on it cause its surface to flex, and the resulting internal friction produces heat which may be responsible for the volcanic activity.

**3. Volcanism on Io**
The first evidence of volcanic activity on Io came from this image, when the curve on Io's limb was identified as a cloud of ash thrown up by an erupting volcano. A second eruption—the bright spot on the terminator— can also be seen. The image was made by Voyager 1 on 8 March 1979.

4A

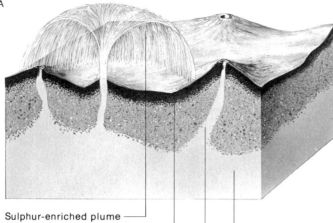

B

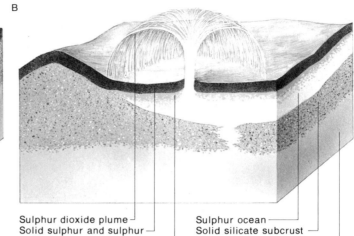

Sulphur-enriched plume
Silicate crust rich in sulphur
Silicate magma chamber
Silicate magma rich in sulphur

Sulphur dioxide plume
Solid sulphur and sulphur
dioxide upper crust
Sulphur aquifer with liquid SO₂

Sulphur ocean
Solid silicate subcrust
Molten silicate interior

**4. Causes of volcanism**
Several explanations of Io's
volcanic activity have been
proposed. The "silicate-
volcanism" model (**A**) involves a
mechanism similar to the volcanic
process on Earth. Silicate magma

is forced up through the crust,
erupting through vents in the
surface. The magma is richer in
sulphur than its terrestrial
equivalent. An alternative model
(**B**) attributes the volcanic
explosions to the effect of molten

sulphur coming into contact with
liquid sulphur dioxide. The
sulphur dioxide is suddenly
decompressed and explodes to
form plumes. According to this
model "oceans" of molten
sulphur lie close beneath the

topmost crust, having been forced
upwards through the molten
silicate interior. These reservoirs
are continually replenished by
sulphur seeping back through the
surface as the ejected material
falls back to the ground.

5

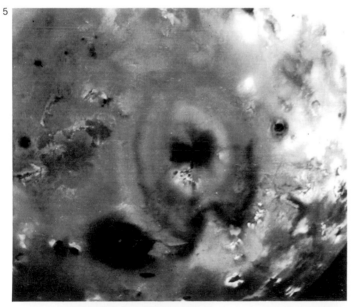

**5. Pele (19°S, 257°W)**
The source of the volcanic cloud
known as Plume 1 was a heart-
shaped volcano, now named Pele.
Its diameter is about 1,200 km at
the widest point

**6. Loki (19°N, 305°W)**
The surface markings in the
region of the eruption known as
Plume 2, whose source has now
been given the name Loki, altered
significantly between the
encounters of Voyager 1 (**A**) and
Voyager 2 (**B**). Volcanic ejecta
appear to have obscured older
patterns, and bright deposits are
visible in several places.

**7. Maasaw Patera (40°S, 341°W)**
The dark features are lava flows
emanating from a caldera.

**8. Gibil (15°S, 295°W)**
This image shows one of the
named volcanic features. Dark
lava flows can be seen in the
upper left corner.

6A

B

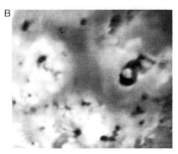

7

8

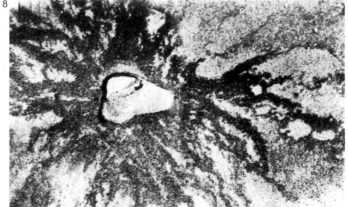

# Photomosaic of Io

**West region**

The surface of Io is active, and so the finer details are subject to change even over short periods of time, but the main features are presumably permanent. The most significant formations are the volcanoes, such as Pele and Loki. Pele was active during the Voyager 1 pass, but inactive when Voyager 2 made its flyby; not far from it is the dark area of Babbar Patera. Loki was active during both passes, and seems to be one of the most violent of the Ionian volcanoes; between it and Pele is the prominent feature known as Hephaestus Patera. Galai and Gibil are other notable objects. The region to the south of Pele and Babbar Patera—the Lerna Regio—has fewer well-defined features and is lighter in color. Ra Patera is another highly volcanic area, with Mazda Catena running alongside it. Also prominent is the structure of Haemus Mons. (The globe appears distorted because different areas are shown from slightly different viewpoints.)

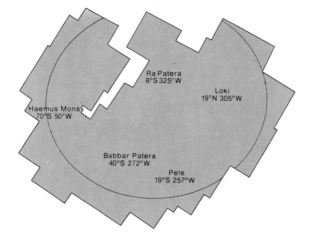

Ra Patera
8°S 325°W

Loki
19°N 305°W

Haemus Mons
70°S 50°W

Babbar Patera
40°S 272°W

Pele
19°S 257°W

Haemus Mons

LERNA REGIO

Ra Patera

Mazda Catena

Loki

Hephaestus Patera

Gibil Patera

Galai Patera

Babbar Patera

Pele

# Map of Io

**Volcanoes**

| | |
|---|---|
| Amirani (Plume 5) | 27°N, 119°W |
| Loki (Plume 2) | 19°N, 305°W |
| Marduk (Plume 7) | 28°S, 210°W |
| Masubi (Plume 8) | 45°S, 53°W |
| Maui (Plume 6) | 19°N, 122°W |
| Pele (Plume 1) | 19°S, 257°W |
| Prometheus (Plume 3) | 3°S, 153°W |
| Surt | 46°N, 336°W |
| Volund (Plume 4) | 22°N, 177°W |

**Paterae**

| | |
|---|---|
| Amaterasu Patera | 38°N, 307°W |
| Asha Patera | 9°N, 226°W |
| Atar Patera | 30°N, 279°W |
| Aten Patera | 48°S, 311°W |
| Babbar Patera | 40°S, 272°W |
| Bochica Patera | 61°S, 22°W |
| Creidne Patera | 52°S, 345°W |
| Culann Patera | 20°S, 150°W |
| Daedelus Patera | 19°N, 275°W |
| Dazhbog Patera | 54°N, 302°W |
| Emakong Patera | 0°, 110°W |
| Fuchi Patera | 28°N, 328°W |
| Galai Patera | 11°S, 289°W |
| Gibil Patera | 15°S, 295°W |
| Heno Patera | 57°S, 312°W |
| Hephaestus Patera | 2°N, 290°W |
| Hiruko Patera | 65°S, 331°W |
| Horus Patera | 10°S, 340°W |
| Inti Patera | 68°S, 349°W |
| Kane Patera | 48°S, 15°W |
| Loki Patera | 13°N, 310°W |
| Maasaw Patera | 40°S, 341°W |
| Mafuike Patera | 15°S, 261°W |
| Malik Patera | 34°S, 128°W |
| Manua Patera | 35°N, 322°W |
| Masaya Patera | 22°S, 350°W |
| Maui Patera | 20°N, 125°W |
| Mihr Patera | 16°S, 306°W |
| Nina Patera | 40°S, 165°W |
| Nusku Patera | 63°S, 7°W |
| Nyambe Patera | 0°, 345°W |
| Ra Patera | 8°S, 325°W |
| Reiden Patera | 14°S, 236°W |
| Ruwa Patera | 0°, 2°W |
| Sengen Patera | 33°S, 304°W |
| Shakuru Patera | 23°N, 267°W |
| Shamash Patera | 36°S, 152°W |
| Svarog Patera | 48°S, 267°W |
| Tohil Patera | 28°S, 157°W |
| Ülgen Patera | 41°S, 288°W |
| Uta Patera | 35°S, 27°W |
| Vahagn Patera | 27°S, 359°W |
| Viracocha Patera | 62°S, 284°W |

**Other features**

| | |
|---|---|
| Apis Tholus | 11°S, 349°W |
| Bactria Regio | 45°S, 125°W |
| Chalybes Regio | 55°N, 85°W |
| Colchis Regio | 10°N, 170°W |
| Dodona Planum | 60°S, 350°W |
| Haemus Mons | 70°S, 50°W |
| Inaehus Tholus | 29°S, 354°W |
| Lerna Regio | 65°S, 300°W |
| Mazda Catena | 8°S, 315°W |
| Media Regio | 0°, 70°W |
| Mycenae Regio | 35°S, 170°W |
| Nemea Planum | 80°S, 270°W |
| Silpium Mons | 62°S, 282°W |
| Tarsus Regio | 30°S, 55°W |

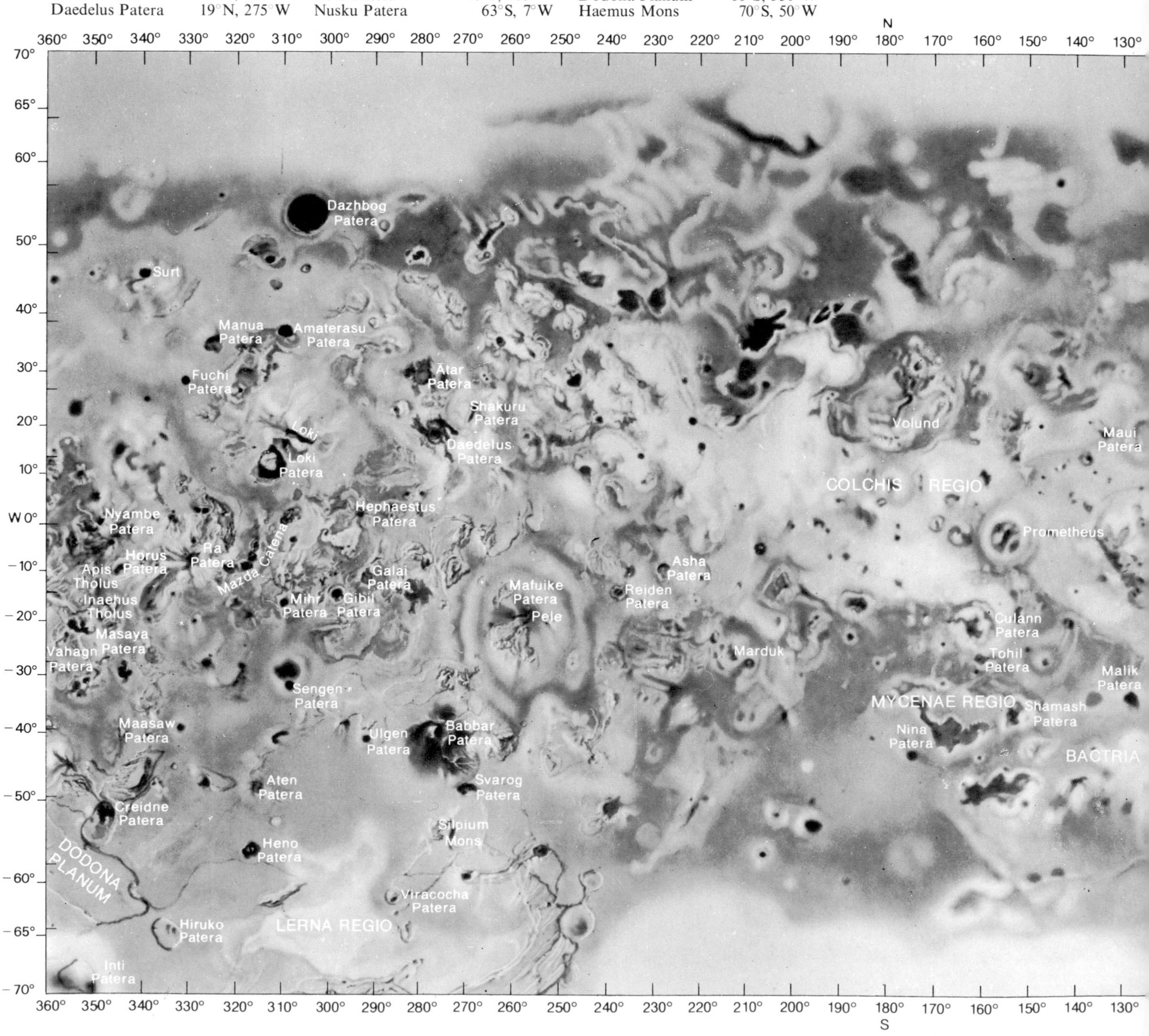

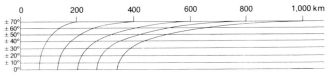

**Mercator Projection**

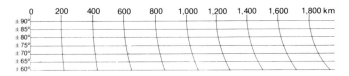

**Polar Stereographic Projection**

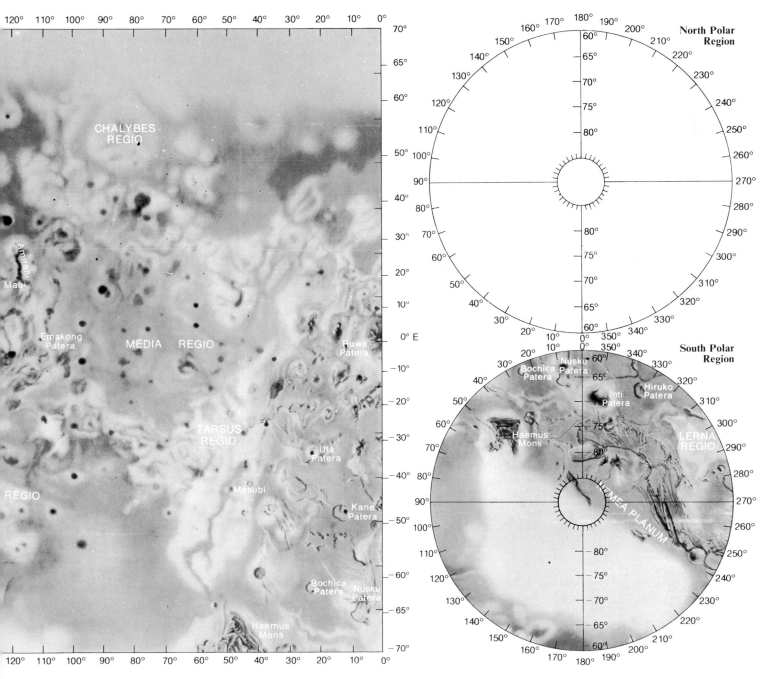

# Europa

With a radius of 1,525 km, Europa is the smallest of the Galilean
satellites, and the only one inferior in size and mass to the Moon.
The overall density is about $3.3 \times 10^3$ kg m$^{-3}$, appreciably less than
that of Io, but still great enough to indicate a relatively large silicate
core. Europa is the most reflective of the Galileans and has an albedo
of 0.64, but in the days before the Voyager mission there was no
reason to suppose that it was markedly different from Io. However, although Voyager 1 did not approach Europa closer than
732,230 km and although the resolution of the pictures of its surface
was much lower than with the other Galileans, they showed Europa
to be totally unlike Io. The general hue was whitish instead of red;
there were no volcanoes, active or dead, and there were practically no
craters. Even more strange was the apparent absence of vertical
relief, as revealed by pictures of the satellite's terminator. Europa is
remarkably smooth, to the extent that it has been compared with a
billiard-ball.

Voyager 2 made a much closer approach, 204,030 km at its
nearest point, and sent back high-quality pictures which revealed
darkish, mottled regions and lighter areas; but the main impression
was of a complex maze of bright and dark lines, criss-crossing each
other. There were linear features up to 40 km wide and thousands of
kilometers long, with narrow ridges having widths of up to 10 km
and lengths of at least 100 km.

New terms have been introduced to describe Europa's terrain:
"linea" (a dark or bright elongated marking, either straight or
curved), "flexus" (a very low curvilinear ridge with a scalloped
pattern), and "macula" (a dark spot, sometimes irregular in shape).
Yet no particular feature stands out prominently to compare with
those on the other Galileans. As far as Europa is concerned one
feature is remarkably like another. The three named "macula"
objects (Thera Macula, Thrace Macula and Tyre Macula) can be
identified without difficulty, but are not very striking.

## Craters

One of the really puzzling aspects is the paucity of craters. On Io
any craters would soon be obliterated by the constant surface
activity, but Europa is quiescent, and there is little doubt that the
visible surface is made up of ice. The three craters identified with
fair certainty are between 18 and 25 km in diameter, but they are
not alike in morphology, one is bowl-shaped, another shallow and
associated with what seems to be a system of dark rays, and the third
is raised up, as though the surrounding surface had subsided for
some reason or other.

One reasonable suggestion was that water had surged upward to
the surface, forming a layer of ice over the silicate interior. If the
layer were as much as 100 km thick, it would be more than adequate
to cover any surface relief; but it may be much thinner than this, so
that in places the silicate base is not far below the visible surface. It
is also possible that there is a relatively thin ice crust lying over
water, or softer ice which could be described as "slush". However,
the surface of Europa is at least several hundreds of millions of
years old and more probably several thousands of millions. The
lack of impact craters still has to be accounted for, assuming that
craters of the type found on Ganymede and Callisto really have
been produced by meteoritic bombardment.

One possible answer is that the crust remained comparatively
warm and slushy until the main bombardment era was over.
Radioactive heating could have been involved, and Europa must
also experience interior tidal forces which heat it in much the same
way as with Io, although since Europa is further from Jupiter the
effect is only about 10 percent as great. Assuming that the outer
crust is rigid, the fractures have presumably been filled in with
material rising from below. It seems that after fracture the pieces of
crust remained in their original positions, in contrast to Ganymede,
where fragments of crust appear to have shifted about relative to
each other.

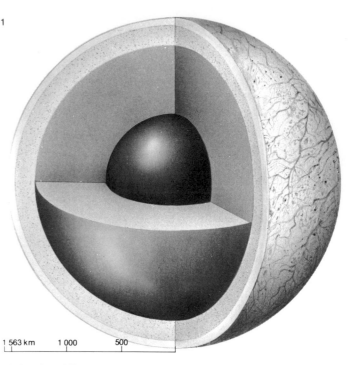

1 563 km    1 000    500

**1. Interior of Europa**
Europa is believed to have a
relatively large silicate core. Its
high albedo suggests that it has
an icy surface. Europa's density
of $3.1 \times 10^3$ kg m$^{-3}$ would result
from a mixture of silicates and
water-ice in the ratio of 9:1 by
weight. A 70 km thick crust of ice
may cover a region of slushy
water-ice, 100 km thick, beneath
which lies the 1,400 km silicate
core. Radioactive decay is likely
to continue to heat the interior,
which may reach temperatures of
around 2,800 K.

**2. Terminator**
Europa's terminator shows the
surface to be extremely smooth.
The absence of impact craters
suggests that the surface may have
been too soft to have retained
impressions made by meteoritic
bombardment. This image, made
by Voyager 2 on 9 July 1979,
reveals bright ridge-like features
rising to about 100 m above the
surface. They are typically 100 km
long and between 5 and 10 km in
width. Broader dark bands are
also apparent. These may be
several thousand kilometers long.

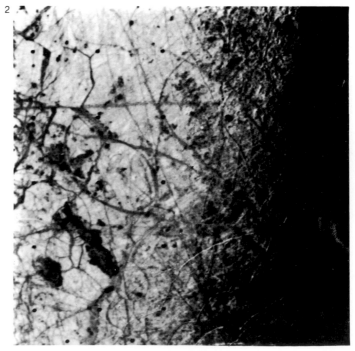

3

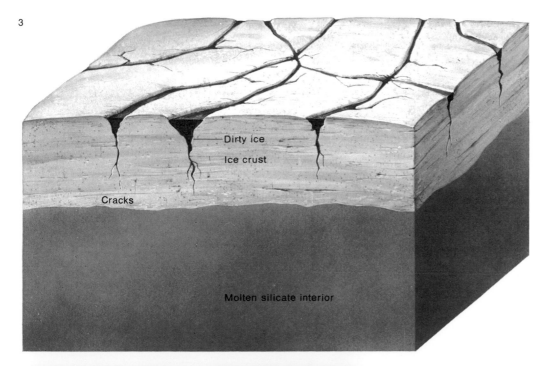

Dirty ice

Ice crust

Cracks

Molten silicate interior

**3. Surface structure**
The markings on Europa's surface suggest that fractures have occurred in the crust and that material from below has been forced upwards to fill the cracks. It is possible that at present there is a 100 km thick layer of water or water-ice separating the crust from the silicate interior, but this layer is not shown in the illustration as considerable uncertainty exists.

4

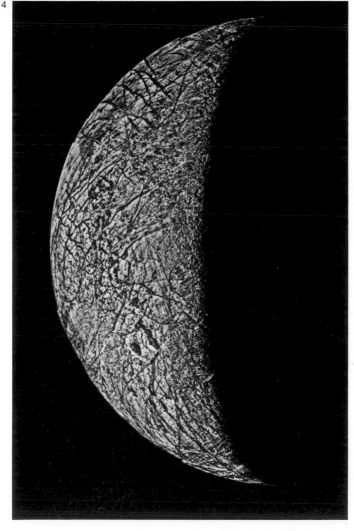

**4. Mosaic of Europa**
Two basic types of terrain are apparent in Europa's equatorial region, one bright, the other mottled and slightly darker. The darker terrain exhibits small depressions which may, in fact, be impact craters.

**5. Cadmus Linea and Belus Linea**
Two of the named "linea" features are shown in this image. Cadmus Linea is the uppermost dark band in the picture, while Belus Linea is the lower of the two nearly parallel bands cutting diagonally across the image from lower left to upper right. The two features intersect at 27°N, 172°W.

**6. Tyre Macula (34°N, 144°W)**
This image shows one of the dark, irregular-shaped features that have been named. (The others are Thera Macula and Thrace Macula.)

5

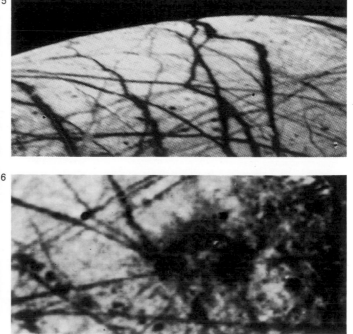

6

# Map of Europa

Adonis Linea
Agenor Linea
Argiope Linea
Asterius Linea
Belus Linea
Cadmus Linea
Libya Linea
Minos Linea
Pelorus Linea
Phineus Linea
Sarpedon Linea
Thasus Linea

Cilicia Flexus
Gortyna Flexus
Sidon Flexus

Thera Macula      45°S, 178°W
Thrace Macula     44°S, 169°W
Tyre Macula       34°N, 144°W

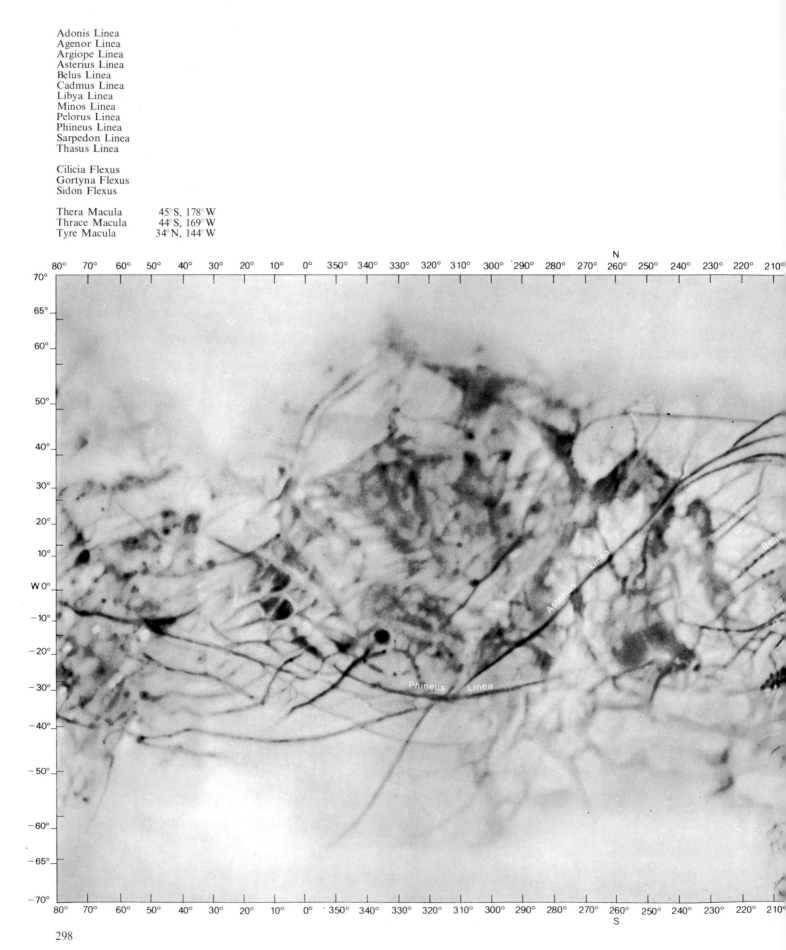

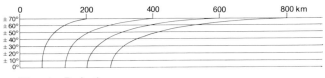

**Mercator Projection**

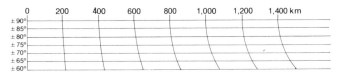

**Polar Stereographic Projection**

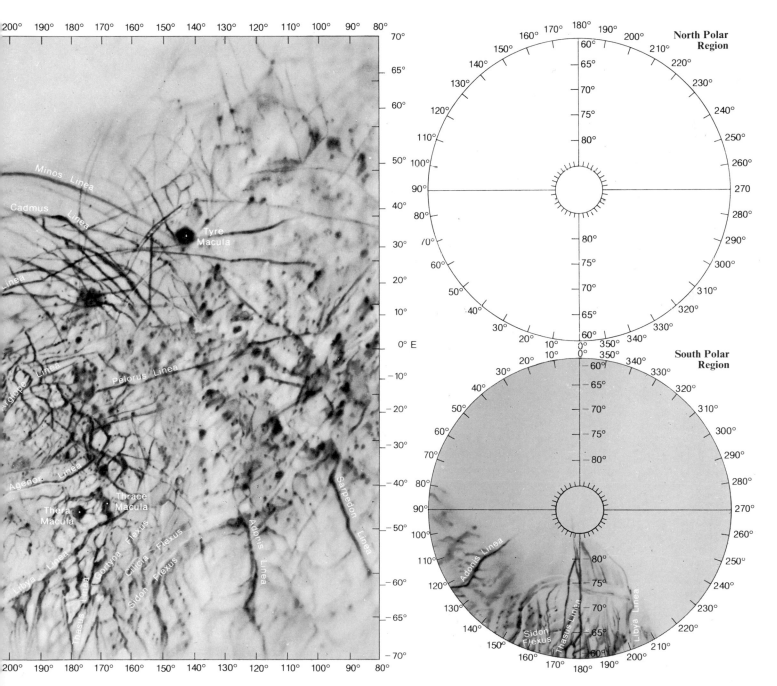

# Ganymede

Ganymede, the third of the principal satellites, is the largest and brightest member of the Jovian family. It has an albedo of 0.4, and a radius of 2,635 km, and is, in fact, one of the largest satellites in the Solar System, rivalled only by Titan in Saturn's system and Triton in Neptune's. Telescopically, Ganymede is always easy to identify, and observers using large telescopes have been able to make out at least one surface feature, the huge circular, darkish region shown in detail by the Voyagers and now named Galileo. The overall density of Ganymede is less than twice that of water, and from this it may be inferred that it consists of about half rock and half water (or ice). The crust is presumably composed of ice mixed with rocky materials, and is darkest in the most ancient regions. It is thought to cover a convective layer of water or soft ice, which in turn surrounds a small silicate core. It is very unlikely that major changes take place there now. There may be a certain amount of radioactive heating, and present-day tidal effects must also contribute to the heating of the satellite. Ganymede moves inside Jupiter's magnetosphere. No trace of atmosphere has yet been detected.

## Surface features
Both Voyagers obtained detailed views of Ganymede. The first probe passed at a distance of 112,030 km, and the second at only 59,530 km. Eighty percent of the total surface has been covered down to a resolution of 5 km or less.

The first impression is of a surface superficially similar to that of the Moon, with craters of various kinds (some of them with central peaks), ray systems, and brighter regions which are presumably younger than the dark terrain. If the craters are mainly of impact origin, it would lead to the assumption that Ganymede was subjected to very intense bombardment in the distant past. The surface features are certainly very old compared with those of Europa, and even more so compared with Io. Yet in some respects the surface is not truly "lunar". The relatively smooth terminator shows that the surface relief is much less marked than that of the Moon, and this is only to be expected if the materials there are largely icy.

Generally speaking there are two main types of geological unit. First there is the ancient, darkish, heavily cratered terrain. Such regions are often roughly polygonal, and may be up to several tens of kilometers across. Secondly there are brighter, presumably younger regions which are characterized by "bundles" of long, parallel grooves. The term "sulcus", meaning a groove or furrow, has been used to describe these features. However, close studies of the Voyager results show that these two types of terrain intermingle, so that the overall picture is one of great complexity. The large stripe-like features are characteristic of Ganymede, and it is these which divide the cratered terrain into isolated polygons up to a thousand kilometers across.

It is likely that Ganymede's surface was originally darkish and composed of soft ice, and that it cooled and froze rapidly during the first few hundred million years after Ganymede was formed. The grooved terrain developed gradually, and the grooves were filled with frozen water; faults are common. (Ganymede is incidentally the only world, apart from the Earth, known to exhibit lateral faulting.) Some of the craters are ghostly, reminiscent of Stadius on the Moon, but others look comparatively fresh, and there are ray systems which undoubtedly represent the youngest of all the features. Around these ray craters the ice is almost white, because it is so much "cleaner" than the rest of the surface.

It is probable that the surface is incapable of supporting much weight. There must be more water than on Europa, and the grooves have been interpreted as valleys between ice extrusions. It has been suggested that the past surface activity, great enough to create "tectonic blocks" (icy versions of the Earth's plates), was due to the slowing-down of Ganymede's axial rotation by the tidal forces that produced the present synchronous orbit.

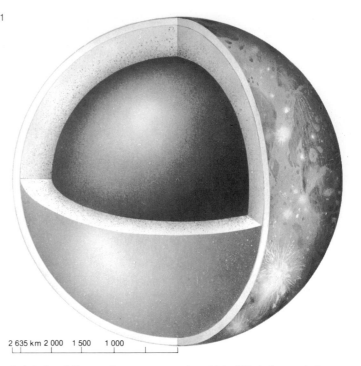

2 635 km  2 000   1 500   1 000

**1. Interior of Ganymede**
Ganymede is believed to have an ice crust less than 100 km thick, with a convecting mantle of water or soft ice between 400 and 800 km thick. Below this there is thought to be a relatively large silicate core of radius 1,800 to 2,200 km. The icy surface has become dirty with age, except where fresh white ice has been ejected as a result of meteoritic impacts. At some stage in its history, Ganymede's surface consisted of "tectonic" blocks of ice which shifted about relative to one another, rather like the Earth's continental plates.

**2. Terminator**
The surface of Ganymede is heavily cratered and scored with a large number of grooves. These are typically between 5 and 10 km in width, and the relative smoothness of the terminator indicates that the surface relief is fairly shallow. This mosaic shows the terminator near the south pole of the satellite.

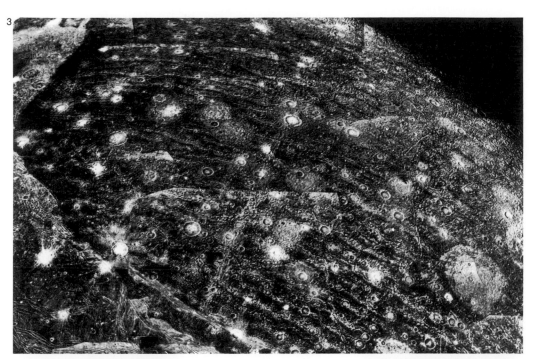

**3. Galileo Regio**
The most important of the large, dark types of terrain, Galileo Regio, covers about one-third of the hemisphere turned permanently away from Jupiter. It is heavily cratered, but its northern part is less dark than the rest of the interior, and may indicate some kind of condensate. Crossing Galileo Regio may be seen a series of parallel, gently curved bright streaks. These could be the result of a vast impact some distance away, but no trace of any impact center has been found, and the cause of the features may, in fact, be internal. Galileo Regio shows comparatively little vertical relief, because of the glacier-like "creep" in a crust composed largely of ice. The region is about 3,200 km in diameter, and probably represents the oldest surface on view on the satellite.

**4. Ray crater (18°S, 192°W)**
The ice surrounding ray craters is almost white; fresh material has been thrown up as a result of a meteoritic impact. This image shows one of the smaller unnamed ray craters; the crater at the lower left is called Eshmun.

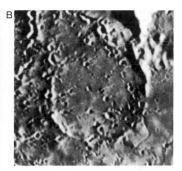

**5. Ghostly crater (20°N, 120°W)**
A large crater (A) has been almost obscured by fresher material, and a smaller, overlapping crater has been formed more recently. A photograph of the ghostly lunar crater Stadius is reproduced for comparison (B).

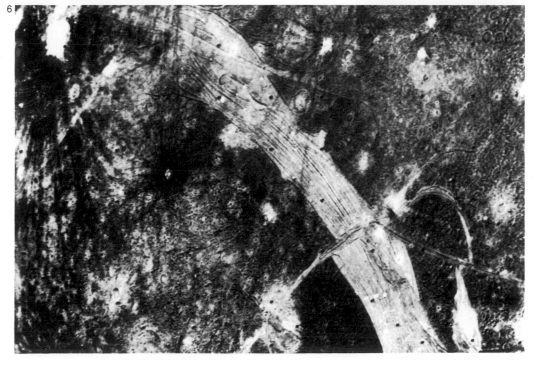

**6. Tiamat Sulcus (3°S, 210°W)**
A bright band of grooved terrain is shown here dividing areas of darker terrain in the Marius Regio. The bright band, Tiamat Sulcus, appears to be fractured by a fault extending from Kishar Sulcus. There is a discrepancy in the number of grooves on either side of the fault: fourteen on the northern side, as compared with twenty on the southern. The width of the grooves also differs. One explanation is that the grooves resulted from fractures that took place at different times on either side of the fault. The large number of craters suggests that the dark areas in the photograph are extremely ancient, while the brighter grooves, which exhibit fewer craters, are likely to have been formed more recently.

# Photomosaics of Ganymede

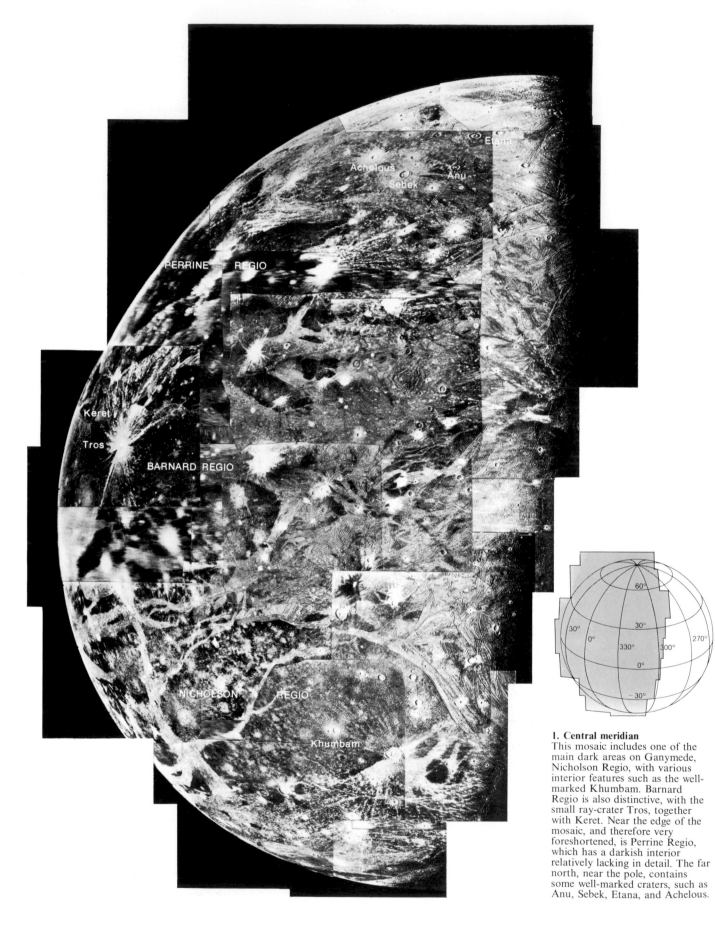

**1. Central meridian**
This mosaic includes one of the main dark areas on Ganymede, Nicholson Regio, with various interior features such as the well-marked Khumbam. Barnard Regio is also distinctive, with the small ray-crater Tros, together with Keret. Near the edge of the mosaic, and therefore very foreshortened, is Perrine Regio, which has a darkish interior relatively lacking in detail. The far north, near the pole, contains some well-marked craters, such as Anu, Sebek, Etana, and Achelous.

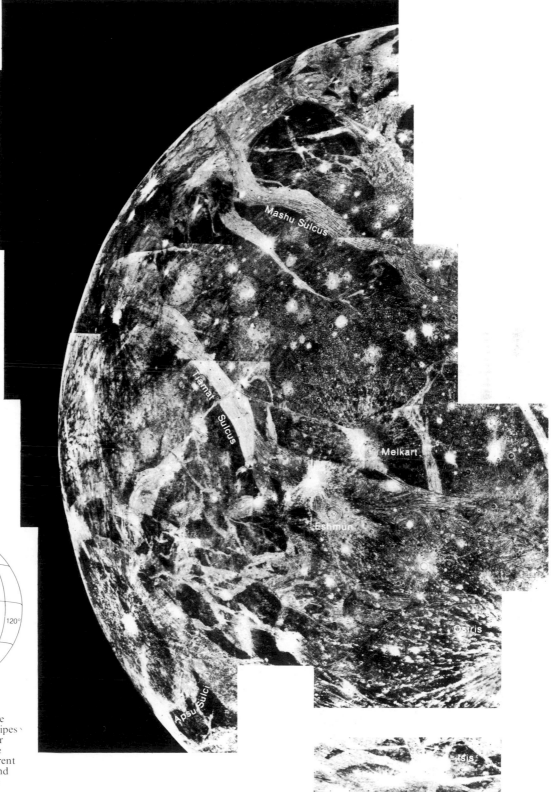

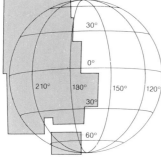

### 2. East region

The eastern part of Ganymede contains many of the light stripes such as Tiamat Sulcus, Kishar Sulcus and Mashu Sulcus; the overall aspect is entirely different from that of the Nicholson and Perrine areas. There are some well-marked craters here and there, notably Melkart and Eshmun, and to the lower right may be seen some of the bright streaks coming from the crater Osiris, which is just off the mosaic to the right. Apsu Sulci may be seen to the lower left.

# Map of Ganymede

**Craters**

| | |
|---|---|
| Achelous | 66°N, 4°W |
| Adad | 62°N, 352°W |
| Adapa | 83°N, 22°W |
| Ammura | 36°N, 337°W |
| Anu | 68°N, 332°W |
| Asshur | 56°N, 325°W |
| Aya | 67°N, 303°W |
| Ba'al | 29°N, 326°W |
| Danel | 4°N, 21°W |
| Diment | 29°N, 346°W |
| Enlil | 52°N, 301°W |
| Eshmun | 22°S, 187°W |
| Etana | 78°N, 310°W |
| Gilgamesh | 58°S, 124°W |
| Gula | 68°N, 1°W |
| Hathor | 70°S, 265°W |
| Isis | 64°S, 197°W |
| Keret | 22°N, 34°W |
| Khumbam | 15°S, 332°W |
| Kishar | 78°N, 330°W |

| | |
|---|---|
| Melkart | 13°S, 182°W |
| Mor | 35°N, 323°W |
| Nabu | 36°S, 2°W |
| Namtar | 49°S, 343°W |
| Nigitsu | 48°S, 308°W |
| Nut | 61°S, 268°W |
| Osiris | 39°S, 161°W |
| Ruti | 15°N, 304°W |
| Sapas | 59°N, 31°W |
| Sebek | 65°N, 348°W |
| Sin | 56°N, 349°W |
| Tanit | 59°N, 32°W |
| Teshub | 2°N, 16°W |
| Tros | 20°N, 28°W |
| Zaqar | 60°N, 31°W |

**Regiones**

| | |
|---|---|
| Barnard Regio | 22°N, 10°W |
| Galileo Regio | 35°N, 145°W |
| Marius Regio | 10°S, 200°W |
| Nicholson Regio | 20°S, 0°W |
| Perrine Regio | 40°N, 30°W |

**Sulci**

| | |
|---|---|
| Anshar Sulcus | 15°N, 200°W |
| Apsu Sulci | 40°S, 230°W |
| Aquarius Sulcus | 50°N, 10°W |
| Dardanus Sulcus | 20°S, 13°W |
| Harpagia Sulci | 0°, 317°W |
| Kishar Sulcus | 15°S, 220°W |
| Mashu Sulcus | 22°N, 200°W |
| Mysia Sulci | 10°N, 340°W |
| Nun Sulci | 50°N, 320°W |
| Philus Sulcus | 37°N, 215°W |
| Phrygia Sulcus | 20°N, 5°W |
| Sicyon Sulcus | 44°N, 3°W |
| Tiamat Sulcus | 3°S, 210°W |
| Uruk Sulcus | 0°, 157°W |

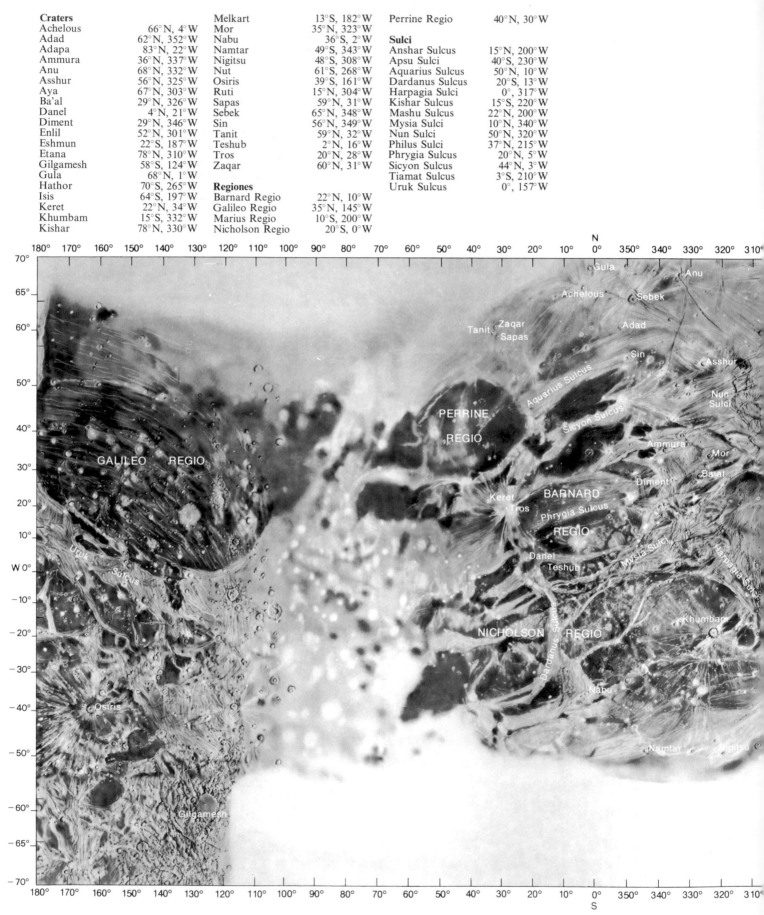

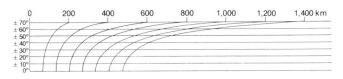

**Mercator Projection**

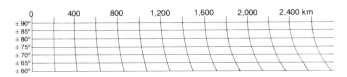

**Polar Stereographic Projection**

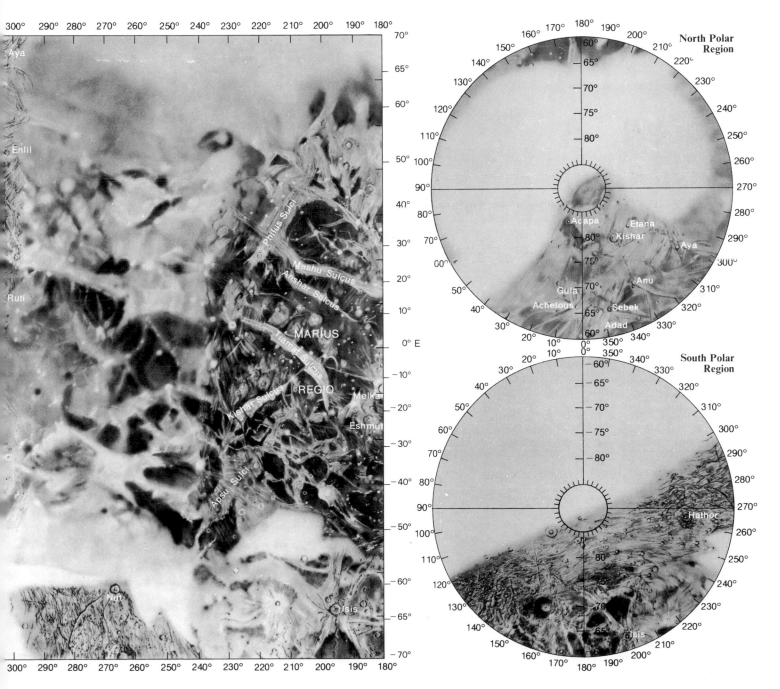

# Callisto

Callisto, the outermost of the Galileans, orbits Jupiter within the extensive magnetosphere, but beyond the main radiation belts. It therefore experiences relatively little effect from the energetic particles or thermal radiation from the planet. It has a radius of about 2,500 km, and at $1.6 \times 10^3$ kg m$^{-3}$ it is the least dense of the Galileans. Indeed, this value is among the lowest of any satellite that has been measured. It is also the least reflective, with an albedo of only 0.2. Callisto therefore appears fainter than its companions, even though in size it is not much inferior to Ganymede. However, since the size and density of Callisto are known to a lower degree of accuracy than those of the other Galileans, there is a correspondingly lower degree of certainty in theoretical models of the satellite's internal structure and geological history based on these figures. Its radius, for example, is known only to within $\pm$ 150 km, as compared with $\pm$ 10 km for Ganymede.

### Interior
The geology of Callisto is apparently much simpler than that of Ganymede. Following the crater-forming period, very little has happened. There is almost certainly a thick ice and rock crust, extending to a depth of up to 300 km, below which lies a convecting water or soft ice mantle which overlies a silicate core. It seems certain from the satellite's density that water accounts for a great deal of Callisto's total bulk. Tidal effects at the present time are much weaker than for Ganymede or the inner Galileans.

### Surface features
Voyager 1 passed Callisto at a distance of 123,950 km and Voyager 2 at 212,510 km; as with Ganymede, some 80 percent of the surface was examined down to a resolution of 5 km or less.

The surface has been likened to "dirty ice", and it is undoubtedly very ancient. It may date back for as much as 4 thousand million years, in which case it seems to be the oldest landscape so far studied in the Solar System. Its surface is saturated with craters; Callisto is the most heavily cratered body known. Some of the craters are associated with ray systems, but the overall aspect is different from that of the Moon, because of the comparative lack of vertical relief; the terminator appears practically smooth, consistent with the picture of an icy crust. It may well be that Callisto now resembles Ganymede in its early period. There is abundant evidence of past internal activity and crustal movement upon Ganymede, but virtually none on Callisto.

There are, however, various large, concentric rings, which are still prominent even though they have been reduced by the expected crustal flow. Much the most striking is Valhalla, which lies slightly north of the satellite's equator. It is a vast structure with a brighter circular region about 600 km in diameter; the outermost of the concentric rings is almost 3,000 km across. It was the first basin to be recognized in the Jovian system, and has been compared with the Mare Orientale on the Moon, the Caloris Basin on Mercury, and Hellas on Mars, though the resemblance may well be no more than superficial. It is distinguished from these, however, by the absence of really high ridges, ring peaks or anything in the nature of a major central depression, indicating that the event which produced Valhalla caused widespread melting and flow, with the forming of shock waves in the crust, after which refreezing took place quickly enough to prevent the concentric shock rings from being destroyed. It is also notable that no ejecta patterns are visible. In the inner region there are fewer craters than elsewhere on Callisto; presumably pre-existing craters were destroyed at the time when Valhalla was formed. In the outer part of the structure, crater frequency is about the same as on the rest of the surface, but there is evidence that some of these craters are older than Valhalla itself, in which case total devastation occurred only for a radius of about 300 km around the central point. There are various other large ringed structures, notably Asgard in the south, but none to rival Valhalla.

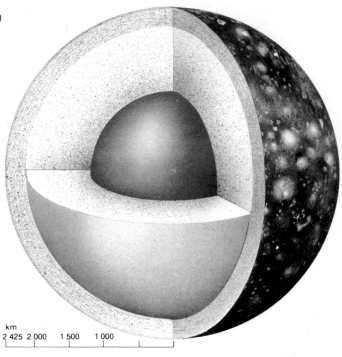

km
2 425  2 000     1 500     1 000

**1. Interior of Callisto**
Callisto is believed to have a thick crust of ice and rock extending to a depth of 200 to 300 km. Below the crust there is thought to be a 1,000 km thick mantle of convecting water or soft ice, similar to that of Ganymede; finally there is a silicate core 1,200 km in radius. Although Callisto is less reflective than the other Galileans, its surface is nevertheless thought to consist mainly of ice, its darker appearance resulting from a greater proportion of impurities.

**2. Terminator**
A view of the terminator near Asgard, one of the most prominent of Callisto's features, shows the surface of the satellite to be fairly smooth, at least in comparison to the Moon. However, a certain amount of relief is evident in this image as a result of Asgard's concentric rings.

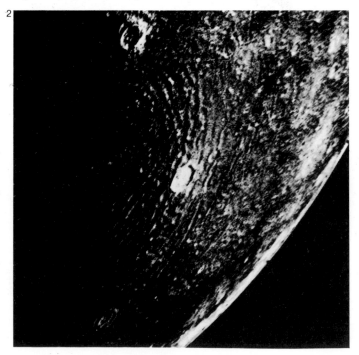

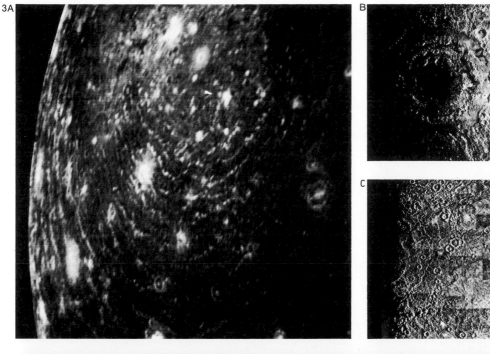

3A

B

C

**3. Valhalla** (10°N, 55°W)
The largest and most striking
feature on Callisto is Valhalla (**A**),
a large impact basin, similar to
Mare Orientale on the Moon (**B**)
and Caloris Basin on Mercury
(**C**). The outer ring of Valhalla is
believed to have been caused by
an immense impact, and it is
thought that the impacting body
may have come close to
penetrating Callisto's crust.

**4. Mosaic of Callisto**
The exaggerated contrast of this
image of Callisto shows up the
extremely dense cratering of the
satellite's surface. Crater
distribution is almost uniform,
but none of the craters are
remarkably high. The giant
feature Asgard is visible near the
limb towards the upper right-
hand corner of the image.

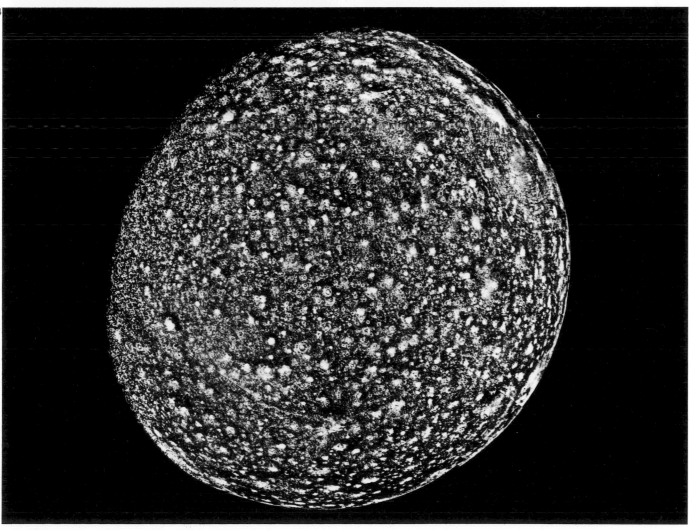

4

# Map of Callisto

| | | | | | | | |
|---|---|---|---|---|---|---|---|
| Adal | 77°N, 79°W | Dia | 73°N, 56°W | Habrok | 77°N, 129°W | Nerivik | 22°S, 55°W |
| Adlinda | 58°S, 20°W | Dryops | 77°N, 29°W | Haki | 26°N, 315°W | Nidi | 66°N, 93°W |
| Ägröi | 42°N, 12°W | Durinn | 66°N, 87°W | Har | 6°N, 357°W | Nori | 46°N, 347°W |
| Akycha | 74°N, 325°W | Egdir | 31°N, 35°W | Hepti | 64°N, 27°W | Nuada | 62°N, 269°W |
| Alfr | 9°S, 222°W | Erlik | 66°N, 358°W | Hodr | 69°N, 87°W | Oski | 56°N, 266°W |
| Ali | 57°N, 58°W | Fadir | 56°N, 15°W | Hoenir | 36°S, 261°W | Ottar | 60°N, 100°W |
| Anarr | 43°N, 3°W | Fili | 65°N, 349°W | Hogni | 14°S, 5°W | Pekko | 17°N, 6°W |
| Aningan | 51°N, 11°W | Finnr | 14°N, 14°W | Igaluk | 5°N, 315°W | Reginn | 42°N, 88°W |
| Asgard | 30°N, 140°W | Freki | 82°N, 10°W | Ivarr | 6°S, 322°W | Rigr | 69°N, 240°W |
| Askr | 53°N, 327°W | Frodi | 69°N, 136°W | Jumo | 62°N, 15°W | Sarakka | 8°S, 53°W |
| Balkr | 27°N, 12°W | Fulla | 74°N, 102°W | Kari | 47°N, 103°W | Seqinek | 55°N, 27°W |
| Bavorr | 48°N, 23°W | Fulnir | 58°N, 37°W | Karl | 56°N, 335°W | Sholmo | 52°N, 18°W |
| Beli | 61°N, 79°W | Geri | 66°N, 353°W | Lodurr | 52°S, 270°W | Sigyn | 33°N, 27°W |
| Bragi | 77°N, 69°W | Gipul Catena | 65°N, 55°W | Loni | 4°S, 215°W | Skoll | 57°N, 317°W |
| Brami | 26°N, 18°W | Gisl | 56°N, 35°W | Losy | 68°N, 329°W | Skuld | 6°N, 37°W |
| Bran | 25°S, 207°W | Gloi | 48°N, 246°W | Mera | 63°N, 73°W | Sudri | 53°N, 137°W |
| Buga | 22°N, 326°W | Goll | 58°N, 323°W | Mimir | 30°N, 54°W | Sumbur | 69°N, 332°W |
| Buri | 43°S, 44°W | Gondul | 59°N, 115°W | Mitsina | 57°N, 97°W | Tindr | 5°S, 355°W |
| Burr | 40°N, 136°W | Grimr | 43°N, 214°W | Modi | 67°N, 115°W | Tornarsuk | 25°N, 130°W |
| Dag | 56°N, 74°W | Gunnr | 64°N, 100°W | Nama | 57°N, 336°W | Tyn | 68°N, 229°W |
| Danr | 61°N, 75°W | Gymir | 61°N, 55°W | Nar | 4°S, 45°W | Valfodr | 3°S, 246°W |

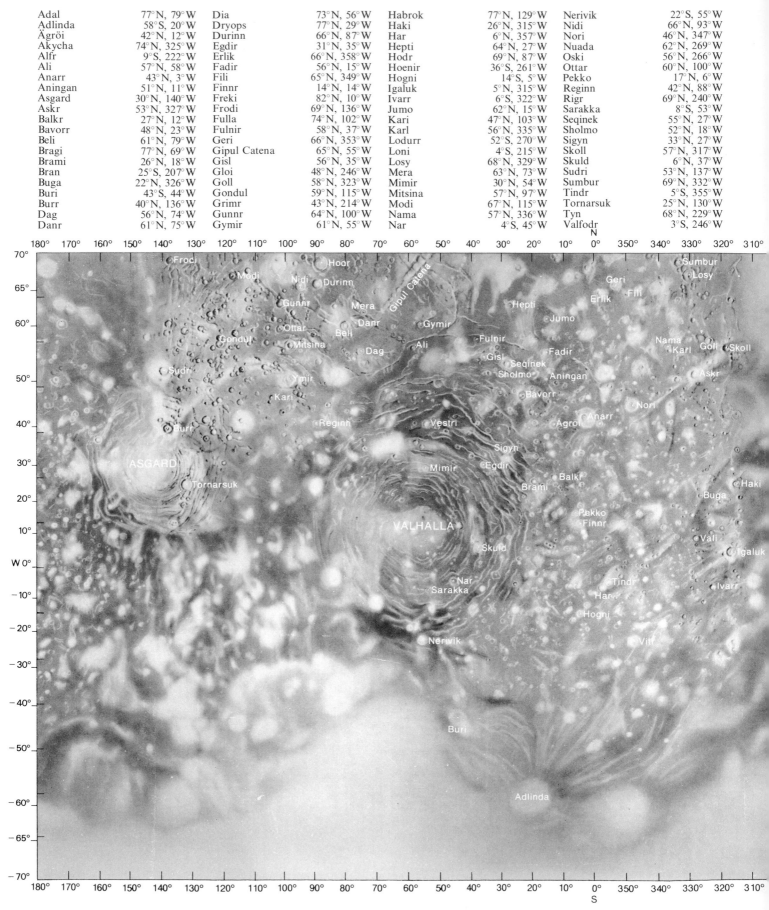

Valhalla          10°N, 55°W
Vali              9°N, 327°W
Vestri            42°N, 54°W
Vitr              23°S, 347°W
Ymir              51°N, 97°W

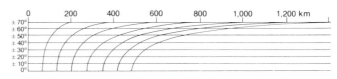

**Mercator Projection**

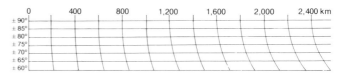

**Polar Stereographic Projection**

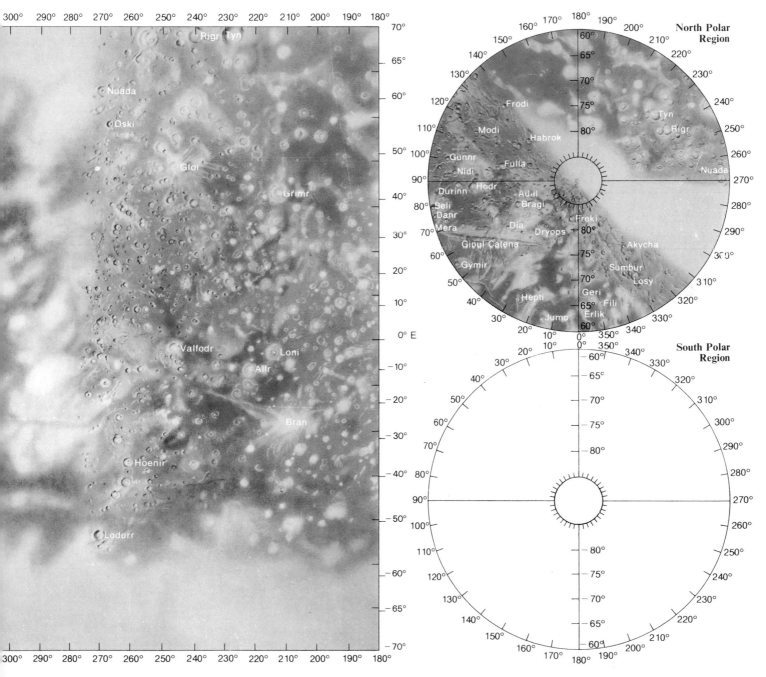

# Photomosaics of Callisto

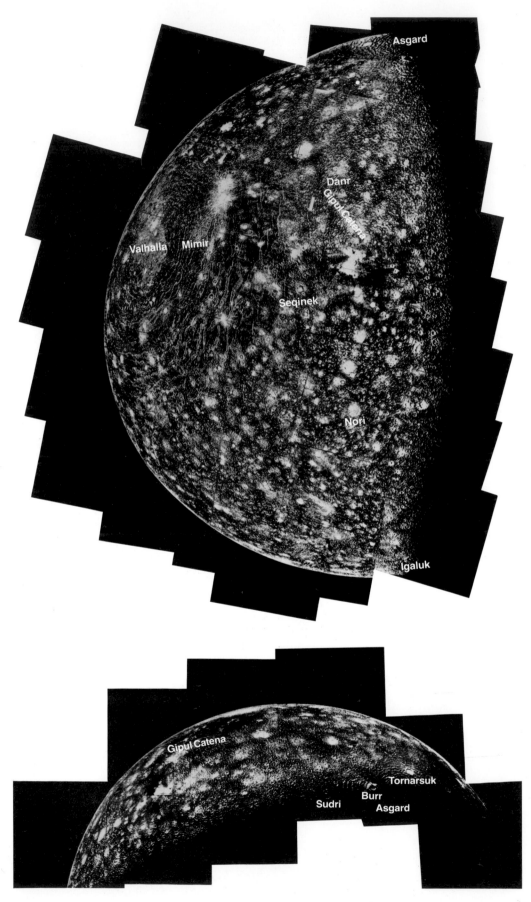

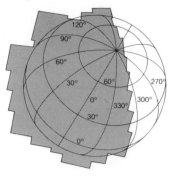

**1. North region**
The whole surface of Callisto is extremely ancient and heavily cratered, but there are two features of predominant importance: the ringed structures of Valhalla and Asgard. Valhalla is shown on this mosaic, to the left, and its complexity is striking. As far as Callisto is concerned, it is just as significant as Mare Orientale on the Moon or the Caloris Basin on Mercury. It is notable that there is no central structure, but there are some well-marked smaller craters such as Mimir, Skuld, Nar and Sarakka. A small part of Asgard is seen at the top of the mosaic; the north pole of Callisto is to the right, above center. There are many identifiable craters (for instance, Nori, Igaluk, Seqinek, Danr), and another feature easily recognized is Gipul Catena, a type of formation which is rather uncommon on Callisto.

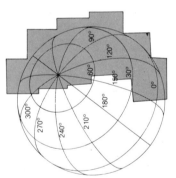

**2. North-west region**
The north pole of Callisto is just off the mosaic to the left, above center. Of special interest is Gipul Catena, very foreshortened on this mosaic but known to be made up of a whole chain of craters. The most important feature is the ringed basin Asgard; part of its wall is shown on the lower right. It is not nearly as large as Valhalla, but is of the same general type; on its borders are two prominent craters, Burr and Tornarsuk. The well-marked crater Sudri is also shown, above Burr and on the edge of the illuminated region. An unnamed crater with a bright nimbus is seen to the upper right of Burr and to the right of Sudri.

# Outer Satellites

Eight known satellites move around Jupiter beyond the orbit of Callisto. All are very faint, and only Himalia is brighter than magnitude 14. Estimates of their radii are very approximate, based on measurements of their observed brightness.

All these discoveries were photographic. For many years they remained unnamed and were referred to by number, based on the order of their discovery. However, unofficial names were used (Satellite VI, for instance, was called "Hestia"), and finally official names were ratified by the International Astronomical Union. Minor Jovian satellites with names ending with the letter "a" (Leda, Himalia, Lysithea and Elara) have direct orbital motion, while those ending in "e" (Ananke, Carme, Pasiphaë and Sinope) have retrograde motion.

## Himalia and Elara
Himalia was discovered by C. D. Perrine, using the Crossley reflector at the Lick Observatory in the United States, on a plate taken on 3 November 1904. At the time Perrine was not certain whether it could be a new satellite: it might equally well have been an asteroid, but several weeks' observation sufficed to show that it was a genuine attendant of Jupiter. Early in the following year he detected an even fainter satellite, Elara. The discovery plate was taken on 2 January, again with the Crossley telescope, although it was not studied until February.

## Pasiphaë and Sinope
The next discovery was made on 27 January 1908, when P. J. Melotte, at Greenwich, photographed a tiny speck of light which proved to be Satellite VIII, Pasiphaë. Meanwhile S. B. Nicholson, with the Crossley reflector, had been continuing the search, and on 21 July 1914 he discovered the ninth satellite, Sinope. Actually, the discovery was somewhat fortuitous. Nicholson had set out to photograph Pasiphaë, and had given the plate an exposure time of $2\frac{1}{2}$ hr; when he developed the plate, he found that he had recorded not only Pasiphaë but also the newcomer. Had the two satellites not been so close together, Sinope would undoubtedly have escaped detection, on this occasion at least.

## Lysithea, Carme, Ananke and Leda
There matters remained for almost a quarter of a century. During the First World War the Hooker reflector at Mount Wilson in California was completed, and subsequently had a long reign as the largest and most powerful telescope in the world. With it, Nicholson made three more discoveries: Lysithea on 6 July 1938, Carme on 30 July of the same year, and Ananke on 29 September 1951. For a time the identity of Ananke was disputed, and there were suggestions that it was identical with Lysithea; a further month was needed to establish that the two were separate. Finally, in 1974, Charles Kowal at Palomar detected Leda, which is only of magnitude 20, and therefore one of the faintest known objects in the Solar System.

## Orbits
Tracking these tiny bodies proved to be very much of a problem. Indeed, Pasiphaë was "lost" after its discovery in 1908, found again in 1922, lost once more until 1938, and again between 1941 and 1955. The main trouble is that the orbits alter from one revolution to another because all the satellites are so far from Jupiter that they are subject to very pronounced solar perturbations.

It is at once evident that these outer satellites fall into two main groups. Leda, Himalia, Lysithea and Elara move around Jupiter at distances of between 11 and 12 million km, while the distances of the rest range between 20 and 24 million km. Moreover, the members of the outermost group—Ananke, Carme, Pasiphaë and Sinope—have retrograde motion. Only two other retrograde satellites are known, Phoebe in Saturn's system and Triton in

Neptune's, and of these only Triton is large. (Since the planet Uranus has an axial inclination of 98°—more than a right angle—and its five satellites move virtually in the equatorial plane, their movements are technically retrograde; but they are not usually reckoned as such, because they move around Uranus in the same sense as that in which Uranus rotates on its axis.)

There is no firm evidence that the outer Jovian satellites are ex-asteroids, but the retrograde motions of the more distant group are rather difficult to account for in any other way. On the other hand, the size distribution of the objects does not lend support to the idea that they are captured asteroids. An alternative suggestion is that they resulted from the break-up of larger bodies, possibly following a collision between asteroids or satellites. About their physical characteristics nothing is known at the moment, and the Voyagers did not provide any information. Himalia is the only one with a diameter of over 150 km; Elara comes next in size, but the rest are true midgets. Leda may be less than 10 km in diameter, making it even smaller than Phobos and Deimos, the tiny satellites of Mars. Seen from Jupiter, it would have an apparent diameter of less than two seconds of arc.

Although far more is known now than would have seemed possible only a year or two ago, there is clearly a great deal more to learn about the satellite system of Jupiter. There have already been many surprises, and it has been claimed that the members of the Jovian family are as interesting as Jupiter itself.

**1. Leda**
The photograph on which Leda was discovered was taken at Hale Observatories on 10 September 1974. Leda is the white spot indicated by an arrow.

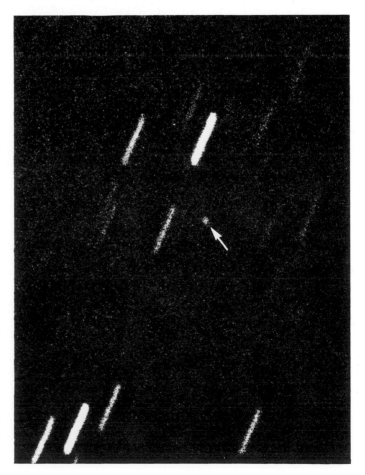

# SATURN

# Characteristics

Saturn, the most remote planet known in ancient times, is sixth in order of distance from the Sun. The ancients named the planet in honor of Jupiter's father, the first ruler of Olympus. Its mass is 95 times greater than that of the Earth—and is greater, in fact, than any other planet apart from Jupiter. It has more than 20 known satellites (*see* pages 354–86), of which one, Titan, is unique among satellites in the Solar System in having a dense atmosphere.

Saturn orbits the Sun at a mean distance of 1,427 million km. Its orbit, like all the planets, is appreciably eccentric: at perihelion it lies at a distance of 1,347 million km, and at aphelion it moves out as far as 1,507 million km. Saturn's orbital period is 10,759.2 Earth days (29.46 years). Although Saturn is so massive, it is much the least dense of all the planets. The outer layers are made up of gas, chiefly hydrogen and helium, the interior is mainly liquid hydrogen, and there is a relatively small, solid core.

As seen with the naked eye, Saturn is slightly yellowish in color. At its most brilliant, it outshines most stars: the maximum magnitude is −0.3, inferior only to Sirius and Canopus. This figure is bettered, however, by Venus, Jupiter and Mars at its brightest. The magnitude is affected by the angle of presentation of the ring system. The rings are more reflective than the planet and when they are edgewise-on to the Earth Saturn is never brighter at opposition than +0.8. Saturn's mean synodic period is 378 days.

## Nomenclature

Through the telescope, Saturn's disc appears yellowish and obviously flattened. As with Jupiter, the dark streaks crossing the planet are known as belts and the bright bands as zones. Since the surface is gaseous, there can be no stable features, but several belts are always to be seen, and change little in latitude. The more stable features have been given a standard nomenclature (opposite).

## Rotation

Like Jupiter, Saturn has a rapid axial rotation and it is differential. The rotational period at the equator (System I) is 10 hr 15 min, while at higher latitudes (System II) it is 10 hr 38 min. The rotational period of System III (based on radio emissions from Saturn) is 10 hr 39.4 min. Again as with Jupiter, because of the liquid nature of Saturn, various definite features have rotation periods of their own, so that they drift about in longitude. Well marked spots on Saturn observable from Earth are rare, however. Saturn's quick spin means it is appreciably flattened at the poles: the equatorial diameter is 120,660 km, the polar only 108,000 km.

Saturn's belts are less prominent than those of Jupiter, partly because Saturn is smaller and further away, and partly because they are genuinely less pronounced. Saturn has a blander appearance because of the greater amount of haze overlying the cloud tops. However, the two main belts, the North Equatorial and the South Equatorial, are always present, and this applies also to the less pronounced North and South Temperate Belts.

There is a strong equatorial current on Saturn, as with Jupiter, but the planets differ in one important respect; the boundaries of different currents closely follow the limits of the belts and zones with Jupiter, but not with Saturn.

## The rings

The glory of Saturn lies in its ring system (*see* pages 330–36), which makes the planet the most beautiful object in the sky. The main rings are, in order of decreasing distance, lettered A, B and C. A and B are separated by the Cassini Division and A is divided by the Enke Division. There are many narrower divisions in the main system, dividing it into thousands of ringlets.

In addition to the main rings, between C and Saturn is Ring D; C and D are separated by the French Division. Outside Ring A is Ring F and the two are separated by the Pioneer Division. Farther out still are the extremely elusive G and E rings.

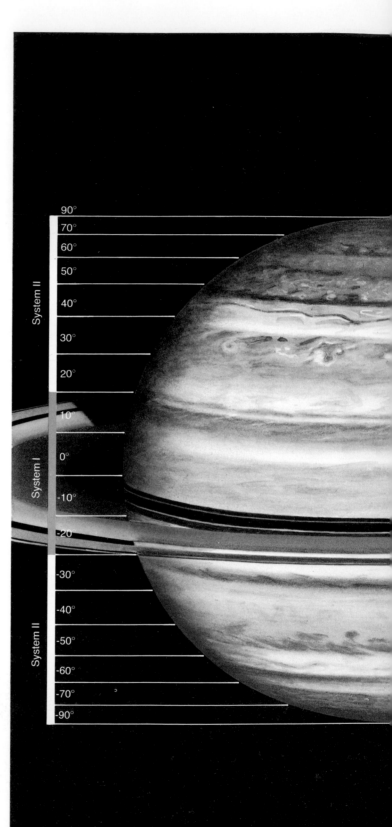

| | |
|---|---|
| 90° | |
| 70° | NPR |
| 60° | |
| 50° | NTZ |
| 40° | |
| | NTB |
| 30° | NTrZ |
| 20° | |
| | NEB |
| 10° | |
| 0° | EZ |
| -10° | |
| -20° | |
| | SEB |
| -30° | STrZ |
| -40° | |
| | STB |
| -50° | |
| -60° | STZ |
| -70° | |
| -90° | SPR |

**North Polar Region (NPR)** Lat. +90° to +55° approx.
The northernmost part of the disc. Its color is variable: sometimes it is comparatively bright; at other times it is comparatively dusky.
**North Temperate Zone (NTZ)** Lat. approx. +70° to +40°
Generally fairly bright; but from Earth few details can be seen.
**North Temperate Belt (NTB)** Lat. +40°
One of the more active belts on the disc, and usually easy to see telescopically except when covered by the rings.
**North Tropical Zone (NTrZ)** Lat. +40° to +20°
A generally fairly bright zone between the two dark belts.
**North Equatorial Belt (NEB)** Lat. +20°
A prominent belt, always easy to see and generally fairly dark. Activity within it can sometimes be observed from Earth.
**Equatorial Zone (EZ)** Lat. +20° to −20°
The brightest part of the planet. Details can be observed in it, and there are occasional white spots. The most prominent example of white spots in the twentieth century was in 1933.
**South Equatorial Belt (SEB)** Lat. −20°
A dark belt, usually about the same intensity as the corresponding belt in the northern hemisphere.
**South Tropical Zone (STrZ)** Lat. −20° to −40°
A generally bright zone. Little detail to be seen telescopically.
**South Temperate Belt (STB)** Lat. −40°
Generally visible when not covered by the rings.
**South Temperate Zone (STZ)** Lat. −40° to −70°
A brightish zone, with little or no visible detail as seen from Earth.
**South Polar Region (SPR)** Lat. approx. −70° to −90°
The southernmost part of the disc. Like the north polar region, somewhat variable in its depth of shading.

**Size comparison**
Saturn's diameter is less than that of Jupiter (142,800 km) but it dwarfs the Earth's (12,756 km).

Jupiter    Saturn    Earth

**Physical data**

| | Saturn |
|---|---|
| Equatorial diameter | 120,000 km |
| Ellipticity | 0.102 |
| Mass | $5.684 \times 10^{26}$ kg |
| Volume (Earth = 1) | 752 |
| Density (water = 1) | 0.70 |
| Surface gravity (Earth = 1) | 1.19 |
| Escape velocity | 35.6 km s$^{-1}$ |
| Equatorial rotation | 10 hr 15 min |
| Axial inclination | 26.73° |
| Albedo | 0.33 |

Cassini Division
Encke Division

F Ring
A Ring
B Ring
C Ring
D Ring

# Observational Background

Saturn must have been known since very early times because, at its brightest, it outshines all stars except Sirius and Canopus. The first recorded observations of Saturn seem to have been made in Mesopotamia in the mid-7th century BC. About 650 BC there is a record that Saturn "entered the Moon", which is presumably a reference to an occultation of the planet. But it was not until July 1610 that Galileo turned his telescope towards Saturn, and saw the planet's disc for the first time.

### The rings: nature and structure

Galileo was puzzled by Saturn. The telescope he used had a magnification of only 32, but it was still powerful enough to show that there was something unusual about the planet's shape. In 1610 the rings were placed at a narrow angle to the Earth, and Galileo could not see them in their true guise; instead he wrote that "the planet Saturn is not one alone, but is composed of three, which almost touch one another and never move nor change with respect to one another. They are arranged in a line parallel to the Zodiac, and the middle one is about three times the size of the lateral ones."

Two years later a fresh surprise awaited him: the attendant globes had disappeared. What had happened, of course, was that the rings had become edgewise-on to the Earth, and Galileo had no hope of seeing them. He then wrote: "Are the two lesser stars consumed after the manner of solar spots? Has Saturn, perhaps, devoured his own children? . . . The unexpected nature of the event, the weakness of my understanding, and the fear of being mistaken, have greatly confounded me."

He never solved the mystery, though when he looked at Saturn again, in 1616, the rings were wider open, and had he been seeing them for the first time he might well have realized the truth. A drawing made by the French astronomer P. Gassendi, in November 1636, also shows what seems to be a ring, but Gassendi too failed to interpret it correctly, and the true explanation was not given until 1659, in Christiaan Huygens' celebrated *Systema Saturnium*. Huygens had begun his telescopic observations in 1655, with a telescope that would bear a magnification of 50, and in his book he gave the solution of an anagram which he had published earlier to ensure priority. The anagram read, in translation: "[The planet] is surrounded by a thin flat ring, nowhere touching [the body of the planet] and inclined to the ecliptic."

Earlier explanations of the rings seem strange today. For example, the French mathematician Gilles de Roberval believed Saturn to be surrounded by a torrid zone giving off vapors, transparent and in small quantity but reflecting sunlight at the edges if of medium density, and producing an elongated appearance if very thick. Another French mathematician, Honoré Fabri, believed that the appearances could be explained by the presence of two large, dark, unreflecting satellites close to the planet and two large bright ones further away. Another theory was proposed in 1658 by Sir Christopher Wren, who had been a professor of astronomy at Oxford before turning to architecture. Wren believed that Saturn had an elliptical corona, meeting the globe in two places and rotating with Saturn once in each sidereal period. Wren never published his theory, because before he was ready to do so he heard of Huygens' solution, and accepted it at once. Others, including Fabri, were less wise. However, further observations showed that Huygens' theory was incontestable, and by 1665 even his most vehement opponents had accepted it.

### Divisions in the rings

In 1675 the Italian astronomer G. D. Cassini, who had become the first Director of the Paris Observatory, found a dark line in Saturn's ring which subsequently proved to be in the nature of a gap, and is named in his honor. Up till then it had been tacitly assumed that the rings must be either solid or liquid, but the discovery of a division cast obvious doubt upon this theory, though the first definite

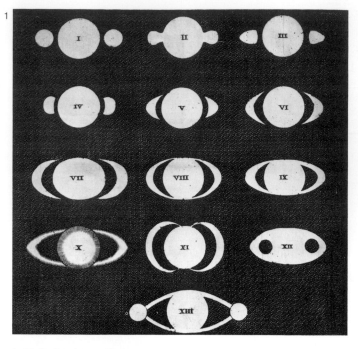

**1. Early drawings of Saturn**
Taken from Huygens' *Systema Saturnium* they are: I Galileo (1610); II Scheiner (1614); III Riccioli (1614 or 1643); IV–VII Hevel (theoretical forms); VIII and IX Riccioli (1648–50); X Divini (1646–8); XI Fontana (1636); XII Biancani (1616); XIII Fontana (1644–5). Some of these drawings, notably IX, had a very ring-like appearance several years before Huygens' theory was widely accepted.

**2. Huygens' ring cycle**
In *Systema Saturnium*, Huygens explained the inclination of Saturn's rings according to the planet's orbital position with respect to the Earth. As a result of Saturn's axial inclination, the rings are at maximum opening only at A and C, and then, Huygens stated, Saturn is at its most brilliant. The tilt of the planet also results in a pronounced seasonal variation, as is experienced on Earth.

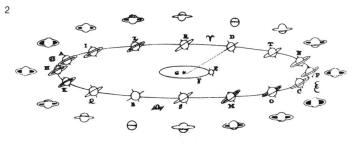

**3. The Cassini Division**
This sketch of Saturn, made in 1676 by J. D. Cassini, shows the division in the ring. This was the first known drawing in which the Division is unmistakably shown, and proves that credit for its discovery must go to Cassini.

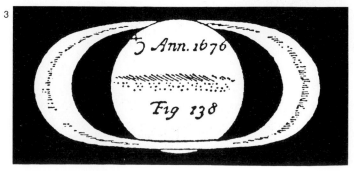

316

challenge to it was delayed until 1705, and was due to J. J. Cassini, who had succeeded his father at the Observatory in Paris.

A division in Ring A was announced by J. F. Encke, Director of the Berlin Observatory, in 1837. It was much less prominent than the Cassini Division, and for some time its nature was regarded as dubious, though there could be no doubt that a feature of some sort existed there. Minor divisions in both bright rings were also reported by various observers, but confirmation was lacking, and it is now known that the many minor divisions which actually exist cannot possibly be seen from Earth.

### The Crêpe Ring
The next major advance came in 1850, with the discovery of Ring C. It was due to W. C. and G. P. Bond, at Harvard, although it seems that the first true interpretation of their observations was made by C. W. Tuttle, an assistant at the Harvard Observatory. Quite independently the ring was discovered by W. R. Dawes in England, and was at once confirmed by W. Lassell. In 1852 Lassell found that the ring was more or less transparent. It is, in fact, not a difficult object and it was suggested that the ring was not observed earlier because it was fainter. This, however, is improbable.

The Bonds regarded the rings as fluid, and this was also the opinion of B. Peirce, who published his theories in 1855. But this posed problems, and in 1855 the University of Cambridge announced that the subject of their Adams Prize Essay would be "to determine the extent to which the stability and appearance of Saturn's rings would be consistent with alternative opinions about their nature—whether they are rigid or fluid, or in part aeriform". The Prize was won in 1857 by James Clerk Maxwell, who showed that neither a solid, a liquid nor a gaseous ring could persist. He concluded, therefore, that the rings were nothing more nor less than swarms of small particles, so close together that at the distance of Saturn they gave the impression of a solid sheet.

Final proof came in 1895 by J. E. Keeler, who showed spectroscopically that the inner sections of the ring system revolve around the planet faster than the outer parts, in agreement with Kepler's Laws. By means of the Doppler effect, Keeler was even able to determine the velocities of the ring particles. His investigation was quickly followed up by W. W. Campbell, who gave the following velocities: inner edge of Ring B, 18.94 km/sec; middle of the bright ring, 17.37 km/sec; outer edge of Ring A, 15.8 km/sec.

### The Kirkwood Gaps
The next important investigation was due to the American astronomer Daniel Kirkwood, in 1866, when he found that there are well-marked gaps in the zone of minor planets or asteroids, which move around the Sun between the orbits of Mars and Jupiter. There are certain "zones of avoidance" in which the revolution periods would be simple fractions of the period of Jupiter (11.75 years); cumulative perturbations would, therefore, drive any asteroids out of this forbidden region. In the following year Kirkwood applied the same theory to Saturn's rings, and concluded that the Cassini Division was due to the cumulative perturbations upon ring particles produced by the inner satellites Mimas, Enceladus, Tethys and Dione. In the Cassini Division, a ring particle would have a period half that of Mimas, one third that of Enceladus, one quarter that of Tethys and one sixth that of Dione, so that it would soon be moved out of the Division.

The space-probe results have since shown, however, that although the effect may play an important role in the production of the Cassini Division, it cannot possibly be the complete answer in view of the many minor divisions now known and, even more significantly, the fact that there are several well-defined ringlets inside the Cassini and Encke divisions.

An occultation of a star (or Saturnian satellite) by the ring system gives valuable information concerning the densities of the various rings. Important observations were made in 1917 by M. A. Ainslie and J. Knight when the star BD +21°1714 was occulted. These confirmed that Ring C is transparent and Ring A partly so.

**6. Changing aspects of the rings** More photographs taken at the Lowell Observatory, between 1921 and 1965, have been chosen to show to full advantage the changing aspect of Saturn's rings—edge-wise on in 1936 and fully open in 1945—and with each ring surface illuminated and tilted towards the Earth.

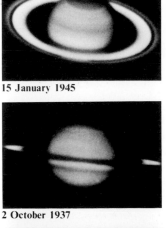

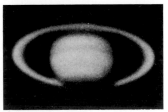

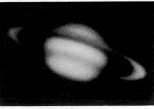

15 January 1945    22 March 1948

2 October 1937    5 August 1965

19 September 1934    15 September 1929

**4. Jeffreys' model of Saturn, 1923** Jeffreys dismissed the belief that the giant planets were very hot and gaseous, and said they were cold and solid, but composed of low-density materials—hydrogen, nitrogen, oxygen and helium.

**5. Wildt's model of Saturn, 1938** Wildt assumed a metallic and rocky core, with a density of 6 gm/cm³. The surrounding layer of frozen water and carbon dioxide (1.5 gm/cm³) was topped by solid hydrogen (0.3 gm/cm³).

# Interior

## Planetary magnetism

There was no definite proof of a magnetic field associated with Saturn until Pioneer 11. It was reasonable to assume that a field should exist, since Jupiter was known to have a powerful magnetic field and in many ways Jupiter and Saturn are of a similar type. With Saturn, however, there is the added complication of the ring system, and it was thought that the rings might prevent energetic electrons from being trapped within a magnetic field.

The Pioneer and Voyager probes have, however, shown that there is indeed a strong magnetic field associated with Saturn. Generally speaking, the mechanism generating a magnetic field may be compared with a dynamo (*see* diagram 1). Some ordinary dynamos contain a permanent magnet, which provides the magnetic field, but it is also possible to construct a "self-exciting" dynamo, which generates its own field, and this probably illustrates the principle underlying the planetary magnetic fields studied so far. The two essential features are the presence of a good electrical conductor and a source of mechanical energy to drive the dynamo.

In the case of a planet, the electrical conductor is believed to be a conducting fluid at or near the planet's core. The force driving the system is still not known, and may involve a combination of several factors, but it certainly involves planetary rotation.

## Interior

It is clear that the origin of Saturn's magnetic field must be closely related to the structure of the interior of the planet. The presence of a magnetic field, as in the case of Jupiter, may indeed be interpreted as evidence that parts of the interior are liquid, though it must be

admitted that there is no direct evidence in either case. Theory indicates that the electrically conducting core of Saturn may be $0.46R_s$ ($R_s$ standing for Saturn's radius), as against $0.7R_J$ (Jupiter's radius) for Jupiter.

Like Jupiter, Saturn consists primarily of the light elements hydrogen and helium, so that these giant worlds are quite unlike the terrestrial planets. The density of Saturn is only 0.7 times that of water: the terrestrial planets—Mercury, Venus, Mars and Earth—have mean densities in the region of five times the density of water. Both Saturn and Jupiter have gravitational fields that are powerful enough to retain the lightest elements, and models of their interiors necessarily depend heavily upon the assumed proportion of hydrogen to helium.

In this respect there is a marked difference between Jupiter and Saturn, since the Voyager observations show that there is much less helium, relatively speaking, in Saturn. Helium accounts for only about 11 percent of the mass of the atmosphere above the clouds of Saturn, as against 19 percent in the case of Jupiter. This indicates that the internal structures of the two planets are likely to be different from each other.

## Internal structure

It is worth comparing the internal structures of Jupiter and Saturn according to the latest theoretical models. The transition from gaseous to liquid hydrogen in Jupiter takes place at a depth of about 1,000 km from the visible surface; the temperature at this level is about 2,000 K, and the pressure is 5,600 atmospheres (that is, 5,600 times the atmospheric pressure at sea level on the Earth's surface). At a depth of 3,000 km the temperature is 5,500 K and the pressure 90,000 atm, so that the hydrogen is highly compressed. At 25,000 km below the cloud tops—one third the radius of Jupiter—the temperature has risen to more than 11,000 K and the pressure to 3,000,000 atm. Under these conditions the hydrogen undergoes a dramatic change; it alters from its "liquid molecular" form to a state in which it is a good electrical conductor, and is then termed "liquid metallic hydrogen". Deeper still the temperatures and pressures rise, reaching about 30,000 K and 100 million atmospheres at the center. The core of Jupiter is rocky, with about 10 to 20 times the mass of the Earth; it is composed mainly of iron and silicate materials.

Saturn has a similar structure. A pressure of 1 atmosphere occurs just below the visible cloud tops, where the temperature is 140 K. The transition from molecular to metallic hydrogen occurs 32,000 km below the clouds, where the temperature is 9,000 K and there is a pressure of 3,000,000 atm. The outer boundary of the rocky, icy core has a temperature of 12,000 K, and the pressure is 8,000,000 atm. The core has a radius that is 16 percent of Saturn's total radius, so it is about the same size as the Earth, but three times as massive. The differences in the internal structures of Jupiter and Saturn may account for the different weather systems in the atmospheres of the two planets. Moreover, the presence of large quantities of liquid hydrogen is critical with regard to the existence, or otherwise, of a magnetic field.

## Magnetic field

The existence of a Saturnian magnetic field was not confirmed until 1979. Saturn is so much more distant than Jupiter and it is not nearly as strong a radio source. The first definite proof came from Pioneer 11, which detected the presence of a magnetic field when it was still 1,440,000 km away from Saturn. There were further interesting results as the Pioneer passed beneath the outer edge of Ring A; the flux of charged particles was abruptly cut off, so that evidently there were well-marked interactions between the rings and the magnetic field.

Further results from the Voyagers have confirmed that the strength of Saturn's magnetic field is 1,000 times greater than the

### 1. The principle of the dynamo

A simple model of a disc dynamo consists of a metal disc rotating in a magnetic field between two permanent magnets (**A**). The field produces a force on the free electrons in the disc, pushing them towards the center. As a result there is a difference in electrical potential between the edge and the center of the disc, which produces a current if the circuit is closed. In a self-exciting

dynamo (**B**) this current is used to drive an electromagnetic coil, which replaces the original permanent magnets. The resulting system generates a magnetic field as long as the disc is kept spinning. The model thus demonstrates how mechanical energy may be converted into magnetic energy: an analogous process is thought to be responsible for creating planetary magnetic fields.

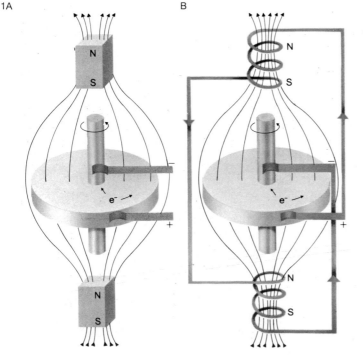

1A

B

Earth's, but about 20 times weaker than Jupiter's. At the cloud tops over Saturn's equator the strength is 0.22 Gauss: on Earth it is 0.3 Gauss at the surface. The magnetic axis is within one degree of the axis of rotation, in sharp contrast to those of the Earth, Jupiter and Mercury, which are all tilted at about 10° with respect to the axis of rotation.

Saturn emits a radio pulse with a period of 10 hr 39.4 min, at frequencies of a few hundred hertz. This period may be regarded as the true rotation period of the planet, and is referred to as System III. The bursts of emission occur at local noon. Such a pulse would be easy enough to explain if Saturn's field were as complex and powerful as that of Jupiter, but it is not. The space-probes have provided few clues about the mechanism or mechanisms that are responsible for the pulse.

It is known, however, that the direction of polarization of the pulse is different in the two hemispheres of Saturn, and it therefore appears that the source is confined to narrow regions of latitude in the polar regions. It is worth noting that these polar latitudes—about 80°N and S—are also the regions where aurorae have been detected. Radio waves were detected from Jupiter in the 1950s, but Saturn is a much weaker source. It is smaller and has lower temperatures than Jupiter, and lacks a violently active satellite such as Io—the innermost Galilean satellite of Jupiter—which interacts extensively with its parent planet.

2

4

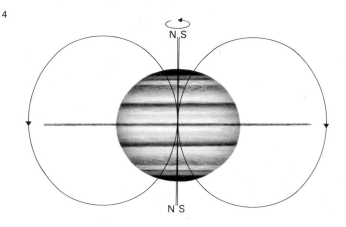

**2. Jupiter's magnetic field**
The magnetic field of Jupiter is inclined to the planet's axis of rotation by an angle of about 11°. Although the field is more complex than those of other planets, it still may be thought of as behaving as if a bar-magnet were embedded at the planet's core. Jupiter's magnetic field is 20 times greater than that of Saturn, which is 1,000 times greater than the magnetic field of the Earth.

**3. Jupiter's interior**
Jupiter is composed almost entirely of gases and liquids, although at present it is believed that there is a small core of rocky materials. The innermost part of the planet around the core is thought to consist of liquid metallic hydrogen, surrounded by a shell of liquid molecular hydrogen. The upper region consists of a deep atmospheric layer of hydrogen and helium.

**4. Saturn's magnetic field**
Saturn's magnetic field is stronger at the north pole (0.69 Gauss) than at the south pole (0.53 Gauss), the center of the field being displaced northward some 2,400 km along the axis of the planet. Saturn's field is unique in that its magnetic axis corresponds almost exactly to the axis of rotation: the angle between the two axes is less than one degree, much less than with other planets.

**5. Saturn's interior**
According to the theory put forward by W. B. Hubbard and his colleagues, Saturn has a rocky core the same size as Earth but three times as massive. This is overlain by a metallic hydrogen zone, which is smaller than Jupiter's because of Saturn's lower mass, gravitational field strength, and internal pressure. Above is molecular hydrogen, then a deep atmosphere.

3

5

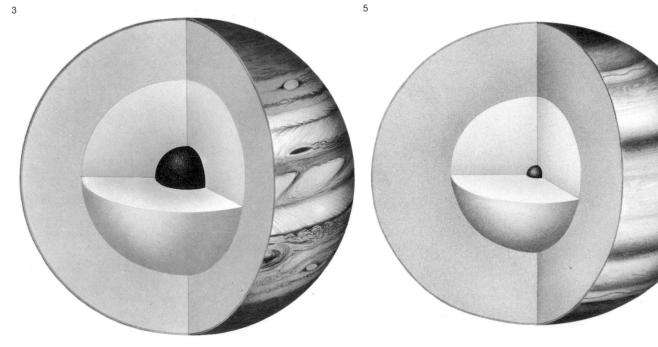

# Magnetosphere

The magnetosphere of Saturn appears to be intermediate between those of the Earth and Jupiter, in terms of both extent and population of trapped energetic particles. The Earth's magnetosphere is of considerable extent, and Jupiter's even more so; indeed, the "magnetic tail" of Jupiter extends beyond the orbit of Saturn, so that at certain times Saturn lies within it. The presence of a magnetic field means that there must also be a magnetosphere, but that of Saturn was not definitely known before the flyby of Pioneer 11. It surrounds the planet like a giant bubble, extending away from the Sun like a wind-sock (see diagram 1).

Beyond the magnetosphere is the region dominated by the solar wind. This is made up of low-energy particles ejected from the solar atmosphere at speeds of up to 400 km/sec and in all directions; it is "gusty" in as much as its strength varies considerably, and this in turn affects planetary magnetospheres.

When the solar wind reaches the boundary of a planetary magnetosphere, it abruptly changes direction to avoid electromagnetic collision. The region in which this takes place is termed the bow shock; with Saturn, the average distance between the bow shock and the planet is 1,800,000 km. The magnetosphere itself lies considerably closer to Saturn, at a mean distance of about 500,000 km, but this distance is not constant, and changes according to the characteristics of the prevailing solar wind. Saturn's largest satellite. Titan—moving around the planet at a mean distance of 1,221,400 km, or $20.3R_s$—lies very close to the edge of the magnetosphere (see diagram 1), so that it is sometimes inside and sometimes outside. It was outside when Voyager 1 made its pass: the spacecraft made five crossings of the bow shock in all, at distances ranging between $26.1R_s$ and only $22.7R_s$. At the time of the Voyager 2 encounter, the magnetosphere was slightly more extensive, and Titan lay inside it, giving an opportunity for measurements of the interactions between Titan itself and the solar wind. Such interactions are the source of a huge, doughnut-shaped torus of hydrogen extending from the orbit of Titan inward as far as the orbit of the second-largest satellite, Rhea. Altogether Titan spends about 20 percent of the time inside Saturn's magnetosphere. The satellite is the source of both neutral and ionized molecules as a result of photodissociation in the upper atmosphere.

The magnetosphere is divided into several definite regions. Within a distance of about 400,000 km of Saturn there is a torus made up of ionized hydrogen and oxygen atoms. The plasma's ions and electrons spiral up and down magnetic field lines and contribute to the local field. At its outer edge some of these ions have been accelerated to high velocities. These indicate temperatures of from 400 million to 500 million degrees K.

Beyond the inner torus is a region of plasma that extends out to about 1,000,000 km, produced by material coming partly from Saturn's outer atmosphere and partly from Titan's. Titan is not

**1. Saturn's magnetosphere**
The region of space in which Saturn's magnetic field is dominant is less extensive than in the case of Jupiter. The region where the solar wind meets the magnetosphere is the bow shock, and this varies with the strength of the solar wind. The region immediately inside the bow shock is the magnetopause, and the whole magnetically active region is enveloped in a magnetosheath. Titan can lie inside or outside the magnetosphere (inset), thus affecting the satellite's interaction with the solar wind and the magnetosphere.

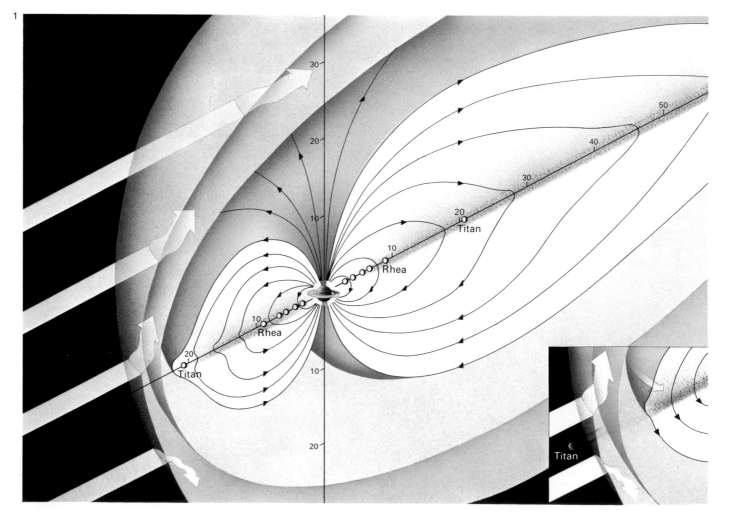

alone in having marked effects upon the magnetosphere; the other large inner satellites—Rhea, Dione, Tethys, Enceladus and Mimas—also play an important role (like their counterparts around Jupiter), although they are devoid of individual atmospheres. For example, protons in the magnetosphere are strongly absorbed by some of the inner satellites as well as the tenuous Ring E. The magnetic tail of Saturn has a diameter of $80R_s$ and is relatively devoid of plasma at high altitudes.

Of great importance are the interactions between the magnetosphere and Saturn's rings. This is, in fact, unique to Saturn since the much less spectacular rings of Jupiter and Uranus cannot have comparable effects. For example, the numbers of electrons fall off sharply at the outer edge of the exterior bright ring, Ring A, because the electrons are absorbed by the ring particles. It is also significant to note that the general magnetic field of Saturn excludes cosmic rays from the inner magnetosphere. (Cosmic rays are atomic nuclei coming from all directions in space, but they are electrically charged and are influenced by magnetic fields.) Accordingly, the region between Ring A and Saturn is almost completely shielded, and is the most radiation-free region in the entire Solar System, excluding the atmospheres and solid globes of planets, asteroids, large satellites and the Sun.

### Aurorae
Aurorae are common phenomena in the Earth's atmosphere: they occur as charged particles emitted by the Sun cascade into the upper atmosphere, producing the glows in the sky more commonly known as polar lights. The Voyagers had confirmed the presence of aurorae in the atmosphere of Jupiter, and subsequently found the same phenomena with Saturn, between latitudes 78° and 80° in each hemisphere. The presence of aurorae was not unexpected, although the displays had not been proved either from Pioneer 11 or from the artificial Earth satellite International Ultraviolet Explorer (IUE).

The brightness of aurorae is generally given in units termed Rayleighs; one Rayleigh is equal to 100,000 photons $cm^{-2}s^{-1}$. On Earth, an aurora with a brilliancy of 1,000 Rayleighs is just visible, and is an almost permanent feature of the polar night sky. The Jovian aurorae were found to be more energetic, reaching a value of 60,000 Rayleighs at ultraviolet wavelengths. The Saturnian aurorae attain from 2,000 to 5,000 Rayleighs, and seem to be associated with the positions of the edge of the "polar hoods" defined by lines of force of the magnetic field. Saturn's aurorae are weaker than Jupiter's by a factor of about 10.

### Radio waves from Saturn
Saturn is not as rich a source of radio emissions as Jupiter, which is one reason why it was so difficult to establish the presence of a magnetosphere before the Pioneer and Voyager encounters. Yet at kilometer wavelengths Saturn is a powerful enough radio source; there is a broad band of emission extending from about 20 KHz to about 1 MHz. (One Hertz—abbreviation Hz—indicates a frequency of one cycle per second; Khz indicates a thousand hertz, and MHz a million hertz.) The maximum intensity occurs between 100 and 500 KHz, and there is a period of 10 hr 39.4 min, which is taken as the System III rotation period for Saturn.

The periodicity of the radio emissions is not easy to explain. A basic question concerning the radio emission is whether the rotational control is caused by a radiation pattern that rotates with the planet rather like a rotating searchlight. Alternatively, there could be a variation with time—like a flashing light. The apparent absence of a phase difference between the inbound and outbound Voyager observations implies that the radiation is emitted simultaneously over a wide range of directions in a particular plane of Saturn's rotation. Saturn differs from Jupiter, whose radio waves are strongly influenced by the inner, large and strongly volcanic Jovian satellite Io. It should be remembered also that the magnetic axis and the rotational axis of Saturn are almost exactly the same. Moreover, the situation changed between the passes of the two Voyagers. Voyager 2 detected the effects of Jupiter's magnetotail when approaching Saturn (*see* diagram 2), and soon afterwards, when Saturn was presumably immersed in the Jovian magnetotail, the kilometric radio emissions from Saturn were not detectable. It would, however, be premature to claim that this apparent shutting-off was due directly to the effects of Jupiter.

There are suggestions of a 2.7-day period in Saturn's radio emissions, and this is the period of one of the larger satellites, Dione, which may or may not be significant. However, Dione, unlike Io in Jupiter's system, is an inert, icy world. Other discrete, low-frequency radio emissions suggest that other satellites may be involved in the generation of radio emissions.

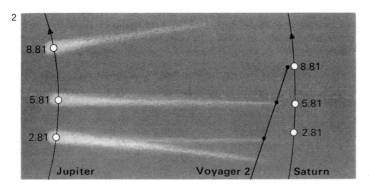

2

8.81

5.81

2.81

Jupiter          Voyager 2          Saturn

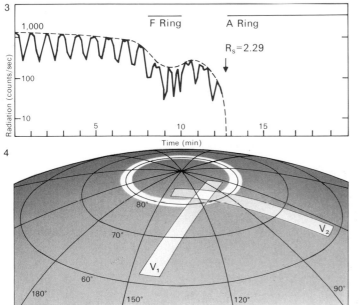

3

1,000

F Ring          A Ring

$R_s = 2.29$

100

Radiation (counts/sec)

10

5          10          15

Time (min)

4

80°

70°

$V_2$

$V_1$

60°

180°          150°          120°          90°

**2. Jupiter's magnetotail**
On its way to Saturn, Voyager 2 encountered the magnetotail of Jupiter in February 1981, even though it was not downstream of the planet until August. This may have been due to a branching filament or the rarefaction of the magnetotail, which was not detected at all in August.

**3. Ring shielding**
Radiation levels were recorded by Pioneer 11 as it approached Saturn. They show a dramatic reduction and then obliteration within the ring system. Electrically charged particles in space are affected by the magnetic field, which in this case acts as a shield.

**4. Radio signals**
Attempts to identify the source of Saturn's radio signals by both Voyager missions have narrowed the area down to a small, high-latitude region (where the two grey paths cross). Emission occurs only when this area crosses the noon meridian.

# Atmosphere I

**Thermal Radiation**

When a body is heated, it radiates energy. If a curve is drawn, plotting the intensity of radiation against wavelength, it is found to have a characteristic form which reaches its peak at a certain wavelength depending upon the temperature of the body: the shorter the wavelength at which the peak occurs, the hotter the body. Radiation of this type is termed thermal, and is easily demonstrated in everyday life: for example, when an electric fire is switched on, the bars start to glow first dull red, then yellow, and eventually white or white tinged with blue. Physicists interpret this kind of thermal radiation in terms of what is called a "black body". For such a black body it is possible to predict the exact intensity-wavelength curve for any given temperature.

Thus the thermal emission of Saturn, at wavelengths ranging from a few microns to centimeters, gives information about the temperatures at different levels in the planet's atmosphere. Radiation emitted from inside the planet passes through the atmosphere of Saturn before it is detected by instruments, and as it does so it is scattered by the cloud particles, so that it is absorbed and subsequently re-emitted. The molecules in the atmosphere absorb the radiation at different discrete wavelengths, so providing information about the chemical composition of Saturn's atmosphere. Extending the analysis beyond the thermal radio wavelengths into the infrared and visible range of the spectrum can also yield information about the variations of temperature and pressure at different levels in Saturn's atmosphere. If, for example, the temperature of the atmosphere increases with altitude gases at higher levels will appear to emit rather than absorb radiation.

Saturn's axis of rotation is inclined at 27° to the perpendicular to the orbit (as against 23°.5 for the Earth), and there are definite seasonal effects in the upper atmosphere during the long Saturnian year. The Sun crossed the equator into Saturn's northern hemisphere in early 1980, but the effects were not immediate; there is a prolonged lag, which explains the relative coldness of the northern hemisphere as recorded by the Voyagers.

Investigations of Saturn are more difficult than of Jupiter, because there are no wavelengths at which the clouds are sufficiently transparent for the lower levels of the atmosphere to be probed. It is possible, however, to produce maps of the thermal radiation at different wavelengths.

At the time of Voyager 2's encounter the temperature was about 10 K lower in the north polar region than in the south. Strong latitudinal temperature gradients were found in the troposphere of the northern hemisphere. These were related to strong westerly jets. The largest longitudinal variations were found in the regions between 32° and 42°N, 50° and 57°N, and between 65°N and the pole. The first two regions indicate relatively weak retrograde jets.

The temperature structure in the troposphere of Saturn is symmetrical with respect to the equator—a surprising result in view of the axial inclination, which would be expected to produce seasonal effects. The probable explanation is to be found in the low level of solar energy in these remote parts of the Solar System: the available energy is only about one hundredth of that received upon the Earth, and Saturn's atmosphere has a very slow response to radiation changes. Perhaps an even more surprising discovery is that there is little connection between Saturn's temperature structure and the visible cloud belts.

**Heat budget**

The total energy budget of a planetary atmosphere is an important parameter relating to the mechanisms of weather systems. Saturn, like Jupiter, has a strong internal heat source. Jupiter radiates about 1.6 times as much energy as it would do if it depended entirely upon what it receives from the Sun. The value for Saturn is also about 1.6. Both Jupiter and Saturn were formed about 4,600 million years ago, mainly from hydrogen and helium. If it is

### Composition of Jupiter's atmosphere above cloud tops

|  | % volume |
|---|---|
| Hydrogen | $\approx 90$ |
| HD | $\approx 1.8 \times 10^{-3}$ |
| Helium | $\approx 4.5$ |
| Methane | $\approx 7 \times 10^{-2}$ |
| Deuterated methane | $\approx 3 \times 10^{-5}$ |
| Ammonia | $\approx 2 \times 10^{-2}$ |
| Ethane | $\approx 10^{-2}$ |
| Acetylene | $\approx 10^{-2}$ |
| Water vapor | $\approx 10^{-4}$ |
| Phosphine | $\approx 10^{-6}$ |
| Carbon monoxide | $\approx 10^{-7}$ |
| Germanium tetrahydride | $\approx 10^{-7}$ |

### Composition of Saturn's atmosphere

|  | % volume |
|---|---|
| Hydrogen | $\approx 94$ |
| Helium | $\approx 6$ |
| Ammonia | $\approx 2 \times 10^{-4}$ |
| Phosphine | $\approx 10^{-6}$ |
| Methane | $\approx 8 \times 10^{-4}$ |
| Ethane | $\approx 5 \times 10^{-6}$ |
| Acetylene | $\approx 2 \times 10^{-8}$ |
| Methylacetylene | $\approx 10^{-10}$ |
| Propane | $\approx 10^{-10}$ |
| HD | $\approx 5 \times 10^{-5}$ |
| $CH_3D$ | $\approx 2 \times 10^{-5}$ |

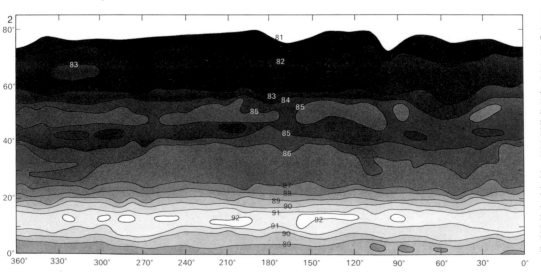

**1. Heat budget**
A comparison of absorbed solar energy and emitted infrared radiation has been plotted for the Earth, Jupiter and Saturn. Values have been averaged for season, time of day and longitude. Both Jupiter and Saturn emit more infrared than sunlight absorbed, and the constancy of emission suggests heat transport across latitude circles deep within the planets. Saturn's seasonality and rings make figures uncertain.

**2. Thermal map**
A map of brightness temperatures within the spectral interval 330 to 400 cm$^{-1}$, corresponding to a region with 150 mb pressure, was plotted during Voyager 2. The tropopause was 10 K cooler in the northern hemisphere (winter).

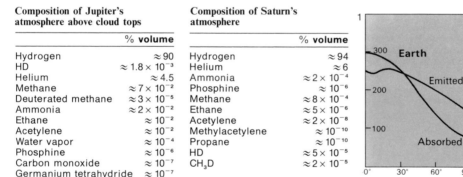

assumed simply that their globes have cooled down since then, the computed value for Jupiter's excess radiation can be accounted for, but with Saturn things are less straightforward: there is more emitted energy than there ought to be.

Since Saturn is much less massive than Jupiter, it has cooled faster, and its present internal heat is less than Jupiter's. About 2,000 million years ago the internal temperature dropped below the critical point where helium condensed on the surface of the fluid core of metallic hydrogen. At this point droplets of helium formed, and fell in what might be described as rain, dropping through the hydrogen fluid and releasing heat by their interactions with it. This seems to account for the discrepancy. In the case of Jupiter, the critical point at which helium condenses is only now being reached in the planet's interior. Extra proof is provided by the fact that there is relatively less helium in Saturn's atmosphere than in Jupiter's.

The internal heat sources in Jupiter and Saturn, in addition to weak solar heating, drive meteorological systems. The internal heating is not likely to be uniform with latitude, as with the terrestrial planets, and, in effect, only differential solar heating operates; with Jupiter and Saturn the net heat loss is greatest at the poles and least at the equator.

### The structure of the atmosphere

The atmospheres of planets are conventionally divided into certain well-defined regions. The lowest is the troposphere. On Earth the temperature above the surface decreases with altitude until it reaches a height of about 15 km. One of the main constituents of the troposphere is water vapor and it is there that weather processes, in particular the formation of clouds, take place. The equivalent region on Saturn is bounded at the upper level by a temperature minimum of about 90 K and a pressure of 100 millibars (mb) (compared with 1,000 mb at sea-level on Earth).

Above the troposphere is a transition region, the tropopause,

above which is the stratosphere. The temperature in the Earth's stratosphere is controlled largely by radiation processes and is relatively high. This temperature inversion is the result of the heating of the upper atmosphere of the ozone layer, which absorbs ultraviolet radiation from the Sun. The same effect occurs on Saturn, but here the cause is to be found in methane.

On Earth, the temperature gradient changes again at an altitude of about 30 km, decreasing with height for a further 30 to 40 km. This is the mesosphere. It is assumed that the same applies to Saturn, but there is not yet enough information to confirm this.

### The ionosphere

Above the mesosphere is the ionosphere, where the density is very low indeed, and the electrical conductivity increases. The name is derived from the higher proportion of atoms and molecules which are ionized, that is stripped of one or more of their electrons: the increased conductivity is due to the consequent presence of greater numbers of free electrons.

The Voyager 1 observations, made at wavelengths of 3.6 and 12 cm, provided the first look at Saturn's ionosphere at high latitudes, around 73°S. The atmosphere in this region is highly rarefied, and the molecules are easily ionized by energetic radiation from the Sun at wavelengths in the ultraviolet region less than 100 nanometers. Energy from the Sun in the ultraviolet part of the spectrum is absorbed by the atoms and molecules in the ionosphere, causing them to lose electrons and become positively charged ions.

The extreme ultraviolet solar radiation which causes ionization originates in the Sun's upper atmosphere and corona. Changes in these parts of the solar atmosphere therefore produce detectable variations of density in the ionosphere. The main constituent of Saturn's ionosphere is ionized hydrogen ($H^+$ ions) as with the Earth and Jupiter. The structure of the ionosphere varies according to the degree of solar activity. The ionosphere of Saturn is modified by a wider range of physical effects than any other similar structure yet studied in the Solar System.

### The composition of the atmosphere

Hydrogen is the most abundant element (about 94 percent) in Saturn's atmosphere, then helium (about 6 percent). The main difference between the atmospheres of Jupiter and Saturn is the lesser quantity of helium in the case of Jupiter. Detailed comparisons of the spectra of Jupiter and Saturn reveal other important differences. In the troposphere, ammonia is much less prominent on Saturn because of the lower temperature, which makes the ammonia condense out at higher levels. Phosphine ($PH_3$) is dominant in the region of the spectrum centered on $1,000 \text{ cm}^{-1}$. On Jupiter it has been found that phosphine demonstrates strong convective processes, bringing it up from lower levels, where the temperature is about 2,000 K, to a region where the temperature is only about 100 K. It is thought that vertical motions on Saturn are more violent than those of Jupiter (*see* page 325).

In the stratosphere, emissions due to $CH_4$ and its photochemical derivatives, acetylene ($C_2H_2$) and ethane ($C_2H_6$), are seen on both planets. The concentrations of these two gases are approximately the same in each case, although there are minor variations due to differences in the pressures and temperatures involved. Traces of propane ($C_3H_8$) and methylacetylene ($C_3H_4$) have been detected in Saturn's atmosphere.

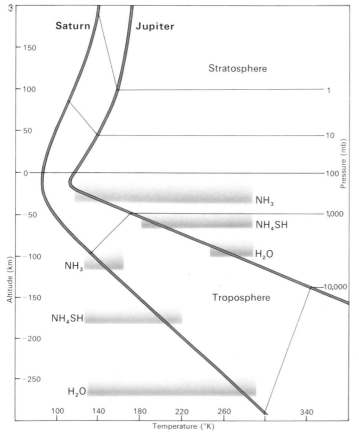

**3. Atmosphere profiles**
Profiles of temperature and pressure in the upper atmospheres of Jupiter and Saturn are based on measurements at infrared and radio wavelengths. The colored bands indicate the altitudes where various cloud layers should form—based on a solar composition of gases. The altitude is shown in kilometers above the 100 mb mark. Temperatures are lower on Saturn because of its greater distance from the Sun, and the atmosphere is more distended than that of Jupiter as a result of Saturn's weaker gravity, so the cloud layers are broader.

# Atmosphere II

One of the basic differences between the visible appearances of Jupiter and Saturn is the apparent scarcity of distinct cloud systems in the atmosphere of Saturn. Features with diameters greater than 1,000 km are at least ten times rarer than on Jupiter. The largest feature observed by the Voyagers is about the size of a Jovian white oval and only about half the size of Jupiter's Great Red Spot.

The lower contrast of features on Saturn is explained in part by lower temperatures and weaker gravity. Since Saturn's atmosphere is colder than Jupiter's, the condensation point of compounds such as ammonia is reached at a higher pressure (greater depth) on Saturn. The pressure at the 150 K level, for example, is at 0.7 bar on Jupiter, but as much as 1.4 bars on Saturn, while the tropopause is at 0.1 bar (that is, 100 millibars) on each planet. Taking into account the difference in gravity, the mass per unit area between the 150 K level and the tropopause is about $2 \times 10^3 \mathrm{kg m}^{-2}$ for Jupiter and about $10^4 \mathrm{kg m}^{-2}$ for Saturn. If a haze of ammonia or other particles was mixed throughout these layers in the same ratios relative to gas, the mass per unit area would be five times greater for Saturn. As a consequence, the colors and contrasts on Saturn would be expected to be greatly reduced in comparison with Jupiter, and this is confirmed by observations. These extensive clouds and hazes may well mask Saturn's dynamic cloud activity.

## Circulation of the atmosphere

The meteorological phenomena that produce the visible cloud structures take place in the troposphere. Saturn, like Jupiter, is an essentially liquid planet with only a very small solid core, so that unlike the Earth, Mars and Venus there is no solid surface to produce observable effects. The driving mechanisms in Saturn's atmosphere are also essentially different from those of the terrestrial planets. For example, there is only a small temperature gradient between the poles and the equator: the difference is a mere 5 K or so, and the same is true of Jupiter, where polar and equatorial temperatures are very much the same. On Earth, the poles are much colder than the equator, and so the transfer of heat from the equator plays an important role in terrestrial meteorology.

A further important difference between Jupiter and Saturn on the one hand and the Earth on the other, is the heat source that drives the weather systems. On Earth, the principal source of energy is the radiation received from the Sun, but both Jupiter and Saturn have additional internal energy, which affects the circulation. Moreover, the fact that both Jupiter and Saturn are rapidly rotating planets also constrains the motions.

Nevertheless, many of the individual features on Saturn's visible surface may be interpreted by analogies with terrestrial weather systems. For example, on Earth air will flow outward from a high-pressure region, producing an anticyclone, which spirals in a clockwise direction in the northern hemisphere and anticlockwise in the southern hemisphere. With low-pressure systems the direction of flow is inward, producing a cyclone, which spirals in a direction opposite to that of an anticyclone in each hemisphere.

The first detailed measurements of the wind patterns of Saturn's clouds, made from the analyses of Voyager images, showed characteristics that differed markedly from those on Jupiter. On Saturn, all the velocities are relative to System III, which is the solid-body rotational period. The broad equatorial jet, with a peak velocity of more than 500 ms$^{-1}$, moves at approximately two thirds of the speed of sound in the region where the cloud temperature is about 100 K. The westerly flow spreads over more than 35 degrees in each hemisphere. There are only three easterly jets, symmetrically placed in each hemisphere at latitudes of about 40, 58 and 70 degrees respectively. As with Jupiter, these easterly jets mark regions of unstable flow, and, moreover, the overall flow pattern seems to be stable over long periods.

Saturn also differs from Jupiter in that the easterly and westerly flow patterns of zonal winds do not closely correlate with the light and dark cloud bands. A further complication arises from the variations in the boundaries of these bands when Saturn is observed at different wavelengths ranging between ultraviolet and red in the solar spectrum, while the zonal velocity profile remains unaltered.

**1. Zonal velocities**
The graph shows the relative zonal velocities of Saturn's mean eastward winds at various latitudes. They are measured relative to the planet's rotation and are plotted alongside an image showing the bands. Negative velocities are winds moving westward. There is a symmetry between northern and southern hemispheres.

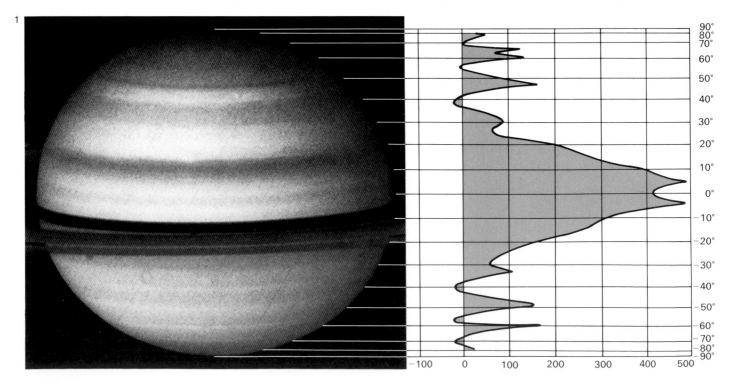

This suggests that the cloud movements observed are simply the average for a vertical layer whose thickness is less than 20 km.

A particularly surprising observation is that the Saturnian winds are completely symmetrical with respect to the equator (*see* diagram 1). This seems to be in complete contrast with the 27-degree inclination of the axis of rotation, and indicates that the weak radiation received from the Sun has only a slight effect upon the observed motions of the clouds. This raises the basic question of how deeply the motions extend into the atmosphere.

One possible explanation may be related to the interior structure of Saturn: the zonal velocity profile may extend, essentially unchanged, deep into the planet. If the liquid interiors of Jupiter and Saturn are adiabatic (without transference of heat) any steady zonal motion will take the form of differentially rotating, concentric cylinders (*see* diagram 2) whose common axis coincides with the planet's axis of rotation. Cylinders that do not reach the metallic core will extend from the top of the adiabatic zone in the north to the corresponding top of the adiabatic zone in the south. The zonal velocity profiles would, therefore, be symmetrical from north to south at the top of this zone, which may well extend to the base of the clouds. Any departures from symmetry would then arise as a result of non-adiabatic conditions within this cloud zone.

If this is true, then the density change at the top of the metallic hydrogen layer will decouple Saturn's northern and southern hemispheres poleward of about latitude 65°. For Jupiter the decoupling will take place at latitude 40° to 45°, because the metallic hydrogen core is larger. In each hemisphere of both planets, there are about three complete cycles of alternating easterly and westerly winds from the equator to the latitudes at which decoupling occurs.

This "deep-atmosphere" model provides a satisfactory explanation of the zonal motions, but the energy budget of the motions must be taken into account. With Jupiter, it has been found that the motions result from energy being transported from eddies into the mean zonal flow. A similar process helps to maintain the Earth's jet-streams, but there is an important difference. On Earth, this energy transfer, averaged over the planet, is only about one thousandth of the total energy flowing through the atmosphere as sunlight and infrared radiation. On Jupiter, the transfer of energy by this mechanism is more than one tenth of the energy flow, so that Jupiter is able to harness the thermal energy flow a hundred times more efficiently than the Earth's atmosphere. This may be due to the fact that on Earth the mid-latitudinal cyclones and anti-cyclones obtain their energy from a horizontal transfer of heat due to the pole–equator temperature difference, while the Jovian eddies obtain half their energy from convection from the interior.

Saturn is likely to behave in a similar way, but since the interior heat provides more than twice the energy from solar heating, the buoyancy is presumably much greater, thereby accounting for the much stronger vertical motion in the clouds of Saturn as compared with those of Jupiter.

In recent years considerable progress has been made in studies of the Earth's meteorology, using numerical models which give detailed representations of the physical processes and surface conditions over the planet. One such model has been extended to the conditions on Jupiter and Saturn (*see* diagram 3). With this type of approach it is possible to vary different parameters so as to examine the response of the atmosphere. The results, generated by a computer, closely resemble the large-scale frequencies of the atmospheres of Jupiter and Saturn, notably the main easterly and westerly jet profiles. It would seem that instabilities in Saturn's atmosphere provide the energy responsible for the overall circulation, creating large-scale waves resembling the belts and zones, while the jets are due to the interactions of propagating waves.

**2. Cylinder model**
This model shows the large-scale flow possible within the molecular fluid envelope of Saturn. There is a unique rotation rate for each cylinder and the zonal winds (diagram 1) may be the surface manifestation of this phenomenon. The tendency of fluids in a rotating body to align with the axis of rotation is likely if Saturn's interior is adiabatic.

**3. Computer model**
This model shows the possible atmospheric circulation on Saturn, assuming that the energy to power such circulation comes from within a narrow atmospheric layer (as on Earth); vertical heat transport is not accounted for. In the model small-scale eddies (**A**) become unstable (**B**), producing zonal jets (**C,D**). These eventually dominate the circulation (**E,F**).

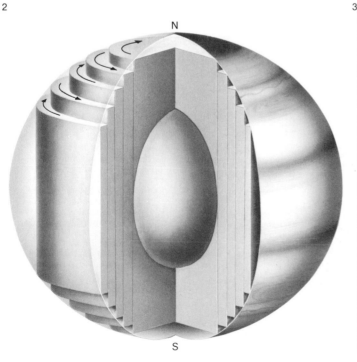

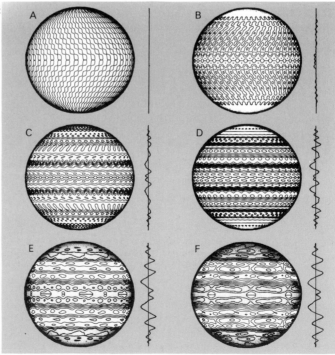

# Atmosphere III

Saturn's greater distance from the Sun as compared with Jupiter results in lower light levels, which, together with the lower contrast and less conspicuous features of the disc, make images of Saturn much more difficult. Voyager 2 images, however, greatly extended our information about the dynamics of the Saturnian atmosphere and provided many more details of the cloud systems.

The greater detail revealed many similarities with the cloud features of Jupiter. Long-lived oval spots and tilted features in east–west shear zones are similar characteristics, as are high-speed jet streams, alternating between eastward and westward directions with increasing latitude. The jet streams are thought to be powered by small-scale eddies as in the Jovian atmosphere. Greater wind speeds and latitudinal spacing of the zonal jets on Saturn are the major differences. Winds are up to four times stronger than on Jupiter. The zonal jets are twice or even four times as wide and bear little relation to the banded cloud structure.

The pattern of jet streams on Saturn occurs at higher latitudes. There is a dominance of eastward jet streams which tends to suggest that the winds are not restricted to the cloud layer but extend down by at least 2,000 km into the atmosphere. They may extend even further. The symmetrical pattern of jet streams in the northern and southern hemispheres is consistent with the cylindrical model of the structure of the atmosphere (*see* page 325).

The extensive rings of Saturn cast a huge shadow on the planet's clouds. Although the rings obscure sunlight from regions near the equator, this does not have any noticeable effect on planetary weather systems. Saturn's atmosphere behaves rather like a huge ocean, and therefore responds very slowly to the weak solar radiation, and over a length of time that is greater than that of the passing ring shadow.

### Northern hemisphere

In the region extending to 7°N, wispy cloud structures have been observed which move very rapidly, the cloud-top winds attaining velocities of up to 500 ms⁻¹. As with Jupiter, there is a minimum zonal velocity at the equator. Wisps of cloud in this region of Saturn's atmosphere are inclined in a way that tends to support this conclusion.

As in the case of Jupiter, there is a wide range of atmospheric cloud systems on Saturn. Several stable symmetric ovals of various colors (white, brown and red) have been observed at several latitudes. Most of the features are located in the anticylonic shear zones. A shearing action is set up by a decrease in wind velocity with increasing latitude. At 27°N, a feature prominent at ultraviolet wavelengths (the UV Spot), and, therefore, higher than the surrounding clouds, was observed throughout both Voyager encounters. Three brown spots are situated at 42°N. Brown spot 1 is 5,000 by 3,300 km. The flow characteristics of the Saturnian brown spots are similar to those of the Jovian white ovals. The

**1. Northern hemisphere**
This Voyager 2 image of the northern hemisphere shows how, at close quarters, the light and dark banding breaks down into a variety of very different smaller features. Several cloud shapes suggest motion, some indicating convection currents from lower levels in Saturn's atmosphere. The ribbon-like feature is as yet unique to this planet.

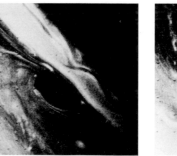

**2. Brown Spot**
The evolution of Brown Spot 1 during two successive rotations of Saturn shows evidence of anticyclonic rotation (clockwise direction in the northern hemisphere) around the edge of the spot. This feature is at latitude 42°.5N and measures some 5,000 km across. These images were taken about two days before Voyager 2's encounter.

**3. Convective features**
There are intermittent eruptions of convective clouds similar to the equatorial plumes on Jupiter. They flow with a westward jet at 39°N and appear to originate in several source regions in this latitude band. Why they are restricted to this region is not known. Individual cloud components of the feature are bright, white, irregular in shape and short-lived. To the south (27°N) is the UV Spot,

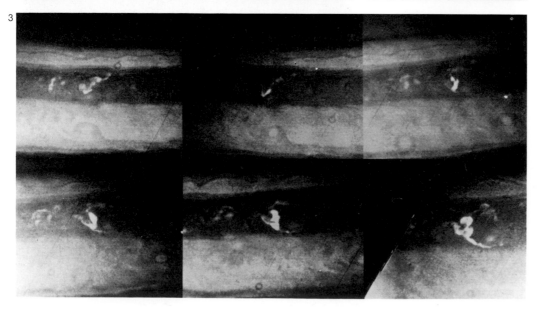

largest stable oval, nicknamed Big Bertha, is a reddish cloud measuring 10,000 by 6,000 km and is situated at 72°N.

In the northern polar regions two hurricane cloud systems have been observed at about 72°N. Each is 250 km in diameter with a distinctive core 60 km in diameter. The overall structure and size of these systems is very similar to those of terrestrial hurricane systems. On Saturn, these convective cloud systems are thought to be driven primarily by energy released by condensation processes in the water ice layer beneath the visible ammonia clouds. The strong upward motions are also influenced by the internal heat source of the planet itself. At the highest level, ammonia rains out, so that condensates are recycled, thus prolonging the lifetime of the cloud disturbance. On Earth, however, such storms are strongly influenced by the availability of a warm, moist surface layer.

Convective features associated with the strong easterly jet at 39°N were prominent in the Voyager 2 observations, and resembled the equatorial plumes and disturbances at higher latitudes in the Jovian atmosphere. Vertical motions associated with these features are much stronger than the Jovian plumes. This is a region which altered its appearance significantly during the nine months between the Voyager encounters. The extensive cloud systems situated between 20° and 35°N during the first Voyager encounter dispersed during the following months, revealing a great many previously unknown cloud details. One feature first appeared as a "cloud knot" with a diameter of 3,000 km. Within 64 hr the shearing action of the jet streams coiled the cloud into a tight oval resembling a figure 6, before the cloud ultimately detached itself from the jet to the north. The cloud oval appeared to have a total lifetime of about two weeks.

Numerous unstable cloud storms seem to originate in this region, where the wind speeds are at a minimum. One system resembles a train of vortices that might be shed by a solid vertical cylinder moving to the west. In the laboratory the characteristics of a flow of liquid are affected by the presence of a solid cylinder in the flow, an effect called a "Taylor column". This type of situation can also be seen on Earth with the air flow over the Canary Islands.

On Saturn a deep convective tower originating in the hotter lower levels of the globe acts as a barrier to the flow, since it is situated in a region of at most a very weak westerly motion. The change in the convective processes causes the tower to pulsate and spawn smaller eddies to the north: these eddies become entrained in adjacent westerly moving jets. This process could provide an almost continuous source of energy for the zonal winds.

The ribbon-like feature at 46°N latitude may be unique to Saturn. This dark, wavy line moves with the peak westerly winds of about 150 ms$^{-1}$ at this latitude. Each crest or trough covers about 5,000 km in an east–west direction. To the north of the ribbon, cyclonic vortices nestle in the troughs, and their filaments spiral towards the center in an anticlockwise direction. To the south, under the crests, are anticylonic vortices which spiral inward in a clockwise direction. A further ribbon-like feature appears at 78°N on Saturn, with a length of 16,000 km; this coincides with the jet further to the northwest.

A complex interaction between two easterly white spots moving at 15 and 20 ms$^{-1}$ respectively, and a brown spot moving at 5 ms$^{-1}$ in a westerly direction, was observed (see illustration 4). In the beginning the two white spots were at approximately the same latitude, and about 10,000 and 15,000 km to the east of the brown spot. All the spots exhibited an anticyclonic circulation. At the beginning of the sequence the furthest white spot (WS1) appeared to be slightly south of the second feature (WS2) and, if they had followed the mean flow, they would have moved closer together as WS2 overtook WS1. However, instead of colliding, WS1 moved further south and went around WS2.

Four days later WS2 passed WS1. While some merging did take place, there appeared to be a connected circulation between the two spots. The dark band stretched out considerably during the next few days as WS2 continued to move in a westerly direction relative to WS1. After six days the band stretched more than 500 km and during the next two rotations of Saturn the dark band became much narrower, while the brown spot approached to within a few thousand kilometers of the two white spots. The band between these spots appeared to be on the verge of disappearing a day later. The interaction between the three spots was a new phenomenon in atmospheric fluid dynamics.

**4. Interaction of spots**
The mutual action of two white spots as they approached Brown Spot 1 was tracked by Voyager 2 over the course of seven days. The first image (**A**) was taken at 8.3 million km from Voyager's closest approach to Saturn.

Frame **B** was taken 3 Saturn rotations later and images **C** to **G** followed at intervals of 4, 2, 5, 1 and 1 rotations respectively. By image **G** the Voyager was only 1.3 million km from the planet. The width of each panel is approximately 16,000 km.

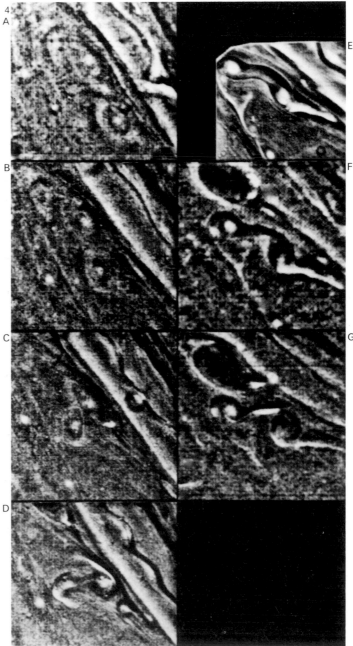

# Atmosphere IV

### Southern hemisphere

The major feature observed in the southern hemisphere of Saturn is the large red spot (Anne's Spot) situated at 55°S, which measures about 5,000 by 3,000 km. The feature was first observed in August 1980, and was tracked throughout both Voyager encounters until September 1981. It resides in a region where there is a westerly flow at about 20 ms⁻¹. The red color of the feature is thought to be the result of the same processes as those operating within the Great Red Spot on Jupiter. Saturn has a large amount of phosphine present in the cloud-top levels, but it is believed that this originates in the deep atmosphere of the planet. When ultraviolet sunlight is incident upon the phosphine, red phosphorus is produced. So the red spot would appear to be another region of strong upward motion. The presence of such a feature in Saturn's atmosphere provides further evidence that red spots are naturally occurring features in the gaseous atmospheres of the giant planets.

The Voyager observations did not permit a detailed study of the southern hemisphere at the resolution that was possible for the northern hemisphere. However, a ribbon-like feature was observed similar to that in the northern hemisphere, which suggests that there may be some symmetry in the cloud morphology as the zonal wind profile indicates.

### Long-lived features

The atmospheres of Jupiter and Saturn both possess examples of cloud systems that seem to persist for months, years, even decades and centuries. Eddies in the oceans and atmosphere of Earth are much less enduring by several orders of magnitude. In the Atlantic Ocean eddies tend to drift in an easterly direction until they merge with the Gulf Stream off the east coast of North America, and their lifetimes are measured in months and sometimes even years. In the atmosphere, the long-lived eddies are generally associated with specific surface features, such as mountain chains, or boundaries between continent and ocean. On Saturn and Jupiter, however, there is no such topography to generate eddies and flow patterns similar to those on Earth.

Two possible explanations have been suggested to account for the long-lived behavior and interactions of cloud features in their planetary atmospheres. One suggestion is that we are observing a "solitary wave", which is a self-containing wave with a single crest instead of a train of crests and troughs. Such a system could produce a flow pattern similar to that observed around Jupiter's Great Red Spot. However, when two solitary waves meet, they simply pass through each other, while observations show that on Jupiter and Saturn the ovals sometimes merge.

An alternative proposal assumes that the east–west flow pattern of the clouds is related to a much deeper system, perhaps a system of rotating cylinders (*see* page 325). It is also assumed that the vortex extends downwards only to the top of the adiabatic zone, and that the stability of the vertical extent of the flow enables the vortices to survive even large-scale perturbations. Large spots may grow by consuming smaller ones, which are transient in behavior.

### Weather and color

An overall understanding of the meteorological processes on Saturn and Jupiter will depend upon determining the depth to which the motions extend, and the individual roles of internal and solar heating in driving those flows.

One of the fundamental problems associated with Jupiter and Saturn is to explain the origin of the colors in their atmosphere, and the apparent differences between the two. The Earth's clouds are composed of water, and particles of water ice that are white. The colorful appearance of the clouds on the two giant planets provides vital evidence of the differences in the chemical composition of their atmospheres. Color is caused by a disturbance to the chemical equilibrium, by charged particles or vertical motion for example.

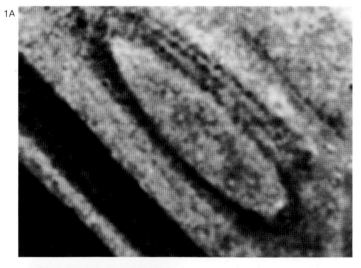

1A

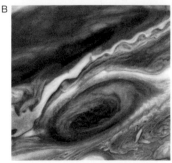

B

**1. Anne's Spot**
In Saturn's southern hemisphere at 55° latitude, Anne's Spot (**A**) is red in color, rather like Jupiter's Great Red Spot. Saturn's feature is about 3,000 km in diameter and it moves in an easterly direction at 30 ms⁻¹. This Voyager 2 image was taken through a green filter two days before encounter. The Great Red Spot on Jupiter (**B**) is a much larger feature. We know it is much colder than its surroundings and its color may come from phosphorus.

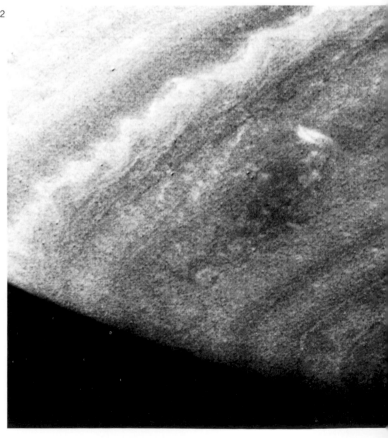

2

Altitude reflects the processes that cause chemical reactions in the first place: there are more charged particles at higher levels of the atmosphere as well as more sunlight. Ammonia, which is known to be a constituent of the Saturnian (and Jovian) atmosphere, condenses into a layer of cloud at a particular level in the troposphere; and it is believed that ammonia clouds constitute the regions of white clouds in the Saturnian and Jovian atmospheres. Saturn's atmosphere is colder than Jupiter's, and there are more extensive clouds of ammonia. Beneath the clouds of ammonia other layers are believed to form, possibly including ammonia hydro-sulphide ($NH_4SH$), water ice and ammonia solution.

### Saturn's pastel colors

As temperatures decrease with increasing height, different colors are seen. On Saturn, however, the colors seem more pastel than those observed on Jupiter. It is possible that diffuse layers of photochemically produced aerosol products are distributed throughout the upper atmosphere of Saturn, forming a thick haze that obscures both features and colors underneath. Consequently, this extra layer of material would reduce the contrast in the colors. Furthermore, it is possible that polymer substances produced in the clouds generate varied colors at these lower temperatures. There may just be a lack of atmospheric colored compounds (chromophores). Chromophores may be generated at a lower rate than the rate of the zonal wind flow so that eddies, for instance, are dispersed more quickly than chromophores can form, which could account for Saturn's bland appearance. The planet's lower temperature than Jupiter's, particularly at cloud-top levels, could mean that chemical reactions are slower, the result of which could be a more chemically homogeneous atmosphere with a more uniform distribution of chromophores. All these factors could contribute to the pastel appearance of Saturn.

Saturn's atmosphere is known to include small but important amounts of hydrocarbons, such as methane, acetylene and other exotic products. The clouds are bathed in ultraviolet sunlight, which is a strong source of energy for chemical reactions. There are also likely to be violent lightning flashes, which, like terrestrial storms, are associated with regions of strong convective activity. These storms extend throughout the clouds, and they provide a second important source of energy to affect the complicated atmospheric chemistry at a variety of levels.

### Origin of the colors

The colorful clouds, which tend to lie in broad longitudinal bands, are thought to be created from a mixture of methane, ammonia and probably sulphur too. Molecules are broken up by ultraviolet sunlight and violent lightning, and then stirred up by the dynamic weather patterns. A possible key to understanding the diversity of colors would be the identification of sulphur-bearing substances, since sulphur and certain of its compounds produce a range of colors from yellow to brown or black depending on the temperature. A suitable candidate is hydrogen sulphide ($H_2S$), from which hydrogen polysulphide ($H_xS_y$) or ammonium polysulphide ($(NH_4)_xS_y$) might form. These chemicals are capable of producing the colors that are to be seen on Saturn, but other compounds—yet to be detected in the planet's atmosphere—may be responsible for some of the colors.

The color of the red spot on Saturn may be related to the presence of phosphine in the atmosphere; for this to be possible, red spots must penetrate more deeply into the atmosphere than other features and they probably represent regions of strong upward motion. It is possible that hydrocarbons such as acetylene and ethane act as scavengers to the phosphine reactions, reducing the amount produced. Variations in the concentrations of these hydrocarbons and therefore of phosphine and other products may help to explain the variety of colors and shades.

**2. Southern hemisphere**
From a distance of 442,000 km Voyager 1 took this wide-angle photograph of the south polar region and the mid-southern latitudes of Saturn. Again the light and dark bands can be seen to be made up of many small-scale features such as waves and eddies. Saturn's equivalent of Jupiter's Red Spot can also be seen. The Voyagers' scrutiny of the southern hemisphere was less detailed than their survey of the northern hemisphere.

**3. Cloud layers**
All the observable features on the Voyager images correspond to clouds of various colors and brightness. Infrared observations show that the bands of color on Saturn distinguish different levels in the planet's cloud structure. Blue clouds have the highest brightness temperature so they must lie at the deepest levels in the atmosphere. We see them only through gaps in higher cloud which permit us to look down. Brown clouds are the next highest, and finally white clouds. If there are any red clouds, as in Jupiter's Great Red Spot, they are at the highest level and represent very cold features as indicated by infrared readings. In the case of Saturn, there is probably a haze layer above the clouds and that this results in muted colors.

3

Haze

$NH_3$

White

$NH_4HS$

Brown

$H_2O$

Blue

# Rings I

Saturn's system of rings is no longer unique, in as much as both Jupiter and Uranus have rings associated with them; but those of Saturn are in a class of their own. In particular they are bright, with a maximum albedo (*see* Glossary) higher than that of Saturn's disc, which is why the planet appears much brighter, as seen from Earth, when the rings are wide open than when they are edgewise-on. Edgewise presentations occur at alternate intervals of 13.75 yr and 15.75 yr. The inequality is due to Saturn's eccentric orbit.

The rings lie in the plane of Saturn's equator, which is tilted at an angle of 28°. During an edgewise presentation the rings are hard to see even with powerful telescopes, because they are thin and because they are barely illuminated by the Sun. There are times when the Sun and the Earth lie to opposite sides of the ring-plane, so that the unilluminated ring-face is presented. The rings lie inside the Roche limit for Saturn; that is, the minimum distance from the planet at which a body with virtually no gravitational cohesion could survive without being disrupted.

## General characteristics

There are three main rings: two of these are bright, and are lettered A (the outer) and B; they are separated by the Cassini Division, named in honor of G. D. Cassini, who discovered it. Closer in to the planet is Ring C, also known as the Crêpe or Dusky Ring, which is semitransparent and is less evident. The main rings together measure about 275,000 km in width.

New rings have been found from instruments carried on the space-probes. Pioneer 11 detected Ring F, 3,600 km beyond the outer edge of Ring A, and the Voyagers added two more; Ring G, which is excessively tenuous and has been seen only in forward-scattering light, and the very broad and also very tenuous Ring E, which extends out into the main satellite system.

Pioneer 11 provided a tantalizing first glimpse of the ring system at close quarters in transmitted rather than the more usual reflected sunlight, but it was the Voyagers that showed many unexpected features. It had been thought that the rings would be of fairly straightforward structure: several well-defined rings separated by gaps which could be attributed to the gravitational effects of inner satellites such as Mimas. The Voyagers showed that there were literally thousands of rings, with minor divisions.

Information about the composition of the ring particles can be derived from the rings' ability to reflect or absorb light at different wavelengths. The A,B and C rings are poor reflectors of sunlight at certain near-infrared wavelengths, which indicates water ice.

The sizes of the particles range from tiny "grains" to blocks several tens of meters in diameter. Observations at radar wavelengths provide information on the size of the particles. The high reflectivity of the A and B rings implies that most particles are at least comparable in size to the radar wavelengths of several centimeters. Radar observations have set the upper size limit of particles, while observations of the scattering of sunlight at visible wavelengths show smaller particles—of the order of a micrometer. The particles are less reflective in blue than in red light, perhaps because of other substances present. Dust containing iron oxide may be one possible source of the reddish color. The gaps are not true gaps; even the Cassini Division was found to contain numerous ringlets.

### 1. The ring system

The D Ring, closest to Saturn, extends from 12,700 km ($1.2R_s$) above the planet to the cloud tops or the upper atmosphere. The C Ring is in fact dozens of ringlets, at least one of which is eccentric. It is comparatively transparent, and extends to $1.53R_s$. The next ring out—the B Ring—has the spokes, and extends outward as far as $1.95R_s$. It is separated from the A Ring by the Cassini Division. The A Ring extends from $2.01R_s$ to $2.26R_s$ and contains the Encke Division. Beyond the classical rings lie the F Ring ($2.33R_s$), G Ring ($2.8R_s$) and the E Ring at $3.5R_s$ extending as far as $3.9R_s$.

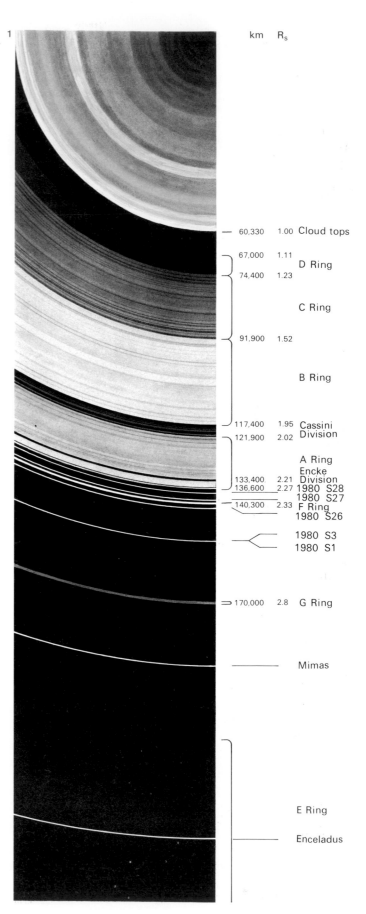

| km | $R_s$ | |
|---|---|---|
| 60,330 | 1.00 | Cloud tops |
| 67,000 | 1.11 | D Ring |
| 74,400 | 1.23 | |
| | | C Ring |
| 91,900 | 1.52 | |
| | | B Ring |
| 117,400 | 1.95 | Cassini Division |
| 121,900 | 2.02 | |
| | | A Ring |
| 133,400 | 2.21 | Encke Division |
| 136,600 | 2.27 | 1980 S28 |
| | | 1980 S27 |
| 140,300 | 2.33 | F Ring |
| | | 1980 S26 |
| | | 1980 S3 |
| | | 1980 S1 |
| 170,000 | 2.8 | G Ring |
| | | Mimas |
| | | E Ring |
| | | Enceladus |

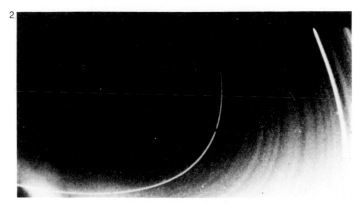

2

## Ring D
It is hardly correct to describe this as a true ring, and it is better termed the D region. It is very thin, though there are various narrow components. There may be no well-defined inner edge, in which case the D region seems to extend right down to the cloud tops of Saturn, though it is also possible that there is a lower limit at a distance of about 6,450 km from the uppermost clouds. The D Ring has been seen only in forward-scattering light, and, due to the intrinsic brightness of Saturn, it could not be seen from Earth.

## Ring C
The Crêpe Ring fills the area between the D region and the inner edge of the brightest ring, B, at 25,000 km above the cloud tops. It is a very complicated, grooved region, with a large number of discrete ringlets which, as seen from Earth, merge together. This was shown by Voyager 1, and even more dramatically by Voyager 2, which provided higher-resolution images. The structure of the C Ring is regularly ordered and cannot be readily associated with gravitational resonances. There are at least two particularly noticeable gaps, the outer of which is about 270 km wide, and is flanked symmetrically by several sharply bordered bands, which are less transparent than their surroundings. This gap is not empty. It contains a narrow ringlet, which is eccentric in shape rather than being perfectly circular. It is about 35 km wide and 100 km from the outer edge of the gap at its periapse (nearest to the planet) and 90 km wide and 50 km from the outer edge of the gap at apapse (furthest from the planet). Eccentric rings came as a great surprise; there are quite a number of them, and they cannot yet be fully explained. The eccentricity may be caused by a small satellite or satellites in the vicinity; by gravitational instabilities within the rings; or by density waves caused by satellite resonances.

**2. D Ring**
This Voyager 2 image of the D Ring was recorded at a range of 250,000 km. It is an extremely faint ring and is composed of several bands where ring particles are concentrated. The limb of Saturn is seen in the upper left, and the planet's shadow cuts diagonally across the image. The D Ring's outer edge is the C Ring's inner border.

3

## Ring B
Ring B is the brightest part of the entire ring system, and is opaque when seen from Earth. It had been expected to be simple in structure, but was instead highly complex. Voyager 1 found hundreds of ringlets and minor divisions, while Voyager 2 increased this number to thousands. The inner part of the B Ring contains the most numerous and transparent gaps. Near the innermost edge the ring tends to be more opaque and to resemble the extremely opaque outer third quarter of the ring. These similar regions have fewer narrow gaps. The most opaque regions have the spokes.

The thickness of the ring system is presumably at its maximum in Ring B, and before the Voyager pass estimates had ranged from a few millimeters to as much as 15–20 km. It now appears that the true value is between 100 and 150 meters, which explains why the rings almost disappear when seen edge-on. The Voyagers also showed that there is a large, rarefied cloud of neutral hydrogen extending to about 60,000 km above and below the ring-plane and beyond the outer edge of Ring A. Water ice in the ring particles is presumably the source; it has been estimated that the density of the hydrogen is 600 atoms per cubic centimeter.

Images of the B Ring are particularly spectacular, though it must be borne in mind that the actual images from Voyager are taken in black and white; the colors are produced by the use of suitable filters. When the images are computerized it is possible to reproduce faithfully the actual colors, though in many cases the colors are either enhanced or modified to make analysis easier.

The particles making up the B Ring are not identical in composition with those of the C Ring or the D region. In particular they are distinctly redder, and the average particle size is probably of the order of several centimeters to a few meters in diameter. The B Ring extends altogether from 25,000 to 54,000 km above the uppermost clouds. It thus extends from the outer boundary of the C Ring to the inner edge of the Cassini Division. The extremely sharp boundary between the B and C rings does not show any gaps.

**3. Reflectivity**
Comparative shots of the dark and illuminated sides of Saturn's rings show a reversing out of the system's components. On the illuminated side, material in the C Ring and the Cassini Division is not apparent, while the A and B rings are bright in forward-scattering light, indicating a significant proportion of very small particles. On the dark side the optically thin C Ring and Cassini are bright as a result of diffuse scattering.

**4. Detail of the B Ring**
This Voyager 2 detail of the B Ring was taken at a distance of less than 800,000 km and it revealed about ten times more ringlets than had previously been suspected in the B Ring. The finest rings to be seen on this image are about 15 km across.

4

# Rings II

## The Cassini Division

The Cassini Division is one of the most prominent features of Saturn's ring system. When the rings are wide open, as they will be in the mid-1980s, a very small telescope will show the Division, and even when the rings are tilted at a less favorable angle the dark line can generally be seen near the ansae, or ring-tips. Before the space-probe encounters, the Cassini Division was assumed to be a true gap, virtually devoid of ring particles. At one time there was even talk of sending Pioneer 11 through it.

The Division is of considerable width, nearly 4,000 km—as wide as the North American continent. Its material is similar to that of the Crêpe Ring; the particles are decidedly less red than those of Ring B and there is a dearth of small particles. Both the Cassini Division and the C Ring exhibit discrete, regularly spaced bands of uniform brightness, together with a variety of narrow, sharp-edged, empty gaps with a radial width of 50–350 km. Some of the gaps contain even narrower, equally sharp-edged ringlets that are quite opaque. These differ in color from the surrounding ring material and resemble more closely the material in the A and B rings.

The origin of the Division has to be considered together with the lesser divisions in the ring system. It is worth noting that there is something strange about the outer edge of Ring B, which is also the boundary of the Cassini Division. Its distance from Saturn is not constant, but varies perceptibly to the extent of about 140 km. Moreover, although it is elliptical, it does not obey Kepler's Laws, because Saturn lies at the center of the ellipse, not at one of the foci.

According to Kepler's Laws, a planet moving around the Sun will travel in an elliptical orbit; the Sun occupies one of the foci of the ellipse, while the other focus is empty. The same applies to a satellite in orbit around a planet, though the situation is complicated by the presence of several large satellites, as with Jupiter in particular and Saturn to a lesser degree. The particles in the B Ring should behave essentially like tiny individual satellites do; each moves around the planet in its own independent path, and should therefore follow a Keplerian orbit. This, however, is not the case. It may be significant that the particles at the outer edge of Ring B have a period which is half that of Mimas, the innermost of the main satellites; the periods are 11.41 hr and 23.14 hr respectively.

## Ring A

The A Ring also contains an important division. The A Ring is not as bright as its inner neighbor, and the difference is quite marked, even from Earth. Ring A is also much less opaque, though the inner edge is as sharply defined as the boundary between Ring B and the Crêpe Ring. In Ring A, as with Ring B, there are large numbers of ringlets and minor gaps, quite apart from the principal division, popularly known as the Encke Division, but now officially called the A Ring Gap.

As Voyager 2 drew away from Saturn there was a period when a bright star, Delta Scorpii, lay on the opposite side of the ring-plane, and was therefore occulted by the rings as seen from the spacecraft. Continuous measurements were made, Delta Scorpii being hidden every time it passed behind a ringlet and reappearing every time its light shone through a gap. From the results it would seem that there are very few clear gaps anywhere in the ring system. The Encke Division, like the Cassini Division, contains several ringlets, at least two of which are eccentric. There are two apparently discontinuous rings inside the Division. The clumpiness of these rings can be seen to orbit as a pattern at the orbital period of the rings at that radius. Both rings vary in brightness and one has a kinked morphology, maybe caused by perturbations created by eccentric satellites. The width of the Division is only 200 km, but even so it is visible with a moderate Earth-based telescope under good conditions. It lies about 3,000 km inside the outer edge of the A Ring. Outward of the A Ring Gap there is a pattern of unresolved ring features. They may represent a sequence of classical resonances converging on one of

**1. Rings of gold**
This Voyager 1 picture was taken shortly after the spacecraft's closest approach to Saturn in November 1980. It clearly shows the Cassini and Encke divisions as well as, at middle left, the opaque nature of the B Ring.

**2. Cassini's Division**
The Cassini Division has been shown to have a very complex structure. From the outer edge inward in this image can be seen a medium-dark ringlet 800 km wide; four brighter ringlets about 500 km wide and separated by dark divisions; and a hardly visible, narrow but bright ringlet

at the inner edge, and little more than 100 km wide. The entire Cassini Division is defined as that area between the two dark ringlets and was photographed by Voyager 1 at a distance of 6 million km from the ring system.

**3. B Ring variations**
Radial slices through the Cassini Division (top) and the B Ring (bottom) show variations of up to 140 km in the width of the gap separating the two, believed to be the result of resonances between the ring material and the satellite Mimas. The presence of an eccentric ringlet can also be detected within the gap.

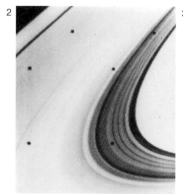
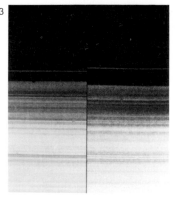

### 4. Resonances
Resonant orbits are a feature of several regions of the Solar System. A resonance occurs between two objects where the orbital period of one is commensurate with the orbital period of the other. A resonant position may be vacant or packed with objects. Kirkwood Gaps in the asteroid belt occur where asteroids would have orbital periods that are simple fractions of the period of Jupiter, which dominates the asteroid zone. The Cassini Division is like a Kirkwood Gap; a particle in the Division has a period about one half of the satellite Mimas. Several gaps in Saturn's A Ring have resonant locations with S10 and S11, and the edges of rings A and B are near resonant locations.

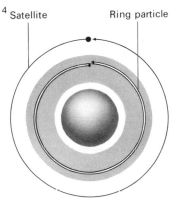

4  Satellite        Ring particle

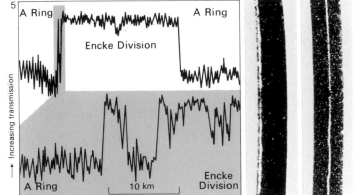

5  A Ring      A Ring

Encke Division

↑ Increasing transmission

A Ring     10 km

Encke Division

6

### 5. Structure of the Encke Division
The graphs show the amount of light recorded from Delta Scorpii as the ring system passed through the line of sight. The Division's structure was revealed, including a feature near the inner edge. The upper graph was averaged to a 3.5 km resolution, the lower 300 m.

### 6. The Encke Doodle
The irregular ringlet in the Encke Division was nicknamed the Encke Doodle. First seen by Voyager 1, it was observed more closely by Voyager 2. There are believed now to be two ringlets: at the center and the inner edge.

the inner satellites. The entire region outside the A Ring Gap is about 25 percent brighter than the region inside. A similar situation is found in a region at the outer edge of the A Ring, several hundreds of kilometers wide, that is set off from the A Ring by an apparently vacant gap tens of kilometers wide. It appears to be about 50 percent brighter in forward-scattering light than the region inside the gap. The outer edge of the A Ring is very sharp, which may be related to the proximity of the satellite Atlas.

### Causes of gaps
Before the Voyager results, it was regarded as virtually certain that the gaps in the ring system were due to the gravitational perturbations of the satellites, particularly Mimas and Enceladus, which were the closest members of the satellite family then known. It was assumed that if a ring particle has a revolution period that is an exact fraction of that of a satellite, it would be subject to cumulative perturbations, and would be moved out of the relevant area in a fairly short time. When ring particles are resonating with a nearby satellite, they begin to swing back and forth, nearer and further from Saturn. The effect is rather like pushing a swing: if the swing is large enough the swinging will increase. Mimas, for example, exerts a gravitational push on the particles in the ring system. This swinging causes particles to collide and certain areas are therefore "cleared out". In other words, the satellites would keep the Cassini and Encke divisions "swept clean", and no ring particle entering these divisions could stay there for long.

There were analogies to be found in the asteroid belt. This lies between the orbits of Mars and Jupiter, and contains many thousands of small worlds, most of them only a few kilometers or tens of kilometers in diameter. There are certain well-defined gaps where few or no asteroids are to be found; they are known as the Kirkwood Gaps. An asteroid moving in a Kirkwood Gap would have a revolution period that was an exact fraction of that of Jupiter, and so it would be unable to stay in such an orbit.

This was all very well as far as Saturn's rings were concerned when it appeared that there were only a few divisions. There are only a few strong resonances, however, and not enough to explain the highly complicated structure of the ring system, which involves thousands of what may be regarded as divisions even though they are by no means devoid of ring particles. There is evidence that satellite perturbations play a role, but there must be other influences as well.

The current theory concerns what may be termed density waves, which are formed by the gravitational effects of the satellites. The Voyager observations show a sequence of smoothly undulating brightness fluctuations in the outer Cassini Division within a band located between about 120,700 and 121,900 km from the center of Saturn. The observed behavior of these variations suggests that they are spiral density waves caused by resonances between local ring particles and Saturn's outermost large satellite, Iapetus. These spread outward from positions where the revolution periods of ring particles are exact fractions of that of one of the satellites, rather as a moving ship produces a wake in a calm sea.

It appears that at the distance at which a ring particle would have a period of one half that of the satellite 1980 S1 there is a series of outward-propagating density waves, with characteristics indicating that there are about 60 grams of ring material per square centimeter of ring area and that the relative velocities of the particles to each other is of the order of one millimeter per second. Other examples could be cited, but this explanation is probably the best available, even though it is far from complete.

Finally, it is too early to say whether the minor ringlets and divisions shown by the Voyagers are permanent or not. They may be no more than transitory, though it seems likely that the Cassini Division is a genuinely permanent feature of the ring system, and the same may well be true of the Encke Division in Ring A.

# Rings III

The main ring system ends at the outer boundary of Ring A, at a distance of 73,000 km from Saturn's cloud tops or about 136,000 km from the center of the planet. There are three more rings further out, lettered F, G and E. They are faint, and without space-probes it is not likely that they would have been detected.

### Ring F
Ring F was detected by particle absorption experiments on board Pioneer 11 in 1979, and caused tremendous interest when Voyager 1 approached Saturn and sent back pictures showing that Ring F was compound. It appeared to be made up of three separate strands, which were intertwined; it was nicknamed the "braided" ring, and seemed to defy all the known laws of dynamics. When Voyager 2 made its pass the aspect of Ring F had changed. There were five separate strands in a region that appeared to have no braiding, and yet the former appearance was seen in another part of the ring. The Delta Scorpii occultation showed that the F Ring consists of at least ten individual strands. The optical depth in the thickest part of the ring is as great as in many parts of the A and B rings. It is, however, apparently restricted to a region less than 3 km wide. The thick feature has a depth of about 100 m.

The explanation is probably to be found in the presence of the two small satellites Prometheus and Pandora, which were discovered by Voyager 1 and are known as the "shepherd" satellites. These move in orbits to either side of Ring F, Pandora between 2,000 and 500 km from the ring and Prometheus 500 km closer in. Both are very small—of the order of 200 km in diameter—and both are presumably made up of ice, but their gravitational pulls are sufficient to confine the F Ring particles to a definite, narrow region. They must also be responsible for the unusual structure of the ring, though precise details are not yet known. It may be that the braiding phenomenon is either temporary, or restricted to one particular part of the ring.

The fact that the F Ring is kept in place by shepherding satellites led to the suggestion that the same might apply to the minor gaps inside the bright rings, so that these rings would contain hundreds or even thousands of tiny shepherds. Voyager 2 was put on a course that took it as close as possible to the rings in order to search for such bodies, but none came to light.

Ring F is not circular: on average it is 140,000 km from the center of Saturn, but this varies over a range of at least 400 km. Again it seems that perturbations by satellites provide a likely explanation.

### Ring G
Next in order of increasing distance from Saturn is Ring G, which is 107,000 km above the cloud tops or 170,000 km from the center of Saturn's globe. Its existence was detected by experiments carried on Pioneer 11, but it was not confirmed until the Voyager passes, and even then it was seen only in forward-scattering light. It is extremely tenuous even when compared with Ring F. It lies between the orbit of Mimas and those of the two co-orbital satellites, Janus and Epimetheus, so that the average revolution period of its particles is 19.9 hr. The distance between Rings F and G is about 30,000 km. The G Ring is optically thin, but it contains sufficient material to endanger passing spacecraft. Voyager 2 passed 2,000 km from it at the time of the ring-plane crossing and it is thought that some of the tiny fragments that make up the ring collided with the spacecraft.

### Ring E
Finally, there is Ring E, which extends from a distance of 147,000 km from the cloud tops at the inner edge as far as 237,000 km at the outer edge. It seems that the inner portion is brighter than the outer, and that the maximum brightness occurs at a distance of 230,000 km from the center of Saturn. The distance of Enceladus from the planet's center is 240,200 km, so that the brightest part of Ring E lies just inside Enceladus' orbit. Whether this is coincidence or not is a matter for debate. Enceladus, with its

**1. Braiding**
The apparently braided components of the F Ring. The ring's complex structure includes strands that appear to cross several times, but no known law predicts what can be seen in this Voyager 1 image. It is thought that this phenomenon must result from the gravitational effects of the shepherd satellites, Prometheus and Pandora.

**2. Structure of the F Ring**
This Voyager 2 image is an almost edge-on view of the F Ring and was taken from a range of 103,000 km. It reveals at least four faint components and one bright one. The total radial extent of these strands is about 500 km. The innermost component is the faintest: it is smooth and does not appear to interact with the other ring components.

**3. Clumping material**
A detail of the bright component of the F Ring reveals another interesting result from Voyager—the presence of "clumps" within the ring structure. High resolution images show up these bright clumps, some of which have very sharply defined edges. These may be individual, relatively large particles. Less well-defined patches may be concentrations of smaller particles. The orbital period of the clumps corresponds to the F Ring's period of approximately 15 hr. Again, these features may be associated with the shepherding satellites.

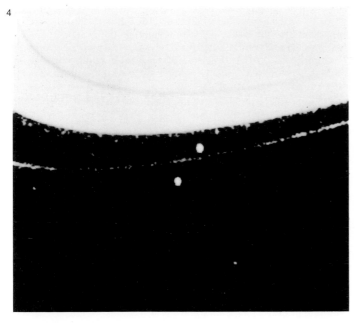

4

partially smooth surface and high albedo, may well be an active world, and there is a possibility that particles from its surface, eroded by bombarding micrometeorites, may constantly enter Ring E and replenish it. It has also been suggested that the periodical venting of material from below the surface of Enceladus, the result of tidal interactions with Dione, may also provide a source of particles for Ring E.

Are there any more rings beyond Ring E? Observations from spacecraft and ground-based measurements suggest that no further rings of comparable optical depth exist further out.

### Origin of the ring system

When Saturn was the only known ringed planet, theories of the origin of the ring system were regarded as peculiar to Saturn alone. The situation today is different. Jupiter has a ring; Uranus has a whole set, though admittedly they are very different from Saturn's—the rings of Saturn are as bright as ice, whereas those of Uranus are as black as coal dust, and the Jovian ring is so obscure that it cannot be seen from Earth. It seems that the origin of Saturn's ring system is different from those of Jupiter or Uranus; certainly the rings make Saturn one of the most beautiful objects in the entire sky.

It has been seen that Saturn's system possesses many more ringlets and divisions than was first thought. There are several examples of clumping material, representing instabilities in the system. It has been suggested that the presence of tiny moonlets could be associated with the production of many narrow features in optically thick regions and broad gaps in optically thin regions.

The first widely accepted theory of Saturn's rings was proposed more than a century ago by the French mathematician Edouard Roche. For Saturn, the Roche limit lies close to the outer edge of the main ring system. In the pre-Space Age all known planetary satellites lay outside the Roche limits for their primaries, but this is no longer true. According to Roche, a large object encountered Saturn at a fairly early stage in the evolution of the Solar System. The body may have been a large meteoroid, an asteroid, or even a moonlet that had been formed close to Saturn. Saturn's pull of gravity would create a shearing force because of differential gravitational attraction on the parts of the body closer to the planet and those parts further away. The result was that the intruder was broken up and the fragments spread around to form the ring system that exists today.

Some of Roche's conclusions have been modified in the light of more recent findings, and in particular it seems that a solid satellite less than about 100 km in diameter will not be disrupted at any distance from the planet. A large moon cannot be disrupted at a greater distance than 0.4 of the planetary radius from the surface of the body. For Saturn this would place the disruption threshold inside the inner edge of the main ring system. It is possible for two particles that differ greatly in size to resist tidal disruption at distances well within the classical limit.

An alternative suggestion is that a single large satellite moving round Saturn was broken up by collision with a wandering object such as an asteroid, so that, again, the fragments would be spread out to form a ring system. However, there is growing support for the idea that the ring particles were never part of a larger body, and that they represent the debris associated with Saturn as the planet evolved from the original solar nebula. Small grains of material can grow by condensation of vapor onto their surfaces and then by local accretion due to collisions.

Saturn's rings are probably the result of several of these processes. It is possible that the obscure ring of Jupiter is a temporary phenomenon, and the rings of Uranus are more Jovian than Saturnian in nature. As far as rings are concerned, Saturn is a planet apart. There can be no doubt that the rings will last for as long as Saturn itself.

**4. Guardian satellites**
The F Ring has two satellite companions, Prometheus and Pandora, one either side of the ring. They may be responsible for some of the peculiar phenomena exhibited by the F Ring.

**5. Shepherd satellites**
Gravitational shepherding of particles within the ring system by moonlets may produce the characteristic banding and gaps. If a moonlet and two particles are in orbit around a planet (**A**), the inner particle moves faster than the moonlet, which travels faster than the outer particle. The inner particle catches up with the moonlet as it overtakes the outer particle (**B**). Gravity draws the particles closer to the moonlet just after they are neck and neck (**C**). The result is that the net gravitational pull of the moonlet lifts the outer particle's orbit and lowers the inner particle's orbit (**D**).

5A

B

C

D

**6. The G Ring**
The G Ring is about 30,000 km beyond the outer edge of the A Ring, which can be seen to the right of the F Ring.

6

# Rings IV

## General characteristics

The radial features, or spokes, detected by Voyager 1 and confirmed by Voyager 2 in the B Ring have proved to be among the most interesting features of Saturn's rings. They were completely unexpected and have not yet been adequately explained. They appear dark in back-scattering light and bright in forward-scattering illumination. Observations indicate that small particles constitute a large proportion of the B Ring and are more visible perhaps because they are elevated above the ring plane within the spokes. Most of the spokes are confined to the central B Ring in a region ranging from 43,000 to 57,000 km above Saturn's cloud tops. There is no sign of them closer to Saturn, across the Crêpe Ring, or in Ring A. Many of them are 10,000 km long by 2,000 km wide. At all times during the Voyager 1 and 2 encounters there were spokes on view in Ring B. In each case the results were combined to make "moving pictures" of the spokes which were remarkably interesting to those trying to explain the phenomenon.

The spokes are essentially radial. They are thought to form radially, in a frame of reference rotating closely to the co-rotational rate of Saturn's magnetic field, and then to follow the differential orbital motion of the individual ring particles. The rotation period of the inner edge of Ring B is 7.93 hr and that of the outer edge 11.41 hr. So, in accordance with Kepler's laws, particles in the inner regions of the ring have shorter periods than those in the outer regions. Consequently, no radial features should exist. There is no apparent reason why they should form, and if they do, they should dissipate after a very short period of time. Yet the spokes in Ring B persist for hours. They are distorted by the rotation, but they do not disappear as quickly as might be expected.

## Formation of spokes

It appears that the magnetic field is responsible for the formation of the spokes and that Keplerian motion is responsible for the dynamics of the particles. Some of the spokes appear to be wedge shaped—with their broadest end towards Saturn—and this shape may reflect the difference between Keplerian and magnetic orbital motion over the time it takes for one spoke feature to be formed. The time scale for the formation of a 6,000 km feature may be as short as five minutes. "Young" spokes are close to the co-rotational rate of the magnetosphere, whereas "older" spokes have a Keplerian rate. These are inevitably distorted and are termed "non-radial" by some astronomers, although this term could be misleading and is not an official categorization.

Some spokes have been tracked as they rotate through 360° or more. It is not clear, however, whether the same spoke pattern is observed throughout that period, or whether a new one is reprinted on top of the old one. It is not thought that the lifetime of any one spoke is very long.

## Dark spokes

Spokes are also seen on the unilluminated side of the B Ring, which is illuminated predominantly by sunlight scattered off Saturn's atmosphere. It is believed that these are dark-side phenomena and not bright spokes shining through optically thin parts of the B Ring. The shapes seen on the dark side suggest that the spokes follow similar morphological behavior as is observed on the illuminated side. The fact that these features must be created on the dark side of the B Ring while others, it is thought, form in Saturn's shadow has implications for the mechanism of spoke formation and casts doubt on the theory that photoionization of small particles levitates them out of the ring-plane. The detection of electrostatic charges (lightning) at radio wavelengths from the ring does suggest that this mechanism is related to the spokes. Voyager 2's detection of discharges was only 10 percent that of Voyager 1, however, indicating that conditions within the ring system are constantly changing.

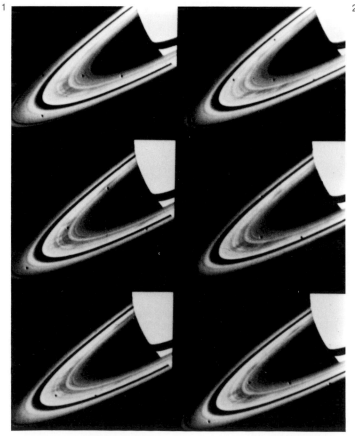

**1. Rotation of spokes**
Time-lapse photography showed not only a relatively short time in which spokes are formed, but also the co-rotation of spokes with the rings. The spokes' ability to do this is believed to be associated with Saturn's magnetic field.

**2. Bright-side spokes**
One of a sequence of Voyager 2 shots of the B Ring shows several spokes with a variety of widths and tilts. The Cassini Division is visible at the left-hand side of the photograph.

**3. Dark-side spokes**
Spokes seen from the underside of Saturn's rings appear bright because they are illuminated by forward-scattering light reflected off the planet's disc.

# Plates

Saturn, Tethys and Dione
Voyager 1
3 November 1980

**Saturn**
**Catalina Observatory (top)**
**11 March 1974**

**Saturn**
**Pioneer 11 (bottom)**
**29 August 1979**

**Saturn**
**Voyager 1 (top)**
**30 October 1980**

**Saturn**
**Voyager 2 (bottom)**
**4 August 1981**

**Saturn**
**Voyager 1 (top)**
**30 October 1980**

**Saturn**
**Voyager 2 (bottom)**
**4 August 1981**

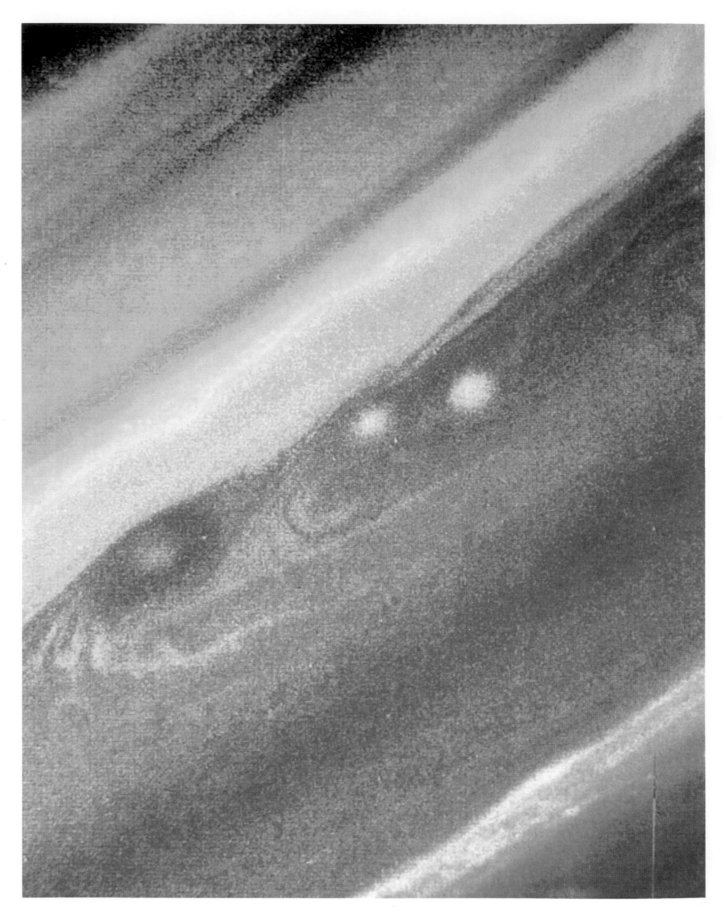

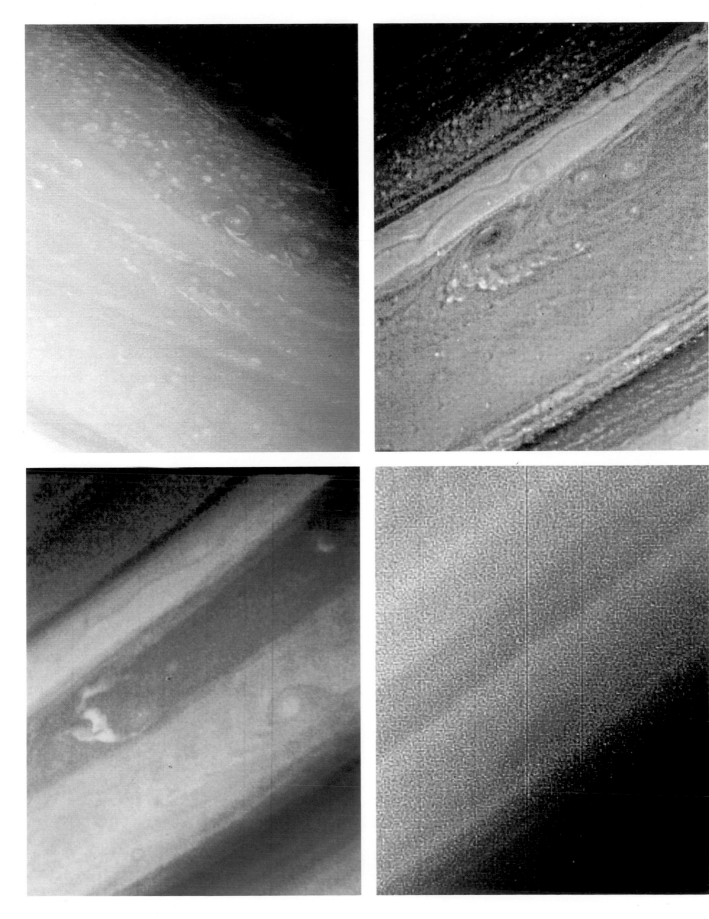

North Polar Region (top left)
Voyager 2
25 August 1981

Northern Hemisphere (top right)
Voyager 2
19 August 1981

Northern Hemisphere (bottom left)
Voyager 1
5 November 1980

Red Oval
Voyager 1
6 November 1980

343

The Rings
Voyager 2 (top)
23 August 1981

The Rings
Voyager 1 (bottom)
12 November 1980

The Rings
Voyager 2 (top)
23 August 1981

The Rings
Voyager 1 (bottom)
12 November 1980

344

Titan
Voyager 1 (top left)
9 November 1980

Voyager 2 (top right)
25 August 1981

Titan limb (bottom)
Voyager 1
12 November 1980

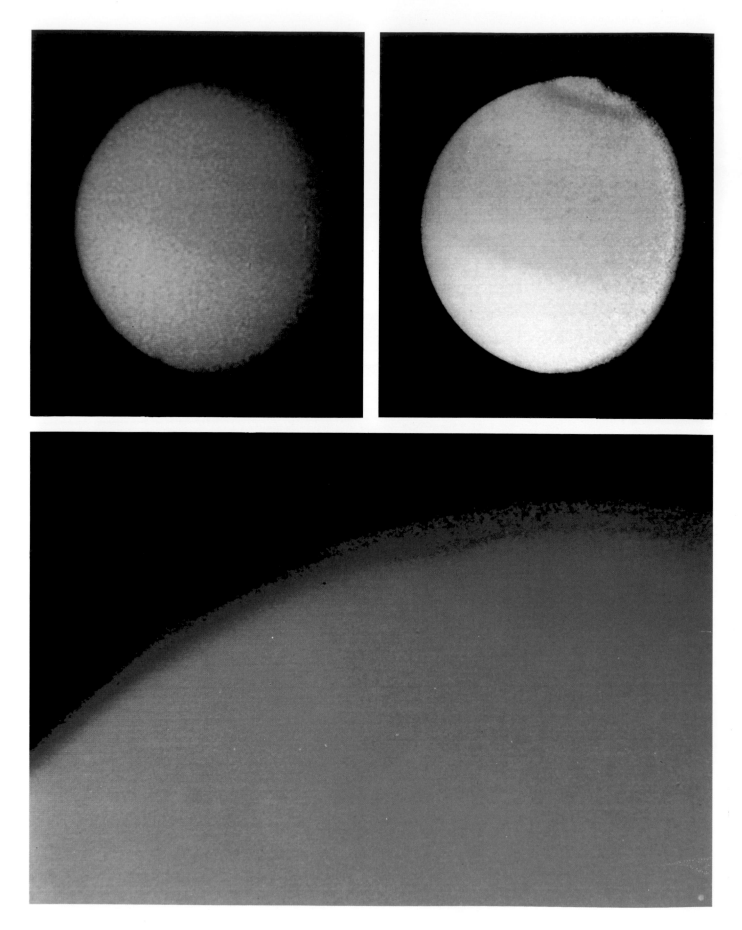

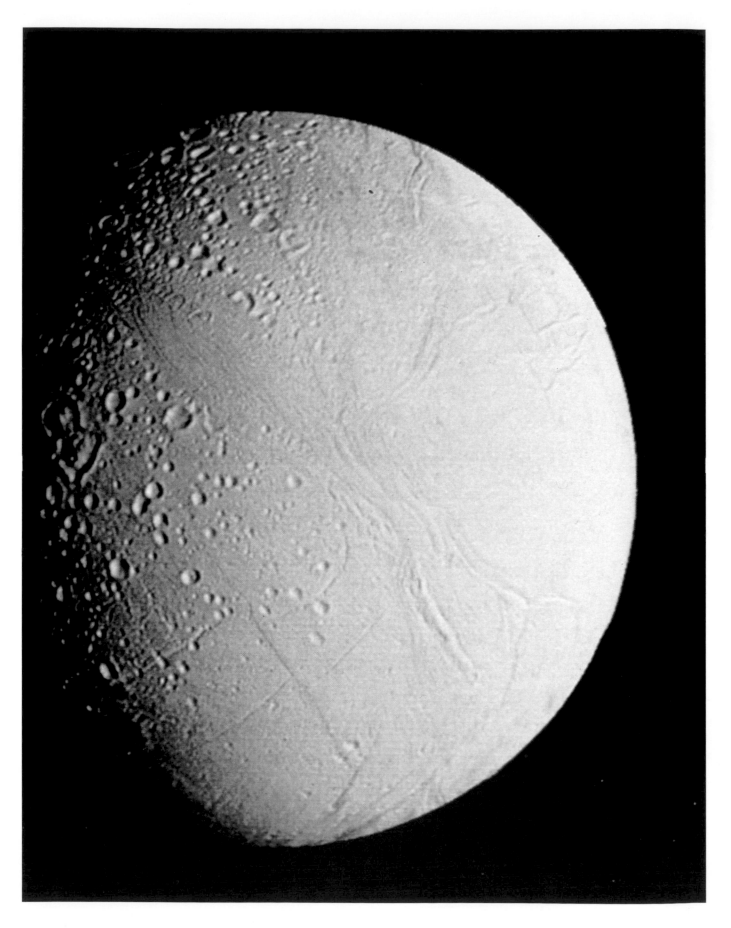

## Notes to Color Plates

**Page 337.** Saturn and two of its moons, Tethys and Dione, from a distance of 13 million km. One of the moons has cast a shadow on the cloud tops just below the rings in the image.

**Page 338.** View of Saturn from Earth in 1974 (top), with the 1.5 m telescope at Catalina Observatory in Arizona. A 1979 Pioneer 11 photograph (bottom) shows light scattered through the ring system.

**Page 339.** The Voyager 1 image (top) is contrast-enhanced and shows Saturn in back-scattered light. The Voyager 2 image, from a distance of 21 million km, is similarly enhanced (bottom).

**Pages 340/341.** Four days after its encounter with Saturn, Voyager 1 looked back on the planet from a distance of more than 5 million km and saw Saturn as a crescent.

**Page 342.** Three spots in Saturn's northern hemisphere are visible in this false-color image from Voyager 2. The largest one is 3,000 km in diameter. To the north of them is a ribbon-like feature.

**Page 343.** Photographs of Saturn's North Polar Region (top left) and Northern Hemisphere (top right) show a variety of features, many of which indicate motion within the clouds. The North Temperate Belt (bottom left) also appears to be active. This image is color-enhanced and shows convective features. Anne's Spot in the southern hemisphere (below right) is a miniature version of Jupiter's Great Red Spot.

**Pages 344/345.** A series of false-color images from Voyager 2 (top) from a range of 2.8 million km were processed to enhance color differences between the rings and give clues about compositional variations. Voyager 1 took dramatic shots of the underside of the rings from a range of more than 700,000 km (bottom).

**Page 346.** Titan from Voyager 1 (top left), photographed three days before the spacecraft passed within 7,000 km of the satellite, showed a slightly brighter southern hemisphere. When Voyager 2 photographed Titan from 2.3 million km there was a dark north polar ring. False-color processing of Titan's limb (bottom) reveals the atmosphere as a blue haze.

**Page 347.** The south polar region of Mimas shows a very heavily cratered terrain.

**Page 348.** A dramatic variety of terrains can be seen on Enceladus, which, viewed from 119,000 km, resembles Ganymede in Jupiter's satellite system.

**Page 349.** A long trench on Tethys winds its way through cratered regions on this moon that is practically all water ice.

**Page 350.** Dione is another moon that is intensely cratered but also exhibits sinuous valleys and bright wispy streaks.

**Page 351.** Voyager 1 was only 70,000 km above the north pole of Rhea when this photograph was taken. Features as small as 2 km across can be seen.

**Page 352.** The surface of Iapetus shows both very bright and very dark areas, the nature of which is still subject to debate.

# Satellites I

Jupiter, Saturn and Uranus each possesses a system of regular satellites—bodies with nearly circular orbits in the primary planet's equatorial plane. But that of Saturn differs from the others. Jupiter has four major attendants and more than a dozen small ones, the outer members of which may be captured asteroids. Uranus has 15 known satellites, all between 550 and 1,800 km in diameter.

Saturn's satellite system, however, is dominated by one large body, Titan, which is larger than any other known satellite apart from Ganymede in Jupiter's system (and possibly Triton in Neptune's). The rest are much smaller. The present total of known satellites is definitely 21, and possibly 23, giving Saturn the most extensive retinue known in the Solar System.

And Saturn's satellites are not all of the same general type. All contain a high percentage of ice, but there are great dissimilarities: Dione, only slightly larger than Tethys, is much more dense; Mimas has one huge crater that is about one third the diameter of the satellite, whereas Enceladus has a surface that is partly smooth; Hyperion has an irregular shape; Iapetus has one bright and one dark hemisphere; and Phoebe has a retrograde motion and may be asteroidal in origin.

The recently discovered satellites have orbits that, while regular, display special dynamical features. They occur as ring shepherds or are co-orbital with the larger satellites. Saturn's system is the first to be studied closely, and, as a consequence, more is known about the dimensions, densities, chemical composition and possible evolution of the satellites.

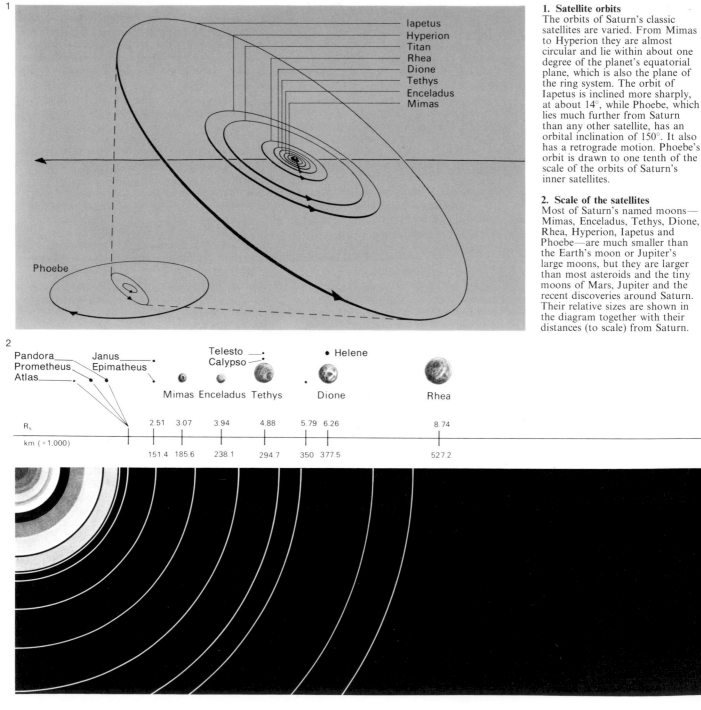

**1. Satellite orbits**
The orbits of Saturn's classic satellites are varied. From Mimas to Hyperion they are almost circular and lie within about one degree of the planet's equatorial plane, which is also the plane of the ring system. The orbit of Iapetus is inclined more sharply, at about 14°, while Phoebe, which lies much further from Saturn than any other satellite, has an orbital inclination of 150°. It also has a retrograde motion. Phoebe's orbit is drawn to one tenth of the scale of the orbits of Saturn's inner satellites.

**2. Scale of the satellites**
Most of Saturn's named moons—Mimas, Enceladus, Tethys, Dione, Rhea, Hyperion, Iapetus and Phoebe—are much smaller than the Earth's moon or Jupiter's large moons, but they are larger than most asteroids and the tiny moons of Mars, Jupiter and the recent discoveries around Saturn. Their relative sizes are shown in the diagram together with their distances (to scale) from Saturn.

The Saturnian Satellites

| No. | Satellite | Discoverer | Year of discovery | Mean distance from Saturn km | Diameter km | Magni-tude | Orbital inclination degrees | Orbital eccentricity | Sidereal period days | Mean synodic period d hr min sec |
|---|---|---|---|---|---|---|---|---|---|---|
| | Atlas | From | | 137,670 | 37 × 34 × 27 | 18.1 | 0.3 | 0.002 | 0.609166 | ? |
| | Prometheus | Voyager | 1980 | 139,353 | 148 × 100 × 68 | 16.5 | 0.0 | 0.004 | 0.613000 | ? |
| | Pandora | photographs | | 141,700 | 110 × 88 × 62 | 16.3 | 0.1 | 0.007 | 0.628541 | ? |
| | Janus | Fountain & Larson | 1978 | 151,422 | 194 × 190 × 154 | 14.5 | 0.1 | 0.009 | 0.694333 | ? |
| | Epimetheus | Fountain & Larson | 1978 | 151,472 | 138 × 110 × 110 | 15.5 | 0.3 | 0.020 | 0.694667 | ? |
| I | Mimas | Herschel | 1789 | 185,600 | 398 | 12.9 | 1.5 | 0.0202 | 0.942422 | 22 37 12.4 |
| II | Enceladus | Herschel | 1789 | 238,100 | 498 | 11.8 | 0.0 | 0.0045 | 1.370218 | 1 8 53 21.9 |
| III | Tethys | Cassini | 1684 | 294,700 | 1,046 | 10.3 | 1.1 | 0.000 | 1.887803 | 1 21 18 54.8 |
| | Telesto | Group led by Smith | 1980 | 294,700 | 30 × 26 × 16 | ? | 2 | ? | ≈ 1.9 | ? |
| | Calypso | | 1980 | 294,700 | 30 × 16 × 16 | ? | 2 | ? | ≈ 1.9 | ? |
| IV | Dione | Cassini | 1684 | 377,500 | 1,120 | 10.4 | 0.2 | 0.0022 | 2.736916 | 2 17 42 9.7 |
| | Helene | Lacques & Lecacheux | 1980 | 378,060 | 36 × 34 × 28 | ? | ? | ? | ≈ 2.7 | ? |
| V | Rhea | Cassini | 1672 | 527,200 | 1,528 | 9.7 | 0.3 | 0.0010 | 4.517503 | 4 12 27 56.2 |
| VI | Titan | Huygens | 1655 | 1,221,600 | 5,150 | 8.3 | 0.3 | 0.0292 | 15.945448 | 15 23 15 31.5 |
| VII | Hyperion | Bond | 1848 | 1,483,000 | 360 × 280 × 236 | 14.2 | 0.6 | 0.1042 | 21.277657 | 21 7 39 5.7 |
| VIII | Iapetus | Cassini | 1671 | 3,560,000 | 1,436 | 10–12 | 14.5 | 0.0283 | 79.33085 | 79 22 4 59 |
| IX | Phoebe | Pickering | 1898 | 12,950,000 | 220 | 16.5 | 150 | 0.1633 | 550.337 | 523 13 — — |

**Several other satellites have been reported, but not yet confirmed.**

Properties of the largest Saturnian satellites

| Satellite | Mean orbital radius km | Mass kg | Density g cm$^{-3}$ | Albedo |
|---|---|---|---|---|
| Mimas | 185,600 | 3.76 × 19$^{19}$ | 1.44 ± 0.18 | 0.7 |
| Enceladus | 238,100 | (7.40 × 10$^{19}$) | 1.16 ± 0.55 | 1.0 |
| Tethys | 294,700 | 6.26 × 10$^{20}$ | 1.21 ± 0.16 | 0.8 |
| Dione | 377,500 | 1.05 × 10$^{21}$ | 1.43 ± 0.06 | 0.5 |
| Rhea | 527,200 | (2.28 × 10$^{21}$) | 1.33 ± 0.09 | 0.6 |
| Titan | 1,221,600 | 1.36 × 10$^{23}$ | 1.88 ± 0.01 | 0.2 |
| Hyperion | 1,483,000 | (1.10 × 10$^{20}$) | ? | 0.3 |
| Iapetus | 3,560,100 | (1.93 × 10$^{21}$) | 1.16 ± 0.09 | 0.5, 0.05 |

Titan
20.25
1,221.6

Hyperion
24.55
1,485

Iapetus
59.02
3,559.1

Phoebe
214.7
12,950

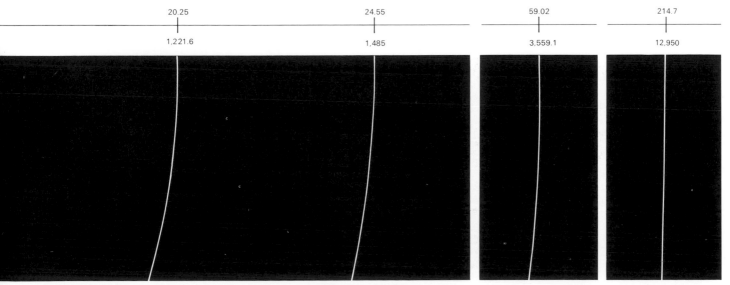

# Satellites II

Titan, the largest of Saturn's satellites, was also the first to be discovered. It was recorded on 25 March 1655 by Christiaan Huygens, and it took him little time to decide that it really was a satellite rather than a star; it moved with Saturn, and Huygens found that its revolution period was 16 days. It is an easy object, with a magnitude just below 8.

### Cassini's discoveries
The next discoveries were made by G. D. Cassini. Like Huygens, Cassini used small-aperture, long-focus refractors, with lenses made by the best instrument-makers of the period such as Campani and Divini. In 1671, using one of Campani's telescopes, he discovered another satellite, much further away from Saturn, and calculated that its revolution period was a little less than 80 days; it was named Iapetus. The true revolution period is 79.3 days, so Cassini's estimate was very accurate. He also found, again correctly, that the orbit is appreciably inclined to the plane of Saturn's equator; in fact by 14.7°. Moreover, he observed that the brilliancy of Iapetus was not constant. When west of the planet it was an easy object, but when east of Saturn it became too faint for him to see with his telescope. He therefore concluded that "one part of his surface is not so capable of reflecting to us the light of the Sun which maketh it visible, as the other part is".

This also was correct, though final proof was not obtained until the Voyager missions. It followed from this observation that the rotation period was captured or synchronous; that is, Iapetus keeps the same hemisphere turned permanently towards Saturn, as the Moon does to Earth. Tidal friction is responsible for this, and all major planetary satellites behave in the same way.

In 1672, when he announced the variability of Iapetus, Cassini discovered another satellite, now known as Rhea, which is of approximately the same magnitude as Iapetus at its best. Then, in March 1684, he made new studies, using an unwieldy aerial telescope with a focal length of more than 40 meters, and discovered two rather fainter satellites, Dione and Tethys. Their magnitudes are between 10 and 10.5, so that they again are not difficult objects; and both have orbits closer to Saturn than that of Titan.

By the end of the seventeenth century five satellites were known, all of which had nearly circular orbits, and of which only Iapetus had an orbit appreciably inclined to Saturn's equatorial plane.

### Herschel's discoveries
On 19 August 1787 William Herschel, using his 6-meter reflector (aperture 46 centimeters), suspected a new inner satellite, but did not follow up his observation until August 1789, by which time he had completed his 12-meter reflector (aperture 122 centimeters). Herschel soon confirmed his original suspicion, and discovered a seventh satellite still nearer to the planet; these new objects were named Enceladus and Mimas.

### The discovery of Hyperion
The first major American observatory was at Harvard. From there studies of Saturn were carried out in 1848, and G. P. Bond detected "a star of the 17th magnitude in the plane of Saturn's ring, between Titan and Japetus" ("Japetus" was the old spelling of "Iapetus"). It was soon confirmed, both by G. P. Bond and his father, W. C. Bond, Director of the Observatory. By mid-October they had found that the revolution period was a little over 21 days, and that the satellite moved practically in the plane of Saturn's equator.

Before the news reached England, William Lassell, using his reflector of 6 meters focal length, had found the satellite. His name for it—Hyperion—was accepted. Lassell rightly concluded that Hyperion is intrinsically fainter than Mimas, but was easier to see because it is so much further from the planet. He also suggested that its brightness was decidedly variable. The modern estimate of its mean magnitude is 14.2, considerably brighter than Bond's original

**1. Christiaan Huygens 1629–95**
Huygens discovered Titan in 1655 with a magnification of 50 on his small-aperture refractor of long focal length. The observation was soon confirmed by other observers. Huygens failed to find any other satellites around Saturn and he correctly concluded that if any other moons existed they must be fainter than Titan.

**2. Herschel's giant telescope**
This 12-meter reflector was by far the largest telescope made at that time, and it was not surpassed until Lord Rosse's 122 cm reflector in 1845. Herschel's telescope was, however, clumsy and unwieldy to use and his discovery of Enceladus and Mimas with it was perhaps its greatest contribution.

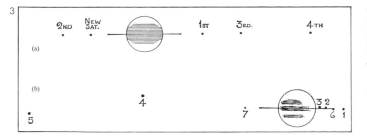

**3. Herschel's sketches**
In 1789 Herschel illustrated the alignment of Saturn's satellites in August (top) and then in October (bottom) to show his new discoveries—Enceladus (6) and Mimas (7). The other satellites drawn are Tethys (1), Dione (2), Rhea (3), Titan (4) and finally Iapetus (5).

**4. Drawings of Titan**
These remarkable drawings were made by B. Lyot, A. Dollfus and their colleagues at the Pic du Midi Observatory. Observations and drawings of Titan's tiny disc were made independently between 1943 and 1950. These drawings represent features that were detected by several observers.

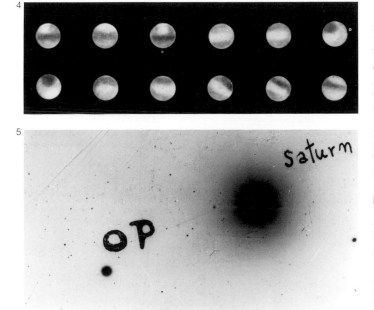

**5. Discovery of Phoebe**
Pickering's discovery of Phoebe from photographs was the first of its kind. Between 16 and 18 August 1898, he exposed four plates for two hours each. The plates were compared and an object—not a star—was found to have moved. Saturn's ninth satellite had been found.

**6. The discovery of Janus**
From photographs taken in December 1966, Dollfus believed that he had found another satellite, which he named Janus. He estimated its period to be 0.815 days and its diameter 300 km. The object shown here is one of the co-orbital satellites later found on the Voyager images.

value. Changes in magnitude are much less easy to detect than with Iapetus because Hyperion is so much fainter, but the Voyager 2 pictures have shown that the satellite is irregular in shape, so that the magnitude may be expected to vary to some extent, and this could explain Lassell's findings.

**Pickering's results**
In 1898 W. H. Pickering, a foremost American planetary observer, began a search for new satellites of Saturn from Arequipa in Peru, the southern station of the Harvard College Observatory. The method used was photographic; the telescope used had an aperture of 61 centimeters. Four plates were exposed for two hours each, and they revealed more than 400,000 stars between them. A new satellite was found to be orbiting well beyond Iapetus and Pickering named it Phoebe. For some time its existence was questioned, but in 1904 E. E. Barnard confirmed it from the Yerkes Observatory. It was found to have retrograde motion and to be smaller than any of the previously discovered satellites. Despite its great elongation from Saturn, which can exceed 34 minutes of arc, it is difficult to see visually. The magnitude is about 16.

Shortly afterwards, Pickering believed that he had discovered another satellite, moving around Saturn at a distance of 1,460,000 km in a period of 20.85 days; this would have placed it between the orbits of Titan and Hyperion. The photographs concerned had been taken in 1899, but Saturn was then in a very rich region of the Milky Way, and positive identification was difficult. Before Pickering's claim could be verified, the satellite apparently vanished! It was named—Themis—and elements of its orbit continued to be published in some astronomical almanacs (such as the *Connaissance des Temps* in France) as recently as 1960, but it has never been confirmed, and it now seems certain that Themis does not exist. Pickering was misled either by a star or, more probably, by an asteroid.

**New small satelites**
When the rings of Saturn are wide open, the fainter satellites, and particularly those close to the planet, are difficult to observe. When the rings are edgewise-on, as happens every 15 to 17 years, conditions are much more favorable. The rings were so placed in 1966, and careful searches were made by A. Dollfus at the high-altitude Pic du Midi Observatory in the Pyrenees. His photographs indicated the existence of an inner satellite, whose distance from Saturn was given as 169,000 km. A few confirmatory observations were obtained, but many observers were sceptical about the existence of the new satellite, named Janus. From Earth, no confirmation could be expected before the next edgewise presentation of the rings, in 1980. Subsequently two small co-orbital satellites were found on Voyager pictures, and one of these was given the name of Janus; it seems that Dollfus observed either the currently-named Janus or else its companion Epimetheus.

Other new, small satellites were found on the Voyager images. Atlas, Prometheus and Pandora are very close-in; Telesto and Calypso are co-orbital with Tethys, and Helene is co-orbital with Dione. Others have been suspected; a Mimas co-orbital (Synott and Terrile, 1982), three more Tethys co-orbitals (Synott and Terrile, 1982) and another Dione co-orbital (Synott, 1982). These have not been positively confirmed, but probably they do exist, and it is also likely that other minor satellites remain to be discovered.

Of the satellites, only Phoebe, much the most distant, has retrograde motion and is probably asteroidal.

# Satellites III

Saturn's greater distance from Earth means there can be no observations comparable with those of Jupiter's Galilean satellites, which show obvious discs even in a small telescope. Unsuccessful efforts were made to record surface details of Titan, the only large satellite in Saturn's retinue, but the surface is permanently concealed below the atmospheric "smog". Even Voyager 1 showed virtually nothing. All that could be said from ground-based observations was that Titan appeared decidedly orange or yellowish in color. Before the Voyager missions nothing was known of the surface features of the other satellites.

There was, however, an important development in 1943–44, when G. P. Kuiper studied Titan spectroscopically, and reported the existence of an atmosphere that was assumed to be composed chiefly of methane. In 1977, L. Trafton reported the existence of molecular hydrogen. Voyager 1 showed the atmosphere of Titan to be mainly nitrogen, with some methane, and a ground pressure 1.6 times that of the Earth at sea-level.

### Transits, eclipses and occultations
These phenomena, so familiar to all observers of the Galilean satellites of Jupiter, are much less easy to follow in Saturn's system. Transits and shadow transits of Titan can be seen and the phenomena may also be detected with Rhea, Dione and Tethys, although telescopes of considerable aperture are needed. The shadow of Rhea on Saturn's disc has, however, been glimpsed with a reflector as small as 15 cm aperture.

Eclipses and occultations of the brighter satellites (Titan, Rhea, Iapetus, Dione and Tethys) are not difficult to observe. Those of Iapetus are the most interesting, because it is so much further from Saturn and because it has an orbit appreciably inclined to Saturn's equatorial plane. An early observation was made in 1889 by E. E. Barnard, with a 30 cm refractor; Iapetus was eclipsed by Ring C. The last opportunities to date occurred in 1977 and 1978, and P. Doherty, using a 41 cm reflector, found that the satellite was never lost while in the shadow of Ring A; and a slight brightening was produced by the Encke Division.

Mutual phenomena are very rare indeed, but can be observed occasionally: thus, in 1921 W. H. Pickering described occultations of Rhea by Titan, and of Dione by Rhea. The only really well-observed eclipse of one satellite by the shadow of another was that of 8 April 1921, when Rhea passed into the shadow of Titan for more than half an hour.

### The Voyagers
During the Voyager missions, great attention was paid to the observation of satellites. Voyager 1 made close-range surveys of Titan, Rhea, Dione and Mimas. Titan was regarded with special interest. Although the surface was found to be masked by clouds, valuable measurements were obtained. Voyager 2 surveyed Iapetus, Hyperion, Tethys and Enceladus.

Dione and Rhea were surveyed by both Voyagers. Voyager 2's closest approach to Rhea was over the satellite's north pole, and close-range pictures were obtained, revealing that the north polar region contained zones with different albedoes. Voyager 1 obtained preliminary images of Tethys, including the first indications of the huge trough now known as Ithaca Chasma, and Voyager 2 extended the survey to a much wider area of the surface. Voyager 2 uncovered the most interesting characteristics of Enceladus—indications that the surface may still be active. It became clear from Voyager 1 that the trailing hemisphere of Iapetus was bright and the leading hemisphere dark; the main charting, however, was done by Voyager 2, and this was also the case with the irregularly shaped Hyperion. Phoebe, unfortunately, was not within useful range of either Voyager, and all that can be said as yet is that the surface is darkish, and the rotation period is so far unique in Saturn's family in being non-synchronous.

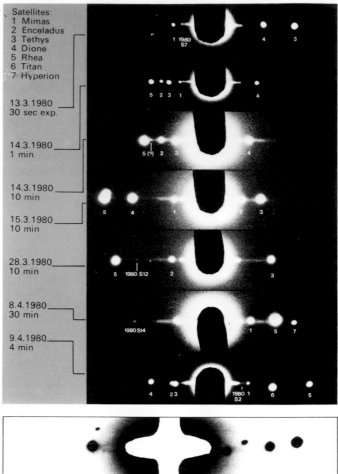

Satellites:
1 Mimas
2 Enceladus
3 Tethys
4 Dione
5 Rhea
6 Titan
7 Hyperion

13.3.1980
30 sec exp.

14.3.1980
1 min

14.3.1980
10 min

15.3.1980
10 min

28.3.1980
10 min

8.4.1980
30 min

9.4.1980
4 min

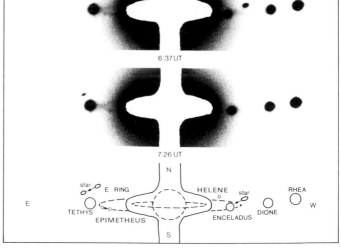

6:37 UT

7:26 UT

**1. Saturn ring-plane crossing**
A series of observations were made during March (**A**) and April (**B**) in 1980 from the Catalina Observatory. They show the satellites in the equatorial plane of Saturn—Mimas (1), Enceladus (2), Tethys (3), Dione (4), Rhea (5), Titan (6) and Hyperion (7).

**2. Satellite phenomena**
The various configurations of the satellites provide interest for observers. The more common examples include shadow transits (**A**) as a satellite passes in front of Saturn, and eclipses (**B**), when the satellite passes behind the planet and in its shadow.

### 3. Voyager 1 flyby of Saturn

The equatorial view of Voyager 1's trajectory shows the space-probe's path in relation to the planet's equatorial plane, and hence the ring system and the orbital plane of the inner satellites. The polar view shows the trajectory in relation to the rings and satellite orbits. The satellites are not drawn to scale and are placed at the point of Voyager's closest approach in light. The smaller diagrams show the flight path as seen from Earth and the trajectory with respect to the south pole and rings.

**Closest approach distances (V1)**

| Satellite | km |
| --- | --- |
| Atlas | 219,000 |
| Prometheus | 300,000 |
| Pandora | 270,000 |
| Janus | 121,000 |
| Epimetheus | 297,000 |
| Mimas | 88,440 |
| Enceladus | 202,040 |
| Tethys | 415,670 |
| Telesto | 237,332 |
| Calypso | 432,295 |
| Helene | 230,000 |
| Dione | 161,520 |
| Rhea | 73,980 |
| Titan | 6,490 |
| Hyperion | 880,440 |
| Iapetus | 2,470,000 |
| Phoebe | 13,537,000 |

### 4. Voyager 2 flyby of Saturn

Similar views—equatorial and polar—show the trajectory of Voyager 2 nine months later. The spacecraft's flight path was considerably revised from original plans in order to take another look at the unexpected and unexplained phenomena that were observed by Voyager 1. Voyager 2 was reprogrammed in flight so that the encounter with Saturn would further explore the results from the first mission. The paths of both Voyager flybys were calculated in each case to include occultations of the Earth, the Sun or stars by Saturn, various satellites and the rings. The small diagram shows Voyager 2's encounter with the outer satellites —Iapetus, Hyperion and Phoebe.

**Closest approach distances (V2)**

| Satellite | km |
| --- | --- |
| Atlas | 287,170 |
| Prometheus | 246,590 |
| Pandora | 107,000 |
| Janus | 147,010 |
| Epimetheus | 222,760 |
| Mimas | 309,990 |
| Enceladus | 87,140 |
| Tethys | 93,000 |
| Telesto | 284,396 |
| Calypso | 153,518 |
| Helene | 318,200 |
| Dione | 502,250 |
| Rhea | 645,280 |
| Titan | 665,960 |
| Hyperion | 470,840 |
| Iapetus | 909,070 |
| Phoebe | 1,473,000 |

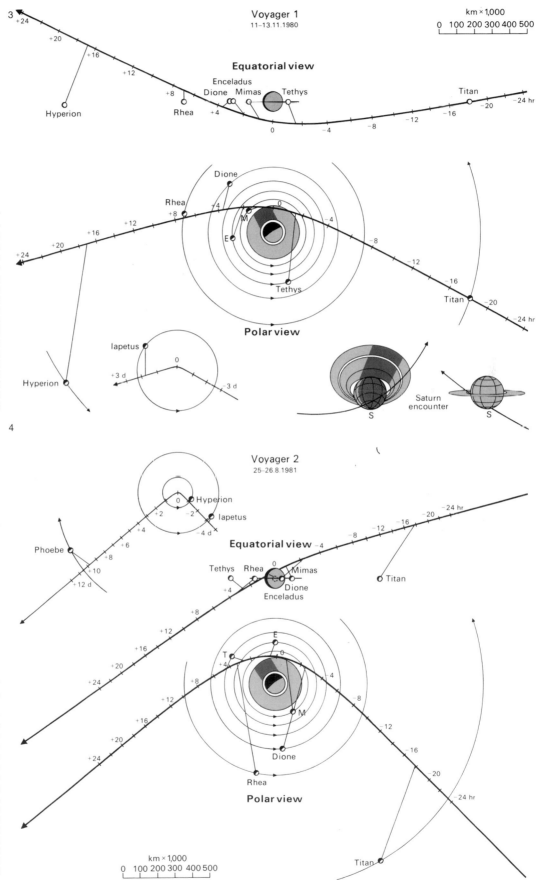

# Small Satellites

The number of known members of Saturn's system of satellites has increased considerably in recent years. Before Pioneer 11 and the Voyager flights nine satellites were known; now we know there are perhaps 23, of which five small ones orbit closer to Saturn than Mimas.

First there is Atlas, which moves just beyond the edge of the A Ring. It was discovered on the Voyager 1 photographs, and is an elongated object; its maximum diameter may be as much as 40 km, and it has an orbital period of 14.5 hr. Its presence may ensure that the outer edge of the A Ring is comparatively well defined.

The next two satellites, Prometheus and Pandora, are considerably larger. They move in orbits lying to either side of Ring F, and they are known as shepherd satellites because their gravitational forces keep the particles of Ring F in a stable orbit.

Further out still, 151,500 km from Saturn and 10,000 km beyond the F Ring—approximately midway between rings F and G—are two very interesting satellites, Janus and Epimetheus. Each is irregular in shape; the former measures about 90 by 40 km, and its companion 100 by 90 km. They give every indication of being fragments of a former single body which broke up, and this is made even more probable by the fact that their orbits are practically identical. In fact, the distance between their orbits is less than the sum of the diameters of the bodies, which leads to an extraordinary state of affairs which has been likened to a game of cosmic musical chairs. At present Janus is slightly closer to the planet, and its revolution period is 16.664 hr as against 16.672 for Epimetheus. This means that it will slowly catch up with its companion, but as they approach each other there will be a mutual interaction such that the inner, faster-moving body will be slightly slowed down while the outer, slower one will be speeded up. The end result is that the two satellites interchange orbits. The interval between two successive encounters is four years. Collision cannot take place, otherwise the two bodies would not continue to exist independently.

In February 1982 S.P. Synnott of the Jet Propulsion Laboratory, Pasadena, announced that the Voyager 2 data had shown the existence of another small inner satellite, moving at about the same distance from Saturn as Mimas and perhaps co-orbital with it, but this has yet to be confirmed. Its diameter is estimated at 10 km.

Further out are more co-orbital situations. Tethys has two, Telesto moving 60° ahead of it and Calypso 60° behind; Synnott suspects a third. This is what is termed a Lagrangian situation; it was first described by the French mathematician Lagrange in 1772. A body moving either 60° ahead of or 60° behind an orbiting satellite will be stable, and will be in no danger of being pulled into destruction. These so-called Lagrangian points had already been proved by the groups of asteroids known as the Trojans, which move in the same orbit about the Sun as Jupiter, oscillating around their stable Lagrangian positions and always keeping at a respectful 60° or so from Jupiter.

Helene, co-orbital with Dione, was the first of the co-orbitals to be detected. It was discovered by the French astronomers P. Lacques and J. Lecacheux, not from the Voyager results but by telescopic observation in 1980, when the ring system was edgewise-on as seen from Earth, and so conditions were ideal for observing small, faint satellites close to the planet. Helene is about 160 km in diameter, which makes it about equal in size to Phoebe. Nothing is known about its surface, but it is probably icy in nature.

There can be little doubt that further small satellites exist, and there may also be satellite-sized bodies embedded in the ring system, but observations from Earth are hardly likely to reveal them. In all probability we must await the next probe to the Saturnian system, Cassini, which is not scheduled to reach the planet until early in the next century. At least we know that Saturn has a very extensive satellite system.

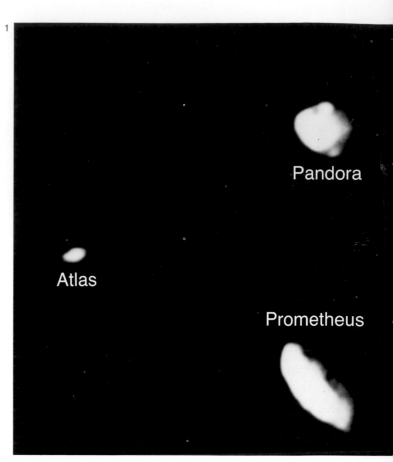

**1. Minor satellites**
Eight minor satellites in the inner part of the Saturnian system were observed by Voyagers 1 and 2. All eight either "guard" a ring or "share" an orbit. From the left and reading top to bottom they are Atlas (A Ring shepherd); Pandora and Prometheus (F Ring shepherds); Epimetheus and Janus (the co-orbitals); Telesto and Calypso (Tethys Lagrangians); and Helene (Dione B). In this montage the images are correct to their relative scale. Several of them clearly show extensive cratering.

**2. Positions of inner satellites**
The positions of the five inner members of Saturn's recently discovered smaller satellites are shown with respect to the ring system. The A Ring shepherd (Atlas) is the smallest satellite of Saturn yet to be discovered. The F Ring shepherds (Prometheus and Pandora) effectively hold the F Ring in position. Finally, the co-orbital moons (Epimetheus and Janus) may well be the fractured halves of what was once a single satellite which was disrupted in the remote past.

Encke Division

A Ring outer edge
Atlas Outer A shepherd

Prometheus inner F shepherd
F ring
Pandora outer F shepherd

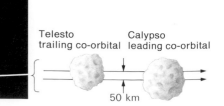

Telesto
trailing co-orbital

Calypso
leading co-orbital

50 km

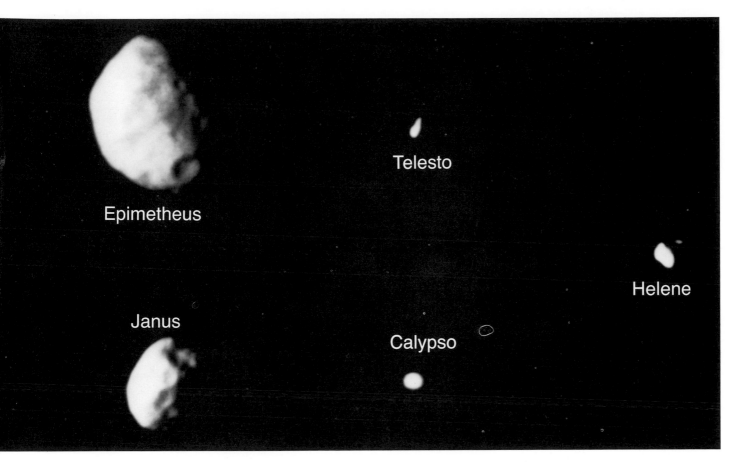

Epimetheus

Telesto

Janus

Calypso

Helene

### 3. Transit of Janus
Two Voyager 1 pictures of the eleventh moon of Saturn's system, Janus, were taken 13 minutes apart and show a narrow shadow moving across the satellite. It is believed that the shadow was of the F Ring, a few thousand km away from the satellite. Janus is a trailing co-orbital and has an irregular shape and a cratered surface. These pictures show the south polar region of the moon and they were taken at a range of 177,000 km with different filters and exposure times.

### 4. Inner F shepherd
Proof of the F Ring's guardian satellite—Prometheus—is in this dramatic photograph of satellite and ring. The presence of moons within the ring system like this was an interesting find. Some scientists believe there may be more as yet undiscovered, but dominating the dynamics of the rings. These irregularly shaped bodies lie well within the Roche limit: ring particles may still be accreting or the satellites could be torn apart in the future, adding material to the rings.

### 5. Orbit sharing
Co-orbital satellites have differential orbital periods which result in gravitational interactions between the two bodies every so often. The innermost co-orbital, with a lesser orbital period, gradually approaches its companion (A) and, in so doing, extracts orbital momentum from it. The added momentum raises it to a higher orbit (B), slowing it down relative to its companion, which is dropped to a lower, faster orbit (C). The two bodies have, therefore, exchanged orbits

(D). This is what is believed to happen with Saturn's co-orbitals Janus and Epimetheus, which lie roughly halfway between Rings F and G. Smaller bodies are also known to co-orbit with larger satellites. They are found at what is known as the Lagrangian points—60° to either side of the larger satellite. These are points of stability so that the smaller bodies survive and are not torn apart by gravitational forces. Tethys has two such companions—Telesto and Calypso. There may well be other such interactions elsewhere.

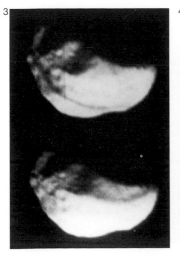

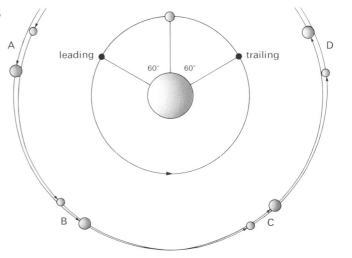

# Mimas

Mimas, the innermost of Saturn's classic satellites, is a faint
telescopic object. It has a diameter of 390 km—a value reliable to
within a range of only 10 km. Practically all our knowledge of its
surface has been drawn from the observations made from Voyager
1, which approached to within a distance of 88,400 km; Voyager 2
passed by at a minimum distance of more than 300,000 km, which
was too far away to obtain large-scale images. Like all the satellites
apart from Phoebe, Mimas has a synchronous or captured
rotation, so it keeps the same hemisphere turned towards Saturn all
the time; the revolution period and the rotation period are the same.
Most of the surface was surveyed from Voyager 1, although the
north polar regions were inaccessible.

### Craters

The most striking feature is a giant crater, Herschel, 130 km
in diameter, nearly centered on the leading hemisphere. The walls
rise on average to a height of 5 km above the floor, which in parts is
10 km deep. An enormous central peak, 20 km by 30 km at its base,
rises 6 km from the crater floor. This feature may have been
generated by a rebound of the floor under the extremely weak
gravity of Mimas. This peak is almost exactly on Mimas' equator.
The crater floor includes considerable detail: adjoining it is a valley
and numerous smaller craters.

The diameter of the crater is one third that of Mimas itself. This
raises interesting questions: the density of Mimas is only 1.2 times
that of water, and for a body of its size the diameter of the crater is
probably near the maximum that can be produced by meteoritic
impact without breaking up the satellite.

All the other craters are a good deal smaller than this giant
feature, and only a few craters larger than 50 km have been
observed. Smaller craters, however, are abundant and in general
more uniformly distributed, and many of them with diameters
greater than 20 km have pronounced central peaks. The region
from 40°W to 260°W has very few craters in the range 20–50 km
and it may have a younger surface on which later processes have
obliterated preexisting large craters. The number of such craters
increases substantially from 260°W to the giant crater at 100°W.

Most of the craters on Mimas are approximately bowl shaped
and much deeper than craters of comparable size on the Moon or
the Galilean satellites. The greater depths are probably due to the
extremely low gravity field of Mimas. Craters are frequently
superimposed upon other craters, while many of the older features
are strongly degraded. No ray craters have been seen on the surface
of Mimas, perhaps as a result of the intrinsic brightness of the
mature surface itself. The apparently uniform distribution of
craters with diameters less than 30 km may indicate an equilibrium
in the production and destruction of the features. The giant crater is
relatively unmodified by superimposed craters, and may well be
younger than the rest of the surface of the satellite.

There are valleys, too, on Mimas and the surface of the satellite is
scored by grooves, which extend to 90 km in length, and are
generally 10 km wide and 1 to 2 km deep. The most conspicuous
grooves trend northwest and west-northwest, rather like the "grid
system" on the Moon. Some are straight and may have formed over
deep-seated fractures or fracture systems. The less regular systems
may actually consist of chains of coalesced craters. The grooves
may have been produced when the giant crater was formed.
Alternatively they may have been developed by tidal interactions as
the body cooled, and then froze. A few local hills are evident in the
trailing hemisphere which are mostly 5 to 10 km across and less
than 1 km high.

The surface of the satellite is icy, and there is every reason to
suppose that ice makes up much of the entire body. The overall
surface of Mimas is consistent with a surface coated with water
frost and bearing scars, which some believe are the result of
bombardment from exterior bodies.

1

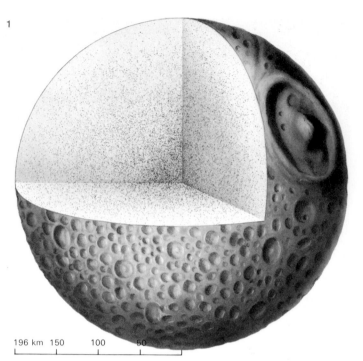

196 km  150       100       50

**1. Interior of Mimas**
Mimas is a regular icy satellite of
low density. Its surface is, at least
in part, covered with water ice,
but little is known of the internal
structure other than that its
composition is at least 60 percent
ice, condensed out from the initial
Saturnian nebula. The remaining
"rocky" material may result from
accretion and bombardment from
external sources.

**2. Terminator**
The terminator of Mimas shows a
heavily cratered surface, which
may represent a record of the
bombardment that occurred in the
Solar System during its early
history, some 4 billion years ago.
Some of the small craters visible
in this photograph, taken at a
range of 129,000 km, are as tiny
as 2 km in diameter. Craters of
this size are abundant on Mimas.

2

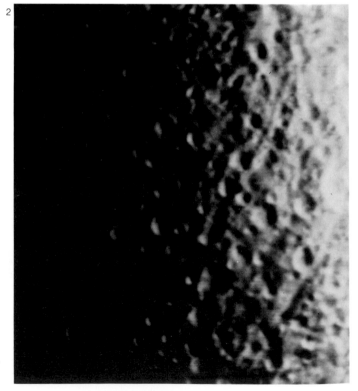

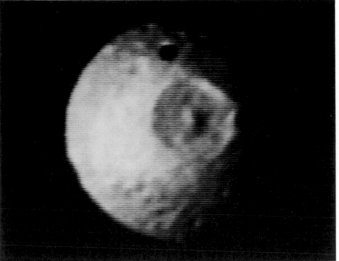

3A

B

## 3. Mimas' giant crater Herschel
This well-formed structure—130 km in diameter—is almost centered on Mimas' leading face. The fact that this feature has a diameter almost one third that of the satellite itself, which can be appreciated from this Voyager 1 image (**A**) taken from a range of about 660,000 km, means that Mimas has one of the largest crater diameter/satellite diameter ratios in the Solar System. The second photograph (**B**), taken closer in at a range of 425,000 km, shows up dramatically the crater walls, which rise to a height of 5 km above the floor, the central peak, prominent in both images, rises to 6 km above the floor.

# Map of Mimas

| | | | |
|---|---|---|---|
| Arthur | 190°W,35°S | Mark | 297°W,28°S |
| Balin | 82°W,22°N | Merlin | 215°W,38°S |
| Ban | 149°W,47°N | Modred | 213°W,5°N |
| Bedivere | 145°W,10°N | Morgan | 240°W,25°N |
| Bors | 165°W,45°N | Pellinore | 128°W,35°N |
| Dynas | 75°W,8°N | Percivale | 171°W,1°S |
| Elaine | 102°W,44°N | Tristram | 26°W,58°S |
| Gaheris | 287°W,46°S | Uther | 244°W,35°S |
| Galahad | 135°W,47°S | | |
| Gareth | 280°W,44°S | Avalon Chasma | 120–160°W,20–57°N |
| Gwynevere | 312°W,12°S | Camelot Chasma | 0–45°W,25–60°S |
| Herschel | 104°W,0° | Oeta Chasma | 105–130°W,10–35°N |
| Igraine | 225°W,40°S | Ossa Chasma | 280–305°W,10–30°S |
| Iseult | 35°W,48°S | Pangea Chasma | 290–340°W,25–55°S |
| Kay | 116°W,46°N | Pelion Chasma | 200–235°W,20–25°S |
| Launcelot | 317°W,10°S | Tintagil Chasma | 190–235°W,43–60°S |
| Lot | 227°W,30°S | | |

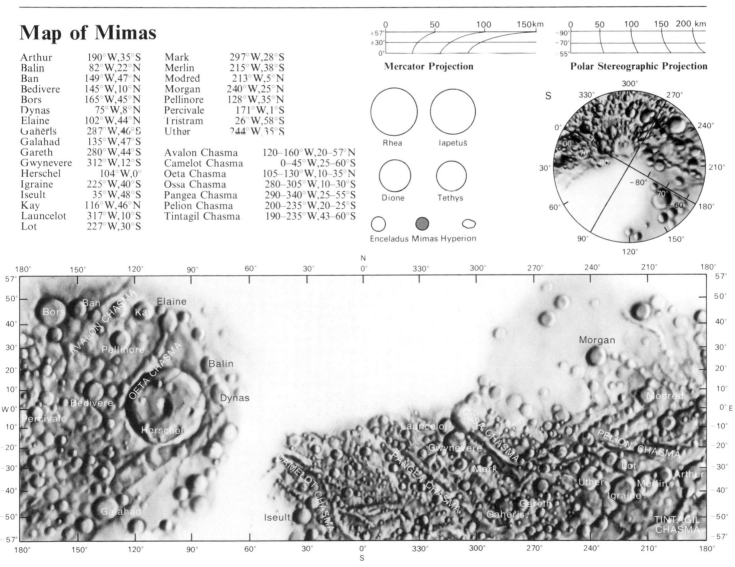

# Enceladus

Enceladus is one of Saturn's smaller satellites: it shows the greatest geological evolution and is the most "youthful". Its diameter is only 500 km, less than half that of Tethys or Dione, and its density is low, only a little greater than that of water. The Voyager 1 images were obtained across a minimum distance of more than 200,000 km, but were detailed enough to show that the surface of Enceladus is different from those of the other icy satellites, for it is smooth. Voyager 2 passed Enceladus at 87,140 km. In general, Enceladus is bright; it is in fact the most reflective body in the entire Solar System, with an albedo not far short of 100 percent.

Six different types of terrain have been found. First there are the heavily cratered plains A and B, thought to be the oldest parts of the surface now on view. There are no giant features, but there are plenty of smaller craters in the 10–30 km diameter range. According to the impact theory of cratering, these formations were produced during the period when Enceladus was being bombarded by assorted debris orbiting Saturn. In region A many of the craters show evidence of collapse, and those with central peaks show very gentle, rounded mountains. In region B similar-sized craters are highly preserved, suggesting that the thermal histories of the two regions are different. The average depth of the craters in region B is greater. Regions C, D and E are of intermediate type, with linear features (valleys and ridges up to 1 km high) cutting through cratered plains. These make up the central part of the visible disc and the bowl-shaped craters are 5–10 km in diameter. But the most remarkable area of Enceladus is the extensive plain F, where there are virtually no craters at all, and the surface of the satellite is dominated by long grooves.

These variations in terrain suggest a complex geological history, during which the surface has been replaced in several stages. The crater plains on Enceladus have a crater density comparable to the least-cratered surfaces seen elsewhere on the Saturnian satellites, such as the smooth plains of Dione's leading hemisphere. The ridged plains, which are in the trailing hemisphere of Enceladus, have a crater density lower by a factor of 50. The present global mean cratering rate is estimated to be equal to the present lunar cratering rate by both asteroids and comets. The plains are thought to be $10^9$ years old, which is only a quarter of the age of the satellite. Consequently a major resurfacing stage must have occurred on Enceladus relatively late in the geological history of this most unusual Saturnian satellite.

Valleys and ridges indicate crustal movements. They may have been formed by faulting accompanied by the extrusion of fluids. In the major region of the ridged plains, the ridges tend to form a concentric pattern near the border of the unit. The possible explanation is that they are pressure ridges formed by convective upwelling and the formation of new crust in the center of the region with compression and folding of the crust along the margins.

One possible cause of the exceptional surface of Enceladus is to be found in its connection with Dione, which is much larger, denser and more massive. Dione's orbital period is twice that of Enceladus, and the gravitational pull of Dione keeps the orbit of Enceladus slightly elliptical. This produces tidal effects, and it may well be that these are sufficient to keep the interior comparatively warm, so that material can be extruded from faults in the surface – presumably in the form of soft ice. Enceladus, unlike the other icy satellites, may be active, a surprising feature for so small a body. Such a situation is not unique, however; the same phenomenon is evident with Jupiter's inner Galilean satellite Io. It is unlikely that there is significant heating of Enceladus from radionuclide decay.

The E Ring of Saturn appears to exhibit a pronounced peak in brightness along the orbit of Enceladus and may consist chiefly of particles that have escaped from the satellite. If the surface of Enceladus were punctured, by meteoritic impact for example, water would outgas, forming supercooled droplets and ice crystals, which would escape from the weak gravitational field of Enceladus.

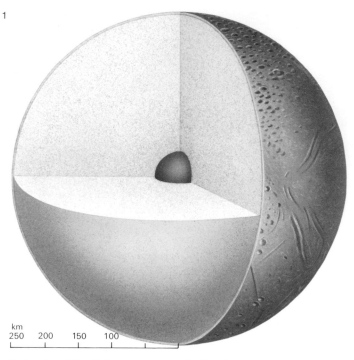

km
250    200    150    100

**1. Interior of Enceladus**
Like most of Saturn's satellites, Enceladus is believed to have a roughly 60:40 ice/rock ratio. It has, however, the most extensive geological history. One theory is that tidal forces exerted by Dione and Saturn produce heating in the satellite's interior. This may in turn result in outgassing of water, and perhaps methane as well, onto the ice-crusted surface.

**2. Terminator**
This view of the terminator of Enceladus shows up craters on the border of a cratered terrain and the ridged plain. At the lefthand edge of the cratered terrain (known as CT1) craters are older than those of the area to the right (CT2), where craters are deep and bowl shaped. These two cratered regions have had different thermal histories.

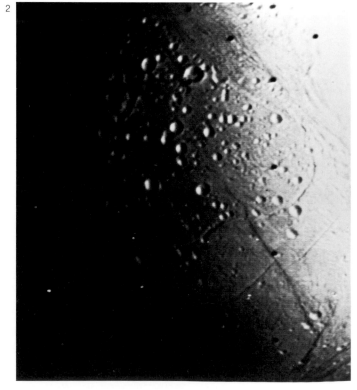

3A

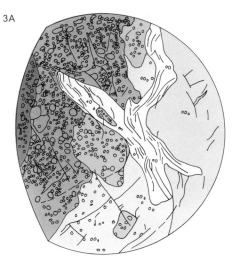

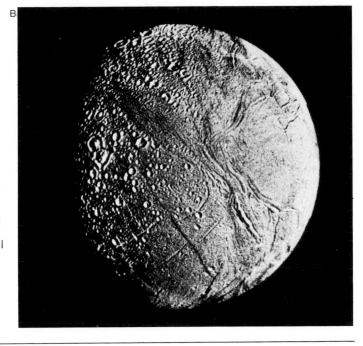

| 1 | Ridged plains |
| 2 | Smooth plains I |
| 3 | Smooth plains II |
| 4 | Cratered plains |
| 5 | Cratered terrain I |
| 6 | Cratered terrain II |

**3. Types of terrain**
The distribution of the six types of terrain found on Enceladus are shown on the diagram (**A**). Cratered terrains are the oldest, possibly from early bombardment of the satellite. The cratered plains are younger but have a range of ages. The ridged plain is the youngest terrain. The Voyager 2 image (**B**) has been enhanced to show up topography.

# Map of Enceladus

| Ali Baba | 11°W,55°N | Bassorah Fossa | 23–345°W,40–50°N |
| Dalilah | 244°W,53°N | Daryabar Fossa | 20–335°W,5–10°N |
| Dunyazad | 200°W,43°N | Isbanir Fossa | 0–350°W,10°S 20°N |
| Julnar | 340°W,54°N | | |
| Shahrazad | 200°W,49°N | Diyar Planitia | 250°W,0° |
| Shahryar | 222°W,58°N | Sarandib Planitia | 300°W,5°N |
| | | Harran Sulci | 210–270°W,5°S–35°N |
| | | Samarkand Sulci | 300–340°W,10°S–75°N |

**Mercator Projection**

**Polar Stereographic Projection**

Rhea  Iapetus

Dione  Tethys

Enceladus Mimas Hyperion

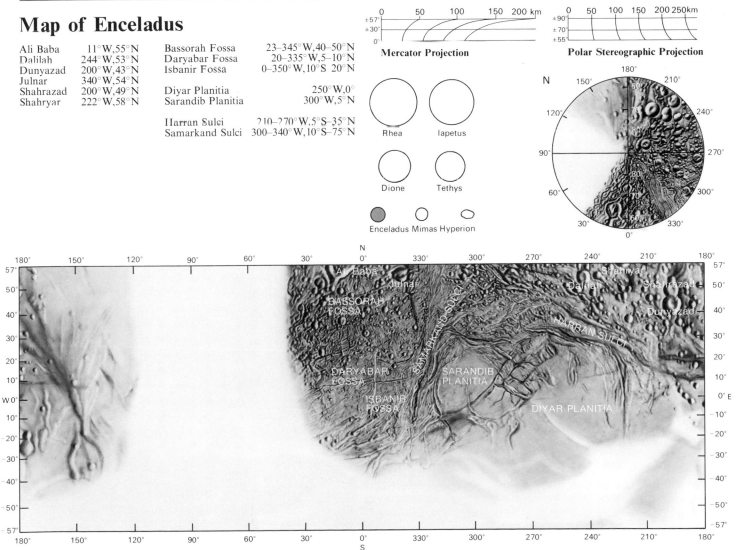

# Tethys

Tethys, the innermost of the larger satellites, has a diameter of 1,050 km. In size it is almost a twin of Dione, but is different in nature, mainly because of its exceptionally low density, which is about the same as that of water. Tethys seems to be made up of almost pure ice, and it has an albedo of 0.8, higher than that of any of the other satellites apart from Enceladus. Very little was revealed on Tethys by Voyager 1 apart from the huge Ithaca Chasma, but Voyager 2 bypassed the satellite at only 93,000 km, enabling much of the surface to be mapped. The Voyagers also discovered two small co-orbital satellites, one 60° ahead of Tethys and the other 60° behind. Both are small and very faint and would not have been detected from Earth.

There are two surface features of paramount importance. One of these is Ithaca Chasma, which is unlike anything else in the Solar System. It is a huge trench, extending around Tethys' globe from near the north pole down to the equator and down to the neighborhood of the south pole. Its average width is 100 km; it is 4–5 km deep, while the slightly raised rim reaches a height of approximately 0.5 km. Its origin is decidedly uncertain. It has been suggested that in its early stages of evolution Tethys consisted of a globe of liquid water with a thin solid crust; if so, the freezing of the interior would have caused expansion, producing the huge trough as a surface crack. Why only one crack was produced instead of a whole system of smaller ones has not been explained. Ithaca Chasma is definitely not a crater-chain of any kind, and it does indicate that there was a catastrophic event upon Tethys in the distant past.

## Craters

The second feature of special interest is an enormous crater, about 400 km in diameter, lying at approximately latitude 30°N and longitude 130°. Its diameter is about 40 percent of that of Tethys itself, and it is larger than the whole of Mimas. It has a central peak, and is the largest crater with a well-developed central peak so far found in the Solar System. The crater floor may have rebounded tens of kilometers above the subdued rim. The surface of the interior follows the curve of the mean radius of Tethys, so that however it may have been formed it is a very ancient feature. It is much less well preserved than the large crater on Mimas, and there have been suggestions that beneath it there is a slightly warmer layer which has caused the crater to become flattened. It is not easy, however, to visualize such a layer in a body made up of pure ice, as Tethys appears to be. Whether it is in any way associated with the origin of Ithaca Chasma is a matter for debate, but that would appear to be rather doubtful.

There is another large crater to the north of the equator, at longitude 250°, with a central peak. Elsewhere there are very heavily cratered regions, with alignments that seem to be very approximately parallel to the overall direction of Ithaca Chasma. The region north of the equator is hilly and densely cratered.

Part of the trailing hemisphere exhibits a less rugged surface, with a few large craters surrounded by plains with a considerably lower crater density than the hilly terrain. The plains were probably produced by a flood of material that erupted from the interior. The flood may have surrounded large, preexisting craters, and overlapped their associated rim deposits. Tethys, then, appears to be an "ice ball" that bears the scars of activity on the surface that happened early in the satellite's history.

It has been tentatively suggested that in the past Tethys suffered such intense bombardment that it was actually broken up, after which its fragments came together again. In this case the whole satellite would be extremely fragile, but the theory is highly speculative, and rests upon no certain evidence. Undoubtedly the two principal features raise interesting questions, and in its way Tethys is as puzzling as some of the other members of Saturn's extensive family of satellites.

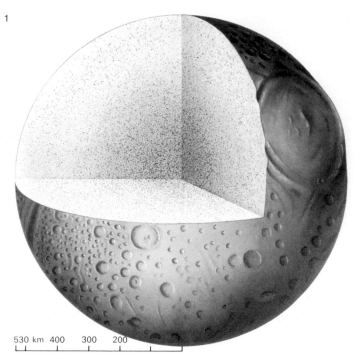

530 km  400  300  200

**1. Interior of Tethys**
Tethys is another low-density satellite thought to be almost entirely composed of water ice. There is evidence of resurfacing despite Tethys' small size. Any melting and surface alteration resulting from internal processes probably occurred after the main cratering. Less-cratered regions may well result from outpouring from the interior.

**2. Ithaca Chasma**
This enormous trench is centered on the Saturn-facing hemisphere of Tethys and extends for more than 270° around the body. It is not indicative of internal activity, but probably formed as the watery interior froze during Tethys' evolution. This would have created a 5 to 10 percent increase in surface area which matches the floor of the rift zone.

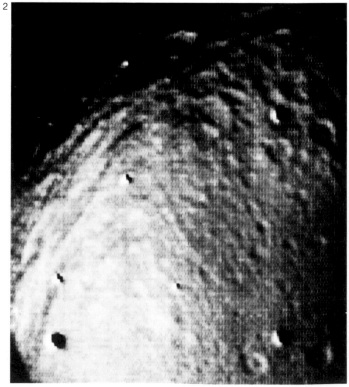

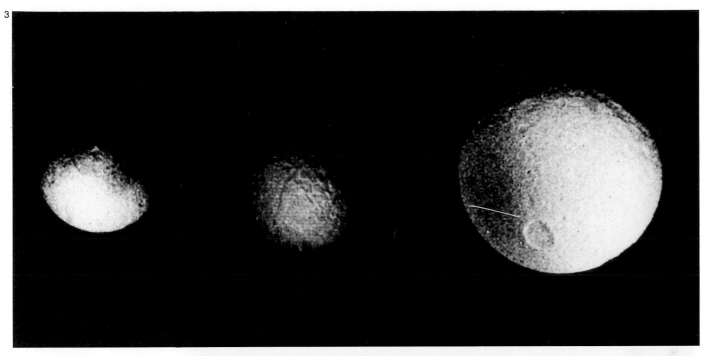

### 3. Giant crater
A series of photographs taken at 4-hourly intervals during Voyager 2's approach of Tethys shows the large, flattened crater—400 km in diameter—that dominates Tethys' other hemisphere. The second large crater can also be seen on the third image.

### 4. Global view of Tethys
The best global view of Tethys (Voyager 2) shows the trailing hemisphere and illustrates a variety of terrains.

### 5. Terminator
This was the highest-resolution image that was taken of Tethys before a spacecraft malfunction prevented further imaging of this satellite. It shows the hilly, heavily cratered region north of the equator and from this image a distribution of craters with diameters as little as 5 km has been determined on Tethys.

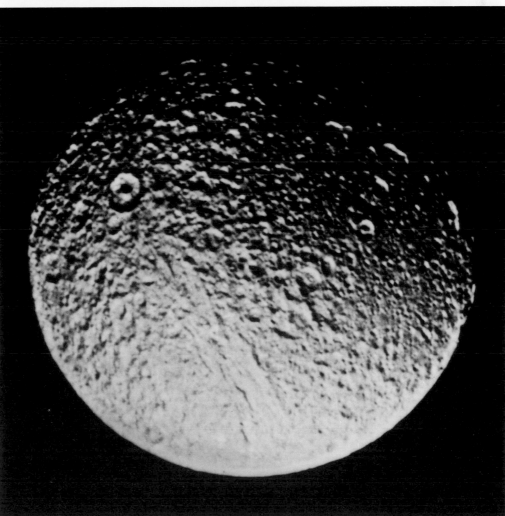

# Map of Tethys

| | | | |
|---|---|---|---|
| Ajax | 285°W,30°S | Ithaca Chasma | 30–340°W,60°S–50°N |
| Anticleia | 38°W,55°N | | |
| Circe | 49°W,8°S | | |
| Elpenor | 268°W,54°N | | |
| Eumaeus | 47°W,27°N | | |
| Eurycleia | 247°W,56°N | | |
| Laertes | 60°W,50°S | | |
| Mentor | 39°W,3°N | | |
| Nestor | 58°W,57°S | | |
| Odysseus | 130°W,30°N | | |
| Penelope | 252°W,10°S | | |
| Phemius | 290°W,12°N | | |
| Polyphemus | 285°W,5°S | | |
| Telemachus | 338°W,56°N | | |

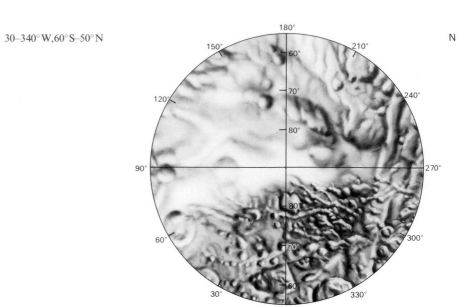

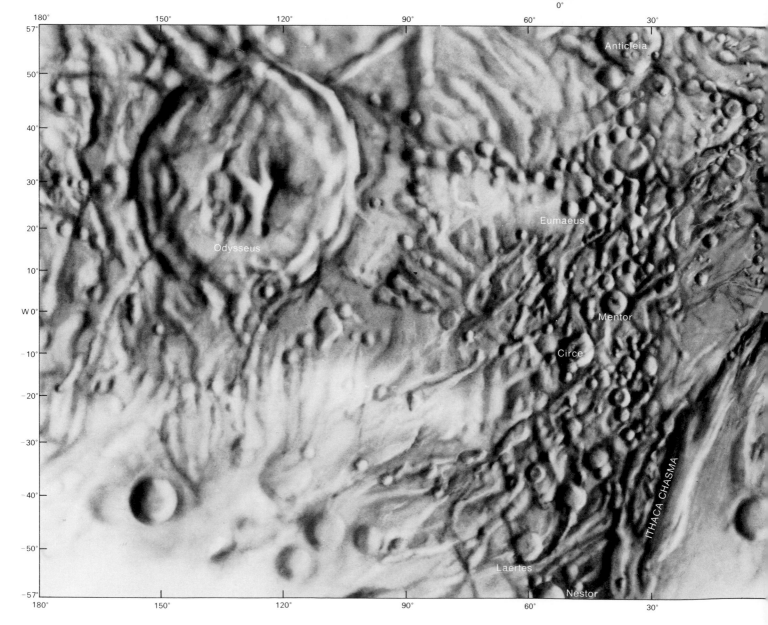

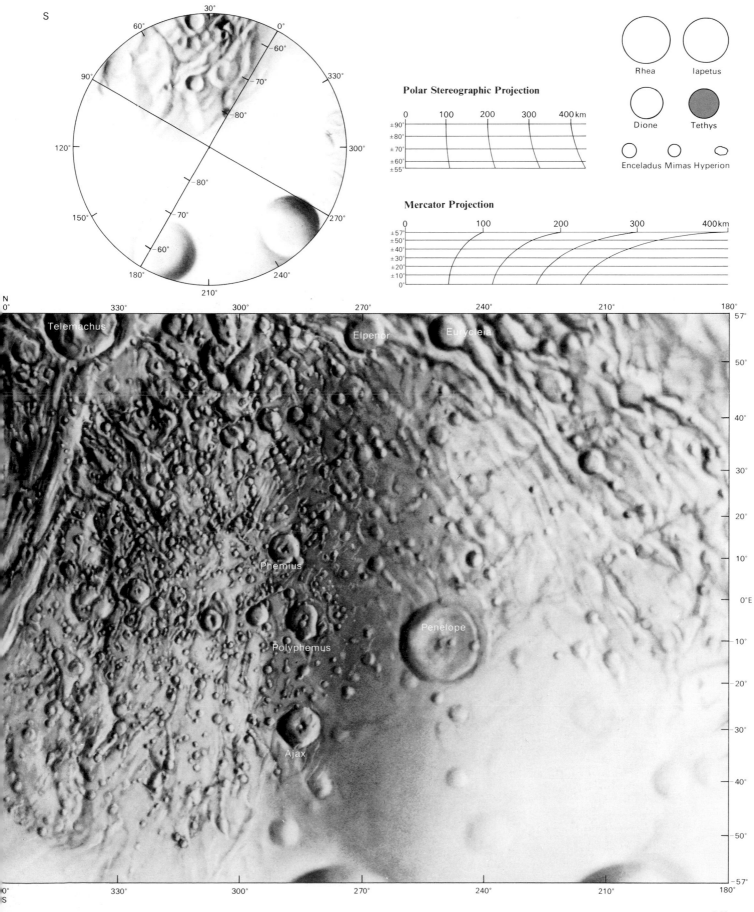

S

30°
60°
0°
90°
-60°
-70°
-80°
330°
120°
-80°
300°
270°
150°
-70°
-60°
180°
240°
210°

**Polar Stereographic Projection**

0    100    200    300    400 km
±90°
±80°
±70°
±60°
±55°

**Mercator Projection**

0    100    200    300    400 km
±57°
±50°
±40°
±30°
±20°
±10°
0°

Rhea    Iapetus

Dione    Tethys

Enceladus  Mimas  Hyperion

N
0°    330°    300°    270°    240°    210°    180°

57°

Telemachus    Elpenor    Eurycleia

50°
40°
30°
20°
10°

Phemius

0°E

Penelope

10°

Polyphemus

20°

30°

Ajax

40°

50°

57°
0°    330°    300°    270°    240°    210°    180°
S

# Dione

Dione is unusual in several respects. Its diameter, 1,120 km, is only slightly greater than that of Tethys, but its albedo is distinctly less, so that as seen from Earth it appears no brighter. Its density, 1.4 times that of water, is considerably greater than for any of the other named satellites apart possibly from Phoebe, and it has been suggested that Dione may affect both the radio emissions from Saturn and the surface of Enceladus, although the latter idea admittedly rests upon fragmentary evidence. The surface features of Dione also differ from those of the other satellites. It was surveyed both by Voyager 1 and, less effectively, by Voyager 2. A considerable portion of the total surface has been studied, although there is a gap between longitudes 80° and about 200°. Some features have been recorded in each polar region. Dione has one small co-orbital satellite, 1980 S6 (Dione B), which moves 60° ahead of it, and another has recently been reported after further studies of the Voyager results.

An important characteristic of Dione is that the brightness of the surface is far from uniform. The trailing hemisphere is comparatively dark, with an albedo of only about 0.3, while the brightest features on the leading hemisphere have an albedo of about 0.6. Only Iapetus, in Saturn's satellite system, displays greater differences in surface brightness.

The most striking feature on Dione is Amata, which lies at the center of a system of bright, wispy features that divide up the dark trailing hemisphere. They appear to be associated with narrow linear troughs and ridges that are extensions of bright lines. These features are possibly the result of internally generated stresses such as those that might have been produced by the freezing of the interior. The bright material, which is probably water ice, may have extruded from inside the satellite. Amata has a maximum diameter of about 240 km and its exact nature is uncertain: it may be an irregular crater, or it may be more in the nature of a basin; images so far have not been clear enough to tell. It is not the focus of a lunar-type ray system, indeed there are virtually no ray-craters on Dione. The wispy features are presumably faults or fractures, and there can be little doubt that Amata is related to their formation, but these features remain something of a mystery.

## Craters

There are not many craters on Dione with diameters of more than 30 km, though a few have 40 km diameters. The larger craters—up to 165 km in diameter—seem to be decidedly shallower than those on Tethys, but some of them have pronounced central peaks. Craters are most numerous on the plain that covers much of the observed part of the region between longitudes 0° and 70°, though the Voyager coverage has not revealed whether or not the crater frequency beyond longitude 70° is equally great. It is thought that this is the most ancient part of Dione's surface: the brighter regions have presumably been resurfaced with a layer of material thick enough to conceal any craters that had been formed there. This may indicate that a former internal heat source kept Dione at least moderately active until a relatively late stage in the satellite's evolution. A high concentration of material undergoing radioactive disintegration may be the heat source. This would be consistent with Dione's density. The satellite is considered by some to be intermediate in type between Enceladus on the one hand, and Mimas and Tethys on the other, although the appreciably greater density of Dione must not be forgotten.

In the southern part of the heavily cratered plain there are a number of broad, low ridges trending in a northeasterly direction, and there are well-marked valleys. Close to the south pole there is a long linear valley, which has a length of about 500 km. The whole of the observed south polar region is thickly cratered; some of these craters have central peaks. In conclusion, Dione shows a very different aspect according to the hemisphere which is being presented to the observer.

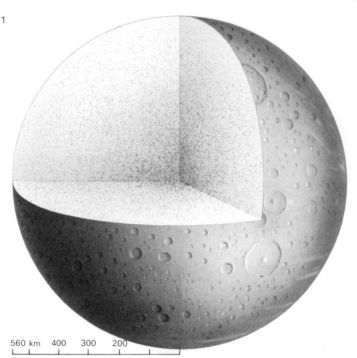

560 km    400    300    200

**1. Interior of Dione**
Dione, though roughly the same size as Tethys, is much denser and is believed to have a higher rock content. If the bright streaks represent an outpouring from the interior, then there was once, and possibly still is, an internal heat source. A relatively high concentration of radiogenic nuclides may be responsible and could also account for the satellite's higher density.

**2. Dione's trailing hemisphere**
From a range of 790,000 km, this Voyager 1 photograph shows the dramatic contrast between light and dark areas of Dione's trailing hemisphere. The bright, wispy features are thought to be surface frost. The bright bands, which are brighter than the brightest features on Jupiter's satellites, may be the result of internal activity at some time during Dione's geological history.

### 3. Dione's bright features
The Saturn-facing hemisphere of Dione is seen here at a range of about 240,000 km (from Voyager 1). The bright radiating features may be the rays of debris thrown out from visible craters. Other bright areas might be ridges and valleys. The irregular valleys are old fault troughs that have been degraded by impacts.

### 4. Craters and faults
The large crater is less than 100 km in diameter and has a well-developed central peak. The valley is probably a fault line.

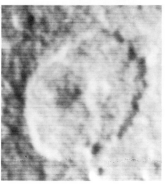

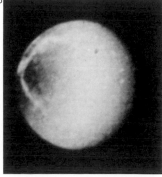

### 5. Crater close-up
This crater—less than 100 km in diameter—was photographed by Voyager 1 from a range of 162,000 km. This feature is part of an area of Dione showing many similar impact craters.

### 6. Dione and Saturn
This Voyager 1 photograph, taken from a range of 377,000 km, shows the satellite against a backdrop of Saturn itself. This view of the anti-Saturn face of Dione shows up well the darker trailing hemisphere and the brighter leading hemisphere.

# Map of Dione

| | | | | |
|---|---|---|---|---|
| Aeneas | 47°W,26°N | Larissa Chasma | 15–65°W,20–48°N | |
| Amata | 287°W,7°N | Latium Chasma | 64–75°W,3–45°N | |
| Anchises | 63°W,35°S | Tibur Chasmata | 60–80°W,48–80°N | |
| Antenor | 8°W,6°S | | | |
| Caieta | 80°W,25°S | Carthage Linea | 310–337°W,10–20°N | |
| Cassandra | 245°W,42°S | Padua Linea | 190–245°W,5°N–40°S | |
| Catillus | 275°W,1°S | Palatine Linea | 285–320°W,10–55°S | |
| Coras | 268°W,3°N | | | |
| Creusa | 78°W,48°N | | | |
| Dido | 15°W,22°S | | | |
| Ilia | 344°W,3°N | | | |
| Italus | 76°W,20°S | | | |
| Latagus | 26°W,16°N | | | |
| Lausus | 23°W,38°N | | | |
| Magus | 24°W,20°N | | | |
| Massicus | 52°W,36°S | | | |
| Remus | 30°W,10°S | | | |
| Ripheus | 29°W,56°S | | | |
| Romulus | 24°W,8°S | | | |
| Sabinus | 190°W,44°S | | | |
| Turnus | 342°W,21°N | | | |

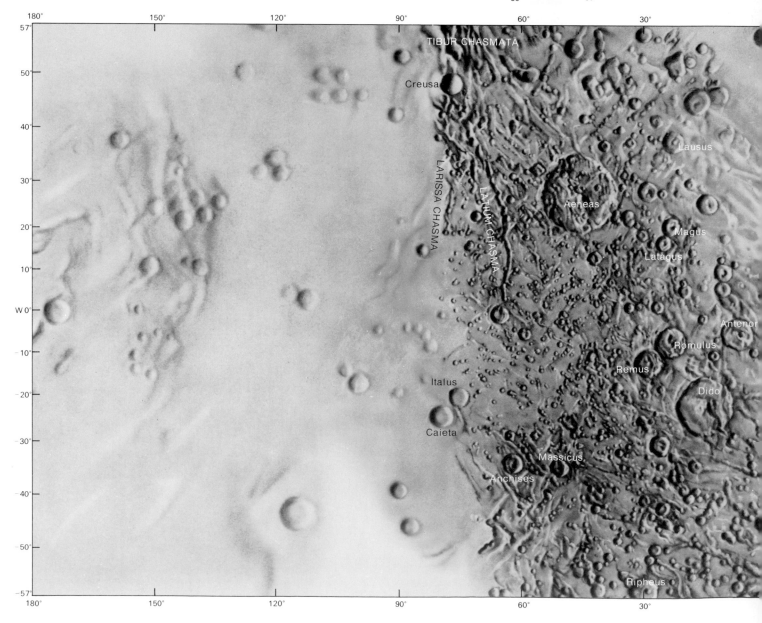

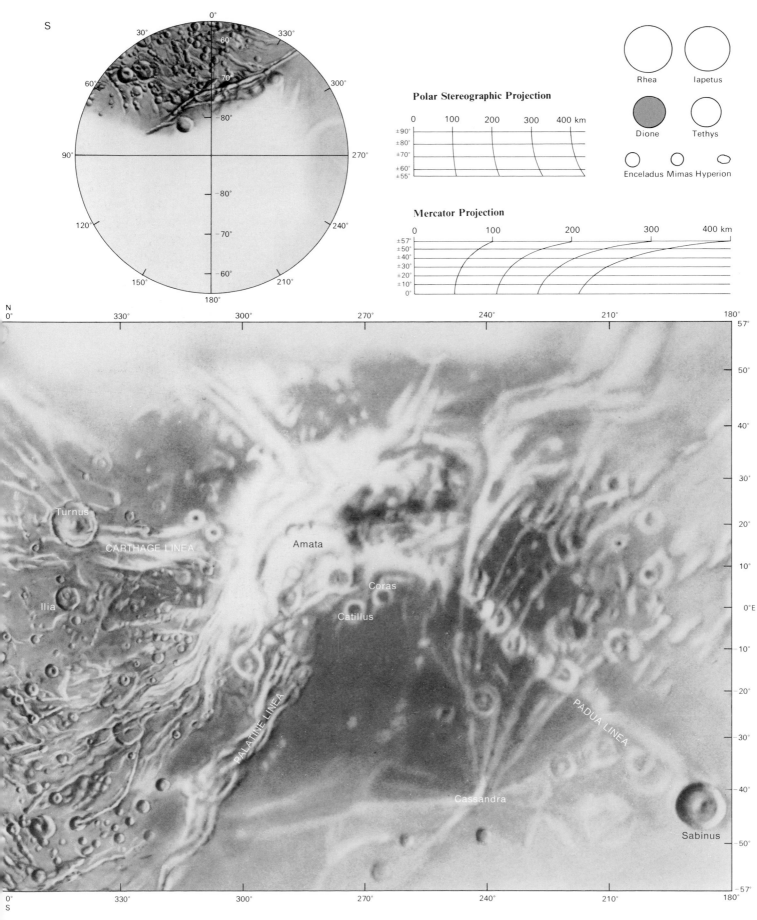

S

30°
60°
90°
120°
150°
180°

0°
−60°
−70°
−80°
330°
300°
270°
240°
210°

−80°
−70°
−60°

Rhea

Iapetus

Dione

Tethys

Enceladus  Mimas  Hyperion

**Polar Stereographic Projection**

0       100       200       300       400 km

±90°
±80°
±70°
±60°
±55°

**Mercator Projection**

0          100          200          300          400 km

±57°
±50°
±40°
±30°
±20°
±10°
0°

N
0°       330°       300°       270°       240°       210°       180°

57°
50°
40°
30°
20°
10°
0°E
10°
20°
30°
40°
50°
57°

Turnus

CARTHAGE LINEA

Amata

Ilia

Coras

Catillus

PALATINE LINEA

PADUA LINEA

Cassandra

Sabinus

0°       330°       300°       270°       240°       210°       180°
S

# Rhea

Rhea is the largest member of Saturn's system of satellites apart from Titan. Its diameter is 1,530 km, slightly greater than that of Iapetus, not slightly less, as was formerly believed. Its density, 1.3 times that of water, is somewhat less than that of Dione but considerably greater than that of Tethys and Iapetus. Voyager 1 bypassed Rhea at a distance of only 73,980 km, and obtained images of greater resolution than for any of the other icy satellites. Voyager 2 made its pass at more than 640,000 km, so it added little to Voyager 1's findings. A large part of Rhea's surface has been surveyed, including more than half the north polar region.

As with Dione and, even more markedly, with Iapetus, there is a pronounced difference between the two hemispheres of the satellite. The leading hemisphere is comparatively uniform and bland, though there is one diffuse feature near 90° which may possibly be a ray-center. This feature is unfortunately situated in an area of Rhea that was not studied at close range, so its precise nature remains uncertain. The trailing hemisphere is darker, and includes some wispy features that are not unlike those of Dione, though they are much less conspicuous; presumably they are the result of the same process, the exact nature of which is as yet uncertain. There is no doubt that the surface of Rhea is icy.

## Craters

The equatorial region, particularly to the north of the equator, is densely cratered, and resembles the rolling cratered highlands of the Earth's moon. It has been commented that if these craters are the result of meteoritic impact, then Rhea may have suffered a terrible battering in the past. Yet the craters are not identical in type to those of some of the other satellites. Many of them are irregular, with a marked tendency to be polygonal in shape. There are no flattened craters. Rhea's low gravity and relatively small diameter permitted rapid cooling and freezing, thus preserving crater forms. The overall impression is that Rhea's outer crust is covered with a layer of what can only be described as rubble, and parts of the surface are extremely ancient even by Solar System standards. Craters follow the usual pattern of distribution, with small craters intruding into larger ones rather than vice versa, and some of the craters have prominent central peaks. Strings and chains of relatively small craters are common. There are few really large craters on Rhea.

## Bright and dark areas

Two areas in the north polar region were better surveyed than any other part of Rhea's surface in as much as Voyager 1 passed right over them at the time of its closest approach. One bright and one dark area whose albedos differ by about 20 percent were found: they are reasonably well defined and the boundary between them lies at approximately longitude 315°. The bright terrain includes a greater number of comparatively large craters, some with fairly bright floors that have relatively little detail. The sharp boundary between the two areas suggests that the difference in crater density reflects differences in the ages of the areas.

According to the impact theory, there were two separate periods of intense bombardment. Between these two there was a period of resurfacing, and the second bombardment produced fewer large craters. In Rhea's case the dark material was extruded from below the surface, covering up the craters that had been produced by the initial bombardment. Localized bright patches seen inside some of the craters could be due to the exposure of comparatively fresh ice on the satellite's surface.

The equatorial region of the satellite includes several areas of low crater density near the equator (60–70° and 300–330°). There is a complex region of narrow linear grooves and troughs in the region from 290° to 330°W and 30° to 40°S. It ends abruptly on the western margin of an area of large craters. Rhea certainly exhibits evidence of a varied geological history.

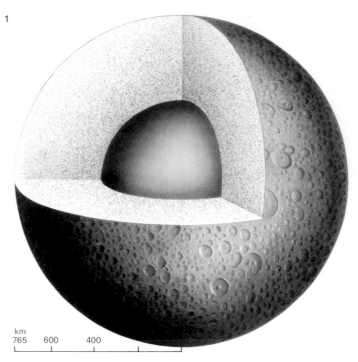

km
765   600      400

**1. Interior of Rhea**
According to some theories, the larger, more dense outer satellites of Saturn, such as Rhea, may have a core. It is believed that Rhea has a roughly 50:50 ice/rock composition, and at some stage in its thermal history rock constituents were differentiated. The core's size is a guesstimate.

**2. Craters**
From the relatively close range of 73,000 km, multiple craters were seen on Rhea by Voyager 1. Many are old and degraded by later impacts or crustal disturbances. Central peaks may have been formed by rebound of the crater floor. Crater diameters are up to 75 km.

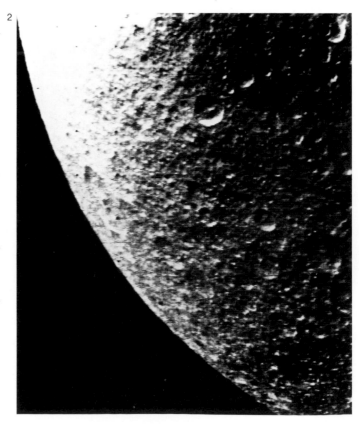

3A

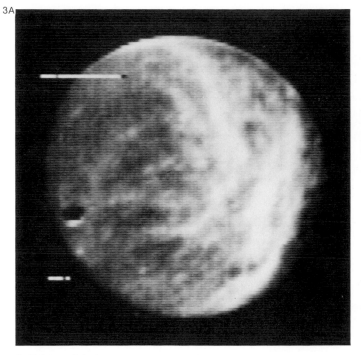

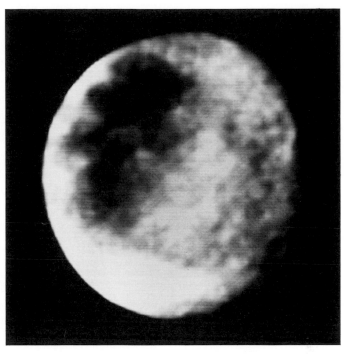

### 3. Bright and dark
Both these Voyager 1 images—taken from 1,925,000 km (**A**) and 2,700,000 km (**B**)—show broad bright areas against a darker background. The entire surface is thought to be covered by water frost and ice: the bright streaks may be pulverized ice particles thrown out from craters.

### 4. North polar region
A mosaic of Voyager 1 images at a distance of about 80,000 km shows the most heavily cratered of Saturn's moons. The largest in this picture is about 300 km in diameter. The lefthand diagram sketches light and dark areas and the right shows new craters (black) and mantled ones.

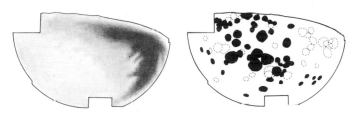

4

# Map of Rhea

| | | | | |
|---|---|---|---|---|
| Aananin | 330°W,39°N | Pedn | 340°W,48°N |
| Adjua | 126°W,46°N | Qat | 347°W,23°S |
| Arunaka | 21°W,14°S | Sholmo | 340°W,13°N |
| Atum | 0°,45°S | Taaroa | 99°W,14°N |
| Bulagat | 14°W,35°S | Thunupa | 15°W,51°N |
| Con | 10°W,24°S | Tika | 87°W,25°N |
| Djuli | 46°W,26°S | Wuraka | 357°W,28°N |
| Faro | 121°W,52°N | Xamba | 347°W,4°N |
| Haik | 27°W,34°S | Yu-ti | 85°W,55°N |
| Haoso | 8°W,9°N | | |
| Heller | 310°W,9°N | Kun Lun Chasma | 275–300°W,37–50°N |
| Iraca | 120°W,45°N | Pu Chou Chasma | 85–115°W,10–35°N |
| Izanagi | 298°W,49°S | | |
| Izanami | 310°W,46°S | | |
| Jumo | 65°W,56°N | | |
| Karora | 16°W,7°N | | |
| Khado | 349°W,45°N | | |
| Kiho | 354°W,10°S | | |
| Kumpara | 321°W,11°N | | |
| Leza | 304°W,19°S | | |
| Lowa | 9°W,45°N | | |
| Manoid | 2°W,33°N | | |
| Melo | 6°W,51°S | | |
| Num | 93°W,23°N | | |

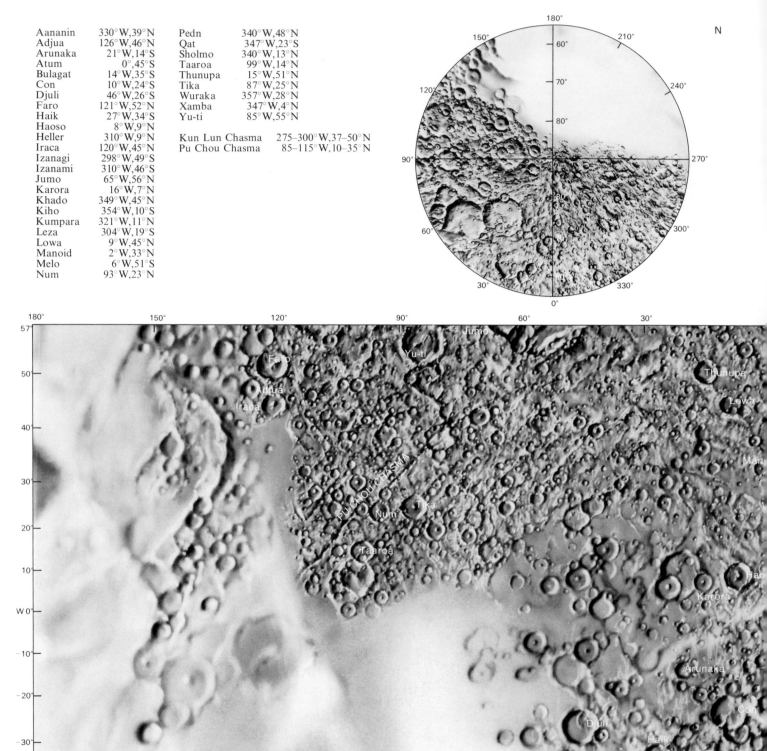

376

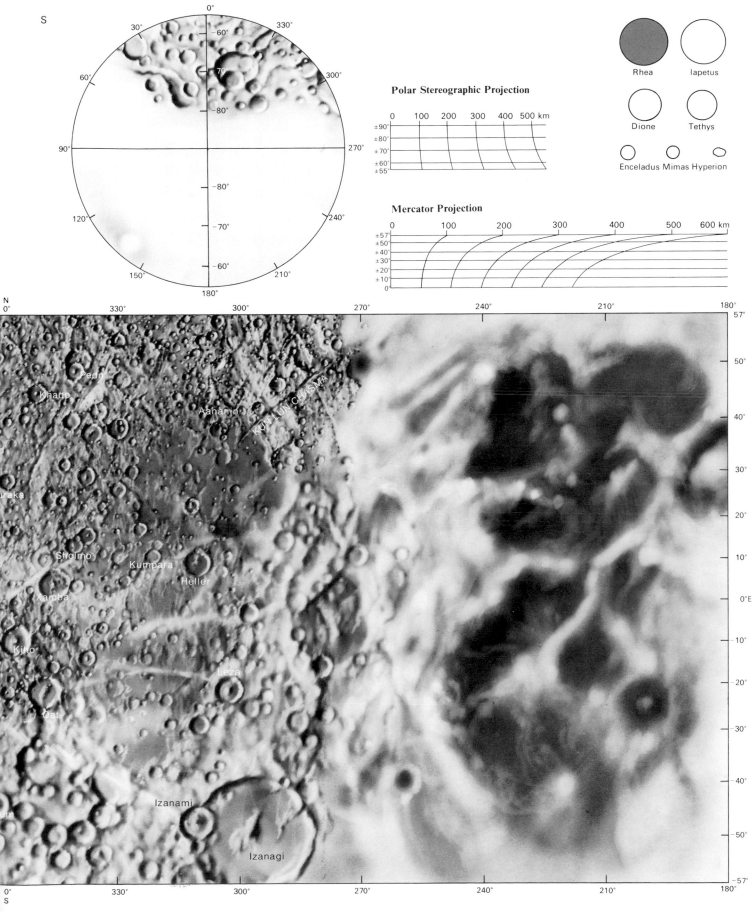

S

0°
30°
330°
60°
-60°
-70°
300°
-80°
90°
270°
-80°
120°
-70°
240°
150°
-60°
210°
180°

Rhea
Iapetus
Dione
Tethys
Enceladus Mimas Hyperion

**Polar Stereographic Projection**

0    100    200    300    400    500 km
±90°
±80°
±70°
±60°
±55°

**Mercator Projection**

0        100        200        300        400        500        600 km
±57°
±50°
±40°
±30°
±20°
±10°
0

N
0°
330°
300°
270°
240°
210°
180°
57°
50°
40°
30°
20°
10°
0°E
-10°
-20°
-30°
-40°
-50°
-57°

Pedn
Khado
Aananin
KUN LUN CHASMA
raka
Sholmo
Kumpara
Heller
Xamba
Kiho
Leza
Oat
Izanami
Izanagi

0°
330°
300°
270°
240°
210°
180°
S

# Titan

Titan, by far the largest of Saturn's satellites, was formerly believed to be the largest satellite in the Solar System. Voyager observations have shown, however, that it is slightly inferior to Ganymede in Jupiter's system. The diameter of Titan is 5,150 km. The mean density of the body is about twice that of water, so that it may be assumed that it is probably composed of equal amounts by mass of rock and ice.

Titan's escape velocity is almost 2.5 km/sec. It is feasible, therefore, that at the distance of the Saturnian system from the Sun, Titan might retain an atmosphere. The first indications of this were given in 1903 by the Spanish astronomer J. Comas Solá, who found that there was considerable limb darkening. Observations of this kind are not easy, and cannot be conclusive, but in the winter of 1943–44 Gerard P. Kuiper showed spectroscopically that Titan has an atmosphere containing methane.

## Constituents of the atmosphere

To the surprise of many investigators, Voyager 1 analyses showed that Titan's atmosphere is made up almost entirely of nitrogen, with less than one percent of methane in the upper part of it. It was found that no surface details were visible at all on the satellite, and the atmosphere was much denser than had been generally believed before the Voyagers. The surface pressure is 1.6 times that on the surface of the Earth, but conditions are very dissimilar; the temperature on Titan's surface is a mere 92 K. However, Titan and the Earth are the only bodies with nitrogen-rich atmospheres.

In addition, Voyager 1 found that there is what may be termed "smog", or aerosols—products of the methane/ammonia chemistry. Acetylene, ethylene, hydrogen, methylacetylene and propane exist there (*see* diagram 1). Carbon dioxide has also been detected and carbon monoxide is suspected but has not yet been confirmed. The hydrocarbon concentrations on Titan are considerably higher than in the case of Jupiter and Saturn. The Titan/Saturn ratio for ethane is about 4, and for acetylene it is about 150; up to the present time ethylene has not been detected on Jupiter or Saturn.

On Titan, about 95 percent of the dissociated methane is irreversibly converted into acetylene and ethane. The constant removal of methane from the upper atmosphere by chemical reactions means that it must be constantly replenished from the surface. The temperature reaches a minimum at a level where the atmospheric pressure is about 200 millibars, and this acts as a "cold trap" for the methane and any other atmospheric constituent whose source is lower down. In effect, the tropopause regulates and maintains a constant supply of methane in Titan's stratosphere.

The atoms and molecules of hydrogen that are produced by photochemical reactions can easily escape from Titan because of the comparatively weak gravitational pull. However, they cannot escape from the pull of Saturn itself, and the result is a doughnut-like torus of hydrogen that remains in the orbit of Titan. Hydrogen is indeed widely distributed throughout the Saturnian system, and appears to form a cloud of uniform density encircling the planet between 8 and 25 $R_s$ near the equatorial plane. Most of the hydrogen seems to be concentrated within 6 $R_s$.

Although molecules of hydrogen cyanide have been observed in interstellar space, this is the first detection of them in a planetary atmosphere. This discovery is of major importance in as much as hydrogen cyanide is a key intermediate between the synthesis of amino acids and the bases present in nucleic acids. Inevitably this has led to the suggestion that Titan may support life, but this would be very improbable in such low temperatures as exist on Titan.

## The color of Titan

Titan's disc appears reddish or orange in color, but the two hemispheres are not identical. The southern hemisphere is relatively uniform in brightness, while the northern hemisphere is darker and redder. At the time of the Voyager 1 pass the north pole

### Composition of Titan's atmosphere

|  | % volume |
| --- | --- |
| Nitrogen | $\approx 94$ |
| Helium | 6 |
| Methane | $10^{-2}$ |
| Ethane | $2 \times 10^{-5}$ |
| Acetylene | $3 \times 10^{-6}$ |
| Propane | $2 \times 10^{-5}$ |
| Diacetylene | $10^{-8} - 10^{-7}$ |
| Methylacetylene | $3 \times 10^{-8}$ |
| Hydrogen cyanide | $2 \times 10^{-7}$ |
| Cyanoacetylene | $10^{-8} - 10^{-7}$ |
| Cyanogen | $10^{-8} - 10^{-7}$ |
| Carbon dioxide | $10^{-10}$ |
| Carbon monoxide | ? |

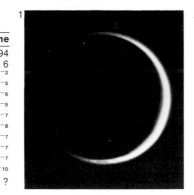

**1. Titan's atmosphere**
Visually, Titan was something of a disappointment, as no surface details were observed by either Voyager. This view of the night side of Titan from a range of 907,000 km shows the satellite's extended atmosphere, revealed by the scattering of sunlight.

**2. Structure of the atmosphere**
Voyager 1 dramatically increased our knowledge of Titan's dense atmosphere. Two previously unknown layers were discovered: an ultraviolet absorption layer that is transparent to light, and below this a haze layer. Beneath the haze is a layer of aerosol particles. Suspended at an altitude of some 200 km, the particles scatter sunlight in the atmosphere and obscure any small-scale contrast of light and dark features that might have been seen on Titan's surface. During the Voyager encounters no breaks were observed in the cloud cover. The aerosol particles are thought to aggregate, forming larger particles, which fall to the surface. Just above the surface are methane clouds and possibly methane rain, although as yet this has not been confirmed. During the Voyager 1 encounter the spacecraft was occulted by the atmosphere and readings obtained indicate a surface temperature of about 95 K and a pressure of 1,500 mb. A thermal inversion, as indicated in the diagram by the thick black line, is anticipated at high altitudes in the atmosphere.

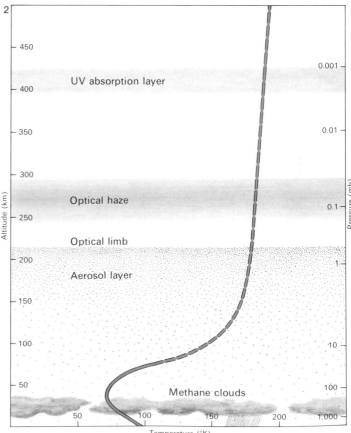

was covered by what appeared to be a dark hood; at the time of
Voyager 2, nine months later, it looked more like a dark collar.
There are certainly substantial differences in the atmospheric
composition at the north pole and the mid-latitude regions; at lower
latitudes there is considerably more acetylene and methylacetylene
than in the polar zones.

The boundary between the two hemispheres lies in Titan's orbital
plane, and this symmetry is almost certainly produced by the
satellite's rotation, which is synchronous. The inclination of Titan's
axis to the perpendicular to the orbital plane is believed to be 5° or
less. Contrast enhancement of images shows up zonal features that
resemble the belts and zones of Jupiter and Saturn.

## Titan's surface

Undoubtedly the surface of Titan is both interesting and varied.
The temperature is near the triple point of methane; that is to say,
methane may exist either as a liquid, a solid or a vapor, as with
water on Earth. There may be a steady rain of methane ice or snow;
there could be cliffs of solid methane; there may be oceans or rivers
of methane, which in the coldest parts may remain rather slushy.
The surface temperature is thought to be about 95 K.
Measurements from Voyager 1 suggest that this varies by about
three degrees only between equator and poles.

## Weather systems

Although no small-scale structures have been observed in the
clouds of Titan, the atmosphere is likely to possess varied weather
systems. In the upper atmosphere, where there are temperature
contrasts of about 20 K, there may be winds as strong as 100 ms⁻¹.
In fact, the rotation of the upper atmosphere is much more rapid
than at lower layers—as on Venus. The circulation of Titan's
atmosphere suggests large-scale features in the troposphere, and
an inter-hemispherical circulation in the stratosphere.

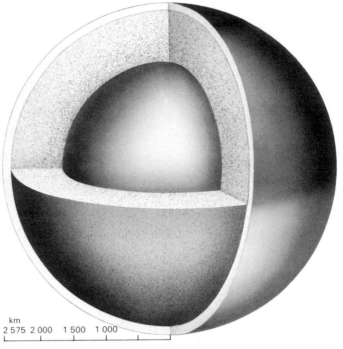

**3. Interior of Titan**
Titan has a density nearly twice
that of water, which suggests that
it is composed roughly of half
rock and half water ice. It has a
solid surface and the absence of
an intrinsic magnetic field means
that there is no liquid electrically
conducting core. One model
suggests that the structure of
Titan is rather like that of
Ganymede in Jupiter's system.
This would mean that there was a
soft ice mantle beneath the crust,
and a relatively large, silicate core
beneath that.

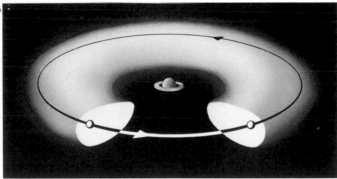

**4. Cross-section through Titan**
Titan's layered atmosphere is
composed of an outer bluish,
translucent haze of organic (that
is, carbon-based) compounds. The
polar hoods are believed to be
concentrations of additional
compounds within this layer.
Beneath it is a relatively clear
zone and then a thick layer of
smog. This again may be
composed of organic compounds
that are characteristically red in
color. This obscures the view
below, but methane clouds may
rain or sleet onto the surface.
Titan's surface is still a matter for
conjecture. It could be solid or
liquid, flat or hilly. Methane may
well play the same role on Titan
as water does on Earth, forming
oceans like those on Earth.

**5. Torus of neutral hydrogen**
A thick disc of neutral hydrogen
atoms surrounds Saturn: this
extends from the orbit of Rhea to
just beyond Titan, giving a width
of about 1,000,000 km. The
hydrogen supply for this torus is
believed to come from Titan.
Atoms are thought to escape from
the upper atmosphere of the
satellite at a rate of about 1 kg/sec⁻¹
to maintain the observed
hydrogen density. The torus,
which may also be composed of
oxygen, co-rotates with the
magnetosphere of Saturn. Titan
has no magnetic field of its own
but interacts with that of the
parent planet. The magnetosphere
flows around the satellite, creating
electrical currents and waves in
the surrounding plasma.

# Iapetus

Iapetus is the outermost of Saturn's larger satellites, moving around the planet at a mean distance of almost 3,560,000 km in the comparatively long period of 79 days, and in an orbit that is appreciably inclined (by 14.7°) to the ring-plane. The diameter is 1,440 km, so that in size it is almost a twin of Rhea, but its density is lower—only slightly greater than that of water—and, therefore, so is its mass. Voyager 1 went nowhere near Iapetus, but Voyager 2 passed by at a distance of 909,000 km, which was close enough to obtain reasonable images, though the surface of Iapetus is not nearly as well known as those of the other icy satellites.

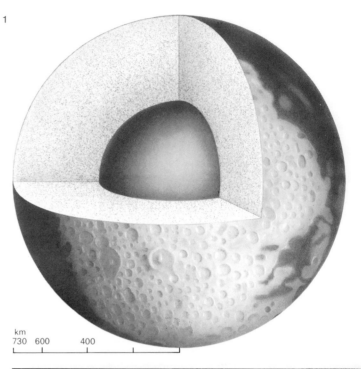

## Light and dark sides

Iapetus is variable in brightness. This was established by G. D. Cassini, who discovered the satellite in 1671. Since it was always accepted that the rotation is synchronous, it followed that either the shape of the globe must be irregular or that the two hemispheres were of unequal albedo. When west of Saturn as seen from Earth, Iapetus is an easy telescopic object; when to the east of Saturn it is markedly fainter.

The Voyager results have shown that the second idea is correct. The leading hemisphere of Iapetus is dark (albedo 0.04–0.05), whereas the trailing edge is bright (albedo 0.5). The demarcation line between the two is not abrupt; there is a transition region with a width of between 200 and 300 km, and the boundary itself is somewhat meandering. It is thought that craters occur in both types of terrain, though unfortunately knowledge about the dark leading hemisphere is incomplete.

The reason for this curious division is still a matter for debate. It is certain that the satellite itself is icy, so that it is the dark material that covers an essentially bright surface rather than vice versa. One suggestion is that the leading hemisphere of Iapetus may have been covered by dark material knocked off the surface of Saturn's outermost satellite, Phoebe, by micrometeoritic impacts; the material would spiral downward towards Saturn and be collected by Iapetus. The symmetry of the dark area supports the idea of an exogenous origin. Phoebe is very small, and never approaches within 7,000,000 km of Iapetus; moreover, investigations carried out by D. Cruikshank from the Mauna Kea Observatory in Hawaii, using UKIRT (the United Kingdom Infra-Red Telescope), have shown that the constituent elements of Phoebe differ in nature from the dark material on the leading hemisphere of Iapetus, making the idea of Phoebe as a source rather implausible.

Another objection, however, to the idea of an extraneous cause is that close to the boundary between the bright and dark areas, at approximately latitude 0° and longitude 330°, there is a ring of dark material, 400 km in diameter. Clearly no such ring could have been produced by material "dusted" onto Iapetus' surface. the nature of this strange feature is still unknown. Also of significance is the fact that some of the craters in the bright trailing hemisphere have dark floors. Some are near the center of the trailing hemisphere, where they would be shielded from impact. This would point to an internal origin of the dark material. There is no evidence of craters on the dark side.

It seems likely that the dark material has extruded from Iapetus' interior. This occurred in the case of the Earth's moon, but the material in that case was lava, and no volcanism of this kind is possible upon an icy world such as Iapetus. The extruded material could be a mixture that includes ammonia, soft ice and a dark substance of some kind or other. It has even been suggested that it might be organic in nature. It is not known how thick the deposit may be; it could be a surface coating a millimeter or two deep, or it could be quite thick. It would seem to be a deposit of some kind and, therefore, younger than the cratered terrain. Both regions are red in color, but there is considerable variation from region to region. The dark red material is similar to that on Callisto in Jupiter's system.

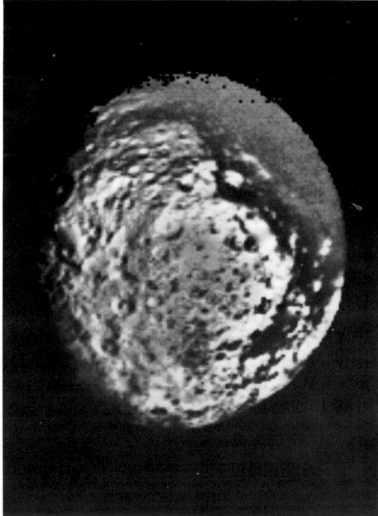

### 1. Interior of Iapetus

Iapetus has a density of about 1.1, so it is almost as low as that of pure water ice, and the body is at least in part covered with ice. There has been considerable speculation about the nature of the dark material and its origin. The low albedo and red color suggest carbonaceous material such as that found in meteorites and asteroids. This may have erupted from the interior in a slurry with ice and ammonia, and may indicate that Iapetus has a differentiated core.

### 2. The dark ring

Voyager 1 only passed within 3.2 million km of Iapetus, but this image of the Saturn-facing hemisphere of the satellite still shows up the dramatic contrast between light and dark regions. Of particular interest is a dark ring extending beyond the dark hemisphere. This feature is approximately 400 km in diameter and has a dark spot in its center. It is probably an impact structure, outlined by dark material that was thrown out from the point where the impacting body struck.

### 3. Bright/dark boundary

A sequence of Voyager 2 images of Iapetus shows the anti-Saturn face of the satellite and the boundary between the bright and dark regions. The dark material of Iapetus' leading hemisphere, which reflects only 4 to 5 percent of the light falling on it, extends into the bright trailing hemisphere (with an albedo of nearly 50 percent) in the region of the equator. The north pole is near the large crater on the terminator. There also appears to be dark

material on crater floors in the bright zone close to the boundary. Such a sharp but complex boundary tends to refute the idea that the dark material originated somewhere out in space and fell to coat Iapetus, and tends to suggest rather that it extruded from the satellite's interior. Neither theory can be confirmed from Voyager data. These images were taken from a range of 1.1 million km and the smallest features that can be seen are about 20 km in diameter.

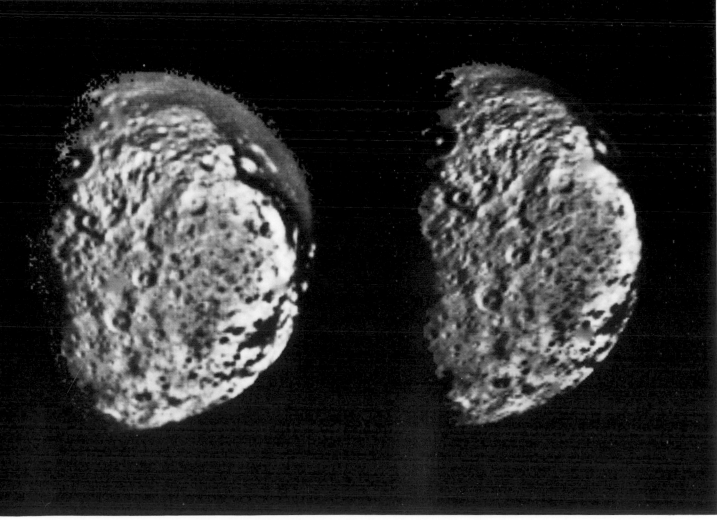

# Map of Iapetus

| | | | | |
|---|---|---|---|---|
| Baligant | 225°W,15°N | | Cassini Regio | 210–340°W,48°S–55°N |
| Basan | 197°W,30°N | | | |
| Charlemagne | 266°W,54°N | | Roncevaux Terra | 130–300°W,30°S–90°N |
| Geboin | 175°W,56°N | | | |
| Grandoyne | 215°W,18°N | | | |
| Hamon | 271°W,10°N | | | |
| Marsilion | 177°W,41°N | | | |
| Ogier | 274°W,42°N | | | |
| Othon | 344°W,24°N | | | |
| Turpin | 0°,43°N | | | |

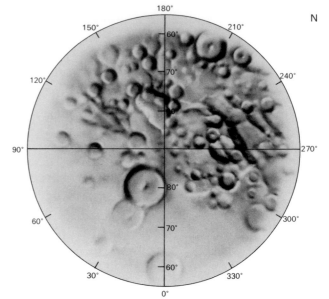

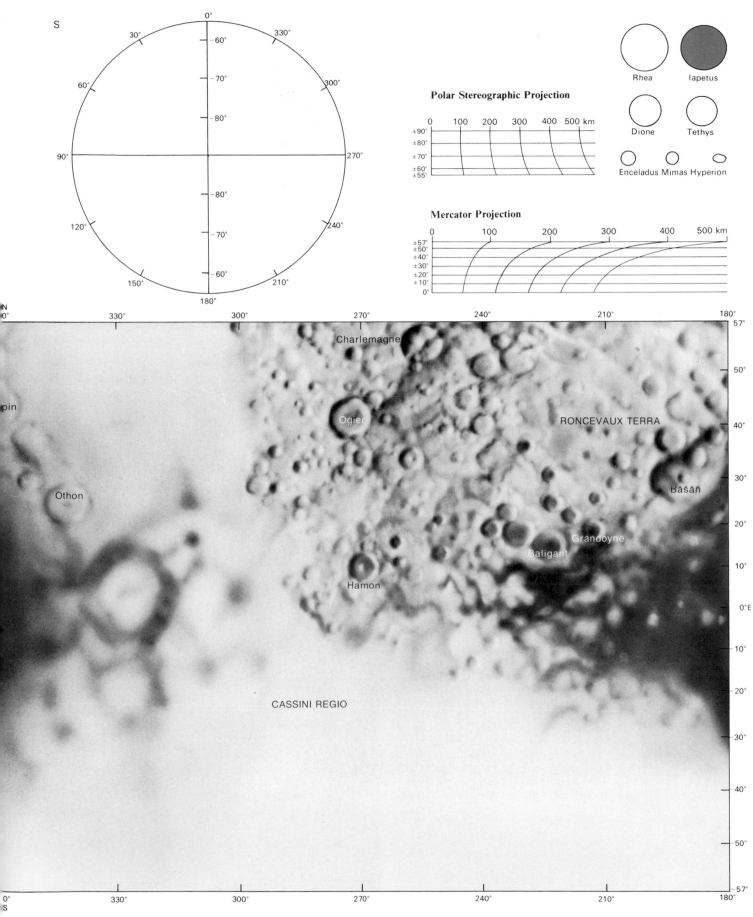

S

0°
30°
330°
−60°
60°
300°
−70°
−80°
90°
270°
−80°
120°
240°
−70°
150°
210°
−60°
180°

Rhea  Iapetus

Polar Stereographic Projection

0  100  200  300  400  500 km
±90°
±80°
±70°
±60°
±55°

Dione  Tethys

Enceladus  Mimas  Hyperion

Mercator Projection

0  100  200  300  400  500 km
±57°
±50°
±40°
±30°
±20°
±10°
0°

N
0°
330°  300°  270°  240°  210°  180°
57°

Charlemagne

pin

50°

Ogier
RONCEVAUX TERRA
40°

30°

Othon
Basan

20°

Grandoyne

Baligant
10°

Hamon

0°E

10°

20°

CASSINI REGIO
30°

40°

50°

S
0°
330°  300°  270°  240°  210°  180°
57°

# Hyperion

Hyperion is one of the smaller satellites of Saturn. Its elliptical orbit lies between those of Titan and Iapetus, but considerably closer to that of Titan. Hyperion was not well surveyed from Voyager 1, which passed by it at a distance of over 880,000 km, but better results were obtained from Voyager 2 at only 470,840 km.

Like all the other satellites, Hyperion provided its quota of surprises. In particular, it is not spherical but irregular in shape, measuring approximately 400 by 250 by 240 km, with angular features and facets as well as rounded features. A body of this size should normally be regular in form, and the strange figure of Hyperion indicates that something unusual may have happened to it in the remote past. Moreover, the long axis is not pointed towards Saturn, as would be expected in a stable configuration; it is at a definite angle. The obvious inference is that Hyperion has suffered collision with another body and has been knocked out of alignment. Eventually it would return to a stable configuration—longest axis turned Saturnward. If this theory is correct, the collision happened a relatively short time ago. Hyperion is a long way from Saturn (almost 1,500,000 km) and tidal interactions are weak, so it would take a long time for a state of equilibrium to be reestablished. Yet what happened to the colliding body? It has been proposed that Hyperion is itself the remnant of a larger satellite that was disrupted at the time of the collision. The rest may have been broken up.

Hyperion has a lower albedo (0.3) than most of the other satellites, and this has been held to be due to material dusted onto it from the surface of the outermost satellite, Phoebe; but if this were so, then the leading hemisphere of Hyperion would be darker than the trailing hemisphere, as with Iapetus, and this does not seem to be the case. There are, however, marked albedo variations (10 to 20 percent) over the whole of the surface. The rotation is not synchronous: there are variations and the current period is 13 days.

## Craters and scarps

Several fairly large craters are to be seen on Hyperion. One of them is at least 120 km in diameter, with about 10 km of relief. There are several other comparatively deep craters of the order of 40 to 50 km in diameter, and the surface is dotted with smaller craters with diameters of 10 km or less. However, the most prominent surface features are the scarps, which are linked to form one long scarp system nearly 300 km in length. It marks the boundary of a feature more than 200 km across which has a broad, rather low, dome-like structure in its center, and may be a crater. Some of the scarps may be as much as 30 km above the mean surface level of the satellite.

Hyperion is not identical in nature to the inner icy satellites, but from its low density it is thought that ice makes up a significant fraction of the composition. Moreover, lines due to water ice have been observed in its reflection spectrum. The relatively low albedo and incomplete ice cover may be the result of extrinsic dark rocky material. If the dark material on the leading side of Iapetus is derived in part from Phoebe, then some of this material could also reach Hyperion and darken and redden the surface. Titan would effectively eliminate any dust ejected by Phoebe that was not swept up by Iapetus or Hyperion but spiralled in further towards Saturn. Thus the surface of the satellites interior to Titan would not be contaminated by the dark material which may affect the surfaces of Iapetus and Hyperion. This method of darkening would require the rotation of Hyperion not to be tidally locked to Saturn, otherwise it should show the same leading–trailing asymmetry of the darkened surface as there is on Iapetus.

**Images of Hyperion**
Voyager 2 images of Hyperion show its irregular shape and areas of light and dark. Impact craters are also clearly visible: their diameters ranging from 120 km (with 10 km of relief) to deep craters with diameters of 40 to 50 km and small features 10 km across. The most prominent features are a series of scarps, some of which are linked.

# Phoebe

The outermost of Saturn's satellite system, Phoebe, was discovered in 1898 and was the first satellite discovery to be made with the aid of photography. Phoebe is smaller than any of the other named satellites, with a diameter of about 160 km, so it is a very faint telescopic object as seen from Earth. Its mean distance from Saturn is more than 10,000,000 km, and it moves in a retrograde orbit with a period of 550.4 days.

Phoebe—alone of the named satellites—was not well surveyed by either Voyager. The minimum distance from Voyager 1 was more than 13,500,000 km, and Voyager 2 came no closer to it than 1,473,000 km. Therefore we have no large-scale pictures of the surface, and Phoebe is less well known than any other member of the Saturnian family. This is unfortunate, because it is unlike any of the remaining satellites, and may be a captured asteroid.

The surface is darkish (with an albedo of about 0.05), but less red than that of Iapetus. The low albedo and color suggest that Phoebe is similar to a class of asteroids that is believed to be common in the outer Solar System. These asteroids are thought to be of primitive composition. This could mean that Phoebe is the first relatively unmodified primitive body in the outer Solar System to be photographed by a spacecraft. A few discrete features can be made out, but nothing definite. The albedo contrasts of these features are as much as 50 percent. It is reasonable to assume that Phoebe is cratered, but proof must await the findings of a future space-probe encounter.

Phoebe, unlike Hyperion, is spherical in form as far as we can tell. From the identified features, it seems that the rotation period is of the order of 9 hr, so Phoebe does not have synchronous rotation, which again may make it unique. More important is the fact that it has a retrograde motion, and the orbit is inclined at 150° to Saturn's equatorial plane.

## Origin of Phoebe

These characteristics have led to the suggestion that Phoebe is not a bona-fide satellite at all, but was captured by Saturn in the remote past, and has not changed since it accreted early in the evolution of the Solar System. It may have been ejected from the inner Solar System by the gravitational field of the growing planet Jupiter. If Phoebe is less heavily cratered than the inner icy satellites, it may be assumed that capture by Saturn took place after the last bombardment of Iapetus, otherwise Phoebe would have been in grave danger of disruption. This, in turn, assumes that craters on the other satellites are indeed mainly of impact origin. Less plausible is the suggestion that Phoebe represents the encrusted nucleus of a gigantic comet. It seems that the mass of Phoebe, slight though it may be by Solar System standards, is too great for the satellite to be cometary in nature.

## The Chiron enigma

The theory that Phoebe is a captured body, formerly moving around the Sun in an independent path, seems less unlikely now than it did a few years ago, because there is positive evidence that asteroidal-type objects do move in this part of the Solar System. The first important discovery was made in 1977 by Charles Kowal, using the Schmidt telescope at the Palomar Observatory in California. During a search for distant comets, Kowal discovered a remarkable object which was found to be moving in an orbit for the most part between those of Saturn and Uranus. It was not a comet, and was assumed to be an asteroid, though the region beyond Saturn is not a region where asteroids are expected to be found. It has been named Chiron, and given an official asteroid number, 2,060. At perihelion its distance from the Sun is 8.5 astronomical units or 1,278 million km, which is just inside Saturn's orbit; at perihelion it recedes to 18.9 astronomical units, or 2,827 million km. Its orbital inclination is 6.9°. The diameter is uncertain because there is no information about the albedo, but it may be about 100 km if its composition is icy. It was believed that it had a fairly low albedo, in which case its diameter could have been as much as 650 km (much larger than Phoebe). In 664 BC it approached Saturn to within a distance of 16,000,000 km, which is not much greater than the distance between Saturn and Phoebe, and it was tempting to believe that the two bodies were of the same type, in which case Phoebe might well have been a latecomer to the Saturnian family.

However, the recent (1989) revelations that Chiron has developed a coma, and may therefore be cometary in nature, cast doubt on this. It is always possible that Chiron has a surface layer which vaporizes near perihelion, in which case it could still be genuinely asteroidal, but whether or not it is in any way similar to Phoebe remains to be seen. At the moment it must be put into a class of its own.

In short, the precise nature of Phoebe is uncertain, but its small size, relatively low albedo, and above all its retrograde motion indicate that it may well be asteroidal.

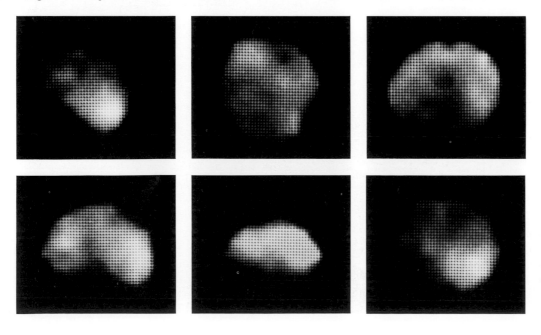

**Images of Phoebe**
Voyager 1 obtained no data on Saturn's most distant satellite. Voyager 2 took a series of images but still from a distance of more than 2.2 million km. The first five frames were at intervals of 70° of longitude. The final one was taken one revolution after the first frame. Even at this distance bright and dark features, with contrasting albedoes of 50 percent, can be seen clearly. The north pole of Phoebe's rotation is at the top of each image.

# OUTER SOLAR SYSTEM

# Uranus

The seventh planet of the Solar System, Uranus, was discovered in 1781 by William Herschel while he was making a telescopic survey of the heavens. He immediately recognized that it was not a star but at first believed it to be a comet. The new planet was later named Uranus, which in Greek mythology was the personification of heaven and ruler of the world. Uranus' great distance makes it a difficult object to study: it is grouped with the giant planets for it is a huge, rapidly rotating, low-density object with an optically thick atmosphere, but Uranus (and also Neptune) has fewer light elements and a greater proportion of methane in its atmosphere than Jupiter and Saturn. Through a telescope Uranus appears as a small bluish-green disc about four seconds of arc in diameter. Not even the largest optical instruments or equipment carried in balloons high in the Earth's atmosphere can detect surface features.

The distance of Uranus from the Sun ranges from 3,004 million kilometers at aphelion to 2,735 million kilometers at perihelion, and it takes 84.01 years to complete its journey around the Sun. The most recent aphelion passage was in 1925 and the next will be in 2009: the most recent perihelion passage was in 1966 and the next will be in 2050. In 1966 Uranus was 2,586 million kilometers from Earth at its closest point.

From Uranus the Sun would be between 1,000 and 1,300 times brighter than full moon seen from Earth. Saturn would be fairly bright when well placed, but Neptune would only just be visible with the naked eye when near opposition.

The axial inclination of Uranus is unique in the Solar System. At an angle of 98°, it makes the planet's rotation technically retrograde, although it is not generally said to be so. The axis is more or less in the plane of the orbit and this means that twice during each orbit of the Sun the axis is at right angles to the direction of the Sun. In between these times one or other of the poles directly faces the Sun, so that the temperature régime there is very unusual indeed. The rotation period is 17.24 hours.

Uranus has an equatorial diameter of 51,118 km; it is therefore very slightly larger than Neptune, though appreciably less massive. The globe is noticeably flattened; the polar diameter is only 49,946 km.

In 1934 R. Wildt proposed that the planet must have a rocky core, overlaid by a thick layer of ice which was in turn overlaid by the atmosphere. In 1951 an alternative model by W.R. Ramsey assumed that the globe was made up largely of methane, ammonia and water. Telescopically there is almost nothing to be seen upon the pale, distinctly greenish disc, but before the Voyager period it was already established that Uranus, unlike the other giant planets, has no strong internal heat source.

There has been much discussion about the reason for the unusual axial tilt. Suggestions that in its early history Uranus was struck by some massive body and literally "knocked sideways" do not seem very plausible, but it is difficult to think of anything better—and it is best to admit that we simply do not know the answer. It is however significant that the satellites move in virtually the equatorial plane.

Five satellites were known before the Voyager 2 mission: Miranda, Ariel, Umbriel, Titania and Oberon, all of which are considerably smaller than our Moon. In 1797 William Herschel announced the discovery of four more satellites, but three of these were certainly faint stars; the fourth may have been Umbriel, but

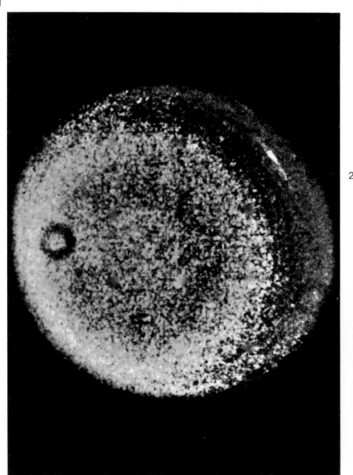

**1. A large bright cloud**
This Voyager 2 false-colour image of Uranus shows a large bright cloud in the atmosphere of the planet at top right.

**2. The crescent Uranus**
Some 1,000,000 km beyond Uranus after the encounter, Voyager 2 transmitted this beautiful image back to Earth.

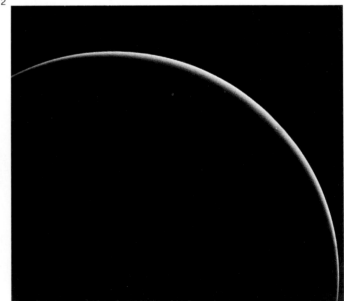

there is considerable doubt. Voyager 2 added ten new, small inner satellites. A system of dark rings was detected in 1977, and subsequently studied in detail by Voyager 2.

### Voyager 2 at Uranus

The only spaceprobe to have by-passed Uranus as yet is Voyager 2, which made its closest approach (80,000 km) on 24 January 1986; Voyager 2 had already sent back data from Jupiter and Saturn, and after leaving Uranus went on to a rendezvous with Neptune in August 1989.

As Voyager drew in, the view was naturally very different from those at Jupiter and Saturn, partly because Uranus is so much less active and partly because the spacecraft was approaching the planet pole-on. It was only a few days before closest encounter that the first definite clouds were seen, so that for the first time the rotation period could be reliably measured.

Radio emissions were first detected five days before closest approach, and at a distance of 27.5 Uranian radii from the planet Voyager crossed the bow shock of the magnetosphere. It was subsequently established that there is a reasonably strong magnetic field, but that the magnetic axis is inclined to the rotational axis by almost 60 degrees—and to make matters even more curious, the magnetic axis does not pass through the centre of the globe; it is perceptibly offset. The magnetic pole facing the Sun (and the Earth) had north magnetic polarity, whereas logically it should be south. It has been suggested, though without proof, that Uranus may be experiencing a "magnetic reversal", or that the dynamo region is closer to the surface than it is on the other planets.

The magnetosphere extends for 590,000 km on the day side and 6,000,000 km on the night side. Ultraviolet observations showed strong emissions on the day side, producing the so-called electro-glow, the origin of which is still somewhat obscure (it may be due to electrons exciting hydrogen molecules in the upper atmosphere).

Two theories of the internal structure were current. It was suggested that there might be a rocky core surrounded by a hot liquid ocean beneath the atmosphere, but it is now thought more likely that there is a dense atmosphere in which gases are mixed with "ices" (that is, substances which would be frozen at the low temperatures at the surface). Hydrogen and hydrogen compounds are dominant in the spectrum; it is methane absorption at red wavelengths which gives Uranus its greenish colour. The planet contains a great deal of water, ammonia and methane; methane freezes at the lowest temperature, and so forms the uppermost cloud-layer. Uranus contains more heavy elements than Jupiter or Saturn, and is in a way intermediate in type between the hydrogen- and helium-rich larger giants on the one hand and the rocky inner planets on the other. Certainly the Jupiter/Saturn pair differs markedly from the Uranus/Neptune pair, but it is also true that Uranus and Neptune are very far from being identical. With its remarkable axial tilt, its lack of a strong internal heat-source, and the 60-degree offset position of its magnetic axis, Uranus is an exceptional world.

Unfortunately, no more space probes to Uranus are scheduled in the foreseeable future, so that we may have a long wait before finding out a great deal more.

---

### 3. Size comparison
Uranus and Neptune are similar in size and are both at least four times larger than the Earth.

| Physical data | |
|---|---|
| | **Uranus** |
| Equatorial diameter | 51,118 km |
| Ellipticity | 0.024 |
| Mass (Earth = 1) | 14.6 |
| Volume (Earth = 1) | 67 |
| Density (water = 1) | 1.27 |
| Surface gravity (Earth = 1) | 1.17 |
| Escape velocity | 22.5 km s$^{-1}$ |
| Equatorial rotation | 17.24 hr |
| Axial inclination | 97.86° |
| Albedo | 0.35 |

3

Uranus    Neptune    Earth

### 4. Changing face of Uranus
The diagram shows the different aspect of Uranus with relation to the Earth during its long revolution period (84 Earth years). The other planets rotate such that their axes of rotation are nearly perpendicular to their axes of revolution. Uranus' axis of rotation lies almost in the plane of its orbit, so that sometimes a pole directly faces the Earth. 21 years later the equatorial region crosses our field of view and 21 years after that we see the other pole. This tilt produces peculiar seasonal effects.

### 5. Interior of Uranus
Uranus is thought to have a rocky core containing metals and silicates, and an icy mantle of methane, ammonia and water. Hydrogen and helium form a "crust" that grades into the planet's atmosphere.

4

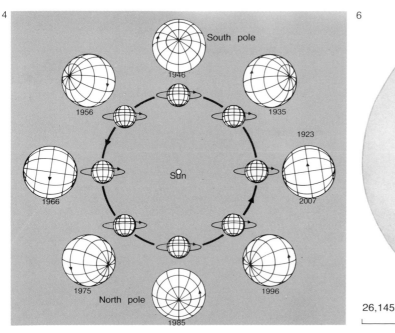

South pole

1946

1956                          1935

1923

Sun

1966                          2007

1975                          1996

North pole

1985

6

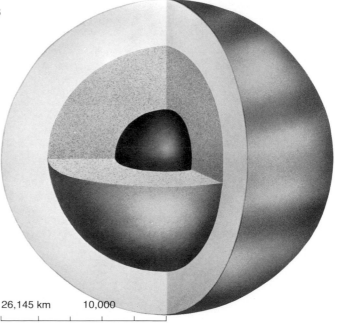

26,145 km          10,000

# Uranus: Rings and Satellites

Uranus was the third planet in the Solar System found to have a ring system. Its discovery came as a complete surprise to astronomers observing the occultation of the star SAO 158687 by Uranus in 1977. Thirty-five minutes before the occultation was due to begin the star "winked" several times; after occultation there were more "winks", symmetrical with the first set, thereby indicating the presence of a ring system—fully confirmed by later occultation results even before the Voyager 2 pass in 1986.

Eleven rings are now known; the system of nomenclature is chaotic, and will no doubt be revised eventually! The rings are thin and narrow, and very dark. Not all are circular, and the outermost ring (the $\epsilon$ ring) is not symmetrical; it is narrowest at its closest to Uranus, and the small satellites Cordelia and Ophelia act as "shepherds" to it. It has been maintained that the $\epsilon$ ring contains dark boulders at least a metre across, and in any case the rings are quite unlike the bright, icy rings of Saturn.

When Voyager 2 took its final picture, after closest approach, it showed several hundreds of very diffuse, nearly transparent bands of microscopic dust surrounding the main rings. There is a definite chance that the rings are young, and may not even be permanent features of the Uranian system.

**The New Satellites**
Miranda, the innermost of the satellites known before the Voyager 2 mission, moves round Uranus at a distance of 129,400 km. New satellites were confidently expected, and in fact ten were found, all closer-in than Miranda. The first of these, Puck, was discovered on 30 December 1985, almost a month before closest approach; it has a diameter of 154 km. Hasty calculations showed that Voyager 2 would pass within half a million kilometres of it on 24 January 1986, and one image was obtained, showing that Puck is roughly spherical, with a low albedo (7 percent). Three craters were recorded, now named Bogle, Lob and Butz.

A second inner satellite (Portia) was found on 3 January 1986, and others followed; all were smaller than Puck, with diameters below 110 km, and all were darkish. Cordelia and Ophelia are the shepherd satellites of the ring; searches for more shepherd satellites, perhaps inside the main ring system, were unsuccessful.

**1. First photograph of the rings**
The first image of the rings of Uranus was obtained by Nicholson *et al* in 1978. Using the 508 cm telescope at Mount Palomar, the scientists scanned the planet at two infrared wavelengths. On one of these images the planet itself appeared darker than the rings and on the other it was brighter. In effect, therefore, the images of the planet were cancelled out and it "disappeared". The rings appear wider than they are because of scattered light.

**2. Structure of the ring system**
Uranus possesses nine narrow rings extending between about 1.6 and 1.96 planetary radii. With increasing distance from Uranus the rings are labelled 6, 5, 4, $\alpha$, $\beta$, $\eta$, $\gamma$, $\delta$ and $\epsilon$. The ring system was revealed during the occultation of the star SAO 158687 by Uranus in 1977. Starlight was recorded before and after its occultation by the planet's disc. Dips in the tracing indicate the presence of nine circular rings. They appear to have well defined edges, which suggests that there are satellites that in effect hold the particles of the rings in position.

**3. Rings and dust**
As Voyager passed by the rings of Uranus, it captured this view of dust and a multitude of rings, backlit by the sun. The bottom ring is the $\epsilon$ ring, thought to consist of boulders at least 1 meter in diameter, and the bright ring above it is the newly discovered 1986 U1R.

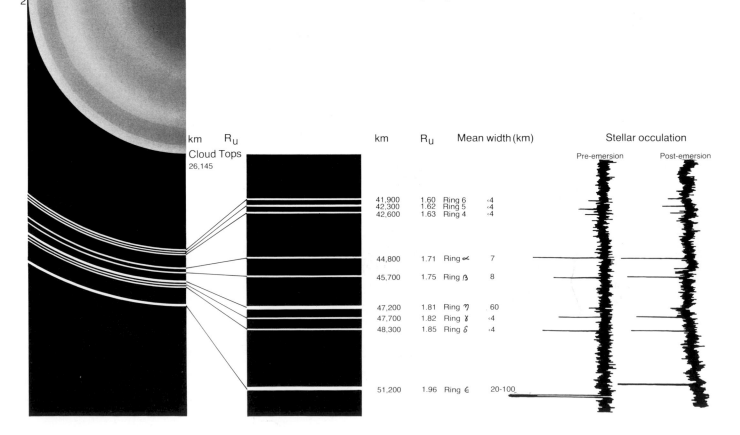

| km | $R_U$ | | Mean width (km) | Stellar occulation |
|---|---|---|---|---|
| Cloud Tops 26,145 | | | | Pre-emersion / Post-emersion |
| 41,900 | 1.60 | Ring 6 | ‹4 | |
| 42,300 | 1.62 | Ring 5 | ‹4 | |
| 42,600 | 1.63 | Ring 4 | ‹4 | |
| 44,800 | 1.71 | Ring $\alpha$ | 7 | |
| 45,700 | 1.75 | Ring $\beta$ | 8 | |
| 47,200 | 1.81 | Ring $\eta$ | 60 | |
| 47,700 | 1.82 | Ring $\gamma$ | ‹4 | |
| 48,300 | 1.85 | Ring $\delta$ | ‹4 | |
| 51,200 | 1.96 | Ring $\epsilon$ | 20-100 | |

**Properties of the Ring System**

| Ring | Distance from Uranus, km | Eccentricity | Inclination degrees | Width km | Optical depth |
|------|--------------------------|--------------|---------------------|----------|---------------|
| 1986 U2R | 37,000–39,500 | 0 | 0 | 2500 | 0.0001–0.001 |
| 6 | 41,850 | 1 | 63 | 1–3 | 0.2 |
| 5 | 42,240 | 1.9 | 52 | 2–3 | 0.5 |
| 4 | 42,580 | 1.1 | 32 | 2 | 0.3 |
| $\alpha$ | 44,730 | 0.8 | 14 | 8–11 | 0.3–0.4 |
| $\beta$ | 45,670 | 0.4 | 5 | 7–11 | 0.2 |
| $\eta$ | 47,180 | 0 | 2 | 2 | 0.1–0.4 |
| $\gamma$ | 47,630 | 0 | 11 | 1–4 | 1.3–2.3 |
| $\delta$ | 48,310 | 0 | 4 | 3–9 | 0.3–1.0 |
| 1986 U1R | 50,040 | 0 | 0 | 1–2 | 0.1` |
| $\epsilon$ | 51,160 | 7.9 | 1 | 22–93 | 0.5–2.1 |

**Satellites of Uranus**

| Name | Discoverer | Year of discovery | Mean distance from Uranus km | Diameter km | Magni- tude | Orbital inclination degrees | Orbital eccentricity | Sidereal period days | Mean synodic period d hr min sec |
|------|-----------|-------------------|------------------------------|-------------|-------------|-----------------------------|----------------------|----------------------|----------------------------------|
| Cordelia | (Voyager 2 images) | | 49,471 | 26 | | | | 0.330 | |
| Ophelia | | | 53,796 | 30 | | | | 0.372 | |
| Bianca | | | 59,173 | 42 | | | | 0.433 | |
| Cressida | | | 61,777 | 62 | | | | 0.463 | |
| Desdemona | | | 62,676 | 54 | | | | 0.475 | |
| Juliet | | | 64,352 | 84 | | | | 0.493 | |
| Portia | | | 66,085 | 108 | | | | 0.513 | |
| Rosalind | | | 69,941 | 54 | | | | 0.558 | |
| Belinda | | | 75,258 | 66 | | | | 0.622 | |
| Puck | | | 86,000 | 154 | | | | 0.762 | |
| Miranda | Kuiper | 1948 | 129,400 | 472 | 16.5 | 0.0 | 0.017 | 1.414 | 1  9 55 31 |
| Ariel | Lassell | 1851 | 191,000 | 1158 | 14.4 | 0.0 | 0.003 | 2.520 | 2 12 29 39 |
| Umbriel | Lassell | 1851 | 266,300 | 1169 | 15.3 | 0.0 | 0.004 | 4.144 | 4  3 28 25.8 |
| Titania | Herschel | 1787 | 435,000 | 1578 | 14.0 | 0.0 | 0.002 | 8.706 | 8 17  0  1.2 |
| Oberon | Herschel | 1787 | 583,500 | 1523 | 14.2 | 0.0 | 0.001 | 13.463 | 13 11 15 36.5 |

# Uranus: The Larger Satellites

All the main satellites were surveyed by Voyager 2, though of course only their day sides were accessible, and our maps are very incomplete—to image the hemispheres which were in darkness at the time of the Voyager encounter, we must wait for another spaceprobe. The satellites were not alike. Miranda has a bewilderingly varied surface; Umbriel has a darkish, rather muted aspect, while the remaining three satellites are brighter and icy, though Ariel shows much more evidence of past tectonic activity than either Titania or Oberon.

### Miranda
Voyager 2 passed Miranda at only 3,000 kilometres, and sent back pictures with a resolution down to 600 metres. There are features of all types, obviously formed at different epochs, which is amazing when we remember how small Miranda is.

One area is apparently ancient and cratered. Another is grooved, with linear valleys and ridges, adjoining a region with curvilinear ridges and troughs. Inverness Corona, nicknamed "the Chevron", is large, dark and rectangular; Arden Corona ("the Race-track") is a huge, grooved area, and there is another grooved enclosure, Elsinore Corona. Craters, ridges, fault valleys, cliffs and scarps abound.

It is difficult to explain this landscape; evidently Miranda has had a varied and troubled history. The surface is of course icy.

### Ariel
Ariel was imaged from 130,000 km, giving a resolution down to 2.4 km. There are craters and ray-centres, but the dominant features are broad, branching valleys with smooth floors such as Korrican

Chasma and Kewpie Chasma. There are also many scarps, cliffs and faults. There must have been considerable activity in the remote past, and the valleys give every indication of having been cut by some liquid—presumably water, since Ariel has an icy surface.

### Umbriel
Umbriel is fainter than Ariel, but has been found to be slightly larger. It was imaged from 537,000 km, so that the resolution is down to only 10 km. The surface is darkish and subdued, but there are prominent craters here and there; one, Skynd, is about 110 km in diameter, and has a bright central peak. The most interesting feature, Wunda, lies near the limb in the picture—this is the satellite's equator, since our view is pole-on. Wunda seems to be a ring about 140 km across, but it is so foreshortened that its nature is uncertain. Obviously, Umbriel's surface is essentially icy, but there is a darker coating. The mean density of the globe is slightly less than that of Ariel, Titania or Oberon.

### Titania
Titania was imaged from a range of 369,000 km. It is only marginally larger than Oberon, but is rather denser and more massive. There is evidence of past tectonic activity; craters are plentiful, with ice-cliffs, valleys and faults. Of the fault valleys, the most impressive is Messina Chasmata, 1,500 km long. One large enclosure, Gertrude, may be more in the nature of a basin than a crater.

### Oberon
The images of Oberon are less detailed than those of the other main satellites; Voyager passed by at 660,000 km, giving a resolution no better than 12 km. Oberon shows less evidence of past tectonic activity than Titania, and inside some of its larger craters, such as Hamlet, Othello and Falstaff, there is dark material—possibly a mixture of ice and carbonaceous material erupted from the interior. One interesting feature is what appears to be a 6-km-high mountain on the edge of the disc; it protrudes from the limb (otherwise it might not be identifiable) but its exact nature remains doubtful.

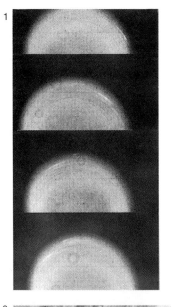

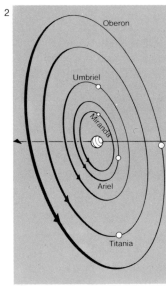

**1. Clouds on Uranus**
This time sequence of images shows the movement of cloud-like features in the atmosphere of the planet.

**2. The satellite orbits**
The five major satellites of Uranus all orbit in approximately the same plane as the planet's equator.

**3. Miranda**
This detail of Miranda shows a region some 250 km across. The large impact crater is about 25 km wide.

**4. Umbriel**
Voyager captured this image of Umbriel from a distance of 537,000 km.

**5. Titania**
This picture of Titania was taken at a distance of 369,000 km.

**6. Oberon**
On encounter day Voyager 2 took this image of Oberon at a distance 660,000 km.

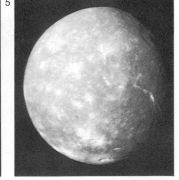

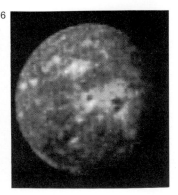

7

**7. Ariel**
At a distance of 130,000 km, this Voyager picture shows a resolution down to 2.4 km. The dominant features are long, smooth-floored valleys, and there are also plenty of craters.

**Satellites of Uranus**

| No. | Name | Discoverer | Year of discovery | Mean distance from Uranus km | Diameter km | Magni-tude | Orbital inclination degrees | Orbital eccentricity | Sidereal period days | Mean synodic period d hr min sec |
|---|---|---|---|---|---|---|---|---|---|---|
| V | Miranda | Kuiper | 1948 | 130,500 | 350 | 16.5 | 0.0 | 0.000 | 1.4135 | 1 9 55 31 |
| I | Ariel | Lassell | 1851 | 191,800 | 1,050 | 14.4 | 0.0 | 0.003 | 2,5204 | 2 12 29 39 |
| II | Umbriel | Lassell | 1851 | 267,200 | 800 | 15.3 | 0.0 | 0.004 | 4.1442 | 4 3 28 25.8 |
| III | Titania | Herschel | 1787 | 438,000 | 1,300 | 14.0 | 0.0 | 0.002 | 0.7050 | 8 17 0 1.2 |
| VI | Titania co-orb. | | | ≈ 438,000 | | | | | | |
| IV | Oberon | Herschel | 1787 | 586,300 | 1,150 | 14.2 | 0.0 | 0.001 | 13.463 | 13 11 15 36.5 |

Titania Co-orbital

Miranda

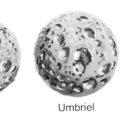

Ariel

Umbriel

Titania

Oberon

| $R_U$ | | 5.13 | 7.54 | 10.5 | | 17.2 | | 23.0 |
|---|---|---|---|---|---|---|---|---|
| km (x 1000) | | 130 | 192 | 267 | | 438 | | 586 |

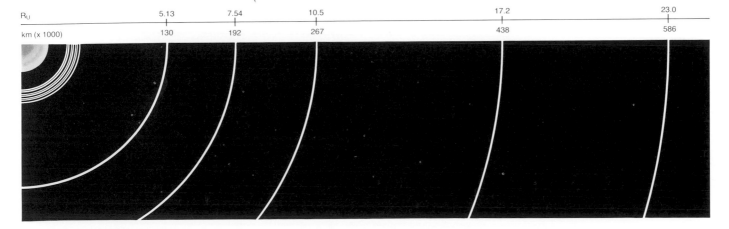

# Neptune

Neptune, the eighth planet of the Solar System, was the first to be discovered by mathematical calculation. Perturbations in the motions of Uranus were studied independently by two mathematicians, John Couch Adams in England and Urbain Le Verrier in France; in 1846 Neptune was visually identified by J. Galle and H. D'Arrest, at Berlin, on the basis of Le Verrier's work.

In size and mass Neptune is very like Uranus. It is slightly smaller (equatorial diameter 50,538 km) but rather more massive. The distance from the Sun ranges between 4,537,000,000 km at aphelion to 4,456,000,000 km at perihelion; the sidereal period is 164.8 years. The rotation period is now known to be 16hr 3m. The axis is inclined by 29 degrees, so that Neptune does not share Uranus' remarkable tilt. The composition is presumably not unlike that of Uranus, with the important exception that Neptune, like Jupiter and Saturn, has a marked internal heat-source. The atmosphere is made up of 85 percent hydrogen, 13 percent helium and 2 percent methane. It is the methane which gives Neptune its characteristic bluish color.

### Voyager 2 passes Neptune
Little surface detail on Neptune can be seen from Earth, and most of our information about the planet is drawn from the results obtained by Voyager 2, which passed Neptune 4,900 km from its cloud-tops on 25 August 1989. Closest approach was over the north pole, which was then in darkness. The encounter was a complete success; all the experiments worked perfectly. Six new inner satellites were discovered, and the existence of rings was confirmed. Detailed images were also obtained of the large satellite Triton. Contact with Voyager 2 is expected to be maintained until about the year 2020, when it will be near the boundary of the heliosphere. In some 296,000 years from now it should pass within 4.3 light years of Sirius, assuming that it has not been destroyed by collision with some wandering object.

### Magnetosphere and Radio Emissions
Radio emissions were detected by Voyager 2, and the existence of a magnetic field was confirmed. The magnetic field is weaker than that of the other giants, with a surface strength of around 1.2 gauss in the southern hemisphere. The magnetic axis is inclined to the rotational axis by 50 degrees, and is offset from the planet's centre by 10,000 km, so that—very surprisingly—Neptune closely resembles Uranus from the magnetic point of view.

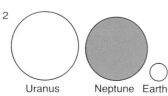

Uranus  Neptune  Earth

**2. Size comparison**
Neptune is slightly smaller in diameter than Uranus but is still four times the size of Earth.

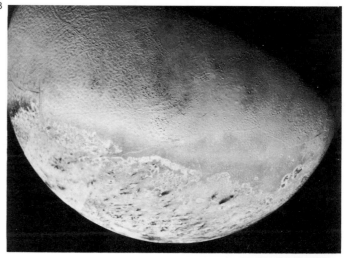

**1. The rings of Neptune**
These two 591-second exposures of the rings of Neptune were taken with a clear filter by the Voyager 2 wide-angle camera on August 26, 1989 from a distance of 280,000 km. The bright glare in the center is due to the over-exposure of the crescent of Uranus.

**3. Triton photomosaic**
This Voyager 2 photomosaic of Triton was assembled from 14 individual frames, and shows the great variety of its surface features.

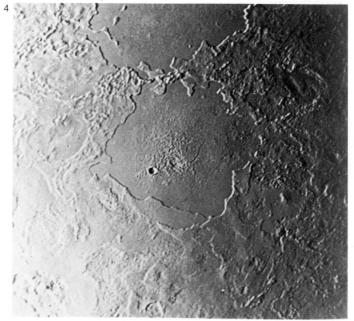

**4. Triton**
This Voyager 2 photograph has a resolution of 900 meters, and is about 500 km across.

**5. Clouds in Neptune's atmosphere**
This image taken from Voyager 2 at a distance of 6,100,000 km clearly shows the Great Dark Spot with its bright white companion.

**Physical data**

| | |
|---|---|
| Equatorial diameter | 50,538 km |
| Polar diameter | 49,600 km |
| Mass | $1.028 \times 10^{23}$ kg |
| Volume (Earth = 1) | 54 |
| Density (water = 1) | 1.77 |
| Surface gravity (Earth = 1) | 1.22 |
| Escape velocity | 24.6 kms$^{-1}$ |
| Equatorial rotation | 16hr 3m |
| Axial inclination | 29.56° |
| Albedo | 0.4 |

## Rings

Ring-arcs had been suspected by observations of stellar occultations, but Voyager showed that there are complete rings. The outer main ring (63,000 km from the planet's centre) is "clumpy", and contains discrete moonlets which may or may not be compacted bodies; there is an inner ring at 53,000 km, and another inner, diffuse ring at 42,000 km. Between the two main rings there is a "plateau" of diffuse material which extends several thousands of kilometres inward, and contains only 10 percent of dust. All the rings are very elusive, and the inner diffuse ring was only just above Voyager's threshold of visibility. Two of the new satellites act as "single shepherds". Searches for additional shepherd satellites were unsuccessful.

## Surface

Neptune is a dynamic planet. On its blue surface the most prominent feature was the Great Dark Spot (GDS), at latitude $-30°$, which in size bears the same relation to Neptune as the Great Red Spot does to Jupiter. Above the GDS are high clouds of methane cirrus, with a clear layer of from 50 to 75 km between the cirrus and the cloud-deck; the clouds change rapidly. There is a

second dark spot further south in latitude, and also a cloud feature nicknamed the "Scooter" because of its relatively rapid rotation. The GDS has a rotation period of just over 18 hours, and the Scooter about 16 hours, so that the wind pattern of the planet can be studied; at the latitude of the GDS there is a 350 ms$^{-1}$ eastward flow, decreasing to zero at latitude $-50°$ and giving way to a more gentle westward wind, again decreasing to zero at the pole. Apparently there is a regular atmospheric cycle. Solar ultraviolet destroys the methane in the high atmosphere, converting it to hydrocarbons such as ethane and acetylene; these descend to the colder, lower stratosphere, where they evaporate and condense. The hydrocarbon ice particles fall back into the warmer troposphere, evaporate, and are converted back to methane. The convective, buoyant methane clouds return methane vapour to the stratosphere, so that the overall methane content remains unchanged.

## Satellites

Of the two satellites known before the Voyager mission, one (Nereid) has a highly eccentric orbit and was poorly imaged, so that little is known about it. One of the new satellites is actually larger than Nereid; it was imaged and found to be dark and irregular in shape, with one large crater. All the new satellites are too close to the planet to be seen from Earth.

Triton has a retrograde orbit. It was found to be smaller and colder than expected (temperature $-236°C$) and has a very thin atmosphere, with a ground pressure no more than 10 microbars, made up of nitrogen with some methane at lower levels. The surface is covered with nitrogen and methane ice, and is varied; the southern hemisphere (sunlit at the time of the Voyager pass) is coated with "pinkish snow", while there is a bluish band near the equator, due presumably to tiny crystals of methane ice. There are few craters, and little surface relief—no more than 100 metres—but many irregular enclosures, and what may be called "frozen lakes". It is thought that around 30 metres below the surface there may be liquid nitrogen. When this percolates upward, and reaches a region where the pressure is about $\frac{1}{10}$ that of the Earth's atmosphere, the liquid "explodes" with a rush of gas and nitrogen ice at an ejection velocity of around 50 ms$^{-1}$. These "ice geysers" produce dark surface streaks, from material blown downwind for up to 75 km. Very probably these ice geysers are active at the present time.

It is likely that Triton is not a true satellite of Neptune, but was captured in the remote past. If so, its original orbit round Neptune would have been eccentric; over a period of 1,000 million years or so, the orbit was forced into the circular form, with tidal flexing and internal heating, so that wide areas were flooded when water-dominated slurries erupted from below. Certainly Triton is quite unlike any other body in the Solar System.

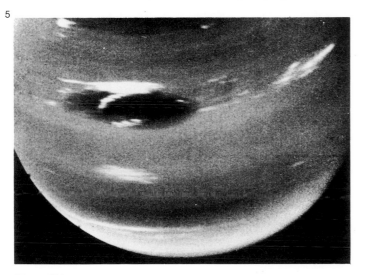

**Rings of Neptune**

| Ring | Distance from centre of Neptune, km | Width km | Dust % |
|---|---|---|---|
| N3A | 42,000 | <50 | 40–60 |
| N2A | 53,000 | <50 | 40–60 |
| "Plateau" | 56,000 | 4,000 | 10 |
| N1A | 63,000 | <50 | 30–60 |

**Satellites of Neptune**

| Satellite | Discoverer | Year of discovery | Mean distance from Neptune km | Mean angular distance from Neptune | Sidereal period days | Mean synodic period d hr min sec | Apparent diameter from Neptune | Orbital eccentricity | Orbital inclination degrees | Diameter km | Magnitude | Mass (Neptune = 1) | Escape velocity kms$^{-1}$ |
|---|---|---|---|---|---|---|---|---|---|---|---|---|---|
| N6 | Voyager 2 | 1989 | 48,200 | | | 0.30 | | Low | 4.5 | 50 | | – | |
| N5 | Voyager 2 | 1989 | 50,000 | | | 0.31 | | Low | <1 | 90 | | – | |
| N3 | Voyager 2 | 1989 | 52,500 | | | 0.33 | | Low | <1 | 140 | | – | |
| N4 | Voyager 2 | 1989 | 62,000 | | | 0.40 | | Low | <1 | 160 | | – | |
| N2 | Voyager 2 | 1989 | 73,600 | | | 0.56 | | Low | <1 | 200 | | – | |
| N1 | Voyager 2 | 1989 | 117,600 | | | 1.12 | | Low | <1 | 420 | | – | |
| Triton | Lassell | 1846 | 353,000 | 16.9" | 5.877 | 5 21 3 29.8 | 1°01' | 0.000 | 159.9 | 2720 | 13.5 | 1/750 | ±2 |
| Nereid | Kuiper | 1949 | 5,560,000 | 4' 23.9" | 359.881 | 362 1 | (av.)19" | 0.749 | 27.2 | 169 | 19 | | |

# Pluto

### Discovery of Pluto

With the discovery of Neptune in 1846, the Solar System was again regarded as complete. However, there were still slight perturbations in the movements of the outer planets that remained to be accounted for, and the possibility of a trans-Neptunian planet could not be ruled out.

Preliminary calculations of the likely position of a new planet were made by Percival Lowell. Unlike John Couch Adams and Urbain Le Verrier before him, Lowell was in a position to undertake a search, and accordingly surveys were made from the Lowell Observatory at Flagstaff. They proved to be negative, and when Lowell died in 1916, his "Planet X" had not been found. Subsequently a search was made by Milton Humason at Mount Wilson on the basis of calculations made by William Pickering, again without result; and for some years the hunt was abandoned.

In 1929, under the direction of Vesto Slipher, astronomers at the Lowell Observatory decided to return to the problem. The Observatory was equipped with a new 33 cm refractor, and a young amateur astronomer, Clyde Tombaugh, was put in charge of the search. During January 1930 Tombaugh took photographic plates of a star-field in the constellation of Gemini, and when he examined them carefully during the following month the discovery of a new planet was made. Tombaugh's method was to photograph the same region of the sky on two nights at times separated by a suitable interval, and then compare the plates by using an ingenious device known as a blink-microscope or blink-comparator. The stars would remain in the same relative positions, but a planet would shift during the interval and its motion would be revealed.

The selected star field was close to the star Delta Geminorum, and the image of the new planet was quite clear, although it was very faint. When the orbit was calculated, the planet's distance from the Sun was found to be greater than that of Neptune. The planet was named Pluto, after a god of the underworld and darkness—a suitable name for the most remote planet currently known in the Solar System.

### Orbit and nature

Yet before long, several disquieting facts were established. First, Pluto is not a giant planet: it is certainly no larger than the Earth, in which case it can hardly exert any measurable influence upon giant planets such as Uranus and Neptune. Second, Pluto's orbit is unusual. Its eccentricity—0.248—is greater than for any other planet, and when near perihelion Pluto comes closer in than Neptune, although a relatively high orbital inclination (17°) means that there is no danger of a collision. Pluto's mean distance from the Sun is 5,900,000,000 km, but this ranges from 7,375,000,000 km at aphelion to 4,425,000,000 km at perihelion. The last perihelion was in 1989. Between 1979 and 1999 Pluto temporarily forfeits its title of "outermost planet". The period of revolution is 248 years.

The small angular diameter means that the size of Pluto is very difficult to measure. For many years the true status of Pluto remained a puzzle. It is also extremely difficult to determine the mass of Pluto.

### Discovery of Charon

One method of measuring Pluto's diameter is to observe the occultation of a star. This method, pioneered by Gordon Taylor of the Royal Greenwich Observatory, had already given good results for some asteroids; but Pluto is so small and so slow moving that occultations by it are very rare. Photographic plates made over the years at the United States Naval Observatory in Flagstaff were carefully studied, and James Christy realized that the image was not symmetrical. It seemed to be elongated and Christy suggested that this might be due to the presence of a close, relatively large satellite. For some time—even after the satellite was given a name, Charon—the independent existence of a companion to Pluto was

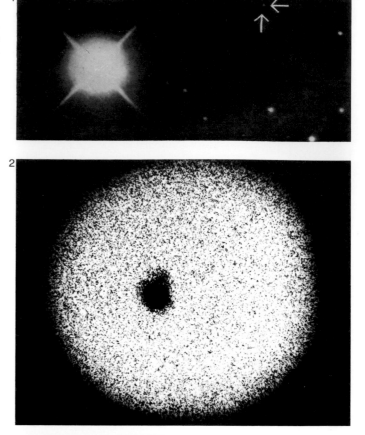

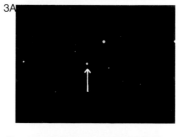

**1. Discovery of Pluto**
Between 2 and 5 March 1930 Clyde Tombaugh noticed the shift of the point of light indicated by the arrows. The overexposed image is of Delta Geminorum, a third-magnitude star. Pluto was thus discovered. It is now of about magnitude 14.

**2. Confirmation of a satellite**
This photograph, taken on 2 July 1978 by Conrad C. Dahn from the US Naval Observatory, provided confirmation of the existence of Pluto's satellite Charon. The bulge just to the right of the top of the image is Charon, whose existence was suspected from earlier images.

**3. Motion of a planet**
Two images taken 24 hr apart show the movement of Pluto.

### Physical data

| | Pluto | Charon |
|---|---|---|
| Discoverer | Tombaugh | Christy |
| Year of discovery | 1930 | 1978 |
| Diameter | 2,445 km | 1,199 km |
| Ellipticity | ? | ? |
| Mass | $6.6 \times 10^{23}$ kg | ? |
| Volume (Earth=1) | 0.01 | ? |
| Density (water=1) | 4.7 | ? |
| Surface gravity (Earth=1) | 0.20 | ? |
| Escape velocity | 7.7 km s$^{-1}$ | ? |
| Equatorial rotation | 6.3 d | 6.3 d |
| Axial inclination | ⩾50°? | ? |
| Albedo | 0.5 | 0.5 |

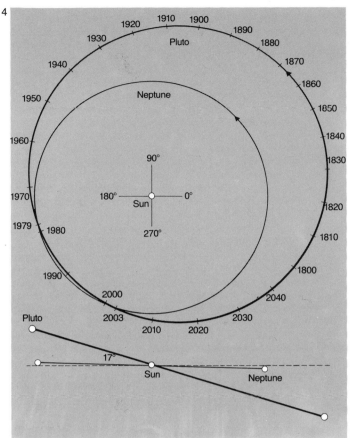

4

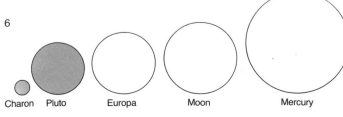

**4. Pluto's orbit**
Pluto has the largest, the most
eccentric and the most highly
inclined orbit of all known planets
in the Solar System. For part of
its orbit Pluto is in fact the eighth
planet from the Sun when it
passes within the orbit of the
planet Neptune.

**5. Charon's orbit**
Charon orbits Pluto at a mean
distance of about 17,000 km in an
almost circular path.

5

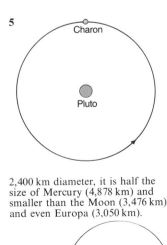

**6. Size comparison**
The revised estimate of Pluto's
diameter means it is the Solar
System's smallest planet. With a

2,400 km diameter, it is half the
size of Mercury (4,878 km) and
smaller than the Moon (3,476 km)
and even Europa (3,050 km).

6

Charon    Pluto    Europa    Moon    Mercury

doubted; but finally French astronomers at the high-altitude
Mauna Kea Observatory in Hawaii, managed to take a picture
which showed the two bodies separately.

As soon as Charon had been found, an estimate of the mass of the
whole system could be derived. It now seems that the diameter of
Pluto is 1,199 km (smaller than the Moon) and that of Charon 2,445
km. The two together have only one-fifth the mass of the Moon.
Moreover, Charon moves round Pluto in a period of 6.3 days: it had
already been established, from brightness variations, that this is
also the rotation period of Pluto itself. The two therefore make up a
"fixed" pair that is unique in the Solar System. The distance
between their surfaces is only 17,000 km.

**Nature of Pluto and Charon**
Because Charon has half the diameter of Pluto, there are grounds
for regarding the pair as a double planet—or, perhaps, a double
asteroid. Yet the two are not alike. In 1988 Pluto occulted a faint
star; just before occultation the star was obviously dimmed, and
from this it has been deduced that Pluto has an extensive if tenuous
atmosphere—at least as deep as Pluto's diameter—whereas
Charon probably has much less.

Probably the atmosphere of Pluto is composed largely of
methane. When Pluto moves out toward aphelion, the temperature
will fall, and the methane may condense out on to the surface, so
that the atmosphere may be a cyclic phenomenon.

From Earth, mutual occultations of Pluto and Charon occur
only for one five-year period every 124 years (half the time that it
takes Pluto and Charon to orbit the Sun). By a very lucky chance,
this period happened during the 1980s; during 1987 and 1988
Charon was periodically hidden by Pluto, and the series of
occultations ends only in October 1990. From this, it has been
possible to deduce that while Pluto is covered in part with methane
frost, the coating of Charon's surface is ordinary water frost. The
movements of the two bodies indicate that Pluto is made up of a
mixture of rock and ice, with rock making up about three-quarters
of the total mass.

In every way Pluto is an enigma. Suggestions that it may once
have been a satellite of Neptune are now largely discounted. It is
unfortunate that none of the current spacecraft will go anywhere
near Pluto and Charon.

**Planet X?**
If Pluto is not the planet for which Lowell was searching, it is likely
that the real Planet X remains to be discovered. After finding Pluto
in 1930, Clyde Tombaugh continued his planet-hunting, and
altogether he examined images including about 90,000,000 stars.
No further planetary objects beyond Neptune were found so that
any new planet was presumably below the limiting magnitude of his
search.

Various attempts have been made to give the position of the
hypothetical Planet X. For example, investigations have been
carried out by J. D. Anderson, of the Jet Propulsion Laboratory in
California, who has studied perturbations of Uranus and Neptune
over past years. He suggests that perturbations were detectable
before 1910, but not since, so that Planet X may have a very
elliptical and inclined orbit, and a period of from 700 to 1,000
years—so that at the moment it is too far away to exert any
measurable perturbations upon Uranus or Neptune.

There is just a chance that clues may be provided by the two
Pioneer and two Voyager spacecraft which are now on their way out
of the Solar System. If any of these craft wanders away from its
predicted path, the approximate position of the perturbing body
might be found. It is a slim chance, but it does exist; and most
astronomers believe that there really is another planet orbiting the
Sun well beyond Neptune and Pluto.

# Comets I

## The nature of comets

Comets have been described as stray members of the Solar System. Spectacular though they may sometimes become, they are not nearly as important as they look, and by planetary standards their masses are very low indeed.

Brilliant comets have caused great alarm in past ages, when they were regarded as forerunners of evil. Even Shakespeare wrote: "When beggars die, there are no comets seen: The heavens themselves blaze forth the death of princes." The Rev. William Whiston, a contemporary of Newton, went as far as to predict that the Earth would eventually be destroyed by collision with a comet. Even in modern times the unease aroused by comets has not entirely died out, and in 1970 a fairly bright comet (Bennett's) caused alarm in Arab countries because it was mistaken for an Israeli war weapon.

In fact, the Earth has on several occasions passed through the tail of a comet without suffering any damage. Even a direct collision with the nucleus of all but the largest comets would cause no more than local devastation, and there is a good chance that the object which impacted in the Tunguska region of Siberia in 1908 was indeed the head of a small comet. Nevertheless it has been seriously proposed that in the past collisions between the Earth and large cometary nuclei may have caused disruption on a global scale.

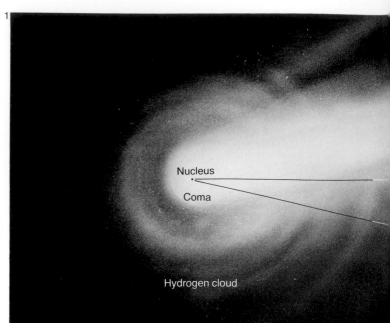

## The orbits of comets

Comets move around the Sun, but most of them do so in extremely elliptical orbits. Because they depend upon reflected sunlight (though they do emit a certain amount of light of their own when near perihelion), comets are visible only when they are relatively close to the Sun and the Earth: few are detected much beyond the orbit of Jupiter. With one exception (Halley's Comet), very brilliant comets have periods so long that they cannot be predicted, and are seen only once in many hundreds or even thousands of years, remaining visible for a few weeks or months before drawing back into the far reaches of the Solar System.

There are, however, many comets with periods of a few years; the shortest period is that of Encke's Comet (3.3 years). Short-period comets are in general too faint to be seen with the naked eye, though their periods and paths are well known.

## The composition of comets

A large comet consists of a nucleus (containing most of the mass), a head or coma, and a tail or tails. The head is made up of small particles, mainly ices, while there are two main types of tails; dusty and gaseous. A gas-tail is generally straight, while a dust-tail is curved. Some comets have one of each. A dust-tail results from dust particles that were impurities in the ices of the nucleus and were released when the ice vaporized. They are usually about one micron in diameter and are believed to be composed of silicates. Gas-tails are made up of ions (for example, $CO^+$, $N_2^+$ and $CO_2^+$) that are blown out behind the comet by the solar wind. These ions are of such a size that they tend to be repelled by the solar wind; thus the tail always points more or less away from the Sun. When a comet has passed perihelion, and is moving outward from the Sun, it travels tail first.

When a comet is far out in the Solar System it has no tail. As it nears the Sun and is heated, a tail develops, only to shrink once more as the comet moves outward after passing perihelion. Many of the smaller comets (including most short-period comets) never develop appreciable tails.

If a comet loses some of its material each time it comes to perihelion, it must be comparatively short-lived; and this has been demonstrated by the fact that several comets have eventually disintegrated. Biela's Comet, which used to have a period of 6.75 years, broke in two at its return in 1845; twin comets came back in

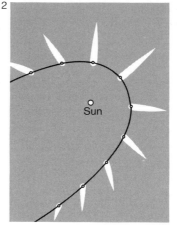

**1. Structure of a comet**
A comet consists of the coma, nucleus, hydrogen cloud and tail. The coma is a spherical envelope of gas and dust surrounding the nucleus. The "icy conglomerate" model of a nucleus envisages several irregular layers of ice around a core and encased by an outer crust. Some comets are surrounded by a huge hydrogen cloud that can extend for millions of km and is believed to be produced by photodissociation.

**2. A comet's tail**
The growth and decay of the tail of a comet as it travels around the Sun is noticeable. The tail always points away from the Sun, and comets can travel tail first.

**3. Comet Morehouse**
Morehouse's Comet of 1908 had a tail with a complex structure that changed as the comet moved around the Sun. It is thought that structural changes within tails of a comet are indicative of various

disturbances within it, but in this case the comet was not bright enough to observe the details of these changes from the Earth. Many comets exhibit multiple tails and/or changes of structure within any one tail.

Tail

1852, but they have not been seen since, and are assumed to no longer exist. There are several other cases of "dead" comets; for instance, Ensor's Comet of 1925 faded out as it approached the Sun and finally vanished. Short-period comets are believed to have lost much tail-forming matter in previous passages around the Sun.

## Cometary nomenclature
Comets are generally known by the names of their discoverer or discoverers, though occasionally by the name of the mathematician who first computed the orbit. Amateur observers have a fine record of comet discoveries; for instance G. E. D. Alcock, a Peterborough schoolmaster, has now discovered three, while the bright comet of 1970 was first seen by the South African amateur Jack Bennett. The record number of comet discoveries is held by the nineteenth-century French astronomer J. L. Pons, who found no less than 37.

When discovered, a comet is given a letter against its year; thus Comet 1982a was the first comet to be found in 1982. The comet is numbered as it passes perihelion; thus the first comet to come to perihelion in 1989 was 1989 I, the second 1989 II, and so on. A periodical comet is distinguished by P; thus Encke's Comet is P/Encke, Halley's Comet P/Halley, and so on.

## Comets and meteor showers
There is a close association between comets and meteor showers; indeed many meteors may be regarded as cometary debris. As a comet moves, it leaves a "dusty trail" after it, and after a sufficient period this trail is spread out all along the orbit, so that we see evidence of it each year as the Earth crosses the orbit. Thus the Perseid meteors, seen between about 27 July and 17 August, move in the path of Comet P/Swift-Tuttle. Biela's lost comet was responsible for a shower of meteors each November, although the stream has become very sparse in recent years.

## Comet families
Many faint short-period comets have their aphelion points at about the same distance from the Sun as the orbit of Jupiter, and are referred to as Jupiter's comet family. Whether or not similar families belong to other giant planets is much less certain, and there can be no doubt that Jupiter, with its strong gravitational pull, is the dominating influence. Comets are of such slight mass that they are easily perturbed, and their orbits may be completely changed by an encounter with a giant planet. There are also a few comets with orbits that are much less eccentric; a good example is Comet P/Schwassmann-Wachmann 1, which has a revolution period of 15 years, and keeps between the orbits of Jupiter and Saturn.

## Origin of comets
It is generally believed that comets are genuine members of the Solar System. According to a theory developed by J. H. Oort in 1950, there is a diffuse cloud or reservoir of gas, dust and comets that is gravitationally part of the Solar System but about 40,000 AU distant from the Sun. The passage of stars close to the cloud disturbs comets, which then fall in towards the center of the Solar System. Oort estimated that there must be about 100 billion comets in the cloud. They probably originated as planetisimals in the solar nebula (*see* pages 10–11). If a comet passes at a suitable distance from a giant planet (usually Jupiter) on its way into the Solar System it may be thrown into a short-period orbit. If it is not so perturbed, it will move back towards the Oort cloud, not to return to the neighborhood of the Sun for a very long time.

There are some astronomers, notably V. Clube and W. Napier of the Royal Observatory, Edinburgh, who believe comets to be of interstellar origin, so that they are captured by the Sun instead of being original members of the Solar System. This remains at present a minority view, but it cannot be discounted; there are many questions about the origin of comets that remain to be answered.

**Data table for periodical comets observed at more than one return**

| Comet | Period yr | Dist. from Sun(AU) min. | max. | Eccent. | Inclin. |
|---|---|---|---|---|---|
| Encke | 3.3 | 0.34 | 4.09 | 0.85 | 12.0 |
| Grigg-Skjellerup | 5.1 | 1.00 | 4.94 | 0.66 | 21.1 |
| Tempel 2 | 5.3 | 1.36 | 4.68 | 0.55 | 12.5 |
| Honda-Mrkós-Pajdusakováá | 5.3 | 0.58 | 5.49 | 0.58 | 13.1 |
| Neujmin 2 | 5.4 | 1.34 | 4.84 | 0.57 | 10.6 |
| Tempel 1 | 5.5 | 1.50 | 4.73 | 0.52 | 10.5 |
| Tuttle-Giacobini-Kresák | 5.6 | 1.15 | 5.13 | 0.63 | 13.6 |
| Tempel-Swift | 5.7 | 1.15 | 5.22 | 0.64 | 5.4 |
| Wirtanen | 5.9 | 1.26 | 5.16 | 0.61 | 12.3 |
| D'Arrest | 6.2 | 1.17 | 5.61 | 0.66 | 16.7 |
| Du Toit-Neujmin-Delporte | 6.3 | 1.68 | 5.15 | 0.51 | 2.9 |
| Di Vico-Swift | 6.3 | 1.62 | 5.21 | 0.52 | 3.6 |
| Pons-Winnecke | 6.3 | 1.25 | 5.61 | 0.64 | 22.3 |
| Forbes | 6.4 | 1.53 | 5.36 | 0.56 | 4.6 |
| Kopff | 6.4 | 1.57 | 5.34 | 0.55 | 4.7 |
| Schwassmann-Wachmann 2 | 6.5 | 2.14 | 4.83 | 0.39 | 3.7 |
| Giacobini-Zinner | 6.5 | 0.99 | 5.98 | · 0.71 | 31.7 |

**4. Comet Arend-Roland**
The comet of 1957 was the best recent example of a comet with a Sun-pointing "tail". This spike is not in fact a tail but meteoritic debris lying along the comet's orbit that is being illuminated.

**5. Brooks' Comet**
Brooks' Comet of 1911 displayed a particularly brilliant coma and a fan-shaped tail that had several rays. The coma shines by means of fluorescent processes when excited gases radiate energy.

4

5

# Comets II

Comet-hunting is extremely time-consuming. A comet that becomes brilliant as it nears the Sun may appear at first only as a dim patch of misty light against the starry background. Comets are also unpredictable: some brighten up rapidly when approaching perihelion, while others do not. A recent case of a comet that promised more than it fulfilled was that of Kohoutek's Comet of 1973. It was discovered by L. Kohoutek at Hamburg Observatory on 7 March when it was still almost 700,000,000 km from the Sun. Few comets are detectable as far away as this, and it was expected that Kohoutek's Comet would become very brilliant during the winter of 1973–4, but in the event it was only dimly visible with the naked eye. However, it was studied by astronauts on Skylab, and found to be surrounded by a huge envelope of rarefied hydrogen.

## Multi-tailed comets

Many comets show both gas and dust tails. There are three types of cometary tail: type I includes tails that are long and straight and often have streamers and knots or other structures within them. Types II and III are curved tails and progressively shorter than Type I. They tend to be fuzzy and have little internal structure. There are also cases of multi-tailed comets. Perhaps the most celebrated of these appeared as long ago as 1744, and is generally known as Comet Chéseaux, although in fact it was seen by the Dutch astronomer Klinkenberg four days before de Chéseaux saw it from Switzerland. There were at least six bright, broad tails, although unfortunately the comet remained bright for only a brief period, and there are few reliable drawings of it.

Probably the most beautiful comet of near-modern times was Donati's of 1858. The main tail was curved, with two shorter, straight ones. During October 1858 the length of the tail reached 80,000,000 km. The period is not known: 2,000 years may be a conservative estimate. The nature of a comet's tail can change over a period of days or even hours. Comet Mrkos in 1957 displayed long streamers of a Type I tail as well as a curved tail of Type II. The extent of each tail varied considerably over a period of 48 hr. The Great Comet of 1811 was one of the most spectacular on record. Its long, straight tail stretched over a distance of about 160,000,000 km, and the diameter of the coma was 2,000,000 km— larger than the Sun. However, it was surpassed in brilliancy by the Great Comet of 1843, whose tail attained a record length of 330,000,000 km, considerably greater than the distance between the Sun and the orbit of Mars.

## Exceptional Orbits

Many of the brilliant comets that can be classed as non-periodical (because their periods are extremely long) pass close to the Sun, and are intensely heated. There are also cases of comets which have failed to survive perihelion; the first known instance was Comet Howard-Kooman-Michels of 1979, which was certainly destroyed.

It is likely that many of the Sun-grazing comets represent the fragments of larger comets that have broken up in the remote past. It follows that when a Sun-grazer appears, we may hope to see others in similar orbits. There are also comets which have very large perihelion distances—such as Schuster's Comet of 1976, which came to perihelion midway between the orbits of Jupiter and Saturn. The most distant comet ever observed was Bowell's Comet of 1982, which was followed out to a distance of more than 1,500,000,000 km, between the orbits of Saturn and Uranus.

## Dead comets

Some small comets look like steller objects, and there are indeed two periodical comets, Neujmin II and Arend-Rigaux, which now show no traces of cometary material. It has been suggested that the Earth-grazing asteroids of the Apollo type are nothing more nor less than comets that have been stripped of all their gaseous components, leaving only their nuclei.

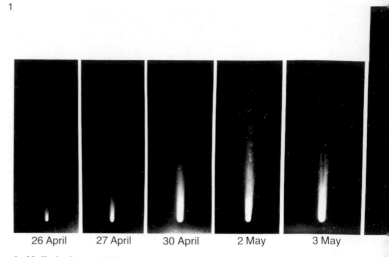

26 April   27 April   30 April   2 May   3 May

**1. Halley's Comet 1910**
A series of photographs clearly shows the growth and decay of the tail of Halley's Comet as the comet approached perihelion and then retreated. The seventh picture in the sequence shows the tail shortly before perihelion passage, and reveals an enormous bright tail. By the time the comet was last seen the tail had disappeared altogether.

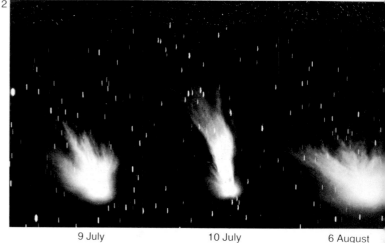

9 July   10 July   6 August

**4. Comet Kohoutek 1974**
Comet Kohoutek displayed a fan-shaped tail that was made up of several streamers. With the naked eye the comet was a disappointing object, but several rocket-borne instruments and telescopes revealed its spectacular form and also the presence of an enormous hydrogen cloud, which extended far into space. The comet will not return for 75,000 years.

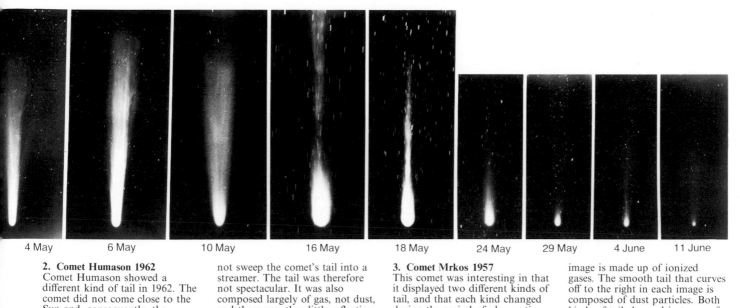

| | | | | | | | | |
|---|---|---|---|---|---|---|---|---|
| 4 May | 6 May | 10 May | 16 May | 18 May | 24 May | 29 May | 4 June | 11 June |

**2. Comet Humason 1962**
Comet Humason showed a different kind of tail in 1962. The comet did not come close to the Sun and, consequently, the extremely weak solar wind could

not sweep the comet's tail into a streamer. The tail was therefore not spectacular. It was also composed largely of gas, not dust, and there was thus little reflection, which normally occurs from dust.

**3. Comet Mrkos 1957**
This comet was interesting in that it displayed two different kinds of tail, and that each kind changed during the period of observation. The straight tail to the left in each

image is made up of ionized gases. The smooth tail that curves off to the right in each image is composed of dust particles. Both kinds of tail changed in terms of shape and extent.

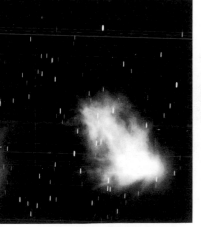

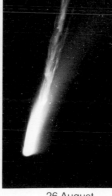

| | | | | |
|---|---|---|---|---|
| 8 August | 22 August | 24 August | 26 August | 27 August |

5A

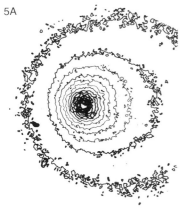

**5. A comet's structure**
The structure of the hydrogen cloud of Comet Kohoutek in 1974 is revealed in a computerized composite image (**A**). The contour lines represent different levels of intensity on the original ultraviolet photographs. The structure of Comet Bennett, a spectacular naked-eye object in 1970, can be seen in an equal density tracing (**B**). The head of the comet is composed of concentric shells which can also be manifested on photographic images as concentric halos, probably the result of rotation of the comet's nucleus.

**6. Collision with the Sun**
A coronagraph aboard a US satellite recorded the collision of a comet with the Sun, the first evidence of such an event. The comet streaked towards the Sun

at great speed (**A**) before impact, after which there was a cometary dust storm (**B**) high above the solar surface. Eleven hours after the comet disintegrated this glow could still be seen.

6A

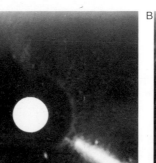

B

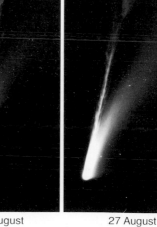

B

# Halley's Comet

Comets were once regarded as atmospheric phenomena. Most European observers agreed with Aristotle that comets were simply "exhalations" of the atmosphere and few measurements of the positions of ancient comets were made as they were thought to move in straight lines rather than in ellipses. The first prediction of the return of a comet was made by Dr Edmond Halley (1656–1742), Britain's second Astronomer Royal. Halley was a younger contemporary and friend of Sir Isaac Newton and consequently was very well acquainted with the latter's work. Newton had demonstrated that a body moves around the Sun in ellipses, but calculations of such orbits were extremely difficult.

In 1682 a bright comet appeared. Halley observed it, and when he computed the orbit he found that it was moving in a path very similar to those of comets seen previously in 1607 and 1531. He concluded that the three objects must be one and the same, and that the comet had a revolution period of approximately 76 years, so that it might be expected to return in 1758. After Halley's death the comet indeed returned on schedule; it was first seen on Christmas night 1758, and passed through perihelion in early 1759.

### Halley's comet in history

Halley's Comet is the only periodical comet that can become a bright naked-eye object. It has been traced back to 240 BC, and it is possible that it may also have been the comet seen in 467 BC. It returned in 87 BC and 12 BC, and since then it has been observed at every perihelion.

As a result of its changing position with respect to the Earth, Halley's Comet is not equally brilliant at every return. It was splendid in 989, and also in 1301 and in 1456; on the latter occasion Pope Calixtus III went as far as to issue a declaration against "the Devil, the Turk and the comet". At other returns it has been much less striking, and unfortunately conditions during the last perihelion passage, that of 1986, were about as unfavorable as possible. The next return of Halley's Comet will be in 2061 but this too will be unfavourable.

Because of perturbations by Jupiter and Saturn, the revolution period of the Comet is not constant, and may be anything between 74 and 78 years. It is associated with two meteor streams, annual Eta Aquarids and the Orionids.

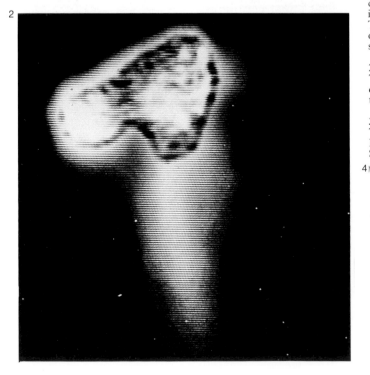

**1. Bayeux Tapestry**
One notable return of Halley's Comet was in 1066, when William of Normandy was preparing to invade Britain. The Bayeux Tapestry shows Harold tottering on his throne while his courtiers stare up at the comet in terror.

**2. The nucleus of Halley's Comet**
This view of Halley's Comet during its 1986 return came from the Soviet probe Vega 2.

**3. Halley's Comet, March 1986**
This telescopic photograph of Halley's Comet was taken in South Africa in mid-March 1986.

**4. Path of Halley's Comet**
This series of drawings shows the relative motions of the Earth and the comet from 15 May 1909 to 29 July 1910.

**5. Halley's Comet 1910**
In 1910 conditions for observing Halley's Comet were better. Here the comet is shown in the morning sky with Venus.

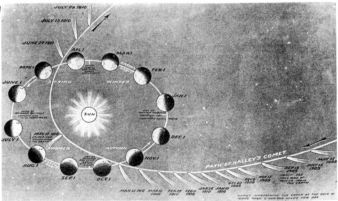

### The 1910 return

At its last visit, in 1910, Halley's Comet was very bright, although it was less spectacular than a comet seen a few weeks earlier—in January 1910—which was known as the Daylight Comet because it was visible even when the Sun was above the horizon. Many people who today claim to have seen Halley's Comet in 1910 really observed the Daylight Comet.

The 1910 return was first detected from Germany on 12 September 1909 by Max Wolf; it was then more than 400,000,000 km from the Sun. The comet remained under observation until June 1911. On 18–19 May of that year it passed directly between the Earth and the Sun, but no trace of it could be seen—confirmation of the extreme flimsiness of cometary material.

### The 1986 return

The return of 1986 was eagerly awaited, even though it was known that the general appearance would be disappointing as a spectacle—at the time of perihelion, the comet was on the far side of the Sun, and was out of view; even at its best it was never prominent, though it was an easy naked-eye object.

Preparations had been made to send spacecraft to encounter the comet. The Americans withdrew, on the grounds of expense, but there were five probes in all; two Japanese (Suisei and Sakigake), two Russian (Vegas 1 and 2) and one European (Giotto). All were successful, and Giotto actually went through the comet's head, obtaining close-range images of the nucleus. Giotto's camera functioned until 14 seconds before the closest approach, when the spacecraft was jolted by the impact of a "dust" particle about the size of a rice grain.

The comet's nucleus was found to measure $15 \times 8 \times 8$ km, and to be dark, so that there was a layer of material overlying the icy nucleus itself. Water ice was found to be the main constituent of the nucleus, with some formaldehyde, carbon dioxide and other volatiles. Dust-jets were active, though only from a small area of the nucleus on the sunward side. The temperature of the sunward side was 47°C, far higher than expected. At each return the comet must lose around 300,000,000 tons of material. The rotation period of the nucleus was found to be 53 hours with respect to the long axis, with a 7.3-day rotational period around this axis.

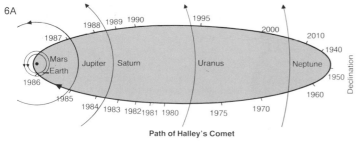

Path of Halley's Comet

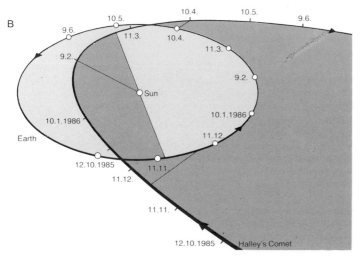

**6. Orbit of Halley's Comet**
The diagram shows the orbital positions of Halley's Comet from 1940 until 2010 (**A**). Each year between 1975 and 1995 is marked which clearly shows how slowly the comet moves when at aphelion and how quickly it moves around the Sun at perihelion. As a result, the comet is close to the Earth for only a small part of the orbital period. Its perihelion passage from 1985–6 can be seen in (**B**) in a perspective drawing.

**7. Halley's Comet, January 1989**
This photograph was taken by Richard West using the 60-inch telescope at La Silla, Chile in January 1989. The CCD image shows the comet as a blurred object near the centre; the hard lines are star trails and the fuzzy line is the trail of a galaxy. The magnitude of the comet was then 26!

# Meteors

## Shooting stars

There are many tiny particles, usually smaller than a grain of sand but sometimes tens of meters in diameter, orbiting the Sun. When beyond the Earth's atmosphere these particles are much too small and faint to be seen and are known collectively as meteoroids, but when, attracted by the Earth's gravity, they enter the upper reaches of the atmosphere and become heated by friction, they destroy themselves in streaks of radiance popularly called shooting stars. The luminous appearance comes not from the tiny particles themselves, but from the effects produced in the atmosphere as they plunge downwards. The molecules of gas are heated, becoming ionized so that they glow and produce a nebula around the dust particle. Meteors may enter the air at velocities of up to 70 km/sec. By measuring the velocity and deceleration of a meteor, its mass can be calculated. Most have a mass of only a few milligrams. It was thought at one time that some meteors had velocities exceeding the escape velocity of the Solar System, but this does not now appear to be the case and virtually all meteors that have been recorded belong to the Solar System.

The first measurements of the heights of meteors were made in 1798 by two German students, Brandes and Benzenberg. They used the method of triangulation—that is, observing the same meteor from two different stations, so that the apparent position against the background of stars was not the same for each observer. They concluded, correctly, that most meteors penetrate to an altitude of about 80 km above the ground before all or most of each particle is vaporized. They end their journeys in the form of fine dust. It is now known that the Earth is also bombarded by even smaller particles about 0.1 mm in diameter, known as micrometeorites, which cannot produce luminous effects, though their tails may be detectable using radar observations.

## Shower and sporadic meteors

Meteors can be of two distinct types: shower or sporadic. Sporadic meteors appear at random from any direction at any moment. The best time to observe them is after midnight, when the darkest region of the sky is in the direction of the Earth's orbital motion. On average an observer will see about six meteors every hour. Each will last for about 30 sec and travel 5° across the sky. The shower meteors are members of streams, each of which is probably associated with a comet. The debris of a short-period comet may extend all around its orbit, and every time the Earth passes through this orbit we see a shower of meteors: thus the August Perseids are the debris of Comet P/Swift-Tuttle, while the October–November Taurids are associated with P/Encke.

Many annual showers have been identified, and some last for several days. The number of meteors observed is measured in terms of the ZHR, which stands for zenithal hourly rate. This is defined as the number of meteors that would be seen by a naked-eye observer during one hour, with the radiant at the zenith or overhead point and under perfect conditions. In practice this situation rarely arises, so that the observed hourly rate will always be rather lower than the theoretical ZHR. Moreover, the ZHR values are not constant for any particular shower. In particular, the Leonids are generally rather sparse, although on occasions the ZHR has been known to rise briefly to something like 60,000.

## Radiants

Because the meteors of a stream are moving through space in parallel paths, individual meteors will appear to radiate from one particular position in the sky. The shower name depends upon the constellation in which the radiant lies; the Perseids are from Perseus, the Lyrids from Lyra and so on.

The Quadrantids of early January come from a position that was once included in the constellation of Quadrans. Quadrans has been deleted from the maps, but the name is preserved in the meteor

**1. Leonids of 1833**
The Leonids of 1833 presented a wonderful spectacle, as seen in this woodcut. It was said of this November shower that "meteors rained down like snowflakes".

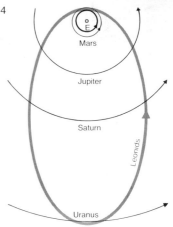

**2. Leonid radiant**
The Leonids were photographed from Kitt Peak in 1966. All the individual meteors appear to radiate from one particular position in the sky, in this case from the direction of the constellation of Leo. The diagram illustrates the principle of the radiant of a meteor shower.

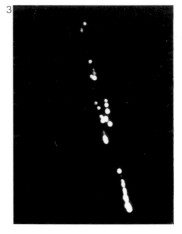

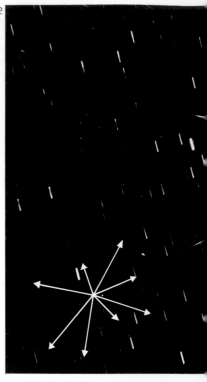

**3. Break-up of a meteor**
This dramatic photograph shows the break-up of a Leonid meteor as it enters the atmosphere of Earth. The luminous effects are not produced by the particles themselves but as a result of the gases of the atmosphere being heated. Molecules become ionized and then they glow, producing nebulosity around the particles.

**4. Orbit of the Leonids**
The orbit of the Leonid meteor stream intersects the orbits of the planets Earth, Mars, Jupiter, Saturn and Uranus. Material is not evenly distributed along the orbit so the intensity of the showers varies from one to the next. Showers sometimes do not occur at all as a result of planetary perturbations.

**Major annual meteor showers**

| Shower | Duration | Max. | ZHR | Associated Comet |
|---|---|---|---|---|
| Quadrantids | 1–6 Jan | 3 Jan | 110 | ? |
| Lyrids | 19–24 Apr | 22 Apr | 12 | Thatcher 1861 I |
| Eta Aquarids | 2–7 May | 4 May | 20 | P/Halley |
| Delta Aquarids | 15 July–15 Aug | 28 July | 35 | ? |
| Perseids | 27 July–17 Aug | 12 Aug | 68 | P/Swift-Tuttle |
| Orionids | 12–16 Oct | 21 Oct | 30 | P/Halley |
| Taurids | 26 Oct–25 Nov | 4 Nov | 12 | P/Encke |
| Leonids | 15–19 Nov | 17 Nov | var. | P/Tempel-Tuttle |
| Geminids | 7–15 Dec | 14 Dec | 58 | ? |

shower (the radiant lies close to Ursa Major). This is a very brief shower, generally lasting for only a few hours, which suggests that the meteors are very bunched up: there is no known associated comet, and it is probable that the comet concerned has long since disintegrated. Generally the richest annual shower is that of the Perseids, though the Geminids may also be spectacular. Again, shower meteors are more plentiful after midnight, when the darkened hemisphere of the Earth is leading and meteors tend to meet it head-on.

## Meteor storms
Occasionally there may be intense displays of meteors. Of these the most famous are due to the Leonids, which are associated with the comet P/Tempel-Tuttle. This comet has a revolution period of 32.9 years and last reached perihelion in 1965. In most years the ZHR is very low, but when the Earth passes through the thickest part of the swarm the results are spectacular. This happens once every 33 years or so. There were brilliant displays in 1799, 1833, and 1866, and again in 1966; the meteor storms of 1899 and 1933 were missed, as the orbit of the stream had been perturbed by Jupiter and Saturn. In 1966 the maximum was very short, but it reached a rate of 60,000 per hour over a period of 40 minutes. The shower occurred during daylight in Europe, although it was well observed from other parts of the world. We may be fairly confident in predicting another major Leonid display in 1999.

The greatest cometary shower of the twentieth century was that associated with Comet P/Giacobini-Zinner. On 9 October 1933, 350 meteors a minute could be seen during a brief period.

## Origin of sporadic meteors
Although sporadic meteors do not come from any particular part of the sky, there is no reason to doubt that they too are cometary debris; either the original streams have been dispersed by the perturbations of the planets, or else the relevant comets have broken up. Sporadic meteors are more common than those from showers, although they are much less easy to observe because they are completely unpredictable.

Very occasionally a meteor can reach magnitude $-15$ or even $-20$, which is brighter than the full Moon. On very rare occasions they may rival even the Sun. In this case the meteor is termed a "fireball". It is possible sometimes to hear the sounds of their passage as they break up into smaller pieces. Some fireballs are due to larger bodies that survive the drop to the ground and are then termed meteorites, but others may come from showers. It is important to note that there is no certain association between meteorites and comets.

## Observing meteors
Until comparatively recent times, meteor observations depended upon visual work, carried out largely by amateurs. The method is to keep a constant watch during the observing spell, obviously concentrating on the region of the radiant. When a meteor is seen, it should be timed, and the duration, magnitude, color (if any) and track should be noted. Photography may be attempted, simply by leaving the camera shutter open; the stars will then appear as trails, and any meteors of sufficient brilliancy will be recorded. Obtaining meteor spectra is very time-consuming, but here too amateurs have carried out invaluable work. It is possible to obtain spectra of bright meteor trails and make a chemical analysis, but direct analysis can be done if larger particles survive to reach the Earth's surface as meteorites.

Today radar is used in meteor research, since a trail left by a falling meteor will produce a radar reflection, and it is even possible to record daytime showers. However, what may be termed "old-fashioned" visual work is still useful, and there are groups of meteor observers scattered all over the world.

**5. Meteor trail**
The Great Bolide meteor trail was photographed in 1923 from an observatory in Prague. The picture clearly shows galaxies and other celestial bodies that formed the background against which the meteor was captured.

**6. Sporadic meteors**
Sporadic meteors can appear at any time and from any direction. While this photograph of Comet Brooks was being taken, a meteor flashed across the line of sight, leaving a trail across the image.

**7. Exploding meteor**
This Andromedid meteor exploded during its descent through the Earth's atmosphere in 1895, providing fine photographic evidence of the fate of meteors on contact with the Earth.

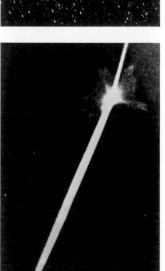

# Meteorites

**Stones from the sky**

In 1807 President Thomas Jefferson of the United States went on record as saying: "I could more easily believe that two Yankee professors would lie than that stones would fall from heaven." Yet even by that time the existence of meteorites had been demonstrated. In April 1803 a shower of "stones" had fallen near the village of L'Aigle in France, and had been examined by the astronomer J. B. Biot, who proved that they were of cosmic origin.

The earliest reports of meteoritic phenomena were recorded on Egyptian papyrus, about 2000 BC. Later, there were records of falls in Crete (1478 BC), Italy (634 BC) and Aegospotamos in Greece (416 BC). The Sacred Stone in Mecca is certainly a meteorite. The oldest meteorite that can be positively dated fell at Ensisheim in Switzerland on 16 November 1492; it is displayed in the church.

Most museums have collections of meteorites. The largest specimen on public display was found by Robert Peary in Greenland in 1897, and is at present in the Hayden Planetarium, New York. However, the largest known meteorite lies where it fell in prehistoric times near Grootfontein, in southern Africa. It is not likely to be moved, since it weighs more than 60 tonnes.

**Types of meteorite**

Traditionally, meteorites are divided into three types: stones (aerolites), stony irons (siderolites) and irons (siderites), although the dividing lines are not always clear-cut. Irons generally contain between four and six percent nickel; stony irons consist of a nickel-iron network enclosing crystals of olivine, whereas stones contain spherical particles known as chondrules, which are fragments of minerals. Chondrules account for about 85 percent of known aerolites. Of special interest are the rare meteorites known as carbonaceous chondrites. Some of these contain "organized elements" that bear a striking resemblance to organic life forms. Some of these elements may result from contamination after the meteorite landed on Earth, but others are believed to be indigenous to the meteorite. Evidence tends to suggest, however, that the latter elements constitute pre-living organic matter. Suggestions made by the Swedish scientist S. Arrhenius in 1908, that terrestrial life might have been brought here by way of a meteorite, have met with little support; the theory raises more difficulties than it solves, though it has recently been revived, in modified form, by Sir Fred Hoyle and Chandra Wickramasinghe.

Meteorites can also be divided into differentiated and undifferentiated meteorites. The most abundant chemical elements found in chondritic meteorites are the same as in the Sun's atmosphere, so the meteorites must be primitive planetary material that has not since been melted. Differentiated meteorites (irons, stony irons and some stones) are the product of melting and the separation of elements. Meteorites are not always easy to identify, but when etched with acid the irons will show the characteristic Widmanstätten patterns, which are not found elsewhere.

**Meteorite falls**

More than 2,000 meteorites have now been located, but only a few have been observed during their descent. Among famous falls that have resulted in meteorite finds are those of the Pribram fireball over Czechoslovakia on 7 April 1959 (recorded as magnitude − 19); Barwell (Leicestershire) on 24 December 1965; and Lost City (Oklahoma) in 1970. The Barwell meteorite broke up during its descent. Like many falls, it was a stony aerolite.

The Allende meteorite fell in Mexico in 1969 and from it many interesting discoveries about meteorites have been made. A chondritic meteorite, Allende does not resemble any terrestrial rock type and radiometric measurement of its age suggests that it dates from the accretion of planets from the solar nebula (*see* pages 10–11). Rocks that are still geologically active are much more fractionated than chrondrites and are younger by about 500 million years.

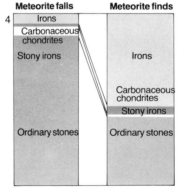

**1. Oschansk fall of 1891**
The Oschansk fall was at Perm in Russia in 1891. An explosion was heard and there followed a rain of incandescent stones. The stones, which fell as splinters, varied in weight between 1 and 300 kg.

**2. Tunguska event**
Several explanations have been proffered to account for an event in Siberia in 1908. An explosion was heard 1,000 km away, animals were killed and trees flattened over a large area. The descending object outshone the Sun and yet no meteoritic fragments have been found anywhere in the area.

**3. Meteorite crater**
Meteor Crater or Coon Butte in Arizona is 1,200 m in diameter, 183 m deep and is surrounded by a wall between 30 and 45 m high. Many meteoritic fragments have been found in the area, and the crater is believed to be about 22,000 years old. It is one of the Earth's youngest impact craters and was formed when an iron mass hit the sedimentary rock at about 11 km a second.

**4. Meteoritic falls and finds**
The diagram compares witnessed falls of the various types of meteorites with those found. Weathering removes many stones even if they survive the fall.

**The largest meteorites**

| Name | Location | Type | Weight tonnes |
|---|---|---|---|
| Hoba West | SW Africa | Iron | 60 |
| Ahnighito (The Tent) | Greenland | Iron | 30.4 |
| Bacuberito | Mexico | Iron | 27 |
| Mbosi | Tanzania | Iron | 26 |
| Agpalik | Greenland | Iron | 20.1 |
| Armanty | Outer Mongolia | Iron | 20 |
| Willamette | Oregon, USA | Iron | 14 |
| Chupaderos | Mexico | Iron | 14 |
| Campo del Cielo | Argentina | Iron | 13 |
| Mundrabilla | Western Australia | Iron | 12 |
| Morito | Mexico | Iron | 11 |
| **In Europe** | | | |
| Magura | Czechoslovakia | Stony | 1.5 |
| Limerick | Ireland | Stony | 0.048 |
| Barwell | England | Stony | 0.046 |

**5. Hoba West meteorite**
Located near Grootfontein in southern Africa, the Hoba West meteorite is the largest ever known. It weighs more than 60 tonnes but there is no crater.

**6. Bacuberito meteorite**
This meteorite was discovered in Mexico in 1871. It is at least 27 tonnes in weight.

**7. Jilin meteorite**
This fell in northeastern China in 1976 and is the largest stony meteorite (1,770 kg) in the world.

**8. Types of meteorite**
Meteorites are basically either stones or irons. Some stones have a high carbon content. They are rare and are known as carbonaceous chondrites (A). Most stones, however, are chondrites and are made up of round particles called chondrules (B). Iron meteorites (C) are up to 90 percent iron. Tektites (D) are small, glass-like objects, often aerodynamically shaped.

**9. Widmanstätten patterns**
When iron meteorites are cut and etched with acid they produce distinctive lines. These are the result of a crystalline metallic structure that forms in unusual conditions. These crystals are unique to iron meteorites.

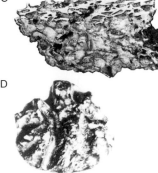

There is no record of death or serious injury due to a falling meteorite, though one or two people have had narrow escapes, and a dog was once killed by a fall in Egypt. There have been two major falls during the present century. The first was that in Tunguska, Siberia, on 30 June 1908. As seen from Kansk, 600 km away, the descending object was said to outshine the Sun, and detonations were heard 1,000 km away. However, the site was not reached by a scientific expedition until 1927, and no meteoritic material was found. Mystery surrounds the nature of the missile: it may have been the head of a small comet, or a meteorite that did not reach the ground before heading off beyond the Earth's atmosphere. The second fall, on 12 February 1947, also took place in Siberia, in the Sikhote-Alin region. This was certainly due to a meteorite that broke up, and more than 100 small craters were identified.

On 10 August 1972 a brilliant object was seen above Utah. It reached its closest point to the ground (58 km) above Montana, and then moved outward again and returned to solar orbit. Its diameter is thought to have been as much as 80 m.

## Meteorite craters
A large meteorite may produce a crater, and various examples are known. The most celebrated is that in Arizona, which is 1,265 m in diameter and 175 m deep, with a slightly raised rim and a perfect bowl shape. It is believed to be about 25,000 years old, and has been preserved because of the suitability of the climate and terrain of the area. The main mass is probably buried under the south wall of the crater. Apparently the meteorite came in at a low angle, but the resulting crater is still circular in form. There is another undoubted meteorite crater at Wolf Creek in Australia, and other examples have been cited; for example, at Waqar in Arabia, and in the Canadian Shield, although it is always difficult to be sure about the origin of a large crater unless there is obvious evidence.

## Origin of meteorites
The origin of meteorites is not known with certainty, but it is probable that there is no difference between a large meteorite and a small asteroid; there may indeed be a complete gradation in size. It is generally thought, therefore, that meteorites come from the asteroid belt. Study of meteorites' trajectories suggest they have orbits similar to Apollo asteroids, and they have probably not always wandered space but were once part of planetary bodies. Their ages are comparable with that of the Earth, and range between 4 and 4.6 thousand million years. More irons are known than stones since irons are much less fragile.

No doubt many meteorites remain to be discovered. Some have been found in Antarctica and they are very well preserved. A systematic search for more specimens has been initiated.

## Tektites
Tektites are small, glassy objects, and have only been found in a few areas. They seem to have been subjected to two periods of heating, and are usually aerodynamically shaped. Four main groups are known; in Australia (Middle-Late Pleistocene Age), Ivory Coast (Lower Pleistocene), Czechoslovakia (Miocene) and parts of the United States (Oligocene). The largest tektite known, with a weight of 3.2 kg, was found in 1932 in Laos.

Tektites are enigmatic objects. They may be meteoritic, but they are quite unlike other types of meteorites, and their restricted distribution indicates that if they came from space some unusual event must have occurred. Suggestions that they may have been ejected by volcanoes on the Moon are generally discounted, but it is quite possible that they originated in terrestrial volcanoes, so that they were sent high into the atmosphere and subsequently fell back to the ground. Many tektites have been found, but systematic searches for them in other parts of the world have been unsuccessful, and at present the mystery remains.

# Comparative Planetology

The Earth is unique among the planets of the Solar System in that it supports living organisms. Man has sought endlessly to understand his own origins and those of the rocky planet upon which he has evolved. A thirst for less tangible knowledge, particularly about the Earth's neighbor planets and, more distantly, the stars themselves, has led to an ever-increasing awareness of both the complexities of the physical nature of objects in the universe and, in starkly objective terms, the Earth's insignificant place in that universe as a whole. This realization has fortunately not led to any decline in man's resolution to solve the puzzles of planetary genesis, evolution and eventual decline.

## Lessons from the lunar landscape
During the last quarter of a century the rapid expansion in our knowledge of the Moon, promoted by the intensive study of photographs of its surface, followed by the actual collection and return of rock samples to Earth, has presented us with new and often unexpected insights into the history of the Earth's nearest natural neighbor. It has, for instance, been shown that between the time of the Moon's creation, some 4,700 million years ago, and a point in time some 700 million years later, its crust formed and was bombarded by huge numbers of rock fragments, many the size of small asteroids. This early period of cratering has become known as the "Great Bombardment".

Hypervelocity impacts disrupted, deformed and melted the lunar surface and, in the absence of active erosion under a mantle of air, the imprint of this epoch of intense activity has remained until this day. As technology has advanced and allowed man to venture farther from the Earth, so it has become apparent that a similar period of bombardment affected Mercury, Mars and most of the satellites of Jupiter and Saturn. Radar mapping of the enigmatic and cloud-covered Venus indicates that there too are circular structures that may well be impact craters.

The Earth, also, undoubtedly experienced the same process, but because the lithosphere has been mobile throughout geological time, any record of such an event has long since been erased. It has been estimated, for example, that a 25 km-diameter crater could only survive for about 500 million years before being obliterated. Before the era of space exploration, geologists were effectively ignorant of this phase of the Earth's evolution. Now some of them have tentatively suggested that early cratering may have been responsible for the onset of mobility and may well have played a major role in the development of the Earth's primitive crust.

Once men landed on the Moon and collected rocks which could be accurately dated by radiometric methods, it became possible to determine the timescale of planetary history. What was surprising was that the heavy bombardment of the Moon had finished by about 4,000 million years ago. Then a period of volcanism ensued, between about 3,800 and 3,200 million years ago, with lavas rising from deep within the Moon and pouring out into the more deeply excavated basins and partially filling them with dark basalts. Once this had occurred, the Earth's satellite had essentially taken on its modern aspect.

## Volcanoes everywhere
Modification of the surfaces of planets largely results from the generation and transfer of heat within the bodies. During the formative years of planetary evolution large impacts produced great amounts of heat, as did the decay of radioactive elements. It is now generally believed that the planets were molten at their creation. Heat was gradually lost to space, flowing through the mantles of the inner planets to their surfaces. Radioactive decay has continued in some cases and the interiors of planets have remained molten throughout geological time. The flow of heat within the interior of the planet continues to have a deformative effect on the surface of the Earth, for example, but on the Moon and Mercury

the outer layers cooled millions of years ago and internal processes have long ceased to modify the surface features of these bodies.

Basalts—the rocks which rose to form the lunar maria—are widespread on Earth and are still generated today, issuing forth from long fissures beneath the Earth's surface below the oceans. From what we have learned of Mercury, Mars and Venus, it seems that volcanic eruptions also characterized these worlds in ancient times. Statistical analysis of impact cratering on Mars, in particular, indicates that a period of intense volcanic activity occurred on the planet between 3,500 and 3,000 million years ago. In the case of Mars, however, it is clear that volcanism continued for much longer than on the Moon, and huge shield volcanoes were formed. While these are in all probability inactive now, there is conjecture that they may still have been active as recently as 500 or 600 million years ago—well into the period of life on Earth.

Lander probes dispatched to both Mars and Venus have made chemical analyses of the rocks they found. On Mars Viking revealed basaltic rocks and soil material rich in iron compounds, while the Soviet Venera craft recently detected both basaltic-type rocks and others more akin to terrestrial graphites. Within the last year there have been reports of an eruption beneath the clouds of Venus. The idea that Venus may still be volcanically active is a very exciting prospect indeed.

The morphology of volcanic features is influenced by the amount and the state of volatiles in the lava and the proportion of iron to silicon. Such characteristics determine the viscosity (resistance to flow) of lava. In general, the greater the viscosity and volatility of lava, the greater will be the effects upon the landscape. On Earth, basalts are iron-rich and are low in volatile content: this produces low relief but extensive volcanic landforms.

Atmospheric conditions also affect the form of volcanic features, as does surface gravity. On Earth, for example, cinder cones have slopes of up to 30°, whereas on the Moon they would be of the order of only one or two degrees. On the Moon there is no atmospheric drag and cinders would travel for a longer time and over a greater distance. As a result, cinder cones on the Moon would have a height less than one-tenth that of a similar cone on the Earth but a diameter at least four times larger. On Mars, where the surface gravity is only about one-third that of the Earth, calderas on the top of large volcanoes are larger than on similar features on Earth. The lesser gravity of Mars, it is thought, may permit the formation of larger collapse features than on Earth.

The basalts that floor the Earth's oceans and underlie the continental crust are far younger than any rocks on the Moon. Any evidence of a prolonged and early volcanic phase on this planet would have been destroyed aeons ago. It may well have occurred, however, and study of neighboring planets throws light on the Earth's formative period. What is certain is that there could never have been basalt-filled basins on Earth like those of the Moon: such basins would have been modified by erosion long before they could become infilled.

## Is plate tectonics unique to the Earth?
On Earth the distribution of volcanic activity can be explained in terms of plate tectonics. According to this theory, the Earth's crust is divided into 15 plates that float in a plastic-like layer beneath: volcanoes are located very often at the boundaries of these plates. Other evidence of tectonic activity that might be viewed from space includes folded mountain ranges, rift valleys, island arcs and oceanic ridges. On other planets there might at first seem to be similar evidence, but certain features are misleading in this respect. The Valles Marineris on Mars, for example, looks rather like the East African Rift Valley on Earth, and the proximity of volcanoes of the Tharsis Ridge is also suggestive of a tectonic origin, but there is a distinct lack of compressional features (such as fold mountains), and the volcanoes are thought more likely to be isolated

magma outpourings rather than subduction zones. There is no evidence of crustal plates: indeed geologists believe that Mars has a rigid crust that is much thicker than the Earth's.

Similarly, on Mercury there is no evidence of the movement of land masses despite a network of escarpments that seem to have deformed craters by means of overthrusting. Neither does the Moon display evidence of tectonic activity: there are no compressional features and, again, this body is thought to have a thick, rigid outer shell. Venus, however, might be a different case. Radar studies have revealed what may be compressional mountains, arcuate ridges and a large trough. Venus' high surface temperatures and Earth-like density may mean that the interior is molten much nearer the surface of the planet—like the Earth, but unlike Mercury, the Moon or Mars.

The latter three bodies may lack tectonic activity because they long ago lost the high internal temperatures necessary for convection in their mantles as a result of their smaller size. In terms of size also, Venus is much more like the Earth. Alternatively, the orginal crust on Earth may have been thinner and therefore able to be torn apart; or the other planets may lack suitable conditions of temperature, pressure and composition beneath their crusts where on Earth low friction enables the crustal plates to glide.

### Climatic change

The Mariner and Viking probes to Mars have contributed greatly to our understanding of planetary climates and changes that have affected them. The discovery of vast networks of channels on Mars, together with smaller anastomosic river courses, implies that, at some time in the past, fluid water was available on the Martian surface and was able to incise it. This epoch appears to have been after the early bombardment but before the onset of major volcanism—that is, very early on in Mars' evolution.

Today conditions on Mars are quite unsuitable for the existence of water on the surface: it is far too cold. Mars' extremely tenuous atmosphere must be quite unlike that which existed when the channels were cut, thus indicating that major changes in the climatic regime must have occurred. Changes, too, are implied by the laminated deposits located near the Martian poles. The stratification of these rocks bears witness to more recent fluctuations in climate, about which we can only speculate. A more enlightened understanding of the origin of these deposits will undoubtedly throw light on the causes of ice ages. Are they related to precessional effects, changes in the output of solar energy, or periods of increased volcanic activity resulting in large volumes of dust being pumped into the atmosphere, so making it opaque to incoming radiation? Climatologists would dearly like to know the answer to this question.

### The outer Solar System

That volcanism still plays an important role in planetary cycles is an established fact. Perhaps less expected was the discovery by the Voyager missions to the outer Solar System of current volcanicity on Io, one of the Galilean moons of Jupiter. Within the confines of the inner Solar System magmas tend to be based on silicates. On Io, however, there are lavas of sulphur. Volcanic activity is so constant there that this little world is resurfacing itself at a very rapid rate, and presents the youngest surface of any solid body so far visited by an exploratory spacecraft.

Also in the outer reaches of the Sun's planetary family are Jovian moons showing fractured and cratered crusts. These fractures and cracks are not necessarily as spectacular as the San Andreas Fault, but they nevertheless bear witness to tectonic movements on a large scale on these distant worlds. They clearly have been (and maybe still are) internally active to some degree.

Our knowledge of the Earth's more distant planetary relatives is still far from good. We currently know much less about the members of the outer Solar System than Earth-like worlds. The large, outer planets evidently accreted from the same cosmic cloud as the Earth and its nearer companions, but exactly how and why they developed into such gaseous giants is still something of a mystery. Puzzling also is the matter of how and exactly when the volatiles that went into the formation of the atmospheres of Venus, Earth and Mars were accreted. Did they originate from cooler, lighter material that fell into the inner regions of the Solar System to be "swept up" by the inner planets? Or were they produced locally from the same matter that generated the cores and mantles of these planets? In the sense that life evolved from these lately-accreted volatile materials, we surely need to find the answers to these questions for therein lie the clues to our origins. At the moment our knowledge is incomplete.

Enough has been learnt about the atmospheres, magnetospheres and satellite systems of Jupiter and Saturn to enable a fairly detailed comparison of these gaseous giants, but several fundamental questions remain unanswered. What is the precise source of heat emanating from both planets? What is responsible for the spectacular colors in Jupiter's atmosphere as observed from Earth? And what is the cause of sporadic radio emissions from these bodies? The Voyagers' observations of the moons of both planets gave us a fascinating new insight into these rock-and-ice worlds, but our understanding of their thermal states and chemical composition is based upon inference rather than established facts.

Similarly, Uranus and Neptune are compared, but again there are yawning gaps in our knowledge. Uranus, alone among the Jovian planets, appears to have no internal heat source. Neptune has a most curious satellite system dominated by the massive Triton which orbits its parent planet in a retrograde motion. And there is even doubt as to the nature of the least-known member of the Solar System—Pluto. Does it in fact constitute a double planet with its satellite Charon, or should Pluto not be considered as a planet in the true sense at all?

### In conclusion

Returning to Earth once more, we need to determine why, of all the terrestrial planets, is ours the most dynamic. The answer probably lies in the fact that the lithosphere is much thinner, certainly than that of the Moon and Mars, and probably of Venus and Mercury too. Little is known, however, of the interior of Venus and this needs to be rectified since in many respects it is Venus that most resembles the Earth.

The size and mean density of both worlds are very similar. Both bodies lie in the inner region of the Solar System and were fashioned from similar materials during the early accretional stage of their evolution. Why then do they have such different atmospheres, and what are the key factors controlling how atmospheres evolve? Is Venus still an active body and does it have continents like those of the Earth? Answers to these questions about Venus have an important bearing on our understanding of how our own planet evolved during its accretionary stages.

Knowledge of the Earth is used to understand the nature of the Earth's neighbors and even more distant members of the Solar System. On the other hand, an understanding of the Earth's early history has benefited greatly over the last two decades from what has been learnt of our close planetary neighbors. No longer should the Earth be viewed apart but as a part of a planetary family.

That family encompasses an extraordinary diversity amongst its members and from one end of the Solar System to the other. Even objects as dissimilar as the inner planets, the outer satellites, asteroids, comets and meteorites are all related in terms of origin and destiny, and they share common characteristics as a result of that relationship. But, the maxim that the more we know, the more there is to know was never more true than when applied to man's search for knowledge within the Solar System.

# Search for Life

Early man believed the Earth to be the supreme body in the universe. To him it was unique—a flat and motionless world, around which the entire heavens revolved every 24 hours. As soon as it became clear that the Earth was a globe and that other worlds existed, speculation about possible life beyond the Earth was inevitable. Such speculation has not diminished throughout the Space Age: indeed, if anything, the question of extraterrestrial life is a more hotly debated topic than ever it was.

Equally inevitably, the Moon was singled out for special attention because it alone is close enough for the Earthbound observer to distinguish surface detail even with the naked eye. It was assumed that the dark areas were seas and the bright areas were land masses. The Greek essayist Plutarch, in his celebrated work *On the Face in the Orb of the Moon*, realized that there were valleys and ravines on the Moon's surface. One of the first science-fiction stories—if not the very first—dates from the second century AD and was the work of a Greek satirist, Lucian of Samosata. In it he described how a party of sailors were caught up in a waterspout while passing through the Strait of Gibraltar, and were hurled upwards so violently that they were deposited upon the Moon, where they found themselves embroiled in a war between the King of the Moon and the King of the Sun.

Lucian did not expect to be taken seriously (although his work was entitled *True History*, he stressed that it was based on fantasy). Yet the idea of lunar inhabitants took hold in much later years. There was, for example, the celebrated *Somnium*, written by no less a person than Johannes Kepler, and published posthumously in 1634. Kepler's hero travels to the Moon, transported by a friendly demon. This story was a mixture of seventeenth-century science and pure fantasy: it is really a defense of the Copernican system, but when Kepler described the various lunar inhabitants he could well have believed that the Earth's satellite was populated.

Later still came some strange accounts by William Herschel, one of the greatest of astronomical observers. He was convinced of the habitability of the Moon, and he even believed that there were inhabitants of the Sun, living in a comfortably warm region below the bright surface. It is true that these were not widely held views at the time, and it is on record that some such comments in a paper of Herschel's were quietly deleted by the then Astronomer Royal, Nevil Maskelyne, prior to publication. But certainly the possibility of extraterrestrial life was taken very seriously.

## The great Moon Hoax

In the 1830s came the celebrated Moon Hoax, perpetrated by a reporter on the New York newspaper *The Sun*. John Herschel, son of William, was in South Africa making the first really important survey of the southern skies. Communications in those days were slow and uncertain, and Richard Alton Lock, *The Sun* reporter, took his chance. He wrote a series of columns in which it was claimed that Herschel's telescope had revealed the Moon in great detail, including crystal mountains, yellow animals and human, or at least quasi-human, inhabitants. There was even a "strange amorphous creature, rolling with great velocity across the pebbly shore". The hoax was soon exposed, but at least some people believed what Lock had written. The *New York Times* declared that the discoveries were both "possible and probable", while one religious sect made inquiries as to the best means of converting the Moon-men to Christianity.

Another serious suggestion came from the astronomer Franz von Paula Gruithuisen, who claimed in 1836 to have detected a lunar city with "dark gigantic ramparts". Nothing is visible in the region of the Moon's surface delineated by Gruithuisen apart from irregular ridges, and his imagination had obviously run riot—not, incidentally, for the first time. He had also suggested that the ashen light of Venus was due to vast forest fires lit by the local inhabitants to celebrate the election of a new Parliament!

## From the Moon to Mars

Observations made during the nineteenth century showed the Moon to be airless and waterless. No life forms could be expected to survive there, and the focus of attention turned to Mars, which then appeared to be much more welcoming than the Moon. At least it had an atmosphere, and the white polar caps were assumed to indicate the presence on the Martian surface of water. A number of suggestions were made as to how to signal to Martians: these included regularly spaced fires across the Sahara Desert, and powerful searchlights. Perhaps the most remarkable idea came from a Frenchman, Charles Cros, who, in the 1870s, wanted to build an enormous glass to focus the rays of the Sun onto the Martian desert, across which he would describe words. And in 1902 a French lady by the name of Guzman offered a substantial prize to the first scientist to make contact with the beings of another world. Mars was excluded from the contest since it was considered to be too easy to call up the Martians! The Guzman Prize remains unclaimed to this day.

Yet it is not so many decades ago that the existence of Martian inhabitants was being championed by several leading astronomers of the day. In 1877 Giovanni Schiaparelli in Milan drew up new maps of Mars and reported regular, artificial-looking features which he called "canali". Percival Lowell, who set up the great observatory in Arizona specially to study Mars, was completely convinced that the canals (as they were inevitably translated) represented a planet-wide irrigation system. In 1909 he wrote: "That Mars is inhabited by beings of some sort or other is as certain as it is uncertain what those beings may be." Lowell's admittedly extreme views were hotly challenged, but it was not until the flight of Mariner 4 in 1965 that the canal controversy was finally settled.

## Life on Earth

As yet there is no proof of life anywhere in the universe except on Earth. It is not even certain how life on Earth began. The majority view is that the origins of life occurred fairly early on in the Earth's history, evolving from inorganic matter. This has been challenged: 80 years ago the Swedish chemist and Nobel Prize winner Svante Arrhenius put forward his "panspermia" theory, in which he suggested that life was brought to Earth by way of a meteor. The theory never attracted much support because it raised more problems than it solved, but it has recently been revived, in a modified form, by Sir Fred Hoyle and Professor Chandra Wickramasinghe. They consider that life was brought to Earth by a comet. In their view, the production of living from non-living matter involves a whole chain of events, each of which is in itself improbable, and they claim that this is hardly likely to have occurred on a small world such as the Earth. It would require a large volume of space; and modern astronomy has revealed the existence of complex organic molecules in interstellar space. Dr Francis Crick, also a Nobel Prize winner and co-discoverer of DNA, goes even further and has proposed what he calls "direct panspermia": this is the idea that life was deliberately planted here by an intelligent agent far away in space.

Uncertainty about the origins of life on Earth makes it even harder to predict whether or not living organisms are likely to be widespread throughout the galaxy or indeed the universe. Will life inevitably appear wherever conditions are suited to it? It had been hoped to answer this question by finding organic matter on Mars, but the Viking soft-landers showed no signs of Martian life in any form, so we are still reduced to little more than speculation.

## Alien life forms

In the search for life, it seems reasonable to confine our attention to life as we know it. All living things in our experience are based on carbon, which has the ability to link up with other atoms to form the large, complex molecules that are the basis of living organisms.

The only other atom that displays anything like the same ability is silicon. Many life forms in science fiction are based on silicon.

It does not seem likely that we will find completely different atom-types in remote parts of the universe. Modern techniques have shown quite conclusively that the most distant objects known to man—galaxies and quasars—are composed of the naturally occurring elements that are found on Earth. Elements heavier than the heaviest naturally occurring element on Earth (uranium) have been produced in the laboratory, but these are not thought suitable for the production of life. The chances of alien life forms seem, therefore, to be slim, and if they do exist, then much of modern science would seem to be wrong.

## Life in the Solar System

Before the Space Age most astronomers believed that, although Lowell's Martians certainly did not exist, the planet probably supported plenty of low-type vegetation, filling the sea beds that were visible as dark areas. One ingenious argument was put forward by Ernst Öpik, an Estonian astronomer working in Northern Ireland. He pointed out that the Martian atmosphere is dust laden, and that unless the dark areas were made up of something that had the ability to push dust out of the way, the whole surface of the planet would soon assume a monotonous ochre hue. We now know that all the dark areas are not basins and that they are not vegetation covered, but at the time Öpik's argument was very convincing.

Venus was thought by many to be more promising. Svante Arrhenius had maintained that Venus probably resembled the Earth during Carboniferous times (about 225,000,000 years ago) so that there would be warm lagoons, luxuriant vegetation, horse-tails and insects such as dragonflies. During the 1950s F. L. Whipple and D. H. Menzel, two of America's most distinguished astronomers, suggested that Venus might be partly or completely covered with water. This would indeed make for conditions not too unlike those on Earth many millions of years ago. It was even suggested that life could develop in the Cytherean (Venusian) oceans as many believe it did on Earth. Mariner 2 in 1965 showed Venus to be an extremely hot and hostile environment.

As far as life is concerned, we can dismiss any worlds without atmospheres and those with extremely tenuous atmospheres. This includes all asteroids, all planetary satellites with the exception of Titan, and Mercury and Pluto. Titan is in a class of its own: all the ingredients for life are there, but temperatures are too low for life forms to have developed. Titan has been called "an Earth in deep freeze", but if the temperatures were higher the atmosphere would probably escape from the satellite.

The giant planets, with their gaseous surfaces and probably mainly liquid interiors, seem singularly unpromising as potential abodes for life. Carl Sagan and others, however, have pointed out that, for example in the case of Jupiter, the outer clouds are bitterly cold, but the core temperature may be well over 10,000° C. There must be a level, it is argued, where temperatures are much the same as on Earth. Could life have developed within Jupiter's globe? Such an idea cannot be ruled out, but it does seem extremely unlikely.

## Terraforming

Life in the Solar System would seem, therefore, to be confined to the Earth, the chances of even primitive life forms occurring elsewhere being extremely slim. Whether life has previously existed and has since died out is a different problem altogether. Life may have started on Venus before the increasing luminosity of the Sun made conditions far too hot. Mars may experience occasional fertile periods, but it could not maintain a reasonably dense atmosphere for long, and life is slow to evolve.

Is it possible to make planets less hostile? The process that has been called "terraforming" is at present little more than fantasy, but in the future it may not seem so outrageous. Venus may be the most suitable candidate. The disadvantages of it as a life-supporting environment are the high temperatures, lack of water, and deadly sulphuric acid clouds. There is plenty of oxygen on Venus locked up in various molecules, but Carl Sagan has suggested that it might be possible to "seed" the atmosphere, thus freeing oxygen from currently unwelcome molecules. This would stem the greenhouse effect, lower the temperatures and transform the Cytherean environment. Mars is less promising, mainly because of its inability to retain an Earth-type atmosphere even if one could be induced. Unless terraforming becomes possible, life on other worlds in the Solar System will always have to exist under completely artificial conditions.

## Planets of other suns

The search for a planet orbiting a star other than the Sun is an important challenge to astronomers and the next logical step, having studied the members of this Solar System. How common are planetary systems throughout the universe, and how likely is it that there might be life forms inhabiting a world within a planetary system elsewhere? The fact that there is no life in this Solar System apart from that on Earth is no argument against the existence of life in other solar systems, of which there must surely be many. A few eminent authorities believe life on Earth to be unique. Some scientists stress that the production of life can only be the result of a whole series of improbable events, so that it is likely to be extremely rare. At the other end of the scale are those who, like Sir Fred Hoyle, believe that the constituents of life lie in deep space and that travelling bodies such as comets could, given sufficient time, deposit living material on any number of worlds whose environment is suitable for its development.

The final proof of the existence of extraterrestrial life would be direct contact. The only possibility at the present is by radio. As early as 1960 the "Project Ozma" experiment was carried out by radio astronomers at Green Bank in West Virginia. They began to "listen" at a selected wavelength to try and pick up rhythmical signals from two near stars which are similar to the Sun—Tau Ceti and Epsilon Eridani. The results were negative, but the experiment was worth trying and further attempts have been made since. Pioneer, launched in 1972 and the first spacecraft to leave the Solar System, bears a plaque that portrays basic information about Pioneer's path from the Solar System and life on Earth.

Accurate studies of a star's motion reveal the presence of planetary companions. A star with an orbiting companion appears to "wobble" as it moves across the sky. This was first observed of Sirius, the brightest star in the sky, in the middle of the nineteenth century. There are several stars, however, which have much less mass than Sirius, that are nearer to the Sun and are now believed to be accompanied. Barnard's Star (5.9 light years distant), which was at first believed to display large planetary perturbations, is now the subject of much doubt and speculation and its data can at best be said to be highly ambiguous. Other possible planetary systems are those of Lalande 21185 (8.2 light years distant), Epsilon Eridani (10.8 light years), 61 Cygni (11.0 light years) and DB + 42°4305 (16.9 light years). Direct detection of possible planetary companions would require extremely sensitive equipment that could blot out the dazzling light of the parent star, thus revealing the much fainter companion or companions.

And in the future? The Earth will certainly not last forever. In about 5,000 million years the Sun will have entered the red giant stage of its stellar evolution: on Earth the oceans will boil and the atmosphere will be stripped off the planet. Whether life on Earth will survive even a fraction of the period between now and then and whether man will explore worlds beyond the Solar System can be no more at present than wild speculation. Either thought presents a challenging yet daunting prospect.

# HISTORY & OBSERVATION

# Astronomers (600 BC – AD 1750)

**THALES** (*c.*624–547 BC). Born in Miletus, the first of the great Greek philosophers. His ideas about the Earth were primitive by modern standards, and he believed the world to be a disc floating on water, but he is notable as having predicted a solar eclipse in 585 BC which put an abrupt end to a war between the Lydians and the Medes. Thales was able to do this by using the Saros cycle: presumably he knew that any eclipse is likely to be followed by another, one Saros cycle later. (One Saros cycle is equal to 18 years 11.3 days.) Thales travelled as far south as Egypt, and is believed to have advised the Greeks to steer, like the Phoenicians, by the Little Bear (Ursa Minor) instead of the Great Bear (Ursa Major).

**ARISTARCHUS** (*c.*310–230 BC). Greek astronomer, born in Samos. He was one of the first to maintain that the Earth is in orbit around the Sun instead of lying at rest in the center of the universe: he also believed that the Earth rotates on its axis. Unfortunately Aristarchus was unable to provide any convincing proof, and he found few followers, but at least he had anticipated the Copernican theory by many centuries. He attempted to measure the relative distances of the Moon and Sun by a method that was perfectly sound in principle: it involved knowing the exact moment when the Moon is at half-phase (dichotomy). In practice this is unworkable because of the rough and jagged appearance of the lunar terminator, and Aristarchus' result was therefore wide of the mark.

**HIPPARCHUS** (second century BC). Greek astronomer, who was born in Nicaea but apparently spent most of his life in Rhodes. He was unquestionably one of the greatest scientists of antiquity, and was a brilliant observer. He drew up a catalog of more than 1,000 stars, determined the length of the tropical year, and discovered the phenomenon of precession, which involved developing what is now known as trigonometry. He believed the Earth to be at the center of the planetary system, but at least he laid a sound foundation upon which his successors could build. His original writings were lost, but his ideas survived by way of Ptolemy's *Almagest*.

**PTOLEMY**, more properly Claudius Ptolemaeus (*c.* AD 150). Ptolemy, the "Prince of Astronomers", was the last great scientist of the classical period. He studied all branches of science, and was the first to draw up a map of the civilized world that was based upon something more than inspired guesswork (even though he did join Scotland to England in a back-to-front position). In his great work the *Almagest* (originally *Syntaxis Mathematike*) he summarized all the work of earlier scientists, and without this book our knowledge of ancient science would be much less detailed, even though the book survived only by way of its Arab translation. Ptolemy also reproduced and extended Hipparchus' star catalog, and made careful studies of the movements of the planets. He perfected the geocentric (Earth-centered) system, always known as the Ptolemaic system even though Ptolemy himself did not invent it. While studying the movements of the Moon, he discovered the phenomenon of evection (*see* page 146). Nothing is known about Ptolemy's life or personality, but periodical attempts to discredit him, and to claim that at best he was a mere copyist, have been notably unsuccessful. Science owes him a great debt.

**COPERNICUS**, more properly Mikołaj Kopernik (1473–1543). Polish cleric and astronomer. Copernicus was born at Toruń on the River Vistula, and studied first in Poland and then in Italy before returning to his native country as Canon of Frombork—a place described by him as "the remotest corner of the earth". At an early stage he became dissatisfied with Ptolemaic theory, and realized that some of its complexities could be removed by taking the Earth away from the center of the planetary system and putting the Sun there instead. He prepared a book, *De Revolutionibus Orbium Coelestium* (On the Revolutions of the Celestial Orbs), but

hesitated about publishing it because he knew that he would meet with strong opposition from the Church. Eventually he was persuaded to publish, but the book appeared only as Copernicus was dying. His fears were well founded; the Church was bitterly hostile. In fact, many of Copernicus' ideas were wrong, and in particular he believed that all planetary orbits must be perfectly circular, but he had taken the essential step.

**TYCHO BRAHE** (1546–1601). Danish astronomer, and certainly the greatest observer of pre-telescopic times. He was of noble birth, and in temperament arrogant and intolerant. At an early stage his attention was drawn to astronomy; in 1572 he observed a brilliant new star (now known to have been a supernova) in the constellation of Cassiopeia and began his observational career. The King of Denmark provided the funds to build an observatory on Hven, an island in the Baltic, and Tycho worked there between 1576 and 1596, compiling a star catalog that was much more accurate than any previously drawn up. He also made careful measurements of the movements of the planets, particularly Mars. He could not accept the Copernican theory, and preferred to believe that while the planets moved around the Sun, the Sun itself moved around the Earth. After quarrelling with the Danish court, he abandoned Hven and went to Prague as Imperial Mathematician to the Holy Roman Emperor. While there, he engaged his last assistant, Johannes Kepler. Tycho died suddenly in 1601, still convinced that the Earth must be at the center of the Solar System.

**GALILEO GALILEI** (1564–1642). Italian astronomer, born at Pisa, who taught mathematics successively at Pisa, Padua and Florence. As soon as he heard of the invention of the telescope, in 1609, Galileo made one for himself, and turned it towards the sky: although he was not the first to use the telescope for astronomical purposes, he was certainly the greatest of the early telescopic observers. From January 1610 he made a series of spectacular discoveries, including the four large satellites of Jupiter, the phases of Venus, the mountains and craters of the Moon, and the myriad stars in the Milky Way; he even saw Saturn's rings, although he could not make out what they were, and believed Saturn to be a triple planet. He was an early convert to the Copernican system, and his observations supported his ideas, about which he made no compromise. He was tactless and outspoken, and aroused the anger of the Church: in 1616 the Papal authorities ordered him to refrain from teaching Copernicanism. A change in the papacy emboldened Galileo to publish his great work *Dialogue on the Two Principal World Systems*, but he had miscalculated the strength of opposition to his theories. In 1633 he was summoned to Rome and forced to make a hollow and completely pointless recantation, after which he was kept in virtual house arrest at his villa for the rest of his life. His books were placed on the Papal Index of prohibited works and not released until the nineteenth century. Quite apart from his astronomical work, Galileo made striking advances in optics and was the real founder of the science of experimental mechanics.

**Johannes KEPLER** (1571–1630). German mathematician and astronomer, born in Württemberg. While studying at Tübingen University he became convinced of the truth of the Copernican system. Kepler was a curious mixture: many of his ideas were rooted in ancient mysticism, but he was also a first-class mathematician. After Tycho Brahe's death all the observations made at Hven fell into Kepler's hands, and he used them to show that the Earth and other planets move around the Sun not in circles but in ellipses—a fact that Tycho himself could never have accepted. Kepler was delicate by constitution and his life was far from happy, but his contributions to science were outstanding, and he wrote that his discoveries would not have been possible but for his faith in the accuracy of Tycho's observations. Kepler's three Laws of Planetary

Motion (the first two published in 1609, the third in 1681) formed the basis of all future work. Law 1 states that the planets move around the Sun in ellipses, the Sun occupying one focus of the ellipse while the other is empty; Law 2 states that the radius vector (that is, the line joining the center of the planet to the center of the Sun) sweeps out equal areas in equal times; and Law 3 provides a link between a planet's revolution period and its distance from the Sun. Kepler's work finally discredited the Ptolemaic system.

**Johann FABRICIUS** (1587–1616). Dutch astronomer, who studied medicine but also helped his father, David Fabricius, in astronomical work. The discovery of sunspots is variously attributed to Fabricius, Galileo and Christoph Scheiner.

**Joannes Baptista RICCIOLI** (1598–1671). Jesuit professor of astronomy in Bologna. His *Almagestum Nova* is a comprehensive account of the astronomy of the time, even though Riccioli never accepted the Copernican system and preferred that of Tycho Brahe. In 1651 he published a map of the Moon (based partly on the observations of his pupil Francesco Grimaldi), which was better than that of Hevelius. Riccioli also introduced a new lunar nomenclature, naming the main craters after eminent persons, and his system is still used, although it has been widely extended. Riccioli had no patience with Galileo's theories—which is why the lunar crater named after Galileo is small and obscure while two important formations bear the names Riccioli and Grimaldi.

**Johann HEVELIUS** (1611–1687). Hevelius is the Latinized form of his name, which is written variously as Hewelcke and Hevel. He was born in Danzig (now the Polish city of Gdańsk) and became a city councillor. He studied in Holland, and then travelled throughout Europe before returning to Danzig in 1639 and devoting all his spare time to astronomy. He set up what was then the finest observatory in Europe, and drew up a catalog of 1,500 stars as well as a catalog of comets. In 1645 he published the first reasonably accurate map of the Moon, though his lunar nomenclature has not survived (for instance, his "Greater Black Lake" is now known as the crater Plato). His observatory was burned down in 1679, but he promptly built another. Hevelius corresponded with, and met, many leading astronomers of the time, including Edmond Halley.

**Jeremiah HORROCKS** (1619–1641). English cleric and astronomer, born in Liverpool. He worked out that a transit of Venus would occur in 1639, and actually observed it. Horrocks showed great promise, but his career was cut short by his early death.

**Giovanni Domenico CASSINI** (1625–1712). Italian astronomer, born at Perinaldo, but spent much of his life in France. Louis XIV invited him to Paris to become the first director of the new observatory there. Cassini was an excellent planetary observer. He discovered four satellites of Saturn (Iapetus, Rhea, Dione and Tethys) as well as the main gap in the ring system which is still known as Cassini's Division. He observed the polar caps of Mars, and improved the theory of the movements of the Moon. By the parallax method he made the first reasonably good estimate of the distance of the Sun, which he gave as 138,000,000 km. Cassini was succeeded in Paris by his son, Jacques.

**Christiaan HUYGENS** (1629–1695). Dutch astronomer, born at The Hague. He turned to astronomy in his early twenties, and made great improvements to the telescope, although he still had to rely upon small-aperture, long-focus refractors, and became the best observer of his day. He discovered the true nature of Saturn's rings; in 1655 he discovered Titan, the largest of Saturn's satellites; and in 1659 he made the first drawing to show any detail on the surface of Mars (the triangular Syrtis Major). He even measured the rotation

period of Mars with fair accuracy. For a time Huygens lived in France, but his Protestant views were unpopular and he returned to Holland. He believed that the stars are suns, and that there must be many inhabited worlds in the universe. His *Cosmotheros*, published posthumously in 1698, summarizes his work, including his studies for the development of a pendulum clock.

**Isaac NEWTON** (1643–1727). Newton was born at Woolsthorpe, Lincolnshire, and graduated from Cambridge University, where he spent most of his life. While the University was closed in 1665 because of the Plague, Newton retired to Woolsthorpe and laid the foundations of his later work, including the theory of gravitation and the splitting-up of sunlight by means of a prism—the forerunner of spectroscopy. He built the first reflecting telescope, presented to the Royal Society in 1671. In 1687 he published his immortal work, the *Principia*, in which he described the theory of gravitation amongst much else: the work has been described as the greatest mental effort ever made by one man. His other great book, *Opticks*, appeared in 1704. Newton received every honor that the scientific world could bestow, and his work was in a class of its own—yet he still believed in alchemy (the method of turning base elements into gold), and was by no means entirely sceptical about astrology. He is buried in Westminster Abbey in London.

**Ole RØMER** (1644–1710). Danish astronomer, born in Aarhus. After spending some years at the Paris Observatory, Rømer returned to Denmark as Director of the Copenhagen Observatory. He devised many astronomical instruments, including the first device for observing transits. By timing the eclipses of Jupiter's satellites, he accurately worked out the velocity of light.

**Robert HOOKE** (1653–1703). Born on the Isle of Wight, Hooke became a leading figure in the Royal Society, and was a brilliant experimenter. He observed the Great Red Spot on Jupiter, and he also made improvements in telescopic optics.

**Edmond HALLEY** (1656–1742). English astronomer. At the age of 19 he published a paper on planetary orbits. He studied at Cambridge University, but left before taking his degree to go to St Helena to draw up the first useful catalog of stars in the southern sky. It was Halley who persuaded Newton to publish the *Principia*—and even paid for it out of his own pocket. In 1705 Halley published his work on comets, in which he claimed that the comet of 1682 had a period of about 76 years and would be seen again in 1758: by then Halley was dead, but the comet was observed on Christmas night 1758 right on cue. Since then it has returned in 1835 and 1910, and was recovered in 1982; it is always known as Halley's Comet. Halley made many other contributions to astronomy: in particular he discovered the proper motions of several bright stars; and in 1715, during a total solar eclipse, he described the chromosphere and the corona of the Sun. In 1720 he was appointed Astronomer Royal in succession to John Flamsteed, and immediately undertook a long series of observations of the movements of the Moon which he completed successfully after 18 years. Halley's pleasant, jovial personality contrasted sharply with the touchiness of many of his contemporaries, including Flamsteed, Hooke and even Newton.

**Mikhail Vasilyevich LOMONOSOV** (1711–1763). Russian astronomer, geologist, meteorologist and grammarian. He studied at St Petersburg (now Leningrad) and then in Germany before returning to St Petersburg as professor of chemistry. He was a pioneer in the kinetic theory of gases, but in astronomy he is notable as having observed the 1761 transit of Venus and deducing that Venus has a dense atmosphere. He was the first Russian champion of the Copernican theory.

# Astronomers (1750-1850)

**Thomas WRIGHT** (1711–1785). English astronomer, born near Durham. He became first a clockmaker's apprentice and then a teacher of mathematics. He suggested that the Galaxy is disc shaped, and claimed, correctly, that Saturn's rings are composed of small independent particles.

**Charles MESSIER** (1730–1817). French astronomer who devoted his whole career to comet hunting. He discovered more than a dozen, though he is best remembered today for his catalog of star clusters and nebulae, completed in 1781 and compiled so that Messier would not mistake such objects for comets.

**Wilhelm HERSCHEL** (1738–1822). Known as William Herschel, he was born in Hanover but came to England as a young man and became a professional organist in Bath. He became interested in astronomy and made his own reflecting telescopes. With one of these, in 1781, he discovered the planet Uranus, becoming famous almost overnight. He was appointed King's Astronomer by George III of England and Hanover, and was able to devote the rest of his life to the post. Herschel was probably the greatest of all observers. Quite apart from his planetary work (including the discovery of two satellites of Saturn and three of Uranus), he discovered thousands of double stars, clusters and nebulae, and drew up the first resonably good picture of the shape of the Galaxy. He was also the best telescope-maker of his time, and his largest instrument, the famous "40-foot reflector", remained unequalled until the middle of the nineteenth century, though admittedly it was clumsy to use. Throughout his observing career Herschel was assisted by his sister Caroline (1750–1848), who was an astronomer in her own right, discovering eight comets as well as a number of star clusters and nebulae. Herschel spent the latter part of his life in Slough, but his old home in Bath, from the garden of which he discovered Uranus, has been preserved and turned into a small museum.

**Anders Johann LEXELL** (1740–1784). Finnish astronomer, although his birthplace, Abö, was then part of Sweden. Lexell became professor of astronomy in St Petersburg: he developed lunar theory, and also studied the movements of comets, notably the comet of 1770, which had been discovered by Messier but is always known as Lexell's Comet (its orbit is not now known). Lexell was also one of the first to show that the object discovered in 1781 by Herschel was indeed a planet (Uranus) and not a comet, as Herschel had at first believed.

**Johann Hieronymus SCHRÖTER** (1745–1816). German astronomer. By profession Schröter was chief magistrate of the town of Lilienthal, near Bremen, but his absorbing interest was astronomy. He set up an excellent observatory, equipping it with the best instruments available. These included a telescope made by William Herschel, with whom Schröter carried on a long correspondence that was uniformly cordial apart from a brief episode concerning Venus. Schröter believed that he had observed high mountains there; Herschel disagreed strongly, but any personal controversy was dispelled by the courtesy of Schröter's reply. Schröter concentrated wholly upon planetary work. He was the first great observer of the Moon: he drew up a map of the markings on Mars, though admittedly he misinterpreted them (he believed them to be atmospheric) and he also observed the other planets. It has been claimed that Schröter's work was reduced in value by his clumsy draughtsmanship, but there can be no doubt that he made invaluable contributions to astronomy. He was president of the so-called "Celestial Police", formed in 1800 at a meeting in Lilienthal to search for a new planet between the orbits of Mars and Jupiter. Sadly, Schröter's observatory was burned down in 1813 by the invading French soldiers, and all his unpublished observations were unfortunately lost.

**Giuseppe PIAZZI** (1746–1826). Italian astronomer, Director of the Observatory of Palermo. In 1801 he discovered the first asteroid, Ceres, while compiling a star catalog. He was not then a member of Schröter's "Celestial Police", although he joined later.

**Johann Elert BODE** (1747–1826). German astronomer, Director of the Berlin Observatory. He published star catalogs and atlases, and for half a century edited the Berlin *Jahrbuch*. He drew attention to a curious relationship linking the distances of the planets from the Sun, and this is still called Bode's Law, though in fact it was first pointed out by Titius of Wittemberg.

**Heinrich OLBERS** (1748–1840). German medical doctor and skilled amateur astronomer. He joined the "Celestial Police", being the first to recover Ceres in 1802 and discovering two more asteroids, Pallas and Vesta. He was a keen observer of comets, and discovered several. He is better known for his cosmological theory known as "Olbers Paradon".

**Pierre Simon LAPLACE** (1749–1827). French mathematician who developed Newton's work. In *Méchanique Céleste* he gave an outline of dynamical astronomy, and in *Système du Monde* he described his so-called nebular hypothesis, according to which the planets were formed from gaseous rings left behind by a shrinking cloud of gas whose central remnant was the Sun. In its original form the nebular hypothesis is untenable, but modern theories are in some ways not dissimilar to it.

**Jean Louis PONS** (1761–1831). The greatest of all comet-hunters. He was French, and his first appointment was at Marseilles Observatory, but as a caretaker. Pons' cometary discoveries, 37 in all, led to his eventual directorship of the Museum Observatory in Florence. In 1818 he discovered the comet now known as Encke's, and in 1819 a periodical comet recovered in 1858.

**Joseph von FRAUNHOFER** (1787–1826). German optician. Fraunhofer's career was unusual: he was orphaned at an early age, and apprenticed to a watchmaker. When the house in which he was lodging collapsed, he was rescued by the Emperor of Bavaria, who befriended him and saw to his education. Fraunhofer became the best optical worker of his time, producing magnificent object-glasses for refracting telescopes, including that for the Dorpat refractor in Estonia which F. G. W. Struve used for his pioneer work on double stars. Fraunhofer invented the diffraction grating, and studied the solar spectrum, mapping 324 dark lines still often called Fraunhofer Lines. He also studied the spectra of the Moon, planets and stars. Sadly, he died while still comparatively young and at the height of his career.

**William Cranch BOND** (1789–1859). One of the most celebrated of early American astronomers. He was born in Maine, and trained as a watchmaker. After a visit to England to see observatories there, he was largely responsible for founding the Harvard College Observatory, becoming its first Director. He discovered Saturn's "Crêpe Ring" and, independently of William Lassell, he discovered Saturn's satellite Hyperion. Bond was a pioneer in astronomical photography, and at the Great Exhibition in 1851 in London, he showed a good daguerreotype of the Moon.

**Heinrich SCHWABE** (1789–1875). Born in Dessau, Schwabe was a German amateur astronomer who searched for an intra-Mercurial planet. He realized that such a planet would be visible only when in transit across the face of the Sun, and so he observed the Sun regularly for many years. No such planet exists, but Schwabe's series of observations led him on to the all-important discovery of the 11-year sunspot cycle.

**Johann Franz ENCKE** (1791–1865). German astronomer, who fought in the Napoleonic Wars, and in 1825 became Director of the Berlin Observatory. Under his direction new star maps were compiled which enabled Johann Galle and Heinrich D'Arrest to confirm optically the discovery of Neptune from calculations submitted by Le Verrier. Encke also contributed to lunar theory, but is probably best remembered for establishing the periodicity of the short-period comet that now bears his name. Encke's Comet has a period of only 3.3 years, still the shortest known. Up to this time only Halley's Comet has been recognized as being periodical.

**Karl Ludwig HENCKE** (1793–1866). German amateur astronomer, postmaster at Driessen. In 1830 he began to search for asteroids, only four of which were then known, and in 1845 he discovered 5 Astraea—a suitable reward for many years of work. He discovered another asteroid, Hebe, in 1847.

**Wilhelm Gotthelf LOHRMANN** (1796–1840). German land surveyor, working in Dresden. He began compiling a large lunar map, and the sections that he completed were excellent, but ill-health prevented him from completing the map. It was, however, used as the basis for Julius Schmidt's large map many years later.

**Wilhelm BEER** (1797–1850) and **Johann Heinrich von MÄDLER** (1794–1874) are best described together. These two German astronomers were responsible for producing the first really good map of the Moon, in 1837, and also the first reasonable chart of Mars. Beer was a wealthy banker; Mädler taught him astronomy, and they observed from Berlin using Beer's 9.5 cm Fraunhofer refractor. Their lunar map, together with a complete description of every named formation, was a materpiece of careful, accurate observation, and was not superseded for almost half a century. In 1840 Mädler left Berlin to go to Dorpat, and carried out little further work on the Solar System.

**William Rutter DAWES** (1799–1868). English amateur astronomer. He became a minister of the Congregational Church in Lancashire, but resigned in 1839 to devote his whole time to astronomy. He discovered Saturn's "Crêpe Ring", independently of Bond, and drew up careful maps of Mars. He also discovered many double stars. Like Edward Barnard, he was renowned for his exceptionally keen eyesight.

**William LASSELL** (1799–1880). English amateur astronomer, a brewer by profession. He built an excellent observatory in Liverpool, and with his large reflector there he discovered Triton, the largest satellite of Neptune, only 17 days after the identification of Neptune itself. He discovered Ariel, the inner satellite of Uranus, and recovered another satellite, Umbriel, which had been glimpsed by Herschel. Lassell was co-discoverer, with Bond, of Saturn's satellite Hyperion. Later he transferred his equipment to Malta and made extensive studies of nebulae, discovering 600.

**Frederik KAISER** (1808–1872). Dutch astronomer, born in Amsterdam. He became Director of the Leiden Observatory, which he reorganized and modernized. He was a skilled planetary observer but concentrated on Mars, charting the surface and measuring the rotation period very accurately.

**Urbain Jean Joseph LE VERRIER** (1811–1877). French astronomer. Independently of John Couch Adams, he set out to search for a trans-Uranian planet. When he had completed his calculations he sent them to Berlin Observatory, where Johann Galle and Heinrich D'Arrest began a search for the new planet, which they found on their first night's observation. Le Verrier also believed in an intra-Mercurial planet, which was even given a name (Vulcan) but which does not exist. Le Verrier became Director of the Paris Observatory, and undertook much valuable work, both theoretical and practical. He was not popular, however, and in 1870 was requested to resign as Director of the Paris Observatory, although he was reinstated when his successor, Charles Delaunay, was drowned in a boating accident.

**Anders Jonas ÅNGSTRÖM** (1814–1874). Swedish astronomer, born at Logdo and educated at Uppsala, becoming Director of Uppsala Observatory in 1843. He was a pioneer astrophysicist, and in 1862 published a list of elements discovered in the Sun: in 1868 he completed the first really large-scale map of the solar spectrum. The Ångström unit of length is named in his honor.

**Daniel KIRKWOOD** (1814–1895). American astronomer, born in Maryland. He became professor of mathematics at Delaware College until 1856, when he moved to the Indiana Observatory. His main work was in connection with comets, meteors and asteroids. As early as 1861 he claimed that meteor streams were the debris of disintegrated comets. He established that there were various gaps in the main asteroid zone in which few bodies move due to the cumulative perturbations of Jupiter. These zones are known as the Kirkwood Gaps.

**John Couch ADAMS** (1819–1892). English astronomer, born at Lidcot in Cornwall. While an undergraduate at Cambridge he determined to tackle the problem of the movements of Uranus in the hope of tracking down a trans-Uranian planet. He completed his work and submitted it to the Astronomer Royal, Sir George Airy, but no immediate action was taken. When Airy heard of similar calculations made in France by Le Verrier he requested James Challis, professor of astronomy at Cambridge, to begin a search, but Challis was far from energetic, and did not identify the new planet—Neptune—until it had been found from Berlin on the basis of Le Verrier's calculations. Nowadays Adams and Le Verrier are recognized as co-discoverers. Adams became Director of the Cambridge Observatory in 1860, and carried out much valuable work, particularly in lunar motions and meteor streams.

**Lewis SWIFT** (1820–1913). American amateur astronomer who specialized in the discovery of comets and nebulae: he discovered 13 comets and 900 nebulae. At the eclipse of 1878 he made a search for the hypothetical planet Vulcan, but it is now certain that the objects he saw were simply stars. Swift is one of the few noted astronomers to have lived through two returns of Halley's Comet.

**Ernst Wilhelm Liebrecht TEMPEL** (1821–1889). German astronomer, who worked successively in Venice, Marseilles and Arcetri. He discovered five asteroids and several comets, including two short-period comets. He also discovered the nebulosity in the star-cluster Pleiades.

**Heinrich Ludwig D'ARREST** (1822–1875). German astronomer, born in Berlin. While D'Arrest was a student there, Encke allowed him to join Johann Galle in the search for Neptune. D'Arrest made many contributions to the study of comets and asteroids, and discovered the periodical comet that bears his name. He also published observations of 2,000 nebulae, many of which he had himself discovered.

**Friedrich Wilhelm Gustav SPÖRER** (1822–1895). German astronomer, for many years on the staff of the Potsdam Observatory. His main work was in connection with the Sun; he confirmed Richard Carrington's results, and discovered the variations in latitude of sunspot zones during the solar cycle (Spörer's Law). He also made extensive studies of the history of solar observation.

# Astronomers (1850-1900)

**Pierre Jules César JANSSEN** (1824–1907). French astronomer, who became a pioneer of astronomical spectroscopy. In 1868 he was the first to observe the spectra of solar prominences without waiting for a total eclipse, and confirmed the existence of the chromosphere beyond any doubt. In 1904 he published his great solar atlas, which contained more than 8,000 photographs. During the German seige of Paris in 1870 he escaped from Paris in order to observe a total eclipse—although ironically he was defeated by cloudy skies. He also believed, erroneously, that he had detected water vapor in the spectrum of Mars. In 1876 he became Director of the Meudon Observatory near Paris, where he remained for life.

**Johann Friedrich Julius SCHMIDT** (1825–1884). German astronomer, who worked in Hamburg and Bonn before going to Greece in 1858 as Director of the Athens Observatory. He specialized in studies of the Moon, and it was his report in 1866 that the small crater Linné had disappeared that drew the attention of astronomers back to lunar work—though it now seems certain that no real change had occurred in Linné. In 1878 Schmidt published an elaborate 1.8 m lunar map, which remained the standard work for many years and was much more detailed than that of Beer and Mädler. Schmidt was also concerned with the orbits of comets and meteors, and he discovered two novae, the recurrent nova T Coronae in 1866 and Q Cygni in 1876.

**Giovanni Battista DONATI** (1826–1873). Italian astronomer, best remembered for his discovery of the beautiful, brilliant comet of 1858. Comets were then believed to shine only by reflected sunlight, but Donati examined the spectrum of Tempel's Comet of 1864 and discovered bright lines due to glowing gases. He became Director of the Museum Observatory in Florence, and was largely responsible for the establishment of the famous observatory at Arcetri.

**Richard Christopher CARRINGTON** (1826–1875). English astronomer, born in London and educated at Cambridge. He set up a private observatory at Redhill in Surrey, where he undertook valuable solar research. He determined the rate of the Sun's rotation, and established that it varies with solar latitude. Independently of Spörer, Carrington discovered the law of sunspot zones, and he was the first to make a visual observation of a solar flare. His health deteriorated in 1865, and he was unable to continue his astronomical work.

**Asaph HALL** (1829–1907). American astronomer, born in Connecticut. He was apprenticed to a carpenter, but then went to Michigan University to study astronomy. In 1862 he joined the staff of the Washington Observatory, and in 1876 used the large refractor there to determine the rotation period of Saturn. He is best remembered, however, for his discovery in 1877 of the two Martian satellites, Phobos and Deimos.

**Maurice LOEWY** (1833–1907) and **Pierre Henri PUISEUX** (1855–1928) worked together at the Observatory of Meudon, producing in 1896 the first good photographic atlas of the Moon—which inevitably superseded the best charts based on observation. Loewy also carried out observations of the asteroid Eros in an attempt to measure the length of the astronomical unit—the distance between the Earth and the Sun.

**Giovanni Virginio SCHIAPARELLI** (1835–1910). Italian astronomer, who graduated from Turin and then went successively to Berlin and Pulkovo before becoming Director of the Brera Observatory in Milan. In 1877 he carried out a careful study of Mars, revising the Martian nomenclature and producing a new map. He recorded regular lines crossing the ochre areas, and he called them "canali", the Italian word for channels. They became notorious as the Martian "canals". In 1879 Schiaparelli re-observed the channels, and claimed that some of them were double. He never committed himself to the theory that they might be artificial, although, in his own words, he was "careful not to oppose the suggestion". Schiaparelli also discovered the connection between the Perseid meteor shower and Comet Swift-Tuttle (1862 III)—a development of tremendous importance. He observed Mercury, producing a map of the surface, although he was wrong in supposing that both Mercury and Venus had synchronous rotations, thereby keeping the same face turned permanently towards the Sun. Schiaparelli was also an authority on the history of astronomy, and continued with this research after deteriorating eyesight forced him to give up observation.

**Joseph Norman LOCKYER** (1836–1920). English astronomer, born in Rugby. His early astronomical work was in connection with Mars, which he drew with great care, but he then turned to spectroscopy, and in 1868 he discovered the method of studying the solar prominences at times of non-eclipse—quite independently of Janssen. In 1869 Lockyer identified lines in the solar spectrum that could not be associated with any element then known. These lines are in fact due to helium, which was not identified on Earth until much later. Lockyer also developed a theory of stellar evolution which involved meteoritic infall, and which is quite erroneous. He was knighted in 1897 and retired to Sidmouth in Devon, where he set up a private observatory.

**Camille FLAMMARION** (1842–1925). French astronomer who studied for the priesthood, but abandoned theology in favor of astronomy. He worked for the Paris Observatory and at the Bureau des Longitudes, but finally set up a private observatory at Juvisy-sur-Orge, where he remained for the rest of his life. He was an energetic popularizer of astronomy, and founded *L'Astronomie*, the periodical of the Société Astronomique de France. He wrote many books, some technical and others for the lay reader, and he was firm in his belief that inhabited worlds must be common in the universe. He made numerous planetary observations, and wrote two books upon the history of Martian research.

**George Howard DARWIN** (1845–1912). English astronomer, son of Charles Darwin. He was born at Downe in Kent, and educated at Cambridge, becoming professor of astronomy there in 1883. He specialized in research into tidal phenomena, and in 1898 outlined his celebrated tidal theory of the origin of the Moon. This was widely accepted for some years even though it has now been rejected by almost all authorities. He was knighted in 1906.

**William Frederick DENNING** (1848–1931). English astronomer, born in Somerset and trained as an accountant. He never held an official scientific post, but was recognized as a foremost authority on meteors: he determined the radiant points of more than 1,100 showers, and was an expert observer. He discovered several comets, re-measured the rotation period of Saturn and made good observations of Mercury, Mars, Venus and Jupiter. It is as a meteor observer that he is best remembered, and the data that he collected has proved to be of immense value.

**Edmund NEISON** (1851–1938). English astronomer. His surname was Nevil, but he preferred to be known as Neison. In 1876 he published an exhaustive book about the Moon, together with a detailed map. Between 1882 and 1910 he was Director of the Natal Observatory in Durban, South Africa.

**Henri Alexandre DESLANDRES** (1853–1948). French astronomer, who was born in Paris and spent most of his working life at the Meudon Observatory, becoming Director in 1908. His main research work was in connection with the Sun and, independently of George Hale, he invented the spectroheliograph.

**Percival LOWELL** (1855–1916). American astronomer, born in Boston of an old New England family. He was educated at Harvard, and spent some years in the Far East on diplomatic missions before returning to America to devote his life to astronomy. He was wealthy and was able to found a private observatory at Flagstaff in Arizona. It was completed in 1894, and is still one of the world's foremost astronomical establishments. Lowell was fascinated by Mars, and carried out careful studies with the 61 cm refractor at Flagstaff. He believed he had observed an extensive canal network, and was quite convinced that it was part of a planet-wide irrigation system built by intelligent Martians—a point of view that aroused strong criticism even at the time. Lowell and his colleagues remained convinced of the existence of Martians, and it is unfortunate that Lowell is now remembered mainly on this score. His Martian canals were optical illusions, but he carried out much good work in other directions. With Vesto Slipher, he measured the rotation period of Uranus by means of spectroscopy; and he made observations of Mercury, Venus and the other members of the Solar System, although again he tended to draw nonexistent linear features. He made contributions to astronomical photography, and he made the calculations that led to the discovery in 1930 of Pluto by Clyde Tombaugh at the Lowell Observatory. Pluto cannot have been responsible for the effects upon Uranus and Neptune which Lowell had used to determine its position, because its mass is too small. The real "Planet X" of Lowell's papers may well remain to be discovered, but it is certainly true that Lowell's work provided the inspiration for the search.

**Edward Emerson BARNARD** (1857–1923). American astronomer, born at Nashville in Tennessee. He was apprenticed to a photographer, but was very interested in astronomy and began to search for comets. At the age of 23 he discovered the great comet of 1881. After graduating from Vanderbilt University he went first to the Lick Observatory and then to Yerkes, earning himself a great reputation as an observer—due partly, though not entirely, to his exceptionally keen eyesight. His main work was in stellar astronomy, but he also carried out careful observations of the planets, producing drawings that were probably the best of their time. Perhaps significantly, he was never able to record the canals on Mars. In 1892 he discovered Amalthea, the fifth satellite of Jupiter. This was the last discovery of a planetary satellite to be made visually; all subsequent discoveries have been by means of photography. In 1919 Barnard published an important catalog of dark nebulae, and he also identified Barnard's Star at a distance of only six light-years—the closest star beyond the Sun apart from the Alpha Centauri trio. In recent years P. van de Kamp and his colleagues at the Sproul Observatory in the United States have claimed that irregularities in the proper motion of Barnard's Star indicate the presence of at least one planet, probably two.

**William Henry PICKERING** (1858–1938). American astronomer, younger brother of Edward Charles Pickering who specialized in stellar astronomy and worked mainly at Harvard, where he was Director of the Observatory. W. H. Pickering, however, was more interested in Solar System studies. He became assistant to his brother in 1887, and for some time was in charge of the southern station of Harvard College Observatory at Arequipa in Peru. He made careful studies of Mars, claiming that the canals crossed the dark areas as well as the ochre tracts, and that where two canals crossed there was usually a dark, circular patch or "oasis". In 1898 he spent some time at Flagstaff with Percival Lowell, and finally proved that the dark areas on Mars are not aqueous. In 1898 Pickering discovered Phoebe, the ninth satellite of Saturn, by

means of photography—the first time that such a discovery had been made this way. From 1900 he directed the Harvard station in Jamaica, concentrating upon lunar work and compiling a good photographic atlas showing each area of the Moon under several different angles of illumination. He was convinced that changes occurred, attributing these either to vegetation or to swarms of insects. Pickering was also interested in the problem of a trans-Neptunian planet, and worked out several possible positions for it. At Mount Wilson in 1919, Milton Humason carried out a brief photographic search on the basis of Pickering's calculations, but failed to find Pluto, even though later examination of the plates showed that the image of Pluto had been recorded twice.

**Maximilian Franz Joseph Cornelius WOLF** (1863–1932), always known as Max Wolf. He was born at Heidelberg in Germany, and educated at Heidelberg and Stockholm. In 1884 he discovered a periodical comet, and then turned his attention to the discovery of asteroids by means of photography, finally discovering well over 200, whose orbits have been well determined. Wolf was, in fact, the pioneer of this method. He went to Königstuhl Observatory in 1893 as Director, and contributed major studies in stellar astronomy, particularly in connection with bright and dark nebulae.

**Andrew Claude de la Cherois CROMMELIN** (1865–1939). Crommelin was of French descent, but was born in Northern Ireland and educated at Cambridge. In 1891 he joined the staff at the Royal Observatory, Greenwich, and became a world authority on the subject of asteroid and cometary orbits. Together with P. H. Cowell, he made accurate calculations of the date of perihelion of Halley's Comet in 1910: their estimated date was in error by only three days. Crommelin also established the identity of a comet observed independently by Pons, Coggia, Winnecke and Forbes; the comet, now known as Crommelin's Comet, has a period of more than 27 years, and returns to perihelion in 1983.

**George Ellery HALE** (1868–1938). American astronomer, born in Chicago. He studied at Harvard and Berlin, specializing in solar research, and in 1892 he invented the spectroheliograph, which enabled the Sun to be photographed in the light of a single element only—a development of exceptional importance (invented independently, at about the same time, by Deslandres in France). In 1897 Hale became Director of the Yerkes Observatory, and in 1905 went to Mount Wilson, where he continued his solar work. He believed in the value of large reflectors, and was responsible for the 152 cm and 254 cm instruments at Mount Wilson, both of which were financed by friendly millionaires. He also planned the 508 cm Palomar reflector, although he did not see its completion in 1948. In solar research Hale's discoveries included the magnetic fields in sunspots (1908) and a tentative theory of their formation (1912). After his premature retirement in 1922 due to bad health, he invented the spectrohelioscope, the visual equivalent of the spectroheliograph. He is best remembered as the instigator of the building of large telescopes, but his solar research was of equal importance.

**Eugenios ANTONIADI** (1870–1944). Greek-born astronomer, who emigrated to France in 1893 and remained there for the rest of his life, becoming a naturalized Frenchman. Between 1893 and 1902 he was assistant to Flammarion at the Juvisy Observatory, and then, in 1909, he joined the staff at Meudon Observatory, where he was able to make full use of the 83 cm refractor. Antoniadi was a planetary specialist, and probably the best of his time. He made maps of Mars, which have proved to be so good that his nomenclature has been essentially retained. He saw no artificial-looking canals, and had no patience with Lowell's theories of intelligent Martians. He made careful studies of the various types of

# Astronomers (1900-1950s)

clouds in the Martian atmosphere, and also drew a map of Mercury, but with less success. He followed Schiaparelli's method of observing Mercury in daylight, when the planet was high in the sky, and believed that he had confirmed Schiaparelli's view that the rotation of Mercury was synchronous—that is, equal to the Mercurian "year". It is now known that the period is 58.6 days. Antoniadi's map of Mercury was not accurate, but this is hardly surprising. He made observations of Venus, confirming that the few visible surface markings are cloudy in nature, and he also made first-class drawings of Jupiter and Saturn.

**Carl Otto LAMPLAND** (1873–1951). American astronomer. Lampland went to Lowell Observatory at Flagstaff in 1903, and remained there. He made extensive visual observations of Mars, but is best known for his work in planetary temperatures and spectra.

**Vesto Melvin SLIPHER** (1875–1970). American astronomer, educated at Indiana University. He joined the staff of Lowell Observatory in 1901, and carried out pioneer work in planetary spectroscopy. He succeeded Lowell as Director of the Observatory in 1917, and it was at his instigation that Clyde Tombaugh was invited to Flagstaff in 1928 to search for the trans-Neptunian planet on the basis of Lowell's work. Quite apart from this, Slipher was the first to show that the spectra of the objects we now term galaxies show red shifts and are receding from the Earth. He also used the Lowell refractor to prove that some nebulae, including that in the Pleiades cluster, shine solely by reflected starlight.

**James Hopwood JEANS** (1877–1946). English astronomer, born in London and educated at Cambridge. He made important advances in the study of stellar evolution, and developed the theory according to which the planets were pulled out of the Sun in a cigar-shaped "tongue" by the action of a passing star—a theory which was widely accepted for some time, but which has now been rejected. Jeans was knighted in 1928, and during the latter part of his life became an energetic popularizer of astronomy.

**Harold BABCOCK** (1882–1968). American astronomer. He graduated from the University of California in 1907 as an electrical engineer, went to Berkeley in 1908, and then became a physicist at the Pasadena Laboratory at the invitation of Hale. Babcock carried out studies of the spectrum of the Sun and, following his official retirement in 1948, he worked on problems of the Sun's general magnetic field. His son, Horace, has continued this work.

**Earl Carl SLIPHER** (1883–1964). American astronomer, brother of Vesto. He too joined the Lowell Observatory, in 1906, and specialized in planetary photography. His beautiful atlas contained some of the best photographs ever taken before the Space Age.

**Harold JEFFREYS** (1891–    ). English astronomer, educated at Newcastle and Cambridge, and who in 1922 became lecturer in geophysics at Cambridge University. He developed Jeans' theory of the origin of the Solar System. During the 1920s he published a series of classic papers which demonstrated that the giant planets are not self-luminous, as had been widely supposed, and thereby laying the foundations of modern research into their nature.

**Karl Wilhelm REINMUTH** (1892–1979). German astronomer, born in Heidelberg. He became assistant to Max Wolf and, like Wolf, made many asteroid discoveries photographically. In 1933 Reinmuth discovered Apollo, which has an Earth-crossing orbit and has given its name to this family of asteroids, and in 1937 he detected Hermes, which brushed past the Earth at a distance less than twice that of the Moon.

**Walter BAADE** (1893–1960). German astronomer, who emigrated to the United States in 1931 and stayed there until 1958, when he returned to Germany. He discovered two exceptional asteroids, Hidalgo in 1920 and Icarus in 1940, but was concerned mainly with stellar research. In a classic paper in 1952 he showed that there had been a serious error in the estimates of the distances of the outer galaxies, and that the universe was at least twice as large as had been perviously believed.

**Bernard LYOT** (1897–1952). French astronomer. Lyot joined the staff of the Meudon Observatory in 1918, subsequently becoming Director. He made photometric and polarimetric measurements of the Moon and planets, developing equipment specially for the purpose. In 1929 he published an important paper on the polarization of the light from Mars, showing that the Martian atmosphere is affected by haze or dust, and that there are irregularities produced by clouds over the south polar cap. From the high-altitude Pic du Midi Observatory in the Pyrenees, Lyot made observations of the Galilean satellites of Jupiter, using the large refractor, and also took planetary photographs which were as good as any previously obtained. Above all, he is remembered as the inventor of the coronograph, which enables the Sun's inner corona to be observed at times of non-eclipse, and for the development of the "Lyot filter" or monochromatic filter, now widely used for studies of solar prominences. Lyot died in 1952 while returning from an eclipse expedition to Africa.

**Gerard Peter KUIPER** (1906–1973). Dutch astronomer. In 1928 he became research assistant astronomer at Leiden, and in 1933 emigrated to the United States, where he remained for the rest of his life; he became an American citizen in 1937. He worked at the Lick Observatory (1933–5), Harvard University (1935–6), the University of Chicago (1936–7) and then McDonald Observatory in Texas (1939–60), latterly as Director. From 1947–9 and 1957–60 he was also Director of the Yerkes Observatory. In 1960 he went to the University of Arizona to found the Lunar and Planetary Laboratory, which remains at the forefront of Solar System research. Kuiper himself was concerned mainly with the Moon and planets, taking part in the compilation of an elaborate lunar photographic atlas, which remained the standard work until the Orbiter flights of the 1960s. He gave the first interpretation of the atmospheres of Jupiter and Saturn, and gave the first proof of carbon dioxide in the atmosphere of Mars. In 1944 he established the presence of an atmosphere on Titan, the largest of Saturn's satellites. He was chief scientist in charge of the Ranger probes to the Moon, and from 1967–70 chairman of the International Astronomical Union sub-committee on Martian nomenclature. He published many books and papers, and was closely concerned with the early phase of lunar and planetary exploration by means of rocket vehicles. The first identified crater on Mercury, revealed by the Mariner 10 probe in March 1974, has been named in his honor.

**Clyde William TOMBAUGH** (1906–    ). Clyde Tombaugh was born on a Kansas farm, and took an interest in astronomy from an early age. Using a telescope that he had made, he carried out observations of Mars and sent them to Slipher, Director of Lowell Observatory. Slipher was impressed and in 1929, when it had been decided to recommence the search for a trans-Neptunian planet, Tombaugh was invited to Flagstaff to take charge of the hunt, using a telescope obtained specially for the purpose. In the following year Tombaugh identified Pluto. Subsequently he continued searching for further planets, and altogether examined the images of 90,000,000 stars. He has also undertaken pioneer work in connection with meteor streams, and has made extensive observations of the planets, particularly Mars. He retired officially some years ago, and is now Professor Emeritus at the University of Las Cruces.

# Solar Missions

| Name | Country | Launch d m yr | Remarks |
|---|---|---|---|
| Pioneer 4 | USA | 3. 3.59 | Studies of solar flares and the Earth's magnetic field. In solar orbit (0.987 × 1.142 AU). |
| Vanguard 3 | USA | 18. 9.59 | Studies of solar X-radiation. |
| Pioneer 5 | USA | 11. 3.60 | Solar orbit (0.8061 × 0.995 AU). Studies of flares and solar wind. |
| OSO 1 | USA | 7. 3.62 | (Orbiting Solar Observatory 1) Entered Earth orbit at 560 km altitude, and sent back data on 75 flares before operation ceased on 6 August 1963. |
| Cosmos 3 | USSR | 24. 4.62 | Studies of solar and cosmic radiation and measurements of the density of the Earth's upper atmosphere. |
| Cosmos 7 | USSR | 28. 7.62 | Monitoring of solar flares during the manned flights Vostok 3 and 4. |
| Explorer 18 | USA | 26.11.63 | IMP (Interplanetary Monitoring Platform 1). In Earth orbit (202,000 × 125,000 km). Monitored solar flares during manned Apollo and Skylab missions. |
| OGO 1 | USA | 4. 9.64 | (Orbiting Geophysical Observatory 1) Studies of Earth–Sun relationships, and the effects of the Sun on the Earth's magnetic field. |
| OSO 2 | USA | 3. 2.65 | General studies of solar activity and measurements of ultraviolet, X-ray and gamma radiation. |
| OSO C | USA | 25. 8.65 | Launch failure. |
| OGO 2 | USA | 14.10.65 | Studies of ultraviolet and X-radiation from the Sun, and effects of the Sun on Earth's magnetic field. |
| Explorer 30 | USA | 18.11.65 | Solrad. Studies of solar radiation in connection with the IYQS (International Year of the Quiet Sun). |
| Pioneer 6 | USA | 16.12.65 | Solar orbit (0.814 × 0.985 AU). Studies of the solar atmosphere. Together with Pioneer 7, studied data from a strip of the solar surface extending half-way around the Sun. |
| OGO 3 | USA | 7. 6.66 | Studies of Earth–Sun relationships, the solar wind, cosmic radiation and the geocorona |
| Pioneer 7 | USA | 17. 8.66 | Solar orbit (1.010 × 1.125 AU). Same program as Pioneer 6. |
| OSO 3 | USA | 8. 3.67 | Studies of general solar activity, with emphasis on flares. |
| Cosmos 166 | USSR | 16. 6.67 | Studies of solar X-radiation. |
| OGO 4 | USA | 28. 7.67 | Studies of Earth–Sun relationships, atmospheric ionization, the Earth's magnetic field and aurorae. |
| OSO 4 | USA | 18.10.67 | Studies of very shortwave ultraviolet radiation from the Sun, and other general solar activity. |
| Pioneer 8 | USA | 13.12.67 | Solar orbit (1.1 × 1.0 AU). General studies of solar radiation. |
| OGO 5 | USA | 4. 3.67 | Studies of general solar activity and the Earth's magnetic field. |
| Explorer 37 | USA | 5. 3.68 | Solrad. General studies of solar activity and radiation. |
| Cosmos 230 | USSR | 6. 7.68 | General solar studies. |
| Pioneer 9 | USA | 8.11.68 | Solar orbit (0.75 × 1.0 AU). General studies of solar activity. |
| HEOS 1 | USA | 5.12.68 | (Heliographic Earth Orbiting Satellite) Earth orbit (418 × 112,440 km). Together with HEOS 2, covered 7 years of the 11-year cycle; studied interplanetary solar particles. |
| Cosmos 262 | USSR | 26.12.68 | Earth orbit (262 × 965 km) Studies of solar ultraviolet and X-radiation. |
| OSO 5 | USA | 22. 1.69 | Studies of flares and solar activity. |
| OGO 6 | USA | 5. 6.69 | Study of solar influence on the Earth's ionosphere and aurorae. |
| OSO 6 | USA | 9. 8.69 | Studies of solar flares, the corona and general solar activity. |
| Pioneer E | USA | 27. 8.69 | Launch failure. |
| Intercosmos 1 | International | 14.10.69 | Launched from Plesetsk. Studies of solar ultraviolet and X-radiation. |
| Azur | West Germany | 8.11.69 | In solar synchronous orbit. Studied flux of solar particles and their effects. |

| Name | Country | Launch d m yr | Remarks |
|---|---|---|---|
| Intercosmos 4 | International | 14.10.70 | Launched from Kapustin Yar. Earth orbit (628 × 250 km). Studies of the Earth's magnetosphere and solar ultraviolet and X-radiation. |
| Explorer 44 | USA | 8. 7.71 | Solrad. Studies of solar radiation. |
| Shinsei SS1 | Japan | 28. 9.71 | Launched from Kagoshima. Earth orbit (1,870 × 870 km). Studies of solar and cosmic rays. Data recorder failed after four months. |
| OSO 7 | USA | 29. 9.71 | Studies of solar flares, the corona and general solar activity. |
| Intercosmos 5 | International | 2.12.71 | Launched from Kapustin Yar. Carried Soviet and Czech equipment for general solar studies. |
| HEOS 2 | USA | 31.1.62 | Originally orbited around the Earth (405 × 240,164 km), later orbit altered (5,442 × 235,589 km). Studies of high-energy solar particles. |
| Cosmos 484 | USSR | 6.4.72 | Solar and cosmic radiation studies. |
| Prognoz 1 | USSR | 14.4.72 | Studies of the solar wind and X-radiation; studies of the Earth's magnetosphere. |
| Prognoz 2 | USSR | 29.6.72 | Same program as Prognoz 1. |
| Intercosmos 7 | International | 30.6.72 | Launched from Kapustin Yar into Earth orbit (267 × 568 km). Controlled by a Russian-Czech-East German team. Studies of solar short-wave radiation. |
| Aeros 1 | West Germany | 16.12.72 | Launched from Vandenberg Air Force Base into Earth orbit (223 × 867 km). Studies of UV radiation. |
| Prognoz 3 | USSR | 15.2.73 | Solar flares, X-radiation and gamma-radiation studies. |
| Intercosmos 9 | International | 19.4.73 | Launched from Kapustin Yar; Earth orbit (202 × 1,551 km). General studies of solar radiation and activity. |
| Skylab | USA | 14.5.73 | Manned space-station carrying 3 successive crews (see pages 424–5). |
| Taiyo (MS-T2) | Japan | 16.2.74 | Launched from Kagoshima into Earth orbit (3,135 × 255 km). Studies of solar UV and soft X-rays. |
| Intercosmos 11 | International | 17.5.74 | Launched from Kapustin Yar; solar X- and UV radiation studies. |
| Explorer 52 | USA | 3.6.74 | Studies of solar wind and general activity. |
| Aeros 2 | West Germany | 16.7.74 | Launched from Vandenberg Air Force Base into Earth orbit (224 × 869 km). General solar studies. |
| Helios 1 | West Germany | 10.12.74 | On 15 March 1975 passed the Sun at 48,000,000 km at a velocity of 238,000 km/h, spinning at 1 revolution per second. Close-range studies of the solar wind and solar surface. |
| Aryabhata | India | 19.4.75 | Launched from Kapustin Yar into Earth orbit (561 × 619 km). Studies of solar neutrons and gamma-radiation. |
| OSO 8 | USA | 21.6.75 | Studies of solar UV radiation and cosmic X-radiation. |
| Prognoz 4 | USSR | 22.12.75 | Solar radiation and geomagnetic field studies associated with the IMS (International Magnetospheric Study). |
| Helios 2 | West Germany | 15.1.76 | Approached the Sun to 45,000,000 km; same program as Helios 1. |
| Intercosmos 16 | International | 27.7.76 | Launched from Pleseck into Earth orbit (465 × 523 km). Carried Russian and Swedish solar equipment. |
| Prognoz 5 | USSR | 25.11.76 | Solar wind, X- and gamma-radiation studies associated with the IMS. |
| Prognoz 6 | USSR | 22.9.77 | Effects of solar X- and gamma-radiation on the Earth's magnetic field and studies of UV, X- and gamma-radiation from the Galaxy. |
| Prognoz 7 | USSR | 30.10.78 | Solar UV and gamma-radiation and studies of the Earth's magnetosphere. |
| Solar Maximum Mission (SMM) | USA | 14.2.80 | Detailed studies of the Sun in all its aspects near the maximum of the solar cycle (see pages 424–5). |

# Lunar Missions

| Name | Country | Launch d m yr | Remarks |
|---|---|---|---|
| Thor-Able 1 | USA | 17. 8.58 | First attempt to reach the Moon. Failed. |
| Pioneer 1 | USA | 11.10.58 | Failed. Sent data for 43 hours. |
| Pioneer 2 | USA | 8.11.58 | Failed to reach the Moon. |
| Pioneer 3 | USA | 6.12.58 | Failed, but provided radiation data. |
| Luna 1 | USSR | 2. 1.59 | Passed Moon at 6,000 km. Went into solar orbit. |
| Pioneer 4 | USA | 3. 3.59 | Passed Moon at 60,000 km. Went into solar orbit. |
| Luna 2 | USSR | 12. 9.59 | First probe to hit the Moon, crash landing at 30°N, 1°W. |
| Luna 3 | USSR | 4.10.59 | Photographed the Moon's far side. |
| Atlas-Able 4 | USA | 26.11.59 | Failed to reach the Moon. |
| Atlas-Able 5 | USA | 25. 9.60 | Failed to reach the Moon. |
| Atlas-Able 5B | USA | 15.12.60 | Failed to reach the Moon. |
| Ranger 3 | USA | 26. 1.62 | Missed Moon by 36,800 km. |
| Ranger 4 | USA | 23. 4.62 | Hit the Moon (at 15°5S, 130°7W), but cameras failed. |
| Ranger 5 | USA | 18.10.62 | Missed Moon by 725 km. In solar orbit. |
| Unnamed | USSR | 4. 1.63 | Probable unsuccessful lunar probe. |
| Luna 4 | USSR | 2. 4.63 | Unsuccessful soft-lander. Missed the Moon by 8,500 km, and entered solar orbit. |
| Ranger 6 | USA | 30. 1.64 | Hit the Moon (at 0°2N, 21°5E), but TV system failed; no data. |
| Ranger 7 | USA | 28. 7.64 | Landed in Mare Nubium at 10°7S, 20°7W. Returned 4,308 photographs. |
| Ranger 8 | USA | 17. 2.65 | Landed in Mare Tranquillitatis at 2°7N, 24°8E. Returned 7,137 photographs. |
| Cosmos 60 | USSR | 12. 3.65 | Probable unsuccessful lunar probe. |
| Ranger 9 | USA | 21. 3.65 | Landed in Alphonsus (12°9S, 2°4W). Returned 5,814 photographs. |
| Luna 5 | USSR | 9. 5.65 | Crashed in Mare Nubium at 31°S, 8°E; failed soft-lander. |
| Luna 6 | USSR | 8. 6.65 | Missed Moon by 161,000 km (11 June) and entered solar orbit. |
| Zond 3 | USSR | 18. 7.65 | Passed Moon at 9,219 km; returned 25 pictures of the far side. Entered solar orbit. |
| Luna 7 | USSR | 4.10.65 | Crashed in Oceanus Procellarum at 9°N, 40°W. Failed soft-lander. |
| Luna 8 | USSR | 3.12.65 | Crashed in Oceanus Procellarum at 9°1N, 63°3W. Failed soft-lander. |
| Luna 9 | USSR | 31. 1.66 | Successful soft-lander in Oceanus Procellarum at 7°1N, 64°4W. 100 kg capsule landed. Photographs returned. |
| Cosmos III | USSR | 1. 3.66 | Probable unsuccessful lunar probe. |
| Luna 10 | USSR | 31. 3.66 | Lunar satellite; minimum distance from Moon 350 km; contact maintained for 460 orbits in two months. |
| Surveyor 1 | USA | 30. 5.66 | Landed at 2°5S, 43°2W, near Flamsteed; returned 11,237 photographs. |
| Explorer 33 | USA | 1. 7.66 | Failed lunar orbiter. |
| Lunar Orbiter 1 | USA | 10. 8.66 | Photographed Moon until 29 August 1966. Impacted at 6°7N, 162°E, on 29 October 1966. |
| Luna 11 | USSR | 24. 8.66 | Minimum distance 159 km. Transmitted until 1 October 1966. |
| Surveyor 2 | USA | 20. 9.66 | Unsuccessful soft-lander; crashed at 5°N, 25°W, near Copernicus. |
| Luna 12 | USSR | 22.10.66 | Transmitted until 19 January 1967. |
| Lunar Orbiter 2 | USA | 6.11.66 | Lunar satellite. Transmitted 422 pictures before impacting at 4°S, 98°E. |
| Luna 13 | USSR | 21.12.66 | Soft-landed at 18°9N, 62°W in Oceanus Procellarum. Transmitted until 27 December 1966. Soil studies. |
| Lunar Orbiter 3 | USA | 4. 2.67 | Lunar satellite, impacted at 14°6N, 91°7W. Returned 307 pictures. |
| Surveyor 3 | USA | 17. 4.67 | Landed at 2°9S, 23°3W in Oceanus Procellarum, 612 km east of Surveyor 1, near Apollo 12 site. Returned 6,315 pictures. Soil physics. |
| Lunar Orbiter 4 | USA | 4. 5.67 | Returned 326 pictures. |
| Surveyor 4 | USA | 14. 7.67 | Failed. Crashed at 0°4N, 1°3W in Sinus Medii. |
| Explorer 35 | USA | 19. 7.67 | Studies of Earth's magnetic field. |
| Lunar Orbiter 5 | USA | 1. 8.67 | Lunar satellite. Controlled impact at 0°, 70°W on 31 January 1968. |
| Surveyor 5 | USA | 8. 9.67 | Landed at 1°4N, 23°2E in Mare Tranquillitatis, 25 km from Apollo 11 site. Returned 18,006 pictures. |
| Surveyor 6 | USA | 7.11.67 | Landed at 0°5N, 1°4W in Sinus Medii. Returned 30,065 pictures. |
| Surveyor 7 | USA | 7. 1.68 | Landed at 40°9S, 11°5W on north rim of Tycho. Returned 21,274 pictures. Soil analyses. |
| Zond 4 | USSR | 2. 3.68 | Lunar probe, but exact purpose unknown. |
| Apollo 6 | USA | 4. 4.68 | Failed to reach the Moon. |
| Luna 14 | USSR | 7. 4.68 | Lunar satellite; minimum distance from Moon 160 km. |
| Zond 5 | USSR | 15. 9.68 | Went round Moon and landed in Indian Ocean, 21 September 1968. |
| Zond 6 | USSR | 10.11.68 | Went round Moon and returned to Earth on 17 November 1968. |
| Apollo 8 | USA | 21.12.68 | Manned orbiter; 10 orbits completed. (Lovell, Borman, Anders.) |
| Apollo 10 | USA | 18. 5.69 | Manned orbiter; went to within 14.9 km of the Moon and tested Lunar Module. (Stafford, Cernan, Young.) |
| Luna 15 | USSR | 13. 7.69 | 52 orbits; crashed at 17°N, 60°E in Mare Crisium, 21 July 1969. |
| Apollo 11 | USA | 16. 7.69 | Manned landing at 0°7N, 23°4E, 20 July 1969, in Mare Tranquillitatis. (Armstrong, Aldrin, Collins.) |
| Zond 7 | USSR | 8. 8.69 | Went round Moon and returned to Earth. |
| Apollo 12 | USA | 14.11.69 | Landed on 19 November 1969 at 3°2S, 23°8W in Oceanus Procellarum. (Conrad, Bean, Gordon.) |
| Apollo 13 | USA | 11. 4.70 | Unsuccessful manned lander; returned 17 April 1970. (Lovell, Haise, Swigert.) |
| Luna 16 | USSR | 12. 9.70 | Landed at 0°7S, 55°3E in Mare Fecunditatis, 0°41S, 56°18E, on 20 September 1970. Returned 100 g of soil. |
| Zond 8 | USSR | 20. 9.70 | Orbited Moon; returned 27 October 1970 |
| Luna 17 | USSR | 10.11.70 | Carried Lunokhod 1 to Moon; landed at 38°3N, 35°W in Mare Imbrium. 17 November 1971, returning over 20,000 pictures. |
| Apollo 14 | USA | 31. 1.71 | Manned lander. Landed at 3°7S, 17°5W in Fra Mauro on 5 February 1971. (Shepard, Mitchell, Rossa.) |
| Apollo 15 | USA | 26. 7.71 | Manned lander; Hadley–Apennines (26°1N, 3°7E). (Scott, Irwin, Worden.) |
| Luna 18 | USSR | 2. 9.71 | Crashed at 3°6N, 56°5E in Mare Fecunditatis after 54 orbits. |
| Luna 19 | USSR | 28. 9.71 | Contact maintained for over a year and 4,000 orbits. |
| Luna 20 | USSR | 14. 2.72 | Landed 21 February 1972 at 3°5N, 56°6E in Mare Fecunditatis 120 km from Luna 16's impact. Returned samples; landed on Earth 25 February 1972. |
| Apollo 16 | USA | 16. 4.72 | Landed at 8°6S, 15°5E in Descartes area on 21 April 1972. (Young, Duke, Mattingley.) |
| Apollo 17 | USA | 7.12.72 | Landed at 21°2N, 30°6E in Taurus–Littrow, 11 December 1972. (Cernan, Schmitt, Evans.) |
| Luna 21 | USSR | 8. 1.73 | Carried Lunokhod 2 to Lemonnier area on 16 January 1973. Lunokhod 3 transmitted until 3 June 1973, returning over 80,000 pictures. |
| Explorer 49 | USA | 10. 6.73 | Radio astronomy from far side. |
| Luna 22 | USSR | 29. 5.74 | Transmitted until 6 November 1975. |
| Luna 23 | USSR | 28.10.74 | Landed in Mare Crisium. Sampling unsuccessful. Transmitted until 9 November 1975. |
| Luna 24 | USSR | 9. 8.76 | Landed at 12°8N, 62°2E in Mare Crisium. Drilled to 2 m. Landed back on Earth 22 August 1976. |

# Planetary Missions

**Mercury and Venus probes**

| Name | Country | Launch d m yr | Remarks |
|------|---------|---------------|---------|
| Venera 1 | USSR | 12. 2.61 | Contact lost at 8,500,000 km. Closest approach about 100,000 km. |
| Mariner 1 | USA | 22. 7.62 | Total failure. |
| Mariner 2 | USA | 26. 8.62 | Flyby. Data transmitted. Closest approach 35,000 km. |
| Zond 1 | USSR | 2. 4.64 | Contact lost in a few weeks. Closest approach about 100,000 km. |
| Venera 2 | USSR | 12.11.65 | In solar orbit. No data received. Closest approach 24,000 km. |
| Venera 3 | USSR | 16.11.65 | Crushed during descent (1 March 1966). No data received. |
| Venera 4 | USSR | 12. 6.67 | Data sent back during 94 min descent (19 October 1967). |
| Mariner 5 | USA | 14. 6.67 | Flyby. Data transmitted. Closest approach 4,000 km. |
| Venera 5 | USSR | 5. 1.69 | Crushed during descent (16 May 1969). Data transmitted. |
| Venera 6 | USSR | 10. 1.69 | Crushed during descent 17 May (1969). Data transmitted. |
| Venera 7 | USSR | 18. 8.70 | Tranmitted for 23 min after landing (15 December 1970). |
| Venera 8 | USSR | 26. 3.72 | Transmitted for 50 min after landing (22 July 1972). |
| Mariner 10 | USA | 3.11.73 | Pictures of upper clouds of Venus; data transmitted. Closest approach 5,800 km (5 February 1974). Went on to three flyby encounters with Mercury (29 March 1974). |
| Venera 9 | USSR | 8. 6.75 | Transmitted for 53 min after landing (21 October 1975). One picture received. Data received from orbiter. |
| Venera 10 | USSR | 14. 6.75 | Transmitted for 65 min after landing (25 October 1975). One picture received. Data received from orbiter. |
| Pioneer–Venus 1 | USA | 20. 5.78 | In orbit. Data sent back. Closest approach 145 km. |
| Pioneer–Venus 2 | USA | 8. 8.78 | Multiprobe. Four probes landed (9 December 1978). Data sent back from probes and "bus". |
| Venera 11 | USSR | 9. 9.78 | Transmitted for 60 min after landing (21 December 1978). Data received from orbiter. |
| Venera 12 | USSR | 14. 9.78 | Transmitted for 60 min after landing (25 December 1978). Data received from orbiter. |
| Venera 13 | USSR | 30.10.81 | Transmitted for 127 min after landing (1 March 1982). Two color pictures received. Soil analysis carried out. Data received from orbiter. |
| Venera 14 | USSR | 4.11.81 | Transmitted for over 60 min after landing (5 March 1982). Two color pictures received. Soil analysis carried out. Data received from orbiter. |
| Venera 15 | USSR | 2. 6.83 | Successful polar orbiter. Radar mapping of surface. |
| Venera 16 | USSR | 7. 6.83 | Successful polar orbiter. Radar mapping of surface. |
| Vega 1 | USSR | 15.12.84 | Passed Venus (11.6.85) en route to Halley's Comet; balloon dropped into Venus' atmosphere. |
| Vega 2 | USSR | 20.12.84 | Passed Venus (15.6.85) en route to Halley's Comet; balloon dropped into Venus' atmosphere. |
| Magellan | USA | 5. 5.89 | Radar mapper; arrival August 1990. |

**Mars probes**

| Name | Country | Launch d m yr | Remarks |
|------|---------|---------------|---------|
| Mars 1 | USSR | 1.11.62 | Contact broken at 105,000,000 km. |
| Mariner 3 | USA | 5.11.64 | Shroud failure. Entered solar orbit, but went nowhere near Mars. |
| Mariner 4 | USA | 28.11.64 | Flyby. Returned 21 pictures of Mars. Now in solar orbit. Contact finally lost on 20 December 1967. Closest approach 9,789 km. |
| Zond 2 | USSR | 30.11.64 | Contact lost. Probable rendezvous with Mars in August 1965. |
| Mariner 6 | USA | 24. 2.69 | Flyby. Flew over Martian equator, returning 75 pictures. Closest approach 3,392 km. Now in solar orbit. |
| Mariner 7 | USA | 27. 3.69 | Flyby. Flew over southern part of Mars, returning 126 pictures. Closest approach 3,504 km. Now in Solar orbit. |
| Mariner 8 | USA | 8. 5.71 | Total failure: fell in the sea. |
| Mars 2 | USSR | 19. 5.71 | In Mars orbit, 2,448 × 24,400 km. Dropped a capsule onto Mars, carrying a Soviet pennant at 44°S, 213°W. |
| Mars 3 | USSR | 28. 5.71 | In Mars orbit, 1,552 × 212,800 km. Lander touched down at 45°S, 158°W, but lost contact after 20 sec: no useful data received. |
| Mariner 9 | USA | 30. 5.71 | In Mars orbit, 1,640 × 16,800 km. Operated from 13 November 1971 to 27 October 1972, returning 7,329 pictures. Closest approach 1,640 km. |
| Mars 4 | USSR | 21. 7.73 | Failed to orbit. Missed Mars by over 2,080 km. Obtained flyby data. |
| Mars 5 | USSR | 25. 7.73 | In Mars orbit. Contact lost. |
| Mars 6 | USSR | 5. 8.73 | Contact lost during landing sequence. Probable landing on Mars at 24°S, 25°W. |
| Mars 7 | USSR | 9. 8.73 | Failed to orbit; missed Mars by 1,280 km. |
| Viking 1 | USA | 20. 8.75 | Landed 20 July 1976, in Chryse (22°N, 47°W). Lander and orbiter returned pictures and data. |
| Viking 2 | USA | 9. 9.75 | Landed, 3 September 1976, in Utopia (48°N, 226°W). Lander and orbiter returned pictures and data. |
| Phobos 1 | USSR | 7. 7.88 | Intended Phobos lander and mapper. Contact lost 29.8.88. |
| Phobos 2 | USSR | 12. 7.88 | Intended Phobos lander and mapper. Contact lost March 1989; a few images obtained. |

**Jupiter and Saturn probes**

| Name | Country | Launch d m yr | Remarks |
|------|---------|---------------|---------|
| Pioneer 10 | USA | 2. 3.72 | Jupiter flyby (3 December 1973). Transmitted pictures and much miscellaneous data. |
| Pioneer 11 | USA | 5. 4.73 | Jupiter flyby (2 December 1974); confirmed results of Pioneer 10. Went on to rendezvous with Saturn (September 1979). |
| Voyager 2 | USA | 20. 8.77 | Four-planet flyby; Jupiter (1979), Saturn (1981). Included pictures and data from Saturn and several satellites, as well as those not fully covered by Voyager 1. Now en route for Uranus (1986) and Neptune (1989). |
| Voyager 1 | USA | 5. 9.77 | Two-planet flyby; Jupiter (5 March 1977), Saturn (November 1980). Included pictures and data from Saturn and several satellites, including Titan. |

# Skylab and Solar Maximum Missions

The largest structure so far to have been placed in space, the American manned orbiting laboratory Skylab, was set into orbit at an altitude of 435 km on 14 May 1973. Crews were ferried to and from the laboratory by means of Apollo-type spacecraft. With the Apollo Command Service Module (CSM) attached, Skylab had an overall length of 36 m and a mass of 90,600 kg. The major component, the orbital workshop, provided a total of 292 m³ of living space, and three crews, each of three astronauts, spent a total of 513 man-days in space, carrying out a wide variety of experiments and observations; solar astronomy occupied about 30 percent of the total time allocation. The astronauts were able to coordinate their experiments closely with the work of ground-based observatories, with whom they maintained contact constantly.

Eight different instruments were mounted on the Apollo Telescope Mount (ATM), a structure which allowed the instruments to be held rigidly and pointed with high accuracy towards the Sun. The basic ATM structure measured 4.4 m tall, weighed 11,090 kg, and was powered by solar panels arranged like windmill sails 31 m in diameter. The instruments comprised a white light coronagraph (which allowed the corona to be studied to a distance of about $6 R_S$), three X-ray instruments, which produced X-ray photographs and spectra, three ultraviolet instruments, which supplied images and spectral data, and two $H_\alpha$ telescopes, permitting direct observation and photography of hydrogen light phenomena and flares. In addition, there was a hand-guided X-ray/ultraviolet solar photography experiment operated through an airlock in the spacecraft's laboratory.

Skylab itself reentered the atmosphere and broke up on 11 July 1979. Ironically, this occurred as a result of the effects of increasing solar activity on the uppermost regions of the Earth's atmosphere. The contribution of Skylab to knowledge of the Sun was immense, including the discovery of important new phenomena such as coronal holes (*see* pages 66–7).

**Solar Maximum Mission (SMM)**
Designed primarily to make a concentrated attack on problems associated with the nature, trigger-mechanism and effects of solar flares, this satellite was launched on 14 February 1980 into a circular orbit at an altitude of 574 km. It contained a battery of carefully matched instruments operating in white light, ultraviolet, X-ray and γ-ray ranges. Other objectives included the study of the evolution of the corona around a maximum period of solar activity, and the accurate measurement over a prolonged period of the total solar output of radiation.

The spacecraft itself has a mass of 2,315 kg and measures 4 m long by 1.2 m wide. It consists of an instrument module, housing all seven solar instruments (see table), together with the Fine Pointing Sun Sensor used to lock the instruments onto the Sun, as well as the supporting spacecraft, which itself comprises three modules of essential subsystems: attitude control, power and communications, and data handling. Two arrays of solar panels supply about 3,000 W of power for the spacecraft's systems, while battery power is also available for the periods when the spacecraft is in the Earth's shadow, or in eclipse.

The unique feature of this mission is the degree of integration that exists between the various instruments, and the degree of flexibility that is built into the mission operations. The investigators associated with each experiment are housed together in the Experimenters' Operations Facility at NASA's Goddard Space Flight Center and meet daily to decide, on the basis of current observations, which active area to concentrate their efforts on during each 24 hr period. In this way the onboard equipment can be used with the greatest efficiency to study such transient phenomena as flares. Solar Max had a protracted career involving one repair in space from the shuttle. It reentered the atmosphere, and it was destroyed, in 1989.

**1. Solar Maximum Mission**
The spacecraft carried seven scientific instruments: (1) X-ray Polychromator, designed to investigate the activity that produces solar plasma with temperatures between 1.5 and 50 million K; (2) Solar Irradiance Monitor, to measure variations in the solar constant (*see* page 25); (3) Coronagraph/Polarimeter, producing images of coronal evolution and coronal transient activity; (4) Hard X-Ray Burst Spectrometer, to investigate the role of energetic electrons in solar flares; (5) Ultraviolet Spectrometer and Polarimeter, to study coronal active regions and flares by observing spectral features in the UV range; (6) γ-Ray Spectrometer, to examine ways in which high-energy particles are produced in solar flares; (7) Hard X-Ray Imaging Spectrometer, to provide information about the position, extension and spectrum of hard X-ray bursts in flares.

**2. Skylab**
Skylab had a length of 25 m and was 6.7 m wide at the main workshop section. While in orbit its instruments were powered by solar cells. (One of the large solar panels shown in this illustration was torn off during launch, and photographs show it with only one in position.)

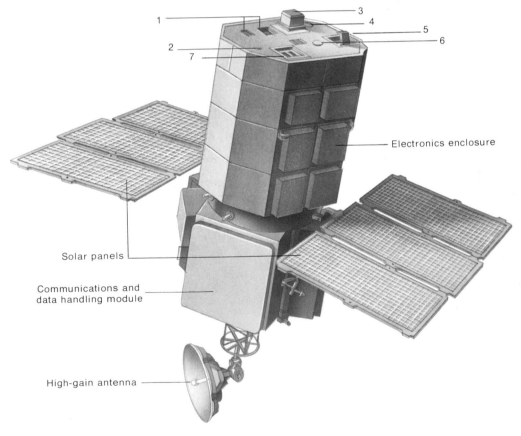

Electronics enclosure

Solar panels

Communications and
data handling module

High-gain antenna

2

1. Modified Apollo Command Module and Service Module
2. Service Propulsion Engine
3. Radiators
4. Attitude Control Jets
5. Crew Station
6. Apollo Telescope Mount
7. Solar Cells
8. Sun Shield
9. Telescope Apertures
10. Oxygen Tank
11. Nitrogen Tank
12. Maneuvering Unit
13. Gravity Substitute Workbench
14. Food Provisions
15. Solar Cells
16. Sleep Restraints
17. Water Containers
18. Aerial
19. Multiple Docking Adapter
20. Alternative Docking Port
21. Atmosphere Interchange Duct
22. Descent Battery Packs

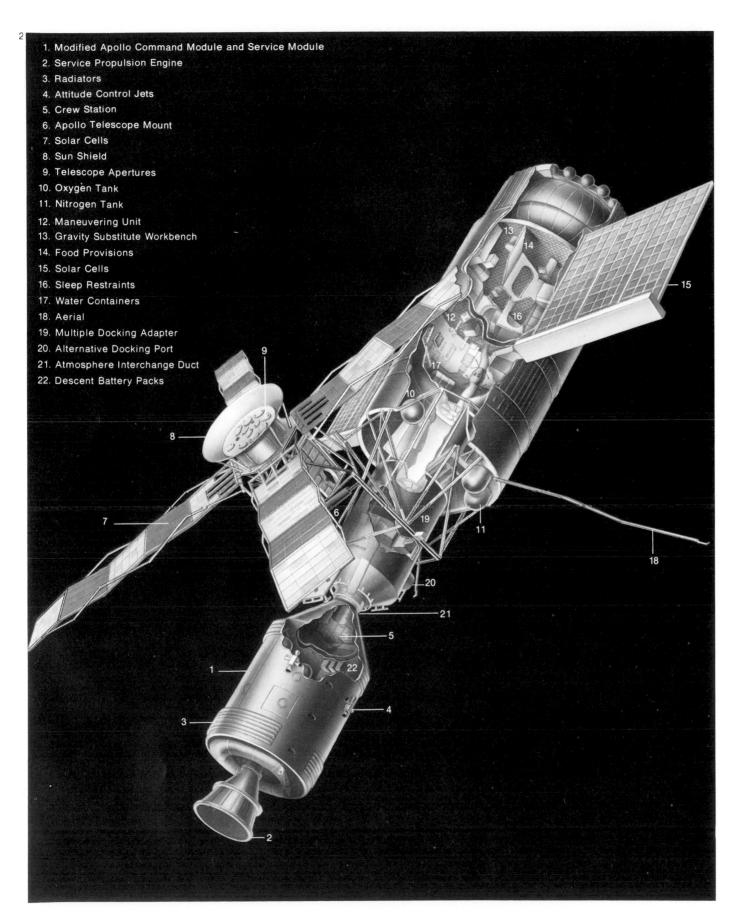

# Unmanned and Manned Lunar Missions

## Soviet lunar missions

The age of space exploration began on 4 October 1957, with the launching of the Soviet satellite Sputnik 1. Less than two years later, on 2 January 1959, the Soviet researchers dispatched their first Moon probe, Luna (or Lunik) 1, which passed within 5,955 km of the Moon. No pictures were obtained, but some useful information was sent back, notably confirmation that the Moon has no detectable overall magnetic field. Next, on 12 September of the same year, came Luna 2, a "hard-lander". The real triumph came with Luna 3, launched on 4 October 1959—exactly two years after Sputnik 1. It went right around the Moon, at a minimum distance of 6,200 km, taking the first photographs of the far side.

The next four Lunas were unsuccessful, but in July 1965 Zond 3 obtained further pictures of the Moon's far side, from a minimum distance of 9,219 km. Luna 9, launched on 31 January 1966, made a successful soft landing on the grey plain of the Oceanus Procellarum and sent back pictures direct from the lunar surface. Lunas 10, 11 and 12 were put into closed paths and sent back useful data as well as pictures; Luna 13, sent up on 21 December 1966, was another successful soft-lander, capable of carrying out soil analyses and other valuable researches. Since then there have been more than a dozen further unmanned vehicles, including Lunas 16 and 20, which obtained samples of lunar material, and brought them back for analysis.

## American unmanned missions

The United States' first real success was with Ranger 7, which hit the Moon on 31 July 1964. Before impact it sent back 4,316 photographs of a region in the Mare Nubium since named the Mare Cognitum or "Known Sea". The Ranger project was followed by two further series of unmanned probes, the Lunar Orbiters and the Surveyors. The Orbiter program consisted of a total of five probes, all of which were launched within a year, starting on 10 August 1966. All worked faultlessly, and even today many of the thousands of photographs sent back await examination. On 30 May 1966 the first vehicle in the Surveyor series was launched. It landed north of Flamsteed and returned over 11,000 splendid photographs.

## The American Apollo missions

The Apollo program was initiated in the early 1960s, with the enthusiastic support of President John F. Kennedy. The culmination of all the preparatory tests was Apollo 11, which was launched on 16 July 1969 from Cape Canaveral. The crew consisted of Neil Armstrong, Edwin Aldrin and Michael Collins; Collins was to remain in the Command Module while Armstrong and Aldrin made the descent to the lunar surface. The actual landing was carried out manually to enable the astronauts to avoid boulders. The Lunar Module touched down in the Mare Tranquillitatis at 0.7°N, 23°E; first Armstrong, then Aldrin, set foot on the bleak landscape of the Mare Tranquillitatis early on 21 July; the two astronauts remained outside the LM for over two hours, setting up the ALSEP or Apollo Lunar Surface Experimental Package. The astronauts also collected 21.75 kg of rock samples for analysis.

Less than four months after the triumph of Apollo 11, Apollo 12 touched down in the Oceanus Procellarum on 19 November 1969; it was a precision landing close to the old unmanned probe Surveyor 3 and the astronauts, Charles Conrad and Alan Bean, set up an improved ALSEP powered by nuclear energy.

There followed four further successful missions. Apollo 14 came down near Fra Mauro, and the astronauts took a "lunar cart" to help in collecting samples. Apollo 15 was brought down in the Hadley-Apennines area and was even more ambitious, as the astronauts (Scott and Irwin) had a special Lunar Roving Vehicle which enabled them to drive around, covering almost 30 km and taking them close to the great chasm known as Hadley Rill. Apollo 16 took astronauts Young and Duke to the highlands of Descartes; this new type of landscape made the results particularly important.

With Apollo 17, of December 1972, there was a new development: Dr. Harrison Schmitt, a professional geologist, was trained as an astronaut and made the journey. His specialized knowledge was extremely useful, and there were some surprises – particularly the discovery of "orange soil", at first thought to be indicative of comparatively recent volcanism, but later found to be due to very ancient, colored glassy particles. The landing was made in the Taurus-Littrow area.

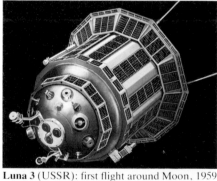

**Luna 3** (USSR): first flight around Moon, 1959

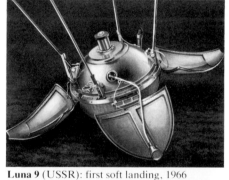

**Luna 9** (USSR): first soft landing, 1966

**Lunokhod 1** (USSR): Moon crawler, 1970

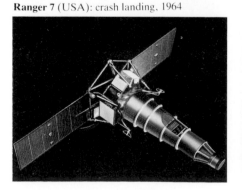

**Ranger 7** (USA): crash landing, 1964

**Lunar Orbiter 5** (USA): photo coverage, 1966

**Surveyor 3** (USA): soft landing, 1967

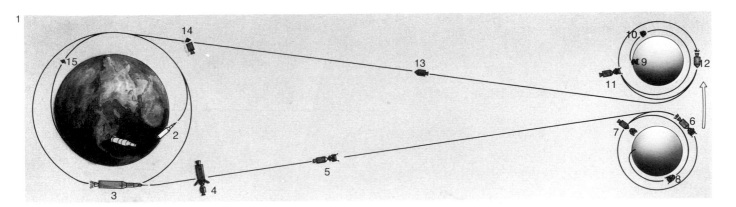

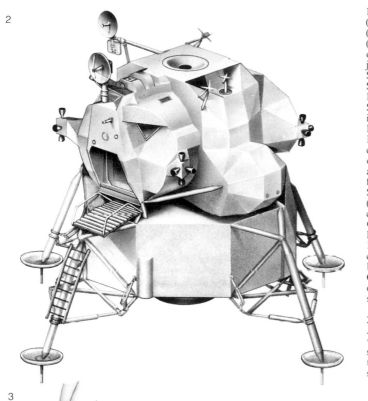

## 1. Flight plan of Apollo 11
(1) Lift-off from Cape Kennedy.
(2) Separation of first stage.
(3) Engine ignition to take rocket out of Earth-orbit and begin journey to the Moon. (4) Separation of Command and Service Module from Lunar Module. (5) Engine ignition of Service Module. (6) Lunar orbit insertion. (7) Separation of Lunar Module from Command Service Module. (8) Lunar Module descent engine ignition. (9) Start of return journey; Lunar Module ascent. (10) Maneuver to prepare for rendezvous with Command Service Module. (11) Docking. (12) Transfer of crew and equipment from Lunar Module to Command Service Module. (13) Return journey. (14) Service Module jettisoned. (15) Re-entry; communication blackout. (16) Final touchdown in the Pacific Ocean. The entire journey lasted 8 days, including the $21\frac{1}{2}$ hours spent on the lunar surface.

## 2. Lunar Module (LM)
All the Apollo Lunar Modules were of the same type. The lower stage served as a launch pad when the astronauts took off from the surface of the Moon.

## 3. Lunar Roving Vehicle (LRV)
LRVs, carried on the last three Apollo missions, could travel at about $15\,\mathrm{km\,hr^{-1}}$ on surfaces that were level.

## 4. Apollo Lunar Surface Experimental Package (ALSEP)
The general layout of an ALSEP is shown here; the equipment is not all drawn to the same scale. (1) ALSEP central station; coordinates the experiments and relays their signals back to Earth. The rod-like feature is an antenna. (2) Radioisotope Thermal Generator (RTG). Approximately 70 W of power for the experiments was provided by radioactive decay. (3) Passive Seismometer, covered by an insulating blanket, monitored small vibrations of the Moon's surface. (4) Lunar Surface Magnetometer measured the magnetic field along three axes. (5) Solar Wind Spectrometer, measuring the energy, density, direction and variations with time of the Solar Wind. (6) Heat Flow Experiment, with two probes placed in holes drilled in the lunar surface. (7) Suprathermal Ion Detector and Cold Cathode Ion Gauge; two experiments studying the lunar atmosphere.

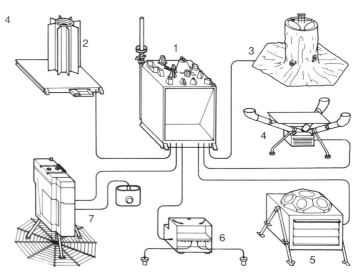

# Mariner 10

Mariner 10's journey to the inner planets represented a pioneering achievement in space technology. Although building on techniques established in previous Mariner missions, Mariner 10 was the first vehicle to make use of the gravity of another planet – Venus – to gain its correct trajectory. It was also the first mission to transmit its television pictures as they were taken. Both the cameras and the more refined sensors incorporated on the vehicle provided a wealth of scientific information.

The design, development and construction of the whole vehicle was completed in $2\frac{1}{2}$ years. The assembly was completed in good time for the launch date in November 1973. The facilities required to maintain the equipment (below) were the propulsion system, the attitude control, temperature control, the power supply and the communications system.

Because Mariner used the gravitational fields of Venus and Mercury at each encounter to provide the main changes in direction and velocity, the propulsion unit was required to make only fine adjustments. The system provided a total velocity change of 119 m per second, and could be started and stopped as often as needed.

As the vehicle continued on its trajectory, it had to maintain a fixed attitude in space, with its Z axis pointing to the Sun, in the direction of the sunshade, and its X and Y axes diagonal to the ecliptic plane, along the axes of the solar panels and magnetometer boom respectively. With this attitude, the sunshade protected the very delicate instruments from the searing heat of the direct solar radiation, and the scientific instruments and the antennae were correctly pointed to make and transmit their measurements.

To maintain this attitude during flight required continuous control by venting minute jets of nitrogen through reaction control jets mounted at the ends of the solar panels and on the antennae and magnetometer outriggers. Two bearings ensured the correct attitude; one on the Sun, controlled by the Sun sensors, which kept the Z axis pointing at the Sun, the other on the bright star Canopus, monitored by the Canopus tracker.

The mainstay of the temperature control was the sunshade, supplemented by multilayer thermal blankets at the top and bottom of the vehicle. Five of the eight sides were also equipped with louvered panels, which gave active temperature control by adjusting the amount of surface able to radiate into space.

The electrical power needed to operate Mariner was generated by the two solar panels carried on outriggers, which kept them clear of the shadow cast by the sunshade. A storage battery provided power for use when the solar cells were not aligned with the Sun, as happened during maneuvers or when the Sun was occulted.

In order to function at all, Mariner had to receive commands from Earth, and transmit data back to Earth. This was achieved by the low-gain antenna and the high-gain antenna respectively; the data was received and transmitted in binary digital form.

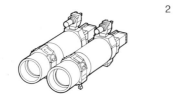

## 1. Television camera
The optical system of the Cassegrain telescope focuses a highly magnified image onto the sensitive screen of a vidicon tube. In front of the shutter is an eight-position filter wheel that enables pictures to be taken at different wavelengths, including ultraviolet. The filter wheel also carries a diagonal mirror that can switch the camera to a wide-angle lens system. Exposure of the videcon screen to light produces a pattern of variable charge, proportional to the light intensity at each point. The pattern is scanned and transmitted in digital form as a series of eight-digit binary numbers (*see* page 15).

## 2. Infrared radiometer
The radiometer measured the temperature of the surface layers of Mercury and Venus by sensing the radiation emitted. It was accurate to within 0.5° C. The radiometer works by measuring the minute changes of temperature in a thermopile placed at the focal point of a telescope. Two separate telescopes and thermopiles measured the warm and cold regions respectively. Both are mounted side by side. Through a motorized mirror they viewed in succession, an internal calibration surface at a known temperature, empty space and the surface of the planet to eliminate inconsistencies.

## 3. Plasma science
The forerunner of the experiments carried on the Voyager missions, the plasma experiment on Mariner 10 studied the properties of the very hot ionized gases that exist in the interplanetary regions. The instrument consists of two plasma detectors, one pointing in the direction of the Earth and the other at a right angle to the first. It analyzed the properties of the solar wind and its interaction with Mercury and Venus. The first encounter revealed the presence of a magnetic field on Mercury strong enough to deflect the solar wind and form a magnetosphere that is similar in its geometry to that of the Earth.

## 4. Magnetometer
The magnetic fields of both Venus and Mercury were measured using the two magnetometers. Each consists of three measuring coils mounted at right angles to one another. Each coil measures the field strength along its own axis. By vector addition of the three components, the external field strength and direction can be reconstructed. Two such magnetometers are used so that by comparing them both, the spacecraft's own magnetic field is eliminated. To ensure the readings are constant, the coils are turned through 180° periodically. In all 100 readings a second were made and transmitted back to Earth.

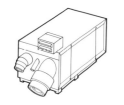

## 5. Charged particle telescope
The charged particle experiment studied the interaction of charged particles with the planet. Together with the plasma science experiment, it confirmed the existence of a magnetic field on Mercury. It is also designed to investigate the solar wind and cosmic rays.

## 6. Airglow spectrometer
The airglow ultraviolet spectrometer detects faint light emissions from gases above the dark hemisphere. Collimation plates admit sufficient light, which is reflected by a concave diffraction grating; this grating focuses a beam as a spectral band, which is then measured.

## 7. Occultation spectrometer
Together with the airglow spectrometer, the occultation spectrometer was used to detect an atmosphere on Mercury. Light in the extreme ultraviolet portion of the electromagnetic spectrum has the property of interacting strongly with most gases. Gas molecules will absorb certain ultraviolet frequencies which correspond to their characteristic wavelengths. This produces dark lines in the ultraviolet spectrum. An irradiated gas is also luminescent, emitting light at these characteristic wavelengths. Both these effects were used to search for an atmosphere. The occultation ultraviolet spectrometer attempts to trace an atmosphere through the absorption of solar ultraviolet. Sunlight enters the spectrometer and is collimated into narrow beams. These beams are reflected by a plane mirror on which a diffraction grating is ruled. The grating splits the six beams into spectral bands. Detectors measure the solar ultraviolet flow at four different wavelengths centered on the absorption bands for neon, helium, argon and krypton. Two infrared beams time the moment of occultation. Measurements could be made only during the first encounter, as this was the one time that an occultation of the Sun occurred.

1    Airglow UV – spectrometer
2    Charged particle telescope
3    High-gain antenna
4    Infra-red radiometer
5    Low-gain antenna
6    Magnetometer (inboard)
7    Magnetometer (outboard)
8    Occultation UV – spectrometer
9    Plasma science
10   Reaction control jets
11   Solar panels
12   Sunshade
13   Television camera

# Venera and Pioneer Venus

## Soviet missions

The Venera program demonstrates the particular interest Russian scientists have in the Earth's "twin planet" Venus. Starting in February 1961 three Veneras were sacrificed before the first data was sent back during the descent of Venera 4 in October 1967. Subsequently, soft landings were achieved with Veneras 7 and 8 and transmission made from the surface. These proved the forerunners of the most spectacular landings on the planet by Veneras 9 and 10 (June 1975) and Veneras 13 and 14 (March 1982). Each was able to transmit television pictures from the surface, the latter two probes using color filters. The achievement cannot be underestimated. The technical challenge of landing the spacecraft and operating them in the hostile environment of Venus was immense. They not only had to withstand the enormous atmospheric pressure, 100 times that on the Earth, but they also had to operate at a temperature of 500° C, high enough to melt lead or zinc. The difficulties of operating a camera to obtain pictures, and, in the case of Veneras 13 and 14, a sampling system to analyze soil samples inside the spacecraft, were truly formidable. Like their American counterparts the Veneras' scientific experiments also concentrated on analyses of the atmosphere. In addition, Veneras 13 and 14 carried French-built gamma-ray detectors and microphones that detected electrical discharges.

## American missions

In comparison to the Russians, American interest in Venus has been minimal; out of a total of 31 missions attempted the United States have been responsible for six. The Mariner program had some success with Mariners 2, 5 and 10; however, with Pioneer Venus in 1978 the Americans achieved the greatest number of experiments ever launched in a single mission. This was achieved by using two spacecraft—an orbiter and a multiprobe. Both utilize the same design to the maximum extent possible to minimize cost. The orbiter carries 12 scientific instruments which concentrated principally on measuring the characteristics of the upper atmosphere of Venus and mapping the surface contours and gravitational field using radar. The multiprobe spacecraft is actually five independent spacecraft consisting of the probe carrier or "bus", one large probe and three smaller probes. In total they contain 18 scientific instruments. The four probes were separated from the bus and successfully transmitted data on atmospheric density and composition of the lower atmosphere during the descent to the surface. The bus itself transmitted data on the upper atmosphere before it was destroyed. One small probe survived the jarring surface impact and continued transmitting for 67 min from the sunlit side of Venus before its circuitry succumbed to the planet's intense heat.

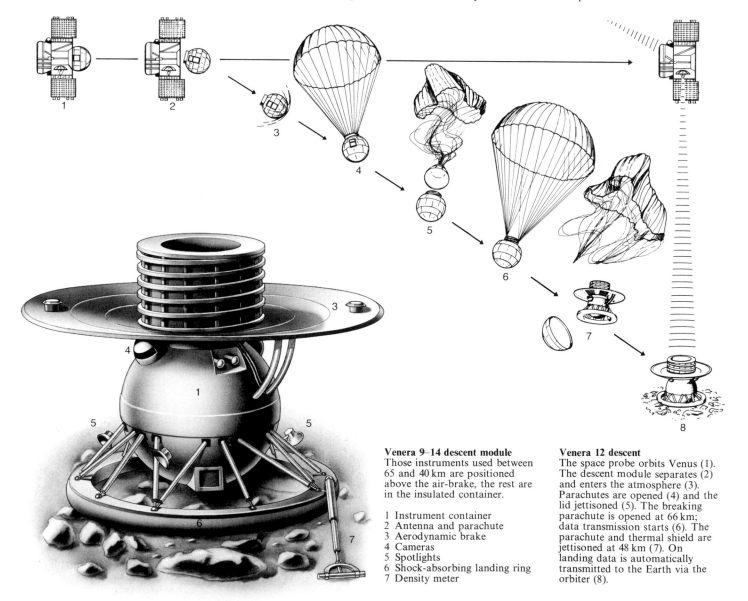

### Venera 9–14 descent module
Those instruments used between 65 and 40 km are positioned above the air-brake, the rest are in the insulated container.

1 Instrument container
2 Antenna and parachute
3 Aerodynamic brake
4 Cameras
5 Spotlights
6 Shock-absorbing landing ring
7 Density meter

### Venera 12 descent
The space probe orbits Venus (1). The descent module separates (2) and enters the atmosphere (3). Parachutes are opened (4) and the lid jettisoned (5). The breaking parachute is opened at 66 km; data transmission starts (6). The parachute and thermal shield are jettisoned at 48 km (7). On landing data is automatically transmitted to the Earth via the orbiter (8).

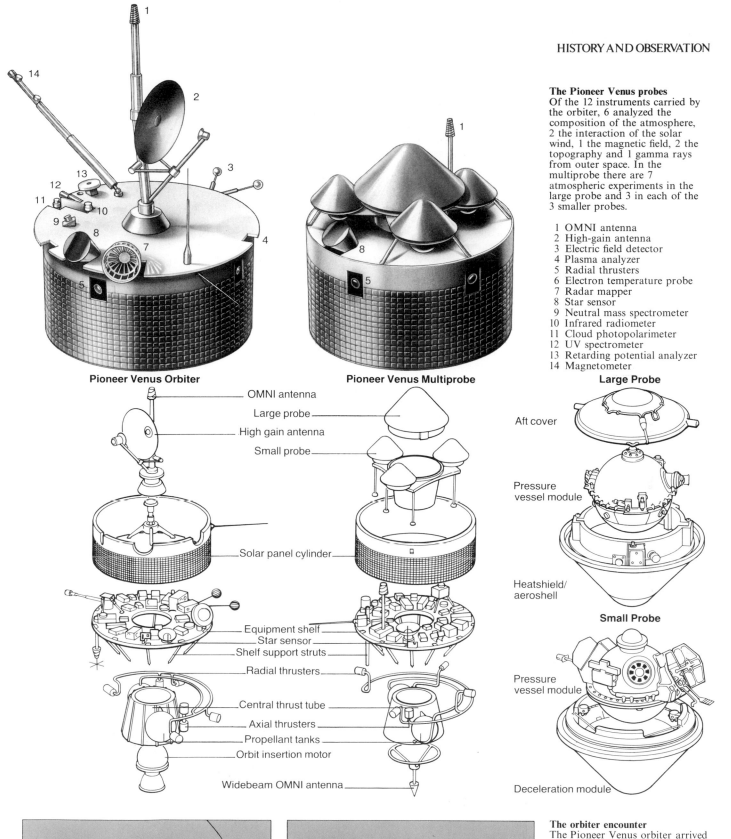

Pioneer Venus Orbiter

Pioneer Venus Multiprobe

OMNI antenna

Large probe

High gain antenna

Small probe

Solar panel cylinder

Equipment shelf
Star sensor
Shelf support struts
Radial thrusters
Central thrust tube
Axial thrusters
Propellant tanks
Orbit insertion motor

Widebeam OMNI antenna

**Large Probe**

Aft cover

Pressure
vessel module

Heatshield/
aeroshell

**Small Probe**

Pressure
vessel module

Deceleration module

**The Pioneer Venus probes**
Of the 12 instruments carried by the orbiter, 6 analyzed the composition of the atmosphere, 2 the interaction of the solar wind, 1 the magnetic field, 2 the topography and 1 gamma rays from outer space. In the multiprobe there are 7 atmospheric experiments in the large probe and 3 in each of the 3 smaller probes.

1  OMNI antenna
2  High-gain antenna
3  Electric field detector
4  Plasma analyzer
5  Radial thrusters
6  Electron temperature probe
7  Radar mapper
8  Star sensor
9  Neutral mass spectrometer
10 Infrared radiometer
11 Cloud photopolarimeter
12 UV spectrometer
13 Retarding potential analyzer
14 Magnetometer

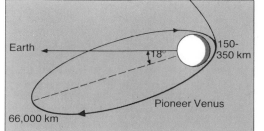

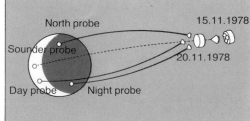

**The orbiter encounter**
The Pioneer Venus orbiter arrived at Venus on 4 December 1978. It entered an eliptical orbit which took it as close as 150 km from the surface.

**The release of the space probes**
The large probe was released on 15 November 1978 and the 3 small probes 5 days later. Descent positions are shown on the day and night hemispheres.

# Mariner 9 and Viking

## Mariner 9

The first phase of direct Martian exploration ended with the probes of 1969. The second phase began with the spacecraft of 1971, of which there were four: two American and two Russian. Their aims were not identical. The United States craft were designed to go into close orbits around Mars and to map the surface as thoroughly as they could, naturally obtaining much miscellaneous information at the same time. The Soviet planners hoped to carry out soft landings, and to pick up transmissions direct from the planet.

In the event, only one of the four vehicles was successful. This was Mariner 9, which more than made up for the failures or near-failures of the other three. Before the Viking program, indeed, Mariner 9 was the source of practically all our reliable information about Mars. On 14 November 1971 its rocket motor was fired for 15 min 23 sec, slowing the spacecraft down and putting it into its circum-Martian path. The initial revolution period was 12 hr 34 min, slightly more than half a sol, but it was then changed to 11 hr 58 min 14 sec, which brought the probe down to a minimum distance of 1,360 km from the surface. Unfortunately Mariner 9's arrival coincided with a major dust-storm which had begun in mid-September. It could not start work until the end of the year.

On 30 December the orbit was altered again, giving a new period of 11 hr 59 min 28 sec. Transmissions continued until 27 October 1972, and altogether Mariner sent back 7,329 pictures, covering practically all of the planet as against the puny 10 percent recorded from Mariners 6 and 7 of 1969. The results were spectacular. The southern hemisphere was shown to be heavily cratered over much of its extent, with abundant craters in the dark regions of Sinus Sabaeus and Tyrrhena Patera, while the northern hemisphere was seen to contain fewer of these thickly cratered areas and more volcanoes. Measurements from Mariner 9 showed that in general the southern hemisphere lies about 2.8 km above the nominal mean radius, with the northern hemisphere lower as well as smoother.

## The Vikings

On the strength of Mariner 9's photographic coverage the United States' most ambitious soft-landing mission was launched in 1975. Both Vikings 1 and 2 consisted of an orbiter and an attached lander. When the spacecraft arrived at Mars in July and August of 1976, each was put in a predetermined orbit around the planet, and the search for a landing place began. Cameras aboard the orbiters were the principal source of information on which the choice of the landing sites was based; important data also came from infrared sensors on the orbiters and from radar observatories on Earth. The sole consideration in the final selection of the sites was the safety of the spacecraft.

On command from the Earth each lander separated from its orbiter. With the help of its retro-engines and parachute it dropped to the surface of Mars. From an initial speed of almost 5 km/sec it decelerated to a final touchdown speed of under 10 km/hr. Both orbiters continued to circle the planet, operating their own scientific instruments and relaying to Earth data transmitted from the landers. On 20 July 1976, the Viking 1 lander came to rest in the Chryse Planitia region of Mars, some 23° north of the equator. Six weeks later the Viking 2 lander settled down in the Utopia Planitia region, some 48° north of the equator. In longitude the two landers are separated by almost exactly 180°, thus placing them on opposite sides of the planet. In addition to the photographic analyses both by the orbiter and lander, the principal investigations concentrated on the emotive question of life on Mars. Each Viking lander analyzed the atmosphere by means of two mass spectrometers. One spectrometer, operating during the descent to the surface, sampled and analyzed the atmospheric gases every five seconds. The second spectrometer operated on the ground. Together with a combined gas chromatograph, the second spectrometer analyzed the surface for organic material alongside three instruments designed to detect the metabolic activities of organisms that might be present.

## Mariner 9

The instruments carried by Mariner 9 included the narrow-angle and wide-angle television cameras mounted on a scan platform that together sent back over 7,000 pictures; an infrared spectrometer that measured gases, particles and temperatures on and above the Martian surface; an ultraviolet spectrometer that identified gases in the upper atmosphere; and an infrared radiometer that measured the diurnal range of temperatures on the surface of the planet.

1 Low-gain antenna
2 Maneuver engine
3 Fuel tank
4 Canopus tracker
5 Propulsion unit
6 Temperature control louvers
7 Infrared spectrometer
8 Narrow-angle TV camera
9 Ultraviolet spectrometer
10 Wide-angle TV camera
11 Infrared radiometer
12 High-gain antenna
13 Acquisition Sun sensors
14 Cruise Sun sensor
15 Medium-gain antenna
16 Solar panel

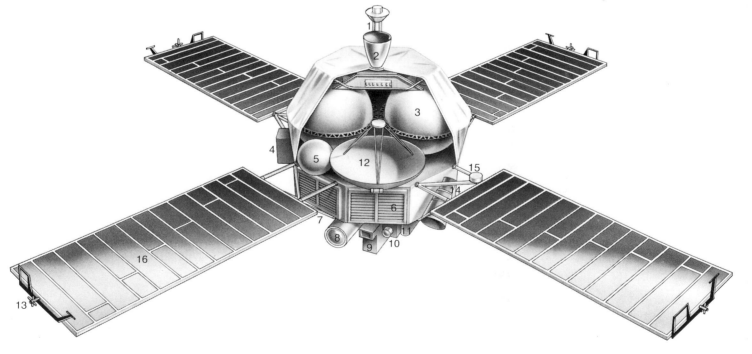

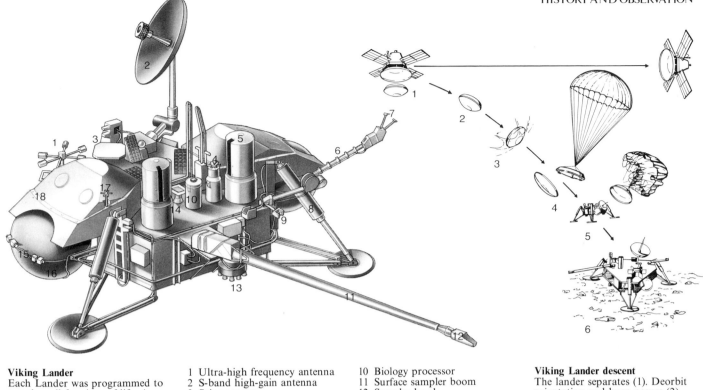

## Viking Lander

Each Lander was programmed to test the soil for signs of life. A mechanical arm scooped up samples and transferred them to an automatic biological laboratory inside the Lander, where three experiments were carried out. No certain proof of life was obtained.

## Viking Orbiter

In addition to relaying information from the Lander back to Earth, the Orbiter sent back pictures that far surpassed those of Mariner 9. It also carried out important analyses with its atmospheric water detector,

surveying water vapor over the entire planet through infrared spectrometers. The results showed that the highest concentration of atmospheric water vapor was at the edge of the north polar cap. Water vapor is a significant ingredient of the atmosphere.

1. Ultra-high frequency antenna
2. S-band high-gain antenna
3. Seismometer
4. Gas chromatograph mass spectrometer processor
5. Television camera
6. Meteorology boom
7. Meteorology sensors
8. Landing shock absorber
9. Magnet cleaning brush

10. Biology processor
11. Surface sampler boom
12. Sampler head
13. Terminal descent engine
14. X-ray fluorescence funnel
15. Roll engine
16. Terminal descent fuel tank
17. S-band low-gain antenna
18. Radioisotope Thermoelectric Generator (RTG)

1. S-band low-gain antenna
2. Stray light sensor
3. Propulsion unit
4. Fuel tank
5. Pressurant tank
6. Relay antenna
7. S- and X-band high-gain antenna

## Viking Lander descent

The lander separates (1). Deorbit orientation and burn occurs (2) before entry into the atmosphere (3). Parachute deploys at 5,791 m (4). Parachute jettisons at 1,402 m (5). On landing (6) foot-pad sensors shut down engines. The Viking Lander then commenced surface analyses.

8. Oxidizer tank
9. Infrared thermal mapper
10. Visual imaging camera
11. Atmosphere water detector
12. Thermal control louvers
13. Attitude control thrusters
14. Solar panel
15. Canopus tracker

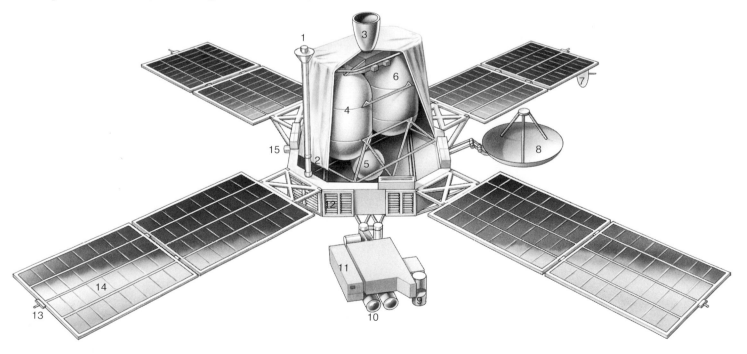

# Voyager

Three vehicles have bypassed Jupiter and Saturn. The most recent, Voyager 1 and Voyager 2, are more advanced in design than their predecessor, Pioneer 11, in a considerable number of respects. In particular, their onboard computer systems are capable of directing more sophisticated experimental equipment, and the vehicles carry a more powerful source of electricity in the form of three Radioisotope Thermoelectric Generators (RTGs). While in flight, the spacecraft is stabilized along three axes, using the Sun and the star Canopus as celestial references. After the first 80 days of flight the 3.66 m parabolic reflector points constantly back to Earth.

Behind the reflector dish is a ten-sided aluminium framework containing the spacecraft's electronics. This framework surrounds a spherical tank containing hydrazine fuel used for maneuvering.

Twelve thrusters control the spacecraft's attitude, while another four are used to make corrections to the trajectory. Only trajectory changes require instructions from the ground; all other functions can be carried out by the onboard computer, in contrast to the Pioneer spacecraft which had be be flown "from the ground". There are three engineering subsystems: the Computer Command Subsystem (CCS), the Flight Data Subsystem (FDS) and the Attitude and Articulation Control Subsystem (AACS).

The entire vehicle weighs only 815 kg and carries equipment for 11 science experiments.

1

**1. CRS**
The Cosmic Ray Detector System is designed to measure the energy spectrum of electrons and cosmic ray nuclei. It studied the composition of Jupiter's and Saturn's radiation belts, as well as the characteristics of energetic particles in the outer Solar System generally. The experiment uses three independent systems: a High-Energy Telescope System (HETS), a Low-Energy Telescope System (LETS) and an Electron Telescope (TET). These enabled the spacecraft to study a wide range of particles that make up cosmic rays.

2

**2. PLS**
The Plasma experiment studied the properties of the very hot ionized gases that exist in the interplanetary regions. The instrument consists of two plasma detectors, one pointing in the direction of the Earth and the other at a right angle to the first. This equipment analyzed the properties of the solar wind and its interaction with first Jupiter and then Saturn. The PLS also studied the properties of Jupiter's and Saturn's magnetospheres.

3
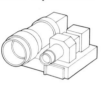

**3. ISS**
The Imaging Science Subsystem consists of two television-type cameras mounted on a scan platform. One of the cameras has a 200 mm wide-angle lens with an aperture of f/3, while the other uses a 1,500 mm f/8.5 lens to produce narrow-angle images. The design is a modified version of the cameras used on previous Mariner vehicles. Both cameras have a range of built-in filters as well as variable shutter speeds and scan rates. The ISS, the IRIS and the PPS instruments were able to view the same region of the planet simultaneously.

4

**4. IRIS**
The Infrared Radiometer Interferometer and Spectrometer measured radiation in two regions of the infrared spectrum, from 2.5 to 50 $\mu$m and from 0.3 to 2.0 $\mu$m. It provided information about the temperatures and pressures at various levels of the planets' atmospheres, as well as about the chemical composition of their clouds. Mounted on a scan platform, the instrument has two fields of view, one using a 0.5m Cassegrain telescope to achieve a narrow, quarter-degree field of view, the other, pointed off the telescope sight, for a wider view.

5
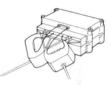

**5. LECP**
The Low-Energy Charged Particle experiment uses two solid-state detector systems mounted on a rotating platform. The two subsystems consist of the Low Energy Particle Telescope (LEPT) and the Low Energy Magnetospheric Particle Analyzer (LEMPA). This equipment studied the planets' magnetospheres and the interaction of charged particles with their satellites. It is also designed to investigate various other interplanetary phenomena such as the solar wind and cosmic rays emanating from sources outside the Solar System.

6

**6. PWS and PRA**
Two separate experiments, the Plasma Wave System and the Planetary Radio Astronomy experiment, share the use of the two long antennas which stretch out at right angles to one another forming a "V". The PWS studied wave-particle interactions and measured electric-field components of plasma waves over a frequency range of 10 Hz to 56 kHz. This experiment is also designed to measure the density of thermal plasma near the planets visited by the space-probe. The PRA experiment detected and analyzed radio signals emitted by the planets; Saturn was expected to be a powerful source of radio waves from Earth-based experiments. The PRA receiver covers two frequency bands, the first in the range of 20.4 kHz to 1,300 kHz, and the second between 2.3 MHz and 40.5 MHz.

7

**7. PPS**
The Photopolarimeter System consists of a 0.2 m telescope fitted with filters and polarization analyzers and is mounted on a scan platform. It covers eight wavelengths in the region between 235 $\mu$m and 750 $\mu$m. The PPS measured gases in planetary atmospheres, examined particles present in the atmospheres, and searched the sky background for interplanetary particles. The PPS instrument is also designed to examine the surface texture and composition of the satellites of Jupiter and Saturn, and the atmosphere of Titan.

8

**8. UVS**
The Ultraviolet Spectrometer covers the wavelength range of 40 $\mu$m to 180 $\mu$m looking at planetary atmospheres and interplanetary space. Its purpose is to study the chemistry of the upper layers of the atmospheres, and to measure how much of the Sun's ultraviolet radiation they absorb during occultation; it also measures ultraviolet emissions from the planetary atmosphere. The instrument collects and channels light through a collimator, which directs a number of narrow parallel beams onto a diffraction grating.

9
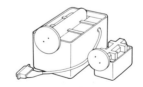

**9. MAG**
The Magnetic Fields Experiment consists of four magnetometers; two are low-field instruments mounted on a 10 m boom away from the field of the spacecraft, while the other two are high-field magnetometers mounted on the body of the spacecraft. Each pair consists of two identical instruments, which makes it possible to eliminate the spacecraft's field from the results. Each magnetometer measures the magnetic component along three perpendicular axes, from which the direction and strength of the field can be determined.

**10. RSS**
The investigations of the Radio Science System are based on the radio equipment which is also used for two-way communications between the Earth and Voyager. For example, the trajectory of the spacecraft can be measured accurately from the radio signals it transmits; analysis of the flight path as it passes near a planet or satellite makes it possible to determine the mass, density and shape of the object in question. The radio signals are also studied at occultations for information about the occulting body's atmosphere and ionosphere.

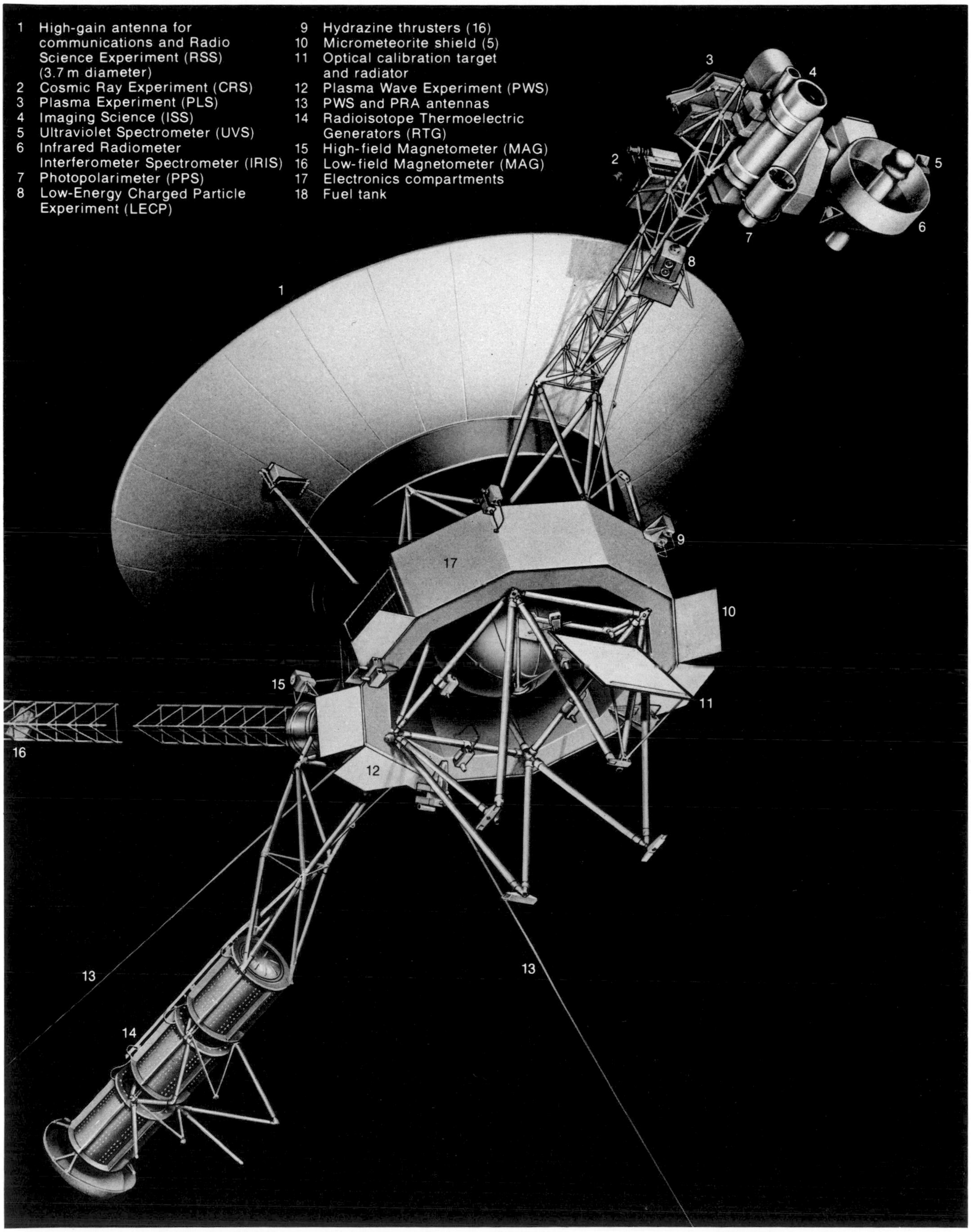

1 High-gain antenna for
  communications and Radio
  Science Experiment (RSS)
  (3.7 m diameter)
2 Cosmic Ray Experiment (CRS)
3 Plasma Experiment (PLS)
4 Imaging Science (ISS)
5 Ultraviolet Spectrometer (UVS)
6 Infrared Radiometer
  Interferometer Spectrometer (IRIS)
7 Photopolarimeter (PPS)
8 Low-Energy Charged Particle
  Experiment (LECP)

9 Hydrazine thrusters (16)
10 Micrometeorite shield (5)
11 Optical calibration target
   and radiator
12 Plasma Wave Experiment (PWS)
13 PWS and PRA antennas
14 Radioisotope Thermoelectric
   Generators (RTG)
15 High-field Magnetometer (MAG)
16 Low-field Magnetometer (MAG)
17 Electronics compartments
18 Fuel tank

# Future Probes to the Planets

So far as space research is concerned, making forecasts is always dangerous—and let it be said that these words are being written in early 1990. Voyager 2 has completed its main task; it and Voyager 1, like Pioneers 10 and 11, are now on their way out of the Solar System. Venus continues to be surveyed by radar orbiters, and the Magellan probe is on its way. So it may not be inappropriate to speculate as to the likely course of events during the 1990s.

First, of course, comes Magellan, which has had a somewhat chequered history. Cutbacks in the NASA budget, plus the *Challenger* disaster, have led to a very long delay, and also to curtailment of the intended program. It was finally launched in May 1989, but will not reach its target until August 1990, and its sole purpose now is to provide better and more complete radar maps of the surface of the planet. The orbit round Venus will have a period of 3.15 hours, and the radar will operate for about 40 minutes per orbit, from altitudes between 250 km and 1,900 km. On each orbit the radar will cover a swath 16,000 km long and 25 km wide. The main program is scheduled to last for one Venus year (that is to say, 243 Earth days) and altogether over 1,850 swaths will be transmitted back to Earth. It is hoped that the resolution will be down to 300 metres, with height variations as little as 100 metres. Once the main program has been completed, Magellan will undertake other tasks concerned with radar mapping, and will continue to operate so long as its power lasts.

At the moment the Russians have not announced further spaceprobes to Venus, and at the meeting of the International Astronomical Union in 1988 they stated that they intended to switch their main attention to Mars, this being a more rewarding target. Unfortunately, as we now know, their two Phobos probes were unsuccessful, but it is planned to send up a Mars Orbiter in the 1990s, and—if all goes well—there should be a "Mars Rover" before the end of the century. Obviously the most attractive project is to obtain samples from the planet's surface, but no date for a sample-and-return probe, either Russian or American, has yet been announced.

Next comes the Galileo probe to Jupiter, which, like Magellan, was launched from the Shuttle. During its flight to Jupiter, the 2½-tonne spacecraft will fly by Venus once and the Earth twice, thereby picking up enough gravitational energy to swing it out to its main target. As it passes through the asteroid belt, it should obtain close-range images of two asteroids, Gaspra and Ida. The whole journey to Jupiter will take much longer than had originally been hoped, because cutbacks have meant using a much less powerful launching system, but Galileo should reach Jupiter in 1995.

The spacecraft consists of an entry probe and an orbiter. The entry probe will be released five months before the rendezvous date, and will descend by parachute into Jupiter's clouds, relaying information about the chemistry, temperature and general conditions. Obviously it will not survive for long, but it should provide invaluable information during the last stages of its career. The orbiter will be put into a closed path round Jupiter, and should continue to operate for almost two years, returning images and data of the planet itself, the satellites and the ring.

Of equal interest is the Cassini probe to Saturn, which has now been definitely accepted even though it is still in the planning stage. Here again there are two parts, this time an orbiter and a Titan probe—remembering that Titan is of special interest inasmuch as it has a dense, nitrogen-rich atmosphere; what the surface conditions are like is still a matter for debate.

Cassini's journey to Saturn will be complicated and protracted. The scheduled launch date is August 1996. The first part of the journey takes the probe out to the asteroid belt and then back for a flyby of Earth in 1998; meantime there should have been an encounter with Asteroid No.66, Maja. After picking up extra energy from the Earth encounter, Cassini will move out to Jupiter, again picking up energy; this should occur in 2000. The arrival at

Saturn should be in October 2002. The initial orbital period will be 100 days, during which a Titan entry probe will be released (December 2002). This probe will plunge into Titan's clouds at a velocity of 6 km per second, and a combination of rocket braking and parachute will, it is hoped, slow the probe down so that its descent to the surface will take three hours, during which time data will be sent to the orbiter and relayed back to Earth. If all goes well, data will also be sent back from the actual surface of Titan.

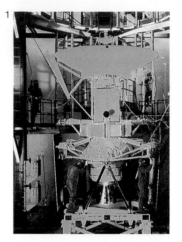

**1–3. The Magellan probe to Venus**
The first image shows the assembly of the Magellan spacecraft, and images 2–3 are artist's impressions of its deployment.

436

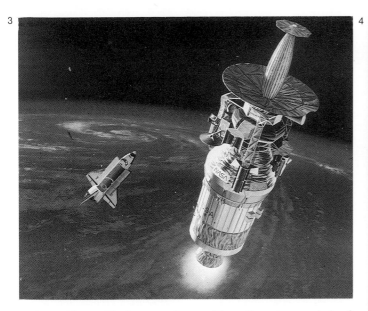

The orbiter will then continue with a four-year period of operations in orbit round Saturn, using a Titan gravity assist for each of the orbits and ending in a near-polar path round Saturn. Continued studies will be made of both Saturn itself and Titan, plus any other satellites which will be within range.

These are the main probes planned for the coming decade, but plans are liable to change quickly, and no doubt there will be other missions as well—quite apart from possible manned expeditions to the Moon; the idea of a Lunar Base by the year 2000 is no longer far-fetched. We live in exciting times, and the direct exploration of the Solar System is well under way.

**4–5 The Galileo probe to Jupiter**
These artist's impressions show the probe in orbit around Jupiter.

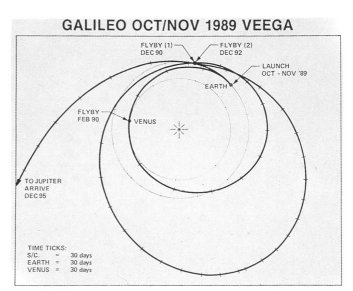

# Observing for Amateurs

Since the start of the Space Age, amateur observation of bodies in the Solar System has undergone a distinct change in character. Previously, cartography of the lunar and planetary surfaces was of paramount importance, and amateur-produced maps were of the greatest value. Today this program has been completed, and the amateur observer of the 1980s must necessarily be more specialized. Considerable areas for research do exist, particularly with regard to time-dependent phenomena, since the various probes cannot monitor the Moon and planets continuously.

It must be admitted, however, that very little can be done with regard to Mercury and the three outer planets, Uranus, Neptune and Pluto. The phase of Mercury is easy to see with a modest telescope, but ordinarily no surface markings will be seen; Uranus shows a pale greenish disc; Neptune is clearly non-stellar, but Pluto appears as nothing more than a very faint star. The only opportunities for useful amateur research are concerned with occultations of stars by the outer planets, which are, unfortunately, very rare. Occultations by asteroids are more frequent, and amateurs can carry out useful work, though in general it will involve photoelectric equipment together with a telescope of at least 200 mm aperture.

## Recording observations

Each observation should be accompanied by the following data: name of observer; aperture and type of telescope; magnification; time (GMT); seeing conditions (1 to 5 on the Antoniadi scale; 1 being perfect and 5 very poor) and any additional comments. Moreover, observational results can be of real use only if combined in an overal program. In Britain, the national observational society is the British Astronomical Association, while there are several analogous societies in Europe and the United States.

Regarding magnification, it is generally held that a telescope will bear a power of 25 per cm of aperture under good conditions. It is important not to use too high a magnification: a small, sharp image is preferable to a larger, slightly blurred one.

An equatorial mounting and clock drive is not absolutely essential, except for photography, but it certainly makes observing much easier and more convenient.

## The Sun

In many ways the Sun is an ideal object for the amateur observer to study. Weather permitting, it is available all day every day, and large telescopes are not required for its observation. A 50 mm aperture telescope will suffice to show sunspots, and apertures in the range of 75 mm to 150 mm are the most useful size for the amateur, refractors being rather more convenient for this purpose than reflectors.

It cannot be stressed too strongly that it is dangerous to look at the Sun – even without a telescope – and on no account should an observer risk looking at the Sun through a telescope, even for the briefest instant. Serious and irreparable eye damage, and quite probably permanent blindness, would be the result.

Dark filters placed over the eyepiece are not safe; if they should crack, there is little chance of the observer removing his eye in time to avoid damage. The Herschel wedge, or solar diagonal, contains an unsilvered glass wedge that passes most of the light and heat straight through and reflects only a small proportion to an eyepiece; even so, a dark filter is required at the eyepiece. Thin sheets of polyester-type material coated with aluminium may be purchased commercially. Placed over the full aperture of the telescope, these are not subject to concentrated heating, but great care has to be taken to ensure that the film is fully effective over its full width, and that it cannot be ruptured or blown away.

Without doubt the safest way to observe the Sun is by the projection method, the focus of the telescope being adjusted to produce a sharp image on a white card or other form of screen held beyond the eyepiece. A card placed round the body of the telescope is necessary to cast a shadow on to the screen so that the projected image is not obscured by direct sunlight; many observers construct a projection box to attach to the rear of the telescope to provide a shaded compartment into which to project the image.

One straightforward type of observation is to make daily records of sunspot activity. The projected image should be drawn to a standard size (a 150 mm diameter disc is most convenient), the positions of sunspots, groups, and faculae being marked in first and details added later, possibly with the aid of a higher magnification. If sunspots are allowed to drift across the field of view, their direction of motion can be used to establish the east–west direction so that the observer may orientate his drawing to allow for the tilt of the solar disc at different times of the year (see pages 32–3).

The Sun may be photographed at prime focus or by eyepiece projection with the camera body attached to the telescope, but care must be taken to avoid overheating the camera itself. Alternatively, the image projected onto a screen may be photographed using a projection box with open sides.

Commercially available $H_\alpha$ filters have opened up a fruitful field for the amateur in the plotting of plages, filaments, prominences and flares. Solar radio astronomy also offers great scope for the amateur. Solar bursts and noise storms lie within the capacity of fairly basic equipment: strong bursts may be picked up even with a simple yagi aerial and a television receiver. Most amateurs use equipment operating at frequencies of a few tens to a few hundreds of MHz, depending on the equipment and type of aerial available.

## The Moon

No valuable research can be carried out until the observer is thoroughly familiar with the Moon in all its aspects. Photographic studies, which require an equatorial mount and clock drive, are helpful even though they have little scientific value. For the sake of gaining experience the observer should also make numerous drawings of features from direct telescopic observations. It is unwise to use too small a scale in a lunar sketch; generally, 1 cm to 12 km is suitable: Plato, for example, would be 8 cm in diameter.

For familiarization a small telescope (150 mm reflector) is adequate, but for serious research a larger aperture—at least 190 mm—is required. Much, however, depends upon the observer's skill.

One of the most important fields of research today is that of TLP (see page 151). There is very strong evidence in favor of their reality (quite apart from N. A. Kozyrev's classic observation of a red event in Alphonsus, in 1958), but knowledge of them is still very incomplete, and although many theories have been put forward the cause and distribution of TLP are still not understood with certainty. Some TLP are reddish, but the color is never striking. One method of detection involves using a "Moon-Blink" device, which consists of a rotating filter, half red and half blue. The suspected formation is viewed through each filter in quick succession, and any red area will then show up as a "winking patch". Yet it is only too easy to be misled. If the TLP is suspected, the first step is to examine other formations in the same area. If these too show the same phenomenon, then it is clear that conditions in the Earth's atmosphere are responsible. Other TLP take the form of localized obscurations, such as a slight blurring of detail over a restricted area. Here again, features near the suspected formation should be examined to make sure that the effect is not due to the terrestrial atmosphere. In addition to the usual observational recordings the area concerned should be carefully noted, and either sketched or (preferably) outlined on a photograph.

Finally, observations of occultations are still required. They can be timed to within an accuracy of 0.1 sec. Occasionally stars appear to fade out instead of disappearing instantaneously; it will generally be found that in such cases the star is a close double.

## Venus

At first sight Venus appears to be singularly unpromising. Generally nothing can be seen apart from the characteristic phase, and any surface markings are very vague and impermanent; it is seldom possible to define them with any accuracy. Moreover, conditions when Venus is a brilliant naked-eye object are generally very inferior, because the planet will be low in the sky. It is better to observe either in daylight, with the Sun above the horizon, or shortly after sunset or before sunrise. Remember that if the Sun is visible, Venus should be found only by using a telescope equipped with accurate setting circles. On no account "sweep around" looking for it, as there is a very real danger that the Sun will enter the field of view – with disastrous results to the observer's eyesight.

The vague markings on Venus should be drawn as carefully as possible, though it will generally be necessary to exaggerate them. Attention should be paid to the time of dichotomy, or exact half-phase; at evening elongations (waning) dichotomy is always early, while at morning elongations (waxing) it is always late—the well known Schröter effect. Undoubtedly it is due to the planet's atmosphere, but the date of dichotomy should be timed if possible, though it often happens that the terminator appears sensibly straight for several successive days.

The Ashen Light, or faint luminosity of the night hemisphere, can be seen only when Venus is a crescent; this does involve observing when the planet is low over the horizon, and it is essential to block out the bright crescent with an occulting device fitted in the eyepiece. Observations of the Ashen Light are probably the most useful that the amateur can make with respect to Venus, but they are very difficult, and it is only too easy to be deceived.

## Mars

Mars is an awkward planet to observe in as much as it comes to opposition only in alternate years, and with telescopes of the size generally used by amateurs useful observations are possible for only a few weeks to either side of opposition. Attention should be paid to the following points: the polar caps, noting their decrease and their outlines, which are often irregular; the dark features, which are in general permanent, but some of which do show variations in size and shape for reasons which are still not completely understood (the Solis Lacus is a good example of this); any clouds which are visible. Some clouds are relatively small, and may be followed from night to night; others are much more extensive. Sometimes, too, Mars experiences global or partially global dust storms, which mask the familiar surface features completely. Careful attention should then be paid to the planet, to check when the dust begins to clear and features become visible.

Because Mars is a relatively small planet, it is necessary to use fairly high magnifications, and small telescopes are not adequate. An aperture of 300 mm is about the minimum for useful work.

## Jupiter and its satellites

Almost any small telescope will show the main belts on Jupiter. For serious work, a 150 mm aperture is probably the minimum requirement for a reflector. With a 200 mm reflector there is ample scope, while with a 300 mm reflector a full program of observation can be carried out.

Jupiter is very obviously flattened at the poles and this must always be taken into account in drawings. It is advisable to have prepared discs printed. The exact size is not important, but should not be less than 5 cm. The phase of Jupiter as seen from Earth is so slight that it may safely be neglected.

Because of Jupiter's quick rotation, the main details should be drawn as quickly as is consistent with accuracy. The finer details can then be filled in, probably by using a higher magnification. It is wise to complete the whole sketch in 15 minutes or less. Drawings of special features may be made at a more leisurely rate.

Probably the most important observational program for the owner of a modest telescope is the timing of surface transits. The moment when a feature is brought to the central meridian may be timed with surprising accuracy—certainly to within one minute—and by using the relevant tables (*see* pages 442–3) the longitude of the feature may then be calculated.

It may happen that no actual transit can be taken, either because of interruption by clouds or because the transit had already occurred before observing began. The experienced observer may then give an estimated time of transit. This will be of lower accuracy, and it is never wise to make estimates for features more than half an hour away from the central meridian.

Only the Galileans are visible with telescopes of the size usually owned by amateurs. The main interest centers on their phenomena: eclipses, occultations, transits and shadow transits. Timings may be made, though these are now chiefly for the observer's interest rather than their scientific value.

## Saturn and its satellites

The rings and satellites of Saturn are beyond the power of binoculars, and a 70 mm refractor is probably the minimum aperture that will give a reasonable view of the rings. With a slightly larger instrument of the kind commonly used by amateurs—a 150 mm reflector, for example—the view will be magnificent when the rings are well displayed, as they will be during most of the 1980s. Serious observation requires a larger aperture, and 200 mm is the lowest limit for a reflector or 80 mm for a refractor.

Saturn is an awkward object to draw properly. Using prepared discs is not convenient in the usual way, because the angle of the ring system varies continuously, and it is true to say that a good drawing of Saturn requires considerable artistic skill. The flattening of the disc should never be neglected, and the size of the disc itself should not be much less than 5 cm.

To make a sketch, begin by putting in the disc and ring outlines. Add the main features, such as the Cassini Division (when the rings are sufficiently wide for it to be seen) and the visible belts. Take great care to put in the shadows correctly – both the shadow of the ring on the globe, and of the globe on the ring. Then change to a higher magnification and put in the finer details, paying special attention to any features visible on the disc. With Jupiter, a drawing has to be completed in a quarter of an hour at most because of the wealth of details and the rapid rotation; with Saturn the observer can be more leisurely, because there will be far fewer visible features on the disc, and any which are seen are likely to be rather ill-defined.

It is useful to make intensity estimates of the various belts, zones and other features, using a scale from 0 (white) to 10 (black). The equatorial zone and Ring B, the brightest of the rings, will generally be of brightness 1 or 1.5 on this scale. Observations of this kind are valuable, because Saturn does show definite variations.

As we have noted, Saturn appears much blander than Jupiter, because of the greater amount of high-altitude haze in its atmosphere. The equatorial belts are always visible with an adequate telescope (except when covered up by the rings), but bright spots are rare. They do occur sometimes, as in 1933, and it is important to study them as intensively as possible, taking timings of their transits across the central meridian of the planet.

Only Titan is visible with a very small telescope. With a 70 mm refractor, Iapetus, at or near western elongation, and Rhea are easy, and Dione and Tethys may also be glimpsed; larger apertures are needed to show the rest. The main work to be done is with regard to the magnitudes of the satellites.

Transits and shadow transits of the satellites are not easy to observe, except with Titan, but they are interesting to watch when they do occur and can be studied with a telescope of at least 200 mm aperture (preferably at least 300 mm). Mutual satellite phenomena are very rare, and so are occultations of satellites by the rings.

# Glossary

**Albedo** The ratio of the amount of light reflected by a body to the amount of light incident on it; a measure of the reflecting power of a body. A perfect reflector would have an albedos of 1. The albedos of the planets are as follows: Mercury 0.06, Venus 0.76, Earth 0.29, Mars 0.16, Jupiter 0.34, Saturn 0.33, Uranus 0.34–0.5, Neptune 0.34–0.5, Pluto 0.5.

**Allotropy** The property in a chemical element of existing in different forms, with distinct physical properties but capable of forming identical chemical compounds. Ozone, for example, is an allotropic form of oxygen.

**Altitude** In astronomy, the angular distance of a celestial body from the horizon. In conjunction with a measurement of AZIMUTH, it describes the position of an object in the sky at a given moment.

**Aphelion** The point or moment of greatest distance from the Sun of an orbiting body such as a planet. The opposite of PERIHELION.

**Astronomical unit** A unit of distance defined by the mean distance of the Earth from the Sun and equal to 149,597,870 km.

**Azimuth** The angular distance along the horizon, measured in an eastward direction, between a point due north and the point at which a vertical line through a celestial object meets the horizon. (This is the normal convention for an observer in the northern hemisphere; other conventions are sometimes followed.) *See also* ALTITUDE.

**Celestial equator** The circle formed by the projection of the Earth's equator onto the surface of the CELESTIAL SPHERE.

**Celestial sphere** An imaginary sphere, centered on the Earth, onto whose surface the stars may be considered, for the purposes of positional measurement and calculation, to be fixed.

**Conjunction** The near or exact alignment of two astronomical bodies in the sky. Also used to describe an alignment between a planet and the Sun as seen from Earth. When the planet passes behind the Sun, the conjunction is called "superior"; in the special case of Mercury or Venus passing between the Sun and the Earth, the conjunction is "inferior".

**Coriolis effect** The apparent deflection of a body moving in a rotating coordinate system. For example, a projectile fired northward from the Earth's equator will appear to be deflected to the east, because the point on the equator from which it is fired will be rotating faster than its target to the north. The Coriolis effect plays an important part in determining the directions of wind and ocean currents.

**Culmination** The maximum altitude of a celestial body above the horizon.

**Declination** The angular distance of a celestial body from the CELESTIAL EQUATOR; one of the two celestial coordinates, roughly equivalent to latitude on the Earth, used to represent the position of a celestial object. *See also* RIGHT ASCENSION.

**Ecliptic** The circle on the CELESTIAL SPHERE defined by the Sun's apparent annual motion against the stellar background. The ecliptic represents the plane in which the Earth orbits the Sun and, because the Earth's rotational axis is tilted, the ecliptic is inclined to the celestial equator at an angle, known as the "obliquity of the ecliptic", which is equal to about $23\frac{1}{2}°$.

**Elongation** The angular distance of a planet from the Sun, or of a satellite from its primary planet.

**Equation of time** The difference between the apparent solar time and the mean time; the value of the equation of time varies throughout the year from about $-14\frac{1}{4}$ min to about $+16\frac{1}{4}$ min.

**First Point of Aries** *See* VERNAL EQUINOX.

**Ion** An atom that is electrically charged as a result of having lost or gained one or more electrons.

**Light-year** A unit of distance defined by the distance travelled by light *in vacuo* in a year, equal to $9.4607 \times 10^{12}$ km or 63,240 ASTRONOMICAL UNITS. In astronomy the more commonly used unit for large distances is the PARSEC, which is equal to 3.2616 light-years.

**Limb** The edge of the visible disc of a celestial body.

**Luminosity** The total amount of energy emitted by a star per unit of time.

**Magnetosphere** The region around a planet within which its magnetic field predominates over the magnetic field of the surrounding interplanetary region.

**Magnitude** A measure of the brightness of a star or other celestial body on a numerical scale which decreases as the brightness increases. The faintest stars visible to the naked eye on a clear night are of magnitude 6; the brightest have a mean magnitude of 1. The "absolute" magnitude of a star is defined as the apparent magnitude it would have if viewed from a standard distance of 10 PARSECS.

**Meridian** A great circle passing through the poles either of the Earth or of the CELESTIAL SPHERE. In astronomical usage, the term usually refers to the "observer's meridian", which passes through the observer's ZENITH.

**Nodes** The points at which two great circles on the CELESTIAL SPHERE intersect; in particular, the points at which the orbit of a body, such as a planet or the Moon, crosses the ECLIPTIC.

**Occultation** The temporary disappearance of one celestial body, usually a star, behind another, usually a planet or moon. A solar eclipse is a particular case of an occultation.

**Opposition** The position of a planet in its orbit when the Earth lies on a direct line between the planet and the Sun. A planet is best placed for observation when it is at opposition.

**Parallax** The apparent change in the position of an object due to an actual change in the position of the observer. Measurement of parallax allows the distances of distant objects to be determined.

**Parsec** A large unit of distance defined as the distance at which a star would have an annual PARALLAX of one second of arc, and equal to $3.0857 \times 10^{13}$ km, 206,265 ASTRONOMICAL UNITS, or 3.2616 LIGHT-YEARS.

**Perihelion** The point or moment of closest approach to the Sun of an orbiting body such as a planet. The opposite of APHELION.

**Perturbations** Irregularities in the orbital motion of a body due to the gravitational influence of other orbiting bodies.

**Phase angle** The angle defined by the position of the Sun, a body, and the Earth, measured at the body.

**Quadrature** The position of the Moon or an outer planet when its ELONGATION is 90°.

**Right ascension** (R.A.) The angle, measured eastward along the CELESTIAL EQUATOR in units of hours, minutes and seconds, between the VERNAL EQUINOX and the point at which the MERIDIAN through a celestial object intersects the celestial equator. Right ascension is roughly equivalent to longitude on the Earth, and in conjunction with one other coordinate, DECLINATION, specifies the exact position of an object in the sky.

**Roche limit** The critical distance from the center of a planet within which gravitational forces would be insufficient to prevent a satellite from being broken up by tidal forces. For a satellite with the same density as the parent planet, the Roche limit lies at 2.4 times the radius of the planet.

**Saros** An interval of 6,583 days (equal to 18 years 11.3 days) after which the Sun, the Moon and the Earth return almost exactly to their previous relative positions. Consequently, the Saros period marks the interval between successive eclipses of similar type and circumstance.

**Sidereal period** The time taken for a body to complete one orbit, as measured against the background of fixed stars. *See also* SYNODIC PERIOD.

**Sidereal time** A system of measurement of time based on the Earth's period of rotation, measured against the background of fixed stars. The sidereal day is taken to begin at the moment at which the VERNAL EQUINOX crosses the observer's MERIDIAN.

**Solar constant** The amount of energy per second that would be received in the form of solar radiation over one square meter of the Earth's surface at the Earth's mean distance from the Sun, if no radiation was absorbed by the atmosphere.

**Solstices** The two points on the ecliptic of maximum or minimum DECLINATION; the times at which the Sun reaches these points along its annual path. The summer solstice (corresponding to the maximum declination) falls around 21 June, the winter solstice (minimum declination) around 21 December.

**Synchrotron radiation** Radiation emitted by electrons travelling in a strong magnetic field at speeds approaching the speed of light.

**Synodic period** The interval between successive CONJUNCTIONS or, more generally, between similar configurations of a celestial body, the Sun and the Earth.

**Terminator** The boundary between the dark and the sunlit hemispheres of a planet or satellite.

**Vernal equinox** The point on the CELESTIAL SPHERE at which the ECLIPTIC crosses the CELESTIAL EQUATOR from south to north (where the direction is defined by the Sun's motion). Also known as the First Point of Aries.

**Zenith** The point on the CELESTIAL SPHERE directly above the observer.

# Tables I

The tables presented on this and the following pages provide standard scientific (background) information against which the texts in the book can be read, as well as practical information for the observer of the Solar System. Specialist publications such as the *Astronomical Ephemeris* and the *American Ephemeris* appear annually and will remain the primary sources for scientific research.

## Using the tables

The units and notation used throughout the book are based on the Système International des unités (SI units), which is currently being introduced universally for scientific and educational purposes. There are seven base units in the system: the *meter* (m), the *kilogram* (kg), the *second* (s), the *ampere* (A), the *Kelvin* (K), the *mole* (mol), and the *candelo* (cd). Other quantities are expressed in units derived from the base units; thus, for example, the unit of force the newton (N) is defined as the force required to give a mass of one kilogram an acceleration of one meter per second squared ($kg\,m\,s^{-2}$).

Some branches of science continue to adhere to a few of the older units, and in one case an editorial concession has had to be made to existing scientific usage: the SI unit of magnetism, the tesla, has been dropped in favor of the more common unit, the Gauss. One tesla is equal to 10,000 Gauss.

For very large and very small numbers, "index notation" has been adopted, so that where appropriate numbers are written as powers of ten. Numbers smaller than one are indicated by negative powers. In addition, a variety of prefixes is used to denote certain multiples of units—listed in Table 1. Table 2 gives the SI equivalents of common imperial units, Table 3 lists a selection of astronomical constants, and Table 4 lists all chemical elements.

Page 440 explains the laws of planetary motion and the orbital elements used to describe the movements of the planets. Tables 5 and 6 summarize the fundamental data of the members of the Solar System (including the Sun and the Moon). Table 5 may be read in conjunction with diagram 3, which illustrates the elements shown in the table. It should be noted, however, that the longitudes are mean rather than "true" values.

Tables 7 to 14 continue the orbital data, including eclipses, elongations and oppositions, while Tables 15 and 16 and 21 to 24 give the positions of the superior planets until the end of the century. Tables 17 and 18 convert the rotational periods of System I and II on Jupiter to degrees of longitude moved after one-minute intervals of time. Tables 19 and 20 can be used to determine the longitudinal drift-rate of individual features on Jupiter, by observing them over a 30-day period.

## Table 1: SI prefixes

| Factor | Name | Prefix Symbol |
|---|---|---|
| $10^{18}$ | exa | E |
| $10^{15}$ | peta | P |
| $10^{12}$ | tera | T |
| $10^{9}$ | giga | G |
| $10^{6}$ | mega | M |
| $10^{3}$ | kilo | k |
| $10^{2}$ | hecto | h |
| $10^{1}$ | deca | da |
| $10^{-1}$ | deci | d |
| $10^{-2}$ | centi | c |
| $10^{-3}$ | milli | m |
| $10^{-6}$ | micro | $\mu$ |
| $10^{-9}$ | nano | n |
| $10^{-12}$ | pico | p |
| $10^{-15}$ | femto | f |
| $10^{-18}$ | atto | a |

## Table 2: SI conversion factors

| | |
|---|---|
| **Length** | |
| 1 in | 25.4 mm |
| 1 mile | 1.609344 km |
| **Volume** | |
| 1 imperial gal | 4.54609 cm³ |
| 1 US gal | 3.78533 liters |
| **Velocity** | |
| 1 ft/s | $0.3048\,m\,s^{-1}$ |
| 1 mile/h | $0.44704\,m\,s^{-1}$ |
| **Mass** | |
| 1 lb | 0.45359237 kg |
| **Force** | |
| 1 pdl | 0.138255 N |
| **Energy (work, heat)** | |
| 1 cal | 4.1868 J |
| **Power** | |
| 1 hp | 745.700 W |
| **Temperature** | |
| °C | = kelvins − 273.15 |
| °F | = $\frac{9}{5}$ (°C) + 32 |

## Table 3: Astronomical and physical constants

| | |
|---|---|
| **Astronomical unit (A.U.)** | $1.49597870 \times 10^{8}$ km |
| **Light-year (l.y.)** | $9.4607 \times 10^{12}$ km = 63,240 A.U. = 0.306660 pc |
| **Parsec (p.c.)** | $30.857 \times 10^{12}$ km = 206,265 A.U. = 3.2616 l.y. |
| **Length of the year** | |
| Tropical (equinox to equinox) | $365^{d}.24219$ |
| Sidereal (fixed star to fixed star) | 365.25636 |
| Anomalistic (apse to apse) | 365.25964 |
| Eclipse (Moon's node to Moon's node) | 346.62003 |
| **Length of the month** | |
| Tropical (equinox to equinox) | $27^{d}.32158$ |
| Sidereal (fixed star to fixed star) | 27.32166 |
| Anomalistic (apse to apse) | 27.55455 |
| Draconic (node to node) | 27.21222 |
| Synodic (New Moon to New Moon) | 29.53059 |
| **Length of day** | |
| Mean solar day | $24^{h}03^{m}56^{s}.555 = 1^{d}.00273791$ mean solar time |
| Mean sidereal day | $23^{h}56^{m}04^{s}.091 = 0^{d}.99726957$ mean solar time |
| Earth's sidereal rotation | $23^{h}56^{m}04^{s}.099 = 0^{d}.99726966$ mean solar time |
| **Speed of light in vacuo (c)** | $2.99792458 \times 10^{5}\,km\,s^{-1}$ |
| **Constant of gravitation** | $6.672 \times 10^{-11}\,kg^{-1}\,m^{3}\,s^{-2}$ |
| **Charge on the electron (e)** | $1.602 \times 10^{-19}$ coulomb |
| **Planck's constant (h)** | $6.624 \times 10^{-34}$ Js |
| **Solar radiation** | |
| Solar constant | $1.37 \times 10^{3}\,J\,m^{-2}\,s^{-1}$ |
| Radiation emitted | $3.86 \times 10^{26}\,J\,s^{-1}$ |
| Visual absolute magnitude ($M_v$) | + 4.79 |
| Effective temperature | 5,780 K |

## Table 4: Chemical elements, their symbols and atomic numbers

| Element | Symbol | No. | Element | Symbol | No. |
|---|---|---|---|---|---|
| actinium | Ac | 89 | mendelevium | Md | 101 |
| aluminium | Al | 13 | mercury | Hg | 80 |
| americium | Am | 95 | molybdenum | Mo | 42 |
| antimony | Sb | 51 | neodymium | Nd | 60 |
| argon | Ar | 18 | neon | Ne | 10 |
| arsenic | As | 33 | neptunium | Np | 93 |
| astatine | At | 85 | nickel | Ni | 28 |
| barium | Ba | 56 | niobium | Nb | 41 |
| berkelium | Bk | 97 | nitrogen | N | 7 |
| beryllium | Be | 4 | nobelium | No | 102 |
| bismuth | Bi | 83 | osmium | Os | 76 |
| boron | B | 5 | oxygen | O | 8 |
| bromine | Br | 35 | palladium | Pd | 46 |
| cadmium | Cd | 48 | phosphorus | P | 15 |
| caesium | Cs | 55 | platinum | Pt | 78 |
| calcium | Ca | 20 | plutonium | Pu | 94 |
| californium | Cf | 98 | polonium | Po | 84 |
| carbon | C | 6 | potassium | K | 19 |
| cerium | Ce | 58 | praeseodymium | Pr | 59 |
| chlorine | Cl | 17 | promethium | Pm | 61 |
| chromium | Cr | 24 | protactinium | Pa | 91 |
| cobalt | Co | 27 | radium | Ra | 88 |
| columbium | Cb | | radon | Rn | 86 |
| copper | Cu | 29 | rhenium | Re | 75 |
| curium | Cm | 96 | rhodium | Rh | 45 |
| dysprosium | Dy | 66 | rubidium | Rb | 37 |
| einsteinium | Es | 99 | ruthenium | Ru | 44 |
| erbium | Er | 68 | samarium | Sm | 62 |
| europium | Eu | 63 | scandium | Sc | 21 |
| fermium | Fm | 100 | selenium | Se | 34 |
| fluorine | F | 9 | silicon | Si | 14 |
| francium | Fr | 87 | silver | Ag | 47 |
| gadolinium | Gd | 64 | sodium | Na | 11 |
| gallium | Ga | 31 | strontium | Sr | 38 |
| germanium | Ge | 32 | sulphur | S | 16 |
| gold | Au | 79 | tantalum | Ta | 73 |
| hafnium | Hf | 72 | technetium | Tc | 43 |
| helium | He | 2 | tellurium | Te | 52 |
| holmium | Ho | 67 | terbium | Tb | 65 |
| hydrogen | H | 1 | thallium | Tl | 81 |
| indium | In | 49 | thorium | Th | 90 |
| iodine | I | 53 | thulium | Tm | 69 |
| iridium | Ir | 77 | tin | Sn | 50 |
| iron | Fe | 26 | titanium | Ti | 22 |
| krypton | Kr | 36 | tungsten | W | 74 |
| lanthanum | La | 57 | uranium | U | 92 |
| lawrencium | Lr | 103 | vanadium | V | 23 |
| lead | Pb | 82 | xenon | Xe | 54 |
| lithium | Li | 3 | ytterbium | Yb | 70 |
| lutetium | Lu | 71 | yttrium | Y | 39 |
| magnesium | Mg | 12 | zinc | Zn | 30 |
| manganese | Mn | 25 | zirconium | Zr | 40 |

# Tables II

## 1. Planetary configurations

An "inferior" planet is one whose orbit lies inside that of the Earth; the others are said to be "superior". A superior planet is in "opposition" ($J_1$) when it lies directly opposite the Sun in the sky; it is in "quadrature" when the angle measured at the Earth between it and the Sun (called the "elongation") is 90° ($J_2$ and $J_4$). When a planet is directly in line with the Sun ($V_1$, $V_3$ and $J_3$) it is said to be in "conjunction". In the case of an inferior planet the conjunction may be either inferior ($V_1$) or superior ($V_3$): maximum elongation is shown at $V_2$.

## 2. Kepler's Laws

According to Kepler's first law, the planets follow elliptical orbits around the Sun, with the Sun at one focus of the ellipse (**A**). An ellipse is an oval curve in which the sum of the distances from any point on the curve to two fixed points or "foci" is constant. Kepler's second law states that the radius vector of a planet sweeps out equal areas in equal times about the focus containing the Sun (**B**). The third law states that, for any planet, the square of its period of revolution is directly proportional to the cube of its mean distance from the Sun (**C**). By "mean distance" what is meant is the semi-major axis of the ellipse, i.e., half of its greatest diameter. Given the mean distance of any orbiting body it is simple to calculate its period, or vice versa.

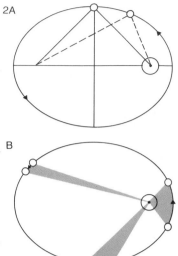

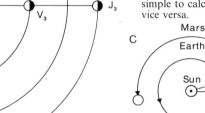

## 3. Orbital elements

Planets travel in elliptical orbits with the Sun at one focus. The dimensions of the ellipse may be described by two elements, the semi-major axis ($a$), and the eccentricity ($e$) as defined by diagram (**A**). Four elements specify fully the orbit as illustrated in diagram (**B**), where the plane of the ecliptic is defined by the Earth's orbit. The angle $i$ is the inclination of the orbit to the plane of the ecliptic; $\Omega$ is the longitude of the ascending node; $\omega$ is the "argument" of the perihelion; and $L$ is the longitude of the planet at a specified moment. ($L$ is given by $\Omega + \omega + v$.) Finally, there is a period (T).

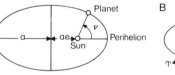

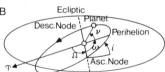

### Table 5: Mean elements of the planetary orbits for epoch 1980 Jan. 1.5 E.T.

| Planet | Mean distance A.U. | Mean distance millions of km $a$ | Eccentricity $e$ | Inclination to ecliptic $i$ ° ′ ″ | of asc. node $\Omega$ ° ′ ″ | of perihelion $\varpi$ ($=\omega+\Omega$) ° ′ ″ | Mean longitude at the epoch $L$ ° ′ ″ | Sidereal period days |
|---|---|---|---|---|---|---|---|---|
| Mercury | 0.3870987 | 57.91 | 0.2056306 | 7 00 15.7 | 48 05 39.2 | 77 08 39.4 | 237 26 09.2 | 87.969 |
| Venus | 0.7233322 | 108.21 | 0.0067826 | 3 23 40.0 | 76 29 59.2 | 131 17 22.7 | 358 08 12.4 | 224.701 |
| Earth | 1.0000000 | 149.60 | 0.0167175 | — — — | — — — | 102 35 47.2 | 100 18 43.2 | 365.256 |
| Mars | 1.5236915 | 227.94 | 0.0933865 | 1 50 59.3 | 49 24 11.6 | 335 41 27.2 | 127 06 26.0 | 686.980 |
| Jupiter | 5.2028039 | 778.34 | 0.0484681 | 1 18 15.2 | 100 14 48.0 | 14 00 01.9 | 147 05 29.8 | 4332.59 |
| Saturn | 9.5388437 | 1,427.01 | 0.0556125 | 2 29 20.9 | 113 28 55.4 | 92 39 22.9 | 165 22 24.3 | 10,759.20 |
| Uranus | 19.181826 | 2,869.6 | 0.0472639 | 0 46 23.5 | 73 53 54.2 | 170 20 10.7 | 227 17 14.5 | 30,684.8 |
| Neptune | 30.058021 | 4,496.7 | 0.0085904 | 1 46 18.8 | 131 33 36.7 | 44 27 01.1. | 260 54 42.6 | 60,190.5 |
| Pluto | 39.44 | 5,900.0 | 0.250 | 17 12 00 | 110 | 223 | | 90,465.0 |

### Table 6: Physical data for the Sun, Moon and planets

| Name | Diameter equatorial km | Diameter polar km | Inclination degrees | Equatorial rotation | Mass kg | Density (water = 1) | Escape velocity km s⁻¹ | Volume (Earth = 1) | Surface gravity (Earth = 1) | Mean vis. opposition Mag. | Albedo |
|---|---|---|---|---|---|---|---|---|---|---|---|
| Sun | 1,392,530 | 1,392,530 | 7.25 | 24.6 d | $1.9891 \times 10^{30}$ | 1.41 | 617.3 | $1.3 \times 10^6$ | 28.0 | − 26.8 | — |
| Moon | 3,476 | 3,476 | 1.53 | 27.32 d | $7.3483 \times 10^{22}$ | 3.34 | 2.37 | 0.02 | 0.165 | − 12.7 | 0.07 |
| Mercury | 4,878 | 4,878 | 0 | 58.65 d | $3.3022 \times 10^{23}$ | 5.43 | 4.25 | 0.06 | 0.377 | 0.0 | 0.06 |
| Venus | 12,104 | 12,104 | 178 | 243 d | $4.8689 \times 10^{24}$ | 5.24 | 10.36 | 0.86 | 0.902 | − 4.4 | 0.76 |
| Earth | 12,756 | 12,714 | 23.44 | 23.93 hr | $5.9742 \times 10^{24}$ | 5.52 | 11.18 | 1.00 | 1.000 | — | 0.29 |
| Mars | 6,794 | 6,759 | 23.59 | 24.62 hr | $6.4191 \times 10^{23}$ | 3.93 | 5.02 | 0.15 | 0.379 | − 2.0 | 0.16 |
| Jupiter | 143,884 | 134,200 | 3.12 | 9.8 hr | $1.899 \times 10^{27}$ | 1.32 | 59.6 | 1,323 | 2.64 | − 2.6 | 0.34 |
| Saturn | 120,536 | 108,000 | 26.73 | 10.6 hr | $5.684 \times 10^{26}$ | 0.70 | 35.6 | 752 | 1.17 | + 0.7 | 0.33 |
| Uranus | 51,118 | 49,946 | 97.86 | 17.24 hr | $8.6978 \times 10^{25}$ | 1.25 | 21.1 | 67 | 0.93 | + 5.5 | 0.34–0.5 |
| Neptune | 50,530 | 49,600 | 29.56 | 16.1 hr | $1.028 \times 10^{26}$ | 1.77 | 24.6 | 57 | 1.22 | + 7.8 | 0.34–0.5 |
| Pluto | 2,445 | 2,445 | 118 | 6.3 d | $6.6 \times 10^{23}$ | 4.7 | Low | <0.01 | Low | + 14.9 | 0.5 |

## Table 7: Solar eclipses 1983–1999

| Date | Area | Type |
|---|---|---|
| 11. 6.1983 | Indian Ocean, E. Indies, Pacific | Total |
| 4.12.1983 | Atlantic, Equatorial Africa | Annular |
| 30. 5.1984 | Pacific, Mexico, USA, Atlantic, N. Africa | Annular |
| 22/3.11.1984 | E. Indies, S. Pacific | Total |
| 19. 5.1985 | Arctic | Partial |
| 12.11.1985 | S. Pacific, Antarctica | Total |
| 9. 4.1986 | Antarctic | Partial |
| 3.10.1986 | N. Atlantic | Total |
| 29. 3.1986 | Argentina, Atlantic, Central Africa, Indian Ocean | Total |
| 23. 9.1987 | USSR, China, Pacific | Annular |
| 18. 3.1988 | Indian Ocean, E. Indies, Pacific | Total |
| 7. 3.1989 | Arctic | Partial |
| 31. 8.1989 | Antarctic | Partial |
| 26. 1.1990 | Antarctic | Annular |
| 22. 7.1990 | Finland, USSR, Pacific | Total |
| 15/16. 1.1991 | Australia, N. Zealand, Pacific | Annular |
| 11. 7.1991 | Pacific, Mexico, Brazil | Total |
| 4/5. 1.1992 | Central Pacific | Annular |
| 30. 6.1992 | S. Atlantic | Total |
| 24.12.1992 | Arctic | Partial |
| 21. 5.1993 | Arctic | Partial |
| 13.11.1993 | Antarctic | Partial |
| 10. 5.1994 | Pacific, Mexico, USA, Canada | Annular |
| 3.11.1994 | Peru, Brazil, S. Atlantic | Total |
| 29. 4.1995 | S. Pacific, Peru, S. Atlantic | Annular |
| 24.10.1995 | Iran, India, E. Indies, Pacific | Total |
| 17. 4.1996 | Antarctic | Partial |
| 12.10.1996 | Arctic | Partial |
| 9. 3.1997 | USSR, Arctic | Total |
| 2. 9.1997 | Antarctic | Partial |
| 26. 2.1998 | Pacific, Atlantic | Total |
| 22. 8.1998 | Indian Ocean, E. Indies, Pacific | Annular |
| 16. 2.1999 | Indian Ocean, Australia, Pacific | Annular |
| 11. 8.1999 | Atlantic, England (Cornwall), France, Turkey, India | Total |

## Table 8: Lunar eclipses 1983–2000

| Date | Magnitude % | Date | Magnitude % | Date | Magnitude % |
|---|---|---|---|---|---|
| 25. 6.1983 | 34 | 9. 2.1990 | Total | 15. 4.1995 | 12 |
| 4. 5.1985 | Total | 6. 8.1990 | 68 | 4. 4.1996 | Total |
| 28.10.1985 | Total | 21.12.1991 | 9 | 27. 9.1996 | Total |
| 24. 4.1986 | Total | 15. 6.1992 | 69 | 24. 3.1997 | 93 |
| 17.10.1986 | Total | 10.12.1992 | Total | 16. 9.1997 | Total |
| 7.10.1987 | 1 | 4. 6.1993 | Total | 28. 7.1999 | 42 |
| 27. 8.1988 | 30 | 29.11.1993 | Total | 21. 1.2000 | Total |
| 20. 2.1989 | Total | 25. 5.1994 | 28 | 16. 7.2000 | Total |
| 17. 8.1989 | Total | | | | |

## Table 9: Elongations of Mercury 1983–2000

| Date (Western) | Date (Eastern) |
|---|---|
| 8. 2./ 8. 6./ 1.10.1983 | 21. 4./19. 8./13.12.1983 |
| 22. 1./19. 5./14. 9.1984 | 3. 4./31. 7./25.11.1984 |
| 3. 1./ 1. 5./28. 8.1985 | 17. 3./14. 7./ 8.11.1985 |
| 13. 4./11. 8./30.11.1986 | 28. 2./25. 6./21.10.1986 |
| 26. 3./25. 7./13.11.1987 | 12. 2./ 7. 6./ 4.10.1987 |
| 8. 3./ 6. 7./26.10.1988 | 26. 1./19. 5./15. 9.1988 |
| 18. 2./18. 6./10.10.1989 | 9. 1./ 1. 5./29. 8./23.12.1989 |
| 1. 2./31. 5./24. 9.1990 | 13. 4./11. 8./ 6.12.1990 |
| 14. 1./12. 5./ 7. 9./27.12.1991 | 27. 3./25. 7./19.11.1991 |
| 23. 4./21. 8./ 9.12.1992 | 9. 3./ 6. 7./31.10.1992 |
| 5. 4./ 4. 8./22.11.1993 | 21. 2./17. 6./14.10.1993 |
| 19. 3./17. 7./ 6.11.1994 | 4. 2./30. 5./26. 9.1994 |
| 1. 3./29. 6./20.10.1995 | 19. 1./12. 5./ 9. 9.1995 |
| 11. 2./10. 6./ 3.10.1996 | 2. 1./23. 5./21. 8./15.12.1996 |
| 24. 1./22. 5./16. 9.1997 | 6. 4./ 4. 8./28.11.1997 |
| 6. 1./ 4. 5./31. 8./20.12.1998 | 20. 3./17. 6./11.11.1998 |
| 16. 4./14. 8./ 2.12.1999 | 3. 3./28. 6./24.10.1999 |
| 28. 3./27. 7./15.11.2000 | 15. 2./ 9. 6./ 6.10.2000 |

## Table 10: Elongations and conjunctions of Venus 1983–2000

| Date | Phenomenon | Date | Phenomenon |
|---|---|---|---|
| 16. 6.1983 | Eastern elongation | 2.11.1991 | Western elongation |
| 25. 8.1983 | Inferior conjunction | 13. 6.1992 | Superior conjunction |
| 4.11.1983 | Western elongation | 19. 1.1993 | Eastern elongation |
| 15. 6.1984 | Superior conjunction | 1. 4.1993 | Inferior conjunction |
| 22. 1.1985 | Eastern elongation | 10. 6.1993 | Western elongation |
| 3. 4.1985 | Inferior conjunction | 17. 1.1994 | Superior conjunction |
| 13. 6.1985 | Western elongation | 25. 8.1994 | Eastern elongation |
| 19. 1.1986 | Superior conjunction | 2.11.1994 | Inferior conjunction |
| 27. 8.1986 | Eastern elongation | 13. 1.1995 | Western elongation |
| 5.11.1986 | Inferior conjunction | 20. 8.1995 | Superior conjunction |
| 15. 1.1987 | Western elongation | 1. 4.1996 | Eastern elongation |
| 23. 8.1987 | Superior conjunction | 10. 6.1996 | Inferior conjunction |
| 3. 4.1988 | Eastern elongation | 19. 8.1996 | Western elongation |
| 13. 6.1988 | Inferior conjunction | 2. 4.1997 | Superior conjunction |
| 22. 8.1988 | Western elongation | 6.11.1997 | Eastern elongation |
| 5. 4.1989 | Superior conjunction | 16. 1.1998 | Inferior conjunction |
| 8.11.1989 | Eastern elongation | 27. 3.1998 | Western elongation |
| 10. 1.1990 | Inferior conjunction | 30.10.1998 | Superior conjunction |
| 30. 3.1990 | Western elongation | 11. 6.1999 | Eastern elongation |
| 1.11.1990 | Superior conjunction | 20. 8.1999 | Inferior conjunction |
| 13. 6.1991 | Eastern elongation | 30.10.1999 | Western elongation |
| 22. 8.1991 | Inferior conjunction | 11. 6.2000 | Superior conjunction |

## Table 11: Oppositions of Mars 1984–1999

| Date | Closest approach to Earth | Apparent diameter sec of arc | Magnitude |
|---|---|---|---|
| 11. 5.1984 | 19. 5.1984 | 17.5 | −1.8 |
| 10. 6.1986 | 16. 6.1986 | 23.1 | −2.4 |
| 28. 9.1988 | 22. 9.1988 | 23.7 | −2.6 |
| 27.11.1990 | 20.11.1990 | 17.9 | −1.7 |
| 7. 1.1993 | 3. 1.1993 | 14.9 | −1.2 |
| 12. 2.1995 | 11. 2.1995 | 13.8 | −1.0 |
| 17. 3.1997 | 20. 3.1997 | 14.2 | −1.1 |
| 24. 4.1999 | 1. 5.1999 | 16.2 | −1.5 |

## Table 12: Oppositions of Jupiter 1983–2000

| Date | Apparent diameter sec of arc | Magnitude |
|---|---|---|
| 27. 5.1983 | 45.5 | −2.1 |
| 29. 6.1984 | 46.8 | −2.2 |
| 4. 8.1985 | 48.5 | −2.3 |
| 10. 9.1986 | 49.6 | −2.4 |
| 18.10.1987 | 49.8 | −2.5 |
| 23.11.1988 | 48.7 | −2.4 |
| 27.12.1989 | 47.2 | −2.3 |
| 28. 1.1991 | 45.7 | −2.1 |
| 28. 2.1992 | 44.6 | −2.0 |
| 30. 3.1993 | 44.2 | −2.0 |
| 30. 4.1994 | 44.5 | −2.0 |
| 1. 6.1995 | 45.6 | −2.1 |
| 4. 7.1996 | 47.0 | −2.2 |
| 9. 8.1997 | 48.6 | −2.4 |
| 16. 9.1998 | 49.7 | −2.5 |
| 23.10.1999 | 49.8 | −2.5 |
| 28.11.2000 | 48.5 | −2.4 |

## Table 13: Oppositions of Saturn 1983–2000

| Date | Magnitude |
|---|---|
| 21. 4.1983 | +0.4 |
| 3. 5.1984 | +0.3 |
| 15. 5.1985 | +0.2 |
| 27. 5.1986 | +0.2 |
| 9. 6.1987 | +0.2 |
| 20. 6.1988 | +0.2 |
| 2. 7.1989 | +0.2 |
| 14. 7.1990 | +0.3 |
| 26. 7.1991 | +0.3 |
| 7. 8.1992 | +0.4 |
| 19. 8.1993 | +0.5 |
| 1. 9.1994 | +0.7 |
| 14. 9.1995 | +0.8 |
| 26. 9.1996 | +0.7 |
| 10.10.1997 | +0.4 |
| 23.10.1998 | +0.2 |
| 6.11.1999 | 0.0 |
| 19.11.2000 | −0.1 |

## Table 14: Opposition of Uranus 1983–2000

| Date | Date | Date |
|---|---|---|
| 29.5.1983 | 24.6.1989 | 21.7.1995 |
| 1.6.1984 | 29.6.1990 | 25.7.1996 |
| 6.6.1985 | 4.7.1991 | 29.7.1997 |
| 11.6.1986 | 7.7.1992 | 3.8.1998 |
| 16.6.1987 | 12.7.1993 | 7.8.1999 |
| 20.6.1988 | 17.7.1994 | 11.8.2000 |

# Tables III

**Table 15: Position of Mars 1983–1992**

| Date | R.A. | Dec | Date | R.A. | Dec | Date | R.A. | Dec | Date | R.A. | Dec | Date | R.A. | Dec |
|---|---|---|---|---|---|---|---|---|---|---|---|---|---|---|
| d mo yr | h m s | ° ′ ″ | d mo yr | h m s | ° ′ ″ | d mo yr | h m s | ° ′ ″ | d mo yr | h m s | ° ′ ″ | d mo yr | h m s | ° ′ ″ |
| 6. 1.83 | 21 34 45 | −15 34 52 | 6.12.84 | 21 12 42 | −17 33 01 | 6.11.86 | 21 19 23 | −17 47 12 | 6.10.88 | 0 17 29 | − 2 30 38 | 6. 9.90 | 4 03 45 | +19 04 17 |
| 16. 1.83 | 22 05 10 | −12 52 00 | 16.12.84 | 21 42 39 | −15 01 13 | 16.11.86 | 21 44 41 | −15 23 44 | 16.10.88 | 0 08 24 | − 2 40 09 | 16. 9.90 | 4 21 38 | +20 01 42 |
| 26. 1.83 | 22 34 55 | − 9 57 05 | 26.12.84 | 22 11 59 | −12 15 43 | 26.11.86 | 22 10 06 | −12 48 53 | 26.10.88 | 0 03 40 | − 2 20 25 | 26. 9.90 | 4 36 19 | +20 47 55 |
| 5. 2.83 | 23 04 02 | − 6 53 41 | **5. 1.85** | 22 40 44 | − 9 19 46 | 6.12.86 | 22 35 33 | −10 04 40 | 5.11.88 | 0 03 52 | − 1 32 46 | 6.10.90 | 4 46 57 | +21 25 24 |
| 15. 2.83 | 23 32 41 | − 3 45 07 | 15. 1.85 | 23 08 56 | − 6 16 43 | 16.12.86 | 23 00 57 | − 7 13 22 | 15.11.88 | 0 08 43 | − 0 20 55 | 16.10.90 | 4 52 34 | +21 56 08 |
| 25. 2.83 | 0 00 57 | − 0 34 44 | 25. 1.85 | 23 36 44 | − 3 09 36 | 26.12.86 | 23 26 17 | − 4 17 36 | 25.11.88 | 0 17 35 | + 1 10 26 | 26.10.90 | 4 52 14 | +22 21 01 |
| 7. 3.83 | 0 29 00 | + 2 34 18 | 4. 2.85 | 0 04 14 | − 0 01 30 | **5. 1.87** | 23 51 35 | − 1 19 34 | 5.12.88 | 0 29 39 | + 2 56 08 | 5.11.90 | 4 45 33 | +22 38 57 |
| 17. 3.83 | 0 56 57 | + 5 39 06 | 14. 2.85 | 0 31 33 | + 3 04 37 | 15. 1.87 | 0 16 56 | + 1 38 17 | 15.12.88 | 0 44 19 | + 4 52 24 | 15.11.90 | 4 33 10 | +22 46 37 |
| 27. 3.83 | 1 24 57 | + 8 36 53 | 24. 2.85 | 0 58 50 | + 6 06 11 | 25. 1.87 | 0 42 21 | + 4 33 29 | 25.12.88 | 1 01 04 | + 6 55 36 | 25.11.90 | 4 17 15 | +22 41 29 |
| 6. 4.83 | 1 53 04 | +11 24 57 | 6. 3.85 | 1 26 11 | + 9 00 33 | 4. 2.87 | 1 07 57 | + 7 24 00 | **4. 1.89** | 1 19 28 | + 9 02 18 | 5.12.90 | 4 01 21 | +22 26 01 |
| 16. 4.83 | 2 21 26 | +14 01 00 | 16. 3.85 | 1 53 42 | +11 45 18 | 14. 2.87 | 1 33 48 | +10 07 33 | 14. 1.89 | 1 39 17 | +11 09 52 | 15.12.90 | 3 48 42 | +22 08 04 |
| 26. 4.83 | 2 50 05 | +16 22 46 | 26. 3.85 | 2 21 29 | +14 18 16 | 24. 2.87 | 1 59 58 | +12 42 00 | 24. 1.89 | 2 00 19 | +13 15 31 | 25.12.90 | 3 41 20 | +21 56 54 |
| 6. 5.83 | 3 19 04 | +18 28 13 | 5. 4.85 | 2 49 35 | +16 37 19 | 6. 3.87 | 2 26 30 | +15 05 29 | 3. 2.89 | 2 22 23 | +15 16 41 | **4. 1.91** | 3 39 49 | +21 58 05 |
| 16. 5.83 | 3 48 22 | +20 15 40 | 15. 4.85 | 3 18 01 | +18 40 34 | 16. 3.87 | 2 53 26 | +17 16 04 | 13. 2.89 | 2 45 25 | +17 11 13 | 14. 1.91 | 3 43 37 | +22 12 18 |
| 26. 5.83 | 4 17 56 | +21 43 39 | 25. 4.85 | 3 46 48 | +20 26 24 | 26. 3.87 | 3 20 48 | +19 11 57 | 23. 2.89 | 3 09 19 | +18 56 50 | 24. 1.91 | 3 51 59 | +22 37 21 |
| 5. 6.83 | 4 47 41 | +22 51 06 | 5. 5.85 | 4 15 52 | +21 53 24 | 5. 4.87 | 3 48 35 | +20 51 42 | 5. 3.89 | 3 34 00 | +20 31 31 | 3. 2.91 | 4 04 04 | +23 09 20 |
| 15. 6.83 | 5 17 31 | +23 37 20 | 15. 5.85 | 4 45 10 | +23 00 27 | 15. 4.87 | 4 16 43 | +22 13 50 | 15. 3.89 | 3 59 23 | +21 53 33 | 13. 2.91 | 4 19 08 | +23 44 05 |
| 25. 6.83 | 5 47 17 | +24 02 07 | 25. 5.85 | 5 14 36 | +23 46 54 | 25. 4.87 | 4 45 08 | +23 17 20 | 25. 3.89 | 4 25 23 | +23 01 14 | 23. 2.91 | 4 36 38 | +24 17 47 |
| 5. 7.83 | 6 16 51 | +24 05 40 | 4. 6.85 | 5 44 02 | +24 12 22 | 5. 5.87 | 5 13 46 | +24 01 24 | 4. 4.89 | 4 51 52 | +23 53 16 | 5. 3.91 | 4 56 03 | +24 46 58 |
| 15. 7.83 | 6 46 04 | +23 48 37 | 14. 6.85 | 6 13 19 | +24 16 57 | 15. 5.87 | 5 42 27 | +24 25 35 | 14. 4.89 | 5 18 45 | +24 28 35 | 15. 3.91 | 5 17 00 | +25 08 43 |
| 25. 7.83 | 7 14 48 | +23 11 58 | 24. 6.85 | 6 42 19 | +24 01 07 | 25. 5.87 | 6 11 04 | +24 29 52 | 24. 4.89 | 5 45 52 | +24 46 27 | 25. 3.91 | 5 39 10 | +25 20 41 |
| 4. 8.83 | 7 42 57 | +22 17 05 | 4. 7.85 | 7 10 55 | +23 25 43 | 4. 6.87 | 6 39 31 | +24 14 29 | 4. 5.89 | 6 13 05 | +24 46 30 | 4. 4.91 | 6 02 16 | +25 21 02 |
| 14. 8.83 | 8 10 26 | +21 05 31 | 14. 7.85 | 7 39 00 | +22 31 55 | 14. 6.87 | 7 07 38 | +23 40 07 | 14. 5.89 | 6 40 18 | +24 28 41 | 14. 4.91 | 6 26 02 | +25 08 25 |
| 24. 8.83 | 8 37 13 | +19 38 04 | 24. 7.85 | 8 06 31 | +21 21 04 | 24. 6.87 | 7 35 19 | +22 47 41 | 24. 5.89 | 7 07 20 | +23 53 20 | 24. 4.91 | 6 50 16 | +24 41 55 |
| 3. 9.83 | 9 03 16 | +17 59 32 | 3. 8.85 | 8 33 24 | +19 54 46 | 4. 7.87 | 8 02 32 | +21 38 22 | 3. 6.89 | 7 34 05 | +23 01 04 | 4. 5.91 | 7 14 45 | +24 01 07 |
| 13. 9.83 | 9 28 37 | +16 08 48 | 13. 8.85 | 8 59 38 | +18 14 41 | 14. 7.87 | 8 29 11 | +20 13 33 | 13. 6.89 | 8 00 31 | +21 52 41 | 14. 5.91 | 7 39 21 | +23 05 56 |
| 23. 9.83 | 9 53 17 | +14 08 49 | 23. 8.85 | 9 25 15 | +16 22 31 | 24. 7.87 | 8 55 17 | +18 34 42 | 23. 6.89 | 8 26 30 | +20 29 21 | 24. 5.91 | 8 03 56 | +21 56 37 |
| 3.10.83 | 10 17 20 | +12 01 21 | 2. 9.85 | 9 50 17 | +14 20 02 | 3. 8.87 | 9 20 49 | +16 43 21 | 3. 7.89 | 8 52 04 | +18 52 12 | 3. 6.91 | 8 28 23 | +20 33 45 |
| 13.10.83 | 10 40 48 | + 9 48 13 | 12. 9.85 | 10 14 46 | +12 08 54 | 13. 8.87 | 9 45 51 | +14 41 07 | 13. 7.89 | 9 17 11 | +17 02 35 | 13. 6.91 | 8 52 38 | +18 58 04 |
| 23.10.83 | 11 03 45 | + 7 31 13 | 22. 9.85 | 10 38 48 | + 9 50 46 | 23. 8.87 | 10 10 24 | +12 29 33 | 23. 7.89 | 9 41 52 | +15 01 55 | 23. 6.91 | 9 16 40 | +17 10 29 |
| 2.11.83 | 11 26 15 | + 5 11 58 | 2.10.85 | 11 02 27 | + 7 27 22 | 2. 9.87 | 10 34 35 | +10 10 11 | 2. 8.89 | 10 06 10 | +12 51 32 | 3. 7.91 | 9 40 27 | +15 12 10 |
| 12.11.83 | 11 48 22 | + 2 52 09 | 12.10.85 | 11 25 48 | + 5 00 10 | 12. 9.87 | 10 58 26 | + 7 44 38 | 12. 8.89 | 10 30 10 | +10 32 54 | 13. 7.91 | 10 04 02 | +13 04 08 |
| 22.11.83 | 12 10 06 | + 0 33 24 | 22.10.85 | 11 48 56 | + 2 30 46 | 22. 9.87 | 11 22 04 | + 5 14 20 | 22. 8.89 | 10 53 54 | + 8 07 30 | 23. 7.91 | 10 27 25 | +10 47 39 |
| 2.12.83 | 12 31 32 | − 1 42 51 | 1.11.85 | 12 11 57 | + 0 00 46 | 2.10.87 | 11 45 36 | + 2 40 51 | 1. 9.89 | 11 17 30 | + 5 36 40 | 2. 8.91 | 10 50 40 | + 8 24 02 |
| 12.12.83 | 12 52 38 | − 3 55 02 | 11.11.85 | 12 34 55 | − 2 28 27 | 12.10.87 | 12 09 06 | + 0 05 43 | 11. 9.89 | 11 41 03 | + 3 01 53 | 12. 8.91 | 11 13 53 | + 5 54 29 |
| 22.12.83 | 13 13 24 | − 6 01 41 | 21.11.85 | 12 57 56 | − 4 55 17 | 22.10.87 | 12 32 41 | − 2 29 36 | 21. 9.89 | 12 04 38 | + 0 24 43 | 22. 8.91 | 11 37 07 | + 3 20 26 |
| **1. 1.84** | 13 33 48 | − 8 01 37 | 1.12.85 | 13 21 3 | − 7 18 13 | 1.11.87 | 12 56 28 | − 5 03 29 | 1.10.89 | 12 28 23 | − 2 13 27 | 1. 9.91 | 12 00 29 | + 0 43 19 |
| 11. 1.84 | 13 53 44 | − 9 53 26 | 11.12.85 | 13 44 20 | − 9 35 53 | 11.11.87 | 13 20 32 | − 7 34 16 | 11.10.89 | 12 52 24 | − 4 50 55 | 11. 9.91 | 12 24 04 | − 1 55 28 |
| 21. 1.84 | 14 13 05 | −11 36 06 | 21.12.85 | 14 07 50 | −11 46 43 | 21.11.87 | 13 44 59 | −10 00 23 | 21.10.89 | 13 16 48 | − 7 26 00 | 21. 9.91 | 12 48 00 | − 4 34 17 |
| 31. 1.84 | 14 31 39 | −13 08 46 | 31.12.85 | 14 31 33 | −13 49 19 | 1.12.87 | 14 09 54 | −12 20 01 | 31.10.89 | 13 41 41 | − 9 56 59 | 1.10.91 | 13 12 22 | − 7 11 23 |
| 10. 2.84 | 14 49 10 | −14 30 35 | **10. 1.86** | 14 55 31 | −15 42 27 | 11.12.87 | 14 35 21 | −14 31 22 | 10.11.89 | 14 07 10 | −12 21 53 | 11.10.91 | 13 37 19 | − 9 45 03 |
| 20. 2.84 | 15 05 16 | −15 41 12 | 20. 1.86 | 15 19 41 | −17 24 48 | 21.12.87 | 15 01 24 | −16 32 40 | 20.11.89 | 14 33 20 | −14 38 41 | 21.10.91 | 14 02 55 | −12 13 10 |
| 1. 3.84 | 15 19 33 | −16 40 26 | 30. 1.86 | 15 44 01 | −18 55 23 | 31.12.87 | 15 28 04 | −18 22 00 | 30.11.89 | 15 00 16 | −16 45 18 | 31.10.91 | 14 29 18 | −14 33 37 |
| 11. 3.84 | 15 31 22 | −17 28 09 | 9. 2.86 | 16 08 26 | −20 13 26 | **10. 1.88** | 15 55 23 | −19 57 37 | 10.12.89 | 15 28 00 | −18 39 26 | 10.11.91 | 14 56 33 | −16 44 06 |
| 21. 3.84 | 15 40 05 | −18 04 41 | 19. 2.86 | 16 32 48 | −21 18 22 | 20. 1.88 | 16 23 18 | −21 17 49 | 20.12.89 | 15 56 34 | −20 18 51 | 20.11.91 | 15 24 43 | −18 42 08 |
| 31. 3.84 | 15 44 51 | −18 30 03 | 1. 3.86 | 16 56 58 | −22 10 10 | 30. 1.88 | 16 51 46 | −22 21 02 | 30.12.89 | 16 25 57 | −21 41 22 | 30.11.91 | 15 53 50 | −20 25 09 |
| 10. 4.84 | 15 44 53 | −18 43 47 | 11. 3.86 | 17 20 46 | −22 49 09 | 9. 2.88 | 17 20 41 | −23 06 03 | **9. 1.90** | 16 56 05 | −22 44 49 | 10.12.91 | 16 23 53 | −21 50 39 |
| 20. 4.84 | 15 39 43 | −18 45 19 | 21. 3.86 | 17 43 56 | −23 16 08 | 19. 2.88 | 17 49 56 | −23 31 55 | 19. 1.90 | 17 26 52 | −23 27 27 | 20.12.91 | 16 54 47 | −22 56 10 |
| 30. 4.84 | 15 29 29 | −18 33 38 | 31. 3.86 | 18 06 14 | −23 32 38 | 29. 2.88 | 18 19 21 | −23 38 06 | 29. 1.90 | 17 58 08 | −23 47 47 | 30.12.91 | 17 26 26 | −23 39 33 |
| 10. 5.84 | 15 15 33 | −18 09 38 | 10. 4.86 | 18 27 21 | −23 40 41 | 10. 3.88 | 18 48 46 | −23 24 34 | 8. 2.90 | 18 29 43 | −23 44 53 | **9. 1.92** | 17 58 41 | −23 59 02 |
| 20. 5.84 | 15 00 32 | −17 38 48 | 20. 4.86 | 18 46 53 | −23 43 08 | 20. 3.88 | 19 18 04 | −22 51 44 | 18. 2.90 | 19 01 25 | −23 18 23 | 19. 1.92 | 18 31 16 | −23 53 29 |
| 30. 5.84 | 14 47 20 | −17 10 09 | 30. 4.86 | 19 04 26 | −23 43 27 | 30. 3.88 | 19 47 02 | −22 00 36 | 28. 2.90 | 19 33 02 | −22 28 31 | 29. 1.92 | 19 03 59 | −23 22 24 |
| 9. 6.84 | 14 38 25 | −16 54 07 | 10. 5.86 | 19 19 27 | −23 45 48 | 9. 4.88 | 20 15 34 | −20 52 30 | 10. 3.90 | 20 04 22 | −21 16 13 | 8. 2.92 | 19 36 38 | −22 26 02 |
| 19. 6.84 | 14 34 53 | −16 57 23 | 20. 5.86 | 19 31 16 | −23 54 58 | 19. 4.88 | 20 43 33 | −19 29 14 | 20. 3.90 | 20 35 19 | −19 42 55 | 18. 2.92 | 20 08 57 | −21 05 28 |
| 29. 6.84 | 14 36 48 | −17 21 15 | 30. 5.86 | 19 39 10 | −24 15 25 | 29. 4.88 | 21 10 52 | −17 53 04 | 30. 3.90 | 21 05 45 | −17 50 32 | 28. 2.92 | 20 40 49 | −19 22 19 |
| 9. 7.84 | 14 43 46 | −18 03 56 | 9. 6.86 | 19 42 19 | −24 50 38 | 9. 5.88 | 21 37 26 | −16 06 22 | 9. 4.90 | 21 35 36 | −15 41 31 | 9. 3.92 | 21 12 07 | −17 18 50 |
| 19. 7.84 | 14 55 05 | −19 01 04 | 19. 6.86 | 19 40 09 | −25 40 44 | 19. 5.88 | 22 03 12 | −14 11 55 | 19. 4.90 | 22 04 53 | −13 18 22 | 19. 3.92 | 21 42 47 | −14 57 46 |
| 29. 7.84 | 15 10 08 | −20 07 35 | 29. 6.86 | 19 32 48 | −26 40 07 | 29. 5.88 | 22 28 03 | −12 12 52 | 29. 4.90 | 22 33 35 | −10 43 55 | 29. 3.92 | 22 12 50 | −12 21 59 |
| 8. 8.84 | 15 28 25 | −21 18 36 | 9. 7.86 | 19 21 26 | −27 37 47 | 8. 6.88 | 22 51 56 | −10 12 05 | 9. 5.90 | 23 01 45 | − 8 01 14 | 8. 4.92 | 22 42 18 | − 9 34 38 |
| 18. 8.84 | 15 49 29 | −22 28 53 | 19. 7.86 | 19 08 49 | −28 20 39 | 18. 6.88 | 23 14 40 | − 8 13 10 | 19. 5.90 | 23 29 28 | − 5 13 06 | 18. 4.92 | 23 11 15 | − 6 39 00 |
| 28. 8.84 | 16 13 00 | −23 33 38 | 29. 7.86 | 18 58 18 | −28 41 08 | 28. 6.88 | 23 36 02 | − 6 19 45 | 29. 5.90 | 23 56 46 | − 2 22 37 | 28. 4.92 | 23 39 48 | − 3 38 10 |
| 7. 9.84 | 16 38 40 | −24 28 27 | 8. 8.86 | 18 52 29 | −28 39 35 | 8. 7.88 | 23 55 46 | − 4 35 09 | 8. 6.90 | 0 23 42 | + 0 27 12 | 8. 5.92 | 0 08 02 | − 0 35 22 |
| 17. 9.84 | 17 06 09 | −25 09 00 | 18. 8.86 | 18 52 47 | −28 20 28 | 18. 7.88 | 0 13 26 | − 3 03 36 | 18. 6.90 | 0 50 21 | + 3 13 42 | 18. 5.92 | 0 36 04 | + 2 26 15 |
| 27. 9.84 | 17 35 09 | −25 31 42 | 28. 8.86 | 18 59 04 | −27 47 58 | 28. 7.88 | 0 28 28 | − 1 48 53 | 28. 6.90 | 1 16 41 | + 5 54 03 | 28. 5.92 | 1 04 01 | + 5 23 51 |
| 7.10.84 | 18 05 21 | −25 33 33 | 7. 9.86 | 19 10 36 | −27 03 43 | 7. 8.88 | 0 40 13 | − 0 54 25 | 8. 7.90 | 1 42 42 | + 8 25 45 | 7. 6.92 | 1 31 58 | + 8 14 32 |
| 17.10.84 | 18 36 22 | −25 12 20 | 17. 9.86 | 19 26 27 | −26 07 13 | 17. 8.88 | 0 47 47 | − 0 24 01 | 18. 7.90 | 2 08 21 | +10 46 47 | 17. 6.92 | 1 59 57 | +10 55 38 |
| 27.10.84 | 19 07 52 | −24 26 54 | 27. 9.86 | 19 45 32 | −24 57 19 | 27. 8.88 | 0 50 23 | − 0 19 04 | 28. 7.90 | 2 33 29 | +12 55 08 | 27. 6.92 | 2 28 02 | +13 24 50 |
| 6.11.84 | 19 39 31 | −23 17 01 | 7.10.86 | 20 07 00 | −23 32 45 | 6. 9.88 | 0 47 39 | − 0 38 15 | 7. 8.90 | 2 57 54 | +14 49 30 | 7. 7.92 | 2 56 13 | +15 39 55 |
| 16.11.84 | 20 11 00 | −21 43 32 | 17.10.86 | 20 30 12 | −21 52 40 | 16. 9.88 | 0 39 55 | − 1 15 44 | 17. 8.90 | 3 21 22 | +16 29 11 | 17. 7.92 | 3 24 26 | +17 39 09 |
| 26.11.84 | 20 42 07 | −19 48 06 | 27.10.86 | 20 54 27 | −19 57 16 | 26. 9.88 | 0 28 56 | − 1 58 16 | 27. 8.90 | 3 43 28 | +17 53 53 | 27. 7.92 | 3 52 37 | +19 21 11 |

## Table 16: Position of Jupiter 1981–1989

| Date (d mo yr) | R.A. (h m s) | Dec (° ′ ″) | Date (d mo yr) | R.A. (h m s) | Dec (° ′ ″) | Date (d mo yr) | R.A. (h m s) | Dec (° ′ ″) | Date (d mo yr) | R.A. (h m s) | Dec (° ′ ″) | Date (d mo yr) | R.A. (h m s) | Dec (° ′ ″) |
|---|---|---|---|---|---|---|---|---|---|---|---|---|---|---|
| 31. 5.80 | 10 17 39 | +11 50 14 | 1. 5.82 | 14 12 10 | −11 46 6 | 31. 3.84 | 18 50 23 | −22 42 49 | 1. 3.86 | 22 17 12 | −11 32 52 | 30. 1.88 | 1 27 20 | + 7 55 19 |
| 10. 6.80 | 10 21 46 | +11 25 22 | 11. 5.82 | 14 7 24 | −11 22 19 | 10. 4.84 | 18 53 37 | −22 39 31 | 11. 3.86 | 22 26 15 | −10 42 5 | 9. 2.88 | 1 32 56 | + 8 30 47 |
| 20. 6.80 | 10 26 41 | +10 55 42 | 21. 5.82 | 14 3 7 | −11 1 15 | 20. 4.84 | 18 55 36 | −22 37 44 | 21. 3.86 | 22 35 6 | − 9 51 34 | 19. 2.88 | 1 39 21 | + 9 10 13 |
| 30. 6.80 | 10 32 19 | +10 21 45 | 31. 5.82 | 13 59 35 | −10 44 24 | 30. 4.84 | 18 56 15 | −22 37 43 | 31. 3.86 | 22 43 40 | − 9 1 55 | 29. 2.88 | 1 46 28 | + 9 52 42 |
| 10. 7.80 | 10 38 32 | + 9 44 2 | 10. 6.82 | 13 57 0 | −10 32 53 | 10. 5.84 | 18 55 33 | −22 39 36 | 10. 4.86 | 22 51 52 | − 8 13 44 | 10. 3.88 | 1 54 10 | +10 37 23 |
| 20. 7.80 | 10 45 15 | + 9 2 59 | 20. 6.82 | 13 55 29 | −10 27 21 | 20. 5.84 | 18 53 32 | −22 43 17 | 20. 4.86 | 22 59 38 | − 7 27 44 | 20. 3.88 | 2 2 21 | +11 23 30 |
| 30. 7.80 | 10 52 22 | + 8 19 5 | 30. 6.82 | 13 55 6 | −10 28 9 | 30. 5.84 | 18 50 18 | −22 48 27 | 30. 4.86 | 23 6 53 | − 6 44 35 | 30. 3.88 | 2 10 57 | +12 10 20 |
| 9. 8.80 | 10 59 49 | + 7 32 49 | 10. 7.82 | 13 55 51 | −10 35 12 | 9. 6.84 | 18 46 2 | −22 54 36 | 10. 5.86 | 23 13 33 | − 6 4 57 | 9. 4.88 | 2 19 52 | +12 57 12 |
| 19. 8.80 | 11 7 31 | + 6 44 39 | 20. 7.82 | 13 57 43 | −10 48 11 | 19. 6.84 | 18 40 59 | −23 1 10 | 20. 5.86 | 23 19 31 | − 5 29 37 | 19. 4.88 | 2 29 2 | +13 43 32 |
| 29. 8.80 | 11 15 23 | + 5 55 6 | 30. 7.82 | 14 0 38 | −11 6 39 | 29. 6.84 | 18 35 31 | −23 7 31 | 30. 5.86 | 23 24 42 | − 4 59 14 | 29. 4.88 | 2 38 21 | +14 28 46 |
| 8. 9.80 | 11 23 20 | + 5 4 38 | 9. 8.82 | 14 4 30 | −11 29 58 | 9. 7.84 | 18 30 0 | −23 13 16 | 9. 6.86 | 23 29 0 | − 4 34 33 | 9. 5.88 | 2 47 45 | +15 12 26 |
| 18. 9.80 | 11 31 20 | + 4 13 46 | 19. 8.82 | 14 9 15 | −11 57 26 | 19. 7.84 | 18 24 50 | −23 18 4 | 19. 6.86 | 23 32 19 | − 4 16 16 | 19. 5.88 | 2 57 10 | +15 54 8 |
| 28. 9.80 | 11 39 17 | + 3 23 4 | 29. 8.82 | 14 14 49 | −12 28 25 | 29. 7.84 | 18 20 22 | −23 21 51 | 29. 6.86 | 23 34 35 | − 4 4 55 | 29. 5.88 | 3 6 31 | +16 33 29 |
| 8.10.80 | 11 47 8 | + 2 33 5 | 8. 9.82 | 14 21 4 | −13 2 8 | 8. 8.84 | 18 16 53 | −23 24 44 | 9. 7.86 | 23 35 42 | − 4 1 0 | 8. 6.88 | 3 15 43 | +17 10 11 |
| 18.10.80 | 11 54 47 | + 1 44 23 | 18. 9.82 | 14 27 57 | −13 37 58 | 18. 8.84 | 18 14 37 | −23 26 52 | 19. 7.86 | 23 35 38 | − 4 4 46 | 18. 6.88 | 3 24 41 | +17 43 59 |
| 28.10.80 | 12 2 9 | + 0 57 37 | 28. 9.82 | 14 35 22 | −14 15 13 | 28. 8.84 | 18 13 40 | −23 28 22 | 29. 7.86 | 23 34 22 | − 4 16 4 | 28. 6.88 | 3 33 19 | +18 14 41 |
| 7.11.80 | 12 9 11 | + 0 13 24 | 8.10.82 | 14 43 15 | −14 53 14 | 7. 9.84 | 18 14 5 | −23 29 21 | 8. 8.86 | 23 31 58 | − 4 34 27 | 8. 7.88 | 3 41 31 | +18 42 9 |
| 17.11.80 | 12 15 46 | − 0 27 36 | 18.10.82 | 14 51 30 | −15 31 28 | 17. 9.84 | 18 15 52 | −23 29 45 | 18. 8.86 | 23 28 33 | − 4 58 51 | 18. 7.88 | 3 49 12 | +19 6 17 |
| 27.11.80 | 12 21 47 | − 1 4 42 | 28.10.82 | 15 0 3 | −16 9 10 | 27. 9.84 | 18 18 56 | −23 29 27 | 28. 8.86 | 23 24 21 | − 5 27 43 | 28. 7.88 | 3 56 14 | +19 26 59 |
| 7.12.80 | 12 27 10 | − 1 37 12 | 7.11.82 | 15 8 49 | −16 46 17 | 7.10.84 | 18 23 14 | −23 28 12 | 7. 9.86 | 23 19 37 | − 5 59 4 | 7. 8.88 | 4 2 30 | +19 44 15 |
| 17.12.80 | 12 31 46 | − 2 4 26 | 17.11.82 | 15 17 43 | −17 21 54 | 17.10.84 | 18 28 37 | −23 25 43 | 17. 9.86 | 23 14 44 | − 6 30 31 | 17. 8.88 | 4 7 51 | +19 58 6 |
| 27.12.80 | 12 35 30 | − 2 25 42 | 27.11.82 | 15 26 39 | −17 55 45 | 27.10.84 | 18 34 60 | −23 21 43 | 27. 9.86 | 23 10 3 | − 6 59 47 | 27. 8.88 | 4 12 11 | +20 8 31 |
| 6. 1.81 | 12 38 14 | − 2 40 27 | 7.12.82 | 15 35 33 | −18 27 26 | 6.11.84 | 18 42 14 | −23 15 51 | 7.10.86 | 23 5 55 | − 7 24 44 | 6. 9.88 | 4 15 21 | +20 15 32 |
| 16. 1.81 | 12 39 54 | − 2 48 8 | 17.12.82 | 15 44 17 | −18 56 41 | 16.11.84 | 18 50 12 | −23 7 51 | 17.10.86 | 23 2 38 | − 7 43 36 | 16. 9.88 | 4 17 14 | +20 19 11 |
| 26. 1.81 | 12 40 25 | − 2 48 28 | 27.12.82 | 15 52 47 | −19 23 15 | 26.11.84 | 18 58 47 | −22 57 30 | 27.10.86 | 23 0 25 | − 7 55 18 | 26. 9.88 | 4 17 45 | +20 19 28 |
| 5. 2.81 | 12 39 45 | − 2 41 24 | 6. 1.83 | 16 0 54 | −19 46 58 | 6.12.84 | 19 7 51 | −22 44 36 | 6.11.86 | 22 59 25 | − 7 59 11 | 6.10.88 | 4 16 51 | +20 16 26 |
| 15. 2.81 | 12 37 57 | − 2 27 11 | 16. 1.83 | 16 8 32 | −20 7 43 | 16.12.84 | 19 17 18 | −22 29 5 | 16.11.86 | 22 59 41 | − 7 55 1 | 16.10.88 | 4 14 34 | +20 10 6 |
| 25. 2.81 | 12 35 6 | − 2 6 35 | 26. 1.83 | 16 15 34 | −20 25 26 | 26.12.84 | 19 27 1 | −22 10 55 | 26.11.86 | 23 1 13 | − 7 43 3 | 26.10.88 | 4 11 2 | +20 0 41 |
| 7. 3.81 | 12 31 22 | − 1 40 48 | 5. 2.83 | 16 21 51 | −20 40 8 | 5. 1.85 | 19 36 54 | −21 50 11 | 6.12.86 | 23 3 57 | − 7 23 39 | 5.11.88 | 4 6 27 | +19 48 32 |
| 17. 3.81 | 12 27 0 | − 1 11 30 | 15. 2.83 | 16 27 17 | −20 51 54 | 15. 1.85 | 19 46 49 | −21 27 5 | 16.12.86 | 23 7 47 | − 6 57 21 | 15.11.88 | 4 1 7 | +19 34 18 |
| 27. 3.81 | 12 22 19 | − 0 40 43 | 25. 2.83 | 16 31 43 | −21 0 47 | 25. 1.85 | 19 56 43 | −21 1 50 | 26.12.86 | 23 12 37 | − 6 24 54 | 25.11.88 | 3 55 28 | +19 18 57 |
| 6. 4.81 | 12 17 38 | − 0 10 37 | 7. 3.83 | 16 35 2 | −21 6 53 | 4. 2.86 | 20 6 28 | −20 34 49 | 5. 1.87 | 23 18 19 | − 5 46 51 | 5.12.88 | 3 49 56 | +19 3 40 |
| 16. 4.81 | 12 13 16 | + 0 16 42 | 17. 3.83 | 16 37 8 | −21 10 19 | 14. 2.86 | 20 15 58 | −20 6 26 | 15. 1.87 | 23 24 46 | − 5 4 1 | 15.12.88 | 3 44 55 | +18 49 48 |
| 26. 4.81 | 12 9 32 | + 0 39 24 | 27. 3.83 | 16 37 56 | −21 11 9 | 24. 2.86 | 20 25 9 | −19 37 10 | 25. 1.87 | 23 31 52 | − 4 17 4 | 25.12.88 | 3 40 48 | +18 38 42 |
| 6. 5.81 | 12 6 39 | + 0 56 9 | 6. 4.83 | 16 37 24 | −21 9 27 | 6. 3.85 | 20 33 55 | −19 7 38 | 4. 2.87 | 23 39 29 | − 3 26 40 | 4. 1.89 | 3 37 50 | +18 31 21 |
| 16. 5.81 | 12 4 45 | + 1 6 3 | 16. 4.83 | 16 35 35 | −21 5 17 | 16. 3.85 | 20 42 10 | −18 38 25 | 14. 2.87 | 23 47 32 | − 2 33 32 | 14. 1.89 | 3 36 12 | +18 28 31 |
| 26. 5.81 | 12 3 57 | + 1 8 42 | 26. 4.83 | 16 32 33 | −20 58 46 | 26. 3.85 | 20 49 49 | −18 10 13 | 24. 2.87 | 23 55 55 | − 1 38 21 | 24. 1.89 | 3 35 58 | +18 30 00 |
| 5. 6.81 | 12 4 14 | + 1 4 6 | 6. 5.83 | 16 28 30 | −20 50 7 | 5. 4.85 | 20 56 46 | −17 43 46 | 6. 3.87 | 0 4 32 | − 0 41 43 | 3. 2.89 | 3 37 9 | +18 37 11 |
| 15. 6.81 | 12 5 37 | + 0 52 29 | 16. 5.83 | 16 23 40 | −20 39 40 | 15. 4.85 | 21 2 56 | −17 19 48 | 16. 3.87 | 0 13 19 | + 0 15 42 | 13. 2.89 | 3 39 40 | +18 48 14 |
| 25. 6.81 | 12 8 0 | + 0 34 19 | 26. 5.83 | 16 18 25 | −20 27 59 | 25. 4.85 | 21 8 12 | −16 59 4 | 26. 3.87 | 0 22 12 | + 1 13 18 | 23. 2.89 | 3 43 25 | +19 3 3 |
| 5. 7.81 | 12 11 20 | + 0 10 11 | 5. 6.83 | 16 13 5 | −20 15 53 | 5. 5.85 | 21 12 28 | −16 42 20 | 5. 4.87 | 0 31 5 | + 2 10 32 | 5. 3.89 | 3 48 19 | +19 20 55 |
| 15. 7.81 | 12 15 32 | − 0 19 18 | 15. 6.83 | 16 8 2 | −20 4 13 | 15. 5.85 | 21 15 40 | −16 30 15 | 15. 4.87 | 0 39 56 | + 3 6 47 | 15. 3.89 | 3 54 13 | +19 41 6 |
| 25. 7.81 | 12 20 29 | − 0 53 27 | 25. 6.83 | 16 3 37 | −19 54 3 | 25. 5.85 | 21 17 41 | −16 23 26 | 25. 4.87 | 0 48 39 | + 4 1 32 | 25. 3.89 | 4 1 0 | +20 2 48 |
| 4. 8.81 | 12 26 6 | − 1 31 39 | 5. 7.83 | 16 0 6 | −19 46 14 | 4. 6.85 | 21 18 28 | −16 22 16 | 5. 5.87 | 0 57 10 | + 4 54 15 | 4. 4.89 | 4 8 32 | +20 25 18 |
| 14. 8.81 | 12 32 19 | − 2 13 15 | 15. 7.83 | 15 57 41 | −19 41 29 | 14. 6.85 | 21 18 0 | −16 26 53 | 15. 5.87 | 1 5 25 | + 5 44 24 | 14. 4.89 | 4 16 44 | +20 47 54 |
| 24. 8.81 | 12 39 1 | − 2 57 35 | 25. 7.83 | 15 56 29 | −19 40 17 | 24. 6.85 | 21 16 17 | −16 37 9 | 25. 5.87 | 1 13 19 | + 6 31 32 | 24. 4.89 | 4 25 27 | +21 10 0 |
| 3. 9.81 | 12 46 8 | − 3 44 5 | 4. 8.83 | 15 56 32 | −19 42 45 | 4. 7.85 | 21 13 25 | −16 52 29 | 4. 6.87 | 1 20 47 | + 7 15 7 | 4. 5.89 | 4 34 37 | +21 31 3 |
| 13. 9.81 | 12 53 36 | − 4 32 9 | 14. 8.83 | 15 57 51 | −19 48 55 | 14. 7.85 | 21 9 31 | −17 11 52 | 14. 6.87 | 1 27 43 | + 7 54 40 | 14. 5.89 | 4 44 8 | +21 50 37 |
| 23. 9.81 | 13 1 21 | − 5 21 10 | 24. 8.83 | 16 0 23 | −19 58 25 | 24. 7.85 | 21 4 52 | −17 34 0 | 24. 6.87 | 1 34 2 | + 8 29 44 | 24. 5.89 | 4 53 54 | +22 8 18 |
| 3.10.81 | 13 9 17 | − 6 10 37 | 3. 9.83 | 16 4 4 | −20 10 49 | 3. 8.85 | 20 59 46 | −17 57 11 | 4. 7.87 | 1 39 36 | + 8 59 48 | 3. 6.89 | 5 3 49 | +22 23 50 |
| 13.10.81 | 13 17 21 | − 6 59 55 | 13. 9.83 | 16 8 48 | −20 25 35 | 13. 8.85 | 20 54 33 | −18 19 47 | 14. 7.87 | 1 44 20 | + 9 24 26 | 13. 6.89 | 5 13 48 | +22 37 2 |
| 23.10.81 | 13 25 28 | − 7 48 32 | 23. 9.83 | 16 14 29 | −20 42 8 | 23. 8.85 | 20 49 39 | −18 40 16 | 24. 7.87 | 1 48 7 | + 9 43 12 | 23. 6.89 | 5 23 47 | +22 47 44 |
| 2.11.81 | 13 33 34 | − 8 35 56 | 3.10.83 | 16 21 2 | −20 59 48 | 2. 9.85 | 20 45 23 | −18 57 18 | 3. 8.87 | 1 50 49 | + 9 55 38 | 3. 7.89 | 5 33 40 | +22 55 56 |
| 12.11.81 | 13 41 33 | − 9 21 34 | 13.10.83 | 16 28 20 | −21 17 59 | 12. 9.85 | 20 42 3 | −19 10 1 | 13. 8.87 | 1 52 21 | +10 1 28 | 13. 7.89 | 5 43 21 | +23 1 40 |
| 22.11.81 | 13 49 21 | −10 4 57 | 23.10.83 | 16 36 19 | −21 36 7 | 22. 9.85 | 20 39 54 | −19 17 53 | 23. 8.87 | 1 52 38 | +10 0 27 | 23. 7.89 | 5 52 45 | +23 5 4 |
| 2.12.81 | 13 56 52 | −10 45 35 | 2.11.83 | 16 44 50 | −21 53 37 | 2.10.85 | 20 39 2 | −19 20 36 | 2. 9.87 | 1 51 40 | + 9 52 33 | 2. 8.89 | 6 1 47 | +23 6 19 |
| 12.12.81 | 14 3 59 | −11 22 58 | 12.11.83 | 16 53 50 | −22 10 1 | 12.10.85 | 20 39 31 | −19 18 11 | 12. 9.87 | 1 49 28 | + 9 38 6 | 12. 8.89 | 6 10 19 | +23 5 42 |
| 22.12.81 | 14 10 37 | −11 56 39 | 22.11.83 | 17 3 11 | −22 24 53 | 22.10.85 | 20 41 20 | −19 10 45 | 22. 9.87 | 1 46 9 | + 9 17 45 | 22. 8.89 | 6 18 16 | +23 3 36 |
| 1. 1.82 | 14 16 37 | −12 26 12 | 2.12.83 | 17 12 47 | −22 37 51 | 1.11.85 | 20 44 25 | −18 58 15 | 2.10.87 | 1 41 56 | + 8 52 40 | 1. 9.89 | 6 25 31 | +23 0 24 |
| 11. 1.82 | 14 21 54 | −12 51 11 | 12.12.83 | 17 22 34 | −22 48 43 | 11.11.85 | 20 48 39 | −18 41 4 | 12.10.87 | 1 37 7 | + 8 24 31 | 11. 9.89 | 6 31 55 | +22 56 35 |
| 21. 1.82 | 14 26 19 | −13 11 15 | 22.12.83 | 17 32 25 | −22 57 14 | 21.11.85 | 20 53 57 | −18 19 19 | 22.10.87 | 1 32 2 | + 7 55 21 | 21. 9.89 | 6 37 23 | +22 52 40 |
| 31. 1.82 | 14 29 45 | −13 26 3 | 1. 1.84 | 17 42 13 | −23 3 21 | 1.12.85 | 21 0 9 | −17 53 14 | 1.11.87 | 1 27 6 | + 7 27 26 | 1.10.89 | 6 41 45 | +22 49 9 |
| 10. 2.82 | 14 32 5 | −13 35 17 | 11. 1.84 | 17 51 52 | −23 7 7 | 11.12.85 | 21 7 8 | −17 23 2 | 11.11.87 | 1 22 40 | + 7 3 1 | 11.10.89 | 6 44 55 | +22 46 32 |
| 20. 2.82 | 14 33 16 | −13 38 48 | 21. 1.84 | 18 1 16 | −23 8 35 | 21.12.85 | 21 14 47 | −16 48 59 | 21.11.87 | 1 19 4 | + 6 43 55 | 21.10.89 | 6 46 46 | +22 45 14 |
| 2. 3.82 | 14 33 13 | −13 36 29 | 31. 1.84 | 18 10 18 | −23 8 1 | 31.12.85 | 21 22 57 | −16 11 22 | 1.12.87 | 1 16 32 | + 6 31 40 | 31.10.89 | 6 47 12 | +22 45 33 |
| 12. 3.82 | 14 31 56 | −13 28 27 | 10. 2.84 | 18 18 52 | −23 5 41 | 10. 1.86 | 21 31 33 | −15 30 33 | 11.12.87 | 1 15 14 | + 6 27 2 | 10.11.89 | 6 46 12 | +22 47 34 |
| 22. 3.82 | 14 29 31 | −13 15 3 | 20. 2.84 | 18 26 49 | −23 1 58 | 20. 1.86 | 21 40 27 | −14 46 52 | 21.12.87 | 1 15 13 | + 6 30 18 | 20.11.89 | 6 43 47 | +22 51 10 |
| 1. 4.82 | 14 26 4 | −12 56 55 | 1. 3.84 | 18 34 5 | −22 57 19 | 30. 1.86 | 21 49 33 | −14 0 48 | 31.12.87 | 1 16 30 | + 6 41 23 | 30.11.89 | 6 40 4 | +22 56 0 |
| 11. 4.82 | 14 21 50 | −12 35 7 | 11. 3.84 | 18 40 30 | −22 52 13 | 9. 2.86 | 21 58 46 | −13 12 45 | 10. 1.88 | 1 19 0 | + 6 59 44 | 10.12.89 | 6 35 17 | +23 1 33 |
| 21. 4.82 | 14 17 6 | −12 10 59 | 21. 3.84 | 18 45 58 | −22 47 13 | 19. 2.86 | 22 8 1 | −12 23 15 | 20. 1.88 | 1 22 39 | + 7 24 38 | 20.12.89 | 6 29 45 | +23 7 10 |

# Tables IV

## Table 17: Movement of the Central Meridian for System I

| m | 0ʰ | 1ʰ | 2ʰ | 3ʰ | 4ʰ | 5ʰ | 6ʰ | 7ʰ | 8ʰ | 9ʰ | 10ʰ | 11ʰ |
|---|---|---|---|---|---|---|---|---|---|---|---|---|
| 0 | 0.0 | 36.6 | 73.2 | 109.7 | 146.3 | 182.9 | 219.5 | 256.1 | 292.7 | 329.2 | 5.8 | 42.4 |
| 1 | 0.6 | 37.2 | 73.8 | 110.4 | 146.9 | 183.5 | 220.1 | 256.7 | 293.3 | 329.8 | 6.4 | 43.0 |
| 2 | 1.2 | 37.8 | 74.4 | 111.0 | 147.5 | 184.1 | 220.7 | 257.3 | 293.9 | 330.5 | 7.0 | 43.6 |
| 3 | 1.8 | 38.4 | 75.0 | 111.6 | 148.2 | 184.7 | 221.3 | 257.9 | 294.5 | 331.1 | 7.6 | 44.2 |
| 4 | 2.4 | 39.0 | 75.6 | 112.2 | 148.8 | 185.3 | 221.9 | 258.5 | 295.1 | 331.7 | 8.3 | 44.8 |
| 5 | 3.0 | 39.6 | 76.2 | 112.8 | 149.4 | 186.0 | 222.5 | 259.1 | 295.7 | 332.3 | 8.9 | 45.4 |
| 6 | 3.7 | 40.2 | 76.8 | 113.4 | 150.0 | 186.6 | 223.1 | 259.7 | 296.3 | 332.9 | 9.5 | 46.1 |
| 7 | 4.3 | 40.8 | 77.4 | 114.0 | 150.6 | 187.2 | 223.8 | 260.3 | 296.9 | 333.5 | 10.1 | 46.7 |
| 8 | 4.9 | 41.5 | 78.0 | 114.6 | 151.2 | 187.8 | 224.4 | 260.9 | 297.5 | 334.1 | 10.7 | 47.3 |
| 9 | 5.5 | 42.1 | 78.6 | 115.2 | 151.8 | 188.4 | 225.0 | 261.6 | 298.1 | 334.7 | 11.3 | 47.9 |
| 10 | 6.1 | 42.7 | 79.3 | 115.8 | 152.4 | 189.0 | 225.6 | 262.2 | 298.7 | 335.3 | 11.9 | 48.5 |
| 11 | 6.7 | 43.3 | 79.9 | 116.5 | 153.0 | 189.6 | 226.2 | 262.8 | 299.4 | 335.9 | 12.5 | 49.1 |
| 12 | 7.3 | 43.9 | 80.5 | 117.1 | 153.6 | 190.2 | 226.8 | 263.4 | 300.0 | 336.5 | 13.1 | 49.7 |
| 13 | 7.9 | 44.5 | 81.1 | 117.7 | 154.3 | 190.8 | 227.4 | 264.0 | 300.6 | 337.2 | 13.7 | 50.3 |
| 14 | 8.5 | 45.1 | 81.7 | 118.3 | 154.9 | 191.4 | 228.0 | 264.6 | 301.2 | 337.8 | 14.3 | 50.9 |
| 15 | 9.1 | 45.7 | 82.3 | 118.9 | 155.5 | 192.1 | 228.6 | 265.2 | 301.8 | 338.4 | 15.0 | 51.5 |
| 16 | 9.8 | 46.3 | 82.9 | 119.5 | 156.1 | 192.7 | 229.2 | 265.8 | 302.4 | 339.0 | 15.6 | 52.1 |
| 17 | 10.4 | 46.9 | 83.5 | 120.1 | 156.7 | 193.3 | 229.9 | 266.4 | 303.0 | 339.6 | 16.2 | 52.8 |
| 18 | 11.0 | 47.6 | 84.1 | 120.7 | 157.3 | 193.9 | 230.5 | 267.0 | 303.6 | 340.2 | 16.8 | 53.4 |
| 19 | 11.6 | 48.2 | 84.7 | 121.3 | 157.9 | 194.5 | 231.1 | 267.7 | 304.2 | 340.8 | 17.4 | 54.0 |
| 20 | 12.2 | 48.8 | 85.4 | 121.9 | 158.5 | 195.1 | 231.7 | 268.3 | 304.8 | 341.4 | 18.0 | 54.6 |
| 21 | 12.8 | 49.4 | 86.0 | 122.5 | 159.1 | 195.7 | 232.3 | 268.9 | 305.5 | 342.0 | 18.6 | 55.2 |
| 22 | 13.4 | 50.0 | 86.6 | 123.2 | 159.7 | 196.3 | 232.9 | 269.5 | 306.1 | 342.6 | 19.2 | 55.8 |
| 23 | 14.0 | 50.6 | 87.2 | 123.8 | 160.3 | 196.9 | 233.5 | 270.1 | 306.7 | 343.3 | 19.8 | 56.4 |
| 24 | 14.6 | 51.2 | 87.8 | 124.4 | 161.0 | 197.5 | 234.1 | 270.7 | 307.3 | 343.9 | 20.4 | 57.0 |
| 25 | 15.2 | 51.8 | 88.4 | 125.0 | 161.6 | 198.1 | 234.7 | 271.3 | 307.9 | 344.5 | 21.1 | 57.6 |
| 26 | 15.9 | 52.4 | 89.0 | 125.6 | 162.2 | 198.8 | 235.3 | 271.9 | 308.5 | 345.1 | 21.7 | 58.2 |
| 27 | 16.5 | 53.0 | 89.6 | 126.2 | 162.8 | 199.4 | 235.9 | 272.5 | 309.1 | 345.7 | 22.3 | 58.9 |
| 28 | 17.1 | 53.7 | 90.2 | 126.8 | 163.4 | 200.0 | 236.6 | 273.1 | 309.7 | 346.3 | 22.9 | 59.5 |
| 29 | 17.7 | 54.3 | 90.8 | 127.4 | 164.0 | 200.6 | 237.2 | 273.7 | 310.3 | 346.9 | 23.5 | 60.1 |
| 30 | 18.3 | 54.9 | 91.5 | 128.0 | 164.6 | 201.2 | 237.8 | 274.4 | 310.9 | 347.5 | 24.1 | 60.7 |
| 31 | 18.9 | 55.5 | 92.1 | 128.6 | 165.2 | 201.8 | 238.4 | 275.0 | 311.6 | 348.1 | 24.7 | 61.3 |
| 32 | 19.5 | 56.1 | 92.7 | 129.3 | 165.8 | 202.4 | 239.0 | 275.6 | 312.2 | 348.7 | 25.3 | 61.9 |
| 33 | 20.1 | 56.7 | 93.3 | 129.9 | 166.4 | 203.0 | 239.6 | 276.2 | 312.8 | 349.4 | 25.9 | 62.5 |
| 34 | 20.7 | 57.3 | 93.9 | 130.5 | 167.1 | 203.6 | 240.2 | 276.8 | 313.4 | 350.0 | 26.5 | 63.1 |
| 35 | 21.3 | 57.9 | 94.5 | 131.1 | 167.7 | 204.2 | 240.8 | 277.4 | 314.0 | 350.6 | 27.2 | 63.7 |
| 36 | 21.9 | 58.5 | 95.1 | 131.7 | 168.3 | 204.9 | 241.4 | 278.0 | 314.6 | 351.2 | 27.8 | 64.3 |
| 37 | 22.6 | 59.1 | 95.7 | 132.3 | 168.9 | 205.5 | 242.0 | 278.6 | 315.2 | 351.8 | 28.4 | 65.0 |
| 38 | 23.2 | 59.7 | 96.3 | 132.9 | 169.5 | 206.1 | 242.7 | 279.2 | 315.8 | 352.4 | 29.0 | 65.6 |
| 39 | 23.8 | 60.4 | 96.9 | 133.5 | 170.1 | 206.7 | 243.3 | 279.8 | 316.4 | 353.0 | 29.6 | 66.2 |
| 40 | 24.4 | 61.0 | 97.6 | 134.1 | 170.7 | 207.3 | 243.9 | 280.5 | 317.0 | 353.6 | 30.2 | 66.8 |
| 41 | 25.0 | 61.6 | 98.2 | 134.7 | 171.3 | 207.9 | 244.5 | 281.1 | 317.6 | 354.2 | 30.8 | 67.4 |
| 42 | 25.6 | 62.2 | 98.8 | 135.4 | 171.9 | 208.5 | 245.1 | 281.7 | 318.3 | 354.8 | 31.4 | 68.0 |
| 43 | 26.2 | 62.8 | 99.4 | 136.0 | 172.5 | 209.1 | 245.7 | 282.3 | 318.9 | 355.4 | 32.0 | 68.6 |
| 44 | 26.8 | 63.4 | 100.0 | 136.6 | 173.2 | 209.7 | 246.3 | 282.9 | 319.5 | 356.1 | 32.6 | 69.2 |
| 45 | 27.4 | 64.0 | 100.6 | 137.2 | 173.8 | 210.3 | 246.9 | 283.5 | 320.1 | 356.7 | 33.2 | 69.8 |
| 46 | 28.0 | 64.6 | 101.2 | 137.8 | 174.4 | 211.0 | 247.5 | 284.1 | 320.7 | 357.3 | 33.9 | 70.4 |
| 47 | 28.7 | 65.2 | 101.8 | 138.4 | 175.0 | 211.6 | 248.1 | 284.7 | 321.3 | 357.9 | 34.5 | 71.0 |
| 48 | 29.3 | 65.8 | 102.4 | 139.0 | 175.6 | 212.2 | 248.8 | 285.3 | 321.9 | 358.5 | 35.1 | 71.7 |
| 49 | 29.9 | 66.5 | 103.0 | 139.6 | 176.2 | 212.8 | 249.4 | 285.9 | 322.5 | 359.1 | 35.7 | 72.3 |
| 50 | 30.5 | 67.1 | 103.6 | 140.2 | 176.8 | 213.4 | 250.0 | 286.6 | 323.1 | 359.7 | 36.3 | 72.9 |
| 51 | 31.1 | 67.7 | 104.3 | 140.8 | 177.4 | 214.0 | 250.6 | 287.2 | 323.7 | 0.3 | 36.9 | 73.5 |
| 52 | 31.7 | 68.3 | 104.9 | 141.4 | 178.0 | 214.6 | 251.2 | 287.8 | 324.4 | 0.9 | 37.5 | 74.1 |
| 53 | 32.3 | 68.9 | 105.5 | 142.1 | 178.6 | 215.2 | 251.8 | 288.4 | 325.0 | 1.5 | 38.1 | 74.7 |
| 54 | 32.9 | 69.5 | 106.1 | 142.7 | 179.2 | 215.8 | 252.4 | 289.0 | 325.6 | 2.2 | 38.7 | 75.3 |
| 55 | 33.5 | 70.1 | 106.7 | 143.3 | 179.9 | 216.4 | 253.0 | 289.6 | 326.2 | 2.8 | 39.3 | 75.9 |
| 56 | 34.1 | 70.7 | 107.3 | 143.9 | 180.5 | 217.0 | 253.6 | 290.2 | 326.8 | 3.4 | 40.0 | 76.5 |
| 57 | 34.8 | 71.3 | 107.9 | 144.5 | 181.1 | 217.7 | 254.2 | 290.8 | 327.4 | 4.0 | 40.6 | 77.1 |
| 58 | 35.4 | 71.9 | 108.5 | 145.1 | 181.7 | 218.3 | 254.8 | 291.4 | 328.0 | 4.6 | 41.2 | 77.8 |
| 59 | 36.0 | 72.6 | 109.1 | 145.7 | 182.3 | 218.9 | 255.5 | 292.0 | 328.6 | 5.2 | 41.8 | 78.4 |
| 60 | 36.6 | 73.2 | 109.7 | 146.3 | 182.9 | 219.5 | 256.1 | 292.7 | 329.2 | 5.8 | 42.4 | 79.0 |

## Table 18: Movement of the Central Meridian for System II

| m | 0ʰ | 1ʰ | 2ʰ | 3ʰ | 4ʰ | 5ʰ | 6ʰ | 7ʰ | 8ʰ | 9ʰ | 10ʰ | 11ʰ |
|---|---|---|---|---|---|---|---|---|---|---|---|---|
| 0 | 0.0 | 36.3 | 72.5 | 108.8 | 145.1 | 181.3 | 217.6 | 253.8 | 290.1 | 326.4 | 2.6 | 38.9 |
| 1 | 0.6 | 36.9 | 73.1 | 109.4 | 145.7 | 181.9 | 218.2 | 254.4 | 290.7 | 327.0 | 3.2 | 39.5 |
| 2 | 1.2 | 37.5 | 73.7 | 110.0 | 146.3 | 182.5 | 218.8 | 255.0 | 291.3 | 327.6 | 3.8 | 40.1 |
| 3 | 1.8 | 38.1 | 74.3 | 110.6 | 146.9 | 183.1 | 219.4 | 255.7 | 291.9 | 328.2 | 4.4 | 40.7 |
| 4 | 2.4 | 38.7 | 74.9 | 111.2 | 147.5 | 183.7 | 220.0 | 256.3 | 292.5 | 328.8 | 5.0 | 41.3 |
| 5 | 3.0 | 39.3 | 75.5 | 111.8 | 148.1 | 184.3 | 220.6 | 256.9 | 293.1 | 329.4 | 5.7 | 41.9 |
| 6 | 3.6 | 39.9 | 76.2 | 112.4 | 148.7 | 184.9 | 221.2 | 257.5 | 293.7 | 330.0 | 6.3 | 42.5 |
| 7 | 4.2 | 40.5 | 76.8 | 113.0 | 149.3 | 185.5 | 221.8 | 258.1 | 294.3 | 330.6 | 6.9 | 43.1 |
| 8 | 4.8 | 41.1 | 77.4 | 113.6 | 149.9 | 186.1 | 222.4 | 258.7 | 294.9 | 331.2 | 7.5 | 43.7 |
| 9 | 5.4 | 41.7 | 78.0 | 114.2 | 150.5 | 186.8 | 223.0 | 259.3 | 295.5 | 331.8 | 8.1 | 44.3 |
| 10 | 6.0 | 42.3 | 78.6 | 114.8 | 151.1 | 187.4 | 223.6 | 259.9 | 296.1 | 332.4 | 8.7 | 44.9 |
| 11 | 6.6 | 42.9 | 79.2 | 115.4 | 151.7 | 188.0 | 224.2 | 260.5 | 296.7 | 333.0 | 9.3 | 45.5 |
| 12 | 7.3 | 43.5 | 79.8 | 116.0 | 152.3 | 188.6 | 224.8 | 261.1 | 297.4 | 333.6 | 9.9 | 46.1 |
| 13 | 7.9 | 44.1 | 80.4 | 116.6 | 152.9 | 189.2 | 225.4 | 261.7 | 298.0 | 334.2 | 10.5 | 46.7 |
| 14 | 8.5 | 44.7 | 81.0 | 117.2 | 153.5 | 189.8 | 226.0 | 262.3 | 298.6 | 334.8 | 11.1 | 47.4 |
| 15 | 9.1 | 45.3 | 81.6 | 117.9 | 154.1 | 190.4 | 226.6 | 262.9 | 299.2 | 335.4 | 11.7 | 48.0 |
| 16 | 9.7 | 45.9 | 82.2 | 118.5 | 154.7 | 191.0 | 227.2 | 263.5 | 299.8 | 336.0 | 12.3 | 48.6 |
| 17 | 10.3 | 46.5 | 82.8 | 119.1 | 155.3 | 191.6 | 227.8 | 264.1 | 300.4 | 336.6 | 12.9 | 49.2 |
| 18 | 10.9 | 47.1 | 83.4 | 119.7 | 155.9 | 192.2 | 228.5 | 264.7 | 301.0 | 337.2 | 13.5 | 49.8 |
| 19 | 11.5 | 47.7 | 84.0 | 120.3 | 156.5 | 192.8 | 229.1 | 265.3 | 301.6 | 337.8 | 14.1 | 50.4 |
| 20 | 12.1 | 48.4 | 84.6 | 120.9 | 157.1 | 193.4 | 229.7 | 265.9 | 302.2 | 338.5 | 14.7 | 51.0 |
| 21 | 12.7 | 49.0 | 85.2 | 121.5 | 157.7 | 194.0 | 230.3 | 266.5 | 302.8 | 339.1 | 15.3 | 51.6 |
| 22 | 13.3 | 49.6 | 85.8 | 122.1 | 158.3 | 194.6 | 230.9 | 267.1 | 303.4 | 339.7 | 15.9 | 52.2 |
| 23 | 13.9 | 50.2 | 86.4 | 122.7 | 159.0 | 195.2 | 231.5 | 267.7 | 304.0 | 340.3 | 16.5 | 52.8 |
| 24 | 14.5 | 50.8 | 87.0 | 123.3 | 159.6 | 195.8 | 232.1 | 268.3 | 304.6 | 340.9 | 17.1 | 53.4 |
| 25 | 15.1 | 51.4 | 87.6 | 123.9 | 160.2 | 196.4 | 232.7 | 268.9 | 305.2 | 341.5 | 17.7 | 54.0 |
| 26 | 15.7 | 52.0 | 88.2 | 124.5 | 160.8 | 197.0 | 233.3 | 269.6 | 305.8 | 342.1 | 18.3 | 54.6 |
| 27 | 16.3 | 52.6 | 88.8 | 125.1 | 161.4 | 197.6 | 233.9 | 270.2 | 306.4 | 342.7 | 18.9 | 55.2 |
| 28 | 16.9 | 53.2 | 89.4 | 125.7 | 162.0 | 198.2 | 234.5 | 270.8 | 307.0 | 343.3 | 19.6 | 55.8 |
| 29 | 17.5 | 53.8 | 90.1 | 126.3 | 162.6 | 198.8 | 235.1 | 271.4 | 307.6 | 343.9 | 20.2 | 56.4 |
| 30 | 18.1 | 54.4 | 90.7 | 126.9 | 163.2 | 199.4 | 235.7 | 272.0 | 308.2 | 344.5 | 20.8 | 57.0 |
| 31 | 18.7 | 55.0 | 91.3 | 127.5 | 163.8 | 200.0 | 236.3 | 272.6 | 308.8 | 345.1 | 21.4 | 57.6 |
| 32 | 19.3 | 55.6 | 91.9 | 128.1 | 164.4 | 200.7 | 236.9 | 273.2 | 309.4 | 345.7 | 22.0 | 58.2 |
| 33 | 19.9 | 56.2 | 92.5 | 128.7 | 165.0 | 201.3 | 237.5 | 273.8 | 310.0 | 346.3 | 22.6 | 58.8 |
| 34 | 20.5 | 56.8 | 93.1 | 129.3 | 165.6 | 201.9 | 238.1 | 274.4 | 310.6 | 346.9 | 23.2 | 59.4 |
| 35 | 21.2 | 57.4 | 93.7 | 129.9 | 166.2 | 202.5 | 238.7 | 275.0 | 311.3 | 347.5 | 23.8 | 60.0 |
| 36 | 21.8 | 58.0 | 94.3 | 130.5 | 166.8 | 203.1 | 239.3 | 275.6 | 311.9 | 348.1 | 24.4 | 60.6 |
| 37 | 22.4 | 58.6 | 94.9 | 131.1 | 167.4 | 203.7 | 239.9 | 276.2 | 312.5 | 348.7 | 25.0 | 61.3 |
| 38 | 23.0 | 59.2 | 95.5 | 131.8 | 168.0 | 204.3 | 240.5 | 276.8 | 313.1 | 349.3 | 25.6 | 61.9 |
| 39 | 23.6 | 59.8 | 96.1 | 132.4 | 168.6 | 204.9 | 241.1 | 277.4 | 313.7 | 349.9 | 26.2 | 62.5 |
| 40 | 24.2 | 60.4 | 96.7 | 133.0 | 169.2 | 205.5 | 241.8 | 278.0 | 314.3 | 350.5 | 26.8 | 63.1 |
| 41 | 24.8 | 61.0 | 97.3 | 133.6 | 169.8 | 206.1 | 242.4 | 278.6 | 314.9 | 351.1 | 27.4 | 63.7 |
| 42 | 25.4 | 61.6 | 97.9 | 134.2 | 170.4 | 206.7 | 243.0 | 279.2 | 315.5 | 351.7 | 28.0 | 64.3 |
| 43 | 26.0 | 62.3 | 98.5 | 134.8 | 171.0 | 207.3 | 243.6 | 279.8 | 316.1 | 352.4 | 28.6 | 64.9 |
| 44 | 26.6 | 62.9 | 99.1 | 135.4 | 171.6 | 207.9 | 244.2 | 280.4 | 316.7 | 353.0 | 29.2 | 65.5 |
| 45 | 27.2 | 63.5 | 99.7 | 136.0 | 172.2 | 208.5 | 244.8 | 281.0 | 317.3 | 353.6 | 29.8 | 66.1 |
| 46 | 27.8 | 64.1 | 100.3 | 136.6 | 172.9 | 209.1 | 245.4 | 281.6 | 317.9 | 354.2 | 30.4 | 66.7 |
| 47 | 28.4 | 64.7 | 100.9 | 137.2 | 173.5 | 209.7 | 246.0 | 282.2 | 318.5 | 354.8 | 31.0 | 67.3 |
| 48 | 29.0 | 65.3 | 101.5 | 137.8 | 174.1 | 210.3 | 246.6 | 282.8 | 319.1 | 355.4 | 31.6 | 67.9 |
| 49 | 29.6 | 65.9 | 102.1 | 138.4 | 174.7 | 210.9 | 247.2 | 283.5 | 319.7 | 356.0 | 32.2 | 68.5 |
| 50 | 30.2 | 66.5 | 102.7 | 139.0 | 175.3 | 211.5 | 247.8 | 284.1 | 320.3 | 356.6 | 32.8 | 69.1 |
| 51 | 30.8 | 67.1 | 103.3 | 139.6 | 175.9 | 212.1 | 248.4 | 284.7 | 320.9 | 357.2 | 33.5 | 69.7 |
| 52 | 31.4 | 67.7 | 104.0 | 140.2 | 176.5 | 212.7 | 249.0 | 285.3 | 321.5 | 357.8 | 34.1 | 70.3 |
| 53 | 32.0 | 68.3 | 104.6 | 140.8 | 177.1 | 213.3 | 249.6 | 285.9 | 322.1 | 358.4 | 34.7 | 70.9 |
| 54 | 32.6 | 68.9 | 105.2 | 141.4 | 177.7 | 213.9 | 250.2 | 286.5 | 322.7 | 359.0 | 35.3 | 71.5 |
| 55 | 33.2 | 69.5 | 105.8 | 142.0 | 178.3 | 214.6 | 250.8 | 287.1 | 323.3 | 359.6 | 35.9 | 72.1 |
| 56 | 33.8 | 70.1 | 106.4 | 142.6 | 178.9 | 215.2 | 251.4 | 287.7 | 323.9 | 0.2 | 36.5 | 72.7 |
| 57 | 34.4 | 70.7 | 107.0 | 143.2 | 179.5 | 215.8 | 252.0 | 288.3 | 324.5 | 0.8 | 37.1 | 73.3 |
| 58 | 35.1 | 71.3 | 107.6 | 143.8 | 180.1 | 216.4 | 252.6 | 288.9 | 325.2 | 1.4 | 37.7 | 73.9 |
| 59 | 35.7 | 71.9 | 108.2 | 144.4 | 180.7 | 217.0 | 253.2 | 289.5 | 325.8 | 2.0 | 38.3 | 74.6 |
| 60 | 36.3 | 72.5 | 108.8 | 145.1 | 181.3 | 217.6 | 253.8 | 290.1 | 326.4 | 2.6 | 38.9 | 75.2 |

## Table 19: Conversion of Change of Longitude in Thirty Days (Long.) to Rotation Period (P) for System I

| Long. 9h 48m | Long. 9h 49m | Long. 9h 50m | Long. 9h 51m | Long. 9h 52m | P |
|---|---|---|---|---|---|
| −112.4 | −67.5 | −22.7 | +21.9 | +66.3 | 0 |
| 111.7 | 66.8 | 22.0 | 22.6 | 67.0 | 1 |
| 110.9 | 66.0 | 21.3 | 23.3 | 67.8 | 2 |
| 110.2 | 65.3 | 20.5 | 24.1 | 68.5 | 3 |
| 109.4 | 64.5 | 19.8 | 24.8 | 69.3 | 4 |
| 108.7 | 63.8 | 19.0 | 25.6 | 70.0 | 5 |
| 107.9 | 63.0 | 18.3 | 26.3 | 70.7 | 6 |
| 107.2 | 62.3 | 17.5 | 27.1 | 71.5 | 7 |
| 106.4 | 61.5 | 16.8 | 27.8 | 72.2 | 8 |
| 105.7 | 60.8 | 16.0 | 28.5 | 73.0 | 9 |
| 104.9 | 60.0 | 15.3 | 29.3 | 73.7 | 10 |
| 104.2 | 59.3 | 14.6 | 30.0 | 74.4 | 11 |
| 103.4 | 58.5 | 13.8 | 30.8 | 75.2 | 12 |
| 102.7 | 57.8 | 13.1 | 31.5 | 75.9 | 13 |
| 101.9 | 57.0 | 12.3 | 32.2 | 76.7 | 14 |
| 101.2 | 56.3 | 11.6 | 33.0 | 77.4 | 15 |
| 100.4 | 55.5 | 10.8 | 33.7 | 78.1 | 16 |
| 99.7 | 54.8 | 10.1 | 34.5 | 78.9 | 17 |
| 98.9 | 54.1 | 9.3 | 35.2 | 79.6 | 18 |
| 98.2 | 53.3 | 8.6 | 36.0 | 80.4 | 19 |
| 97.4 | 52.6 | 7.9 | 36.7 | 81.1 | 20 |
| 96.7 | 51.8 | 7.1 | 37.4 | 81.8 | 21 |
| 95.9 | 51.1 | 6.4 | 38.2 | 82.6 | 22 |
| 95.2 | 50.3 | 5.6 | 38.9 | 83.3 | 23 |
| 94.4 | 49.6 | 4.9 | 39.7 | 84.0 | 24 |
| 93.7 | 48.8 | 4.1 | 40.4 | 84.8 | 25 |
| 92.9 | 48.1 | 3.4 | 41.1 | 85.5 | 26 |
| 92.2 | 47.3 | 2.6 | 41.9 | 86.3 | 27 |
| 91.4 | 46.6 | 1.9 | 42.6 | 87.0 | 28 |
| 90.7 | 45.8 | 1.2 | 43.4 | 87.7 | 29 |
| 89.9 | 45.1 | −0.4 | 44.1 | 88.5 | 30 |
| 89.2 | 44.4 | +0.3 | 44.8 | +89.2 | 31 |
| 88.4 | 43.6 | 1.1 | 45.6 | | 32 |
| 87.7 | 42.9 | 1.8 | 46.3 | | 33 |
| 86.9 | 42.1 | 2.5 | 47.1 | | 34 |
| 86.2 | 41.4 | 3.3 | 47.8 | | 35 |
| 85.4 | 40.6 | 4.0 | 48.5 | | 36 |
| 84.7 | 39.9 | 4.8 | 49.3 | | 37 |
| 83.9 | 39.1 | 5.5 | 50.0 | | 38 |
| 83.2 | 38.4 | 6.3 | 50.8 | | 39 |
| 82.5 | 37.6 | 7.0 | 51.5 | | 40 |
| 81.7 | 36.9 | 7.8 | 52.2 | | 41 |
| 81.0 | 36.2 | 8.5 | 53.0 | | 42 |
| 80.2 | 35.4 | 9.2 | 53.7 | | 43 |
| 79.5 | 34.7 | 10.0 | 54.5 | | 44 |
| 78.7 | 33.9 | 10.7 | 55.2 | | 45 |
| 78.0 | 33.2 | 11.5 | 56.0 | | 46 |
| 77.2 | 32.4 | 12.2 | 56.7 | | 47 |
| 76.5 | 31.7 | 12.9 | 57.4 | | 48 |
| 75.7 | 30.9 | 13.7 | 58.2 | | 49 |
| 75.0 | 30.2 | 14.4 | 58.9 | | 50 |
| 74.2 | 29.4 | 15.2 | 59.7 | | 51 |
| 73.5 | 28.7 | 15.9 | 60.4 | | 52 |
| 72.7 | 28.0 | 16.7 | 61.1 | | 53 |
| 72.0 | 27.2 | 17.4 | 61.9 | | 54 |
| 71.2 | 26.5 | 18.1 | 62.6 | | 55 |
| 70.5 | 25.7 | 18.9 | 63.4 | | 56 |
| 69.7 | 25.0 | 19.6 | 64.1 | | 57 |
| 69.0 | 24.2 | 20.4 | 64.8 | | 58 |
| 68.2 | 23.5 | 21.1 | 65.6 | | 59 |
| 67.5 | 22.7 | 21.9 | 66.3 | | 60 |
| −66.8 | −22.0 | +22.6 | +67.0 | | |

## Table 20: Conversion of Change of Longitude in Thirty Days (Long.) to Rotation Period (P) for System II

| Long. 9h 52m | Long. 9h 53m | Long. 9h 54m | Long. 9h 55m | Long. 9h 56m | Long. 9h 57m | Long. 9h 58m | Long. 9h 59m | P |
|---|---|---|---|---|---|---|---|---|
| −162.6 | −118.3 | −74.1 | −30.1 | +13.7 | +57.4 | +101.0 | +144.4 | 0 |
| 161.9 | 117.6 | 73.4 | 29.4 | 14.5 | 58.2 | 101.7 | 145.1 | 1 |
| 161.1 | 116.8 | 72.7 | 28.7 | 15.2 | 58.9 | 102.4 | 145.9 | 2 |
| 160.4 | 116.1 | 71.9 | 27.9 | 15.9 | 59.6 | 103.2 | 146.6 | 3 |
| 159.6 | 115.3 | 71.2 | 27.2 | 16.6 | 60.3 | 103.9 | 147.3 | 4 |
| 158.9 | 114.6 | 70.5 | 26.5 | 17.4 | 61.1 | 104.6 | 148.0 | 5 |
| 158.2 | 113.9 | 69.7 | 25.7 | 18.1 | 61.8 | 105.3 | 148.8 | 6 |
| 157.4 | 113.1 | 69.0 | 25.0 | 18.8 | 62.5 | 106.1 | 149.5 | 7 |
| 156.7 | 112.4 | 68.3 | 24.3 | 19.6 | 63.3 | 106.8 | 150.2 | 8 |
| 155.9 | 111.7 | 67.5 | 23.5 | 20.3 | 64.0 | 107.5 | 150.9 | 9 |
| 155.2 | 110.9 | 66.8 | 22.8 | 21.0 | 64.7 | 108.2 | 151.6 | 10 |
| 154.5 | 110.2 | 66.1 | 22.1 | 21.7 | 65.4 | 109.0 | 152.4 | 11 |
| 153.7 | 109.4 | 65.3 | 21.3 | 22.5 | 66.2 | 109.7 | 153.1 | 12 |
| 153.0 | 108.7 | 64.6 | 20.6 | 23.2 | 66.9 | 110.4 | 153.8 | 13 |
| 152.2 | 108.0 | 63.9 | 19.9 | 23.9 | 67.6 | 111.1 | 154.5 | 14 |
| 151.5 | 107.2 | 63.1 | 19.2 | 24.7 | 68.3 | 111.9 | 155.2 | 15 |
| 150.8 | 106.5 | 62.4 | 18.4 | 25.4 | 69.1 | 112.6 | 156.0 | 16 |
| 150.0 | 105.8 | 61.7 | 17.7 | 26.1 | 69.8 | 113.3 | 156.7 | 17 |
| 149.3 | 105.0 | 60.9 | 17.0 | 26.9 | 70.5 | 114.0 | 157.4 | 18 |
| 148.5 | 104.3 | 60.2 | 16.2 | 27.6 | 71.2 | 114.8 | 158.1 | 19 |
| 147.8 | 103.6 | 59.5 | 15.5 | 28.3 | 72.0 | 115.5 | 158.9 | 20 |
| 147.1 | 102.8 | 58.7 | 14.8 | 29.0 | 72.7 | 116.2 | 159.6 | 21 |
| 146.3 | 102.1 | 58.0 | 14.0 | 29.8 | 73.4 | 116.9 | 160.3 | 22 |
| 145.6 | 101.3 | 57.2 | 13.3 | 30.5 | 74.2 | 117.7 | 161.0 | 23 |
| 144.9 | 100.6 | 56.5 | 12.6 | 31.2 | 74.9 | 118.4 | 161.7 | 24 |
| 144.1 | 99.9 | 55.8 | 11.8 | 32.0 | 75.6 | 119.1 | 162.5 | 25 |
| 143.4 | 99.1 | 55.0 | 11.1 | 32.7 | 76.3 | 119.8 | 163.2 | 26 |
| 142.6 | 98.4 | 54.3 | 10.4 | 33.4 | 77.1 | 120.6 | 163.9 | 27 |
| 141.9 | 97.7 | 53.6 | 9.6 | 34.1 | 77.8 | 121.3 | 164.6 | 28 |
| 141.2 | 96.9 | 52.8 | 8.9 | 34.9 | 78.5 | 122.0 | 165.4 | 29 |
| 140.4 | 96.2 | 52.1 | 8.2 | 35.6 | 79.2 | 122.7 | 166.1 | 30 |
| 139.7 | 95.5 | 51.4 | 7.5 | 36.3 | 80.0 | 123.4 | 166.8 | 31 |
| 138.9 | 94.7 | 50.6 | 6.7 | 37.1 | 80.7 | 124.2 | 167.5 | 32 |
| 138.2 | 94.0 | 49.9 | 6.0 | 37.8 | 81.4 | 124.9 | 168.2 | 33 |
| 137.5 | 93.3 | 49.2 | 5.3 | 38.5 | 82.1 | 125.6 | 169.0 | 34 |
| 136.7 | 92.5 | 48.4 | 4.5 | 39.2 | 82.9 | 126.3 | 169.7 | 35 |
| 136.0 | 91.8 | 47.7 | 3.8 | 40.0 | 83.6 | 127.1 | 170.4 | 36 |
| 135.3 | 91.0 | 47.0 | 3.1 | 40.7 | 84.3 | 127.8 | 171.1 | 37 |
| 134.5 | 90.3 | 46.2 | 2.3 | 41.4 | 85.0 | 128.5 | 171.8 | 38 |
| 133.8 | 89.6 | 45.5 | 1.6 | 42.2 | 85.8 | 129.2 | 172.6 | 39 |
| 133.0 | 88.9 | 44.8 | 0.9 | 42.9 | 86.5 | 130.0 | 173.3 | 40 |
| 132.3 | 88.1 | 44.0 | −0.1 | 43.6 | 87.2 | 130.7 | 174.0 | 41 |
| 131.6 | 87.4 | 43.3 | +0.6 | 44.3 | 87.9 | 131.4 | 174.7 | 42 |
| 130.8 | 86.6 | 42.6 | 1.3 | 45.1 | 88.7 | 132.1 | 175.4 | 43 |
| 130.1 | 85.9 | 41.9 | 2.0 | 45.8 | 89.4 | 132.9 | 176.2 | 44 |
| 129.4 | 85.2 | 41.1 | 2.8 | 46.5 | 90.1 | 133.6 | 176.9 | 45 |
| 128.6 | 84.4 | 40.4 | 3.5 | 47.2 | 90.8 | 134.3 | 177.6 | 46 |
| 127.9 | 83.7 | 39.7 | 4.2 | 48.0 | 91.6 | 135.0 | 178.3 | 47 |
| 127.1 | 83.0 | 38.9 | 5.0 | 48.7 | 92.3 | 135.7 | 179.0 | 48 |
| 126.4 | 82.2 | 38.2 | 5.7 | 49.4 | 93.0 | 136.5 | 179.8 | 49 |
| 125.7 | 81.5 | 37.5 | 6.4 | 50.2 | 93.7 | 137.2 | 180.5 | 50 |
| 124.9 | 80.7 | 36.7 | 7.2 | 50.9 | 94.5 | 137.9 | 181.2 | 51 |
| 124.2 | 80.0 | 36.0 | 7.9 | 51.6 | 95.2 | 138.6 | 181.9 | 52 |
| 123.4 | 79.3 | 35.3 | 8.6 | 52.3 | 95.9 | 139.4 | 182.6 | 53 |
| 122.7 | 78.5 | 34.5 | 9.3 | 53.1 | 96.6 | 140.1 | 183.4 | 54 |
| 122.0 | 77.8 | 33.8 | 10.1 | 53.8 | 97.4 | 140.8 | 184.1 | 55 |
| 121.2 | 77.1 | 33.1 | 10.8 | 54.5 | 98.1 | 141.5 | 184.8 | 56 |
| 120.5 | 76.3 | 32.3 | 11.5 | 55.3 | 98.8 | 142.2 | 185.5 | 57 |
| 119.8 | 75.6 | 31.6 | 12.3 | 56.0 | 99.5 | 143.0 | 186.2 | 58 |
| 119.0 | 74.9 | 30.9 | 13.0 | 56.7 | 100.3 | 143.7 | 187.0 | 59 |
| 118.3 | 74.1 | 30.1 | 13.7 | 57.4 | 101.0 | 144.4 | 187.7 | 60 |
| −117.6 | −73.4 | −29.4 | +14.5 | +58.2 | +101.7 | +145.1 | +188.4 | |

# Tables V

**Table 21: Position of Saturn 1982–1992**

| Date d mo yr | R.A. h m s | Dec ° ′ ″ | Date d mo yr | R.A. h m s | Dec ° ′ ″ | Date d mo yr | R.A. h m s | Dec ° ′ ″ | Date d mo yr | R.A. h m s | Dec ° ′ ″ | Date d mo yr | R.A. h m s | Dec ° ′ ″ |
|---|---|---|---|---|---|---|---|---|---|---|---|---|---|---|
| 28.10.82 | 13 40 46 | − 8 2 20 | 27. 9.84 | 14 48 29 | −14 0 32 | 28. 8.86 | 16 6 57 | −19 7 43 | 28. 7.88 | 17 46 8 | −22 20 20 | 28. 6.90 | 19 40 13 | −21 20 41 |
| 7.11.82 | 13 45 18 | − 8 27 38 | 7.10.84 | 14 52 35 | −14 20 17 | 7. 9.86 | 16 8 38 | −19 14 32 | 7. 8.88 | 17 44 16 | −22 21 13 | 8. 7.90 | 19 37 12 | −21 28 37 |
| 17.11.82 | 13 49 43 | − 8 51 39 | 17.10.84 | 14 56 58 | −14 40 33 | 17. 9.86 | 16 10 55 | −19 22 46 | 17. 8.88 | 17 43 1 | −22 22 22 | 18. 7.90 | 19 34 5 | −21 36 38 |
| 27.11.82 | 13 53 56 | − 9 13 59 | 27.10.84 | 15 1 52 | −15 0 60 | 27. 9.86 | 16 13 46 | −19 32 11 | 27. 8.88 | 17 42 26 | −22 23 49 | 28. 7.90 | 19 31 0 | −21 44 19 |
| 7.12.82 | 13 57 53 | − 9 34 14 | 6.11.84 | 15 6 15 | −15 21 15 | 7.10.86 | 16 17 8 | −19 42 31 | 6. 9.88 | 17 42 32 | −22 25 34 | 7. 8.90 | 19 28 8 | −21 51 22 |
| 17.12.82 | 14 1 29 | − 9 52 5 | 16.11.84 | 15 11 2 | −15 40 57 | 17.10.86 | 16 20 57 | −19 53 28 | 16. 9.88 | 17 43 21 | −22 27 35 | 17. 8.90 | 19 25 36 | −21 57 30 |
| 27.12.82 | 14 4 41 | −10 7 11 | 26.11.84 | 15 15 47 | −15 59 49 | 27.10.86 | 16 25 10 | −20 4 44 | 26. 9.88 | 17 44 51 | −22 29 48 | 27. 8.90 | 19 23 32 | −22 2 32 |
| 6. 1.83 | 14 7 23 | −10 19 18 | 6.12.84 | 15 20 27 | −16 17 32 | 6.11.86 | 16 29 41 | −20 16 4 | 6.10.88 | 17 47 0 | −22 32 6 | 6. 9.90 | 19 22 3 | −22 6 19 |
| 16. 1.83 | 14 9 32 | −10 28 10 | 16.12.84 | 15 24 57 | −16 33 51 | 16.11.86 | 16 34 27 | −20 27 11 | 16.10.88 | 17 49 47 | −22 34 21 | 16. 9.90 | 19 21 13 | −22 8 48 |
| 26. 1.83 | 14 11 5 | −10 33 37 | 26.12.84 | 15 29 12 | −16 48 32 | 26.11.86 | 16 39 22 | −20 37 53 | 26.10.88 | 17 53 6 | −22 36 25 | 26. 9.90 | 19 21 4 | −22 9 53 |
| 5. 2.83 | 14 11 59 | −10 35 36 | **5. 1.85** | 15 33 7 | −17 1 21 | 6.12.86 | 16 44 23 | −20 47 55 | 5.11.88 | 17 56 55 | −22 38 10 | 6.10.90 | 19 21 37 | −22 9 35 |
| 15. 2.83 | 14 12 12 | −10 34 3 | 15. 1.85 | 15 36 37 | −17 12 12 | 16.12.86 | 16 49 24 | −20 57 10 | 15.11.88 | 18 1 8 | −22 39 26 | 16.10.90 | 19 22 53 | −22 7 53 |
| 25. 2.83 | 14 11 46 | −10 29 7 | 25. 1.85 | 15 39 38 | −17 20 54 | 26.12.86 | 16 54 20 | −21 5 28 | 25.11.88 | 18 5 42 | −22 40 8 | 26.10.90 | 19 24 49 | −22 4 45 |
| 7. 3.83 | 14 10 40 | −10 21 2 | 4. 2.85 | 15 42 5 | −17 27 23 | **5. 1.87** | 16 59 6 | −21 12 45 | 5.12.88 | 18 10 32 | −22 40 9 | 5.11.90 | 19 27 22 | −22 0 12 |
| 17. 3.83 | 14 9 0 | −10 10 9 | 14. 2.85 | 15 43 55 | −17 31 36 | 15. 1.87 | 17 3 37 | −21 18 56 | 15.12.88 | 18 15 33 | −22 39 25 | 15.11.90 | 19 30 30 | −21 54 17 |
| 27. 3.83 | 14 6 50 | − 9 56 59 | 24. 2.85 | 15 45 6 | −17 33 30 | 25. 1.87 | 17 7 48 | −21 24 4 | 25.12.88 | 18 20 39 | −22 37 56 | 25.11.90 | 19 34 8 | −21 46 58 |
| 6. 4.83 | 14 4 17 | − 9 42 13 | 6. 3.85 | 15 45 35 | −17 33 9 | 4. 2.87 | 17 11 35 | −21 28 7 | **4. 1.89** | 18 25 46 | −22 35 44 | 5.12.90 | 19 38 11 | −21 38 21 |
| 16. 4.83 | 14 1 29 | − 9 26 33 | 16. 3.85 | 15 45 23 | −17 30 36 | 14. 2.87 | 17 14 52 | −21 31 9 | 14. 1.89 | 18 30 49 | −22 32 51 | 15.12.90 | 19 42 36 | −21 28 30 |
| 26. 4.83 | 13 58 36 | − 9 10 50 | 26. 3.85 | 15 44 29 | −17 25 59 | 24. 2.87 | 17 17 36 | −21 33 14 | 24. 1.89 | 18 35 42 | −22 29 24 | 25.12.90 | 19 47 17 | −21 17 30 |
| 6. 5.83 | 13 55 45 | − 8 55 54 | 5. 4.85 | 15 42 58 | −17 19 31 | 6. 3.87 | 17 19 42 | −21 34 26 | 3. 2.89 | 18 40 21 | −22 25 32 | **4. 1.91** | 19 52 9 | −21 5 30 |
| 16. 5.83 | 13 53 7 | − 8 42 30 | 15. 4.85 | 15 40 53 | −17 11 29 | 16. 3.87 | 17 21 9 | −21 34 53 | 13. 2.89 | 18 44 40 | −22 21 25 | 14. 1.91 | 19 57 8 | −20 52 40 |
| 26. 5.83 | 13 50 48 | − 8 31 22 | 25. 4.85 | 15 38 22 | −17 2 12 | 26. 3.87 | 17 21 53 | −21 34 36 | 23. 2.89 | 18 48 35 | −22 17 16 | 24. 1.91 | 20 2 7 | −20 39 13 |
| 5. 6.83 | 13 48 56 | − 8 23 2 | 5. 5.85 | 15 35 32 | −16 52 8 | 5. 4.87 | 17 21 54 | −21 33 42 | 5. 3.89 | 18 52 3 | −22 13 16 | 3. 2.91 | 20 7 4 | −20 25 24 |
| 15. 6.83 | 13 47 35 | − 8 17 56 | 15. 5.85 | 15 32 32 | −16 41 46 | 15. 4.87 | 17 21 13 | −21 32 14 | 15. 3.89 | 18 54 57 | −22 9 38 | 13. 2.91 | 20 11 52 | −20 11 29 |
| 25. 6.83 | 13 46 49 | − 8 16 19 | 25. 5.85 | 15 29 30 | −16 31 37 | 25. 4.87 | 17 19 52 | −21 30 16 | 25. 3.89 | 18 57 16 | −22 6 38 | 23. 2.91 | 20 16 26 | −19 57 46 |
| 5. 7.83 | 13 46 40 | − 8 18 16 | 4. 6.85 | 15 26 37 | −16 22 16 | 5. 5.87 | 17 17 55 | −21 27 50 | 4. 4.89 | 18 58 56 | −22 4 24 | 5. 3.91 | 20 20 44 | −19 44 36 |
| 15. 7.83 | 13 47 8 | − 8 23 46 | 14. 6.85 | 15 24 1 | −16 14 13 | 15. 5.87 | 17 15 28 | −21 25 2 | 14. 4.89 | 18 59 55 | −22 3 5 | 15. 3.91 | 20 24 40 | −19 52 17 |
| 25. 7.83 | 13 48 12 | − 8 32 39 | 24. 6.85 | 15 21 49 | −16 7 55 | 25. 5.87 | 17 12 37 | −21 21 55 | 24. 4.89 | 19 0 12 | −22 2 47 | 25. 3.91 | 20 28 10 | −19 21 9 |
| 4. 8.83 | 13 49 52 | − 8 44 40 | 4. 7.85 | 15 20 8 | −16 3 45 | 4. 6.87 | 17 9 33 | −21 18 39 | 4. 5.89 | 18 59 47 | −22 3 33 | 4. 4.91 | 20 31 10 | −19 11 34 |
| 14. 8.83 | 13 52 5 | − 8 59 34 | 14. 7.85 | 15 19 1 | −16 1 57 | 14. 6.87 | 17 6 23 | −21 15 22 | 14. 5.89 | 18 58 41 | −22 5 20 | 14. 4.91 | 20 33 37 | −19 3 48 |
| 24. 8.83 | 13 54 48 | − 9 16 58 | 24. 7.85 | 15 18 31 | −16 2 41 | 24. 6.87 | 17 3 18 | −21 12 15 | 24. 5.89 | 18 56 58 | −22 8 1 | 24. 4.91 | 20 35 29 | −18 58 8 |
| 3. 9.83 | 13 57 59 | − 9 36 29 | 3. 8.85 | 15 18 41 | −16 5 58 | 4. 7.87 | 17 0 26 | −21 9 33 | 3. 6.89 | 18 54 42 | −22 11 28 | 4. 5.91 | 20 36 43 | −18 54 47 |
| 13. 9.83 | 14 1 33 | − 9 57 44 | 13. 8.85 | 15 19 30 | −16 11 42 | 14. 7.87 | 16 57 57 | −21 7 28 | 13. 6.89 | 18 52 0 | −22 15 28 | 14. 5.91 | 20 37 17 | −18 53 52 |
| 23. 9.83 | 14 5 28 | −10 20 17 | 23. 8.85 | 15 20 56 | −16 19 47 | 24. 7.87 | 16 55 56 | −21 6 11 | 23. 6.89 | 18 49 0 | −22 19 46 | 24. 5.91 | 20 37 10 | −18 55 27 |
| 3.10.83 | 14 9 40 | −10 43 44 | 2. 9.85 | 15 22 59 | −16 29 57 | 3. 8.87 | 16 54 31 | −21 5 53 | 3. 7.89 | 18 45 51 | −22 24 9 | 3. 6.91 | 20 36 25 | −18 59 26 |
| 13.10.83 | 14 14 5 | −11 7 41 | 12. 9.85 | 15 25 36 | −16 41 56 | 13. 8.87 | 16 53 43 | −21 6 40 | 13. 7.89 | 18 42 42 | −22 28 25 | 13. 6.91 | 20 35 3 | −19 5 40 |
| 23.10.83 | 14 18 39 | −11 31 42 | 22. 9.85 | 15 28 44 | −16 55 25 | 23. 8.87 | 16 53 37 | −21 8 34 | 23. 7.89 | 18 39 42 | −22 32 24 | 23. 6.91 | 20 33 7 | −19 13 51 |
| 2.11.83 | 14 23 18 | −11 55 25 | 2.10.85 | 15 32 19 | −17 10 4 | 2. 9.87 | 16 54 12 | −21 11 34 | 2. 8.89 | 18 37 2 | −22 35 60 | 3. 7.91 | 20 30 42 | −19 23 34 |
| 12.11.83 | 14 27 57 | −12 18 27 | 12.10.85 | 15 36 17 | −17 25 34 | 12. 9.87 | 16 55 27 | −21 15 35 | 12. 8.89 | 18 34 47 | −22 39 7 | 13. 7.91 | 20 27 57 | −19 34 20 |
| 22.11.83 | 14 32 32 | −12 40 24 | 22.10.85 | 15 40 36 | −17 41 33 | 22. 9.87 | 16 57 22 | −21 20 29 | 22. 8.89 | 18 35 7 | −22 41 44 | 23. 7.91 | 20 24 58 | −19 45 37 |
| 2.12.83 | 14 36 59 | −13 0 58 | 1.11.85 | 15 45 10 | −17 57 41 | 2.10.87 | 16 59 54 | −21 26 5 | 1. 9.89 | 18 32 3 | −22 43 50 | 2. 8.91 | 20 21 54 | −19 56 50 |
| 12.12.83 | 14 41 13 | −13 19 50 | 11.11.85 | 15 49 56 | −18 13 40 | 12.10.87 | 17 3 0 | −21 32 10 | 11. 9.89 | 18 31 43 | −22 45 24 | 12. 8.91 | 20 18 56 | −20 7 27 |
| 22.12.83 | 14 45 9 | −13 36 39 | 21.11.85 | 15 54 48 | −18 29 13 | 22.10.87 | 17 6 37 | −21 38 33 | 21. 9.89 | 18 32 1 | −22 46 25 | 22. 8.91 | 20 16 12 | −20 17 0 |
| **1. 1.84** | 14 48 43 | −13 51 13 | 1.12.85 | 15 59 42 | −18 44 2 | 1.11.87 | 17 10 39 | −21 44 59 | 1.10.89 | 18 33 4 | −22 46 51 | 1. 9.91 | 20 13 51 | −20 25 6 |
| 11. 1.84 | 14 51 50 | −14 3 17 | 11.12.85 | 16 4 34 | −18 57 55 | 11.11.87 | 17 15 4 | −21 51 15 | 11.10.89 | 18 34 48 | −22 46 38 | 11. 9.91 | 20 11 59 | −20 31 26 |
| 21. 1.84 | 14 54 25 | −14 12 38 | 21.12.85 | 16 9 19 | −19 10 39 | 21.11.87 | 17 19 46 | −21 57 11 | 21.10.89 | 18 37 10 | −22 45 44 | 21. 9.91 | 20 10 42 | −20 35 47 |
| 31. 1.84 | 14 56 25 | −14 19 10 | 31.12.85 | 16 13 50 | −19 22 5 | 1.12.87 | 17 24 41 | −22 2 36 | 31.10.89 | 18 40 8 | −22 44 5 | 1.10.91 | 20 10 5 | −20 38 1 |
| 10. 2.84 | 14 57 47 | −14 22 45 | **10. 1.86** | 16 18 5 | −19 32 6 | 11.12.87 | 17 29 44 | −22 7 21 | 10.11.89 | 18 43 38 | −22 41 36 | 11.10.91 | 20 10 9 | −20 38 6 |
| 20. 2.84 | 14 58 29 | −14 23 24 | 20. 1.86 | 16 21 57 | −19 40 35 | 21.12.87 | 17 34 50 | −22 11 23 | 20.11.89 | 18 47 36 | −22 38 15 | 21.10.91 | 20 10 55 | −20 35 59 |
| 1. 3.84 | 14 58 30 | −14 21 9 | 30. 1.86 | 16 25 21 | −19 47 30 | 31.12.87 | 17 39 54 | −22 14 36 | 30.11.89 | 18 51 58 | −22 33 59 | 31.10.91 | 20 12 21 | −20 31 43 |
| 11. 3.84 | 14 57 51 | −14 16 6 | 9. 2.86 | 16 28 15 | −19 52 48 | **10. 1.88** | 17 44 51 | −22 17 1 | 10.12.89 | 18 56 37 | −22 28 48 | 10.11.91 | 20 14 27 | −20 25 21 |
| 21. 3.84 | 14 56 33 | −14 8 31 | 19. 2.86 | 16 30 33 | −19 56 30 | 20. 1.88 | 17 49 35 | −22 18 38 | 20.12.89 | 19 1 30 | −22 22 43 | 20.11.91 | 20 17 9 | −20 16 57 |
| 31. 3.84 | 14 54 41 | −13 58 45 | 1. 3.86 | 16 32 12 | −19 58 37 | 30. 1.88 | 17 54 2 | −22 19 33 | 30.12.89 | 19 6 32 | −22 15 47 | 30.11.91 | 20 20 23 | −20 6 38 |
| 10. 4.84 | 14 52 20 | −13 47 12 | 11. 3.86 | 16 33 9 | −19 59 12 | 9. 2.88 | 17 58 7 | −22 19 51 | **9. 1.90** | 19 11 37 | −22 8 7 | 10.12.91 | 20 24 5 | −19 54 32 |
| 20. 4.84 | 14 49 39 | −13 34 28 | 21. 3.86 | 16 33 25 | −19 58 18 | 19. 2.88 | 18 1 46 | −22 19 39 | 19. 1.90 | 19 16 40 | −21 59 51 | 20.12.91 | 20 28 11 | −19 40 48 |
| 30. 4.84 | 14 46 45 | −13 21 10 | 31. 3.86 | 16 32 58 | −19 55 60 | 29. 2.88 | 18 4 53 | −22 19 6 | 29. 1.90 | 19 21 38 | −21 51 9 | 30.12.91 | 20 32 36 | −19 25 38 |
| 10. 5.84 | 14 43 47 | −13 7 60 | 10. 4.86 | 16 31 50 | −19 52 25 | 10. 3.88 | 18 7 25 | −22 18 22 | 8. 2.90 | 19 26 23 | −21 42 13 | **9. 1.92** | 20 37 15 | −19 9 13 |
| 20. 5.84 | 14 40 56 | −12 55 39 | 20. 4.86 | 16 30 6 | −19 47 40 | 20. 3.88 | 18 9 18 | −22 17 34 | 18. 2.90 | 19 30 52 | −21 33 20 | 19. 1.92 | 20 42 3 | −18 51 50 |
| 30. 5.84 | 14 38 18 | −12 44 46 | 30. 4.86 | 16 27 49 | −19 41 59 | 30. 3.88 | 18 10 31 | −22 16 50 | 28. 2.90 | 19 35 1 | −21 24 45 | 29. 1.92 | 20 46 56 | −18 33 44 |
| 9. 6.84 | 14 36 3 | −12 35 57 | 10. 5.86 | 16 25 8 | −19 35 32 | 9. 4.88 | 18 11 1 | −22 16 16 | 10. 3.90 | 19 38 44 | −21 16 44 | 8. 2.92 | 20 51 48 | −18 15 14 |
| 19. 6.84 | 14 34 16 | −12 29 40 | 20. 5.86 | 16 22 9 | −19 28 40 | 19. 4.88 | 18 10 49 | −22 15 56 | 20. 3.90 | 19 41 58 | −21 9 34 | 18. 2.92 | 20 56 34 | −17 56 40 |
| 29. 6.84 | 14 33 1 | −12 26 14 | 30. 5.86 | 16 19 3 | −19 21 41 | 29. 4.88 | 18 9 54 | −22 15 52 | 30. 3.90 | 19 44 39 | −21 3 32 | 28. 2.92 | 21 1 12 | −17 38 23 |
| 9. 7.84 | 14 32 23 | −12 25 52 | 9. 6.86 | 16 15 59 | −19 14 57 | 9. 5.88 | 18 8 21 | −22 16 2 | 9. 4.90 | 19 46 44 | −20 58 53 | 9, 3.92 | 21 5 35 | −17 20 45 |
| 19. 7.84 | 14 32 22 | −12 28 38 | 19. 6.86 | 16 13 6 | −19 8 52 | 19. 5.88 | 18 6 14 | −22 16 23 | 19. 4.90 | 19 48 11 | −20 55 48 | 19. 3.92 | 21 9 40 | −17 4 9 |
| 29. 7.84 | 14 32 59 | −12 34 26 | 29. 6.86 | 16 10 32 | −19 3 49 | 29. 5.88 | 18 3 38 | −22 16 53 | 29. 4.90 | 19 48 56 | −20 54 27 | 29. 3.92 | 21 13 23 | −16 48 57 |
| 8. 8.84 | 14 34 14 | −12 43 11 | 9. 7.86 | 16 8 25 | −19 0 5 | 8. 6.88 | 18 0 41 | −22 17 25 | 9. 5.90 | 19 49 1 | −20 54 55 | 8. 4.92 | 21 16 40 | −16 35 33 |
| 18. 8.84 | 14 36 4 | −12 54 35 | 19. 7.86 | 16 6 51 | −18 57 57 | 18. 6.88 | 17 57 33 | −22 17 52 | 19. 5.90 | 19 48 25 | −20 57 10 | 18. 4.92 | 21 19 27 | −16 24 16 |
| 28. 8.84 | 14 38 28 | −13 8 22 | 29. 7.86 | 16 5 53 | −18 57 37 | 28. 6.88 | 17 54 22 | −22 18 30 | 29. 5.90 | 19 47 10 | −21 1 7 | 28. 4.92 | 21 21 42 | −16 15 26 |
| 7. 9.84 | 14 41 22 | −13 24 12 | 8. 8.86 | 16 5 34 | −18 59 8 | 8. 7.88 | 17 51 18 | −22 19 3 | 8. 6.90 | 19 45 19 | −21 6 33 | 8. 5.92 | 21 23 21 | −16 9 19 |
| 17. 9.84 | 14 44 44 | −13 41 43 | 18. 8.86 | 16 5 56 | −19 2 32 | 18. 7.88 | 17 48 31 | −22 19 38 | 18. 6.90 | 19 42 58 | −21 13 11 | 18. 5.92 | 21 24 23 | −16 6 8 |

**Table 22: Position of Uranus 1983–1992**

| Date | R.A. | Dec | Date | R.A. | Dec | Date | R.A. | Dec | Date | R.A. | Dec | Date | R.A. | Dec |
|---|---|---|---|---|---|---|---|---|---|---|---|---|---|---|
| d mo yr | h m s | ° ′ ″ | d mo yr | h m s | ° ′ ″ | d mo yr | h m s | ° ′ ″ | d mo yr | h m s | ° ′ ″ | d mo yr | h m s | ° ′ ″ |
| 6. 1.83 | 16 21 36 | −21 24 13 | 6.12.84 | 16 49 45 | −22 27 30 | 6.11.86 | 17 17 51 | −23 11 43 | 6.10.88 | 17 48 50 | −23 37 59 | 6. 9.90 | 18 24 36 | −23 38 52 |
| 16. 1.83 | 16 23 42 | −21 29 08 | 16.12.84 | 16 52 23 | −22 31 55 | 16.11.86 | 17 20 11 | −23 14 03 | 16.10.88 | 17 50 07 | −23 38 17 | 16. 9.90 | 18 24 28 | −23 38 52 |
| 26. 1.83 | 16 25 32 | −21 33 21 | 26.12.84 | 16 54 57 | −22 36 05 | 26.11.86 | 17 22 40 | −23 16 24 | 26.10.88 | 17 51 44 | −23 38 37 | 26. 9.90 | 18 24 42 | −23 38 39 |
| 5. 2.83 | 16 27 04 | −21 36 50 | **5. 1.85** | 16 57 25 | −22 39 56 | 6.12.86 | 17 25 16 | −23 18 42 | 5.11.88 | 17 53 37 | −23 38 57 | 6.10.90 | 18 25 18 | −23 38 12 |
| 15. 2.83 | 16 28 17 | −21 39 33 | 15. 1.85 | 16 59 43 | −22 43 24 | 16.12.86 | 17 27 54 | −23 20 54 | 15.11.88 | 17 55 46 | −23 39 14 | 16.10.90 | 18 26 16 | −23 37 31 |
| 25. 2.83 | 16 29 08 | −21 41 27 | 25. 1.85 | 17 01 49 | −22 46 28 | 26.12.86 | 17 30 32 | −23 22 56 | 25.11.88 | 17 58 06 | −23 39 26 | 26.10.90 | 18 27 34 | −23 36 36 |
| 7. 3.83 | 16 29 37 | −21 42 32 | 4. 2.85 | 17 03 39 | −22 49 04 | **5. 1.87** | 17 33 07 | −23 24 46 | 5.12.88 | 18 00 36 | −23 39 31 | 5.11.90 | 18 29 12 | −23 35 27 |
| 17. 3.83 | 16 29 43 | −21 42 47 | 14. 2.85 | 17 05 11 | −22 51 12 | 15. 1.87 | 17 35 34 | −23 26 23 | 15.12.88 | 18 03 11 | −23 39 27 | 15.11.90 | 18 31 06 | −23 34 02 |
| 27. 3.83 | 16 29 26 | −21 42 14 | 24. 2.85 | 17 06 22 | −22 52 51 | 25. 1.87 | 17 37 52 | −23 27 47 | 25.12.88 | 18 05 49 | −23 39 15 | 25.11.90 | 18 33 14 | −23 32 22 |
| 6. 4.83 | 16 28 48 | −21 40 54 | 6. 3.85 | 17 07 12 | −22 54 01 | 4. 2.87 | 17 39 56 | −23 28 58 | **4. 1.89** | 18 08 26 | −23 38 54 | 5.12.90 | 18 35 35 | −23 30 27 |
| 16. 4.83 | 16 27 50 | −21 38 52 | 16. 3.85 | 17 07 40 | −22 54 42 | 14. 2.87 | 17 41 45 | −23 29 57 | 14. 1.89 | 18 11 00 | −23 38 26 | 15.12.90 | 18 38 03 | −23 28 18 |
| 26. 4.83 | 16 26 36 | −21 36 11 | 26. 3.85 | 17 07 44 | −22 54 54 | 24. 2.87 | 17 43 15 | −23 30 44 | 24. 1.89 | 18 13 26 | −23 37 52 | 25.12.90 | 18 40 37 | −23 25 56 |
| 6. 5.83 | 16 25 07 | −21 32 57 | 5. 4.85 | 17 07 26 | −22 54 38 | 6. 3.87 | 17 44 25 | −23 31 20 | 3. 2.89 | 18 15 41 | −23 37 14 | **4. 1.91** | 18 43 14 | −23 23 24 |
| 16. 5.83 | 16 23 29 | −21 29 18 | 15. 4.85 | 17 06 47 | −22 53 54 | 16. 3.87 | 17 45 13 | −23 31 48 | 13. 2.89 | 18 17 43 | −23 36 35 | 14. 1.91 | 18 45 49 | −23 20 46 |
| 26. 5.83 | 16 21 45 | −21 25 22 | 25. 4.85 | 17 05 48 | −22 52 45 | 26. 3.87 | 17 45 39 | −23 32 06 | 23. 2.89 | 18 19 29 | −23 36 00 | 24. 1.91 | 18 48 19 | −23 18 05 |
| 5. 6.83 | 16 20 00 | −21 21 21 | 5. 5.85 | 17 04 32 | −22 51 12 | 5. 4.87 | 17 45 42 | −23 32 17 | 5. 3.89 | 18 20 57 | −23 35 29 | 3. 2.91 | 18 50 42 | −23 15 26 |
| 15. 6.83 | 16 18 18 | −21 17 22 | 15. 5.85 | 17 03 02 | −22 49 18 | 15. 4.87 | 17 45 22 | −23 32 20 | 15. 3.89 | 18 22 05 | −23 35 07 | 13. 2.91 | 18 52 54 | −23 12 53 |
| 25. 6.83 | 16 16 43 | −21 13 38 | 25. 5.85 | 17 01 23 | −22 47 07 | 25. 4.87 | 17 44 41 | −23 32 15 | 25. 3.89 | 18 22 50 | −23 34 55 | 23. 2.91 | 18 54 53 | −23 10 33 |
| 5. 7.83 | 16 15 20 | −21 10 20 | 4. 6.85 | 16 59 37 | −22 44 44 | 5. 5.87 | 17 43 41 | −23 32 01 | 4. 4.89 | 18 23 14 | −23 34 55 | 5. 3.91 | 18 56 36 | −23 08 31 |
| 15. 7.83 | 16 14 11 | −21 07 35 | 14. 6.85 | 16 57 51 | −22 42 15 | 15. 5.87 | 17 42 24 | −23 31 38 | 14. 4.89 | 18 23 15 | −23 35 08 | 15. 3.91 | 18 58 00 | −23 06 50 |
| 25. 7.83 | 16 13 19 | −21 05 33 | 24. 6.85 | 16 56 08 | −22 39 46 | 25. 5.87 | 17 40 53 | −23 31 07 | 24. 4.89 | 18 22 54 | −23 35 32 | 25. 3.91 | 18 59 04 | −23 05 35 |
| 4. 8.83 | 16 12 47 | −21 04 19 | 4. 7.85 | 16 54 33 | −22 37 24 | 4. 6.87 | 17 39 13 | −23 30 27 | 4. 5.89 | 18 22 11 | −23 36 07 | 4. 4.91 | 18 59 47 | −23 04 50 |
| 14. 8.83 | 16 12 35 | −21 03 59 | 14. 7.85 | 16 53 08 | −22 35 17 | 14. 6.87 | 17 37 28 | −23 29 38 | 14. 5.89 | 18 21 10 | −23 36 50 | 14. 4.91 | 19 00 09 | −23 04 36 |
| 24. 8.83 | 16 12 46 | −21 04 33 | 24. 7.85 | 16 51 59 | −22 33 31 | 24. 6.87 | 17 35 41 | −23 28 44 | 24. 5.89 | 18 19 53 | −23 37 39 | 24. 4.91 | 19 00 07 | −23 04 53 |
| 3. 9.83 | 16 13 17 | −21 06 03 | 3. 8.85 | 16 51 07 | −22 32 11 | 4. 7.87 | 17 33 58 | −23 27 47 | 3. 6.89 | 18 18 22 | −23 38 29 | 4. 5.91 | 18 59 45 | −23 05 41 |
| 13. 9.83 | 16 14 10 | −21 08 26 | 13. 8.85 | 16 50 35 | −22 31 23 | 14. 7.87 | 17 32 22 | −23 26 50 | 13. 6.89 | 18 16 42 | −23 39 19 | 14. 5.91 | 18 59 01 | −23 06 57 |
| 23. 9.83 | 16 15 24 | −21 11 38 | 23. 8.85 | 16 50 24 | −22 31 09 | 24. 7.87 | 17 30 58 | −23 25 56 | 23. 6.89 | 18 14 57 | −23 40 06 | 24. 5.91 | 18 58 00 | −23 08 36 |
| 3.10.83 | 16 16 56 | −21 15 34 | 2. 9.85 | 16 50 34 | −22 31 32 | 3. 8.87 | 17 29 48 | −23 25 10 | 3. 7.89 | 18 13 10 | −23 40 46 | 3. 6.91 | 18 56 42 | −23 10 34 |
| 13.10.83 | 16 18 46 | −21 20 08 | 12. 9.85 | 16 51 07 | −22 32 31 | 13. 8.87 | 17 28 57 | −23 24 34 | 13. 7.89 | 18 11 28 | −23 41 20 | 13. 6.91 | 18 55 12 | −23 12 45 |
| 23.10.83 | 16 20 50 | −21 25 12 | 22. 9.85 | 16 52 02 | 22 34 05 | 23. 8.87 | 17 28 25 | −23 24 12 | 23. 7.89 | 18 09 53 | −23 41 46 | 23. 6.91 | 18 53 33 | −23 15 02 |
| 2.11.83 | 16 23 07 | −21 30 38 | 2.10.85 | 16 53 17 | −22 36 10 | 2. 9.87 | 17 28 15 | −23 24 05 | 2. 8.89 | 18 08 29 | −23 42 05 | 3. 7.91 | 18 51 49 | −23 17 19 |
| 12.11.83 | 16 25 33 | −21 36 18 | 12.10.85 | 16 54 52 | −22 38 43 | 12. 9.87 | 17 28 27 | −23 24 15 | 12. 8.89 | 18 07 21 | −23 42 18 | 13. 7.91 | 18 50 04 | −23 19 31 |
| 22.11.83 | 16 28 06 | −21 42 03 | 22.10.85 | 16 56 43 | −22 41 39 | 22. 9.87 | 17 29 01 | −23 24 41 | 22. 8.89 | 18 06 30 | −23 42 25 | 23. 7.91 | 18 48 23 | −23 21 31 |
| 2.12.83 | 16 30 43 | −21 47 46 | 1.11.85 | 16 58 50 | −22 44 51 | 2.10.87 | 17 29 57 | −23 25 22 | 1. 9.89 | 18 06 00 | −23 42 27 | 2. 8.91 | 18 46 49 | −23 23 17 |
| 12.12.83 | 16 33 19 | −21 53 21 | 11.11.85 | 17 01 09 | −22 48 14 | 12.10.87 | 17 31 14 | −23 26 17 | 11. 9.89 | 18 05 51 | −23 42 25 | 12. 8.91 | 18 45 28 | −23 24 46 |
| 22.12.83 | 16 35 53 | −21 58 39 | 21.11.85 | 17 03 37 | −22 51 43 | 22.10.87 | 17 32 50 | −23 27 22 | 21. 9.89 | 18 06 04 | −23 42 20 | 22. 8.91 | 18 44 21 | −23 25 53 |
| **1. 1.84** | 16 38 20 | −22 03 36 | 1.12.85 | 17 06 12 | 22 55 12 | 1.11.87 | 17 34 13 | −23 28 31 | 1.10.89 | 18 06 30 | −23 42 12 | 1. 0.01 | 18 43 32 | 23 26 40 |
| 11. 1.84 | 16 40 38 | −22 08 06 | 11.12.85 | 17 08 50 | −22 58 35 | 11.11.87 | 17 36 51 | −23 29 50 | 11.10.89 | 18 07 37 | −23 41 59 | 11. 9.91 | 18 43 03 | −23 27 04 |
| 21. 1.84 | 16 42 44 | −22 12 05 | 21.12.85 | 17 11 28 | −23 01 49 | 21.11.87 | 17 39 11 | −23 31 06 | 21.10.89 | 18 08 55 | −23 41 40 | 21. 9.91 | 18 42 56 | −23 27 05 |
| 31. 1.84 | 16 44 34 | −22 15 30 | 31.12.85 | 17 14 03 | −23 04 49 | 1.12.87 | 17 41 40 | −23 32 19 | 31.10.89 | 18 10 32 | −23 41 16 | 1.10.91 | 18 43 10 | −23 26 42 |
| 10. 2.84 | 16 46 06 | −22 18 19 | **10. 1.86** | 17 16 31 | −23 07 33 | 11.12.87 | 17 44 16 | −23 33 26 | 10.11.89 | 18 12 26 | −23 40 43 | 11.10.91 | 18 43 47 | −23 25 57 |
| 20. 2.84 | 16 47 19 | −22 20 30 | 20. 1.86 | 17 18 49 | −23 10 00 | 21.12.87 | 17 46 54 | −23 34 26 | 20.11.89 | 18 14 34 | −23 40 01 | 21.10.91 | 18 44 46 | −23 24 49 |
| 1. 3.84 | 16 48 09 | −22 22 03 | 30. 1.86 | 17 20 54 | −23 12 07 | 31.12.87 | 17 49 32 | −23 35 16 | 30.11.89 | 18 16 55 | −23 39 09 | 31.10.91 | 18 46 04 | −23 23 18 |
| 11. 3.84 | 16 48 38 | −22 22 55 | 9. 2.86 | 17 22 43 | −23 13 54 | **10. 1.88** | 17 52 06 | −23 35 57 | 10.12.89 | 18 19 24 | −23 38 06 | 10.11.91 | 18 47 42 | −23 21 23 |
| 21. 3.84 | 16 48 43 | −22 23 08 | 19. 2.86 | 17 24 14 | −23 15 21 | 20. 1.88 | 17 54 33 | −23 36 27 | 20.12.89 | 18 21 59 | −23 36 53 | 20.11.91 | 18 49 36 | −23 19 07 |
| 31. 3.84 | 16 48 25 | −22 22 44 | 1. 3.86 | 17 25 25 | −23 16 28 | 30. 1.88 | 17 56 50 | −23 36 50 | 30.12.89 | 18 24 36 | −23 35 30 | 30.11.91 | 18 51 44 | −23 16 30 |
| 10. 4.84 | 16 47 47 | −22 21 41 | 11. 3.86 | 17 26 15 | −23 17 17 | 9. 2.88 | 17 58 53 | −23 37 06 | **9. 1.90** | 18 27 13 | −23 34 00 | 10.12.91 | 18 54 04 | −23 13 33 |
| 20. 4.84 | 16 46 48 | −22 20 05 | 21. 3.86 | 17 26 41 | −23 17 46 | 19. 2.88 | 18 00 41 | −23 37 16 | 19. 1.90 | 18 29 45 | −23 32 24 | 20.12.91 | 18 56 32 | −23 10 19 |
| 30. 4.84 | 16 45 33 | −22 17 58 | 31. 3.86 | 17 26 45 | −23 17 57 | 29. 2.88 | 18 02 10 | −23 37 24 | 29. 1.90 | 18 32 09 | −23 30 46 | 30.12.91 | 18 59 05 | −23 06 51 |
| 10. 5.84 | 16 44 04 | −22 15 23 | 10. 4.86 | 17 26 26 | −23 17 50 | 10. 3.88 | 18 03 19 | −23 37 30 | 8. 2.90 | 18 34 23 | −23 29 10 | **9. 1.92** | 19 01 40 | −23 03 12 |
| 20. 5.84 | 16 42 25 | −22 12 28 | 20. 4.86 | 17 25 46 | −23 17 26 | 20. 3.88 | 18 04 06 | −23 37 37 | 18. 2.90 | 18 36 24 | −23 27 40 | 19. 1.92 | 19 04 13 | −22 59 29 |
| 30. 5.84 | 16 40 40 | −22 09 18 | 30. 4.86 | 17 24 46 | −23 16 44 | 30. 3.88 | 18 04 31 | −23 37 46 | 28. 2.90 | 18 38 08 | −23 26 19 | 29. 1.92 | 19 06 42 | −22 55 45 |
| 9. 6.84 | 16 38 55 | −22 06 02 | 10. 5.86 | 17 23 29 | −23 15 46 | 9. 4.88 | 18 04 32 | −23 37 57 | 10. 3.90 | 18 39 34 | −23 25 13 | 8. 2.92 | 19 09 03 | −22 52 07 |
| 19. 6.84 | 16 37 12 | −22 02 48 | 20. 5.86 | 17 21 59 | −23 14 33 | 19. 4.88 | 18 04 12 | −23 38 11 | 20. 3.90 | 18 40 40 | −23 24 23 | 18. 2.92 | 19 11 13 | −22 48 42 |
| 29. 6.84 | 16 35 37 | −21 59 45 | 30. 5.86 | 17 20 20 | −23 13 07 | 29. 4.88 | 18 03 30 | −23 38 26 | 30. 3.90 | 18 41 25 | −23 23 54 | 28. 2.92 | 19 13 10 | −22 45 34 |
| 9. 7.84 | 16 34 13 | −21 57 01 | 9. 6.86 | 17 18 34 | −23 11 32 | 9. 5.88 | 18 02 30 | −23 38 40 | 9. 4.90 | 18 41 47 | −23 23 47 | 9. 3.92 | 19 14 51 | −22 42 52 |
| 19. 7.84 | 16 33 04 | −21 54 46 | 19. 6.86 | 17 16 47 | −23 09 50 | 19. 5.88 | 18 01 12 | −23 38 53 | 19. 4.90 | 18 41 47 | −23 24 01 | 19. 3.92 | 19 16 13 | −22 40 39 |
| 29. 7.84 | 16 32 12 | −21 53 04 | 29. 6.86 | 17 15 04 | −23 08 07 | 29. 5.88 | 17 59 42 | −23 39 03 | 29. 4.90 | 18 41 25 | −23 24 37 | 29. 3.92 | 19 17 16 | −22 39 01 |
| 8. 8.84 | 16 31 39 | −21 52 03 | 9. 7.86 | 17 13 28 | −23 06 27 | 8. 6.88 | 17 58 01 | −23 39 08 | 9. 5.90 | 18 40 42 | −23 25 32 | 8. 4.92 | 19 17 57 | −22 38 02 |
| 18. 8.84 | 16 31 28 | −21 51 46 | 19. 7.86 | 17 12 04 | −23 04 57 | 18. 6.88 | 17 56 16 | −23 39 07 | 19. 5.90 | 18 39 41 | −23 26 44 | 18. 4.92 | 19 18 17 | −22 37 43 |
| 28. 8.84 | 16 31 38 | −21 52 15 | 29. 7.86 | 17 10 55 | −23 03 40 | 28. 6.88 | 17 54 29 | −23 39 01 | 29. 5.90 | 18 38 23 | −23 28 07 | 28. 4.92 | 19 18 15 | −22 38 04 |
| 7. 9.84 | 16 32 11 | −21 53 30 | 8. 8.86 | 17 10 03 | −23 02 43 | 8. 7.88 | 17 52 46 | −23 38 50 | 8. 6.90 | 18 36 53 | −23 29 38 | 8. 5.92 | 19 17 51 | −22 39 04 |
| 17. 9.84 | 16 33 05 | −21 55 29 | 18. 8.86 | 17 09 31 | −23 02 07 | 18. 7.88 | 17 51 11 | −23 38 34 | 18. 6.90 | 18 35 13 | −23 31 12 | 18. 5.92 | 19 17 08 | −22 40 40 |
| 27. 9.84 | 16 34 19 | −21 58 08 | 28. 8.86 | 17 09 20 | −23 01 57 | 28. 7.88 | 17 49 47 | −23 38 17 | 28. 6.90 | 18 33 28 | −23 32 45 | 28. 5.92 | 19 16 06 | −22 42 47 |
| 7.10.84 | 16 35 53 | −22 01 23 | 7. 9.86 | 17 09 31 | −23 02 14 | 7. 8.88 | 17 48 38 | −23 38 01 | 8. 7.90 | 18 31 42 | −23 34 11 | 7. 6.92 | 19 14 48 | −22 45 18 |
| 17.10.84 | 16 37 43 | −22 05 08 | 17. 9.86 | 17 10 05 | −23 02 57 | 17. 8.88 | 17 47 47 | −23 37 47 | 18. 7.90 | 18 30 00 | −23 35 29 | 17. 6.92 | 19 13 19 | −22 48 07 |
| 27.10.84 | 16 39 49 | −22 09 17 | 27. 9.86 | 17 11 00 | −23 04 04 | 27. 8.88 | 17 47 15 | −23 37 37 | 28. 7.90 | 18 28 26 | −23 36 36 | 27. 6.92 | 19 11 40 | −22 51 06 |
| 6.11.84 | 16 42 07 | −22 13 43 | 7.10.86 | 17 12 16 | −23 05 35 | 6. 9.88 | 17 47 06 | −23 37 33 | 7. 8.90 | 18 27 03 | −23 37 30 | 7. 7.92 | 19 09 57 | −22 54 07 |
| 16.11.84 | 16 44 34 | −22 18 18 | 17.10.86 | 17 13 52 | −23 07 24 | 16. 9.88 | 17 47 18 | −23 37 35 | 17. 8.90 | 18 25 56 | −23 38 11 | 17. 7.92 | 19 08 13 | −22 57 01 |
| 26.11.84 | 16 47 08 | −22 22 56 | 27.10.86 | 17 15 44 | −23 09 28 | 26. 9.88 | 17 47 53 | −23 37 44 | 27. 8.90 | 18 25 06 | −23 38 38 | 27. 7.92 | 19 06 33 | −22 59 43 |

# Tables VI

**Table 23: Position of Neptune 1983–1992**

| Date d mo yr | R.A. h m s | Dec ° ′ ″ | Date d mo yr | R.A. h m s | Dec ° ′ ″ | Date d mo yr | R.A. h m s | Dec ° ′ ″ | Date d mo yr | R.A. h m s | Dec ° ′ ″ | Date d mo yr | R.A. h m s | Dec ° ′ ″ |
|---|---|---|---|---|---|---|---|---|---|---|---|---|---|---|
| 6. 1.83 | 17 48 54 | −22 12 26 | 6.12.84 | 18 02 12 | −22 19 27 | 6.11.86 | 18 16 17 | −22 21 34 | 6.10.88 | 18 32 26 | −22 17 20 | 6. 9.90 | 18 51 21 | −22 03 54 |
| 16. 1.83 | 17 50 26 | −22 12 44 | 16.12.84 | 18 03 48 | −22 19 29 | 16.11.86 | 18 17 31 | −22 21 24 | 16.10.88 | 18 32 58 | −22 17 19 | 16. 9.90 | 18 51 03 | −22 04 37 |
| 26. 1.83 | 17 51 52 | −22 12 55 | 26.12.84 | 18 05 27 | −22 19 24 | 26.11.86 | 18 18 54 | −22 21 04 | 26.10.88 | 18 33 43 | −22 17 06 | 26. 9.90 | 18 50 59 | −22 05 04 |
| 5. 2.83 | 17 53 09 | −22 12 57 | **5. 1.85** | 18 07 04 | −22 19 11 | 6.12.86 | 18 20 24 | −22 20 36 | 5.11.88 | 18 34 41 | −22 16 40 | 6.10.90 | 18 51 10 | −22 05 15 |
| 15. 2.83 | 17 54 17 | −22 12 53 | 15. 1.85 | 18 08 38 | −22 18 52 | 16.12.86 | 18 21 59 | −22 19 58 | 15.11.88 | 18 35 50 | −22 16 00 | 16.10.90 | 18 51 34 | −22 05 07 |
| 25. 2.83 | 17 55 13 | −22 12 43 | 25. 1.85 | 18 10 08 | −22 18 27 | 26.12.86 | 18 23 37 | −22 19 12 | 25.11.88 | 18 37 09 | −22 15 08 | 26.10.90 | 18 52 13 | −22 04 40 |
| 7. 3.83 | 17 55 55 | −22 12 29 | 4. 2.85 | 18 11 29 | −22 17 57 | **5. 1.87** | 18 25 15 | −22 18 18 | 5.12.88 | 18 38 36 | −22 14 04 | 5.11.90 | 18 53 05 | −22 03 55 |
| 17. 3.83 | 17 56 25 | −22 12 12 | 14. 2.85 | 18 12 42 | −22 17 24 | 15. 1.87 | 18 26 51 | −22 17 19 | 15.12.88 | 18 40 08 | −22 12 48 | 15.11.90 | 18 54 08 | −22 02 52 |
| 27. 3.83 | 17 56 39 | −22 11 53 | 24. 2.85 | 18 13 44 | −22 16 51 | 25. 1.87 | 18 28 23 | −22 16 15 | 25.12.88 | 18 41 45 | −22 11 22 | 25.11.90 | 18 55 22 | −22 01 32 |
| 6. 4.83 | 17 56 39 | −22 11 33 | 6. 3.85 | 18 14 33 | −22 16 19 | 4. 2.87 | 18 29 48 | −22 15 10 | **4. 1.89** | 18 43 23 | −22 09 48 | 5.12.90 | 18 56 45 | −21 59 55 |
| 16. 4.83 | 17 56 26 | −22 11 12 | 16. 3.85 | 18 15 09 | −22 15 49 | 14. 2.87 | 18 31 05 | −22 14 05 | 14. 1.89 | 18 45 00 | −22 08 08 | 15.12.90 | 18 58 15 | −21 58 03 |
| 26. 4.83 | 17 55 58 | −22 10 52 | 26. 3.85 | 18 15 31 | −22 15 23 | 24. 2.87 | 18 32 12 | −22 13 04 | 24. 1.89 | 18 46 34 | −22 06 24 | 25.12.90 | 18 59 50 | −21 55 59 |
| 6. 5.83 | 17 55 19 | −22 10 33 | 5. 4.85 | 18 15 38 | −22 15 02 | 6. 3.87 | 18 33 08 | −22 12 07 | 3. 2.89 | 18 48 02 | −22 04 41 | **4. 1.91** | 19 01 27 | −21 53 45 |
| 16. 5.83 | 17 54 29 | −22 10 15 | 15. 4.85 | 18 15 31 | −22 14 47 | 16. 3.87 | 18 33 50 | −22 11 19 | 13. 2.89 | 18 49 23 | −22 03 00 | 14. 1.91 | 19 03 04 | −21 51 24 |
| 26. 5.83 | 17 53 31 | −22 09 58 | 25. 4.85 | 18 15 10 | −22 14 38 | 26. 3.87 | 18 34 19 | −22 10 40 | 23. 2.89 | 18 50 36 | −22 01 26 | 24. 1.91 | 19 04 39 | −21 49 01 |
| 5. 6.83 | 17 52 26 | −22 09 43 | 5. 5.85 | 18 14 37 | −22 14 36 | 5. 4.87 | 18 34 33 | −22 10 12 | 5. 3.89 | 18 51 36 | −22 00 02 | 3. 2.91 | 19 06 10 | −21 46 38 |
| 15. 6.83 | 17 51 18 | −22 09 30 | 15. 5.85 | 18 13 52 | −22 14 39 | 15. 4.87 | 18 34 33 | −22 09 57 | 15. 3.89 | 18 52 25 | −21 58 49 | 13. 2.91 | 19 07 35 | −21 44 19 |
| 25. 6.83 | 17 50 08 | −22 09 19 | 25. 5.85 | 18 12 58 | −22 14 48 | 25. 4.87 | 18 34 19 | −22 09 54 | 25. 3.89 | 18 53 00 | −21 57 52 | 23. 2.91 | 19 08 51 | −21 42 10 |
| 5. 7.83 | 17 49 00 | −22 09 11 | 4. 6.85 | 18 11 56 | −22 15 02 | 5. 5.87 | 18 33 52 | −22 10 03 | 4. 4.89 | 18 53 21 | −21 57 12 | 5. 3.91 | 19 09 57 | −21 40 13 |
| 15. 7.83 | 17 47 55 | −22 09 07 | 14. 6.85 | 18 10 49 | −22 15 19 | 15. 5.87 | 18 33 13 | −22 10 23 | 14. 4.89 | 18 53 28 | −21 56 50 | 15. 3.91 | 19 10 51 | −21 38 33 |
| 25. 7.83 | 17 46 57 | −22 09 08 | 24. 6.85 | 18 09 39 | −22 15 39 | 25. 5.87 | 18 32 23 | −22 10 54 | 24. 4.89 | 18 53 21 | −21 56 46 | 25. 3.91 | 19 11 33 | −21 37 12 |
| 4. 8.83 | 17 46 07 | −22 09 13 | 4. 7.85 | 18 08 30 | −22 16 02 | 4. 6.87 | 18 31 25 | −22 11 33 | 4. 5.89 | 18 53 00 | −21 57 01 | 4. 4.91 | 19 12 01 | −21 36 13 |
| 14. 8.83 | 17 45 27 | −22 09 24 | 14. 7.85 | 18 07 23 | −22 16 26 | 14. 6.87 | 18 30 20 | −22 12 19 | 14. 5.89 | 18 52 27 | −21 57 33 | 14. 4.91 | 19 12 15 | −21 35 38 |
| 24. 8.83 | 17 45 00 | −22 09 41 | 24. 7.85 | 18 06 21 | −22 16 52 | 24. 6.87 | 18 29 12 | −22 13 10 | 24. 5.89 | 18 51 43 | −21 58 20 | 24. 4.91 | 19 12 15 | −21 35 28 |
| 3. 9.83 | 17 44 46 | −22 10 04 | 3. 8.85 | 18 05 27 | −22 17 19 | 4. 7.87 | 18 28 02 | −22 14 04 | 3. 6.89 | 18 50 49 | −21 59 22 | 4. 5.91 | 19 12 01 | −21 35 42 |
| 13. 9.83 | 17 44 45 | −22 10 34 | 13. 8.85 | 18 04 42 | −22 17 46 | 14. 7.87 | 18 26 53 | −22 14 58 | 13. 6.89 | 18 49 47 | −22 00 34 | 14. 5.91 | 19 11 34 | −21 36 20 |
| 23. 9.83 | 17 44 59 | −22 11 08 | 23. 8.85 | 18 04 08 | −22 18 14 | 24. 7.87 | 18 25 49 | −22 15 52 | 23. 6.89 | 18 48 40 | −22 01 53 | 24. 5.91 | 19 10 55 | −21 37 19 |
| 3.10.83 | 17 45 26 | −22 11 47 | 2. 9.85 | 18 03 47 | −22 18 43 | 3. 8.87 | 18 24 50 | −22 16 44 | 3. 7.89 | 18 47 31 | −22 03 17 | 3. 6.91 | 19 10 06 | −21 38 38 |
| 13.10.83 | 17 46 08 | −22 12 29 | 12. 9.85 | 18 03 40 | −22 19 10 | 13. 8.87 | 18 24 00 | −22 17 33 | 13. 7.89 | 18 46 22 | −22 04 43 | 13. 6.91 | 19 09 08 | −21 40 12 |
| 23.10.83 | 17 47 02 | −22 13 12 | 22. 9.85 | 18 03 46 | −22 19 38 | 23. 8.87 | 18 23 21 | −22 18 17 | 23. 7.89 | 18 45 15 | −22 06 08 | 23. 6.91 | 19 08 04 | −21 41 57 |
| 2.11.83 | 17 48 08 | −22 13 56 | 2.10.85 | 18 04 07 | −22 20 03 | 2. 9.87 | 18 22 53 | −22 18 57 | 2. 8.89 | 18 44 13 | −22 07 28 | 3. 7.91 | 19 06 56 | −21 43 51 |
| 12.11.83 | 17 49 24 | −22 14 37 | 12.10.85 | 18 04 42 | −22 20 26 | 12. 9.87 | 18 22 39 | −22 19 30 | 12. 8.89 | 18 43 19 | −22 08 42 | 13. 7.91 | 19 05 47 | −21 45 47 |
| 22.11.83 | 17 50 48 | −22 15 14 | 22.10.85 | 18 05 30 | −22 20 46 | 22. 9.87 | 18 22 39 | −22 19 56 | 22. 8.89 | 18 42 34 | −22 09 48 | 23. 7.91 | 19 04 39 | −21 47 42 |
| 2.12.83 | 17 52 19 | −22 15 47 | 1.11.85 | 18 06 31 | −22 21 00 | 2.10.87 | 18 22 53 | −22 20 14 | 1. 9.89 | 18 42 01 | −22 10 43 | 2. 8.91 | 19 03 35 | −21 49 33 |
| 12.12.83 | 17 53 55 | −22 16 13 | 11.11.85 | 18 07 42 | −22 21 09 | 12.10.87 | 18 23 21 | −22 20 23 | 11. 9.89 | 18 41 40 | −22 11 27 | 12. 8.91 | 19 02 37 | −21 51 16 |
| 22.12.83 | 17 55 33 | −22 16 31 | 21.11.85 | 18 09 03 | −22 21 11 | 22.10.87 | 18 24 02 | −22 20 24 | 21. 9.89 | 18 41 32 | −22 11 58 | 22. 8.91 | 19 01 47 | −21 52 46 |
| **1. 1.84** | 17 57 11 | −22 16 43 | 1.12.85 | 18 10 32 | −22 21 05 | 1.11.87 | 18 24 57 | −22 20 15 | 1.10.89 | 18 41 39 | −22 12 15 | 1. 9.91 | 19 01 08 | −21 54 03 |
| 11. 1.84 | 17 58 46 | −22 16 47 | 11.12.85 | 18 12 05 | −22 20 51 | 11.11.87 | 18 26 03 | −22 19 54 | 11.10.89 | 18 42 00 | −22 12 17 | 11. 9.91 | 19 00 41 | −21 55 03 |
| 21. 1.84 | 18 00 17 | −22 16 43 | 21.12.85 | 18 13 43 | −22 20 29 | 21.11.87 | 18 27 20 | −22 19 24 | 21.10.89 | 18 42 35 | −22 12 04 | 21. 9.91 | 19 00 27 | −21 55 45 |
| 31. 1.84 | 18 01 41 | −22 16 32 | 31.12.85 | 18 15 21 | −22 20 00 | 1.12.87 | 18 28 45 | −22 18 42 | 31.10.89 | 18 43 24 | −22 11 36 | 1.10.91 | 19 00 26 | −21 56 07 |
| 10. 2.84 | 18 02 56 | −22 16 17 | **10. 1.86** | 18 16 58 | −22 19 23 | 11.12.87 | 18 30 16 | −22 17 50 | 10.11.89 | 18 44 25 | −22 10 52 | 11.10.91 | 19 00 41 | −21 56 08 |
| 20. 2.84 | 18 04 00 | −22 15 58 | 20. 1.86 | 18 18 31 | −22 18 41 | 21.12.87 | 18 31 52 | −22 16 48 | 20.11.89 | 18 45 36 | −22 09 53 | 21.10.91 | 19 01 09 | −21 55 48 |
| 1. 3.84 | 18 04 53 | −22 15 36 | 30. 1.86 | 18 19 58 | −22 17 56 | 31.12.87 | 18 33 30 | −22 15 37 | 30.11.89 | 18 46 57 | −22 08 38 | 31.10.91 | 19 01 51 | −21 55 06 |
| 11. 3.84 | 18 05 32 | −22 15 13 | 9. 2.86 | 18 21 18 | −22 17 08 | **10. 1.88** | 18 35 08 | −22 14 21 | 10.12.89 | 18 48 26 | −22 07 11 | 10.11.91 | 19 02 45 | −21 54 04 |
| 21. 3.84 | 18 05 58 | −22 14 50 | 19. 2.86 | 18 22 28 | −22 16 20 | 20. 1.88 | 18 36 43 | −22 12 59 | 20.12.89 | 18 49 59 | −22 05 31 | 20.11.91 | 19 03 52 | −21 52 41 |
| 31. 3.84 | 18 06 09 | −22 14 29 | 1. 3.86 | 18 23 26 | −22 15 34 | 30. 1.88 | 18 38 13 | −22 11 35 | 30.12.89 | 18 51 36 | −22 03 40 | 30.11.91 | 19 05 08 | −21 50 58 |
| 10. 4.84 | 18 06 05 | −22 14 11 | 11. 3.86 | 18 24 12 | −22 14 53 | 9. 2.88 | 18 39 36 | −22 10 11 | **9. 1.90** | 18 53 14 | −22 01 43 | 10.12.91 | 19 06 33 | −21 48 58 |
| 20. 4.84 | 18 05 48 | −22 13 55 | 21. 3.86 | 18 24 44 | −22 14 18 | 19. 2.88 | 18 40 51 | −22 08 51 | 19. 1.90 | 18 54 50 | −21 59 40 | 20.12.91 | 19 08 04 | −21 46 42 |
| 30. 4.84 | 18 05 18 | −22 13 44 | 31. 3.86 | 18 25 02 | −22 13 50 | 29. 2.88 | 18 41 55 | −22 07 37 | 29. 1.90 | 18 56 23 | −21 57 37 | 30.12.91 | 19 09 39 | −21 44 14 |
| 10. 5.84 | 18 04 36 | −22 13 35 | 10. 4.86 | 18 25 06 | −22 13 31 | 10. 3.88 | 18 42 47 | −22 06 32 | 8. 2.90 | 18 57 49 | −21 55 35 | **9. 1.92** | 19 11 16 | −21 41 37 |
| 20. 5.84 | 18 03 43 | −22 13 30 | 20. 4.86 | 18 24 56 | −22 13 21 | 20. 3.88 | 18 43 26 | −22 05 38 | 18. 2.90 | 18 59 08 | −21 53 39 | 19. 1.92 | 19 12 53 | −21 38 54 |
| 30. 5.84 | 18 02 43 | −22 13 28 | 30. 4.86 | 18 24 32 | −22 13 20 | 30. 3.88 | 18 43 51 | −22 04 58 | 28. 2.90 | 19 00 17 | −21 51 53 | 29. 1.92 | 19 14 27 | −21 36 10 |
| 9. 6.84 | 18 01 37 | −22 13 29 | 10. 5.86 | 18 23 55 | −22 13 29 | 9. 4.88 | 18 44 02 | −22 04 32 | 10. 3.90 | 19 01 15 | −21 50 20 | 8. 2.92 | 19 15 56 | −21 33 29 |
| 19. 6.84 | 18 00 28 | −22 13 33 | 20. 5.86 | 18 23 08 | −22 13 46 | 19. 4.88 | 18 43 58 | −22 04 21 | 20. 3.90 | 19 02 00 | −21 49 02 | 18. 2.92 | 19 17 19 | −21 30 55 |
| 29. 6.84 | 17 59 19 | −22 13 38 | 30. 5.86 | 18 22 12 | −22 14 09 | 29. 4.88 | 18 43 41 | −22 04 27 | 30. 3.90 | 19 02 32 | −21 48 03 | 28. 2.92 | 19 18 32 | −21 28 33 |
| 9. 7.84 | 17 58 11 | −22 13 47 | 9. 6.86 | 18 21 08 | −22 14 39 | 9. 5.88 | 18 43 10 | −22 04 46 | 9. 4.90 | 19 02 49 | −21 47 25 | 9. 3.92 | 19 19 35 | −21 26 28 |
| 19. 7.84 | 17 57 08 | −22 13 58 | 19. 6.86 | 18 20 01 | −22 15 13 | 19. 5.88 | 18 42 28 | −22 05 20 | 19. 4.90 | 19 02 53 | −21 47 08 | 19. 3.92 | 19 20 26 | −21 24 42 |
| 29. 7.84 | 17 56 11 | −22 14 12 | 29. 6.86 | 18 18 51 | −22 15 50 | 29. 5.88 | 18 41 36 | −22 06 06 | 29. 4.90 | 19 02 42 | −21 47 12 | 29. 3.92 | 19 21 04 | −21 23 20 |
| 8. 8.84 | 17 55 24 | −22 14 29 | 9. 7.86 | 18 17 42 | −22 16 29 | 8. 6.88 | 18 40 36 | −22 07 01 | 9. 5.90 | 19 02 18 | −21 47 38 | 8. 4.92 | 19 21 29 | −21 22 23 |
| 18. 8.84 | 17 54 47 | −22 14 50 | 19. 7.86 | 18 16 36 | −22 17 08 | 18. 6.88 | 18 39 31 | −22 08 04 | 19. 5.90 | 19 01 42 | −21 48 23 | 18. 4.92 | 19 21 39 | −21 21 53 |
| 28. 8.84 | 17 54 23 | −22 15 13 | 29. 7.86 | 18 15 36 | −22 17 48 | 28. 6.88 | 18 38 22 | −22 09 11 | 29. 5.90 | 19 00 55 | −21 49 26 | 28. 4.92 | 19 21 36 | −21 21 51 |
| 7. 9.84 | 17 54 12 | −22 15 39 | 8. 8.86 | 18 14 43 | −22 18 26 | 8. 7.88 | 18 37 12 | −22 10 21 | 8. 6.90 | 18 59 59 | −21 50 43 | 8. 5.92 | 19 21 18 | −21 22 17 |
| 17. 9.84 | 17 54 15 | −22 16 08 | 18. 8.86 | 18 14 01 | −22 19 02 | 18. 7.88 | 18 36 04 | −22 11 31 | 18. 6.90 | 18 58 56 | −21 52 12 | 18. 5.92 | 19 20 48 | −21 23 08 |
| 27. 9.84 | 17 54 32 | −22 16 40 | 28. 8.86 | 18 13 31 | −22 19 37 | 28. 7.88 | 18 35 01 | −22 12 39 | 28. 6.90 | 18 57 49 | −21 53 49 | 28. 5.92 | 19 20 07 | −21 24 22 |
| 7.10.84 | 17 55 03 | −22 17 11 | 7. 9.86 | 18 13 13 | −22 20 08 | 7. 8.88 | 18 34 05 | −22 13 42 | 8. 7.90 | 18 56 39 | −21 55 29 | 7. 6.92 | 19 19 15 | −21 25 57 |
| 17.10.84 | 17 55 48 | −22 17 43 | 17. 9.86 | 18 13 09 | −22 20 36 | 17. 8.88 | 18 33 17 | −22 14 40 | 18. 7.90 | 18 55 31 | −21 57 10 | 17. 6.92 | 19 18 16 | −21 27 48 |
| 27.10.84 | 17 56 45 | −22 18 12 | 27. 9.86 | 18 13 19 | −22 21 00 | 27. 8.88 | 18 32 41 | −22 15 31 | 28. 7.90 | 18 54 25 | −21 58 48 | 27. 6.92 | 19 17 11 | −21 29 50 |
| 6.11.84 | 17 57 54 | −22 18 39 | 7.10.86 | 18 13 44 | −22 21 18 | 6. 9.88 | 18 32 17 | −22 16 13 | 7. 8.90 | 18 53 25 | −22 00 20 | 7. 7.92 | 19 16 03 | −21 32 00 |
| 16.11.84 | 17 59 13 | −22 19 01 | 17.10.86 | 18 14 22 | −22 21 31 | 16. 9.88 | 18 32 06 | −22 16 46 | 17. 8.90 | 18 52 33 | −22 01 43 | 17. 7.92 | 19 14 54 | −21 34 12 |
| 26.11.84 | 18 00 39 | −22 19 17 | 27.10.86 | 18 15 13 | −22 21 36 | 26. 9.88 | 18 32 09 | −22 17 09 | 27. 8.90 | 18 51 51 | −22 02 55 | 27. 7.92 | 19 13 47 | −21 36 21 |

**Table 24: Position of Pluto 1983–1992**

| Date | R.A. | Dec | Date | R.A. | Dec | Date | R.A. | Dec | Date | R.A. | Dec | Date | R.A. | Dec |
|---|---|---|---|---|---|---|---|---|---|---|---|---|---|---|
| d mo yr | h m s | ° ′ ″ | d mo yr | h m s | ° ′ ″ | d mo yr | h m s | ° ′ ″ | d mo yr | h m s | ° ′ ″ | d mo yr | h m s | ° ′ ″ |
| 6. 1.83 | 14 12 42 | +4 35 07 | 6.12.84 | 14 27 38 | +2 41 39 | 6.11.86 | 14 41 09 | +1 02 24 | 6.10.88 | 14 54 38 | −0 28 12 | 6. 9.90 | 15 09 25 | −1 57 55 |
| 16. 1.83 | 14 13 17 | +4 37 46 | 16.12.84 | 14 28 50 | +2 39 14 | 16.11.86 | 14 42 38 | +0 55 47 | 16.10.88 | 14 55 58 | −0 37 01 | 16. 9.90 | 15 10 16 | −2 06 33 |
| 26. 1.83 | 14 13 40 | +4 41 44 | 26.12.84 | 14 29 54 | +2 38 16 | 26.11.86 | 14 44 04 | +0 50 17 | 26.10.88 | 14 57 23 | −0 45 20 | 26. 9.90 | 15 11 17 | −2 15 21 |
| 5. 2.83 | 14 13 49 | +4 46 52 | **5. 1.85** | 14 30 47 | +2 38 46 | 6.12.86 | 14 45 27 | +0 46 01 | 5.11.88 | 14 58 52 | −0 52 57 | 6.10.90 | 15 12 27 | −2 24 07 |
| 15. 2.83 | 14 13 46 | +4 52 58 | 15. 1.85 | 14 31 29 | +2 40 40 | 16.12.86 | 14 46 43 | +0 43 04 | 15.11.88 | 15 00 21 | −0 59 43 | 16.10.90 | 15 13 45 | −2 32 40 |
| 25. 2.83 | 14 13 29 | +4 59 49 | 25. 1.85 | 14 31 59 | +2 43 53 | 26.12.86 | 14 47 52 | +0 41 29 | 25.11.88 | 15 01 50 | −1 05 28 | 26.10.90 | 15 15 09 | −2 40 47 |
| 7. 3.83 | 14 13 01 | +5 07 08 | 4. 2.85 | 14 32 15 | +2 48 18 | **5. 1.87** | 14 48 51 | +0 41 17 | 5.12.88 | 15 03 15 | −1 10 05 | 5.11.90 | 15 16 36 | −2 48 19 |
| 17. 3.83 | 14 12 23 | +5 14 38 | 14. 2.85 | 14 32 18 | +2 53 44 | 15. 1.87 | 14 49 39 | +0 42 29 | 15.12.88 | 15 04 35 | −1 13 30 | 15.11.90 | 15 18 06 | −2 55 06 |
| 27. 3.83 | 14 11 35 | +5 22 02 | 24. 2.85 | 14 32 08 | +2 59 59 | 25. 1.87 | 14 50 16 | +0 44 58 | 25.12.88 | 15 05 49 | −1 15 36 | 25.11.90 | 15 19 36 | −3 00 58 |
| 6. 4.83 | 14 10 41 | +5 29 02 | 6. 3.85 | 14 31 46 | +3 06 48 | 4. 2.87 | 14 50 39 | +0 48 38 | **4. 1.89** | 15 06 53 | −1 16 24 | 5.12.90 | 15 21 04 | −3 05 49 |
| 16. 4.83 | 14 09 42 | +5 35 22 | 16. 3.85 | 14 31 12 | +3 13 54 | 14. 2.87 | 14 50 49 | +0 53 22 | 14. 1.89 | 15 07 48 | −1 15 53 | 15.12.90 | 15 22 28 | −3 09 33 |
| 26. 4.83 | 14 08 41 | +5 40 47 | 26. 3.85 | 14 30 29 | +3 21 03 | 24. 2.87 | 14 50 46 | +0 58 57 | 24. 1.89 | 15 08 31 | −1 14 06 | 25.12.90 | 15 23 46 | −3 12 06 |
| 6. 5.83 | 14 07 40 | +5 45 03 | 5. 4.85 | 14 29 38 | +3 27 55 | 6. 3.87 | 14 50 29 | +1 05 11 | 3. 2.89 | 15 09 01 | −1 11 09 | **4. 1.91** | 15 24 55 | −3 13 24 |
| 16. 5.83 | 14 06 42 | +5 48 02 | 15. 4.85 | 14 28 41 | +3 34 16 | 16. 3.87 | 14 50 01 | +1 11 48 | 13. 2.89 | 15 09 16 | −1 07 09 | 14. 1.91 | 15 25 55 | −3 13 29 |
| 26. 5.83 | 14 05 49 | +5 49 35 | 25. 4.85 | 14 27 40 | +3 39 50 | 26. 3.87 | 14 49 22 | +1 18 33 | 23. 2.89 | 15 09 21 | −1 02 16 | 24. 1.91 | 15 26 44 | −3 12 21 |
| 5. 6.83 | 14 05 02 | +5 49 39 | 5. 5.85 | 14 26 38 | +3 44 24 | 5. 4.87 | 14 48 34 | +1 25 11 | 5. 3.89 | 15 09 11 | −0 56 42 | 3. 2.91 | 15 27 21 | −3 10 06 |
| 15. 6.83 | 14 04 24 | +5 48 11 | 15. 5.85 | 14 25 38 | +3 47 46 | 15. 4.87 | 14 47 40 | +1 31 24 | 15. 3.89 | 15 08 49 | −0 50 39 | 13. 2.91 | 15 27 45 | −3 06 49 |
| 25. 6.83 | 14 03 56 | +5 45 14 | 25. 5.85 | 14 24 41 | +3 49 49 | 25. 4.87 | 14 46 40 | +1 36 59 | 25. 3.89 | 15 08 15 | −0 44 23 | 23. 2.91 | 15 27 55 | −3 02 38 |
| 5. 7.83 | 14 03 39 | +5 40 51 | 4. 6.85 | 14 23 50 | +3 50 26 | 5. 5.87 | 14 45 38 | +1 41 42 | 4. 4.89 | 15 07 31 | −0 38 08 | 5. 3.91 | 15 27 52 | −2 57 46 |
| 15. 7.83 | 14 03 33 | +5 35 08 | 14. 6.85 | 14 23 07 | +3 49 36 | 15. 5.87 | 14 44 36 | +1 45 22 | 14. 4.89 | 15 06 39 | −0 32 09 | 15. 3.91 | 15 27 36 | −2 52 22 |
| 25. 7.83 | 14 03 40 | +5 28 15 | 24. 6.85 | 14 22 33 | +3 47 18 | 25. 5.87 | 14 43 37 | +1 47 48 | 24. 4.89 | 15 05 41 | −0 26 41 | 25. 3.91 | 15 27 07 | −2 46 39 |
| 4. 8.83 | 14 04 00 | +5 20 21 | 4. 7.85 | 14 22 09 | +3 43 35 | 4. 6.87 | 14 42 42 | +1 48 55 | 4. 5.89 | 15 04 40 | −0 21 57 | 4. 4.91 | 15 26 28 | −2 40 53 |
| 14. 8.83 | 14 04 31 | +5 11 36 | 14. 7.85 | 14 21 57 | +3 38 32 | 14. 6.87 | 14 41 54 | +1 48 38 | 14. 5.89 | 15 03 37 | −0 18 09 | 14. 4.91 | 15 25 40 | −2 35 16 |
| 24. 8.83 | 14 05 13 | +5 02 14 | 24. 7.85 | 14 21 58 | +3 32 17 | 24. 6.87 | 14 41 14 | +1 46 58 | 24. 5.89 | 15 02 35 | −0 15 25 | 24. 4.91 | 15 24 44 | −2 30 03 |
| 3. 9.83 | 14 06 07 | +4 52 26 | 3. 8.85 | 14 22 10 | +3 24 57 | 4. 7.87 | 14 40 45 | +1 43 54 | 3. 6.89 | 15 01 37 | −0 13 55 | 4. 5.91 | 15 23 44 | −2 25 25 |
| 13. 9.83 | 14 07 09 | +4 42 26 | 13. 8.85 | 14 22 35 | +3 16 45 | 14. 7.87 | 14 40 26 | +1 39 32 | 13. 6.89 | 15 00 45 | −0 13 42 | 14. 5.91 | 15 22 41 | −2 21 35 |
| 23. 9.83 | 14 08 20 | +4 32 28 | 23. 8.85 | 14 23 12 | +3 07 50 | 24. 7.87 | 14 40 20 | +1 33 57 | 23. 6.89 | 15 00 00 | −0 14 48 | 24. 5.91 | 15 21 38 | −2 18 42 |
| 3.10.83 | 14 09 38 | +4 22 44 | 2. 9.85 | 14 24 00 | +2 58 26 | 3. 8.87 | 14 40 26 | +1 27 16 | 3. 7.89 | 14 59 25 | −0 17 14 | 3. 6.91 | 15 20 37 | −2 16 55 |
| 13.10.83 | 14 11 01 | +4 13 28 | 12. 9.85 | 14 24 58 | +2 48 44 | 13. 8.87 | 14 40 44 | +1 19 40 | 13. 7.89 | 14 59 00 | −0 20 57 | 13. 6.91 | 15 19 41 | −2 16 18 |
| 23.10.83 | 14 12 27 | +4 04 51 | 22. 9.85 | 14 26 06 | +2 38 58 | 23. 8.87 | 14 41 15 | +1 11 18 | 23. 7.89 | 14 58 47 | 0 25 52 | 23. 6.91 | 15 18 52 | −2 16 55 |
| 2.11.83 | 14 13 54 | +3 57 06 | 2.10.85 | 14 27 20 | +2 29 21 | 2. 9.87 | 14 41 58 | +1 02 23 | 2. 8.89 | 14 58 47 | −0 31 52 | 3. 7.91 | 15 18 11 | −2 18 47 |
| 12.11.83 | 14 15 21 | +3 50 22 | 12.10.85 | 14 28 41 | +2 20 05 | 12. 9.87 | 14 42 51 | +0 53 06 | 12. 8.89 | 14 58 59 | −0 38 49 | 13. 7.91 | 15 17 41 | −2 21 52 |
| 22.11.83 | 14 16 46 | +3 44 50 | 22.10.85 | 14 30 07 | +2 11 23 | 22. 9.87 | 14 43 54 | +0 43 39 | 22. 8.89 | 14 59 23 | −0 46 33 | 23. 7.91 | 15 17 21 | −2 26 07 |
| 2.12.83 | 14 18 06 | +3 40 37 | 1.11.85 | 14 31 34 | +2 03 26 | 2.10.87 | 14 45 06 | +0 34 16 | 1. 9.89 | 15 00 00 | −0 54 54 | 2. 8.91 | 15 17 14 | −2 31 26 |
| 12.12.83 | 14 19 19 | +3 37 47 | 11.11.85 | 14 33 03 | +1 56 25 | 12.10.87 | 14 46 24 | +0 25 09 | 11. 9.89 | 15 00 48 | −1 03 41 | 12. 8.91 | 15 17 20 | −2 37 42 |
| 22.12.83 | 14 20 25 | +3 36 26 | 21.11.85 | 14 34 29 | +1 50 30 | 22.10.87 | 14 47 48 | +0 16 29 | 21. 9.89 | 15 01 47 | −1 12 40 | 22. 8.91 | 15 17 38 | −2 44 46 |
| 1. 1.84 | 14 21 21 | +3 36 32 | 1.12.85 | 14 35 53 | +1 45 48 | 1.11.87 | 14 49 14 | +0 08 29 | 1.10.89 | 15 02 55 | −1 21 42 | 1. 9.91 | 15 18 08 | −2 52 28 |
| 11. 1.84 | 14 22 05 | +3 38 07 | 11.12.85 | 14 37 11 | +1 42 25 | 11.11.87 | 14 50 45 | +0 01 19 | 11.10.89 | 15 04 11 | −1 30 34 | 11. 9.91 | 15 18 51 | −3 00 39 |
| 21. 1.84 | 14 22 38 | +3 41 03 | 21.12.85 | 14 38 21 | +1 40 26 | 21.11.87 | 14 52 13 | −0 04 53 | 21.10.89 | 15 05 33 | −1 39 03 | 21. 9.91 | 15 19 45 | −3 09 07 |
| 31. 1.84 | 14 22 57 | +3 45 16 | 31.12.85 | 14 39 22 | +1 39 52 | 1.12.87 | 14 53 39 | −0 09 57 | 31.10.89 | 15 06 59 | −1 46 59 | 1.10.91 | 15 20 48 | −3 17 40 |
| **10. 2.84** | 14 23 04 | +3 50 34 | **10. 1.86** | 14 40 13 | +1 40 42 | 11.12.87 | 14 55 01 | 0 13 47 | 10.11.89 | 15 08 28 | −1 54 12 | 11.10.91 | 15 22 01 | −3 26 08 |
| 20. 2.84 | 14 22 57 | +3 56 46 | 20. 1.86 | 14 40 53 | +1 42 55 | 21.12.87 | 14 56 16 | −0 16 19 | 20.11.89 | 15 09 58 | −2 00 31 | 21.10.91 | 15 23 21 | −3 34 20 |
| 1. 3.84 | 14 22 38 | +4 03 36 | 30. 1.86 | 14 41 19 | +1 46 23 | 31.12.87 | 14 57 23 | −0 17 30 | 30.11.89 | 15 11 26 | −2 05 50 | 31.10.91 | 15 24 46 | −3 42 03 |
| 11. 3.84 | 14 22 07 | +4 10 51 | 9. 2.86 | 14 41 32 | +1 50 58 | **10. 1.88** | 14 58 19 | −0 17 19 | 10.12.89 | 15 12 51 | −2 10 00 | 10.11.91 | 15 26 15 | −3 49 10 |
| 21. 3.84 | 14 21 26 | +4 18 11 | 19. 2.86 | 14 41 32 | +1 56 30 | 20. 1.88 | 14 59 05 | −0 15 50 | 20.12.89 | 15 14 10 | −2 12 58 | 20.11.91 | 15 27 45 | −3 55 29 |
| 31. 3.84 | 14 20 37 | +4 25 20 | 1. 3.86 | 14 41 19 | +2 02 45 | 30. 1.88 | 14 59 38 | −0 13 05 | 30.12.89 | 15 15 22 | −2 14 41 | 30.11.91 | 15 29 15 | −4 00 55 |
| 10. 4.84 | 14 19 41 | +4 32 02 | 11. 3.86 | 14 40 54 | +2 09 29 | 9. 2.88 | 14 59 58 | −0 09 14 | **9. 1.90** | 15 16 24 | −2 15 06 | 10.12.91 | 15 30 42 | −4 05 18 |
| 20. 4.84 | 14 18 41 | +4 37 59 | 21. 3.86 | 14 40 18 | +2 16 27 | 19. 2.88 | 15 00 05 | −0 04 25 | 19. 1.90 | 15 17 16 | −2 14 16 | 20.12.91 | 15 32 05 | −4 08 35 |
| 30. 4.84 | 14 17 42 | +4 42 59 | 31. 3.86 | 14 39 32 | +2 23 21 | 29. 2.88 | 14 59 58 | +0 01 11 | 29. 1.90 | 15 17 56 | −2 12 14 | 30.12.91 | 15 33 21 | −4 10 42 |
| 10. 5.84 | 14 16 39 | +4 46 49 | 10. 4.86 | 14 38 39 | +2 29 55 | 10. 3.88 | 14 59 39 | +0 07 20 | 8. 2.90 | 15 18 23 | −2 09 06 | **9. 1.92** | 15 34 29 | −4 11 38 |
| 20. 5.84 | 14 15 41 | +4 49 19 | 20. 4.86 | 14 37 40 | +2 35 53 | 20. 3.88 | 14 59 08 | +0 13 48 | 18. 2.90 | 15 18 36 | −2 05 00 | 19. 1.92 | 15 35 27 | −4 11 23 |
| 30. 5.84 | 14 14 49 | +4 50 25 | 30. 4.86 | 14 36 39 | +2 41 02 | 30. 3.88 | 14 58 27 | +0 20 19 | 28. 2.90 | 15 18 36 | −2 00 06 | 29. 1.92 | 15 36 13 | −4 09 59 |
| 9. 6.84 | 14 14 04 | +4 50 01 | 10. 5.86 | 14 35 37 | +2 45 09 | 9. 4.88 | 14 57 37 | +0 26 39 | 10. 3.90 | 15 18 23 | −1 54 35 | 8. 2.92 | 15 36 46 | −4 07 32 |
| 19. 6.84 | 14 13 28 | +4 48 07 | 20. 5.86 | 14 34 37 | +2 48 03 | 19. 4.88 | 14 56 40 | +0 32 30 | 20. 3.90 | 15 17 58 | −1 48 42 | 18. 2.92 | 15 37 07 | −4 04 09 |
| 29. 6.84 | 14 13 02 | +4 44 47 | 30. 5.86 | 14 33 41 | +2 49 38 | 29. 4.88 | 14 55 40 | +0 37 40 | 30. 3.90 | 15 17 21 | −1 42 39 | 28. 2.92 | 15 37 14 | −3 59 57 |
| 9. 7.84 | 14 12 47 | +4 40 03 | 9. 6.86 | 14 32 51 | +2 49 49 | 9. 5.88 | 14 54 37 | +0 41 56 | 9. 4.90 | 15 16 35 | −1 36 43 | 9. 3.92 | 15 37 07 | −3 55 07 |
| 19. 7.84 | 14 12 45 | +4 34 03 | 19. 6.86 | 14 32 10 | +2 48 32 | 19. 5.88 | 14 53 35 | +0 45 08 | 19. 4.90 | 15 15 41 | −1 31 06 | 19. 3.92 | 15 36 44 | −3 49 52 |
| 29. 7.84 | 14 12 55 | +4 26 56 | 29. 6.86 | 14 31 38 | +2 45 51 | 29. 5.88 | 14 52 37 | +0 47 06 | 29. 4.90 | 15 14 42 | −1 26 02 | 29. 3.92 | 15 36 16 | −3 44 24 |
| 8. 8.84 | 14 13 17 | +4 18 51 | 9. 7.86 | 14 31 17 | +2 41 47 | 8. 6.88 | 14 51 43 | +0 47 46 | 9. 5.90 | 15 13 40 | −1 21 45 | 8. 4.92 | 15 35 34 | −3 38 55 |
| 18. 8.84 | 14 13 51 | +4 10 00 | 19. 7.86 | 14 31 08 | +2 36 27 | 18. 6.88 | 14 50 56 | +0 47 04 | 19. 5.90 | 15 12 37 | −1 18 24 | 18. 4.92 | 15 34 44 | −3 33 40 |
| 28. 8.84 | 14 14 36 | +4 00 36 | 29. 7.86 | 14 31 11 | +2 29 58 | 28. 6.88 | 14 50 19 | +0 45 00 | 29. 5.90 | 15 11 36 | −1 16 09 | 28. 4.92 | 15 33 47 | −3 28 50 |
| 7. 9.84 | 14 15 32 | +3 50 50 | 8. 8.86 | 14 31 27 | +2 22 29 | 8. 7.88 | 14 49 52 | +0 41 36 | 8. 6.90 | 15 10 39 | −1 15 06 | 8. 5.92 | 15 32 43 | −3 24 39 |
| 17. 9.84 | 14 16 37 | +3 40 56 | 18. 8.86 | 14 31 55 | +2 14 11 | 18. 7.88 | 14 49 36 | +0 36 56 | 18. 6.90 | 15 09 48 | −1 15 19 | 18. 5.92 | 15 31 42 | −3 21 17 |
| 27. 9.84 | 14 17 50 | +3 31 07 | 28. 8.86 | 14 32 34 | +2 05 15 | 28. 7.88 | 14 49 33 | +0 31 07 | 28. 6.90 | 15 09 05 | −1 16 49 | 28. 5.92 | 15 30 39 | −3 18 52 |
| 7.10.84 | 14 19 09 | +3 21 36 | 7. 9.86 | 14 33 25 | +1 55 58 | 7. 8.88 | 14 49 42 | +0 24 18 | 8. 7.90 | 15 08 32 | −1 19 35 | 7. 6.92 | 15 29 39 | −3 17 32 |
| 17.10.84 | 14 20 33 | +3 12 36 | 17. 9.86 | 14 34 26 | +1 46 18 | 17. 8.88 | 14 50 03 | +0 16 36 | 18. 7.90 | 15 08 10 | −1 23 35 | 17. 6.92 | 15 28 45 | −3 17 21 |
| 27.10.84 | 14 22 00 | +3 04 18 | 27. 9.86 | 14 35 35 | +1 36 42 | 27. 8.88 | 14 50 37 | +0 08 13 | 28. 7.90 | 15 08 00 | −1 28 43 | 27. 6.92 | 15 27 57 | −3 18 21 |
| 6.11.84 | 14 23 28 | +2 56 54 | 7.10.86 | 14 36 52 | +1 27 19 | 6. 9.88 | 14 51 22 | −0 00 39 | 7. 8.90 | 15 08 02 | −1 34 52 | 7. 7.92 | 15 27 11 | −3 20 34 |
| 16.11.84 | 14 24 55 | +2 50 35 | 17.10.86 | 14 38 15 | +1 18 20 | 16. 9.88 | 14 52 18 | −0 09 48 | 17. 8.90 | 15 08 17 | −1 41 53 | 17. 7.92 | 15 26 51 | −3 23 56 |
| 26.11.84 | 14 26 19 | +2 45 27 | 27.10.86 | 14 39 41 | +1 09 58 | 26. 9.88 | 14 53 24 | −0 19 04 | 27. 8.90 | 15 08 45 | −1 49 38 | 27. 7.92 | 15 26 34 | −3 28 23 |

# Bibliography

The literature of planetary astronomy is now so vast that it is impossible to give more than a few of the books particularly relevant to the subject.

Baker, V. R., *The Channels of Mars* (Adam Hilger, 1982)

Brandt, J. C., and Chapman, R. S., *Introduction to Comets* (Cambridge University Press, 1981)

Carr, M. H., *The Surface of Mars* (Yale University Press, 1982)

Cooper, H. S., *Imaging Saturn* (Holt, Reinhard and Wilson, 1982)

Cunningham, C., *Introduction to Asteroids* (Willmann-Bell, 1988)

Elliot, J., and Kerr, R., *Rings* (Massachusetts Institute of Technology Press, 1984)

French, B., *The Moon Book* (Penguin Books, 1977)

Gehrels, T., and Matthews, M. S., (ed.) *Saturn* (University of Arizona Press, 1984)

Grewing, M., Praderite, F., and Reinhard, R., (ed.) *Exploration of Halley's Comet* (Springer Verlag, 1986)

Glass, B., *Introduction to Planetary Geology* (Cambridge University Press, 1982)

Greeley, R., *Planetary Landscapes* (Allen and Unwin, 1987)

Guest, J., and Greeley, R., *Geology on the Moon* (Wykeham Press, 1977)

Hoyt, W., *Planets X and Pluto* (University of Arizona Press, 1980)

Hunt, G., (ed.) *Uranus and the Outer Planets* (Cambridge University Press, 1982)

Hunt, G., and Moore, P., *The Planet Venus* (Faber and Faber, 1982)

Hunt, G., and Moore, P., *Atlas of Uranus* (Cambridge University Press, 1989)

Hunten, D. M., Colin, L., Donahue, T. M., and Moroz, V. I., *Venus* (University of Arizona Press, 1983)

Kondratyev, K., and Hunt, G., *Weather and Climate on the Planets* (Pergamon Press, 1982)

Kowal, C. T., *Asteroids* (Ellis Horwood and John Wylie, 1989)

Kronk, G., *Comets* (Enslow Press, 1984)

Moore, P., *The Planet Neptune* (Ellis Horwood and John Wylie, 1989)

Morrison, D., and Samz, J., *Voyager to Jupiter* (NASA, 1980)

Morton, V., *Halley's Comet, 1785–1984* (Greenwood Press, 1984)

Schultze, P. H., *Moon Morphology* (University of Texas Press, 1976)

Sheehan, W., *Planets and Perception* (University of Arizona Press, 1988)

Strom, R., *Mercury, the Elusive Planet* (Cambridge University Press, 1987)

Tombaugh, C., and Moore, P., *Out of the Darkness: the Planet Pluto* (Stackpole Books, 1980)

various. *The Moon; a New Appraisal* (Royal Society of London, 1975)

Vilas, F., Chapman, C., and Matthews, M. S., (ed.) *Mercury* (University of Arizona Press, 1988)

# Index

453

# Acknowledgements

The publishers gratefully acknowledge the assistance of the following:
R.M. Batson of the US Geological Survey
Professor R. Booth of Onsala Space Observatory, Sweden
Dr J.C. Brown of the University of Glasgow
Gilead Cooper (editor)
The late Charles A. Cross, cartographer of Mercury and Moon maps
Cathy Gill (proof reader)
Peter Gill (researcher)
Dr. John E. Guest of the Mill Hill Observatory, University of London
H.M. Nautical Almanac Office of the Royal Greenwich Observatory, Herstmonceux, which supplied the computer print-out for Tables 15–24
Sean Keogh (art editor)
Nicolas Law (researcher)
Dr. C.T. Pillinger of the University of Cambridge
A.O. Pickersgill (researcher)

**Photographic credits**
NASA 54 (bottom). 55. 63. 81 (6). 82 (2). 83 (3, 4, 5, 6, 7). 84 (2A, B). 85 (3, 4, 5, 6, 7, 8). 86. 87. 88. 89. 106 (1). 107 (2, 6). 108 (3). 109 (4A, B, 5A, B). 111 (1). 113 (left-top, -center, -bottom). 122. 123. 124. 125. 126. 128 (top). 135 (4C). 203 (bottom). 213 (5, 6, 7). 215 (4, 5). 220–21 (1). 220 (2A). 222 (3, 4, 5, 6). 223 (1, 2, 3, 4, 5, 6). 224 (1, 3). 225 (4, 5, 6, 8). 228 (1, 2, 3). 229 (4, 5, 6). 328–9 (2). 334 (3). 361 (3, 4). 374 (2). 375 (3A, B). 381 (2).
NASA/Ames Research Center: 338 (bottom).
NASA/Ames Research Center/Jet Propulsion Laboratory/US Geological Survey: 114 (top, center).
NASA/Harvard College Observatory/E.J. Schmahl, UMd: 56–7.
NASA/Jet Propulsion Laboratory: 108 (2). 113 (right-top, -center, -bottom) Michael Kobrick. 127. 128 (bottom). 217 (4, 5). 218 (1, 2, 3, 4, 5). 219 (8A, B, 9). 220 (2B). 221 (4). 222 (2). 223 (1). 224 (2). 225 (7). 226 (1). 227 (2, 4, 5, 6, 7). 229 (7). 240 (1, 2). 241 (6). 257 (3). 259 (4A, B). 262. 263 (3, 4, 5). 264. 273. 275–88. 271 (5, 6). 272 (1, 2). 290 (3). 291 (5, 6A, B, 7, 8). 292–3. 296 (2). 297 (4, 5, 6). 300 (2). 301 (3, 4, 5A, 6). 302. 303. 306 (2). 307 (3A, C, 4). 310. 324 (1). 326 (1, 2, 3). 327 (4). 328 (1A). 331 (2, 3, 4). 332–3. 335 (4, 6). 339 (bottom). 340–41. 342. 343 (top-left, -right). 344–5 (top). 346 (top right). 348. 349. 351. 352. 360–61 (1). 362 (2). 363 (3A, B). 364 (2, 3A, B). 366 (2). 367 (3, 4, 5). 370 (2). 371 (3, 4, 5, 6). 375 (4). 378 (1). 380–81 (3). 385. 388. 392. 393. 394. 395. 436. 437.
NASA/Lyndon B. Johnson Space Center: 152 (2A, B, C, D, 3A). 153 (4A, B, C, D, 5A). 158 (3). 208 (top, bottom).
NASA/Picturepoint: 193. 194. 297.
NASA/Space Graphics: 334 (1, 2). 336 (1). 337. 339 (top). 343 (bottom-left, -right). 344 (bottom left). 346 (top-left, -bottom). 347. 350.
NASA/Woodmansterne Ltd: 195. 196. 197. 200. 203 (top).

Royal Astronomical Society: 6 (left, center-top) Hale Observatories. 21 (5). 74. 245 (4) Heidelburg Observatory. 316 (1) Angelo Hornak. 392 (2A) McDonald Observatory. 395 McDonald Observatory/Kuiper. 398 (3) Yerkes Observatory. 399 (4) Armagh Observatory. 400 (2) US Naval Observatory. 400–401 (2) US Naval Observatory. 403 (5) Helwan Observatory, Egypt. 404 (2) Kitt Peak Observatory. 405 (5) Ondřejov Observatory, Prague; (6) Lick Observatory; (7) Cambridge Observatory.

p. 6 (center) US Naval Observatory
p. 6 (right) Hale Observatories
p. 7 (top) Hale Observatories
p. 7 (bottom) US Naval Observatory
p. 20 (2A, B, C) Ann Ronan Picture Library
p. 21 (4) Ann Ronan Picture Library
p. 23 (3) Kitt Peak National Observatory
p. 25 (3) Max-Planck-Institut Für Radio Astronomy
p. 25 (5A, B, C, D) Solar Physics Group, American Science & Engineering
p. 26 (1) Royal Greenwich Observatory
p. 26 (2) CSIRO Solar Observatory, Culgoora, N.S.W.
p. 27 (4) Courtesy of R. Leighton, Mount Wilson and Palomar Observatories, Carnegie Institution of Washington
p. 27 (5) Observatoire du Pic-du-Midi
p. 28 (1) Association of Universities for Research in Astronomy, Inc., Sacramento Park Observatory
p. 28 (2) Sacramento Peak Observatory, Air Force Cambridge Research Laboratories
p. 30 (1) Lick Observatory
p. 30 (2A, B) Mount Wilson and Palomar Observatories, Carnegie Institute of Washington
p. 31 (2C, D) Mount Wilson and Palomar Observatories, Carnegie Institute of Washington
p. 31 (3) CSIRO Solar Observatory, Culgoora, N.S.W.
p. 34 (1A, B, C) The Aerospace Corporation, San Fernando Observatory
34 (2) CSIRO Solar Observatory, Culgoora, N.S.W.
p. 32 (1) Kitt Peak National Observatory
p. 33 (6A, B, C) Plate 3.25 from the book *Sunspots* by R.J. Bray and R.E. Loughmead (Dover Publications, Inc., NY, 1979)
p. 36 (1A, B) Astronomical Institute of Czechoslovakian Academy of Sciences, Observatory Ondřejov
p. 37 (3) Royal Greenwich Observatory
p. 38 (1A, B, C) Solar Physics Group, American Science & Engineering
p. 38–9 (2A, B) Big Bear Solar Observatory, Caltech